T0190266

Lecture Notes in Computer Science   10362

Commenced Publication in 1973
Founding and Former Series Editors:
Gerhard Goos, Juris Hartmanis, and Jan van Leeuwen

More information about this series at http://www.springer.com/series/7409

De-Shuang Huang · Kang-Hyun Jo
Juan Carlos Figueroa-García (Eds.)

Intelligent Computing Theories and Application

13th International Conference, ICIC 2017
Liverpool, UK, August 7–10, 2017
Proceedings, Part II

 Springer

Editors

De-Shuang Huang
Tongji University
Shanghai
China

Kang-Hyun Jo
University of Ulsan
Ulsan
Korea (Republic of)

Juan Carlos Figueroa-García
Universidad Distrital Francisco José de
  Caldas
Bogotá
Colombia

ISSN 0302-9743 ISSN 1611-3349 (electronic)
Lecture Notes in Computer Science
ISBN 978-3-319-63311-4 ISBN 978-3-319-63312-1 (eBook)
DOI 10.1007/978-3-319-63312-1

Library of Congress Control Number: 2017946067

LNCS Sublibrary: SL3 – Information Systems and Applications, incl. Internet/Web, and HCI

Printed on acid-free paper

This Springer imprint is published by Springer Nature
The registered company is Springer International Publishing AG
The registered company address is: Gewerbestrasse 11, 6330 Cham, Switzerland

Preface

The International Conference on Intelligent Computing (ICIC) was started to provide an annual forum dedicated to the emerging and challenging topics in artificial intelligence, machine learning, pattern recognition, bioinformatics, and computational biology. It aims to bring together researchers and practitioners from both academia and industry to share ideas, problems, and solutions related to the multifaceted aspects of intelligent computing.

ICIC 2017, held in Liverpool, UK, August 7–10, 2017, constituted the 13th International Conference on Intelligent Computing. It built upon the success of ICIC 2016, ICIC 2015, ICIC 2014, ICIC 2013, ICIC 2012, ICIC 2011, ICIC 2010, ICIC 2009, ICIC 2008, ICIC 2007, ICIC 2006, and ICIC 2005 that were held in Lanzhou, Fuzhou, Taiyuan, Nanning, Huangshan, Zhengzhou, Changsha, China, Ulsan, Korea, Shanghai, Qingdao, Kunming, and Hefei, China, respectively.

This year, the conference concentrated mainly on the theories and methodologies as well as the emerging applications of intelligent computing. Its aim was to unify the picture of contemporary intelligent computing techniques as an integral concept that highlights the trends in advanced computational intelligence and bridges theoretical research with applications. Therefore, the theme for this conference was "Advanced Intelligent Computing Technology and Applications." Papers focused on this theme were solicited, addressing theories, methodologies, and applications in science and technology.

ICIC 2017 received 612 submissions from 21 countries and regions. All papers went through a rigorous peer-review procedure and each paper received at least three review reports. Based on the review reports, the Program Committee finally selected 212 high-quality papers for presentation at ICIC 2017, included in three volumes of proceedings published by Springer: two volumes of *Lecture Notes in Computer Science* (LNCS) and one volume of *Lecture Notes in Artificial Intelligence* (LNAI).

This volume of *Lecture Notes in Computer Science* (LNCS) includes 74 papers.

The organizers of ICIC 2017, including Tongji University and Liverpool John Moores University, UK, made an enormous effort to ensure the success of the conference. We hereby would like to thank the members of the Program Committee and the referees for their collective effort in reviewing and soliciting the papers. We would like to thank Alfred Hofmann, executive editor at Springer, for his frank and helpful advice and guidance throughout and for his continuous support in publishing the proceedings. In particular, we would like to thank all the authors for contributing their papers. Without the high-quality submissions from the authors, the success of the conference would not have been possible. Finally, we are especially grateful to the IEEE Computational Intelligence Society, the International Neural Network Society, and the National Science Foundation of China for their sponsorship.

May 2017

De-Shuang Huang
Kang-Hyun Jo
Juan Carlos Figueroa

ICIC 2017 Organization

General Co-chairs

De-Shuang Huang, China
Abir Hussain, UK

Program Committee Co-chairs

Kang-Hyun Jo, Korea
M. Michael Gromiha, India

Organizing Committee Co-chairs

Dhiya Al-Jumeily, UK
Tom Dowson, UK

Award Committee Co-chairs

Vitoantonio Bevilacqua, Italy
Phalguni Gupta, India

Tutorial Co-chairs

Juan Carlos Figueroa, Colombia
Valeriya Gribova, Russia

Publication Co-chairs

Kyungsook Han, Korea
Laurent Heutte, France

Workshop/Special Session Chair

Wenzheng Bao, China

Special Issue Chair

Ling Wang, China

International Liaison Chair

Prashan Premaratne, Australia

Publicity Co-chairs

Chun-Hou Zheng, China
Jair Cervantes Canales, Mexico
Huiyu Zhou, China

Exhibition Chair

Lin Zhu, China

Program Committee

Khalid Aamir	Yannis Goulermas	Chunmei Liu
Mohd Helmy Abd Wahab	Michael Gromiha	Shuo Liu
Abbas Amini	Fei Han	Xingwen Liu
Vasily Aristarkhov	Kyungsook Han	Xiwei Liu
Waqas Haider	Tianyong Hao	Yunxia Liu
Khan Bangyal	Wei-Chiang Hong	Ahmad Lotfi
Shuui Bi	Yuexian Hou	Jungang Lou
Hongmin Cai	Saiful Islam	Yonggang Lu
Jair Cervantes	Chuleerat Jaruskulchai	Yingqin Luo
Pei-Chann Chang	Kang-Hyun Jo	Jinwen Ma
Chen Chen	Dah-Jing Jwo	Xiandong Meng
Shih-Hsin Chen	Seeja K.R.	Kang Ning
Weidong Chen	Sungshin Kim	Ben Niu
Wen-Sheng Chen	Yong-Guk Kim	Gaoxiang Ouyang
Xiyuan Chen	Yoshinori Kuno	Francesco Pappalardo
Jieren Cheng	Takashi Kuremoto	Young B. Park
Michal Choras	Xuguang Lan	Eros Pasero
Angelo Ciaramella	Xinyi Le	Marzio Pennisi
Jose Alfredo Costa	Choong Ho Lee	Prashan Premaratne
Guojun Dai	Xiujuan Lei	Yuhua Qian
Yanrui Ding	Bo Li	Daowen Qiu
Ji-Xiang Du	Guoliang Li	Jiangning Song
Pufeng Du	Kang Li	Stefano Squartini
Jianbo Fan	Ming Li	Zhixi Su
Paul Fergus	Qiaotian Li	Shiliang Sun
Juan Carlos	Shuai Li	Zhan-Li Sun
Figueroa-Garcia	Honghuang Lin	Jijun Tang
Liang Gao	Bin Liu	Joaquin Torres-Sospedra
Dunwei Gong	Bingqiang Liu	Antonio Uva

Bing Wang
Jim Jing-Yan Wang
Shitong Wang
Xuesong Wang
Yong Wang
Yuanzhe Wang
Ze Wang
Wei Wei
Ka-Chun Wong
Hongjie Wu

QingXiang Wu
Yan Wu
Junfeng Xia
Shunren Xia
Xinzheng Xu
Wen Yu
Junqi Zhang
Rui Zhang
Shihua Zhang
Xiang Zhang

Zhihua Zhang
Dongbin Zhao
Xiaoguang Zhao
Huiru Zheng
Huiyu Zhou
Yongquan Zhou
Shanfeng Zhu
Quan Zou
Lc Zhang

Additional Reviewers

Honghong Cheng
Jieting Wang
Xinyan Liang
Feijiang Li
Furong Lu
Mohd Farhan Md Fudzee
Husniza Husni
Rozaida Ghazali
Zarul Zaaba
Shuzlina Abdul-Rahman
Sasalak Tongkaw
Wan Hussain Wan Ishak
Mohamad Farhan
    Mohamad Mohsin
Masoud Mohammadian
Mario Diván
Nureize Arbaiy
Mohd Shamrie Sainin
Francesco Masulli
Nooraini Yusoff
Aida Mustapha
Nur Azzah Abu Bakar
Anang Hudaya Muhamad
    Amin
Sazalinsyah Razali
Norita Md Norwawi
Nasir Sulaiman
Ru-Ze Liang
Wan Hasrulnizzam
    Wan Mahmood
Azizul Azhar Ramli

Qian Guo
Mingjie Tang
Fang Jin
Muhammad Imran
Barnypok
Weizhi Li
Ri Hanafi
Jieyi Zhao
Zhichao Jiang
Y-H. Taguchi
Chenhao Xu
Shaofu Yang
Jian Xiao
Haoyong Yu
Prabakaran R.
Anusuya S.
A. Mary Thangakani
Fatih Adiguzel
Shanwen Zhu
Yuheng Wang
Xixun Zhu
Ambuj Srivastava
Xing He
Nagarajan Raju
Kumar Yugandhar
Sakthivel Ramasamy
Anoosha Paruchuri
Sherlyn
Ramakrishnan
Chandrasekaran
Akila Ranjith

Harihar Balasubramanian
Vimala A.
Farheen Siddiqui
Sameena Naaz
Parul Agarwal
Soniya Balram
Jaya Sudha
Deepa Anand
Yun Xue
Santo Motta
Nengfu Xie
Lei Liu
Nengfu Xie
Xiangyuan Lan
Rushi Lan
Jie Qin
Tomasz Andrysiak
Rafal Kozik
Adam Schmidt
Zhiqiang Liu
Liang Xiaobo
Zhiwei Feng
Xia Li
Chunxia Zhang
Jiaoyun Yang
Yi Xiong
Wenjun Shen
Xihao Hu
Xiangli Zhang
Xiaoyong Bian
Yan Xu

Lu-Chen Weng
Guan Xiao
Jianchao Fan
Ka-Chun Wong
Fei Wang
Beeno Cheong
Wei-Jie Yu
Meng Zhang
Jiao Zhang
Zhouhua Peng
Shankai Yan
Yueming Lyu
Xiaoping Wang
Cheng Liu
Jiecong Lin
Shan Gao
Xiu-Jun Gong
Lijun Quan
Haiou Li
Junyi Chen
Cheng Chen
Ukyo Liu
Weizhi Li
Xi Yuga
Bo Chen
Binbin Pan
Yaran Chen
Xiangtao Li
Fei Guo
Leyi Wei
Yungang Xu
Bin Qin
Yuanpeng Zhang
Wenlong Hang
Huo Xuan
Feng Cong
Hongguo Zhao
Xiongtao Zhang
Li Liu
Baohua Wang
Jianwei Yang
Naveed Anwer Butt
Seung Hoan Choi
Farid Garcia-Lamont
Sergio Ruiz

Josue Vicente Cervantes
  Bazan
Abd Ur Rehman
Ta Zhou
Wentao Fan
Xin Liu
Qing Lei
Ziyi Chen
Yewang Chen
Yan Chen
Jialin Peng
Sihai Yang
Hong-Bo Zhang
Xingguang Pan
Biqi Wang
Arturo
Lisbeth Rodríguez
Asdrubal Lopez-Chau
Xiaoli Li
Yuriy Orlov
Ken Wing-Kin Sung
Zhou Hufeng
Gonghong Wei
De-Shuang Huang
Dongbo Bu
Dariusz Plewczyński
Jialiang Yang
Musa Mhlanga
Shanshan Tuo
Dawei Li
Chee Hong Wong
Zhang Zhizhuo
Xiang Jinhai
Wada Youichiro
Shuai Cheng Li
Xuequn Shang
Ralf Jauch
Limsoon Wong
Edwin Cheung
Filippo Castiglione
Ping Zhang
Hong Zhu
Tianming Liang
Jiancong Fan
Guanying Wang

Shumei Zhang
Raymond Bond
Ginestra Bianconi
Antonino Staiano
Dong Shan
Xiangying Jiang
Hongjie Jia
Xiaopeng Hua
Sixiao Wei
Shumei Zhang
Francesco Camastra
Weikuan Jia
Huajuan Huang
Hui Li
Xinzheng Xu
Yan Qin
Jiayin Zhou
Xin Gao
Antonio Maratea
Wu Qingxiang
Xiaowei Zhang
Cunlu Xu
Yong Lin
Haizhou Wu
Ruxin Zhao
Alessio Ferone
Zheheng Jiang
Kun Zhang
Fei Chu
Chunyu Yang
Wei Dai
Gang Li
Guo Yanen
Hao Xin
Zhang Zanchao
Yonggang Wang
Yajun Zhang
Guimin Lin
Pengfei Li
Antony Lam
Hironobu Fujiyoshi
Yoshinori Kobayashi
Jin-Soo Kim
Erchao Li
Miao Rong

Yiping Liu
Nobutaka Shimada
Hae-Chul Choi
Irfan Mehmood
Yuyan Han
Jianhua Zhang
Biao Xu
Meirong Chen
Kazunori Onoguchi
Gongjing Chen
Yuntao Wei
Rina Su
Lu Xingjia
Casimiro Aday
   Curbelo Montanez
Mohammed Khalaf
Atsushi Yamashita
Hotaka Takizawa
Yasushi Mae
Hisato Fukuda
Shaohua Li
Keight Robert
Raghad Al-Shabandar
Xie Zhijun
Abir Hussain
Morihiro Hayashida
Ziding Zhang
Mengyao Li
Shen Chong
Wei Wang
Basma Abdulaimma
Hoshang Kolivand
Ala Al Kafri
Abbas, Rounaq
Kaihui Bian
Jiangning Song
Haya Alaskar
Francis, Hulya
Chuang Wu
Xinhua Tang
Hong Zhang
Sheng Zou
Cuci Quinstina
Bingbo Cui
Hongjie Wu
Wenrui Zhao

Xuan Wang
Junjun Jiang
Qiyan Sun
Zhuangguo Miao
Yuan Xu
Dengxin Dai
Zhiwu Huang
Xuetao Zhang
Shaoyi Du
Jihua Zhu
Bingxiang Xu
Zhixuan Wei
Hanfu Wang
Prashan Premaratne
Alessandro Naddeo
Jun Zhou
Bing Feng
Carlo Bianca
Raúl Montoliu
German Martín
   Mendoza-Silva
Ximo Torres
Bulent Ayhan
Qiang Liu
Xinle Liu
Jair Ccrvantes
Junming Zhang
Linting Guan
Guodong Zhao
Tao Yang
Zhenmin Zhang
Tao Wu
Yong Chen
Chien-Lung Chan
Liang-Chih Yu
Chin-Sheng Yang
Chin-Yuan Fan
Si-Woong Lee
Zhiwei Ji
Ke Zeng
Filipe Saraiva
Mario Malcangi
Francesco Ferracuti
José Alfredo Costa
Jacob Schrum
Shangxuan Tian

Yitian Zhao
Yonghuai Liu
Julien Leroy
Pierre Marighetto
Jheng-Long Wu
Falcon Rafael
Haytham Fayek
Yong Xu
Kozou Abdellalı
Damiano Rossetti
Daniele Ferretti
Chengdong Li
Zhen Ni
Pavel Osinenko
Bakkiyaraj Ashok
Hector Menendez
Pandiri Venkatesh
Davendra Donald
Ming-Wei Li
Leo Chen
Xiaoli Wei
Chien-Yuan Lai
Yongquan Zhou
Ma Xiaotu
Yannan Bin
Wei-Chiang Hong
Seongho Kim
Xiang Zhang
Geethan Mendiz
Brendan Halloran
Kai Xu
Di Zhang
Sheng Ding
Jun Li
Lei Wang
Jin Gu
Wang, Xiaowo
Hao Lin
Yao Nie
Vincenzo Randazzo
Yujie Cai
Derek Wang
Gang Li
Fuhai Li
Junfeng Luo
Wei Lan

Chaowang Lan
Lasker Ershad Ali
Shuyi Zhang
Pengbo Bo
Abhijit Kundu
Jingkai Yang
Shamim Reza
Lee Kai Wah
Anass Nouri
Arridhana Ciptadi
Lun Li
Cheeyong Kim
Helen Hong
Guangchun Gao
Weilin Deng
Taeho Jo
Jiazhou Chen
Enhong Zhuo
Carlo Bianca
Gai-Ge Wang
Kwanggi Kim
Jiangling Song
Xiandong Xu
Guohui Zhang
Long Wen
Kunkun Peng
Qiqi Duan
Hui Wang
Shiying Sun
Yu Yongjia
Jianbo Fan
Lishuang Shen
Liming Xie
Geethan Mendiz
Ying Bi
Hong Wang
Juntao Liu
Yang Li
Enfeng Qi
Xiangyu Liu
Jia Liu
Hou Yingnan
Weihua Deng
Chuanfeng Li
Fuyi Li
Tomasz Talaska

Nathan Cannon
Bo Gao
Ting Yu
Kang Li
Li Zhang
Xingjia Lu
Qingfeng Li
Yong-Guk Kim
Kunikazu Kobayashi
Takashi Kuremoto
Shingo Mabu
Takaomi Hirata
Tuozhong Yao
Lvzhou Li
Fuchun Liu
Shenggen Zheng
Jingkai Yang
Chen Li
Sungshin Kim
Hansoo Lee
Jungwon Yu
Yongsheng Dong
Zhihua Cui
Duyu Liu
Timothy L. Bailey
Hongda Mao
Cai Xiao
Xin Bai
Qishui Zhong
Yansong Deng
Giansalvo Cirrincione
Shi Zhenwei
Shuhui Wang
Junning Gao
Ziye Wang
Hui Li
Chao Wu
Fuyuan Cao
Rui Hong
Wahyono Wahyono
Rafal Kozik
Yihua Zhou
Wenyan Wang
Juan Figueroa
Artem Lenskiy
Qiang Yan

Wenlong Xu
Yi Gu
Yi Kou
Chenhui Qiu
Tianyu Yang
Xiaoyi Yu
Rongjing Hu
Yuanyuan Liu
Zhang Haitao
Libing Shen
Dongliang Yu
Yanwu Zeng
Yuanyuan Wang
Giulia Russo
Emilio Mastriani
Di Tang
Yan Jiang
Hong-Guan Liu
Liyao Ma
Panpan Du
Hongguan Liu
Chenbin Liu
Dangdang Shao
Yunze Yang
Yang Chen
Yang Li
Heye Zhang
Austin Brockmeier
Hui Li
Xi Yang
Yan Wang
Lin Bai
Laksono Kurnianggoro
Ajmal Shahbaz
Yan Zhang
Xiaoyang Wang
Bo Wang
Sajjad Ahmed
Ke Zeng
Liang Mao
Yuan You
Qiuyang Liu
Shaojie He
Zehui Cao
Zhongpu Xia
Bin Ye

Zhi-Yu Shen
Alexander Filonenko
Zhenhu Liang
Jing Li
Yang Yang
Wenjun Xu
Yongjia Yu
Jyotsna Wassan
Vibha Patel
Saiful Islam
Ekram Khan

Angelo Ciaramella
Junyu Chen
Ziang Dong
Jingjing Fei
Meng Lei
Xuanfang Fei
Bing Zeng
Taifeng Li
Yan Wu
Paul Fergus
Jiulun Cai

Jingying Huang
Francesco Pappalardo
Xiaoguang Zhao
Qingfang Meng
Mohamed Alloghani
Ruihao Li
Zhengyu Yang
Xin Xie
Yifan Wu
Hong Zeng

Contents – Part II

Blind Source Separation

Intelligent Fault Diagnosis

Machine Learning

Knowledge Discovery and Data Mining

Gene Expression Array Analysis

Systems Biology

Modeling, Simulation, and Optimization of Biological Systems

Intelligent Computing in Computational Biology

Computational Genomics

Computational Proteomics

Gene Regulation Modeling and Analysis

SNPs and Haplotype Analysis

Protein-Protein Interaction Prediction

Protein Structure and Function Prediction

Next-Gen Sequencing and Metagenomics

Structure Prediction and Folding

Biomarker Discovery

Applications of Machine Learning Techniques to Computational Proteomics, Genomics, and Biological Sequence Analysis

Biomedical Image Analysis

Human-Machine Interaction: Shaping Tools Which Will Shape Us

Protein and Gene Bioinformatics: Analysis, Algorithms and Applications

Special Session on Computer Vision based Navigation

Neural Networks: Theory and Application

Pattern Recognition

Face Recognition via Domain Adaptation and Manifold Distance Metric Learning

Bo Li[1,2,3(✉)], Ping-Ping Zheng[1,2], Jin Liu[4], and Xiao-Long Zhang[1,2]

[1] School of Computer Science of Technology,
Wuhan University of Science of Technology, Wuhan 430065, Hubei, China
liberol@126.com
[2] Hubei Province Key Laboratory of Intelligent Information Processing
and Real-Time Industrial System, Wuhan 430065, Hubei, China
[3] School of Electronics and Information Engineering,
Tongji University, Shanghai 201804, China
[4] State Key Laboratory of Software Engineering, Computer School,
Wuhan University, Wuhan, China

Abstract. A novel approach for face recognition via domain adaptation and manifold distance metric learning is presented in this paper. Recently, unconstrained face recognition is becoming a research hot in computer vision. For the non-independent and identically distributed data set, the maximum mean discrepancy algorithm in domain adaption learning is used to represent the difference between the training set and the test set. At the same time, assume that the same type of face data are distributed on the same manifold and the different types of face data are distributed on different manifolds, the face image set is used to model multiple manifolds and the distance between affine hulls is used to represent the distance between manifolds. At last, a projection matrix will be explored by maximizing the distance between manifolds and minimizing the difference between the training set and test set. A large number of experimental results on different face data sets show the efficiency of the proposed method.

Keywords: Unconstrained face recognition · Maximum mean discrepancy · Domain adaptation · Affine hull · Manifold distance

1 Introduction

In daily life, people always identify a person by the facial information including the face images, which makes it possible to use computer for face recognition. Humans can recognize a person by their own cognitive ability when their age changes. But it is a very difficult task for computer to accurately distinguish different human faces. Over the past decades, face recognition has aroused great interests and has been widely used in many fields such as public security, information security, and human-computer interaction.

Face recognition is a hot topic in the field of machine learning. Traditional machine learning approaches assume that the training data and test data are subject to the same probability distribution, i.e. independent and identical distribution (IID) [1]. But this assumption is not always the case in real-world data. In fact, non IID data are widely

© Springer International Publishing AG 2017
D.-S. Huang et al. (Eds.): ICIC 2017, Part II, LNCS 10362, pp. 3–10, 2017.
DOI: 10.1007/978-3-319-63312-1_1

used in many applications. Recently, domain adaptation learning [2] is proposed to overcome the problem that target domain and source domain are not IID. Under such circumstances, knowledge learned from the source data can be well exploited to the target data by considering the distribution difference between them. Thus the distributions of training data and test data are not required to be the same in domain adaption learning. Because domain adaption uses data from different source areas to explore data in the target tasks, it can effectively solve the problem caused by the inconsistent distribution of training data and test data.

Until to now, a large number of face recognition approaches which based on single shot images have been proposed [3–5]. These approaches have already achieved satisfactory performance under controlled conditions. However, most of these approaches are not reliable in real-world applications. Face images in the real environment are affected by a lot of unconstrained conditions such as illumination, facial expression, posture, and shelter. These interference factors result in the more difference between individuals, which makes it more difficult for unconstrained face recognition. Compared with a single image, an image set is able to provide more useful information to cover the variations of a person's facial appearance. Hence, more discriminatory information can be learned from image set to model the facial appearances. It is the image sets with all kinds variations as illumination, pose, expression and occlusion etc. that are easily led to differences between the training set and the test set when applying them for face recognition. In other words, the training set and the test set may be non IID. Thus domain adaptation can be used to solve the problem. In this paper, we can use the maximum mean discrepancy (MMD) [6] to measure the different empirical distributions of the source domain and target domain. In order to solve the problem of low recognition accuracy caused by interference factors, we can also take advantage of the information of image set, where affine hull [7] can be introduced to build a multiply manifold distance metric model. The affine hull can well utilize the information existing in both the within-class data and the between-class data. On the basis of MMD and affine hull, a low dimensional projection space will be explored where the difference between training set and test set is minimum and the distance between the manifolds is maximum. So the face recognition accuracy of our approach will be improved.

2 The Model of DA-MDML

In this section, we present a domain adaptation and manifold distance metric learning (DA-MDML) method in the following.

In unconstrained face recognition, some labeled training images are defined as the source domain and test set is defined as the target domain where the labeled data is very limited. The distribution of the training samples and test samples are different, but we can reduce the difference between the training samples and test samples by a transformation.

Let $X_s = \left\{x_1^s, x_2^s, \cdots, x_n^s\right\}$ be the source set of n persons and $X_t = \left\{x_1^t, x_2^t, \cdots, x_m^t\right\}$ be the target set of m persons. We search for a projection space W in which the distributions of the source set and target set are as similar as possible. And the projection space W is a d-dimensional projection matrix. In particular, the MMD can be used to measure the distance between two distributions. The distance between the source domain and the target domain is expressed as:

$$D_d(W^T X_s, W^T X_t) = \left\| \frac{1}{n} \sum_{i=1}^{n} \Phi(W^T x_i^s) - \frac{1}{m} \sum_{j=1}^{m} \Phi(W^T x_j^t) \right\|_H \tag{1}$$

The unconstrained human face images are influenced by many interference factors and they have the characteristics of low dimensional manifold distribution. In order to reduce the influence of these factors, face images of one person can be taken to establish an affine hull based manifold model. Thus the distances between multiply manifolds can be transformed to the distances between different affine hulls.

Most existing works such as [9, 10] design the manifold-to-manifold distance metric as the dissimilarity between the most similar parts of two image sets. The distance between manifolds M_s^i and M_s^j is

$$D(M_s^i, M_s^j) = D(h_s^i, h_s^j) \tag{2}$$

By using affine hulls to characterize the multiple manifolds, thus the distance metric of multi-manifold can be represented as:

$$D_m^2(W^T X_s, W^T X_t) = \sum_{i,j} \left\| W^T(\varepsilon_i + \lambda_i U_i) - W^T(\varepsilon_j + \lambda_j U_j) \right\|^2 \tag{3}$$

In the proposed method, we naturally integrate MMD and manifold distance metric into one function as:

$$D^2 = \mu D_m^2(W^T X_s, W^T X_t) / (1 - \mu) D_d^2(W^T X_s, W^T X_t) \tag{4}$$

In order to minimize the distance between the source set and the target set and maximize the multi-manifolds distance of the constructed metric model, we need to find a projection space W using the following objective function:

$$W = \underset{W^T W - I}{\arg\max} \left\{ \frac{\mu \sum_{i,j} \left\| W^T(\varepsilon_i + \lambda_i U_i) - W^T(\varepsilon_j + \lambda_j U_j) \right\|_H^2}{(1-\mu) \left\| \frac{1}{n} \sum_{i=1}^{n} \Phi(W^T x_i^s) - \frac{1}{m} \sum_{j=1}^{m} \Phi(W^T x_j^t) \right\|_H^2} \right\} \tag{5}$$

where $\mu(0 < \mu < 1)$ is adjustable parameters. μ is used to quantify the contributions of the difference between the source set and the target set and the multi-manifolds distance to the low dimensional projection space.

In order to find the projection space W for the constructed model, we need to solve the above objective function. The expression of MMD is:

$$D^2(W^T X_s, W^T X_t) = \left\| \frac{1}{n} \sum_{i=1}^{n} \Phi(W^T x_i^s) - \frac{1}{m} \sum_{j=1}^{m} \Phi(W^T x_j^t) \right\|_H^2 = \frac{1}{n^2} \sum_{i,j=1}^{n} \exp\left(-\frac{(x_i^s - x_j^s)^T W W^T (x_i^s - x_j^s)}{\sigma} \right)$$
$$+ \frac{1}{m^2} \sum_{i,j=1}^{m} \exp\left(-\frac{(x_i^t - x_j^t)^T W W^T (x_i^t - x_j^t)}{\sigma} \right) - \frac{2}{mn} \sum_{i,j=1}^{n,m} \exp\left(-\frac{(x_i^s - x_j^t)^T W W^T (x_i^s - x_j^t)}{\sigma} \right) \tag{6}$$

Equation (6) can also be computed efficiently in matrix form as:

$$D^2\left(W^T X_s, W^T X_t \right) = Tr(K_W L) \tag{7}$$

where $K_W = \begin{bmatrix} K_{s,s} & K_{s,t} \\ K_{t,s} & K_{t,t} \end{bmatrix} \in R^{(n+m) \times (n+m)}$ and

$$L(i,j) = \begin{cases} 1/n^2, \text{if } x_i, x_j \in X_s \\ 1/m^2, \text{if } x_i, x_j \in X_t \\ -1/mn, \text{otherwise} \end{cases}, \; L(i,j) \in R^{(n+m) \times (n+m)}$$

Researches point out that Gauss kernel can provide an effective RKHS embeddings, which makes the consistency metric of the different areas distribution distance to be easily implemented [12, 13]. In this paper, the Gauss kernel function is used in the reproducing kernel function of Hilbert space mapping. The Gauss kernel function can be expressed as:

$$k_\sigma(x, z) = \exp\left(-\frac{1}{2\sigma^2} \|x - z\|^2 \right) \tag{8}$$

where $x, z \in X$ and σ refers to the bandwidth that needs to be obtained through experiments.

In the projection space W, the affine hull based multi-manifolds distance metric is stated as Eq. (3), which can also be transformed to:

$$D_M^2(W^T X_s, W^T X_t) = (W^T D_{MD})(W^T D_{MD})^T = W^T D_{MD} D_{MD}^T W = W^T S_b W \tag{9}$$

where D_{MD} is the distance matrix between affine hulls, $S_b = D_{MD} D_{MD}^T$.

In sum, the original objective function can also be expressed as:

$$W = \underset{W^T W - I}{\arg \max} \left\{ \mu W^T S_b W / (1 - \mu)(K_w L) \right\} \tag{10}$$

According to the local fisher discriminant analysis, Eq. (10) can be converted to the problem of solving the following generalized eigen-decomposition.

$$S_b \beta = \lambda \frac{1 - \mu}{\mu} S_d \beta \tag{11}$$

where $S_d = \frac{\partial(K_w L)}{\partial W}$ is the metric matrix between source set and the target set.

Solving Eq. (11), a set of eigenvalues $\lambda_1, \lambda_2, \cdots, \lambda_d$ and the corresponding eigenvectors $\beta_1, \beta_2, \cdots, \beta_d$ can be obtained. Sorting the eigenvalues as descending order $\lambda_1 > \lambda_2 > \cdots > \lambda_d$, thus the eigenvectors associated to the sorting eigen values will be composed of the low dimensional projection subspace as $W = (\beta_1, \beta_2, \cdots, \beta_d)$.

The original high-dimensional face data is projected into the low dimensional projection space W, where the discriminant features can be extracted. At last, KNN classifier is used to identify the unknown face image in low-dimensional space W.

3   Experimental Results and Conclusion

In order to validate the efficiency of the proposed approach, we evaluate the approach by carrying out a number of face recognition experiments on five widely-used data sets. Details of the experiments and results are described below in Table 1. For all these data sets, the data class labels in source domain are known. Data in the source domain are used as training set, and those not labeled data in target domain are used as the test set. Table 2 shows the five data sets in details and the division of test set and training set in the experiment. In order to ensure the test results are objective, the data set is divided by the approach of random partition method.

Table 1. Data sets for face recognition

Data set	Samples	Train/test	Dimensions	Classes
PIE	170	85/85	1024	68
ORL	10	5/5	4096	40
AR	14	7/7	2000	120
AR_60 × 60	26	13/13	3600	119
AR_60 × 60_glasses	20	10/10	3600	119

We compare DM-MDML with other related approaches. These approaches are LDA, R_LDA [14], UDP [15] and DIP [16].

According to the characteristics of this approach, we need to set two parameters. There is an unknown parameter that is the width of the Gauss kernel function, i.e. σ. We use gradient descent algorithm to determine the parameter σ in Gauss kernel function [17], which are stated in Table 2.

Table 2. The value of σ^2 in the Gauss kernel function

Data set	Value of σ^2
PIE	3150
ORL	350
AR	400
AR_60 × 60_glasses	600
AR_60 × 60	3600/3650

There is another parameter in the affine hull. In Eq. (3), λ_i is an unknown parameter related to x_i. We use the Euclidean distance of x_i to μ_i to determine the λ_i. And then normalized the value of λ_i by making $\sum_i \lambda_i = 1$ and $\lambda_i \in (0,1)$.

After setting the parameters, the proposed method are carried out to extract features. At last, KNN classifier is used to recognize the unknown images. The experimental results on above mentioned face sets are reported as follows.

Conclusions can be drawn from Table 3. In data sets as PIE and ORL, the recognition accuracy of the proposed DA-MDML is significantly improved compared with LDA and UDP. In AR data set, the recognition accuracy of DA-MDML has a slighter advantage than R_LDA. Comparison with experimental results on other data sets, DA-MDML has obvious advantages over R_LDA.

Table 3. Recognition accuracy of different methods on different data sets

Data set	Approach			
	LDA	UDP	R_LDA	DA-MDML
PIE	0.9510	0.9486	0.9477	0.9688
ORL	0.9750	0.9250	0.9550	0.9875
AR	0.9631	0.9607	0.9893	0.9899

We can draw the following conclusions from Table 4: With the increasing of the interference factors, the correct accuracy of face recognition is declining. Moreover, the recognition accuracy of DA-MDML is significantly improved compared with UDP and LDA. However, compared with R_LDA, the recognition accuracy of DA-MDML weakly beats the recognition rate ofR_LDA.

Table 4. Recognition accuracy of different methods on AR data subsets

Data set	Approach			
	LDA	UDP	R_LDA	DA-MDML
AR	0.9631	0.9607	0.9893	0.9899
AR_60 × 60_glasses	0.9538	0.9697	0.9807	0.9882
AR_60 × 60	0.9179	0.9573	0.9683	0.9703

The experiments are shown in Table 5 by using two domain adaption methods to these face data sets. On the one hand, the recognition accuracy of DA-MDML is higher than DIP. On the other hand, by comparing the test results of the three subsets of AR data set, we can find that the gap of recognition accuracy between DA-MDML and DIP becomes larger when the interference factors are more. It can be seen that the proposed approach is effective.

Table 5. Recognition accuracy of different domain adaption learning in different data sets

Data set	Approach	
	DIP	DA-MDML
PIE	0.9506	0.9688
ORL	0.9650	0.9875
AR	0.9726	0.9899
AR_60 × 60_glasses	0.9489	0.9882
AR_60 × 60	0.9276	0.9703

4 Conclusion

In unconstrained face recognition, in order to find the most important elements and structures of face data effectively, we remove the influence of interference factors, and then reveal the simple structure behind the complex data. The original redundant data set with a large number of interference factors are projected into a subspace which can retain the important characteristic data, and then the face recognition is carried out. The traditional approach does not take into account the problem of non IID and more interference factors. In this paper, the DA-MDML is proposed to solve these problems. We use the MMD approach to calculate the distribution distance between the training set and test set, and introduce the affine hull-based manifold distance metric to calculate the distance between classes. Our approach retains the more useful feature to judge the category of unconstrained face image. The feasibility and stability of this approach can be proved by the experimental results.

Acknowledgments. This work was supported in part by the grants of Natural Science Foundation of China (61572381, 61273303, 61472280 and 61602349) and Post-doctoral Science Foundation of China (2016M601646).

References

1. Pan, S.J., Yang, Q.: A survey on transfer learning. IEEE Trans. Knowl. Data Eng. **22**(10), 1345–1359 (2010)
2. Pan, S.J., Tsang, I.W., Kwok, J.T., Yang, Q.: Domain adaptation via transfer component analysis. IEEE Trans. Neural Netw. **22**(2), 199–210 (2011)
3. He, X., Yan, S., Hu, Y., Niyogi, P., Zhang, H.J.: Face recognition using laplacian faces. IEEE Trans. PAMI **27**, 328–340 (2005)
4. Belkin, M., Niyogi, P.: Laplacian eigen maps and spectral techniques for embedding and clustering. In: International Conference on Neural Information Processing Systems: Natural and Synthetic, vol. 14, pp. 585–591. MIT Press (2001)
5. Yan, S., Xu, D., Zhang, B., Zhang, H.J., Yang, Q., Lin, S.: Graph embedding and extension: a general framework for dimensionality reduction. IEEE Trans. Pattern Anal. Mach. Intell. **29**(1), 40–51 (2007)
6. Gretton, A., Borgwardt, K.M., Rasch, M.J., Schölkopf, B., Smola, A.: A kernel two-sample test. J. Mach. Learn. Res. **13**(1), 723–773 (2012)

7. Huang, L., Lu, J., Tan, Y.P.: Multi-manifold metric learning for face recognition based on image sets. J. Vis. Commun. Image Represent. **25**(7), 1774–1783 (2014)
8. Kim, T.K., Kittler, J., Cipolla, R.: Discriminative learning and recognition of image set classes using canonical correlations. IEEE Trans. Pattern Anal. Mach. Intell. **29**(6), 1005–1018 (2007)
9. Wang, R., Shan, S., Chen, X., and Gao, W.: Manifold-manifold distance with application to face recognition based on image set, pp. 1–8 (2008)
10. Pan, J., Manocha, D.: Bi-level locality sensitive hashing for k-nearest neighbor computation. In: International Conference on Data Engineering, vol. 41, no. 4, pp. 378–389 (2012)
11. Steinwart, I.: On the influence of the kernel on the consistency of support vector machines. J. Mach. Learn. Res. **2**(1), 67–93 (2002)
12. Sriperumbudur, B.K., Gretton, A., Fukumizu, K., Lkopf, B., Lanckriet, G.R.G.: Hilbert space embeddings and metrics on probability measures. J. Mach. Learn. Res. **11**, 1517–1561 (2009)
13. Balakrishnama, S., Ganapathiraju, A.: Linear discriminate analysis. Institute for Signal and Information Processing, Mississippi State University (1998)
14. Lu, J., Plataniotis, K.N., Venetsanopoulos, A.N.: Regularization studies of linear discriminant analysis in small sample size scenarios with application to face recognition. Pattern Recogn. Lett. **26**(2), 181–191 (2005)
15. Wang, F., Zhao, B., Zhang, C.: Unsupervised large margin discriminative projection. IEEE Trans. Neural Netw. **22**(9), 1446–1456 (2011)
16. Baktashmotlagh, M., Harandi, M.T., Lovell, B.C., Salzmann, M.: Unsupervised domain adaptation by domain invariant projection. In: IEEE International Conference on Computer Vision, pp. 769–776. IEEE (2013)
17. Berry, R.A., Gallager, R.G.: Communication over fading channels with delay constraints. IEEE Trans. Inf. Theory **48**(5), 1135–1149 (2002)

Image Processing

Generalized Cubic Hermite Interpolation Based on Perturbed *Padé* Approximation

Le Zou[1,2,3], Liang-Tu Song[1,2], Xiao-Feng Wang[3(✉)],
and Yan-Ping Chen[3]

[1] Institute of Intelligent Machines, Hefei Institutes of Physical Science,
Chinese Academy of Sciences, 1130, Hefei 230031, China
[2] University of Science and Technology of China, Hefei 230027, China
[3] Key Lab of Network and Intelligent Information Processing,
Department of Computer Science and Technology,
Hefei University, Hefei 230601, China
xfwang@iim.ac.cn

Abstract. Generalized cubic Hermite interpolation was constructed by using perturbed *Padé* approximation in this paper. We generalize our method to the $2n + 1$ times Hermite interpolation of $n + 1$ points and study its barycentric form. Numerical example is given to show the effectiveness of our method. Finally, we further generalize the proposed method to generalized cubic Hermite interpolation based on perturbed Chebyshev-*Padé* approximation.

Keywords: Perturbed *Padé* approximation · Chebyshev-*Padé* approximation · Cubic hermite interpolation · Barycentric form

1 Introduction

The interpolation method plays an important pole in numerical analysis. The interpolation have been widely applied to the numerical approximation, and graphics image processing, image processing. Some scholars have been studying their application in image processing, numerical integration, engineering technology, curves/surface construction. The Hermite interpolation problem is a special polynomial interpolation, and has been widely studied. It is well known that the Hermite interpolation in the Chebychev nodes succeeds where Lagrange interpolation failed, that is, for every continuous function the Hermite interpolation polynomials in the Chebychev nodes converge to the function [1]. The classical formula given by Hermite interpolation takes prefixed values as well as its consecutive derivatives at some fixed points. Several researchers have studied this problem and have obtained interesting results concerning the convergence to the interpolant function [1, 2]. Corless studied Hermite interpolant occur naturally in the context of the numerical solution of initial value problems for ordinary differential equations [2]. Xie studied rational cubic Hermite interpolation spline and its approximation properties [3]. *Padé* approximation is an effective tool for rational approximation. When the function has a convergent power series in the interval [0, 1], Khodier proposed perturbation *Padé* approximation method, the accuracy of the introduced approximation increases as the order increases [4]. Based on *Padé* approximation and Lagrange

© Springer International Publishing AG 2017
D.-S. Huang et al. (Eds.): ICIC 2017, Part II, LNCS 10362, pp. 13–21, 2017.
DOI: 10.1007/978-3-319-63312-1_2

interpolation, Zhao studied generalize blending rational interpolation [5]. Zhao also constructed the barycentric rational Hermite interpolant [6]. Zou studied generalize barycentric Lagrange rational interpolation [7]. Then, Zhang studied the general frames of barycentric blending rational interpolation [8].

Our contribution in this paper is to obtain a new type of cubic Hermite interpolation which combine cubic Hermite interpolation with perturbation *Padé* approximation, when the function has a convergent power series in the interval [0, 1]. The organization of the paper is as follows. In Sect. 2, we solve generalized cubic Hermite interpolation on equally spaced nodes based on perturbed *Padé* approximation. As an application of the preceding results, we give examples to show the effectiveness of our algorithm. In Sect. 3, we generalize our method to solve the $2n + 1$ times Hermite interpolation problem with $n + 1$ points and give its barycentric case. In Sect. 4 we point out that the generalize cubic Hermite interpolation and the case of $2n + 1$ times cubic Hermite interpolation of $n + 1$ points based on perturbed *Padé* approximation can be generalized to construct generalize cubic Hermite interpolation based on Chebyshev-*Padé* approximation.

2 Generalized Cubic Hermite Interpolation Based on Perturbed *Padé* Approximation

2.1 The Construction of Generalized Cubic Hermite Interpolation Based on Perturbed Padé Approximation

In this section we propose and solve generalized cubic Hermite interpolation based on perturbation *Padé* approximation.

Assume that $f(x)$ is a function which has the convergent power series form at $x = x_k$,

$$f(x) = \sum_{i=0}^{\infty} c_i^{(k)} (x - x_k) \quad x \in [0, 1], c_0^{(k)} \neq 0, (k = 0, 1, \cdots, n). \tag{1}$$

The (m, n) type perturbed *Padé* approximation $R_{m,n}^{(k)}(x)$ of the function $f(x)$ at $x = x_k$ is a rational function of the form [8]

$$R_{m,n}^{(k)}(x) = \frac{a_0 + a_1 x + a_2 x^2 + \cdots + a_m x^m}{b_0 + b_1 x + b_2 x^2 + \cdots + b_n x^n} \tag{2}$$

where a_i, b_j can be defined with the method in the paper [8]. Without loss of generality, we restrict our description to $(1, 1)$ type perturbed *Padé* approximation.

We can get $(1, 1)$ type perturbed *Padé* approximation from [4]

$$R_{1,1}^{(k)}(x) = \frac{c_0^{(k)} + [c_1^{(k)} + c_0^{(k)}(\varepsilon^{(k)} - c_2^{(k)})/c_1^{(k)}](x - x_k)}{1 + [(\varepsilon^{(k)} - c_2^{(k)})/c_1^{(k)}](x - x_k)}. \tag{3}$$

where $\varepsilon^{(k)}$ are the perturbation parameters and

$$\varepsilon^{(k)} = \frac{(x - x_k)[c_n^{(k)} c_2^{(k)} (x - x_k)^{n-2} + \sum_{i=0}^{n-3} (c_{i+2}^{(k)} c_2^{(k)} - c_{i+3}^{(k)} c_1^{(k)})(x - x_k)^i]}{\sum_{i=0}^{n-1} c_{i+1}^{(k)} (x - x_k)^i}. \tag{4}$$

Note that if we put $\varepsilon^{(k)} = 0$ we get the classical *Padé* approximation $R_{1,1}(x)$ Then $R_{1,1}^{(k)}(x)$ agrees with the power series of $f(x)$ to a certain order n for all $x \in [0, 1]$ with the perturbation parameter $\varepsilon^{(k)}$.

Assume that $f(x)$ is a function which has the convergent power series form at $x = x_k$.

Lemma 1. If $f(x)$ has the convergent power series form (1) and the perturbed parameters $\varepsilon^{(k)}$ are determined from (4), then the truncation error induced by using the perturbed *Padé* approximation $R_{1,1}^{(k)}(x)$ is of order n + 1.

Now we will get the perturbed *Padé* approximation $R_{1,1}^{(k)}(x)$ of derivatives. Differentiating the Eq. (1), we can get

$$f'(x) = \sum_{j=0}^{\infty} \frac{(1+j)!}{j!} c_{1+j}^{(k)} (x - x_k)^j = \sum_{j=0}^{\infty} \tilde{c}_j^{(k)} (x - x_k)^j. \tag{5}$$

where $\tilde{c}_j^{(k)} = \frac{(1+j)!}{j!} c_{1+j}^{(k)}$.

From [4] we can get perturbed *Padé* approximation of $f'(x)$ at point $x = x_k$

$$\tilde{R}_{1,1}^{(k)}(x) = \frac{\tilde{c}_0^{(k)} + [\tilde{c}_1^{(k)} + \tilde{c}_0^{(k)} (\tilde{\varepsilon}^{(k)} - \tilde{c}_2^{(k)}) / \tilde{c}_1^{(k)}](x - x_k)}{1 + [(\tilde{\varepsilon}^{(k)} - \tilde{c}_2^{(k)}) / \tilde{c}_1^{(k)}](x - x_k)}. \tag{6}$$

Given the points x_0, x_1, and its function $f^{(s)}(x_i) = f_i^{(s)}, s = 0, 1; i = 0, 1$. We can construct a generalized cubic Hermite interpolation based on perturbed *Padé* approximation

$$\bar{H}_3(x) = R_{1,1}^{(0)}(x)(1 - 2\frac{x - x_0}{x_0 - x_1})\left(\frac{x - x_1}{x_0 - x_1}\right)^2 + R_{1,1}^{(1)}(x)(1 - 2\frac{x - x_1}{x_1 - x_0})\left(\frac{x - x_0}{x_1 - x_0}\right)^2$$
$$+ \tilde{R}_{1,1}^{(0)}(x)(x - x_0)\left(\frac{x - x_1}{x_0 - x_1}\right)^2 + \tilde{R}_{1,1}^{(1)}(x)(x - x_1)\left(\frac{x - x_0}{x_1 - x_0}\right)^2. \tag{7}$$

where $R_{1,1}^{(0)}(x)$ and $\tilde{R}_{1,1}^{(0)}(x)$ are the $(1, 1)$ type perturbed *Padé* approximation of $f(x)$ at x_0 and the $(1, 1)$ type perturbed *Padé* approximation of $f'(x)$ at x_0. $R_{1,1}^{(1)}(x)$ and $\tilde{R}_{1,1}^{(1)}(x)$ are the $(1, 1)$ type perturbed *Padé* approximation of $f(x)$ at x_1 and the $(1, 1)$ type perturbed *Padé* approximation of $f'(x)$ at x_1 respectively.

From [4], we know that

$$R_{1,1}^{(0)}(x_0) = f(x_0),\tag{8}$$

$$\tilde{R}_{1,1}^{(0)\prime}(x_0) = f'(x_0),\tag{9}$$

$$R_{1,1}^{(1)}(x_1) = f(x_1),\tag{10}$$

$$\tilde{R}_{1,1}^{(1)\prime}(x_1) = f'(x_1),\tag{11}$$

So we have

$$\bar{H}_3(x_i) = f(x_i), H_3'(x_i) = f'(x_i), i = 0, 1.\tag{12}$$

That is to say, generalized cubic Hermite interpolation based on perturbed *Padé* approximation satisfies interpolation conditions.

We can select the perturbed parameters $\varepsilon^{(k)}, \tilde{\varepsilon}^{(k)}(k = 0, 1)$ which are determined from (4), so that the truncation error induced by using the perturbed *Padé* approximation $R_{1,1}^{(k)}(x)$, $\tilde{R}_{1,1}^{(k)}(x)$ is of order n + 1.

2.2   Numerical Example

In this Section, we will give two examples to show the effectiveness of our method.

Example 1. Given $f(x) = e^x$, $x_0 = 0, x_1 = 1$, then we can get the former power series of the function $f(x)$ at point $x = x_0, x = x_1$:

$$f(x) = 1 + x + \frac{1}{2}x^2 + \frac{1}{6}x^3 + \frac{1}{24}x^4 + \frac{1}{120}x^5 + \cdots,$$

$$f(x) = e + e(x - 1) + \frac{e}{2}(x - 1)^2 + \frac{e}{6}(x - 1)^3 + \frac{e}{24}(x - 1)^4 + \cdots.$$

From Eq. (3), we can get [9]

$$R_{1,1}^{(0)}(x) = \frac{1 + (\varepsilon^{(0)} + 0.5)x}{1 + (\varepsilon^{(0)} - 0.5)x}$$

$$R_{1,1}^{(1)}(x) = \frac{e + (e + \varepsilon^{(1)} - e/2)(x - 1)}{1 + (\varepsilon^{(1)} - e)/e(x - 1)}$$

where $\varepsilon^{(k)}(k = 0, 1)$ are the perturbed parameters given in Eq. (4)

We can get the derivative of $f(x)$

$$f'(x) = 1 + x + \frac{1}{2}x^2 + \frac{1}{6}x^3 + \frac{1}{24}x^4 + \frac{1}{120}x^5 + \cdots,$$

and

$$f'(x) = e + e(x-1) + \frac{e}{2}(x-1)^2 + \frac{e}{6}(x-1)^3 + \frac{e}{24}(x-1)^4 + \cdots.$$

So we can get

$$\tilde{R}_{1,1}^{(0)}(x) = \frac{1 + (\tilde{\varepsilon}^{(0)} + 0.5)x}{1 + (\tilde{\varepsilon}^{(0)} - 0.5)x}$$

$$\tilde{R}_{1,1}^{(1)}(x) = \frac{e + (e + \tilde{\varepsilon}^{(1)} - e/2)(x-1)}{1 + (\tilde{\varepsilon}^{(1)} - e)/e(x-1)}$$

where $\tilde{\varepsilon}^{(k)}(k = 0, 1)$ are the perturbed parameters given in Eq. (4)

By using the constructed method in Sect. 2.1, we can get generalized cubic Hermite interpolation based on perturbed *Padé* approximation

$$\bar{H}_3(x) = R_{1,1}^{(0)}(x)(1+2x)(x-1)^2 + R_{1,1}^{(1)}(x)(2(x-1))x^2 +$$
$$\tilde{R}_{1,1}^{(0)}(x)x(x-1)^2 + \tilde{R}_{1,1}^{(1)}(x)(x-1)x^2$$

It is easy to verify that

$$\bar{H}_3(x) = f(x)$$

and

$$H_3(x) = f'(x).$$

Example 2. Given $f(x) = x^2 + \sin x$, $x_0 = 0, x_1 = \pi/4$, then we can get the former power series of the function $f(x)$ at point $x = x_0, x = x_1$:

$$f(x) = x + x^2 - \frac{1}{3!}x^3 + \frac{1}{5!}x^5 + \cdots$$

$$f(x) = \frac{\sqrt{2}}{2} + \frac{\pi^2}{16} + (\frac{\sqrt{2}}{2} + \frac{\pi}{2})(x - \frac{\pi}{4}) + (\frac{\sqrt{2}}{4} - 1)(x - \frac{\pi}{4})^2 + \frac{\sqrt{2}}{12}(x - \frac{\pi}{2})^3 + \cdots.$$

From Eq. (3), we can get [9]

$$R_{1,1}^{(0)}(x) = \frac{1}{1 + (\varepsilon^{(0)} - 1)x}$$

$$R_{1,1}^{(1)}(x) = \frac{\frac{\sqrt{2}}{2} + \frac{\pi^2}{16} + [\frac{\sqrt{2}}{2} + \frac{\pi}{2} + (\frac{\sqrt{2}}{2} + \frac{\pi^2}{16})(\varepsilon^{(1)} - (\frac{\sqrt{2}}{4} - 1))/(\frac{\sqrt{2}}{2} + \frac{\pi}{2})](x - \frac{\pi}{4})}{1 + (\varepsilon^{(1)} - (\frac{\sqrt{2}}{4} - 1))/(\frac{\sqrt{2}}{2} + \frac{\pi}{2})(x - \frac{\pi}{4})}$$

where $\varepsilon^{(k)}(k = 0, 1)$ is the perturbed parameter given in Eq. (4)

The absolute maximum error obtained by using the perturbed *Padé* approximation with $n = 3$ is 81E-4 instead of 16E-2 for classical *Padé* approximation. Absolute maximum errors obtained for other values of n at point $x = x_0$ are given in Table 1.

Table 1. Absolute maximum error of perturbed *Padé* approximation

n	Absolute maximum error	
	Example 1	Example 2
5	16E-04	20E-05
7	28E-06	27E-07
9	30E-08	25E-09
11	23E-10	16E-11
13	12E-12	76E-14
15	51E-16	28E-16

We can get the derivative of $f(x)$,

$$f'(x) = 1 + 2x - \frac{1}{2}x^2 + \frac{1}{24}x^4 + \cdots,$$

and

$$f'(x) = \frac{\pi}{2} + \frac{\sqrt{2}}{2} + (2 - \frac{\sqrt{2}}{2})(x - \frac{\pi}{4}) - \frac{\sqrt{2}}{4}(x - \frac{\pi}{4})^2 + \frac{\sqrt{2}}{12}(x - \frac{\pi}{4})^3 + \cdots$$

So we can get

$$\tilde{R}_{1,1}^{(0)}(x) = \frac{1 + (\frac{\tilde{\varepsilon}^{(0)}}{2} + \frac{9}{4})x}{1 + (\frac{\tilde{\varepsilon}^{(0)}}{2} + \frac{1}{4})x}$$

$$\tilde{R}_{1,1}^{(1)}(x) = \frac{\frac{\pi + \sqrt{2}}{2} + [\frac{2 - \sqrt{2}}{2} + (\frac{\pi + \sqrt{2}}{2} + \tilde{\varepsilon}^{(1)} + \frac{\sqrt{2}}{4})/\frac{2 - \sqrt{2}}{2}](x - \frac{\pi}{4})}{1 + (\tilde{\varepsilon}^{(1)} + \frac{\sqrt{2}}{4})/\frac{2 - \sqrt{2}}{2}(x - \frac{\pi}{4})}$$

where $\tilde{\varepsilon}^{(k)} (k = 0, 1)$ are the perturbed parameters given in Eq. (4)

By using the proposed method in Sect. 2.1, we can get generalized cubic Hermite interpolation based on perturbed *Padé* approximation

$$\bar{H}_3(x) = R_{1,1}^{(0)}(x)(1 + \frac{8x}{\pi})(\frac{4x - \pi}{\pi})^2 + R_{1,1}^{(1)}(x)(1 - \frac{8x - 2\pi}{\pi})(\frac{4x}{\pi})^2 +$$

$$\tilde{R}_{1,1}^{(0)}(x)x(\frac{4x - \pi}{\pi})^2 + \tilde{R}_{1,1}^{(1)}(x)(x - \frac{\pi}{4})(\frac{4x}{\pi})^2$$

It is easy to verify that

$$\bar{H}_3(x) = f(x)$$

and

$$\bar{H}_3(x) = f'(x).$$

As can be seen from Table 1, we can select the perturbed parameters $\varepsilon^{(k)}, \tilde{\varepsilon}^{(k)} (k = 0, 1)$ which are determined from (4), so that the truncation error induced by using the perturbed *Padé* approximation $R_{1,1}^{(k)}(x)$ and $\tilde{R}_{1,1}^{(k)}(x)$ is of order $n + 1$.

From the above numerical examples, we can see our method gives higher accuracy. Generally, image interpolation can be widely used in digital image processing field, it is important to construct an efficient interpolation function to deal with image processing. Because image interpolation is sensitive to the image contours or edges, which are the high frequency information regions, interpolation kernel function should be more accurately approximate to the ideal interpolation (the Sinc function) in space domain and frequency domain. Hence the rational interpolating kernel is a good choice to process the image edge and texture regions. The Hermite rational interpolation can better preserve the geometric regularity around the color edges and thus generate interpolant images with higher visual quality.

3 $2n + 1$ Times Hermite Interpolation of $n + 1$ Point Based on Perturbed *Padé* Approximation

3.1   Construction of $2n + 1$ Times Hermite Interpolation of $n + 1$ Point Based on Perturbed *Padé* Approximation

Further, we can extend our method to $2n + 1$ times Hermite interpolation of $n + 1$ point based on perturbed *Padé* approximation.

Given $x_0 < x_1 < \cdots < x_n, f^{(s)}(x_i) = f_i^{(s)}$, $s = 0, 1; i =, 1, 2, \cdots, n$, we can get the formula from [9]

$$\tilde{H}(x) = \sum_{k=0}^{n} (R_{1,1}^{(k)}(x)(1 - 2(x - x_k)l_k'(x))l_k^2(x)$$
$$+ \sum_{k=0}^{n} \tilde{R}_{1,1}^{(k)}(x)(x - x_k)l_k^2(x)). \tag{13}$$

where $l_k(x)$ is the Lagrange basic function based on $x_i(i = 0, 1, 2, \cdots, n)$. $R_{1,1}^{(k)}(x)$ is perturbed *Padé* approximation of $f(x)$ at point $x = x_k$. $\tilde{R}_{1,1}^{(k)}(x)$ is perturbed *Padé* approximation of $f'(x)$ at point $x = x_k$.

Similar to the method in Sect. 2, we can get

$$\tilde{H}(x_i) = f(x_i), \tilde{H}'(x_i) = f'(x_i), i = 0, 1, 2, \cdots, n. \tag{14}$$

3.2   The Barycentric Form

Generally, the barycentric rational interpolations have more accuracy than the polynomial interpolation in computation, and the barycentric rational interpolation has more advantage than polynomial interpolation, for example, easy to calculate, the information concerning the existence and location of poles of the interpolation detecting the unattainable points of the interpolation, good numerical stability.

Then, the following Hermite interpolating polynomial [10, 11] can be obtained.

$$P(x) = l^2(x) \sum_{k=0}^{n} \frac{\omega_k^2}{x - x_k} \left(\left(\frac{1}{x - x_k} - \mu_k \omega_k \right) f_k + f'_k \right). \tag{15}$$

With $\omega_k = (l'(x_k))^{-1}, \mu_k = l''(x_k)$, we can get the barycentric version of (15)

$$P(x) = \frac{\sum_{k=0}^{n} \frac{\omega_k^2}{x - x_k} \left(\left(\frac{1}{x - x_k} - \mu_k \omega_k \right) f_k + f'_k \right)}{\sum_{k=0}^{n} \frac{\omega_k^2}{x - x_k} \left(\frac{1}{x - x_k} - \mu_k \omega_k \right)}. \tag{16}$$

Then, we can get $2n + 1$ times barycentric Hermite interpolation of $n + 1$ point based on perturbed *Padé* approximation

$$P(x) = \frac{\sum_{k=0}^{n} \frac{\omega_k^2}{x - x_k} \left(\left(\frac{1}{x - x_k} - \mu_k \omega_k \right) R_{1,1}^{(k)}(x) + \tilde{R}_{1,1}^{(k)}(x) \right)}{\sum_{k=0}^{n} \frac{\omega_k^2}{x - x_k} \left(\frac{1}{x - x_k} - \mu_k \omega_k \right)} \tag{17}$$

Similarly, we can select the perturbed parameters $\varepsilon^{(k)}, \tilde{\varepsilon}^{(k)} (k = 0, 1)$ which are determined from (4), so that the truncation error induced by using the perturbed *Padé* approximation $R_{1,1}^{(k)}(x)$, $\tilde{R}_{1,1}^{(k)}(x)$ is of order n + 1.

4   Conclusion

In this paper, we propose a method for computing the generalized cubic Hermite interpolation based on perturbed *Padé* approximation. We generalize the method to $2n + 1$ times Hermite interpolation of $n + 1$ points based on perturbed *Padé* approximation and its barycentric form.

We can also construct new kinds of generalized cubic Hermite blending rational interpolation based on *Padé* approximation and *Padé*-type approximation with the proposed method. If the Chebyshev series of the given function is given, we can construct the perturbation of Chebyshev-*Padé* approximation [12, 13]. By using Chebyshev-*Padé* instead of the perturbed *Padé* approximation, we can construct a new generalized cubic Hermite blending rational interpolation and $2n + 1$ times Hermite interpolation of $n + 1$ point and its barycentric case.

The Hermite rational interpolation can better preserve the geometric regularity around the color edges and thus generate interpolated images with higher visual quality, so we will construct a novel image magnification scheme based on the proposed method in future.

Acknowledgements. This work is Supported by the grant of Anhui Provincial Natural Science Foundation, No.1508085QF116, the grant of the National Natural Science Foundation of China, Nos. 61672204, 61272024, the grant of Support Key Project for Excellent Young Talent in College of Anhui Province, No.gxyqZD2016269, the grant of Support Project for Excellent Young Talent in College of Anhui Province (X.F. Wang), the Science Research Major Foundation of Education Department of Anhui Province, Nos.KJ2015A206, KJ2016A603, Training Object for Academic Leader of Hefei University, No.2014dtr08, Key Constructive Discipline of Hefei University, No.2016xk05.

References

1. Berriochoa, E., Cachafeiro, A.: Algorithms for solving hermite interpolation problems using the fast fourier transform. J. Comput. Appl. Math. **235**, 882–894 (2010)
2. Corless, R.M., Shakoori, A., Aruliah, D., Gonzalez, Vega L.: Barycentric hermite interpolants for event location in initial-value problems. J. Numer. Anal. Ind. Appl. Math. **3**, 1–18 (2008)
3. Xie, J., Tan, J.Q., Li, S.F.: Rational cubic hermite interpolation spline and its approximation properties applications. Chin. J. Eng. Math. **28**, 385–392 (2011)
4. Khodier, A.M.M.: Perturbed *Padé* approximation with high accuracy. Appl. Math. Comput. **148**, 753–757 (2004)
5. Zhang, N., Zhao, Q.J.: Generalized lagrange blending rational interpolation based on *Padé* approximation. J. Anhui Univ. Sci. Tech. **30**, 68–72 (2010)
6. Zhao, Q.J., Hao, Y.P., Yin, Z.X., Zhang, Y.W.: Best barycentric rational hermite interpolation. In: 2010 International Conference on Intelligent System Design and Engineering Application, pp. 417–419 (2010)
7. Zou, L., Li, C.W.: Barycentric lagrange blending rational interpolation based on *padé* approximation. In: The 3rd International Conference on Computational and Information Sciences, pp. 1124–1127 (2011)
8. Zhang, Y.G.: General interpolation formulae for barycentric blending interpolation. Anal. Theory Appl. **32**(1), 52–64 (2016)
9. Khodier, A.M.M.: Perturbed *Padé* Approximation. Int. J. Comput. Math. **74**, 247–253 (2000)
10. Henrici, P.: Essentials of Numerical Analysis with Pocket Calculator Demonstrations. Wiley, New York (1982)
11. Schneider, C., Werner, W.: Hermite interpolation: the barycentric approach computing. Numer. Algorithms **46**, 35–51 (1991)
12. Khodier, A.M.M.: Perturbed chebyshev rational approximation. Intern. J. Comput. Math. **3** (80), 1199–1204 (2003)
13. Tan, J.Q., Wu, G.H.: Perturbed Chebyshev-*Padé* approximation. J. Hefei. Univ. Tech. **30**, 112–116 (2007)

Automatic License Plate Recognition Using Local Binary Pattern and Histogram Matching

Ashutosh Kumar Bachchan, Apurba Gorai, and Phalguni Gupta[(✉)]

National Institute of Technical Teachers' Training & Research, Kolkata 700106, India
ashutoshbachhan@gmail.com, apurbagorai@gmail.com,
phalgunigupta@nitttrkol.ac.in

Abstract. This paper proposes new real time license plate recognition (LPR) system that is capable of motion tracking and recognition of license plate. The best frame taken from the video has been chosen which is found to be about 4 m apart from camera position. For further processing, lower half section of vehicle image has been cropped of sized (450×140) while tracking. Local Binary Pattern (LBP) and histogram matching technique are used to detect license plate. Due to the robustness of LBP features, this method can adaptively deal with various changes such as rotation, scaling, and illumination in the license plate. Segmentation of the plate region into disjoint characters has been done with bounding box technique with some modifications. Recognition has been done by calculating histogram features. Minimum distance classifier has been used for features matching. The system is tested on more than 300 images and it gives 96.14% detection and 89.35% of recognition accuracy. This system is designed to recognize license plate of small, medium as well as large vehicles. It is also capable to detect single line and two line license plates format.

Keywords: Local binary pattern · License plate recognition · Optical character recognition

1 Introduction

The rapid increase of urban and national road networks over the last three to four decades emerges the need of efficient monitoring and management of road traffic. Information about vehicles in complex traffic conditions can be obtained by recognizing the numbers shown in the license plate of the vehicle. Automatic License Plate Recognition (ALPR) system is used to recognize the characters on license plate which has many applications such as violating traffic rules, empty space for car parking [7], collecting toll automatically [4], controlling security in restricted area [5], measurement of vehicle speed, search for stolen vehicle and cost estimation for traffic congestion.

Extensive studies have been done on license plate recognition. In [16, 17], edge/boundary information based license plate detection using Sobel operator has been used. Character segmentation is done by horizontal and vertical projection of binary image to get initial and final point. Template matching has been used for character recognition.

© Springer International Publishing AG 2017
D.-S. Huang et al. (Eds.): ICIC 2017, Part II, LNCS 10362, pp. 22–34, 2017.
DOI: 10.1007/978-3-319-63312-1_3

Du et al. has suggested feature extraction based License Plate detection and classifier like neural network, fuzzy base recognition system [6]. In [8], Gabor Transform technique has been used to detect the license plate. It provides rough idea about plate boundary. Further, binary split tree technique is used for vector quantization in order to character segmentation and OCR for character recognition. In [9], features like color, shape and size of the characters are extracted using Scale Invariant Feature Transform (SIFT) for license plate detection. It has applied windowing Hough transform to draw rectangle over each character for segmentation. Finally, optical character recognition (OCR) algorithm has been used for recognition to extract text data.

Stroke Width Transform (SWT) which calculates stroke width of each pixel to locate plate area, has been used in [10]. Further, Canny edge detection has been performed to highlight the edges. The pixels width equals stroke width combine as license plate character. They have used connected component analysis (CCA) [21] and neural network classifier for segmentation and recognition respectively [10]. In [11], an ALPR system has been discussed for Bangladeshi license plate with chain code and neural network.

In [12], low-computation advanced LBP (ALBP) operator has been proposed for feature extraction, and applied for the recognition of Chinese license plate. As local feature extraction operator, ALBP produces a feature excursion for characters of barycenter departure. To recognize characters of barycenter, Gabor filters have been used. In [13], statistical and structural type hybrid method is used for license plate recognition. In [18], fuzzy logic based classifier is used to extract texture, color features to differentiate color of car and license plate. For any input image, each pixel is classified if it belongs to the license plate based on the generated fuzzy logic. Artificial neural network (ANN) is used for character recognition.

In [19], different character features like contour crossing counts, directional counts, and peripheral background area are used. The classification is realized by a support vector machine (SVM). In [20], Haar like features and cascade AdaBoost technique have been used for plate localization, while PCA and SVM are used for character recognition. In [21], HSI model is used to detect license plate. Mean and standard deviation is use to detect green and yellow color, while saturation and intensity are used to detect white and yellow. They have used connected component analysis (CCA) and feed forward ANN for segmentation and recognition of characters respectively.

However, there are several challenges in designing any ALPR system with respect to the accuracy. Some of them are as follows:

- Poor resolution, usually because the plate is too far away but sometimes resulting from the use of a low-quality camera.
- Blurred images, particularly motion blur.
- Poor lighting and low contrast due to overexposure, reflection or shadows that changes the pixel intensity value with time.
- In some country everybody may not follow the standard license plate format assigned by the government. Size and writing style vary from one state to other state. Logo of the motor companies is frequently found on the license plate, which makes difficult for the recognition. Some of the Indian license plates with variations in shape and scripts are shown in Fig. 1.

Fig. 1. License plates in India

- Occlusion and dirt on the plate.
- Sometimes, it is difficult to read license plates that are different at the front and the back because of towed trailers, campers, etc.
- Change of lane by the vehicle against the camera's angle of view during license plate reading.

Some examples are shown in Fig. 2.

This paper proposes an efficient ALPR system. Frame differencing technique is used for background subtraction and new technique is used for vehicle tracking. Texture features are extracted using LBP descriptor for detection of license plate. A modified bounding rectangle based technique is used for segmentation and KNN classifier is used for character recognition. It has also proposed an efficient training set in the context of Indian license plate.

Rest of the paper has been organized as follows. Section 2 presents the proposed system; it includes vehicle detection and motion tracking, license plate detection, character segmentation and recognition. Performance of the system has been analyzed in the next section. Conclusions and future works are presented in the last section.

2   Proposed System

Any typical ALPR system generally uses a traffic camera network, which processes captured traffic video on-site and transmits the extracted parameters in real time. Position of the camera may be fixed, movable with the vehicle, or camera designed for special task. ALPR system mainly consists of five components. Such of the typical ALPR system is shown in Fig. 3.

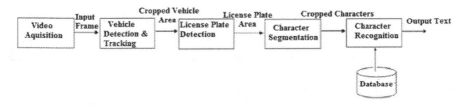

Fig. 3.   System block diagram

We have designed hardware setup with consideration of fixed camera position. We have used Moto G 3G (13 megapixels) camera having specifications, frame Rate = 30, height = 1080, width = 1920, video Format = MP4, pan angle = 7, 10 degree and tilt angle = 0, 3°. The camera is mounted at the height of 800 mm. This setup has the ability to transmit images in real time. To collect data, video recording has been done in two time intervals. The whole process can be performed either in real time video streaming from an operational center or in already stored video materials.

2.1 Vehicle Detection

In this section frame differencing technique has been used for the subtraction of background model. Preprocessing is done to convert the color image into the gray scale image. Gray scale image of two consecutive frames are subtracted in real time to get foreground image. Image is binarized by setting the threshold value. Presence of any static objects in the scenes and bushes affects the threshold value. Selection of threshold value is very difficult because it depends upon the environmental turbulences like illumination of light, movement of tree's leaf due to wind blowing. Background segmented image is shown in Fig. 4.

RGB Image   Grey Image   Binary Image

Fig. 4. Preprocessed images

2.2 Motion Tracking

In this section a new method has been proposed to track an object. It is on pixel intensity characteristics. A motion tracking algorithm has been is discussed in [1] which store each pixel and its neighbors in the past and update with time to know whether those pixels are from background model or not. Probability density function (PDF) of each pixel of background model and the next frame is compared to subtract background which adapts quickly with time. Mean shift [25] and calm shift [26] are the two most common approaches used for the object tracking. In the mean shift algorithm, mask contains the local maximum values of the data distributed in space. In the other one, window size is adjusted itself while tracking the object. The limitation of these two approaches performs poorly when the object is occluded. Color information is used to overcome the shadow due to wall, trees and by vehicle itself [2]. Subudhi et al. have presented spatial (Markov random field) and temporal (change detection mask) segmentation techniques to subtract background and track moving object [3].

In foreground image where the object is lying, intensity of the pixels are changing with time. To track the vehicle, we have drawn a rectangular box around it. In the foreground image white pixels are counted. Foreground image contains pixel intensity of the moving target along with the movement of tree's leaf, bushes et al. with less values. We have removed these values by a predefined threshold value and sum of the white pixels intensity values is calculated along every single row and column to the entire image.

Horizontal and vertical histograms are drawn with this sum data along horizontal and vertical direction. Histogram is smoothened using mean filtering and normalized. Four coordinates of rectangle around moving object can be found from intersection of two histograms. To obtain position of the vehicle along horizontal direction in the frame, difference of the two consecutive row values is calculated. If the difference is less than a threshold value, it is the coordinate value of initial position of moving target along horizontal direction and if difference is greater than the threshold value, it is coordinate of final position along horizontal direction. Similarly to get the exact location of the vehicle along the vertical direction, difference of the two columns is calculated. If the difference is less than the threshold value, it is coordinate of initial position of moving target along vertical direction and if difference is greater than the predefined threshold value, it is coordinate of final position.

It is also possible to capture other moving objects in the camera frame like pedestrians along footpath, bicycles. These unwanted moving objects create problems when they are ahead of the moving target in which we are interested. To remove these unwanted objects moving ahead of the moving target, coordinate value of final position is modified. To get the exact location along horizontal direction, we set coordinate value of final position to final gray value of the frame size. Now difference of the new coordinate position and initial coordinate position is calculated. If the difference of final gray value and coordinate of initial position is less than the predefined threshold, it is coordinate of initial position of moving target and if it is greater than the threshold value, it is coordinate of final position along horizontal direction. Further the difference (area of histogram) which is maximum is our desired moving object because difference (area of histogram) of moving pedestrians and other moving object is less compare to moving vehicle. Similarly, we apply same process to get coordinates value of initial and final position along vertical direction. Let x_1, x_2 are the initial and final position respectively along horizontal direction and y_1, y_2 are the initial and final position respectively along vertical direction. Histogram plot for both directions is shown in Fig. 5 (a) and (d)

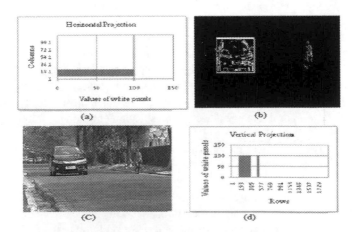

Fig. 5. (a) Horizontal projection to get x_1 and x_2 (b) combined horizontal and vertical projection (c) grey Image (d) vertical image to get y_1 and y_2

respectively, and from intersection of these two histograms we get four coordinates (x_1, y_1), (x_1, y_2), (x_2, y_1) and (x_2, y_2). In Fig. 5 (b) and (c), it is observed that apart from the target object, movement of the pedestrians is also found. It can be analyzed easily also in Fig. 5 (d) in the histogram plot that pedestrians moving ahead of the target. It can be easily observed our system could not track that moving pedestrians in Fig. 5 (b). This kind of false acceptance has been removed if one considers the area acquired by the moving target should be maximum.

Therefore, x_1 and x_2 is calculated in such a way that difference of x_1 and x_2 is maximum along horizontal direction and difference of y_1 and y_2 is maximum along vertical direction. By using these four coordinates, a rectangle has been drawn outside boundary of the vehicle for tracking. Tracked vehicle image has been shown in Fig. 6. By using these four coordinates, vehicle's image has been cropped from the running frame for further processing. We considered only vehicle area. Again, it is known visually license plate position in the front side of the vehicle, exists in the lower half portion. So one should take interest that region only for further processing, and it takes less computational time. Further by using those four coordinates a rectangle has been drawn in the lower half of the portion of the vehicle image and that area has been cropped for final processing to detect license plate, which is our region of interest (ROI), Coordinates which are used to crop the lower half portion are:

$$(y_1 + ((y_2 - y_1)/4.0), x_1 + ((x_2 - x_1)/2.0), ((3 \times (y_2 - y_1))/4.0), (x_2 - x_1)/2))$$

where, x_1 and x_2 are coordinate values of initial and final position along horizontal axis. y_1 and y_2 are coordinate values of initial and final position along vertical axis.

Fig. 6. (a) Original image (b) tracked vehicle image

Different stages of the background subtraction motion tracking and a cropped vehicle image are shown in Fig. 7. For this case, video has been taken at the pan angle of 12°

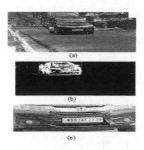

Fig. 7. (a) Input frames (b) cropped vehicle image (c) ROI

and tilt angle of 3°. Original video frames are size of (1920 × 1080). Then ROI from the original is cropped with sized (450 × 140) for detection of license plate.

3   License Plate Detection

A license plate is the unique identification of any vehicle. License plate detection is the most important part of ALPR system. Several approaches based on edge detection and morphological operation have been proposed [5, 11, 16, 17]. However, it is very difficult to obtain the character edges in a license plate due to environmental challenges. Color feature extraction using fuzzy logic [18] and HSI model [21] has been discussed. Haar like features calculated to extract plate area [14, 20]. Texture feature extraction using SHIFT [9] and ALBP [12] are used for plate area detection.

Different features like color, texture, shape, character can be extracted from the license plate. One of the feature extraction methods is LBP [24]. It is possible to describe the texture and shape of a digital image using LBP [27]. This method is quit robust against shape, rotation of the license plate and illumination of light [22, 23]. Features can be extracted by dividing an image into several regions as shown in Fig. 8.

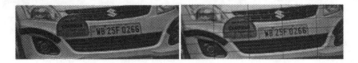

Fig. 8. A preprocessed image divided into 42 regions

Three neighborhood examples have been used to define a texture and calculate a local binary pattern (LBP). A 3 × 3 neighborhood LBP operator is shown in Fig. 9. After that, histogram of each cell is calculated and normalization is done which gives feature vector for entire window.

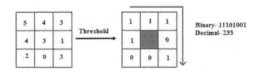

Fig. 9. LBP operator using 3 × 3 neighborhood

The LBP operator can be defined as:

$$LBP\left(x_c, y_c\right) = \sum_{p=0}^{p-1} 2^p\, s\left(i_p - i_c\right)\ s(x) = \begin{cases} 1 & if\ x \geq 0 \\ 0 & else \end{cases}$$

where i_c corresponds to the gray value of the center pixel (x_c, y_c) of a local neighborhood and i_p to the gray values of P equally spaced pixels on a circle of radius R. By extending the neighborhood, one can collect larger-scale texture features.

The extracted image of sized (450 × 140) is taken as input image for detection of license plate. Histogram equalization is performed to increase contrast of the image. To extract the region of interest (ROI) that contains license plate area Local Binary Pattern (LBP) descriptor is used. LBP operator is robust against monotonic gray scale transformations as shown in Fig. 10. Any license plate has alphanumeric characters with high roughness as compared to its background, so this technique can be used for license plate extraction. LBP calculates the pattern along with a block (square shape). Block diagram of the proposed system is shown in Fig. 11.

Fig. 10. (a) Gray images (b) LBP images (c) their histogram

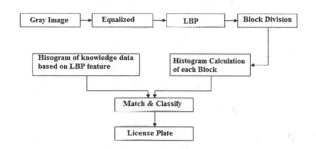

Fig. 11. Block diagram for license plate detection

The LBP image of sized (450 × 140) is divided in to 42 overlapping blocks of 60 × 80 pixels. Normalized histogram of each of 42 blocks is obtained. From the hardware setup some license plate has been are considered as knowledge samples. Their normalized histogram based on LBP features has been obtained and stored as knowledge data. So we have two sets of histogram. Let $h = (h_1, h_2 \ldots\ldots h_{42})$ be the normalized histogram of 42 blocks and $H = (H_1, H_2 \ldots\ldots H_{50})$ for 50 knowledge samples. Normalized histogram of each of 42 blocks is compared with the normalized histogram of each of knowledge data. Intersection method is used to compare the histogram. The metric $(D\ (h_1, H_1, H_2 \ldots\ldots H_{50}))$ is used to compare histogram $(h_1, H_1, H_2 \ldots\ldots H_{50})$ where

$$D\left(h_1,\ H_1,\ H_2 \ldots\ldots H_{50}\right) = \sum_{1}^{42} \min(h_1 H_1 H_2 \ldots\ldots H_{50})$$

D value is ranging from 0 to 1. The D is close to 0, when the features of two blocks are same. When the features of the two blocks are different, the D is close to 1. We have obtained 50 sets of compared values for each of 42 blocks. Again average value is of each comparison has been calculated. The block which has maximum average value is less than the predefined threshold value, results the desired block contained license plate.

Result of the extracted license plate from the ROI of sized (450 × 140) is shown in Fig. 12(a) and (b).

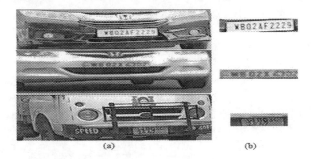

Fig. 12. (a) ROI (b) detected license plate images

4   Character Segmentation

Minimum bounding rectangle (MBR) sometimes called bounding box or envelop is an expression of the maximum area of a two dimensional objects (point, line area) within its 2-D (x, y) coordinate system. This method is frequently used to represent general position of geographic features or data set for display. License plate segmentation has been done based on prior knowledge of characters. Initially, extracted license plate which is gray scale image is binarized by applying adaptive threshold. To draw rectangle on the characters, two points (P_1 and P_2) are chosen such that P_1 gives minimum of x_1 and y_1 coordinates and P_2 is the maximum of x_2 and y_2. Plate sized (128 × 28) is normalized and minimum area 150 and maximum area 500 are considered to draw bounding boxes. The problem with this method is that characters are segmented in increasing order of area of the characters. In license plate area of all the characters may not be same. So segmented characters are not in sequence as per license plate. This problem is solved by analyzing the coordinate values of P_1 and P_2. It is found that the gap between the positions of the two characters is fixed. To crop the characters as per the license plate format order coordinate value P_1 is divided by the gap size and then rounded off the fractional value. It makes the coordinates of P_1 and P_2 arranged according to sequence of license plate. Some examples of segmented characters are shown in Fig. 13. License plate is sized of (128 × 28) has been cropped in characters sized of (12 × 20). It is observed that skewed and rotated license plates can be cropped easily.

Fig. 13. (a) Gray image (b) binary image (c) segmented characters

5   Character Recognition

Character recognition is the final step in ALPR system which recognizes each alpha-numeric character. For training data more than 300 alphanumeric characters has been taken. A database has been created with consideration of license plate format in India. In India the new license plate format is as:

$$S_1 \, S_2 \, D_1 \, D_2 \, A_1 \, A_2 \, N_1 \, N_2 \, N_3 \, N_4$$

where $S_1 \, S_2$ are the two alpha characters used for state code; $D_1 \, D_2$ are the two digits representing district code; and $N_1 \, N_2 \, N_3 \, N_4$ are the four digits of the license plate number and A_1, A_2 are two alpha numeric characters. An example of standard license plates is shown in Fig. 14. In this paper these segmented characters are used for matching of the two test images. All characters are resized of (12 × 20). Their normalized vertical and horizontal histogram used as feature vectors. Features of each segmented images are compared with those of the corresponding type of alpha numeric trained characters. K-nearest neighbor method is used for matching.

$$S_1 \, S_2 \, D_1 \, D_2 A_1 A_2 N_1 N_2 N_3 N_4$$

(a)   (b)

Fig. 14. License plate format (a) two line (b) one line

Let P and Q be the normalized histogram feature vectors of the characters obtained of test image and of the trained images respectively. That is,

$$P = \sum_{i=1}^{10} p_i \qquad Q = \sum_{j=1}^{36} q_j$$

where p_i and q_j are the feature i in P and that of j in Q respectively. P is considered to be maximum word length 10 and Q is (A....Z) and (0 ... 9) alphanumeric characters. If $\| \, p_i - q_j \, \|$ and $\| \, p_i\text{-} \, q_j \, \|$ are the Bhattacharya distance between p_i and its first nearest-neighbor q_j and that between p_i and it's second nearest-neighbor of q_k respectively, then p_i is matched to q_j if $D_B(p, q) < \text{T}$ where T is threshold, $D_B(p, q) = -\ln(BC(p, q))$ where $BC(p, q) = \sum_{x \in X} \sqrt{P(x)q(x)}$ is Bhattacharyya coefficient.

6   Experimental Results

Performance of the proposed system has been analyzed on the data captured along the road side and is shown in Table 1. This database contains 300 images of small and medium size cars, small lorry, and three wheelers. Images are captured in two different sessions with different illumination of light and clutter. It is observed from the table that

Table 1. System performance

Image set	Camera position		Number of vehicles						
	Pan angle °	Tilt angle°	Small		Medium		Large		Total
			One line	Two line	One line	Two line	One line	Two line	
I₁ (12 Noon)	7	0	88	12	72	–	–	8	180
Detection			85	11	69	–	–	7	
Recognition			85	9	68	–	–	6	
I₂ (4 PM)	12	3	70	10	38	–	–	2	120
Detection			67	9	36	–	–	2	
Recognition			67	7	36	–	–	2	
Overall detection accuracy (%)			96.14		Overall accuracy (%)				89.35

the proposed system performs with detection accuracy of 96.14% and detection accuracy of 89.35%. Recognition accuracy is less compared to that of detection. Fonts of some of the characters create some problems at the time of the recognition. Distinguishing the difference between pair of characters like B and 8, 0 and 8, A and 4, 3 and B, I and 1 is found to be difficult. Clampers, different writing style, presence of company logo, occlusion also affect system recognition. The proposed system is rotation invariant. It can detect single and double line written format on license plate for small, medium and large size vehicles. Further, the proposed system has been compared with some well known procedures used for detection and recognition. Results are given in Table 2. The proposed system is simple and can provide better distinctive performance compared to [5, 9, 11, 16].

Table 2. Performance comparison of some typical ALPR systems [License Plate Detection (LPD) and character recognition (CR)]

Methods	Procedures		LPD rate	CR rate	Total rate	Plate format
	LPD	CR				
[5] 805 images	Vertical & horizontal projection	Hidden markov model	97.7%	96.7%	92.9%	Vietnamese plates
[9] 120 images	SIFT	OCR	91%	90%	70%	Jordanian plates
[11] 300 images	Sobel filter, morphology	Neural network	84%	80%	NA	Bangladeshi plates
[12] 497 images	ALBP	Gabor filter	98.3%	99.20	97.4%	Chinese plates
[16] 420 image	Sobel filter, morphology	Template matching	93.4%	87.8%	NA	NA
Our system 300 images	LBP	Vertical & horizontal projection	96.14%	89.35%	93.37%	Indian plates

7   Conclusions

This paper has proposed an efficient license plate recognition system which is robust to illumination, shape and format of license plate, rotation. An approach to make system fast lower half portion which contains license plate has been cropped while tracking. Texture features of the license plate have been calculated using LBP. Recognition is done using K-NN classifier. This system has been tested at different time with different illumination of light. The proposed system performs with accuracy of 96.14% for detection and 89.35% recognition respectively. It is seen that illumination of light and angle of rotation of camera does not affect too much on the system performance. The proposed system detects letters of single and double line on license plate.

References

1. Barnich, O., Droogenbroeck, M.V.: Vibe: a universal background subtraction algorithm for video sequences. IN: IEEE Transactions on Image Processing, pp. 1709–1724 (2011)
2. Elgammal, A., Harwood, D., Davis, L.: Non-parametric Model for Background Subtraction. In: Vernon, D. (ed.) ECCV 2000. LNCS, vol. 1843, pp. 751–767. Springer, Heidelberg (2000). doi:10.1007/3-540-45053-X_48
3. Subudhi, B.N., Nanda, P.K., Ghosh, A.: A change information based fast algorithm for video object detection and tracking. In: IEEE Transactions on Circuits and Systems for Video Technology, pp. 993–1004 (2011)
4. Chang, J.-K., Ryoo, S., Lim, H.: Real-time vehicle tracking mechanism with license plate recognition from road images. J. Supercomput. **65**, 353–364 (2011). Springer
5. Duan, T.D., Duc, D.A., Du, T.L.H.: Combining hough transform and contour algorithm for detecting vehicles' license-plates. In: Proceedings International Symposium Intelligent Multimedia Video Speech Process, pp. 747–750 (2004)
6. Du, S., Ibrahim, M., Shehata, M., Badawy, W.: Automatic license plate recognition (ALPR): a state-of-the-art review. In: IEEE Transactions on Circuits and Systems for Video Technology, pp. 311–325 (2013)
7. Sirithinaphong, T., Chamnongthai, K.: The recognition of car license plate for automatic parking system. In: Proceedings 5th International Symposium on Signal Processing and its Applications, pp. 455–457 (1998)
8. Kahraman, F., Kurt, B., Gökmen, M.: License plate character segmentation based on the gabor transform and vector quantization. In: Yazıcı, A., Şener, C. (eds.) ISCIS 2003. LNCS, vol. 2869, pp. 381–388. Springer, Heidelberg (2003). doi:10.1007/978-3-540-39737-3_48
9. Yousef, K.M.A., Al-Tabanjah, M., Hudaib, E., Ikrai, M.: SIFT based automatic number plate recognition. In: 6th International Conference on Information and Communication Systems (ICICS), Amman, pp. 124–129, April 2015
10. Gorovyi, I.M., Smirnov, I.O.: Robust number plate detector based on stroke width transform and neural network. In: IEEE Conference on Signal Processing Symposium, Debe, pp. 1–4, June 2015
11. Ghosh, A.K., Sharma, K.D., Islam, Md.N., Biswas, S., Akter, S.: Automatic License Plate Recognition (ALPR) for bangladeshi vehicles. In: Global Journal of Computer Science and Technology, pp. 68–73, December 2011

12. Wang, Y., Zhang, H., Fang, X., Guo, J.: Low-resolution chinese character recognition of vehicle license plate based on ALBP and gabor filters. In: Seventh International Conference on Advances in Pattern Recognition, ICAPR 2009, Beijing, pp. 302–305, February 2009

13. Pan, X., Ye, X., Zhang, S.: A hybrid method for robust car plate character recognition. In: IEEE International Conference on Systems, Man and Cybernetics, pp. 4733–4737 (2004)

14. Zhang, H., et al.: Learning-based license plate detection using global and local features. In: International Conference on Pattern Recognition, Hong Kong, China, pp. 1102–1105 (2006)

15. Wang, Y.R., Lin, W.H., Horng, S.J.: A sliding window technique for efficient license plate localization based on discrete wavelet transform. Expert Syst. Appl. **38**, 3142–3146 (2011)

16. Kanayama, K., Fujikawa, Y., Fujimoto, K., Horino, M.: Development of vehicle-license number recognition system using real-time image processing and its application to travel-time measurement. In: Proceedings IEEE Conference on Vehicular Technology, pp. 798–804, May 1991

17. Zhang, S., Zhang, M., Ye, X.: Car plate character extraction under complicated environment. In: Proceedings IEEE International Conference on System Man Cybern, vol. 5, pp. 4722–4726 October 2004

18. Nijhuis, J.A.G., Brugge, M.H.T., Helmholt, K.A., Pluim, J.P.W., Spaanenburg, L., Venema, R.S., Westenberg, M.A.: Car license plate recognition with neural networks and fuzzy logic. In: Proceedings IEEE International Conference on Neural Network, pp. 2232–223,   6 December 1995

19. Wang, S.Z., Lee, H.J.: A cascade framework for a real-time statistical plate recognition system. IEEE Trans. Inf. Forensics Secur., 267–282 (2007)

20. Jia, W., Zhang, H., He, X.: Region-based license plate detection. J. Netw. Comput. Applicat. **30**(4), 1324–1333 (2007)

21. Deb, K., Jo, K.-H.: A vehicle license plate detection method for intelligent transportation system applications. Int. J. Cyber Syst. **40**(8), 689–705 (2009)

22. O'Connor, B., Roy, K.: Facial recognition using modified local binary pattern and random forest. Int. J. Artif. Intell. Appl. (IJAIA) **4**(6), 25–33 (2013)

23. Nigam, A., Krishna, V., Bendale, A., Gupta, P.: Iris recognition using block local binary patterns and relational measures. In: IEEE International Joint Conference on Biometric, pp. 1–6 (2014)

24. Ojala, T., Pietikainen, M.: A comparative study of texture measures with classification based on feature distributions. Pattern Recogn. **29**(1), 51–59 (1996)

25. Comaniciu, D., Meer, P.: Mean shift: a robust approach toward feature space analysis. In: IEEE Transaction on Pattern Analysis and Machine Intelligence, pp. 1–9, May 2002

26. Ali, G., Nouar, O., Raphael, C.: Tracking system using camshift and feature points. In: 14th European signal processing conference, Florence, Italy, 4–8 September 2006

27. Gorai, A., Pal, R., Gupta, P.: Document fraud detection by ink analysis using texture features and histogram matching. In: IJCNN, pp. 4512–4517 (2016)

Leg Ulcer Long Term Analysis

Eros Pasero[✉] and Cristina Castagneri

DET, Politecnico di Torino, c. Duca d. Abruzzi 24, 10129 Turin, Italy
{eros.pasero,cristina.castaggneri}@polito.it

Abstract. Ulcers on legs and feet usually require long-term clinical treatment and follow-up. To facilitate the monitoring, we propose a fully automatic and low-cost method for ulcers detection and analysis. The ulcer segmentation is performed using an automatic processing based on pixel's classification into background or not background classes. Features used to perform the classification are the values of three channels that define each pixel in the RGB color map and in the HSV color map.

We tested the algorithm on a dataset of 92 images, acquired from 14 different patients. The segmentation performances were evaluated in terms of overlap, recall and precision, by comparing the automatic segmentation with the manually one. The results show good average values of overlap, recall and precision.

Then, a Self-Organizing Map (SOM) was used for tissue classification. The SOM was trained in order to identify six colorimetric classes associated to different type of tissues.

Keywords: Segmentation · Wound healing · SOM · Tissue classification · Colour image processing · Skin lesion · Leg ulcers · Tissue composition

1 Introduction

In the last years, the problem of skin ulcers (venous, arterial, diabetic and pressure) is becoming increasingly important, especially in relation to the increase in the aging population and thus the prevalence of chronic disabling diseases. Skin ulcers cause immobility or marked reduction in individual autonomy in many cases; they also cause a worsening of the quality of life of those affected and have a considerable negative psychological impact on patients.

Nowadays, chronic ulcers, particularly vascular leg ulcers, represent a significant public health problem and a major economic burden to the healthcare system. An automatic measurement of a wounds heling status would reduce workloads, save money, give a more objective assessment of wound healing and provide better care for the patient.

The healing process of an ulcer consist of three main phases. The first one, the inflammation phase, is a protective response of the body characterized by the activation of the immune system. In the second phase, tissue formation begins with the formation of granulation tissue. At last, the remodeling occurs with healing and re-epithelialization. In clinical practice, the rate of change of ulcers area is one of the main way to quantify

© Springer International Publishing AG 2017
D.-S. Huang et al. (Eds.): ICIC 2017, Part II, LNCS 10362, pp. 35–44, 2017.
DOI: 10.1007/978-3-319-63312-1_4

progress in wound healing and the quantification of wound healing rate is critical in assessing the efficacy of treatments [1].

At present, the diagnosis of skin ulcers is carried out mainly on the visual assessment of pathological skin and on the evaluation of macroscopic features. Therefore, diagnosis is highly dependent on the observer and on his visual perception [2]. Furthermore, the only visual assessment of the lesion is not able to provide quantitative information on the state of the ulcer.

Several studies [3–6] had proven that the quantification of tissue lesion features might be of essential importance in clinical practice. In the last decade, significant amounts of works have dealt with the problem of objective analysis of skin ulcers, implementing different methods and approaches [7]. Studies have addressed the problem mainly on two sides: on one hand, the development of segmentation algorithms for automatic wound detection [7, 9, 10] and, on the other hand, the implementations of methods for an objective characterization of the composition of ulcerated tissues [11, 13, 14]. Different techniques have been proposed to process wound images as K-nearest neighbors (KNN) algorithms [12], neural networks [15, 21], Bayesian classifier [16] and clustering algorithms [17].

In the last years, segmentation [22] of skin ulcers using camera images has turned out to be an emergent area in this field [18]; however, in some proposed methods, the use of a smartphone camera must be associated with the use of other complex devices, difficult to use in practice [19, 20].

Therefore, there is a need to develop image segmentation methods that will work with images acquired in regular clinical condition. In this work, we propose a fully automatic and adaptive method for automatic ulcers segmentation and for tissue classification. The algorithm analyzes the images of the patient's lesions acquired only with a smartphone camera, without using other external device for the acquisition.

2 Materials and Methods

2.1 Image Dataset

The dataset includes 47 images from 14 patients followed from March 2016 to February 2017 (Table 1).

Table 1. Image dataset

Patient ID	Number of acquired images	Patient ID	Number of acquired images
1	9	8	8
2	8	9	8
3	7	10	8
4	8	11	8
5	8	12	8
6	8	13	8
7	8	14	8

Images were acquired through smartphone camera. A "ruler" was used for the study (Fig. 1) as a reference to make the conversion from pixel to cm^2 (the green area of the ruler defines a region of 3 cm^2).

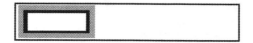

Fig. 1. Ruler representation. The region inside the green rectangle identifies an area of 3 cm^2. (Color figure online)

2.2 Images Acquisition

The ulcer images (Fig. 2) were acquired using a smartphone camera, with a resolution of 1836 × 3264 pixels. The acquisition protocol is the following:

(1) The wound must be placed as close as possible to the camera.
(2) The wound must be placed close to the center of the image.
(3) The ruler must be placed in contact with the skin and at the same plane of the picture.
(4) The flash is not allowed.
(5) Take the photograph.

Images are encoded in *uint8* format, so that each pixel value lies in the range [0, 255].

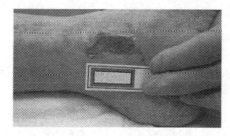

Fig. 2. Example of image acquired correctly.

3 Segmentation Algorithm

3.1 Ruler Identification

Firstly, a color deconvolution filter is applied to the image (Fig. 3A, B) and the green area of the ruler is found according to the following condition on the three RGB components resulting after deconvolution:

$$R < 150 \, and \, G > 200 \, and \, B < 150 \tag{1}$$

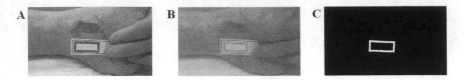

Fig. 3. Ruler identification process. Original image (A), image after color deconvolution (B), ruler mask (C).

A binary mask is built, where all pixels are blacks except those that meet the criteria (1) and the ROI with the greatest dimension is extracted from the resulting mask (Fig. 3C). Finally, the number of pixels within the region ($PIXEL_{ruler}$) is counted and the factor K is defined to make the conversion from pixel (n_{pixel}) to cm^2 using the formulas (2) and (3):

$$PIXEL_{RULER}: 3 \text{ cm}^2 = n_{pixel}: x_{cm^2} \tag{2}$$

$$x_{cm^2} = K \cdot n_{pixel}, \text{ with } K = \frac{3 \text{ cm}^2}{pixel_{RULER}} \tag{3}$$

3.2 Preprocessing

In the first step of preprocessing, the software identifies the centroid of the white portion of the ruler (Fig. 4B). A square of 1 cm^2 centered in that point is identified. In that area, the average of the three RGB channels is calculated ($ruler_{mean}$). Then, the brightness of the image is increased using the ChV factor (4):

$$ChV = \frac{255}{ruler_{mean}} \tag{4}$$

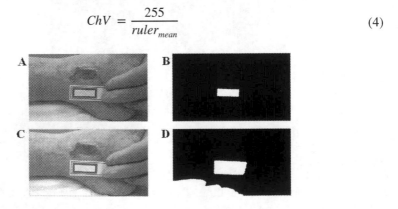

Fig. 4. Original image (A), ruler white mask (B), preprocessed image (C), *exclusion_mask* (D).

A median filter, with a dimension equal to 20, is applied in order to make the image more homogeneous [23] and to improve the segmentation performance. The result of the preprocessing is shown in Fig. 4C. Finally, a binary mask (*exclusion_mask*) is built to exclude those pixels belonging to the ruler and the *white pixels* in the image (Fig. 4D).

All pixels that meet the following criteria (5) on the RGB color vector are classified as *"white pixels"*:

$$R > 250 \, and \, G > 250 \, and \, B > 250 \tag{5}$$

3.3   Ulcer Detection

The segmentation algorithm is based on pixel classification into *background* or *not background* classes. Features used to perform the classification are the values of the three channels that define each pixel in the RGB color map and in the HSV color map.

In the RGB color map, the [R G B] vector defines a point in three-dimensional space in which the three components may vary between 0 and 255. Therefore, the algorithm evaluates the Euclidean distance between the RGB components of two pixels to assess their similarity. The algorithm consists of three main steps.

Step 1. The RGB preprocessed image (Fig. 4C) is converted in the HSV color map (Fig. 5A). A mask is defined on the edge of the image (thickness equal to 120 pixels) and the HSV colors of each pixel belonging to that region is defined as *'background colors'* (Fig. 5B).

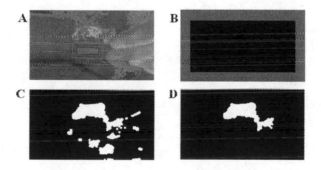

Fig. 5. HSV image (A), background region (red region) (B), *mask1* (C), selected ROI (*ROI1*) (D). (Color figure online)

The binary mask (*mask1*) is built in order to mark pixels classified as *not background*. To compute the classification, the algorithm scans all the pixels of the image: for each one the distances between the HSV vector of the pixel and the HSV vectors of *background colors* are calculated. If the minimum distance of those found is greater than 10, the pixel is classified as not background and its value is set to 1 in the *mask1* (Fig. 5C). A morphological opening on the binary image *mask1* is performed with a disk-shaped structuring element, with radius pair to 50 pixel. In order to identify the ROI (Region of Interest) that may contain the ulcer (ROI1), the greatest region of *mask1* is selected (Fig. 5D).

Step 2. The bounding box (*BOX1*) of the ROI identified at previous step is drawn (Fig. 6A) and then 100 pixel dilated on each side (Fig. 6B). The algorithm identifies the edge of the *BOX1* and draws its edge with a thickness of 40 pixels. The RGB colors of those pixels are considered as the new *background colors* (Fig. 6C).

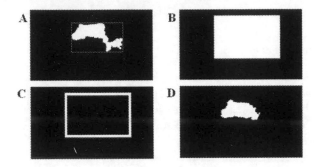

Fig. 6. Bounding box of *ROI1* (A), dilated bounding box (B), edge of *BOX1* (C), *ROI2* (D).

We scan all the pixels belonging to *BOX1*, calculating the distances of each RGB vector with all RGB vectors of *background colors*.

After the scanning of the pixels, we perform a morphological opening with a disk-shaped structuring element, with radius pair to 25 pixel. Then we check if there is a ROI inside *BOX1* completely detached from the edge. If there is more than one ROI with no pixels in common with the edge of the box, we choose the ROI with the greatest dimension. Instead, if there is no ROI that meets this specification, the operation is repeated by increasing the threshold value of 1.5 and proceed iteratively in this way until it locates at least one ROI (*ROI2*) that complies with the specification (Fig. 6D).

Step 3. It repeats the same procedure as in Step 2 starting from the *ROI2* identified (Fig. 6D), dilating the new bounding box of 50 pixels (Fig. 7A, B), using a thickness for the edge of 20 pixels (Fig. 7C) and a structural element of radius 10 pixels. Therefore, the *mask3* is obtained (Fig. 7D) that defines the final automatic segmentation.

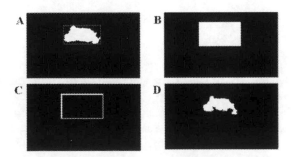

Fig. 7. *ROI2* (A) and *BOX2* (A).

Figure 8 shows the results of automatic segmentation (Fig. 8B), compared with the manual segmentation (Fig. 8A).

Fig. 8. Manual segmentation (A) and result of the automatic segmentation (B).

4 Segmentation Results

The performance were analyzed in terms of overlap, recall and precision (Table 2) [24].

Table 2. Overlap, recall and precision indices calculated for a 47 image subaset.

Image	Overlap	Recall	Precision	Image	Overlap	Recall	Precision
1	0,88	0,91	0,97	25	0,68	0,81	0,80
2	0.79	0,96	0,81	26	0,76	0,89	0,84
3	0,79	0,95	0,80	27	0,47	1,00	0,48
4	0,56	0,93	0,58	28	0,77	0,92	1,00
5	0,68	0,81	0,80	29	0,80	0,96	0,83
6	0,49	1,00	0,49	30	0,76	0,79	0,95
7	0,63	0,87	0,70	31	0,86	0,92	0,92
8	0,70	0,90	0,70	32	0,84	0,88	0,95
9	0,76	0,94	0,75	33	0,75	0,80	0,92
10	0,74	0,93	0,78	34	0,82	0,83	0,85
11	0,85	0,94	0,90	35	0,77	0,93	1,00
12	0,83	0,84	0,98	36	0,81	0,87	0,92
13	0,83	0,88	0,94	37	0,79	0,89	0,87
14	0,71	0,99	0,71	38	0,38	0,78	0,42
15	0,76	0,95	0,79	39	0,85	0,99	0,88
16	0,43	0,82	0,47	40	0,46	0,92	0,48
17	0.68	0,75	0,88	41	0,77	0,95	0,80
18	0,86	0,88	0,97	42	0,77	0,78	0,98
19	0,90	0,92	0,97	43	0,64	0,71	0,86
20	0,83	0,89	0,93	44	0,71	0,74	0,94
21	0,69	0,97	0,70	45	0,86	0,97	0,88
22	0,84	0,95	0,88	46	0,80	0,84	0,94
23	0,84	0,91	0,91	47	0,60	0,92	0,64
24	0,82	0,89	0,91				

Overlap quantifies the agreement between the automatic segmentation and the reference (manual segmentation). This measure is computed as the number of pixels in the intersection of segmentation and reference, divided by the number of pixels in the union of segmentation and reference. Its value goes from 0 to 1 (perfect segmentation).

Recall is a statistical measure of completeness; compared to overlap, this measure only detects underestimation of the volume. It is defined as the number of pixels in the intersection of segmentation and reference, divided by the number of voxels in the reference alone.

Precision is a statistical measure of exactness; it is an index of overestimation of the area. It is defined as the number of pixels in the intersection of segmentation and reference, divided by the number of pixels in the segmentation alone.

5　Tissues Classification

In order to perform the tissue classification, a classifier is built to identify six color classes in the images (Fig. 9): yellow (class associated to fibrotic tissue), red (class associated to granulation tissue), dark red (class associated to mixed fibrotic and granulation tissue), pink (class associated to re-epithelization tissue), green (class associated to infection) and black (class associated to necrotic tissue).

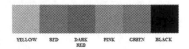

YELLOW　　RED　　DARK RED　　PINK　　GREEN　　BLACK

Fig. 9. Color classes. (Color figure online)

A SOM [25] is built, having output layer with dimensions 10×10, hexagonal layer topology function, neighborhood size equal to 3 and Euclidean neuron distance function. The network was trained with a dataset composed of 500 pixels per class, extracted from the image dataset, on the advice of a medical expert.

After the training, each of the six set of pixel was used as input for the SOM separately. A neuron of the SOM is labelled with a certain class if the number of victories for that class is equal to or higher than 50 [26].

Finally, the SOM is fed with original images pixels and each pixel is assigned to the class of the label in the correspondent neuron (Fig. 10).

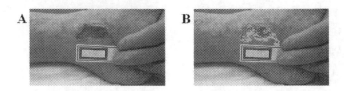

Fig. 10. Example of tissue classification.

6   Conclusion

We have presented a method for the analysis of ulcers from wound segmentation to tissue classification. The segmentation results are enough good but we believe we can improve our method and make it more reliable and accurate, to provide reliable and repeatable objective results in support of medical experts. More patients and more data will be acquired in next months by a medical team which will work with our group. Therefore we will have more knowledge base, especially to focus on tissue classification, and to validate the segmentation algorithm on other image sets.

Acknowledgment.   This project was partly funded by Italian MIUR OPLON project.
Authors would like to thank Dr. Franco Ribero for his precious help and the ulcer images.

References

1. Jessup, R.L.: What is the best method for assessing the rate of wound healing? a comparison of 3 mathematical formulas. Adv. Skin Wound Care **19**, 138–147 (2006)
2. Stremitzer, S., Wild, T., Hoelzenbein, T.: How precise is the evaluation of chronic wounds by health care professionals? Int. Wound J. **4**(2), 156–161 (2007)
3. Fauzi, M.F.A., Khansa, I., Khansa, I., Catignani, K., Gordillo, G., Sen, C.K., Gurcan, M.N.: Computerized segmentation and measurement of chronic wound images. Comput. Biol. Med. **60**, 74–85 (2015)
4. Jagannath, S.K., Poral, N.: Cost-effective wound monitoring system for diabetic patients using smartphone. Int. J. Adv. Res. Trend Eng. Technol. **22**(2), 449–453 (2016)
5. Seixas, J.L. Jr., Barbon, S. Jr., Mantovani, R.G.: Pattern recognition of lower member skin ulcers in medical images with machine learning algorithms. In: 28th International Symposium on Computer-Based Medical Systems, pp. 50–53. IEEE (2015)
6. Wannous, H., Lucas, Y., Treuiller, S., Albouy, B.: A complete 3D wound assessment tool for accurate tissue classification and measurement.In: 15th IEEE International Conference on Image Processing (ICIP), pp. 2928–2931. IEEE (2008)
7. Papazoglou, E.S., Zubkov, L., Mao, X., Neidrauer, M., Rannou, N., Weingarten, M.S.: Image analysis of chronic wounds for determining the surface area. Wound Repair Regen. **18**(4), 349–358 (2010)
8. Kolesnik, M., Fexa, A.: Multi-dimensional color histograms for segmentation of wounds in images. In: Kamel, M., Campilho, A. (eds.) ICIAR 2005. LNCS, vol. 3656, pp. 1014–1022. Springer, Heidelberg (2005). doi:10.1007/11559573_123
9. Hariprasad, R., Sharmila, N.: Foot ulcer detection using image processing technique. Int. J. Comput. Technol. **3**(3), 120–123 (2016)
10. Song, B., Sacan, A.: Automated wound identification system based on image segmentation and artificial neural networks. In: International Conference Bioinformatics and Biomedicine, pp. 1–4. IEEE (2012)
11. Galushka, M., Zheng, H., Patterson, D., Bradley, L.: Case-based tissue classification for monitoring leg ulcer healing. In: 18th Symposium on Computer-Based Medical Systems. IEEE (2005)
12. Zheng, H., Bradley, L., Patterson, D., Galushka, M. Winder, J.: New protocol for leg ulcer tissue classification from colour images. In: 26th Annual International Conference of IEEE EMBS, pp. 1389–1392. IEEE (2004)

13. Hani, A.F.M., Arshad, L. Malik, A.S., Jamil, A., Boon, F.Y.B.: Assessment of chronic ulcers using digital imaging. In: 2011 National Postgraduate Conference, pp. 1–5. IEEE (2011)
14. Dorileo, E., Frade, M., Rangayyan, R., Azevedo Marques, P.: Segmentation and analysis of the tissue composition of dermatological ulcers. In: Canadian Conference on Electrical and Computer Engineering, pp. 1–4. IEEE (2010)
15. Pinero, B.A., Serrano, C., Acha, J.I.: Segmentation of burn images using the L*u*v space and classification of their depths by color and texture information. J. Biomed. Opt. **4684**, 1508–1515 (2002)
16. Veredas, F., Mesa, H., Morente, L.: binary tissue classification on wound images with neural networks and bayesian classifiers. IEEE Trans. Med. Imag. **29**(2), 410–427 (2010)
17. Bhelonde, A., Didolkar, N.M., Jangale, S., Kulkarni, N.L.: Flexible wound assessment system for diabetic patient using android smarphhone. In: Green Computing and Internet of Things (ICGCIoT) International Conference, pp. 466–469. IEEE (2015)
18. Sungkrityayan, K., Swarupa, P., Sambu, S.R., Adithya, K.: Segmenting skin ulcers based on thresholding and watershed segmentation. In: International Conference on Communications and Signal Processing (ICCSP), pp. 1679–1683. IEEE (2015)
19. Bochko, V., Valisuo, P., Harju, T., Alander, J.: Lower extremity ulcer image segmentation of visual and near-infrared imagery. Skin Res. Technol. **16**, 190–197 (2010)
20. Geetha, C., Sathish, K.: Wound assessment for diabetic patients using matlab application in smartphone. Int. J. Adv. Res. Trend Eng. Technol. **3**(24), 255–263 (2016)
21. Pasero, E., Raimondo, G., Ruffa, S.: MULP: a multi-layer perceptron application to long-term, out-of-sample time series prediction. In: Zhang, L., Lu, B.-L., Kwok, J. (eds.) ISNN 2010. LNCS, vol. 6064, pp. 566–575. Springer, Heidelberg (2010). doi:10.1007/978-3-642-13318-3_70
22. Min, H., Wang, X., Huang, D.S., Jia, W.: A novel dual minimization based level set method for image segmentation. Neurocomputing **214**, 910–926 (2013)
23. Yadav, M.K., Manohar, D.D., Mukherjee, G., Chakraborty, C.: Segmentation of chronic wound areas by clustering techniques using selected color space. J. Med. Imag. Health Inf. **3**(1), 22–29 (2013)
24. Giannini, V., Vignati, A., Morra, L., Persano, D., Brizzi, D., Carbonaro, L., Bert, A., Sardanelli, F., Regge,D.: A fully automatic algorithm for segmentation of the breasts in DCE-MR images. In: 2010 Annual International Conference on Engineering in Medicine and Biology Society (EMBC), pp. 3146–3149. IEEE (2010)
25. Zhang, X., Jiao, L., Liu, F., Bo, L., Gong, M.: Spectral clustering ensemble applied to SAR image segmentation. IEEE Trans. Geosci. **46**(7), 2126–2136 (2008)
26. Rosati, S., Giannini, V., Castagneri, C., Regge, D., Balestra, G.: Dataset homogeneity assessment for a prostate cancer CAD system. In: International Symposium on Medical Measurements and Applications (MeMeA), IEEE (2016)

Virtual Reality and Human-Computer Interaction

Assessing Learners' Reasoning Using Eye Tracking and a Sequence Alignment Method

Asma Ben Khedher[(⊠)], Imène Jraidi, and Claude Frasson

University of Montreal, Montreal, QC, Canada
{benkheda, jraidiim, frasson}@iro.umontreal.ca

Abstract. In this paper we aim to assess students' reasoning in a clinical problem-solving task. We propose to use students' eye movements to measure the scan path followed while resolving medical cases, and a sequence alignment method, namely, the pattern searching algorithm to evaluate their analytical reasoning. Experimental data were gathered from 15 participants using an eye tracker. We present by using gaze data that the proposed approach can be reliably applied to eye movement sequence comparison. Our results have implications for improving novice clinicians' reasoning abilities in particular and enhancing students' learning outcomes in general.

Keywords: Eye movements · Sequence alignment · Reasoning · Scan path similarity

1 Introduction

Sensing technology has shown considerable promising results in analyzing learners' behavior and improving their learning experience [1–3]. Particularly, eye tracking sensors have gained much popularity due to their ease of use, high sensitivity and especially non-intrusiveness [4, 5]. Eye tracking systems are used to track users' eye movements and assess their visual behavior. They are being increasingly used in learning [6, 7] where it is important to evaluate the students' learning progression in order to provide the adequate feedback when needed [8, 9]. Indeed, assessing the students' outcomes is not the only interest within computer-based learning environments; how they learn is also a fundamental issue. Yet, it is very laborious to investigate if the students are reasoning well while solving problems, considering the technical and methodological challenges due to the complexity of the reasoning process, as well as the wide range of factors that impact the students' behavior. The use of eye tracking as a mean to understand the students' progress could be particularly beneficial in identifying the link between the task-relevant information and the learners' knowledge state [10].

Eye tracking systems provide different metrics which can be classified into static metrics and dynamic metrics [11]. Static metrics are measured from the observed data, such as fixation counts within a particular area of interest (AOI). Dynamic metrics are measured according to spatiotemporal dimensions, such as the scan path, which is a series of fixations and saccades representing the visual trajectory of a user's eye

© Springer International Publishing AG 2017
D.-S. Huang et al. (Eds.): ICIC 2017, Part II, LNCS 10362, pp. 47–57, 2017.
DOI: 10.1007/978-3-319-63312-1_5

movements. The latter metrics provide richer information as they reveal the dynamics of the visual behavior.

In this paper, we propose to use eye gaze data to model the students' reasoning process. In particular, we use the scan path to represent the sequential visual path a student follows while resolving medical cases. We are interested in assessing the analytical reasoning process the learners use to yield a general conclusion from existing observations.

An experimental protocol was conducted in order to record the learners' eye movements while interacting with our virtual environment, trying to identify for different medical cases the matching diagnosis and treatment. We describe the methodology used to measure the students' scan paths and evaluate their analytical reasoning using a sequence alignment algorithm.

The rest of this paper is structured as follows: Sect. 2 presents some related work, Sect. 3 presents the algorithm we will use. In Sect. 4, we describe our learning environment and the experimental protocol. Section 5 discusses the obtained results and Sect. 6 concludes and presents future work.

2   State of the Art

Eye tracking is a method that is being increasingly employed in many research domains including visualization [12], activity recognition [13] and affect detection [14]. It is also widely used in learning [15–18] due to its close relation to human cognition and brain activity [19–21]. Tai et al. (2006) used eye tracking in a problem solving environment in order to investigate whether there were differences in students' visual behavior. The results showed that gaze data enabled to discriminate between students with different degrees of expertise in chemistry [22]. Ben Khedher et al. (2016) explored eye movements in order to predict learners' level of performance during an interaction with a narrative-centered learning game. They assessed the relationship between eye movements' metrics and learners' performance, and achieved a good accuracy in discriminating between two groups of learners in term of scores in multiple choice quizzes [23].

Although most of the current work on eye tracking use fixations and saccades, there is a growing interest in using the scan path as an alternative metric as it enables to monitor moment-to-moment changes in the individual's attention and focus [24, 25]. For example, Glady and his colleagues (2013) used the scan path to analyze the visual trajectories of children and adults when solving analogy problems [26]. Susac et al. (2014) also investigated the scan path metric to analyze the students' eye movement patterns while solving mathematical equations [27]. They have found that expert students perform a well-defined and organized visual strategy as compared to non-expert students who have more fixations on the presented solutions.

In this paper, we seek to assess the students' reasoning process while resolving clinical cases. First, we propose to use the scan path to model the students' analytical reasoning. Second, we discriminate between learners in a way that identifies those who follow a correct analytical reasoning using a sequence alignment algorithm namely pattern searching.

We are not the first to apply an alignment algorithm to eye movements. In a previous work, Cristino et al. (2010) used the ScanMatch method which is designed to compare two scan paths based on the Needleman-Wunch alignment algorithm [15]. In their work, the authors proposed to use this method within a web-based visual search task.

To our knowledge, this is the first work that uses students' scan path in conjunction with the pattern searching algorithm for cognitive task assessment.

3   Sequence Alignment

In the human genetic system, millions of DNA sequences need to be aligned in order to discover the relationships between them. It is difficult for biologists to discover if a DNA is conserved from one species to another or how many similar residues genes share. Sequence alignment algorithms were developed in order to solve all these issues. They are used, among others, in bioinformatics to compare DNA sequences and to represent the evolutionary events that can occur from one sequence to another.

Aligning two sequences consists in finding if they are homologous. A similarity score between both sequences is computed by measuring the number of editing operations (deletion, insertion or mutation) required to transform one sequence to another. For example, let us consider two sequences, Seq1 and Seq2, to be aligned (Fig. 1). Three possible situations can occur, namely, match, mismatch and gap. A match is a perfect alignment with the same elements (characters) in Seq1 and Seq2 (A-A, B-B and Y-Y). A mismatch requires a mutation since the characters are different (G-O, A-D and R-K). And a gap means an insertion or a deletion in one of the two sequences: a gap in Seq1 leads to an insertion in this sequence (_-L and _-D) and a gap in Seq2 leads to a deletion (T-_ and D-_).

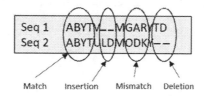

Fig. 1.   Editing operations for aligning two sequences

Pattern Searching
Pattern searching is an alignment algorithm that seeks to find a particular pattern in a DNA sequence [28, 29]. It consists in searching, through a given sequence, for the possible conserved regions. That is, given a small pattern P of size m and a sequence T of size n, aligning the two sequences consists in finding all the approximate occurrences of P in T. Suppose T "CTCAGGT" and P "CAT"; we want to search for all the occurrences of P in T (with a maximum of k errors). The first step of the algorithm is to create a distance-based matrix using gap, match and mismatch metrics (see Fig. 2).

The first row is initialized with zeros; the remainder cells are filled using the global alignment recursion relation shown in (1).

$$V(0, j) = 0$$
$$V(i, 0) = i$$
$$V(i,j) = \min \begin{cases} V(i-1, j) + d \\ V(i,j-1) + d \\ V(i-1,j-1) + \delta(i, j) \end{cases}$$
$$\text{Where } \delta(i, j) = m \text{ if } ti = pj \text{ else } r.$$

(1)

V(i, j) is the distance between i and j, and m, r and d are respectively the values for match, mismatch and gap.

Once the matrix is filled, the algorithm selects in the last row all the cells with a score less or equal to k. Then, from that cell, it backtracks through the array until the first row to extract the possible alignment paths. Finally, the last step is to compute the alignment score as shown in (2), where matches are rewarded with δ, mismatches with μ and gaps with σ.

$$\text{Score} = \delta(\#\text{matches}) + \mu(\#\text{mismatches}) + \sigma(\#\text{gaps})$$

(2)

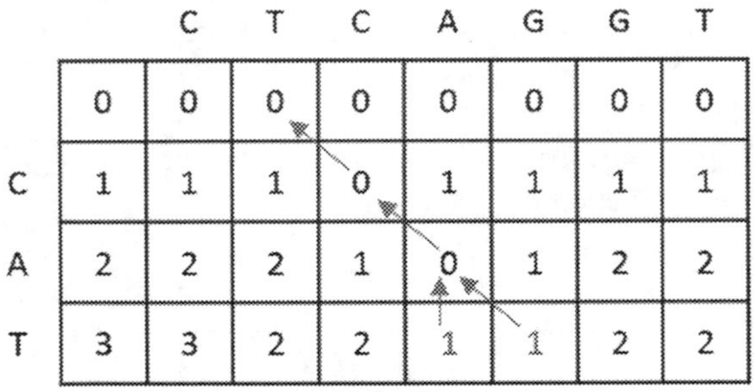

	C	T	C	A	G	G	T
	0	0	0	0	0	0	0	0
C	1	1	1	0	1	1	1	1
A	2	2	2	1	0	1	2	2
T	3	3	2	2	1	1	2	2

Fig. 2. Substitution matrix with the two optimal alignments paths (k = 1)

4   Experimental Protocol

An experimental protocol was conducted in order to record the students' eye movement activity while interacting with a virtual learning environment called AMNESIA. This system, which was specially designed for the experiment, is a medical serious game that assesses the cognitive abilities of novice medical students through clinical decision-making; the curriculum underlying the environment was designed by a medical professional and approved by a doctor. The game features a virtual hospital where the user plays the role of a doctor who was mistakenly diagnosed with amnesia and found himself trapped within the hospital. The player has to prove that he does not suffer from this disease, by resolving in a first step some cognitive tasks (such as memory and logic tests) to prove his cognitive abilities. Then, in a second step, he has to demonstrate his clinical skills by resolving medical cases.

4.1 Medical Cases

The game includes six medical cases. For each case, the students are asked to find out the correct diagnosis and the appropriate treatment. To this end, the players are instructed to analyze a series of observations including the patient's demographic information, antecedents, symptoms and clinical data.

The different diseases to be identified are respectively flu, bacterial pneumonia, measles, ebola, mumps and whooping cough. For each diagnosis and treatment, different response alternatives are given. For the diagnoses, there is only one correct answer, which is the current disease. The students are given three attempts to find out the correct answer. Once the diagnosis is established, the student has to identify the right treatment, there are up to three possible correct responses, and the students are given three attempts to respond. Between the attempts, the student can re-analyze the patient in order to correct his answer.

4.2 Participant and Apparatus

Fifteen participants (8 males) aged between 20 and 27 (M = 21.8 ± 2.73) were recruited for this research. Participants were all undergraduate medicine students at the University of Montreal. Upon arrival at the laboratory, participants were briefed about the experimental objectives and familiarized with the material. They were asked to sign a consent form and placed in front of the eye tracker. They were informed that free head movements were allowed.

A stationary eye tracking system (Tobii Tx300) with a sampling rate of 300 Hz was used. The eye tracker with the infrared sensors and the camera were integrated within a 23-inch computer monitor (1920 × 1080) resolution where the game was displayed. The monitor was placed at a distance of approximately 65 cm from the participants' eyes. The gaze of each participant was calibrated with a 9-point calibration grid. The calibration step, which consists in evaluating the quality of the measured gaze points, is important since it influences the reliability of the data.

At the end of the game session, participants were shown with the medical cases they were resolving and prompted to recall their reasoning process. They were asked to self-report their visual sequence (i.e. the order of the elements they looked at). Participants were also asked to give their opinion about the environment design and usability in order to have feedbacks for potential corrections.

5 Results

The aim of this work is to demonstrate how the pattern searching sequence alignment algorithm can be applied to eye movements to assess students' analytical reasoning while resolving medical cases. First, we describe how we modeled participants' reasoning from the eye gaze data. Then we present how we used our sequence alignment method to evaluate the students' reasoning.

5.1 Students' Reasoning Sequence

The first phase of our approach is to represent the steps the students use to resolve the medical cases as a sequence. To that end, we used the recorded eye movements to extract, for each medical case, the sequential visual path the students followed to yield their conclusions from the existing observations. These scan paths were measured according to six areas of interest (AOI) that were defined in our environment, namely (I) Information, (A) Antecedent, (S) Symptoms, (N) aNalysis, (D) Diagnosis and (T) Treatment (see Fig. 3). First, we identify the AOI on which the learner focused with a fixation duration above 250 ms; this threshold was chosen according to previous studies suggesting that eye fixations should last over 200 ms to be considered as meaningful [30]. Then a scan path is recoded as a sequence of characters (e.g. "IIASSSSSANNDDDTT"). Each character corresponds to a visited AOI, and redundant characters refer to areas that have been visited many times successively. Finally, redundancies are removed to maintain only one occurrence of each visited area (e.g. "IASANDT").

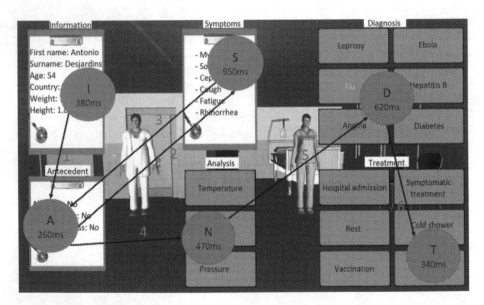

Fig. 3. An example of a scan path over the six AOIs

5.2 Reference Sequence

The second phase of our approach will be to assess whether the student is following a correct reasoning. Hence, we needed to model this correct reasoning in order to check whether the student's followed scan path matches the correct sequence. To this end, we used the hypothetico-deductive clinical reasoning shown in Fig. 4 (adapted from [31]) in order to represent the correct sequence, which will be called hereafter *reference sequence*. In this analytical process, the clinician generates hypotheses from initial clues (patient's information and symptoms) followed by further clinical data (analysis, radiography, medical antecedents, etc.) in order to confirm or negate these hypotheses until a correct diagnosis and treatment are found. From this presented diagram, we draw a sequential representation of the different steps a student needs to perform in order to have a correct reasoning resulting in the following scan path: I → S → A N → D → T, which yields the following sequence: "ISANDT".

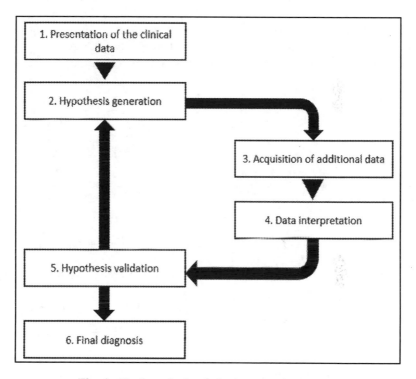

Fig. 4. The hypothetico-deductive reasoning process

5.3   Sequence Comparison

The pattern searching algorithm consists in searching for a particular pattern within a long sequence. In our case, the pattern is the reference sequence; the goal is to align it with the student's reasoning sequence in order to find out if they match in a specific position.

As previously described, the algorithm uses a substitution matrix based on an edit distance, while trying to obtain a low alignment score, which means a strong similarity between the two sequences, implying that the learner followed the correct reasoning process.

Figure 5 describes the execution of the alignment algorithm for two different participants. The scoring metrics were set as follows: 0 for match and 1 for mismatch and gap, since we aim to reduce the distance between the sequences. After filling the matrix, the algorithm searches in the last row for the cell with the smallest score. Then, from that cell, it backtracks until the first row of the array to identify the alignment path. The left part of the figure presents the alignment of a participant who followed a good reasoning process, as the obtained alignment score is zero. This means that there was no error in the alignment of the sequences and that the participant's sequence perfectly matches the reference sequence. In other words, this implies that the reference pattern is entirely included in the participant's sequence. The right part of the figure shows the alignment of a second sequence. In this example, the obtained alignment score is equal to 3, representing the three errors identified when aligning the two sequences: 2 gaps and 1 mismatch. This means that the student's reasoning process deviated from the reference sequence and hence did not match the correct reasoning.

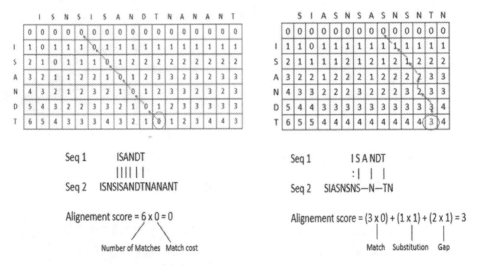

Fig. 5. Two examples of sequence alignment with the corresponding substitution matrix. (Left) a perfect match between the sequences. (Right) an alignment with three errors. "|" means a match, ":" a mismatch and "—" a gap.

To sum up, our approach involves:

1. Representing the learner's reasoning process as a sequence using the scan path followed while resolving the medical cases.
2. Using a reference sequence representing the correct analytical reasoning as a point of comparison.
3. Comparing both sequences using the pattern searching algorithm and computing an alignment score.

The smaller the alignment score (i.e. the distance), the closer the reasoning is to the correct process. Conversely the greater the score, the less accurate the reasoning is. Obviously, this method could be generalized and replicated within a different problem-solving context and different areas of interest.

6   Conclusion

In this paper, we described our approach for comparing two scan paths generated from students' eye movements while interacting with a learning serious game. The goal was to evaluate the analytical reasoning process of novice medical students. An experimental protocol was conducted while 15 participants were resolving six medical cases. They had to identify for each case the correct diagnosis and treatment starting from the patients' information, symptoms and clinical data. An eye tracker was used to record participants' eye movements. The obtained visual scan path was then extracted and compared to a reference pattern using a sequence alignment method, namely, the pattern searching algorithm to compute the similarity between the sequences.

This approach is particularly useful for learning environments and more generally for human-computer interaction applications seeking to understand and monitor the user's decision-making process within problem-solving tasks.

As our future work, we plan to study further physiological variables such as electroencephalography to adapt the system's help strategies according to both the visual attention and the learner's mental state.

Acknowledgment. We acknowledge SSHRC (Social Science and Human Research Council) through the LEADS project and NSERC (National Science and Engineering Research Council) for funding this research. Thanks to Issam Tanoubi from the University of Montreal for his collaboration on the experimental design.

References

1. Jraidi, I., Chaouachi, M., Frasson, C.: A dynamic multimodal approach for assessing learners' interaction experience. In: Proceedings of the 15th International Conference on Multimodal Interaction, pp. 271–278. ACM (2013)
2. Jraidi, I., Frasson, C.: Student's uncertainty modeling through a multimodal sensor-based approach. Educ. Technol. Soc. **16**(1), 219–230 (2013)
3. Jraidi, I., Chaouachi, M., Frasson, C.: A hierarchical probabilistic framework for recognizing learners' interaction experience trends and emotions. Adv. Hum.-Comput. Interact. **6**, 1–6 (2014)
4. Chandra, S., et al.: Eye tracking based human computer interaction: applications and their uses. In: International Conference on Man and Machine Interfacing (MAMI), pp. 1–5. IEEE (2015)
5. Yeo, H.-S., Lee, B.-G., Lim, H.: Hand tracking and gesture recognition system for human-computer interaction using low-cost hardware. Multimed. Tools Appl. **74**(8), 2687–2715 (2015)
6. Taub, M., Azevedo, R.: Using eye-tracking to determine the impact of prior knowledge on self-regulated learning with an adaptive hypermedia-learning environment. In: Micarelli, A., Stamper, J., Panourgia, K. (eds.) ITS 2016. LNCS, vol. 9684, pp. 34–47. Springer, Cham (2016). doi:10.1007/978-3-319-39583-8_4
7. Wang, C.-Y., Tsai, M.-J., Tsai, C.-C.: Multimedia recipe reading: predicting learning outcomes and diagnosing cooking interest using eye-tracking measures. Comput. Hum. Behav. **62**, 9–18 (2016)

8. Jraidi, I., Chalfoun, P., Frasson, C.: Implicit strategies for intelligent tutoring systems. In: Cerri, S.A., Clancey, W.J., Papadourakis, G., Panourgia, K. (eds.) ITS 2012. LNCS, vol. 7315, pp. 1–10. Springer, Heidelberg (2012). doi:10.1007/978-3-642-30950-2_1

9. Chaouachi, M., Jraidi, I., Frasson, C.: MENTOR: a physiologically controlled tutoring system. In: Ricci, F., Bontcheva, K., Conlan, O., Lawless, S. (eds.) UMAP 2015. LNCS, vol. 9146, pp. 56–67. Springer, Cham (2015). doi:10.1007/978-3-319-20267-9_5

10. Ben Khedher, A., Jraidi, I., Frasson, C.: Learners' performance tracking using eye gaze data. In: 1st International Workshop on Supporting Dynamic Cognitive, Affective, and Metacognitive Processes (SD-CAM), Part of the 13th International Conference on Intelligent Tutoring Systems, pp. 15–24 (2016)

11. Takeuchi, H., Habuchi, Y.: A quantitative method for analyzing scan path data obtained by eye tracker. In: IEEE Symposium on Computational Intelligence and Data Mining, pp. 283–286. IEEE (2007)

12. Toker, D., Conati, C.: Eye tracking to understand user differences in visualization processing with highlighting interventions. In: Dimitrova, V., Kuflik, T., Chin, D., Ricci, F., Dolog, P., Houben, G.-J. (eds.) UMAP 2014. LNCS, vol. 8538, pp. 219–230. Springer, Cham (2014). doi:10.1007/978-3-319-08786-3_19

13. Courtemanche, F., et al.: Activity recognition using eye-gaze movements and traditional interactions. Interact. Comput. 23(3), 202–213 (2011)

14. Jaques, N., Conati, C., Harley, J.M., Azevedo, R.: Predicting affect from gaze data during interaction with an intelligent tutoring system. In: Trausan-Matu, S., Boyer, K.E., Crosby, M., Panourgia, K. (eds.) ITS 2014. LNCS, vol. 8474, pp. 29–38. Springer, Cham (2014). doi:10.1007/978-3-319-07221-0_4

15. Cristino, F., et al.: ScanMatch: a novel method for comparing fixation sequences. Behav. Res. Methods 42(3), 692–700 (2010)

16. Ghali, R., Frasson, C., Ouellet, S.: Towards real time detection of learners' need of help in serious games. In: 29th International Flairs Conference, pp. 154–157 (2016)

17. Kardan, S., Conati, C.: Exploring gaze data for determining user learning with an interactive simulation. In: Masthoff, J., Mobasher, B., Desmarais, M.C., Nkambou, R. (eds.) UMAP 2012. LNCS, vol. 7379, pp. 126–138. Springer, Heidelberg (2012). doi:10.1007/978-3-642-31454-4_11

18. Martínez-Gómez, P., Aizawa, A.: Recognition of understanding level and language skill using measurements of reading behavior. In: Proceedings of the 19th International Conference on Intelligent User Interfaces, pp. 95–104. ACM (2014)

19. Kiili, K., Ketamo, H., Kickmeier-Rust, M.D.: Evaluating the usefulness of eye tracking in game-based learning. Int. J. Serious Games 1(2), 51–65 (2014)

20. Tsai, M.-J., et al.: Visual attention for solving multiple-choice science problem: an eye-tracking analysis. Comput. Educ. 58(1), 375–385 (2012)

21. Voisin, S., et al.: Investigating the association of eye gaze pattern and diagnostic error in mammography. In: Proceedings of SPIE (Vol. 867302), Medical Imaging 2013: Image Perception, Observer Performance, and Technology Assessment, pp. 867302–867308 (2013)

22. Tai, R.H., Loehr, J.F., Brigham, F.J.: An exploration of the use of eye-gaze tracking to study problem-solving on standardized science assessments. Int. J. Res. Method Educ. 29(2), 185–208 (2006)

23. Ben Khedher, A., Frasson, C.: Predicting user learning performance from eye movements during interaction with a serious game. In: EdMedia: World Conference on Educational Media and Technology, Association for the Advancement of Computing in Education (AACE), pp. 1504–1511 (2016)

24. Cutrell, E., Guan, Z.: What are you looking for?: an eye-tracking study of information usage in web search. In: Proceedings of the SIGCHI Conference on Human Factors in Computing Systems, pp. 407–416. ACM (2007)
25. Hayhoe, M., Ballard, D.: Eye movements in natural behavior. Trends Cogn. Sci. **9**(4), 188–194 (2005)
26. Glady, Y., Thibaut, J.-P., French, R.M.: Visual strategies in analogical reasoning development: a new method for classifying scanpaths. In: Proceedings of the 35th Annual Meeting of the Cognitive Science Society, pp. 2398–2403 (2013)
27. Susac, A., et al.: Eye movements reveal students' strategies in simple equation solving. Int. J. Sci. Math. Educ. **12**(3), 555–577 (2014)
28. Lange, J., Wyrwicz, L.S., Vriend, G.: KMAD: knowledge-based multiple sequence alignment for intrinsically disordered proteins. Bioinformatics **32**(6), 932–936 (2016)
29. Zou, Q., et al.: HAlign: fast multiple similar DNA/RNA sequence alignment based on the centre star strategy. Bioinformatics **31**(15), 2475–2481 (2015)
30. Liversedge, S.P., Findlay, J.M.: Saccadic eye movements and cognition. Trends Cogn. Sci. **4**(1), 6–14 (2000)
31. Nendaz, M., et al.: Le raisonnement clinique: données issues de la recherche et implications pour l'enseignement. Pédagogie médicale **6**(4), 235–254 (2005)

Healthcare Informatics Theory
and Methods

An Intelligent Systems Approach to Primary Headache Diagnosis

Robert Keight[1]([✉]), Ahmed J. Aljaaf[1], Dhiya Al-Jumeily[1], Abir Jaafar Hussain[1], Aynur Özge[2], and Conor Mallucci[3]

[1] Applied Computing Research Group, The Faculty of Engineering and Technology, Liverpool John Moores University, Byrom Street, Liverpool L3 3AF, UK
R.Keight@2015.ljmu.ac.uk, A.J.Kaky@2013.ljmu.ac.uk,
{D.Aljumeily,A.Hussain}@ljmu.ac.uk
[2] Mersin University School of Medicine, Mersin, Turkey
aozge@mersin.edu.tr
[3] Department of Neurosurgery, The Royal Liverpool Childrens NHS Trust,
Alder Hey, Liverpool, UK
cmallucci@me.com

Abstract. In this study, the problem of primary headache diagnosis is considered, referring to multiple frames of reference, including the complexity characteristics of living systems, the limitation of human information processing, the enduring nature of headache throughout history, and the potential for intelligent systems paradigms to both broaden and deepen the scope of such diagnostic solutions. In particular, the use of machine learning is recruited for this study, for which a dataset of 836 primary headache cases is evaluated, originating from two medical centres located in Turkey. Five primary headache classes were derived from the data obtained, namely Tension Type Headache (TTH), Chronic Tension Type Headache (CTTH), Migraine with Aura (MwA), Migraine without Aura (MwoA), followed by Trigeminal Autonomic Cephalalgia (TAC). A total of 9 machine learning based classifiers, ranging from linear to non-linear ensembles, in addition to 1 random baseline procedure, were evaluated within a supervised learning setting, yielding highest performance outcomes of AUC 0.985, sensitivity 1, and specificity 0.966. The study concludes that modern computing platforms represent a promising setting through which to realise intelligent solutions, enabling the space of analytical operations needed to drive forward diagnostic capability in the primary headache domain and beyond.

Keywords: Cephalalgia · Headache diagnosis · Artificial intelligence

1 Introduction

Over the centuries, the phenomenon of headache, formally termed Cephalalgia, has remained an enduring burden for societies and healthcare systems [1, 2]. With a current global prevalence estimate of 50% [3], an estimated of economic burden set at around $500 million per annum in the UK, and from $5.6 to $17.2 billion in the United States [4], headache conditions account for one of the most common complaints within primary and

© Springer International Publishing AG 2017
D.-S. Huang et al. (Eds.): ICIC 2017, Part II, LNCS 10362, pp. 61–72, 2017.
DOI: 10.1007/978-3-319-63312-1_6

neurological care settings [5]. Principally, headache may be defined as pain occurring anywhere in the region of the head or neck [6], though the boundary of such conditions remains incompletely defined. Today's accepted international standard is provided by the International Classification of Headache Disorders (ICHD) criteria [7], first introduced in 1988, which describes both headache entities and diagnostic criteria. Such conditions are currently dichotomised into two broad categories, primary (benign), and secondary (serious), where the former conditions are relatively common and the latter constitute only around 10% of all known cases [8]. In this study, we focus on demonstrating a potential pathway for the advancement of headache diagnosis for the primary conditions, proposing an intelligent systems methodology based on supervised classification. A consideration of 5 primary headache types is undertaken, comprising Tension Type Headache (TTH), Chronic Tension Type Headache (CTTH), Migraine with Aura (MwA), Migraine without Aura(MwoA), followed by Trigeminal Autonomic Cephalalgia (TAC). Accordingly, the rest of this paper is therefore organised as follows. In Sect. 2 a review of prior works is undertaken, Sect. 3 introduces machine learning as a diagnostic solution; the approach and methodology taken are covered in Sect. 4, followed by Sects. 5 and 6 in which the results of exploratory and classification analysis are presented. Finally, the discussion and conclusions for this study are presented in Sects. 7 and 8 respectively, completing the paper.

2   Review of Prior Works

At the intersection of the domains of primary headache diagnostics and intelligent systems, a range of works are seen to exist in the literature. Intuitionistic fuzzy sets are considered in [9, 10]. The studies use an interval-valued intuitionistic fuzzy (IVIFS) weighted arithmetic average operator, with a min-max composition rule to determine the final diagnosis of disease. The authors concluded that the IVIFS method is viable for medical diagnosis. An Artificial Neural Network approach is considered in [11]. The study considers four types of primary headache disorder over a sample of 2,177 patients. Results reported a sensitivity and specificity of 0.93 and 0.91 for tension type headache, 0.99 and 0.94 for migraine without aura, 1.0 and 0.98 for migraine with aura, and 1.0 and 0.96 for medication overuse headache. A manually defined Decision Tree approach was considered in [12], in which the criteria published by the International Headache Society (IHS) are compared in terms of the performance of ad-hoc criteria. There search found a 10% error rate yielded by use of the IHS diagnostic criteria. Celik et al. address migraine, tension-type, cluster, and other primary headaches using an Artificial Immune System (AIS) algorithm in [13]. A sample of 850 patients was considered, reporting 94% classification accuracy, with the best result reaching 99.65%. In another study, the same authors investigate the use of the k-means algorithm in [14], restricting the scope of the study to migraines with and without aura. The work uses the data of 288 students for use with cluster based analysis, the authors conclude in closing discussion that the final system performance is not acceptable when compared with results obtained from a neurologist. Tezel and Kose apply an artificial immune approach, using a Clonal Selection Algorithm [15] to analyse a sample of 150 headache patients and 65 healthy subjects, reporting a highest classification accuracy of 92%.

3   Machine Learning as a Diagnostic Solution

Diagnostic tasks can be considered a form of information intensive problem, demanding processing pathways capable of maximising the utility of available input. Existing research within healthcare domains [16] and beyond [17, 18], has demonstrated the potential of intelligent systems approaches where such information challenges are presented. The space of primary headaches, seated in biological substrates, forms a complex and demanding diagnostic domain with a high-information factor. Human agency grounded procedures, built around conventional technological settings, may not be optimal for such extensive information tasks. Machine learning offers the capacity to address such information work-loads, without the interference expected of thought-based processing artefacts, namely biologically bounded attention, memory, and decision making [19, 20]; intra and inter-observer variability [21], and limited, localised availability. Consequently, the integration of computationally driven intelligent agency therefore promises to efficiently shift the boundaries in diagnostic accuracy in headache, allowing an arbitrary number of simulta-neous patient features to be analysed without necessitating a team of human experts. Such properties offer a good response to the problem of headache, given both its complexity and prevalence in populations, in addition to guiding our understanding of headache itself.

4   Methodology

4.1   Dataset

The dataset used in this study represents 836 applicable primary headache cases, collected from two medical institutions located in Turkey; namely Cerrahpasa Medical Faculty in Istanbul, and Mersin University Medical Facility, Mersin. The ground truth label for each case was determined by expert opinion, which is assumed in this study to represent a reasonable proxy for the true diagnosis. Diagnoses in the original dataset span a total of 8 primary headache types, which were subsequently reduced to 5 distinct classes through consolidation of the Trigeminal Autonomic Cephalalgias, which indi-vidually lacked sufficient case quantities to be useful. The final headache classes consid-ered therefore comprise: Tension type headache (TTH), Chronic Tension Type Head-ache (CTTH), Migraine with aura (MwA), Migraine without aura (MwoA), followed by Trigeminal Autonomic Cephalalgia (TAC). The TAC category originally existed as 4 distinctly labelled subsets, namely Cluster Headache (CH), Paroxysmal Hemicrania (PHem), Hemicrania continua (HC), and Short-lasting unilateral neuralgiform headache with conjunctival injection and tearing (SUNCT).

4.2   Features

A total of 65 features are considered in this study, which can be summarised according to 9 broad categories. Firstly, the patient's socio-demographics were extracted, revealing 4 features, namely age, gender, smoking status, and smoking duration. Subsequently, Headache characteristics themselves are considered using 6 distinct features, comprising

Headache Onset (months), Headache Frequency (days/months), Headache Character-istic (throbbing, pressing, stabbing, dull, lightening), Headache duration (hours), Head-ache localisation (Bilateral generalised, Bi-temporal, Calvarial, Facial Pain, Frontal, Occipital, Periocular, Secondary generalised, and Unilateral), followed by Headache intensity (Visual Analogue Scale, VAS). In conjunction with the former attributes, further features comprised of Precipitant factors (8 features), Accompanying symptoms (11 features), Psychological condition (5 features), in addition to examinations, namely Fundoscopy (1 feature) and Neurological (1 feature). The patient's medical history was also included (21 features), followed by their family medical history(8 features).

4.3 Preprocessing

Prior to the application of analytical operations, the raw dataset, originating as a result of human input, was scrutinised to ensure consistency with the expected domain of each feature, 2 cases were dropped due to the presence of malformed information. Subse-quently, all human readable character representations were subject to mapping to discrete integer enumerations, yielding a numerical encoding over all features, as is required to provide suitable input for the machine learning phase of analysis. Missing values at this point were substituted for dummy values, such that the number of cases could be maintained; appropriately values were chosen for each feature domain so as not to introduce conflicts. It was later found that this approach did not unduly impact classifier performances. The numerical representation of the problem space was then subjected to z-score normalisation, yielding a consistent scale and location of values over all features. As a result, the original dataset used for this study was applied to the analytical phase in a unified numerically encoded form.

4.4 Supervised Learning Formulation

A numerically encoded representation derived from the original domain data is used in this study as an operational substrate upon which a machine learning problem may be established. In particular, a supervised learning problem is formulated, where data may be interpreted as a series of numerically encoded examples, of the form: $x_{i,1}, \dots x_{i,p}, y_i$, where x represents a feature vector with parameters 1 through p, and $y \in 1 \dots 5$ is a ground truth value. Each value of y may be considered an index pointing to one of the 5 classes of headache considered in this study. Classifier models may be formed through repeated exposure to such x, y pairs, following which the fitted model may be used to produce estimates of y, given x alone, to produce $\hat{y}$. Such a functional mapping may be formalised as $f : \bar{x} \rightarrow y \in \{1, \dots, 5\}$, where the learning element of the algorithm searches by induc-tion a hypothesis space F for a suitable $\hat{f} \in$ F, with the aim of optimal generalisation performance. In effect, once trained, the knowledge synthesised by a model may be used to classify future inputs for which outputs are not known.

4.5   Model Architectures

The series of models considered in this study are listed in Table 1, accompanied by respective architecture and hyperparameter values. In order to investigate which model types may operate effectively with the dataset under consideration, a variety of architectures have been posed, spanning multiple learning paradigms and complexity classes. The use of a diverse model space ensures that the potential information utility of the primary headache data may be profiled while highlighting the strengths and weakness of the member algorithms within this domain. Each model may be thought of as a hypothesis concerning which structural elements of the dataset may be conducive to learning a classification rule. An uninformed model, ROM, is used as a random baseline, while the LNN, LDA, and SVM models represent linear model classes, followed by

Table 1. Models.

No.	Model	Description	Architecture	Training algorithm	Role
1	ROM	Random Oracle Model	Pseudo-random Number Generator	N/A	RAND
2	LNN	Linear Neural Network	35 Units, Linear Activations	Batch training with weight and bias learning rules	LB
3	LDA	Linear Discriminant Analysis	Linear Combination	Between class maximisation via closed form equation [23]	LB
4	SVM	Support Vector Machine	Matrix Kernel	Quadratic Optimisation	LB
5	KNN	K-Nearest Neighbor	5 Neighbours	Instance Induction	WNLB
6	TREEC	Decision Tree Classifier	Decision Tree	Tree induction using information gain criterion	TC
7	RFC	Random Forest Classifier	Decision Tree Ensemble with 200 trees	Random feature subset bagging	TC
8	LEVNN	Levenberg Neural Network	Units: 55-55-5	Levenberg Marquardt	TC
9	H1	Hybrid Stacked Generaliser 1	Learners: LEVNN 55-55-5, SVM, KNN 25,10,3; Combiner: RFC 200 trees	Hybrid	TC
10	H2	Hybrid Stacked Generaliser 2	Learners: KNN 25,20,15,12,10,3; Combiner: RFC 200 trees	Hybrid	TC

Key: RAND = Random Baseline, LB = Linear Baseline
WNLB = Weak Non-linear Baseline, TC = Test Classifier

KNN and TREEC which serve as weak non-linear learners, with the remaining models acting as experimental models, capable of powerful non-linear approximation. In particular, two hybrid ensemble models are introduced, H1 and H2, such that weaknesses of individual models may be overcome while maintaining their strengths [22].

4.6 Simulation Procedure

In order to obtain empirical results for the proposed case study, the dataset previously discussed was used for both the training and testing of the classifiers under study, with 70% holdout assigned to the training set and 30% to the test set, respectively. Feature space mapping via Principal Component Analysis (PCA) was subsequently estimated over the training set and the mapping applied to both the training and test feature vectors, such that noise components and the dimensionality were reduced. The first 55 principal components were used. Simulations were then subjected to 50 repeated trials, such that average responses could be obtained and the effect of outlying realisations attenuated.

4.7 Performance Evaluation

To objectively measure the capability of the classifiers under study, a framework of performance metrics is introduced. A total of 7 scalar metric calculations are listed in Table 2, including sensitivity and specificity, which are applied subsequent to Receiver Operating Curve (ROC) analysis. ROC analysis is a widely accepted performance evaluation technique in the medical domain [24], comprising a parametric curve in two-dimensional ROC space, formed from true and false positive rate measurements. Importantly, in contrast to a simple accuracy calculation, ROC analysis is agnostic to imbalances in class representation, providing robustness of evaluation. In this study, the minimum Euclidean distance from ideal was used to select a final threshold from the ROC analysis. The limit of random guessing performance is noted to fall along the diagonal running from the South-West to the North-East corners of the unit square.

Table 2. Performance metrics.

Metric	Abbr.	Computation	Range
Area Under Curve	AUC	$0 \leq \text{area(ROC)} \leq 1$	[0,1]
Sensitivity	SEN	TP/(TP + FN)	[0,1]
Specificity	SPEC	TN/(TN + FP)	[0,1]
Precision	PRE	TP/(TP + FP)	[0,1]
F1 Score	F1	$2 \cdot (\text{PRE/RC})/(\text{PRE} + \text{RC})$	[0,1]
Youden's J Statistic	J	Sensitivity + Specificity - 1	[−1,1]
Accuracy	ACC	(TP + TN)/(TP + FN + TN + FP)	[0,1]

Key: TP = True Positive Count, TN = True Negative Count
PRE = Precision, RC = Recall, ROC = Receiver Operating Characteristic

5 Exploratory Analysis

Following the derivation of a feature matrix and prior to primary classification analysis, graphical domain representations of the data were constructed, encoding the problem in multiple perceptually accessible forms. Both Principal Component Analysis (PCA) and t-Distributed Stochastic Neighbourhood Embedding(tSNE) [25] techniques were applied, with results listed in Figs. 1 and 2. Jointly, the two graphical views reveal distinctive areas of clustering for TTH, MwoA, and CTTH, with TAC and MwA appearing more diffuse and less coherent. The tSNE domain plot appears to capture more sharply the essence of such clustering, showing both areas of separability and conflation between headache types within the two-dimensional summary space. The visual domain suggests that the data does in fact carry distinct structures and that classification has a justifiable basis.

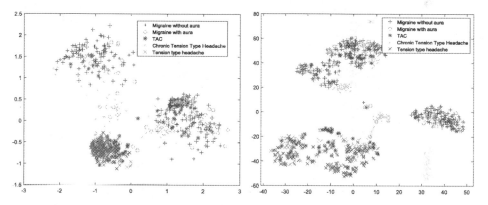

Fig. 1. Principal component analysis plot   **Fig. 2.** t-SNE plot

6 Results

The test results from the experimental procedure undertaken in this study are presented as follows. Table 3 lists the results from each classifier in terms of the 7 scalar perform-ance metrics considered; Figure 3 shows plots resulting from ROC domain analysis, allowing visual inspection of the classifier responses over each model and class, Fig. 4 provides a visual summary of the AUC values obtained from the former.

Table 3. Results (Test Holdout).

Model	Class	Sensitivity	Specificity	Precision	F1	J	Accuracy	AUC
ROM	TTH	0.51	0.61	0.352	0.417	0.12	0.581	0.496
	CTTH	0.615	0.362	0.151	0.242	−0.023	0.401	0.419
	MwA	0.333	0.852	0.148	0.205	0.185	0.814	0.482
	MwoA	0.803	0.376	0.457	0.582	0.179	0.545	0.556
	TAC	0.643	0.575	0.122	0.205	0.218	0.581	0.583
LNN	TTH	0.938	0.966	0.918	0.928	0.904	0.958	0.979
	CTTH	0.923	0.865	0.558	0.696	0.788	0.874	0.93
	MwA	0.667	0.832	0.235	0.348	0.499	0.82	0.757
	MwoA	0.955	0.9	0.865	0.908	0.855	0.922	0.953
	TAC	0.857	0.771	0.255	0.393	0.628	0.778	0.822
LDA	TTH	0.959	0.932	0.855	0.904	0.891	0.94	0.969
	CTTH	0.923	0.83	0.5	0.649	0.753	0.844	0.924
	MwA	0.833	0.652	0.156	0.263	0.485	0.665	0.783
	MwoA	0.97	0.901	0.865	0.914	0.871	0.928	0.954
	TAC	0.857	0.784	0.267	0.407	0.641	0.79	0.822
SVM	TTH	0.939	0.915	0.821	0.876	0.854	0.922	0.97
	CTTH	0.885	0.936	0.719	0.793	0.821	0.928	0.956
	MwA	0.75	0.626	0.134	0.228	0.376	0.635	0.753
	MwoA	0.939	0.901	0.861	0.899	0.84	0.916	0.942
	TAC	0.643	0.863	0.3	0.409	0.506	0.844	0.793
KNN	TTH	1	0.898	0.803	0.891	0.898	0.928	0.969
	CTTH	0.731	0.95	0.731	0.731	0.681	0.916	0.882
	MwA	0.75	0.703	0.164	0.269	0.453	0.707	0.772
	MwoA	0.985	0.901	0.867	0.922	0.886	0.934	0.95
	TAC	0.643	0.81	0.237	0.346	0.453	0.796	0.761
TREEC	TTH	0.735	0.898	0.75	0.742	0.633	0.85	0.878
	CTTH	0.615	0.73	0.296	0.4	0.346	0.713	0.717
	MwA	0.417	0.6	0.0746	0.127	0.0167	0.587	0.535
	MwoA	0.682	0.851	0.75	0.714	0.533	0.784	0.811
	TAC	0.571	0.699	0.148	0.235	0.271	0.689	0.652
RFC	TTH	0.959	0.924	0.839	0.895	0.883	0.934	0.984
	CTTH	0.846	0.872	0.55	0.667	0.718	0.868	0.907
	MwA	0.75	0.781	0.209	0.327	0.531	0.778	0.734
	MwoA	0.955	0.891	0.851	0.9	0.846	0.916	0.957
	TAC	0.714	0.771	0.222	0.339	0.486	0.766	0.803
LEVNN	TTH	0.98	0.941	0.873	0.923	0.92	0.952	0.979
	CTTH	0.923	0.929	0.706	0.8	0.852	0.928	0.94
	MwA	0.667	0.858	0.267	0.381	0.525	0.844	0.81
	MwoA	0.985	0.881	0.844	0.909	0.866	0.922	0.96
	TAC	0.786	0.712	0.2	0.319	0.498	0.719	0.818
H1	TTH	0.959	0.949	0.887	0.922	0.908	0.952	0.985
	CTTH	0.962	0.894	0.625	0.758	0.855	0.904	0.964
	MwA	0.667	0.845	0.25	0.364	0.512	0.832	0.789
	MwoA	0.939	0.921	0.886	0.912	0.86	0.928	0.959
	TAC	0.857	0.732	0.226	0.358	0.589	0.743	0.874
H2	TTH	0.959	0.924	0.839	0.895	0.883	0.934	0.983
	CTTH	0.885	0.894	0.605	0.719	0.778	0.892	0.942
	MwA	0.75	0.626	0.134	0.228	0.376	0.635	0.797
	MwoA	1	0.891	0.857	0.923	0.891	0.934	0.95
	TAC	0.857	0.752	0.24	0.375	0.609	0.76	0.88

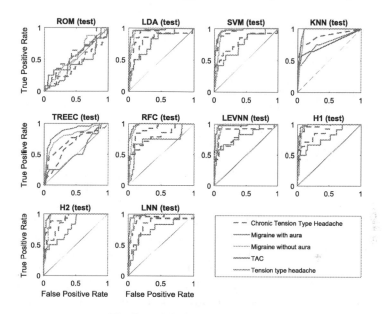

Fig. 3. ROC plots for the test set

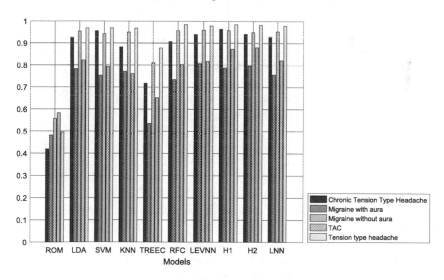

Fig. 4. AUC plots for the test set

7 Discussion

From examination of the 5 primary headache classes, evaluated over the 9 models and 1 random baseline, it is apparent that TTH, CTTH, and MwoA have yielded viable results over nearly all of the informed classifiers, with exception of the decision tree learner, achieving highest AUC values of 0.985, 0.96, 0.964, respectively. Additionally, it can be observed that all three of these classes consistently yield AUC values greater than 0.9, with the exception of evaluation over the decision tree and KNN models. The classes TAC and MwA have yielded a significantly lower band of performance, with AUCs of 0.88 and 0.81 respectively, while following the same relative cross-model performance pattern as the other classes. On observation of the ROC plots, the similarity in the performance profile, with a few exceptions, can be confirmed. All models appear to exhibit a similar behaviour over the classes, with the exception of the decision tree, the KNN model, and to a lesser extent the RFC model. The highest sensitivities were achieved by the KNN model with respect to the TTH class, followed by the H2 ensemble model for the MwoA class, both of which achieved a perfect result in terms of this metric. Sensitivities for the same were 0.898 and 0.891 respectively, indicating that such model-class combinations are a viable prospect. Such a result can be filtered by the ROC geometry, showing that the KNN model, although equally sensitive when compared to the H2 model, shows a less stable performance profile overall in relation to the decision threshold. The highest overall accuracy outcome was obtained by the LNN mode over the TTH class, yielding a value of 0.958. Moreover, the other performance outcomes obtained by this model were also comparable to the more powerful non-linear classifiers, demonstrating the adequacy of a linear model complexity within the considered data scenario. Finally, it is observed that the non-linear models offered little advantage over the linear model classes, though all informed models significantly outperformed the ROM random baseline.

8   Conclusion and Future Work

To investigate primary headache diagnosis within the intelligent systems framework, an experimental procedure was undertaken in this study, considering a dataset of over 800 patient cases originating from two medical institutions in Turkey. Five primary headache classes were defined as targets in the study, over which 9 machine learning based classifiers were simulated. Results showed that a highest AUC of 0.985 could be obtained, followed by a highest sensitivity of 1 and a best specificity of 0.958. The most clearly discriminated classes were found to be the TTH, MwoA, and CTTH respectively, with both linear and non-linear model classes demonstrating significant capability. Less accurate results were obtained for classes TAC and MwA, though it was found that data availability for these headache types comprised significantly fewer cases than the other three targets, leading to a hypothesis that an increase in data may equalise the result distribution, which is to be considered in future work. It was also found that there was little indication of the presence of non-linear components within the dataset, as is demonstrated by the close similarity in performance between the non-linear and linear model types.

In aggregate, the results demonstrate that intelligent systems represent a promising approach for primary headache diagnosis, which in conjunction with wider technological ecosystems likely hold significant potential to disrupt conventional models of diagnostic delivery. In particular, our approach may be positioned as a component of personalised medicine. In future work, we aim to overcome some of the key limitations inherent of the former case study. In particular, the assumption of a single headache diagnosis should be extended to allow for multiple simultaneous diagnoses, since multiple headache types are known to coexist in patients [26]. Additionally, it is understood that the ground truth labels used, the diagnosis provided by expert opinion, has the potential itself to be inaccurate; an investigation of this issue is warranted. The use of deep learning algorithms is a further avenue of interest, since it was apparent that a change in feature representation via PCA caused a significant shift in results, where deep network architectures are known to internally facilitate such representational mappings [27]. Finally, the scope of the study should be expanded to include signals from physiological domains, such as EEG and MEG. The analysis of such biologically grounded information channels may allow for the discovery and operationalisation of otherwise hidden neurophysiological markers, providing a more intimate vantage point of the living system and therefore permitting new classes of diagnostic solution.

References

1. Magiorkinis, E., Diamantis, A., Mitsikostas, D.D., Androutsos, G.: Headaches in antiquity and during the early scientific era. J. Neurol. **256**(8), 1215–1220 (2009)
2. Diamond, S., Franklin, M.A.: Headache Through the Ages. Professional Communications, New York (2005)
3. Steiner, T.J., Stovner, L.J., Al Jumah, M., Birbeck, G.L., Gururaj, G., Jensen, R., Katsarava, Z., Queiroz, L.P., Scher, A.I., Tekle-Haimanot, R., et al.: Improving quality in population surveys of headache prevalence, burden and cost key methodological considerations. J. Headache Pain **14**(1), 87 (2013)
4. Clarke, C., MacMillan, L., Sondhi, S., Wells, N.: Economic and social impact of migraine. QJM **89**(1), 77–84 (1996)
5. Lipton, R.B., Stewart, W.F., Simon, D.: Medical consultation for migraine: results from the American migraine study. Headache: J. Head Face Pain **38**(2), 87–96 (1998)
6. La Jolla Donald, J., et al.: Wolf's Headache and Other Head Pain. Oxford University Press, New York (2001)
7. Headache Classification Committee of the International Headache Society (HIS): The international classification of headache disorders, (beta version). Cephalalgia, vol. 33(9), pp. 629–808 (2013)
8. Martin, V.T.: The diagnostic evaluation of secondary headache disorders. Headache J. Head Face Pain **51**(2), 346–352 (2011)
9. Ahn, J.Y., Han, K.S., Oh, S.Y., Lee, C.D.: An application of interval-valued intuitionistic fuzzy sets for medical diagnosis of headache. Int. J. Innov. Comput. Inf. Control **7**(5), 2755–2762 (2011)
10. Ahn, J.Y., Park, J.H.: Headache diagnosis method using an aggregate operator. CSAM (Communications for Statistical Applications and Methods) 19(3), pp. 359–365 (2012)
11. Mendes, K.B., Fiuza, R.M., Teresinha, M., Steiner, A.: Diagnosis of headache using artificial neural networks. J. Comput. Sci. **10**(7), 172–178 (2010)
12. Andrew, M.E., Penzien, D.B., Rains, J.C., Knowlton, G.E., McAnulty, R.D.: Development of a computer application for headache diagnosis: The headache diagnostic system. Int. J. Bio-Med. Comput. **31**(1), 17–24 (1992)

13. Celik, U., Yurtay, N., Koc, E.R., Tepe, N., Gulluoglu, H., Ertas, M.: Diagnostic accuracy comparison of artificial immune algorithms for primary headaches. Comput. Math. Methods Med. **2015**, 8 (2015)
14. Celik, U., Yurtay, N., Yurtay, Y.: Headache diagnosis with k-means algorithm. Glob. J. Technol. **1**, 1074–1081 (2012)
15. Tezel, G., Kose, U.: Headache disease diagnosis by using the clonal selection algorithm. In: 6th International Advanced Technologies Symposium, pp. 144–148 (2011)
16. Hussain, A.J., Fergus, P., Al-Askar, H., Al-Jumeily, D., Jager, F.: Dynamic neural network architecture inspired by the immune algorithm to predict preterm deliveries in pregnant women. Neurocomputing **151**, 963–974 (2015)
17. Al-Jumeily, D., Hussain, A., Fergus, P.: Using adaptive neural networks to provide self-healing autonomic software. Int. J. Space-Based Situated Comput. **5**(3), 129–140 (2015)
18. Ghazali, R., Hussain, A.J., Al-Jumeily, D., Lisboa, P.: Time series prediction using dynamic ridge polynomial neural networks. In: 2009 Second International Conference on Developments in eSystems Engineering (DESE), pp. 354–363. IEEE (2009)
19. Klingberg, T.: Limitations in information processing in the human brain: neuroimaging of dual task performance and working memory tasks. Prog. Brain Res. **126**, 95–102 (2000)
20. Simon, H.A.: Models of Man: Social and Rational. Wiley, New York (1957)
21. Crombie, D., Cross, K., Fleming, D.: The problem of diagnostic variability in general practice. J. Epidemiol. Community Health **46**(4), 447–454 (1992)
22. Dietterich, T.G.: Ensemble learning. In: The Handbook of Brain Theory and Neural Networks, vol. 2, pp. 110–125 (2002)
23. Balakrishnama, S., Ganapathiraju, A.: Linear discriminant analysis-a brief tutorial. Institute for Signal and information Processing, vol. 18 (1998)
24. Zweig, M.H., Campbell, G.: Receiver-operating characteristic (ROC) plots: a fundamental evaluation tool in clinical medicine. Clin. Chem. **39**(4), 561–577 (1993)
25. Van der Maaten, L., Hinton, G.: Visualizing data using t-SNE. J. Mach. Learn. Res. **9**, 2579–2605 (2008)
26. Lipton, R.B., Diamond, S., Reed, M., Diamond, M.L., Stewart, W.F.: Migraine diagnosis and treatment: results from the American migraine study II. Headache J. Head Face Pain **41**(7), 638–645 (2001)
27. Bengio, Y., Courville, A., Vincent, P.: Representation learning: A review and new perspectives. IEEE Trans. Pattern Anal. Mach. Intell. **35**(8), 1798–1828 (2013)

Genetic Algorithms

Benchmarking and Evaluating MATLAB Derivative-Free Optimisers for Single-Objective Applications

Lin Li[1], Yi Chen[1], Qunfeng Liu[1], Jasmina Lazic[2], Wuqiao Luo[3], and Yun Li[1,3(✉)]

[1] School of Computer Science and Network Security,
Dongguan University of Technology, Dongguan 523808, China
{lilin0214,chenyi,liuqf}@dgut.edu.cn, Yun.Li@ieee.org
[2] MathWorks, Matrix House, Cambridge CB4 0HH, UK
Jasmina.Lazic@mathworks.co.uk
[3] School of Engineering, University of Glasgow, Glasgow G12 8LT, UK
w.luo.1@research.gla.ac.uk

Abstract. MATLAB® builds in a number of derivative-free optimisers (DFOs), conveniently providing tools beyond conventional optimisation means. However, with the increase of available DFOs and being compounded by the fact that DFOs are often problem dependent and parameter sensitive, it has become challenging to determine which one would be most suited to the application at hand, but there exist no comparisons on MATLAB DFOs so far. In order to help engineers use MATLAB for their applications without needing to learn DFOs in detail, this paper evaluates the performance of all seven DFOs in MATLAB and sets out an amalgamated benchmark of multiple benchmarks. The DFOs include four heuristic algorithms - simulated annealing, particle swarm optimization (PSO), the genetic algorithm (GA), and the genetic algorithm with elitism (GAe), and three direct search algorithms - Nelder-Mead's simplex search, pattern search (PS) and Powell's conjugate search. The five benchmarks presented in this paper exceed those that have been reported in the literature. Four benchmark problems widely adopted in assessing evolutionary algorithms are employed. Under MATLAB's default settings, it is found that the numerical optimisers Powell is the aggregative best on the unimodal Quadratic Problem, PSO on the lower dimensional Scaffer Problem, PS on the lower dimensional Composition Problem, while the extra-numerical genotype GAe is the best on the Varying Landscape Problem and on the other two higher dimensional problems. Overall, the GAe offers the highest performance, followed by PSO and Powell. The amalgamated benchmark quantifies the advantage and robustness of heuristic and population-based optimisers (GAe and PSO), especially on multimodal problems.

Keywords: Evolutionary algorithms · Heuristic search · Direct search methods · Derivative-free optimisation · Benchmarking

© Springer International Publishing AG 2017
D.-S. Huang et al. (Eds.): ICIC 2017, Part II, LNCS 10362, pp. 75–88, 2017.
DOI: 10.1007/978-3-319-63312-1_7

1 Introduction

During the past four decades, many heuristic and other derivative-free optimisers (DFOs) [1, 2] have been developed. They have been successfully applied to a wide range of real-world problems and are made available in MATLAB®, which is used daily by a significant number of practising engineers and scientists. With the development of an increasing number of DFOs beyond conventional optimisation means, it has become challenging to choose the most suitable ones for an engineer's application at hand. To provide a practical review of these tools, this paper conducts and reports their benchmarking tests. Further, a scoring system for benchmarking DFOs/EAs is developed, such that a MATLAB user can easily use it without needing to learn the DFOs in depth when contemplating selecting one of MATLAB's DFOs for his/her application.

At present, there exist the following DFO algorithms [1–3]:

Heuristics (a posteriori):

(1) Random search;
(2) Bayesian optimization;
(3) Data-based Online Nonlinear Extremumseeker (DONE);
(4) **Simulated annealing (SA), a special case being heuristic hill-climbing;**
(5) **Particle swarm optimization (PSO);**
(6) **Genetic algorithm (GA);**
(7) Cuckoo search;
(8) Other evolution algorithms (such as Evolution Strategy, Evolutionary Programming, Ant Colony Optimisation, etc.).

Direct Search (a priori):

(9)   **Nelder-Mead Simplex method (Simplex);**
(10) **Pattern search (PS);**
(11) **Powell's conjugate search (COBYLA, UOBYQA, NEWUOA, BOBYQA and LINCOA);**
(12) Coordinate descent and adaptive coordinate descent;
(13) Multilevel Coordinate Search (MCS) algorithm.

Among these, the bolded algorithms (4)–(6), (9)–(10) are available as functions in MATLAB, and (11) is available as an open-source.m file [4]. We shall test these six algorithms, together with a special version of the GA, its elitist variant GAe. Benchmarks used in this paper are consistent with but exceed those used in measuring conventional optimisation algorithms reported in [5, 6].

This paper is not about DFO algorithm or their comparison studies per se. It is about helping engineers who wish to use MATLAB's DFOs to choose one for his/her application at hand, and about providing an amalgamated selection criterion for this purpose. Section 2 of this paper presents the aforementioned benchmarks. Section 3 presents the test functions adopted. Benchmarking results are reported and analysed in Sect. 4. Conclusions are drawn in Sect. 5.

2 Uniform Benchmarks for Testing

For an optimisation problem with a single objective, suppose that its objective function (performance index, cost function for minimisation, or fitness function for maximisation) is:

$$f(x) : X \rightarrow F \tag{1}$$

where $X \subseteq \mathbf{R}^n$ spans the entire search or possible solution space in n dimensions, $x \in X$ represents the collective variables or parameters of n dimensions to be optimised, $F \subseteq \mathbf{R}^m$ represents the m dimensional space of all possible values of m objective, and $f \in F$.

We prefer benchmark test functions have known theoretical solutions, including the minimum or maximum value of the objective function:

$$f_0 = Max\, or\, Min\{f(x)|x \in X\} \in F \tag{2}$$

where $x_0 \in X$ satisfying:

$$f(x_0) = f_0 \tag{3}$$

Normalised benchmarks used in this paper are defined in [5]. We choose five benchmarks, where four have been used in [4], as follow.

2.1 Optimality

Optimality represents the relative figure of merit of an objective value found $\hat{f}_0$ [5]:

$$\text{Optimality}|_a = 1 - \frac{\|f_0 - \hat{f}_0\|_a}{\|\bar{f} - \underline{f}\|_a} \in [0, 1] \tag{4}$$

where $\bar{f}$ and $\underline{f}$ are the lower and upper bounds of f, which are known from proper test functions.

2.2 Accuracy

As the highest optimality reached could be from a local optimum, the benchmark Accuracy needs to be used, which represents the relative closeness of a solution found $\hat{x}_0$, to the theoretical solution, x_0. This may be particularly useful if the solution space is noisy, there exist multiple optima or 'niching' is used. It is defined as [5]:

$$\text{Accuracy} = 1 - \frac{\|x_0 - \hat{x}_0\|}{\|\bar{x} - \underline{x}\|} \in [0, 1] \tag{5}$$

where $\underline{x}$ is the lower bounds of x and $\bar{x}$ is the upper bounds, $[\underline{x}\ \bar{x}]$ represent the search range, which all should be known from a proper test function.

2.3   Convergence

Reach Time. The most frequently used measure of convergence is the number of function evaluations. The stop condition is usually set to when there is little change, less than 1e-6. But it could convergence very quickly ending at local extreme point. Hence the optimality is also evaluated in convergence as reach-time [5]:

$$\text{Reach time}|_b = C^b \tag{6}$$

where C^b represent the total number of function evaluations conducted by which the optimality of the best individual first reaches $b \in [0, 1]$.

NP Time. To estimate the order of the polynomial $C^{0.99999}$ may be plotted against the number of parameters being optimized, n, as revised in:

$$NP - \text{time}(n) = C^{0.999999}(n) \tag{7}$$

Total Number of Evaluations. This is limited to:

$$N = \min\{C^{0.999999}(n), 400mn^2\} \tag{8}$$

which implies that a benchmark test should terminate either when the goal has been reached or $20n$ generations with a population size of $20n \times m$ have been evolved, i.e., the theoretical maximum number of function evaluations will not exceed $400mn^2$. Without loss of considerable generality, in this paper, only single-objective tests are considered, i.e., $m = 1$.

2.4   Optimizer Overhead

Alternative to or in addition to the 'total number of evaluations', the 'total CPU time' may be used in a benchmark test. More quantitatively, the optimizer overhead may be calculated by [1]:

$$\text{Optimiser overhead} = \frac{\text{Total time taken} - T_{\text{PFE}}}{T_{\text{PFE}}} \tag{9}$$

where T_{PFE} is the time taken for pure function evaluations.

2.5   Sensitivity

When the values of optimal parameters found are perturbed, the optimality may well change. This affects the robustness of the solution and hence can impact on an engineering design. To measure how much 'small' relative change in the designed parameters (solution found) will lead to relative change in the quality (objective value found), sensitivity is defined as the ratio between these changes [5]:

$$\text{Sensitivity} = \lim_{\|\Delta x\| \to 0} \left. \frac{\|\Delta f\| / \|\bar{f} - f\|}{\|\Delta x\| / \|\bar{x} - x\|} \right|_{x=\hat{x}_0} = \lim_{\|\Delta x\| \to 0} \left. \frac{\|\Delta f\|}{\|\Delta x\|} \right|_{x=\hat{x}_0} \frac{\|\bar{x} - x\|}{\|\bar{f} - f\|} \tag{10}$$

$$\approx \frac{1 - \text{Optimality}}{1 - \text{Accuracy}} \tag{11}$$

3 Benchmark Test Functions

3.1 Experimental Setup

In the experiments, we investigate the performance of the MATLAB (R2016b) heuristics SA, PSO, GA, and GAe, and direct-search optimisers Simplex, PS, and Powell.

Each experiment is repeated 10 times to obtain a mean value of each benchmark. All starting points of search are randomly generated, i.e., for SA, Simplex, PS and Powell, different runs start from the same initial. For PSO, GA and GAe, different runs start from the same initial population. In all runs, we kept the same stopping criteria as given by (8), for a fair comparison.

In order to obtain T_{PFE} for each tested function, all the benchmark functions (as in Table 2) are run $400n^2$ times and the time is taken as the average value from 10 runs in MATLAB.

3.2 Algorithmic Settings

The parameters are not manually set, but are taken default settings in MATLAB unless otherwise stated. Because it is not to ask the user to tune or test the MATLAB function, but to help them use it to solve their own, practising problem. For the Simplex, reflection coefficient is 1, the expansion coefficient is 2, the contraction coefficient is 0.5 and the shrink coefficient is 0.5. For GA, the population size is set as 10n, where n is the dimension number. The max generation times is 20n, so that the maximum function evaluation is $400n^2$ as described in Sect. 2 about reach-time. The crossover function is using scattered which creates a random binary vector and selects the genes from two parents based on that binary vector. Crossover fraction is set as 0.8. Probability rate of mutation is 0.01. For GA with elitism, 5 out of 100 population are guaranteed to survive to the next generation. For SA, the initial temperature is set to 100, and the temperature is lowered following function of $T = T_0 \times 0.95^{50}$. For PSO, swarm size is 10n, and maximum iteration is set as 40n so that the maximum function evaluations is $400n^2$.

3.3 Benchmark Functions

For comparison, four non-linear functions used in [7, 8] are used here in Table 1. It is a benchmark function and not a benchmark data set that is normally used in

Table 1. Tested functions.

	Name of function	Test function	Search Space	Minimum/Maximum
Unimodal	Quadric [11, 12]	$\min f_1(x) = \sum_{i=1}^{D} ix_i^4$	$[-1.28, 1.28]^D$	$D = 10,$ $y \in [0, 147.64]$ $D = 30,$ $y \in [0, 1248.2]$
Multimodal	n-D Varying Landscape [9, 10]	$\max f_2(x) = \sum_{i=1}^{n} f(x_i) = \sum_{i=1}^{n} \sin(x_i) sin^{2m}\left(\frac{ix_i^2}{\pi}\right)$	$x \in [0, \pi]^D$	$D = 10,$ $y \in [0, 9.65]$ $D = 30,$ $y \in [0, 29.63]$
	Scaffer's F6 Function [7, 8]	$\min f_3(x) = g(x_1, x_2) + g(x_2, x_3) + \ldots + g(x_{D-1}, x_D) + g(x_D, x_1)$ $g(x, y) = 0.5 + \frac{\sin^2(\sqrt{x^2+y^2} - 0.5)}{(1 + 0.001(x^2 + y^2))^2}$	$x \in [100, 100]^D$	$D = 10,$ $y \in [0, 9.98]$ $D = 30,$ $y \in [0, 29.93]$
	Composition Functions [7, 8]	$\min f_4 = 0.3 \times f_{41} + 0.3 \times f_{42} + 0.4 \times f_{43}$ $f_{41} = 418.9829 \times D - \sum_{i=1}^{D} g(z_i)$ $m = mod(|z_i|, 500)$ $g(z_i) = \begin{cases} z_i \sin(|z_i|^{1/2}) & if\ |z_i| \le 500 \\ (500 - mod(z_i, 500))\sin(\sqrt{|500 - mod(z_i, 500)|}) - \frac{(z_i-500)^2}{10000D} & if\ z_i > 500 \\ (mod(|z_i|, 500) - 500)\sin(\sqrt{|mod(|z_i|, 500) - 500|}) - \frac{(z_i+500)^2}{10000D} & if\ z_i < -500 \end{cases}$ $f_{42} = \sum_{i=1}^{D}(x_i^2 - 10\cos(2\pi x_i) + 10)$ $f_{43} = \sum_{i=1}^{D}(10^6)^{\frac{i-1}{D-1}} x_i^2$	$x \in [100, 100]^D$	$D = 10,$ $y \in [0, 5.54e8]$ $D = 30,$ $y \in [0, 1.22e9]$

benchmarking EAs/DFOs. Also, benchmarking through function evaluation is closer to the real world that a MATLB user or a design engineer finds himself/herself in, where simulation results (which functional evaluation mimics) instead to data sets are used. All the functions are tested in 10 and 30 dimensions. Difference types of tested function are chosen for a comprehensive comparison. The first function is Quadric function, which is a unimodal function. The rest three functions are multimodal ones. The n-D Varying Landscape problem, which is also the only maximization problem in the four tested functions, that was introduced by Michalewicz [9] and further studied by Renders and Bersini [10].

The objective function $f_2(x)$ is, in effect, de-coupled in every dimension represented by $f_i(x_i) = \sin(x_i)\sin^{2m}\left(\frac{ix_i^2}{\pi}\right)$. Every such member function is independent. The larger the product mn is, the sharper the landscape becomes. There are $n!$ local maxima within the search space $[0, \pi]^n$. The theoretical benchmark solution to this n-dimensional optimization problem may be obtained by maximizing n independent uni-dimensional functions, $f_i(x_i)$, the fact of which is however unknown to an optimization algorithm being tested.

The third is Scaffer F_6 function with minimum at zero. And the last one is a composition function of Schwefel's function, Rastrigin's function and High Conditioned Elliptic Function. The composition rate is 0.3, 0.3 and 0.4 respectively.

4 Benchmarking Results an Analysis

4.1 Compare Among Algorithms

The results reported in this section are summarized in Tables 2, 3, 4, 5, 6, 7, 8 and 9, grouped by the tested function. All the theoretical values are given at the bottom of each table for comparison. Performance scores are given by sorting every algorithm's results on each indicator. For example, if the mean optimality result of PSO is the best among all 7 algorithms, the PSO is given a score of 7 on Optimality. Then, five sets of scores for all benchmarks of every algorithm are added to the last column in each result table.

Table 2. Powell best on the unimodal problem (10-D Quadric)

Algorithms	Mean optimality	Mean accuracy	FEs or Reach time	Optimizer overhead	Mean sensitivity	Score
SA	98.99%	86.19%	5766	696.24	7.0152	10
PSO	**100%**	99.96%	6220	**8.23**	3.19e-07	28
GA	99.69%	81.92%	13256	11.16	0.0171	12
GAe	100%	99.86%	10520	25.85	9.83e-06	20
Simplex	99.35%	87.32%	8871	8.61	4.3963	16
PS	**100%**	99.99%	**2967**	38.41	**0**	29
Powell	**100%**	**100%**	3422	9.24	**0**	**32**

Table 3. Powell best on the unimodal problem (30-D Quadric)

Algorithms	Mean optimality	Mean accuracy	FEs or Reach time	Optimizer overhead	Mean sensitivity	Score
SA	99.38%	87.77%	30709	253.78	0.0505	11
PSO	**100%**	99.91%	19860	5.3973	1.6983e-08	28
GA	**100%**	99.71%	87627	11.34	1.5395e-06	18
GAe	**100%**	99.74%	72840	11.04	1.4795e-06	22
Simplex	96.03%	77.82%	93779	4.8576	0.1751	10
PS	**100%**	99.99%	**5049**	12.27	**0**	29
Powell	**100%**	**100%**	23900	**3.8695**	**0**	**33**

Table 4. GAe best on the low-dimensional multimodal Varying Landscape Problem (10-D)

Algorithms	Mean optimality	Mean accuracy	FEs or Reach time	Optimizer overhead	Mean sensitivity	Score
SA	19.64%	73.99%	**40000**	593.75	3.1204	12
PSO	82.54%	80.76%	**40000**	**7.34**	0.8637	27
GA	91.36%	85.62%	**40000**	22.65	0.5829	27
GAe	**95.29%**	**86.73%**	**40000**	16.19	0.3549	**31**
Simplex	26.06%	74.41%	**40000**	6.988	**0.3522**	26
PS	79.05%	77.39%	**40000**	34.38	1.0199	19
Powell	70.56%	73.64%	**40000**	8.6780	0.9532	19

Table 5. GAe best on the high-dimensional multimodal Varying Landscape Problem (30-D)

Algorithms	Mean optimality	Mean accuracy	FEs or Reach time	Optimizer overhead	Mean sensitivity	Score
SA	14.16%	73.73%	**360000**	170.19	3.31	14
PSO	66.17%	77.23%	**360000**	4.4801	1.4810	22
GA	88.75%	84.31%	**360000**	8.8706	0.7023	27
GAe	**92.61%**	**86.69%**	**360000**	7.0496	**0.552**	**31**
Simplex	13.39%	70.84%	**360000**	3.6048	0.1765	22
PS	72.98%	75.16%	**360000**	10.16	0.9703	21
Powell	69.31%	73.72%	**360000**	**3.3626**	0.7372	24

Table 6. PSO best on the low-dimensional multimodal Scaffer Problem (10-D)

Algorithms	Mean optimality	Mean accuracy	FEs or Reach time	Optimizer overhead	Mean sensitivity	Score
SA	55.44%	71.96%	**40000**	1457.1	1.6269	18
PSO	80.63%	84.96%	**40000**	**16.88**	**1.37**	**29**
GA	80.71%	91.14%	**40000**	53.49	2.3755	22
GAe	**88.74%**	**95.26%**	**40000**	35.32	2.3669	28
Simplex	53.59%	72.22%	**40000**	17.05	1.6962	20
PS	83.75%	93.29%	**40000**	87.63	2.5063	22
Powell	53.98%	72.32%	**40000**	21.165	1.6913	23

Table 7. GAe best on the high-dimensional multimodal Scaffer Problem (30-D)

Algorithms	Mean optimality	Mean accuracy	FEs or Reach time	Optimizer overhead	Mean sensitivity	Score
SA	53.14%	71.97%	**360000**	462.84	1.6812	17
PSO	63.50%	77.05%	**360000**	11.83	1.6162	25
GA	77.94%	88.74%	**360000**	23.10	2.0218	22
GAe	**83.68%**	**91.07%**	**360000**	17.12	1.8275	**27**
Simplex	52.93%	70.78%	**360000**	9.3363	1.6161	22
PS	83.24%	92.99%	**360000**	28.25	2.5613	23
Powell	53.15%	70.73%	**360000**	**8.6047**	**1.6059**	25

Quadratic function f_1. The first test function is a unimodal function with only one minimum point in the search area $[-1.28 \ 1.28]^D$. The search results are shown in Tables 2, 3 and Fig. 1. The algorithms are running for different dimension of 10 and 30. All seven tested algorithms had optimality up to 0.99 in 10-D. In lower dimension of 10, PSO and Powell [4] showed better performance with lower optimization overhead and better accuracy. In 30-dimensional problem, the Powell had the best score. SA has the most optimizer overhead for both 10-D and 30-D tests. Figure 1 shows convergence trace for each algorithm. For f_1, all the algorithms show a fast and relatively uniformed speed.

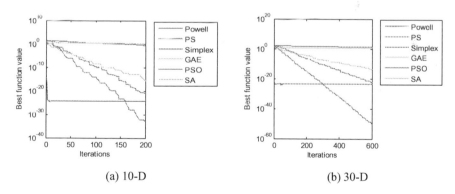

(a) 10-D   (b) 30-D

Fig. 1. Convergence traces of tested algorithms on unimodal f_1 minimisation

n-D Varying Landscape function f_2. The problem described in second function has multiple extremes in $[0, \pi]^D$. The objective is to maximise the function value, which is different from the other three tested function.

From Tables 4 and 5, NM Simplex and SA performances are not satisfactory. It stalled at local maximum point. GA and GAe, on the other hand, showed steady performance, though it took much more optimizer overhead to reach the maximum point than NM Simplex. None of the tested algorithms can reach to 0.9999 optimality in N times of function evaluations. GA and GAe have provided relatively better results than other. Figure 2 shows convergence trace for each algorithm. For f_2, GA, GAe and PSO can get closer to global extremum with more iterations.

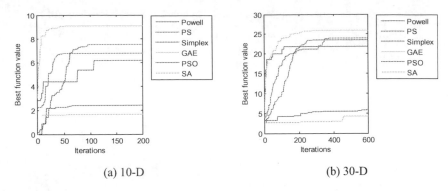

(a) 10-D   (b) 30-D

Fig. 2. Convergence traces of tested algorithms on multimodal f_2 maximisation

Scaffer's F6 function f_3. Like the first and second functions, GA, PS and PSO show consistently better performance than other three algorithms. Figure 3 shows convergence track for each algorithm. For f_3, PSO has a slower convergence speed than GAe, but can reach to the same level of extreme point.

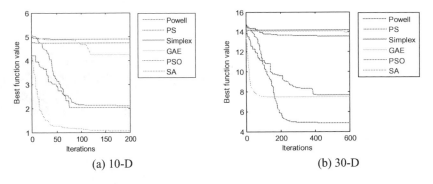

(a) 10-D   (b) 30-D

Fig. 3. Convergence traces of tested algorithms on multimodal f_3 minimisation

However, all the algorithms show high overhead at this tested function, which suggested that optimizer overhead is not only related to algorithms itself, but also related to the fitness function.

Composition function f_4. The composition function is complex which should be expected more optimizer overhead. However, except for SA, the overhead of other six algorithms are less than 20. The optimality all can reach to 0.99, except Simplex, while the accuracy in Simplex is 65.83% in 30-D. Overall, all the seven algorithms can locate the global extreme in the last tested function (Fig. 4).

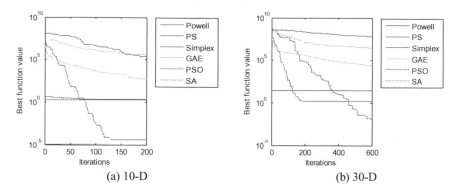

Fig. 4. Convergence traces of tested algorithms on multimodal f_4 minimisation

4.2 Compare Among Functions

Because f_1 is a unimodal function. All the algorithms are performing well. In 30-D, the optimizer overhead of Powell is only 3.8695. For multimodal functions, GAe, PSO and PS methods have an overall better performance than others.

From the theoretical time of pure running of fixed function evaluations, f_4 took most long time and f_3 is the shortest. However, the optimizer overhead of all the algorithms in f_3 is the biggest while small in f_1 and f_4. Moreover, for NM Simplex, the overhead is relatively stable in both 10-D and 30-D, while the optimizer overhead decreased at 30-D for other methods. It is make sense that because GA and PSO are imitation of group activating in nature, which depends on big population to keep diversity. When the dimension increases, the population size is also increased.

4.3 Evaluations of Benchmarks

The mean optimality and mean accuracy represent relative closeness to the theoretical value. It would lost meaning when the range for x and y become very big. The relative value would be too small to show difference between compared algorithms.

The minimum value for function is 0, however, the maximum value is increased exponentially with dimension of n. In this case, the benchmark of optimality can't reflect the distant of searching result to theoretical value. When the distance between f_{max} and f_{min} is too big, the optimality could be very close to 1 even the absolute distance between $\hat{f_0}$ and f_0. As it is shown in Tables 8 and 9. All the optimality can reach to 0.99, but the Simplex.

Though reach-Time or N evaluations of functions also has disadvantages, it gave a convergence speed reflected both convergence and optimality. But because of large range of f in f_4, PSO, GA Powell and PS stop before it reaches to its best solutions. Though the optimizer overhead in this test is very small, it is not the actual overhead time because algorithms stop early. For example, in Table 9, the times for function evaluation of Powell is only 1856, compare to N = 360000. The results could be better

Table 8. GAe best on the low-dimensional multimodal Composition Problem (10-D)

Algorithms	Mean optimality	Mean accuracy	FEs or Reach time	Optimizer overhead	Mean sensitivity	Score
SA	99.89%	79.49%	27019	241.09	0.0054	10
PSO	99.99%	95.65%	6560	**2.6737**	5.20e-04	25
GA	99.99%	91.66%	20100	8.3736	0.0002	19
GAe	**100%**	99.98%	13625	12.64	**0**	**26**
Simplex	99.60%	65.86%	10816	2.9003	1.531	11
PS	**100%**	**99.99%**	**2910**	13.05	1	**26**
Powell	**100%**	99.83%	4033	3.7823	1.0017	25

Table 9. GAe best on the high-dimensional multimodal Composition Problem (30-D)

Algorithms	Mean optimality	Mean accuracy	FEs or Reach time	Optimizer overhead	Mean sensitivity	Score
SA	99.77%	79.87%	275690	96.18	1.0897	9
PSO	**100%**	99.96%	45780	1.9398	**4.003e-04**	26
GA	99.99%	93.26%	35700	0.3925	0.0128	21
GAe	**100%**	99.79%	32460	**0.3219**	0.0084	**30**
Simplex	96.76%	65.83%	360000	1.7590	1.4805	8
PS	**100%**	**99.99%**	25341	4.8462	1	26
Powell	**100%**	99.67%	**1856**	1.5973	1.003	26

if the search continue. However, PSO stops search based on the reach time setting. Thus, it compromised its performance to fit the testing standard.

Benchmarks of optimization, accuracy, convergence, optimizer overhead and sensitivity have shown some advantages for a uniformed standard to compare. Its application still has limits (Table 10).

Table 10. Amalgamated rankings of heuristic and direct search DFOs

Algorithm rankings		Optimality	Accuracy	FEs or Reach time	Optimizer Overhead	Sensitivity/ Robustness	Overall score & rank
Heuristics	SA	6	6	5	7	7	101
	PSO	3	3	3	3	**1**	210
	GA	5	4	6	5	4	168
	GAe	**1**	2	4	4	2	**215**
Direct search	Simplex	7	7	6	2	5	135
	PS	2	**1**	**1**	6	6	195
	Powell	3	5	2	**1**	3	207

5 Conclusions

Using 4 commonly adopted benchmark testing problems, we have compared all 7 MATLAB DFO functions. The PS and Powell have shown good performance for unimodal problems, but have not delivered satisfactory performance for multimodal and high dimension problems. The GA and GAe have shown overall consistently good performance on all benchmark problems, with the second highest robustness (second lowest sensitivity). The PSO has offered the highest convergence speed and relatively lower optimizer overhead, though the optimality and accuracy were not as good as the GA and GAe.

Further, the direct-search numerical optimiser Powell is seen to be the best on the Quadratic Problem, the heuristic numerical PSO algorithm on the lower dimensional Scaffer Problem, the direct-search numerical optimiser PS on the lower dimensional Composition Problem, and the extra-numerical genotype evolutionary algorithm GAe the best on the Varying Landscape Problem and on the other two higher dimensional problems. Overall, the GAe offers the highest performance, followed by PSO and Powell's optimisation methods. It is surprising to see that, while it has reached the top accuracy (implying the global optimum), the numerical PS with a floating-point resolution has not exceeded the optimality delivered by the GAe which has only a resolution of genotype encoding. This confirms the advantage of a population-based optimiser.

Five benchmarks and a scoring system for benchmarking DFOs/EAs have been developed in this paper, which provide a selection criterion for engineers who wish easily to choose one MATLAB DFO most suitable to their application at hand. Meanwhile, the benchmarking technique developed in this paper is not restricted by the engineers' background and does not require them to learn these algorithms pre-requisite.

References

1. Lo, V.M.: Heuristic algorithm for task assignment in distributed systems. IEEE Trans. Comput. **37**(11), 1384–1397 (1988)
2. A. R. Conn, K. Scheinberg and L. N. Vicente. *Introduction to Derivative-Free Optimization*, SIAM (2009)
3. Powell, M.J.D.: Direct search algorithms for optimisation calculations. Acta Numer. **7**, 287–336 (1998)
4. Powell on MathWorks: https://cn.mathworks.com/matlabcentral/fileexchange/15072-unconstrained-optimization-using-powell/content/powell.m. Accessed 29 Mar 2017
5. Feng, W., Brune, T., Chan, L., Chowdhury, M., Kuek, C.K., Li, Y.: Benchmarks for testing evolutionary algorithms. In: Asia-Pacific Conference on Control and Measurement, pp. 134–138 (1998)
6. Luo, W., Li, Y.: Benchmarking heuristic search and optimisation algorithms in matlab. In: 22th International Conference on Automation & Computing, Colchester city, UK, 7 September 2016

7. Chen, Q., Liu, B., Zhang, Q., Liang, J., Suganthan, P., Qu, B.: Problem definitions and evaluation criteria for CEC 2015 special session on bound constrained single-objective computationally expensive numerical optimization. In: 2015 IEEE Congress on Evolutionary Computation, Sendai, Japan, 25 May (2015)

8. Liang, J., Qu, B., Suganthan, P.: Problem definitions and evaluation criteria for the CEC 2014 special session and competition on single objective real-parameter numerical optimization. In: 2013 IEEE Congress on Evolutionary Computation, Cancun, Mexico, 21 June (2013)

9. Michelewicz, Z.: Genetic algorithm + data structure = evolutionary programs, vol. 1, p. 996. Springer-Verlag, New York (1996)

10. Renders, J.-M., Bersini, H.: Hybridizing genetic algorithms with hill-climbing methods for global optimisation: two possible ways. In: 1994 IEEE World Congress on Computational Intelligence, Florida, USA, 26 June 1994

11. Zhan, Z.-H., Zhang, J., Li, Y., Chung, H.S.-H.: Adaptive particle swarm optimisation. IEEE Trans. Syst. Man Cybernet. Part B: Cybernet. **39**(6), 1362–1381 (2009)

12. Yao, X., Liu, Y., Lin, G.: Evolutionary programming made faster. IEEE Trans. Evol. Comput. **3**(2), 82–102 (1999)

Blind Source Separation

Dependent Source Separation with Nonparametric Non-Gaussianity Measure

Fasong Wang[1(✉)], Li Jiang[1], and Rui Li[2]

[1] School of Information Engineering,
Zhengzhou University, Zhengzhou 450001, China
{iefswang, ieljiang}@zzu.edu.cn
[2] School of Sciences, Henan University of Technology, Zhengzhou, China
liruilyric@126.com

Abstract. Separating statistically dependent source signals from their linear mixtures is a challenging problem in signal processing society. Firstly, we show that maximization of the non-Gaussianity (NG) measure among the separated signals can realize dependent source signals separation. Then, based on cumulative distribution function (CDF) instead of traditional probability density function (PDF), the NG measure is defined by utilizing statistical distances between different distributions. After that, the CDF based objective function is estimated by utilizing nonparametric order statistics (OS). At last, by consulting the stochastic gradient rule of constrained optimization problem, the efficiently nonparametric dependent sources separation algorithm is derived and termed as nonpNG. Simulation results demonstrate the validity of the proposed statistically dependent sources separation algorithm.

Keywords: Blind source separation (BSS) · Dependent component analysis (DCA) · Order statistics (OS) · Non-Gaussianity (NG) measure

1 Introduction

Blind separation of statistically independent linear instantaneous superposition of underlying hidden source signals is well done by the so-called independent component analysis (ICA) method, which has attracted considerable attention in the signal processing and neural network fields during the last two decades and several efficient algorithms have been proposed [1, 2].

Despite the success of using various ICA algorithms in many applications of blind source separation (BSS) problem, the primary statistically independent assumption of ICA may not hold for some real-world situations, especially in biomedical signal processing and image processing [3], and consequently, facing these cases, the ICA based BSS approaches would not give ideal performance. Among many extensions of the basic ICA model, several researchers have studied the situation where the source signals are not statistically independent. We call these models dependent component analysis (DCA) model generally. The first extended ICA model to DCA is the multidimensional independent component analysis (MICA) model [3]. Instead, MICA assumes that the source signals can be divided into couples, triplets, or in general *i*-tuples, such that the

© Springer International Publishing AG 2017
D.-S. Huang et al. (Eds.): ICIC 2017, Part II, LNCS 10362, pp. 91–101, 2017.
DOI: 10.1007/978-3-319-63312-1_8

source signals inside a given i-tuple may be dependent on each other, but dependencies among different i-tuples are not allowed. Based on this important extension of the ICA model, there have emerged lots of DCA models and corresponding algorithms, such as independent subspace analysis [4], variance dependent BSS [5], topographic ICA [6], tree-dependent component analysis [7], subband decomposition ICA [8], statistic measure aided methods such as maximum Non-Gaussianity (NG) [9], spectral decomposition [10], time-frequency method [11] and bounded component analysis [12, 13].

Meanwhile, Cardoso has shown that strong relationship exists among mutual information (MI), correlation and NG of source estimations [14]. Following this theory result, instead of resorting minimization the MI, by maximizing the NG, one can get the desired dependent source signals theoretically.

In this paper, based on the generalization of the central limit theorem (CLT) to special dependent variables, we will try to track DCA problem by maximization NG measure, which is defined in terms of cumulative distribution function (CDF) rather than the wide used probability density function (PDF), whose main advantage lies the direct estimation by means of the nonparametric order statistics (OS).

The remaining of the paper is structured as follows. In Sect. 2, the model and theory fundamentals are described. In Sect. 3, the proposed nonpNG algorithm for dependent source signals is developed. In Sect. 4, some simulation results are demonstrated to show the efficiency of the proposed algorithm. Section 5 concludes the paper.

2 Model and Theory Fundamentals

Generally, the noiseless instantaneous BSS problem can be modelled as follows,

$$\mathbf{x}(t) = \mathbf{A}\mathbf{s}(t) + \mathbf{n}(t) \tag{1}$$

where $\mathbf{s}(t) = [s_1(t), \cdots, s_N(t)]^T$ is an unknown source vector. Matrix $\mathbf{A} = [a_{ij}] \in \mathbf{R}^{M \times N}$ is an unknown full column rank mixing matrix and $M = N \cdot \mathbf{x}(t) = [x_1(t), \cdots, x_M(t)]^T$ are the observed mixtures, also called as sensor outputs. $\mathbf{n}(t) = [n_1(t), \cdots, n_N(t)]^T$ is a vector of additive noise which is assumed to be zero in this paper.

The task of BSS lies in estimating the mixing matrix $\mathbf{A}$ or its pseudoinverse separating (unmixing) matrix $\mathbf{W} = \mathbf{A}^\dagger$, then to estimate the original source signals $\mathbf{s}(t)$, given only a finite number of observation data $\{\mathbf{x}(t), t = 1, \cdots, T\}$. Recall that scaling and permutation ambiguities cannot be resolved in BSS without some a *priori* knowledge about the source signals or the mixing process [1, 2].

Most of the methods of BSS perform a spatial decorrelation preprocessing over $\mathbf{x}(t)$ to obtain the decorrelated observations $\mathbf{z}(t) = [z_1(t), \cdots, z_N(t)]^T$. So, the global mixture is expressed as

$$\mathbf{z}(t) = \mathbf{V}\mathbf{s}(t)$$

where **V** is an unknown orthogonal matrix.

After separation algorithms are used to the preprocessed data $\mathbf{z}(t)$, one can find a linear unitary transformation **B** and get the estimation of the source signals $\mathbf{y}(t)$ as,

$$\mathbf{y}(t) = \mathbf{B}\mathbf{z}(t) \tag{2}$$

In order to guarantee the maximum NG to work, the theoretical foundation to be satisfied of source signals should be based on those that allow us to generalize the CLT to special dependent variables. An enormous amount of works has been published since long time ago to establish different sufficient conditions for this generalization [15, 16].

Definition 1. The maximum NG method consists of searching for the linear combinations of mixtures that give source estimates with maximum non-Gaussian distributions restricting the space of search to the unit-variance signals space. More specifically, sources are estimated through relation $\mathbf{y}(t) = \mathbf{B}\mathbf{z}(t)$ over the space of invertible separating matrices **B** providing signals $y_1(t), \cdots, y_N(t)$ with unit-variances (which is equivalent to imposing the covariance matrix $\mathbf{R}_{yy} = \mathbf{B}\mathbf{R}_{zz}\mathbf{B}^T$ to have ones in its main diagonal).

The NG measure is defined as,

$$d(y, g) = \left(\int |p(y) - p_y(y)|^2 dy \right)^{1/2} \tag{3}$$

where the integral is defined in the Lebesgue sense and is taken on all the range of variable y, and $p(y)$ is the normalized Gaussian PDF. So, Eq. (3) is the square of the distance between functions $p(y)$ and $p_y(y)$ in L^2 space.

In this paper, we consider the NG measure using the concept of CDF instead of traditional PDF. Let us call F_y and F_g the CDFs to be analyzed and its equivalent one, respectively. Then, the NG measure based on CDF is defined as follows,

$$d(F_y, F_g) = \left(\int_{-\infty}^{\infty} |F_y(x) - F_g(x)|^2 dx \right)^{1/2} \tag{4}$$

As done in [16], we show that the previous definition hold the following distance property in order to corroborate their feasibility as DCA objective function,

$$\begin{aligned} d(F_y, F_g) &= 0 \Leftrightarrow c = 2, \\ d(F_y, F_g) &> 0, \forall c \neq 2 \end{aligned} \tag{5}$$

where c is the Gaussianity parameter in the generalized Gaussian density (GGD) function [17]. Consequently, the distance measure is the measure of NG since it offers a global minimum when the analyzed distribution is Gaussian and the function grown monotonically according to the distribution shifts far away from the Gaussian (see Fig. 1 for detail).

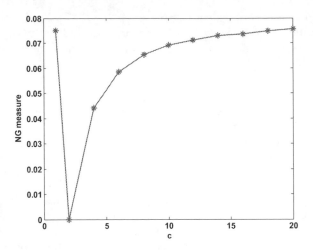

Fig. 1. NG measure based on CDF's expressions of the generalized family Gaussian.

For the property Eq. (5), the close relationship between CDF F and its inverse Q can be generalized as

$$D(Q_{y_i}, Q_g) = 0 \Leftrightarrow c = 2$$
$$D(Q_{y_i}, Q_g) > 0, \forall c \neq 2 \qquad (6)$$

As a result,

$$D(Q_{y_i}, Q_g) = \max_{x \in R} |Q_{y_i}(x) - Q_g(x)| \qquad (7)$$

is also an appropriate NG measure. The estimation of Q_{y_i} can be performed very robustly in a simple practical way by using the set of OS $y_{i(1)} < y_{i(2)} < \cdots < y_{i(n)}$. Eventually, as shown in [16], the efficient estimation of NG measure (7) can be represented as,

$$\hat{D}(Q_{y_i}, Q_g) = \left| y_{i(k)} - y_{i(l)} + 2Q_g\left(\frac{l}{n}\right) \right|$$
$$\text{with } \begin{cases} k, l = 80\%n, 20\%n, \quad \text{or} \\ k, l = n, 1. \end{cases} \qquad (8)$$

3   Proposed NG Based DCA Algorithm–NonpNG

From Definition 1, NG measure and the CLT, we can get that the proposed NG measure will present a local maximum at any output channel if $\mathbf{b}_i$ is forced to be unitary. Therefore, a multistage procedure should be applied to obtain a different component at each output channel: the NG measure is maximized at each output channel successively under the constriction that the vector $\mathbf{b}_i$ has to be orthonormal to the previously obtained vectors. The orthogonality among the separation vectors implies that the extracted components are decorrelated.

Taking into account Eq. (2) in vector form,

$$y_i(t) = \mathbf{b}_i^T \mathbf{z}(t)$$

where $\mathbf{b}_i^T$ is the i-row of the separation matrix $\mathbf{B}$, the goal is to update $\mathbf{b}_i$ at each stage by optimizing an objective function $J(\mathbf{b}_i)$. We take

$$J(\mathbf{b}_i) = D(Q_{y_i}, Q_g)$$

and it will be optimized by the stochastic gradient rule of the constrained optimization method [16],

$$\mathbf{b}_i(t+1) = \mathbf{b}_i(t) + \mu \nabla J(\mathbf{b}_i(t))|_{\mathbf{b}_i(t)},$$
$$\text{subject to:} \tag{9}$$
$$\mathbf{b}_i \text{ is orthonormal to } [\mathbf{b}_1, \cdots, \mathbf{b}_{i-1}]$$

The gradient of $J(\mathbf{b}_i)$ in Eq. (9) is,

$$\nabla J(\mathbf{b}_i(t))|_{\mathbf{b}_i(t)} = S \frac{d(y_{i(k)} - y_{i(l)})}{d\mathbf{b}_i}\bigg|_{\mathbf{b}_i(t)} \tag{10}$$

where $S = \text{sign}\left(y_{i(k)} - y_{i(l)} + 2Q_g\left(\frac{l}{n}\right)\right)\big|_{\mathbf{b}_i(t)}$.

Applying the chain rule we have,

$$\frac{d(y_{i(k)} - y_{i(l)})}{d\mathbf{b}_i}\bigg|_{\mathbf{b}_i(t)} = \mathbf{z}\left(\frac{dy_{i(k)}}{dy_i}\bigg|_{\mathbf{b}_i(t)} - \frac{dy_{i(l)}}{dy_i}\bigg|_{\mathbf{b}_i(t)}\right) \tag{11}$$

After each iteration procedure, $\mathbf{b}_i$ must be normalized and projected over the orthonormal subspace $\mathbf{C}_{i-1}$ to the vectors obtained at every previous stage. Let us denote that the $\mathbf{C}_{i-1}$ expression is,

$$\mathbf{C}_{i-1} = \mathbf{I} - (\mathbf{B}_{i-1}\mathbf{B}_{i-1}^H)^{-1}\mathbf{B}_{i-1}^H \tag{12}$$

where $\mathbf{B}_{i-1} = [\mathbf{b}_1, \cdots, \mathbf{b}_{i-1}]$.

4   Simulation Results

Firstly, we will compare the performance results of our proposed nonparametric NG measure based DCA algorithm, which is called nonpNG algorithm, against the results obtained through the application of some classical BSS/ICA methods, such as FastICA,

AMUSE, Pearson-opt, SANG, SOBI, JADE-opt (all these methods are fully reviewed in [1, 2] and the ICALAB Matlab software package [18]) by separating real world signals.

The four source signals are extracted from the real world photo and the waveforms of them are shown in Fig. 2(a) (The data have been sphered to zero mean and unit variance). These source signals are from different pixels of the photo which aren't far away, therefore, they are correlated. The source signals' correlation coefficients are shown in Table 1.

Table 1. The correlation coefficients between different source signals

	Source1	Source2	Source3	Source4
Source1	1.0000	0.2719	0.2286	0.3568
Source2	0.2719	1.0000	0.4048	0.4226
Source3	0.2286	0.4048	1.0000	0.3640
Source4	0.3568	0.4226	0.3640	1.0000

The mixing matrices $\mathbf{A}$ is a 4×4 random mixing matrix in which the elements are $N(0, 1)$ distributed. The mixtures are displayed in Fig. 2(b).

At the same convergent conditions, the performance of the proposed nonpNG algorithm is compared along the statistically criteria, which is measured using an index called cross-talking error E defined by [2] as,

$$E = \frac{1}{N} \sum_{i=1}^{N} \left(\sum_{j=1}^{N} \frac{|p_{ij}|}{\max_k |p_{ik}|} - 1 \right) + \frac{1}{N} \sum_{j=1}^{N} \left(\sum_{i=1}^{N} \frac{|p_{ij}|}{\max_k |p_{kj}|} - 1 \right) \tag{13}$$

where $\{p_{ij}\}$ is the entries of $\mathbf{P} = \mathbf{WA} = \mathbf{BVA}$ which is the performance matrix.

After using the proposed nonpNG algorithm to the mixtures, the separated dependent source signals are shown in Fig. 2(c), where the order and amplitudes of the estimated signals do not correspond to the sources for the ambiguities.

From Fig. 2, one can obtain that the proposed nonpNG algorithm can separate statistically dependent source signals from their mixtures properly, even though there exists amplitude and sign ambiguous. Other applications will be investigated using the proposed algorithm, such as in data mining [19].

Another statistical performance, or accuracy, was measured by three index defined in the BSSEVAL toolbox [20] to evaluate the extracted source signals. These criteria are called the signal-to-distortion ratio (SDR), the signal-to-interference ratio (SIR) and the signal-to-artifacts ratio (SAR), defined respectively as follows,

$$SDR = 10 \log_{10} \frac{\left\| s_{\text{target}} \right\|^2}{\left\| e_{\text{interf}} + e_{\text{noise}} + e_{\text{artif}} \right\|^2} \tag{14}$$

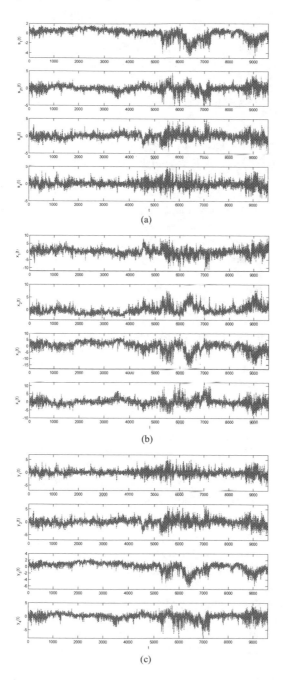

Fig. 2. The source signals with data size 9600 are shown in (a), mixed signals by an random matrix which elements are normalized Gaussian distribution are shown in (b), the separated signals using proposed nonpNG algorithm are shown in (c).

$$\text{SIR} = 10 \log_{10} \frac{\left\|s_{\text{target}}\right\|^2}{\left\|e_{\text{interf}}\right\|^2} \tag{15}$$

$$\text{SAR} = 10 \log_{10} \frac{\left\|s_{\text{target}} + e_{\text{interf}} + e_{\text{noise}}\right\|^2}{\left\|e_{\text{artif}}\right\|^2} \tag{16}$$

where s_{target} is an allowed deformation of the target source signals $s_i(t)$, e_{interf} is an allowed deformation of the source signals which belongs the interference of the unwanted sources, e_{noise} is an allowed deformation of the perturbation noise (but not the source signals), and e_{artif} is an artifact term that may correspond to artifacts of the separation algorithm. Therefore, the estimated source $\hat{s}(t)$ can be decomposed as follows,

$$\hat{s}(t) = s_{\text{target}} + e_{\text{interf}} + e_{\text{noise}} + e_{\text{artif}} \tag{17}$$

According to [20], both SIR and SAR measure local performance. SDR is a global performance index. In this section, we will give all of the results of SDR, SIR and SAR for overall comparisons.

Then, we will demonstrate the performance of the proposed algorithm for the real world MRI images selected from databases of [21] which is shown in Fig. 3(a). The correlation coefficients of source signals are shown in Table 2.

Table 2. The correlation coefficients between four different MRI images

	Source1	Source2	Source3	Source4
Source1	1.0000	0.9178	0.8542	0.8530
Source2	0.9178	1.0000	0.8838	0.8811
Source3	0.8542	0.8838	1.0000	0.9882
Source4	0.8530	0.8811	0.9882	1.0000

The input mixed signals of the algorithm are generated by mixing the four MRI source signals with a random mixing matrix in which the elements are distributed with $N(0, 1)$, the mixtures are shown in Fig. 3(b). After convergence, the extracted signals are shown in Fig. 3(c).

The average results of the performance criteria evaluated by SIR, SDR, SAR and ISR over 10 experiments are shown in Table 3.

Then, we will analyze performance variations of the proposed nonpNG algorithm with different sample data size. The source signals are extracted from the same photo as the first simulation. The algorithm has been applied for data sizes from 512 to 8960, with a step size equals to 256. The accuracy is calculated averaging a total of 200 separation cases for each data size by Eq. (13). The separation performance index E versus sample data size is depicted in Fig. 4.

From Fig. 4 one can conclude that the performance of the proposed nonpNG algorithm is affected by the sample data size apparently. As the sample data size

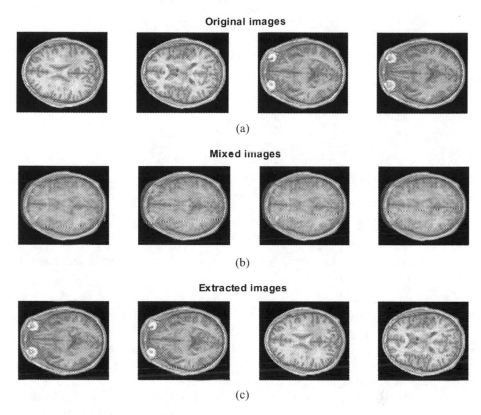

Fig. 3. (a) The 4 source MRI images, (b) The 4 mixed MRI images signals, (c) The 4 extracted MRI images using the proposed algorithm.

increasing, the performance becomes more and better generally. We can also get that when the sample data size T exceeds 1700, the proposed nonpNG algorithm can make an ideal separation result for dependent source signals with the average performance index E below 0.15. In additional, as expected, the separation performance is affected by the correlation coefficients among the source signals.

Table 3. Average SIR, SDR, SAR and ISR for different MRI source signals using the proposed algorithm

Performance index	Source signals			
	Source1	Source2	Source3	Source4
SIR	243.01	236.14	229.53	233.26
SDR	259.11	269.17	252.00	254.74
SAR	242.37	236.47	229.56	233.89
ISR	249.70	252.18	255.74	253.05

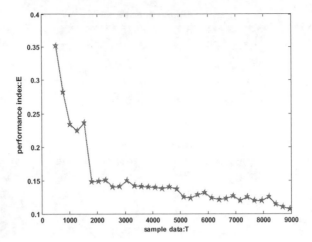

Fig. 4. Separation performance index E versus sample data size T for the four source signals using the proposed nonpNG algorithm.

5   Conclusions and Discussion

In this paper, we show that maximization of the CDF aided NG measure can realize the efficient source separation, which can not only separate the statistically independent but also dependent source signals. The NG measure is defined as the statistical distances between CDF rather than the traditional PDF based approach. Moreover, using non-parametric OS to approximate NG measure makes the proposed nonpNG algorithm has a wide range of applications. Simulation results show that the proposed nonpNG algorithm is able to separate the dependent signals and yield ideal performance with medium data size.

Acknowledgments. This research is financially supported by the National Natural Science Foundation of China (No. 61401401, 61402421, 61571401) and the China Postdoctoral Science Foundation (No. 2015T80779, 2014M561998).

References

1. Comon, P., Jutten, C.: Handbook of Blind Source Separation: Independent Component Analysis and Applications. Elsevier, Oxford (2010)
2. Cichocki, A., Amari, S.: Adaptive Blind Signal and Image Processing: Learning Algorithms and Applications. Wiley, New York (2002)
3. Cardoso, J.F.: Multidimensional independent component analysis. In: Proceedings of International Conference on Acoustics, Speech, and Signal Processing, Seattle, WA, USA, pp. 1941–1944. IEEE (1998)
4. Szabo, Z., Poczos, B., Lorincz, A.: Separation theorem for independent subspace analysis and its consequences. Pattern Recogn. **45**(4), 1782–1791 (2012)

5. Kawanabe, M., Muller, K.R.: Estimating functions for blind separation when sources have variance dependencies. J. Mach. Learn. Res. **6**, 453–482 (2005)
6. Hyvarinen, A., Hoyer, P.O., Inki, M.: Topographic independent component analysis. Neural Comput. **13**(7), 1527–1558 (2001)
7. Bach, F.R., Jordan, M.I.: Kernel independent component analysis. J. Mach. Learn. Res. **3**, 1–48 (2002)
8. Zhang, K., Chan, L.W.: An adaptive method for subband decomposition ICA. Neural Comput. **18**(1), 191–223 (2006)
9. Caiafa, C.: On the conditions for valid objective functions in blind separation of independent and dependent sources. EURASIP J. Adv. Sig. Process. **2012**, 255 (2012)
10. Aghabozorgi, M.R., Doost-Hoseini, A.M.: Blind separation of jointly stationary correlated sources. Sig. Process. **84**(2), 317–325 (2004)
11. Abrard, F., Deville, Y.: A time-frequency blind signal separation method applicable to underdetermined mixtures of dependent sources. Sig. Process. **85**(7), 1389–1403 (2005)
12. Inan, H.A., Erdogan, A.T.: A convolutive bounded component analysis framework for potentially nonstationary independent and/or dependent sources. IEEE Trans. Sig. Process. **63**(1), 18–30 (2015)
13. Inan, H.A., Erdogan, A.T.: Convolutive bounded component analysis algorithms for independent and dependent source separation. IEEE Trans. Neural Netw. Learn. Syst. **26**(4), 697–708 (2015)
14. Cardoso, J.F.: Dependence, correlation and gaussianity in independent component analysis. J. Mach. Learn. Res. **4**, 1177–1203 (2003)
15. Caiafa, C.F., Proto, A.N.: Separation of statistically dependent sources using an L^2 distance non-Gaussianity measure. Sig. Process. **86**(11), 3404–3420 (2006)
16. Blanco, Y., Zazo, S.: New Gaussianity measures based on order statistics: application to ICA. Neurocomputing **51**, 303–320 (2003)
17. Wang, F., Li, H., Li, R.: Unified nonparametric and parametric ICA algorithm for hybrid source signals and stability analysis. Int. J. Innov. Comput. Inf. Control **4**(4), 933–942 (2008)
18. Cichocki, A., Amari, S., Siwek, K., et al.: ICALAB toolboxes (2007). http://www.bsp.brain.riken.jp/ICALAB
19. Wang, F., Li, H., Li, R.: Data mining with independent component analysis. In: Proceedings of the Sixth World Congress on Intelligent Control and Automation, vol. 2, pp. 6043–6047 (2006)
20. Vincent, E., Gribonval, R., Fevotte, C.: Performance measurement in blind audio source separation. IEEE Trans. Audio Speech Lang. Process. **14**(4), 1462–1469 (2006)
21. The Whole Brain Atlas. http://www.med.harvard.edu/AANLIB/home.html

Intelligent Fault Diagnosis

Convolutional Neural Network Based Bearing Fault Diagnosis

Duy-Tang Hoang[1] and Hee-Jun Kang[2(✉)]

[1] Graduate School of Electrical Engineering, University of Ulsan, Ulsan 680-749, South Korea
hoang.duy.tang@gmail.com
[2] School of Electrical Engineering, University of Ulsan, Ulsan 680-749, South Korea
hjkang@ulsan.ac.kr

Abstract. In this paper, we propose a new bearing fault diagnosis method without the feature extraction, based on Convolutional Neural Network (CNN). The 1-D vibration signal is converted to 2-D data called vibration image. Then, the vibration images are fed into the CNN for bearing fault classification. Experiments are carried out with bearing data from the Case Western Reserve University Bearing Fault Database and its result are compared with the results of other methods to show the effectiveness of the proposed algorithm.

Keywords: Convolutional neural network · Deep learning · Bearing fault diagnosis · Signal based fault diagnosis

1 Introduction

Vibration analysis is widely employed in industrial applications. The fault vibration signal generated by the interaction between a damaged area and a rolling surface occurred regardless of the defect type. Therefore, vibration analysis can be employed to detect all types of faults, either localized or distributed. General signal based intelligent diagnosis methodology includes three steps as follows: signal acquisition; feature extraction; and fault recognition [1]. To extract representative features from the complex and non-stationary noisy signal, numerous vibration signal processing approaches have been developed such as statistical analysis, Fourier transform, wavelet transform, empirical mode decomposition (EMD). The feature set obtained from the extraction step often have a high dimension. To enhance the performance of the diagnosis system, dimension reduction methods are used such as principle component analysis (PCA), sequential feature selection (SFS). In the last step, machine learning algorithms such as artificial neural network (ANN), support vector machine (SVM), k-nearest neighbor (k-NN) are exploited to classify the faults.

Performance of machine learning based classification mainly depends on the feature extraction step. But there is no standard for extracting features because of the requirement of expert knowledge. Thus, for every specific fault diagnosis task, feature extractor must be redesigned manually.

© Springer International Publishing AG 2017
D.-S. Huang et al. (Eds.): ICIC 2017, Part II, LNCS 10362, pp. 105–111, 2017.
DOI: 10.1007/978-3-319-63312-1_9

Deep learning is a branch of machine learning based on algorithms that attempt to model high level abstractions of data. CNN is a deep learning algorithm with hierarchical neural networks whose convolutional layers alternate with sub-sampling layers, following with a full connection layer. A CNN primarily mimics the human visual system, which can efficiently recognize the parterns and structures in a visual scenery [2]. As a result, nowadays CNNs are successfully applied in many areas relating to image processing such as face recognition, object recognition, hand written recognition, video analysis, etc.

Recently, 2-D representations of signals have been exploited in various studies of fault diagnosis [3], where time-domain signals are converted to 2-D gray level images. In order to extract the texture information from the converted images, 2-D feature extraction methods are used such as gray-level co-ocurrence matrix (GLCM) [4], global neighborhood structure (GNS) map [5], etc. By exploiting the texture information, the 2-D image based fault diagnosis can achieve high accuracy, but the performance still depends mainly on handcrafted feature extraction.

Motivated by the efficient performance of CNN in image classification and the ability of representing signals in 2-D data, this paper proposes a bearing fault diagnosis algorithm using CNN which do not require any feature extractor. Gray-level images converted from the 1-D vibration signal are fed into CNN for classification. Experiments are carried out with the bearing data from the Case Western Reserve University Bearing Fault Database [6]. Comparison with other machine learning based fault diagnosis approaches (ANN, SVM, k-NN) is carried out to show the effectiveness of the proposed method.

This paper is organized as follows. The proposed fault diagnosis method is explained in detail in Sect. 2. Section 3 describes the implementations and performances. Finally, conclusions are drawn in Sect. 4.

2   The Proposed Bearing Fault Diagnosis Method

To exploit the efficient of CNN in image classification, the vibration signals are converted to gray images. Then vibration images are feed into CNN for classification. The block diagram of the proposed method is shown in Fig. 1.

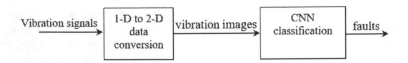

Fig. 1. Block diagram of CNN based bearing fault diagnosis

2.1   Vibration 2-D Gray Image Construction

In the data conversion process, the amplitude of each sample in the vibration signal is normalized to range from 0 to 1. And the normalized amplitude of each sample becomes intensity of the corresponding pixel of the M × N corresponding image. The conversion between normalized amplitude of sample and corresponding pixel can be described as the following equation [3].

$$P[i,j] = A[(i-1) * M + j] \tag{1}$$

where $i = 1{:}N; j = 1{:}M$; $P[i,j]$ is intension of corresponding pixel (i,j) in the $M \times N$ vibration gray image. A[.] is the normalized amplitude of the sample in the vibration signal. Number of pixel in the vibration image equals to number of sample in the vibration signal.

2.2 Image Classification with CNN

Typical CNN consists of four types of layers: convolution layer, sub-sampling layer, full connection layer and output layer. The network layers are arranged in a feed-forward structure: each convolution layer is followed by a sub-sampling layer. The last sub-sampling layer is followed by a full connection layer, which finally followed by the output layer.

At convolutional layer, the previous layer feature maps are convolved with learnable kernels and put through the activation function to form the output feature map. Each kernel is used at every position of the input. CNN exploits sparse connectivity by making the kernel smaller than the input and enforcing a local connectivity pattern among neurons of adjacent layers. Each output map may combine convolution with multiple inputs maps. However, for each output map, the input maps are convolved with distinct kernels. Each kernel is used at every position of the input. The parameter sharing used by the convolution operation means that rather than learning a separate set of parameters for every location, we learn only one set.

Each convolution layer is followed by a sub-sampling layer. A sub-sampling layer produces down-sampled versions of the input maps, progressively reduces the spatial size of the representation. That helps to decrease the number of parameters and computation in the network. Moreover, sub-sampling layer makes the representation become invariant with a small translation of the input. If there are N input maps, there will be exactly N output maps, although the output maps will be smaller.

The full connection layer is a traditional feed-forward neural network, neurons in this layer have full connections to all activations in the previous layer. The purpose of the full connection layer is to use the features from previous layer for classifying the input image into various classes. The final layer in a CNN is output layer, using the softmax as the activate function.

With three architectural ideals: local receptive fields, weight sharing and sub-sampling, CNN has many strength [7, 8]. First, feature extraction and classification are integrated into one structure and fully adaptive. Second, the network extracts 2-D image features at increasing dyadic scales. Third, it is relatively invariant to geometric, local distortions in the image.

3   Experimental Implementation

3.1   Test-Bed Specification

The bearing data were obtained from the Case Western Reserve University Bearing Fault Database [6]. Motivation of this choice is the fact that a public database which accessible to the research community allows a fair comparison of the performance of the proposed algorithms.

Vibration signals of six operating conditions as following: normal condition, inner race fault, ball fault, outer race fault at position 6, 3 and 12 o'clock position.

3.2   Experiment 1: Fault Diagnosis with CNN

At first, the vibration signals are split into non-overlapping segments. The length of the segments is chosen based on two criterions: (i) long enough to capture localized features of the signal and (ii) as short as possible to reduce the computation time. In this experiment, the length of each segment is selected as 441 samples. Signals from fan end are considered. 270 samples/condition are generated, thus for six conditions we have $270 \times 6 = 1620$ segments.

In the next step, vibration images are constructed by the method described in Sect. 2.1. Since the length of each vibration signal segment is 441, we select the size of vibration image $M = 21, N = 21$ pixel, ($M \times N = 21 \times 21 = 441$). By that way, we obtained an image set includes 1620 gray images with size 21×21 pixel. Then the image set is split into two sets: training set (1080 images) and test set (540 images). The vibration images for six bearing conditions are shown in Fig. 2.

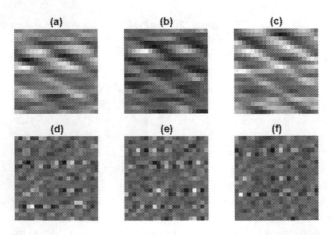

Fig. 2. Vibration gray image of bearing under six conditions: (a)-Normal, (b)-Inner race, (c)-Ball, (d)-Outer Race 6 o'clock, (e)-Outer Race 3 o'clock, (f)-Outer Race 12 o'clock

The configuration of CNN classification is as follow:

- first layer: convolution layer, 30 kernels, each kernel has size 6×6, stride step 1, ReLU activate function;
- second layer: sub-sampling layer with filter size 2×2, stride step 1;
- third layer: convolution layer, 60 kernels, each kernel has size 3×3, stride step 1, ReLU activate function;
- fourth layer: sub-sampling layer with filter size 2×2, stride step 1;
- full connection layer;
- output layer: six output (corresponding to 6 conditions need to be classified).

3.3 Experiment 2: Fault Diagnosis with Conventional Machine Learning

To make the comparison, we consider a machine learning based fault diagnosis approach. In first step, the vibration signals are split into non-overlapping segments as the same way used by Experiment 1 in Sect. 3.2.

In feature extraction step, we consider statistical features and Wavelet Packet analysis. Statistical features are obtained from both time domain and frequency domain. Ten statistical features in the time domain: root mean square (RMS), square root the amplitude (SRA), kurtosis value (KV), skewness value (SV), peak-to-peak value (PPV), crest factor (CF), impulse factor (IF), margin factor (MF), shape factor (SF), and kurtosis factor (KF). Three statistical features in frequency domain: frequency center (FC), RMS frequency (RMSF) and root variance frequency (RVF). Signal from both end (drive end and fan end) are considered. The total number of statistical features is $(10 + 3) \times 2 = 26$ features, i.e., ten statistical features from time domain, three statistical features from frequency domain, taken at both the drive end fan end of the motor of the test-bed.

In this experiment, wavelet packet analysis is used to extract features from the time-frequency domain. The analysis procedure in [9] is used. The mother wavelet is Daubechies 4, and refining is done down to the fourth decomposition level. With a tree depth of 4, 16 final leaves were obtained and consequently, $16 \times 2 = 32$ features were taken for both drive end and fan end.

After the feature extraction step, we obtained a feature set with size 1620×58, i.e., 1620 samples, each sample has $26 + 32 = 58$ features.

In next step, to reduce the dimension of the feature set, we use the sequential forward selection (SFS) [10]. SFS starts with an empty set and then test each candidate together with the already-selected features. After applying SFS algorithm, we obtain a reduced feature set with size 1620×4. In the classification step, artificial neural network (ANN), support vector machine (SVM) and k-nearest neighbor (k-NN) are used.

3.4 Comparison and Analysis

In this part, we make a comparison between the classification results of two above experiments. The first experiment was our proposed fault diagnosis method using CNN. The second experiment used conventional machine learning based fault diagnosis includes following step: feature extraction, dimensional reduction, machine learning classification.

The first classifier, CNN used vibration images to dignose the bearing faults. For the next three classifiers, the reduced feature set with only 4 features was used to train ANN, SVM and k-NN. The last three classifiers are ANN, SVM, and k-NN which trained by the original feature set with 58 features. In all case, after being trained with the same training set (1080 samples, 180 samples for each class), all classifiers were fed the same test set (540 samples, 90 samples for each class) to evaluate the performance.

In this comparison, four standard criteria are used to evaluation the experiment results [11]: accuracy (acc), specificity (tnr), fallout (fpr) and miss rate (fnr).

From the Table 1, we can see some important points are:

- the classifiers ANN, SVM, k-NN wihout using SFS, which used full feature set (58 features) don't have good performance;
- using SFS, the performace of those classifiers are all enhanced by using the reduced feature set (4 features);
- the CNN classifier using 2-D gray level image has very good performance (100% accuracy), does not use feature extractor and feature selection.

Table 1. Classification results

Classifier		acc	fnr	fpr	fnr
	CNN	100	1	0	0
Use SFS to reduce feature set dimension	ANN	100	1	0	0
	SVM	100	1	0	0
	k-NN	99.44	0.9934	0	0.0066
Use all features to classify	ANN	98.90	0.9868	0	0.0132
	SVM	96.11	0.9740	0	0.0260
	k-NN	95.56	0.9551	0.0417	0.0449

4   Conclusion

In this study, we proposed a novel approach based on CNN to classify the faults of rolling element. By converting 1-D vibration signal to 2-D images and exploiting CNN image classification technique, the proposed method has two main strong points:

- firstly, do not require feature extraction and feature selection process which have big effect on the classification accuracy and usually require the expert knowledge about the system.
- secondly, gives high accurate classification of bearing faults, validated through the experiments with real data.

Acknowledgments. This research was supported by Basic Science Research Program through the National Research Foundation of Korea (NRF) funded by the Ministry of Education (NRF-2016R1D1A3B03930496).

References

1. Prieto, M.D., Cirrincione, G., Espinosa, A.G., Ortega, J.A., Henao, H.: Bearing fault detection by a novel condition-monitoring scheme based on statistical-time features and neural networks. IEEE Trans. Industr. Electron. **60**(8), 3398–3407 (2013)
2. Kiranyaz, S., Ince, T., Gabbouj, M.: Real-time patient-specific ECG classification by 1-D convolutional neural networks. IEEE Trans. Biomed. Eng. **63**(3), 664–675 (2016)
3. Nguyen, D., Kang, M., Kim, C.H., Kim, J.: Highly reliable state monitoring system for induction motors using dominant features in a 2-dimension vibration signal. New Rev. Hypermedia Multimedia **13**(3–4), 245–258 (2013)
4. Jang, W.C., Park, Y.H., Kang, M.S., Kim, J.M.: Mechainical fault classification of an induction motor using texture analysis. J. Korea Soc. Comput. Inf. **18**(12), 12–19 (2013)
5. Uddin, J., Kang, M., Nguyen, D.V., Kim, J.-M.: Reliable fault classification of induction motors using texture feature extraction and a multiclass support vector machine. Mathe. Probl. Eng. **2014**, 9 (2014). doi:10.1155/2014/814593. Article ID 814593
6. Loparo, K.A.: Bearing data center. http://www.eecs.case.edu/laboratory/bearing. Case Western Reserve University. Accessed 2016
7. Phung, S.L., Bouzerdoum, A.: MATLAB library for convolutional neural networks. ICT Research Institute, Visual and Audio Signal Processing Laboratory, University of Wollongong, Technical report (2009)
8. Phung, S.L., Bouzerdoum, A.: A pyramidal neural network for visual pattern recognition. IEEE Trans. Neural Networks **27**(1), 329–343 (2007)
9. Rauber, T.W., de Assis Boldt, F., Varejão, F.M.: Heterogeneous feature models and feature selection applied to bearing fault diagnosis. IEEE Trans. Industr. Electron. **62**(1), 637–646 (2015)
10. Guyon, I., Elisseeff, A.: An introduction to variable and feature selection. J. Mach. Learn. Res. **3**, 1157–1182 (2003)
11. Powers, D.M.: Evaluation: from precision, recall and F-measure to ROC, informedness, markedness and correlation (2011)

Machine Learning

A Performance Evaluation of Systematic Analysis for Combining Multi-class Models for Sickle Cell Disorder Data Sets

Mohammed Khalaf[1,2(✉)], Abir Jaafar Hussain[1], Dhiya Al-Jumeily[1], Robert Keight[1], Russell Keenan[3], Ala S. Al Kafri[1], Carl Chalmers[1], Paul Fergus[1], and Ibrahim Olatunji Idowu[1]

[1] Faculty of Engineering and Technology, Liverpool John Moores University, Byrom Street, Liverpool L3 3AF, UK
M.I.Khalaf@2014.ljmu.ac.uk,
{a.hussain,d.aljumeily,c.chalmers,p.fergus}@ljmu.ac.uk,
R.Keight@2015.ljmu.ac.uk, a.s.alkafri@2015.ljmu.ac.uk
[2] Ministery of Higher Education and Scientific Research, Bagdad, Al-Rusafa Region, Iraq
[3] Liverpool Paediatric Haemophilia Centre, Haematology Treatment Centre, Alder Hey Children's Hospital, Eaton Road, West Derby, Liverpool L12 2AP, UK
Russell.keenan@alderhey.nhs.uk

Abstract. Machine learning approach is considered as a field of science aiming specifically to extract knowledge from the data sets. The main aim of this study is to provide a sophisticate model to difference applications of machine learning models for medically related problems. We attempt for classifying the amount of medications for each patient with Sickle Cell disorder. We present a new technique to combine two classifiers between the Levenberg-Marquartdt training algorithm and the k-nearest neighbours algorithm. In this paper, we introduce multi-class label classification problem in order to obtain training and testing methods for each models along with other performance evaluations. In machine learning, the models utilise a training sets in association with building a classifier that provide a reliable classification. This research discusses different aspects of machine learning approaches for the classification of biomedical data. We are mainly focus on the multi-class label classification problem where many number of classes are available in the data sets. Results have indicated that for the machine learning models tested, the combination classifiers were found to yield considerably better results over the range of performance measures that been selected for this research.

Keywords: Machine-learning classifiers · Sickle cell disorder · SCD date sets · Accuracy · Performance evaluation

© Springer International Publishing AG 2017
D.-S. Huang et al. (Eds.): ICIC 2017, Part II, LNCS 10362, pp. 115–121, 2017.
DOI: 10.1007/978-3-319-63312-1_10

1   Introduction

Sickle cell disorder (SCD) is considered a common genetic disease due to the abnormality of the red blood cell (RBCs). The World health Organisation (WHO), showed that, there are around 7 million born every year feel pain from the congenital anomaly [1]. With regards to SCD, the main symptoms effects on individual's health are headaches, decrease in heart bit, and difficulties in breath.

In the case of SCD, recent research has shown the beneficial effects of a drug called hydroxycarbamide in modifying the disease phenotype [2]. The clinical management of this disease modifying therapy is difficult and time consuming for clinical staff. In this research, we present the utilisation of Machine learning models for classifying the Sickle Cell disorder data sets. The growth of clinical information has played a significant role in medical domain [3–5]. It is indicted that machine learning algorithms practically with combining multi-class models produce a great improvement with clinical data sets and have helped in acquiring high accuracy [6].

The reminder of this paper is organized as follows. Section 2 will illustrate the model descriptions. The methodology will then be introduced in Sect. 3, followed by the presentation of our results in Sect. 4. Finally, in Sect. 5 we discuss our conclusions.

2   Models Description

K-nearest neighbours algorithm (KNNC) is commonly applied for regressing and classification, which used in pattern recognition and statistical estimation as a non-parametric technique. This type of algorithm can classify the data sets by taking the majority vote of its neighbours, and the object will be assigned to the class most common amongst its K nearest neighbours measured by a distance function. If $K = 1$, then the case is simply assigned to the class of its nearest neighbour [7].

Multilayer perceptron can be trained using the Levenberg-Marquartdt training algorithm (LEVNN) [8, 9] which is an approximation to Newton's method of least squares optimisation. Consider an error function E(W) as shown in Eq (1) that required to be minimised in association with the parameter vector W. In this case, the Newton's method is defined $\nabla^2 E(\underline{w})$ is refer to the Hessian matrix and $\nabla E(\underline{w})$ is refer to the gradient.

$$\Delta\underline{w} = -[\nabla^2 E(\underline{w})]^{-1}\nabla E(\underline{w}) \tag{1}$$

Figure 1 demonstrates the block diagram for combining two or more models. This study concentrate on combining the final classification results gained using N different kind of features sets $(f_i^1(x), \ldots, f_i^N(x))$. In order to construct both classifiers, we require training the model through using feature sets for each classifier. As illustrated in Fig. 1, x refers to specific input, each model m^n produces it is own output $y^n = (y^n(1), \ldots y^n(Z), \ldots y^n(z))$ T, where z is considered the class label, while $y^n(m)$ corresponds to the probability of c^n. Each classifier i generates L approximations to the probabilities $f_j^N(x), j = 1, \ldots, L$. As shown in the block diagram, z correspond to the final target class label.

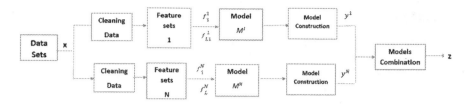

Fig. 1. The combination procedure

The main idea of combined two algorithms is obtain a better result instead of using one algorithm. We attempt for melding outcomes from weak learners classifiers into a high-quality classifier. In order to combine two classifiers, we used Stacked technique to run our experiments. First, the data sets is divided into training, validation, and testing sets using the hold out technique.

Support vector machines (SVM) are considered supervised learning with associated machine learning classifiers, which are able to analyse data sets used for regression and classification analysis. The overall idea of SVM was established into the most commonly used form (soft margin classification) by Cortes and Vapnik in [10], the method itself inventing from the ideas inherent of statistical learning theory [11]. Given a training data sets containing instance-label pairs $\{(x_1, y_1), \ldots, (x_N, y_N)\}$ where $x_i \in R^d$ and $y_i \in \{-1, +1\}$, SVM solves the following optimization problem as shown in Eq. (2):

$$\min_{x, b, \xi_i} \frac{1}{2} \|w\|^2 + C \sum_{i=1}^{N} \xi_i$$
$$\text{subject to } y_i (\langle \emptyset(x_i), w \rangle + b) - 1 + \xi_i \geq 0,$$
$$\xi_i \geq 0, i = 1 \ldots, N. \tag{2}$$

3   Methodology

The proposed model includes of three stages; data collection, data computation, and outcomes evaluation. In the framework, the process starts by taking blood test from patients by specialist nurse at hospital. Clinicians will then provide the amount medication dosage to the patient according to their blood test result. All these process occurs at the haematology laboratory at Alder Hey children hospital. The information are provided to the researcher in order to analyse sickle cell disorder data sets. To this extend, machine learning algorithm which are belong artificial intelligence can be used to exploit these facts by analysing the SCD data sets [12]. To apply machine learning, we need first to clean the data sets in case there is some missing data. Afterward, we select various type of machine learning models for the purpose of evaluating the data sets. We used hold technique that can divide the data sets into three method: training, validation, and testing phases. Eventually, we use a number of evaluation performance to measure the accuracy of our proposed solution. We used sensitivity, Specificity, Area under Curve, Accuracy, and Receiver Operating Characteristic Curve to evaluate the outcomes.

3.1 Data Collection

The real data sets are collected within a five-year period for our experiments. Our date sets comprises 12 features considered crucial values for classifying the sickle disorder. There are 12 features have high impact on the blood test outcomes. In order to collect a huge amount of data sets, the alder hey children trust hospital in Liverpool has supported this research with real datasets for gaining better services and accuracy. The total amount of datasets dataset involved 1186 sample, with a single target value for the amount of hydroxycarbamide dosage in milligrams. The target of medication dosage was divided into 6 bins, starts at classes 1 through 6.

3.2 Experimental Setup

In this part, we cover the design of the test domain that utilised in this study and the algorithms tested the structure of each classifiers, and lastly the performance evaluation techniques used to find out the results of the machine learning algorithms for the whole data set that used in this experiment.

Our model evaluation framework consists training and testing diagnostics, comprising five important performance evaluations. It is involved sensitivity (True positive), specificity (true negative), precision level, F-score (F1), J statistic (J), and accuracy. In addition, the models were characterised utilising ROC figures and the AUC figures, where the classification ability across operating method was ascertained.

4 Results

The results from our experiments are showed in Tables 1 and 2, listing the outcomes for the supervised learning techniques, through applied the training phase and testing phase of the models. We also provide very high performance visualisations through using the receiver operating characteristic curve (ROC) as shown in Figs. 2 and 3 and the use of the area under the curve (AUC) as illustrated in Figs. 4 and 5.

Table 1. Classifiers performance with average of six classes (Training)

Model	Sensitivity	Specificity	Precision	F1	J	Accuracy	AUC
ROM	0.493	0.542	0.178	0.253	0.035	0.543	0.487
KNNC	1	1	1	1	1	1	1
KNNC-LVMNN	0.99	0.996	0.985	0.988	0.987	0.996	0.999
SVM	0.837	0.850	0.5575	0.657	0.687	0.847	0.899
LNN	0.845	0.828	0.523	0.639	0.673	0.833	0.870

Table 2. Classifiers performance with average of six classes (Testing).

Model	Sensitivity	Specificity	Precision	F1	J	Accuracy	AUC
ROM	0.548	0.547	0.196	0.2681	0.094	0.538	0.524
KNNC	0.745	0.761	0.363	0.476	0.506	0.767	0.808
KNNC-LVMNN	0.98	0.986	0.919	0.948	0.967	0.986	0.991
SVM	0.862	0.838	0.499	0.616	0.701	0.837	0.888
LNN	0.830167	0.844	0.526	0.637	0.674	0.838	0.848

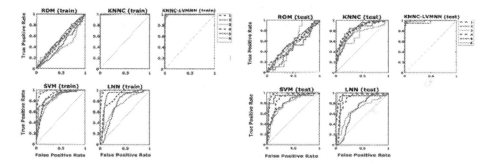

Fig. 2. ROC curve (Train) for classifiers   **Fig. 3.** ROC curve (Testing) for classifiers

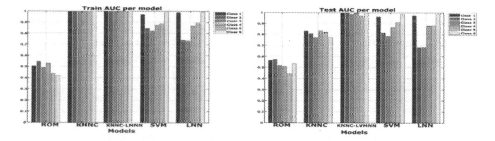

Fig. 4. Train AUC per model   **Fig. 5.** Test AUC per model

The results we obtained from the empirical study indicated that the selected data sets exhibits crucial non-linear relationships, producing great outcomes after applied difference type of machine learning approaches. One of the machine learning classifiers, the combined classifiers under study, the KNNC-LVMNN outperformed the other models, indicating capability for fitting the training for the data sets as well as in generalising to unseen instances. The AUCs obtained for the KNNC-LVMNN classifier during training were 0.999, in comparison to 0.991 over the test sample. Classes 1 to 6 were found to show excellent performance and consistent generalisation test sets for this model. Classes 1 and 6 were found to show reasonably consistent performance representation between the test and train sets for this approach. It is clear that, the reasonable performance obtained for the KNNC-LVMNN classifier, in contrast with the poor performance of the other Machine Learning approaches, could point to a detrimental effect caused by presence of feedbacks from the outputs in the classification setting.

The training sets of the LNN classifier model generated sets produced values between 0.870, while the testing 0.848 as expected. The LNN classifier was incapable to be learnt specifically the non-liner mechanisms. It yields weak classification outcomes against the other models. Random oracles model was unable to offer modest results; indicating by compare the significance of model provide accuracy values between 0.487, while the testing 0.524.

The main reason behind combined classifiers outperformed other types of machine learning is due to the type of data and its distribution. We note that each application and datasets presents different challenges and diverse relationships among the variables. LEVNN and KNNC is superior models that provide a great accuracy and performance with difference data sets

5   Conclusion

In this study, we have conducted an empirical investigation into the use of machine learning models for the classification of SCD effective dosage levels. This research has presented various architectures of machine learning for examining the medical data sets acquired from sickle disorder patients in contrast with traditional medical solutions. Our study sought to investigate the effectiveness of machine learning approaches when posed in the direct classification setting, examining if the machine learning models could positively influence model training and testing. It was found empirical study, involving the use of medical data sets and comparator models for instance, LNN and ROM, and Levenberg-Marquartdt training algorithm (LVMNN), the k-nearest neighbours algorithm (KNNC) although capable of providing some degree of fitting and generalisation, are suboptimal in the classification setting within our experiment. The results obtained from a range of models during our experiments have shown that the combined classifiers LVMNN and KNNC produced significantly better outcomes over the other range of classifiers. We note however that the quantity of data available at the time of this study may limit the generality of these findings, it is suggested that further work is needed to validate our findings with the use of a sufficiently large data sample. Additionally, we consider for future work the use of global optimisation algorithms such as genetic optimisation to explore more comprehensively the space of possible recurrent network architectures.

References

1. Weatherall, D.J.: The importance of micromapping the gene frequencies for the common inherited disorders of haemoglobin. Br. J. Haematol. **149**, 635–637 (2010)
2. Kosaryan, M., Karami, H., Zafari, M., Yaghobi, N.: Report on patients with non transfusion-dependent β-thalassemia major being treated with hydroxyurea attending the Thalassemia Research Center, Sari, Mazandaran Province, Islamic Republic of Iran in 2013. Hemoglobin **38**, 115–118 (2014)
3. Al-Jumeily, D., Hussain, A., Fergus, P.: Using adaptive neural networks to provide self-healing autonomic software. Int. J. Space Based Situated Comput. **5**, 129–140 (2015)

4. Khalaf, M., et al.: Training neural networks as experimental models: classifying biomedical datasets for sickle cell disease. In: Huang, D.-S., Bevilacqua, V., Premaratne, P. (eds.) ICIC 2016. LNCS, vol. 9771, pp. 784–795. Springer, Cham (2016). doi:10.1007/978-3-319-42291-6_78

5. Al-Jumeily, D., Iram, S., Vialatte, F.-B., Fergus, P., Hussain, A.: A novel method of early diagnosis of Alzheimer's disease based on EEG signals. Sci. World J. **2015**, 11 (2015). Article ID: 931387. http://dx.doi.org/10.1155/2015/931387

6. Khalaf, M., Hussain, A.J., Keight, R., Al-Jumeily, D., Fergus, P., Keenan, R., Tso, P.: Machine learning approaches to the application of disease modifying therapy for sickle cell using classification models. Neurocomputing **228**, 154–164 (2017)

7. Ionescu, R.T., Popescu, M.: Knowledge Transfer between Computer Vision and Text Mining. Similarity-Based Learning Approaches. ACVPR. Springer, Cham (2016). doi: 10.1007/978-3-319-30367-3

8. Marquardt, D.W.: An algorithm for least-squares estimation of nonlinear parameters. J. Soc. Ind. Appl. Math. **11**, 431–441 (1963)

9. Hagan, M.T., Menhaj, M.B.: Training feedforward networks with the Marquardt algorithm. IEEE Trans. Neural Netw. **5**, 989–993 (1994)

10. Cortes, C., Vapnik, V.: Support-vector networks. Mach. Learn. **20**, 273–297 (1995)

11. Vapnik, V.: The Nature of Statistical Learning Theory. Springer Science & Business Media, New York (2013)

12. Al Kafri, A.S., Sudirman, S., Hussain, A.J., Fergus, P., Al-Jumeily, D., Al-Jumaily, M., Al-Askar, H.: A framework on a computer assisted and systematic methodology for detection of chronic lower back pain using artificial intelligence and computer graphics technologies. In: Huang, D.-S., Bevilacqua, V., Premaratne, P. (eds.) ICIC 2016. LNCS, vol. 9771, pp. 843–854. Springer, Cham (2016). doi:10.1007/978-3-319-42291-6_83

Knowledge Discovery and Data Mining

Feature Selection Based on Density Peak Clustering Using Information Distance Measure

Jie Cai, Shilong Chao, Sheng Yang[✉], Shulin Wang, and Jiawei Luo

College of Computer Science and Electronic Engineering,
Hunan University, Changsha, China
Yangsh0506@sina.com

Abstract. Feature selection is one of the most important data preprocessing techniques in data mining and machine learning. A new feature selection method based on density peak clustering is proposed. The new method applies an information distance between features as clustering distance metric, and uses the density peak clustering method for feature clustering. The representative feature of each cluster is selected to generate the final result. The method can avoid selecting the irrelevant representative feature from one cluster, where most features are irrelevant to class label. The comparison experiments on ten datasets show that the feature selection results of the proposed method exhibit improved classification accuracies for different classifiers.

Keywords: Feature selection · Clustering · Density peak · Information distance

1 Introduction

Feature selection is the main pre-processing technique in data mining and machine learning [1, 2]. Most of feature selection methods are produced in supervised learning environment [3–6]. Clustering is a common data analysis method in unsupervised learning environment. Its purpose is to divide data according to the degree of dependence between them, thus facilitating accurate analysis and extraction of rules or patterns hidden in the data. Based on this principle, clustering methods can be used to aggregate features with high dependencies (high redundancy). Many feature selection methods are based on feature clustering [7–12].

Many clustering based feature selection methods using information theory have been presented. Au and Chan propose an ACA method [13], which selects an information measure R to measure the correlation between features, clusters features using a K-means-like clustering method. The FAST [14] method proposed by Song uses the hierarchical clustering method. This method takes each minimum spanning tree as a cluster and uses the symmetric uncertainty SU as a metric. FAST is suitable for high dimensional data. Liu proposes a clustering feature selection method MFC [15] based on minimum spanning tree. Unlike FAST, MFC uses the information distance metric VI. Feature selection using the information diversity between features has been a hotspot.

© Springer International Publishing AG 2017
D.-S. Huang et al. (Eds.): ICIC 2017, Part II, LNCS 10362, pp. 125–131, 2017.
DOI: 10.1007/978-3-319-63312-1_11

A new feature selection method based on density peak clustering using information distance is proposed in this paper. Irrelevant clusters are often generated in feature clustering, and they are composed of features irrelevant to the class label. First, features that are irrelevant to the class label are deleted by introducing a feature deletion coefficient to avoid generating irrelevant clusters. Then, by using an information distance measure, the remaining features are clustered by density peak clustering method, where the features with strong redundancy are clustered together. Finally, the representative feature in each cluster is selected to form the final feature subset.

2   Feature Density Peak Clustering Using Information Distance

Information theory is a mathematical discipline founded by Shannon, which has been gradually perfected in theory and widely used in the fields of communication, information, control and so on. Information distance [16, 17] is a measure of difference based on information theory. The most commonly used information theoretic metrics are listed in literature [17]. Here, the information distance measure $d(A, B)$ is selected for two random variables A and B.

$$d(A, B) = H(A) + H(B) - 2I(A; B) = H(A|B) + H(B|A) \qquad (1)$$

In formula (1), $H(A)$ is the entropy of A, $H(A|B)$ is the conditional entropy of A given B, and $I(A; B)$ is the mutual information between A and B. As a distance metric, $d(A, B)$ satisfies non-negativity, symmetry, trigonometric inequality, which have been proved in the literature [17]. Thus, a feature space can be built based on $d(A, B)$, and the features can be clustered in this space.

The density peak clustering method [18] is used for feature clustering in this study. The kernel function is used in the calculation of the local density, and we choose the Gaussian kernel as the kernel function in the experiment. The features with large local density ρ and distance to a high local density point δ are taken as the cluster centers. After the cluster centers are determined, all the noncentral features can be mapped to different clusters according to nearest neighbor principle.

3   The Proposed Feature Selection Method

Based on above analysis, we propose a feature selection method based on Density Peak Clustering using Information Distance (DPCID). The DPCID method is divided into three phases: the irrelevant feature deletion, the density peak clustering and the representative feature selection.

Irrelevant features often exist in the original feature set, which have smaller mutual information to class label C. If the irrelevant features are clustered into one cluster or more, it is improper to select the representative feature from them. Therefore, the irrelevant features are deleted firstly in DPCID by a manageable parameter, feature deletion coefficient, $\alpha, \alpha \in [0, 1)$. After irrelevant features are deleted, the left features are clustered into K clusters by density peak clustering method. Finally, the feature is selected from each cluster whose mutual information with the class label C is the largest. Assume that the original feature set is $F = \{f_1, f_2, \ldots, f_n\}$, d_{ij} represents the information distance between f_i and f_j. The procedure in detail of DPCID is as following.

```
Method: DPCID
Input: F-Initial feature set
       K-The number of target features selected
       α-Feature deletion coefficient
Output: S-The selected subset of features
   For each fᵢ ∈ F

     MI[i] := I(fᵢ; C);

   F:=sort(F) by MI of fᵢ ∈ F descendingly
   Delete the last α*n features in F;
   For each fᵢ ∈ F, fⱼ ∈ F

      Calculate dᵢⱼ;

   Determine the truncation distance d_c;

   For each fᵢ ∈ F
      Calculate ρᵢ and δᵢ;
      N[i]:=nearest neighbor feature;
   For i=1:K
      Center[i] := arg max{ρⱼ * δⱼ};
                   fⱼ∈F

      F := F / {fⱼ};

   For each fᵢ ∈ F
      Cluster[i] := N[i];
   For each cluster
      S := S ∪ {f} which maximizes I(f; C);
   Return S;
```

The time complexity of DPCID mainly depends on the calculation of the information distance and the density peak clustering. If the dataset includes m samples and n features, then the time complexity of the calculation of the information distance is $O(mn^2)$. The time complexity of the density peak clustering is $O(n^2)$. Overall, the complexity of the DPCID method is $O(mn^2)$, which is a polynomial time complexity.

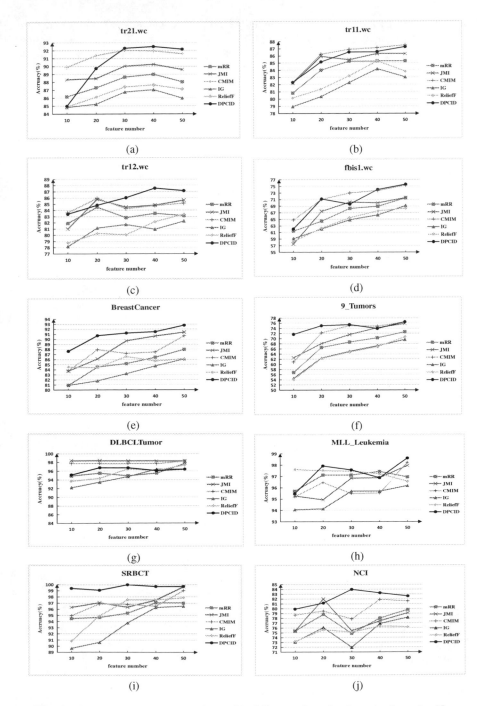

Fig. 1. Average accuracy comparison with different selected subsets by four classifiers

4 Experiments

To test the performance of DPCID, four text classification datasets and six microarray datasets are selected to validate the classification accuracy. All data are discretized by MDL [19] method. We carry out experimental comparisons with five representative feature selection methods. They are JMI [20], CMIM [4], IG, ReliefF [21] and mRR [22], where mRR is clustering based feature selection method using a conditional mutual information distance measure. Furthermore, the four widely used Classifiers: NBC, SVM, KNN1 and C4.5 are employed to generate average classification accuracy on the different feature selection result by ten times 10-fold cross validation. The experiment is implemented on the experimental platform built on Weka 3.7 and Matlab 2014a.

Figure 1 is the average accuracy comparison with different selected subsets by four classifiers. As can be seen, the classification accuracy of DPCID is better than other feature selection methods for most datasets, except DLBCLTumor dataset. The classification accuracy increases with the increase of the number of selected features for tr11.wc, fbis1.wc, BreastCancer, 9_Tumors, MLL_Leukemia. However, the classification accuracy begins to be smooth for tr21.wc, tr12.wc, DLBCLTumor, SRBCT and NCI, when the number of selected features increases to a certain extent. The peak values of most accuracy curves exist when the number of selected features is not less than 20.

Tables 1 and 2 are the best classification accuracies of the six feature selection methods focusing on NBC and SVM. For DPCID, the feature deletion coefficient and the number of selected features are shown in brackets when the classification accuracy is the best. CMIM, JMI, and mRR have their own advantages in different datasets, but the effect of ReliefF is slightly worse. The classification accuracies of DPCID method on most of the datasets are better than JMI, CMIM, IG, mRR and ReliefF, which indicates that the DPCID method can select a feature subset with satisfactory classification ability.

Table 1. Best classification accuracy of features selected by six methods (NBC)

Dataset	DPCID	JMI	CMIM	IG	ReliefF	mRR	unselect
tr21.wc	**92.23**(0.8,20)	89.29	91.96	86.59	87.96	88.03	80.33
tr11.wc	**88.20**(0.8,30)	86.71	87.68	84.59	85.41	86.17	86.23
tr12.wc	**89.78**(0.1,50)	87.54	88.17	84.59	85.61	86.49	87.10
fbis1.wc	73.65(0.8,50)	68.05	**74.98**	69.94	67.39	69.28	70.69
BreastCancer	**97.94**(0,50)	95.76	96.31	90.36	90.92	91.75	92.58
9_Tumors	86.67(0.4,50)	84.39	85.67	76.54	78.92	79.68	**88.67**
DLBCLTumor	97.40(0,20)	**100**	**100**	98.57	98.57	**100**	94.94
MLL_Leukemia	**100**(0.1,30)	98.61	98.61	97.22	98.75	97.22	98.75
SRBCT	**100**(0.1,20)	**100**	**100**	98.60	98.60	98.60	**100**
NCI	**90.16**(0,40)	83.6	83.6	79.81	78.17	80.53	88.5246

Table 2. Best classification accuracy of features selected by six methods (SVM)

Dataset	DPCID	JMI	CMIM	IG	ReliefF	mRR	unselect
tr21.wc	**92.07**(0.8,30)	91.37	91.66	88.37	88.21	91.39	81.64
tr11.wc	87.44(0.8,50)	87.68	**88.4**	84.71	85.93	86.07	83.09
tr12.wc	85.30(0.3,50)	84.66	85.62	82.13	83.04	82.61	**86.26**
fbis1.wc	**78.80**(0.7,40)	75.48	78.76	70.07	69.84	73.69	68.94
BreastCancer	**94.85**(0.2,50)	94.16	92.78	90.92	93.51	90.92	**94.85**
9_Tumors	**68.33**(0.6,50)	65.64	67.51	63.01	61.68	65.22	55.33
DLBCLTumor	98.70(0.6,20)	**100**	**100**	**100**	**100**	**100**	98.83
MLL_Leukemia	**100**(0.1,20)	98.61	**100**	98.61	98.75	**100**	98.75
SRBCT	**100**(0,20)	**100**	98.60	98.60	**100**	**100**	**100**

5 Conclusions

This study proposes a feature selection method called DPCID, which is based on density peak clustering. In this method, the diversity between features is emphasized, and an information distance measure is used to cluster features. The clustering based feature selection is sensitive to the cluster that is composed of the features irrelevant to class label, and that results in a poor feature selection. Therefore, these irrelevant features are deleted before feature clustering in DPCID. DPCID not only ensures the diversity between the features in the final feature subset, but also guarantees the great relevance with the class. The experimental results show that the proposed method performs better than other feature selection methods on ten datasets, and the DPCID method is feasible. Provided that the parameters of DPCID is set properly, the actual effect of the DPCID method can be further improved.

Acknowledgment. The authors would like to acknowledge the assistance provided by National Natural Science Foundation of China (Grant no. 61572180, no. 61472467 and no. 61672011).

References

1. Avrim, L.B., Langley, P.: Selection of relevant features and examples in machine learning. Artif. Intell. **97**(1), 245–271 (1997)
2. Liu, H., Motoda, H.: Feature Selection for Knowledge Discovery and Data Mining. Kluwer Academic Publishers, Boston (1998)
3. Peng, H., Long, F., Ding, C.: Feature selection based on mutual information: criteria of max-dependency, max-relevance, and min-redundancy. IEEE Trans. Pattern Anal. Mach. Intell. **27**(8), 1226–1238 (2005)
4. Fleuret, F.: Fast binary feature selection with conditional mutual information. J. Mach. Learn. Res. **5**, 1531–1555 (2004)
5. Hall, M.A.: Correlation-based feature selection for discrete and numeric class machine learning. In: Proceedings of the 7th International Conference on Machine Learning, pp. 359–366 (2000)

6. Yu, L., Liu, H.: Efficient feature selection via analysis of relevance and redundancy. J. Mach. Learn. Res. **5**, 1205–1224 (2004)

7. Mitra, P., Murthy, C.: Unsupervised feature selection using similarity. IEEE Trans. Pattern Anal. Mach. Intell. **24**(3), 301–312 (2002)

8. Ienco, D., Meo, R.: Exploration and reduction of the feature space by hierarchical clustering. In: Proceedings of the 2008 SIAM Conference on Data Mining, Atlanta, Georgia, USA, pp. 577–587 (2008)

9. Witten, D., Tibshirni, R.: A framework for feature selection in clustering. J. Am. Stat. Associ. **105**, 713–726 (2010)

10. Liu, H., Wu, X., Zhang, S.: Feature selection using hierarchical feature clustering. In: Proceedings of the 20th ACM International Conference on Information and Knowledge Management, Glasgow, United Kingdom, pp. 979–984 (2011)

11. Zhao, X., Deng, W., Shi, Y.: Feature selection with attributes clustering by maximal information coefficient. Procedia Comput. Sci. **17**(2), 70–79 (2013)

12. Bandyopadhyay, S., Bhadra, T., Mitra, P., et al.: Integration of dense subgraph finding with feature clustering for unsupervised feature selection. Pattern Recogn. Lett. **40**(1), 104–112 (2014)

13. Au, W.H., Chan, K.C., Wong, A.K., Wang, Y.: Attribute clustering for grouping, selection, and classification of gene expression data. IEEE/ACM Trans. Comput. Biol. Bioinf. **2**(2), 83–101 (2005)

14. Song, Q., Ni, J., Wang, G.: A fast clustering-based feature subset selection algorithm for high-dimensional data. IEEE Trans. Knowl. Data Eng. **25**(1), 1–14 (2013)

15. Liu, Q., Zhang, J., Xiao J., Zhu, H., Zhao, Q.: A supervised feature selection algorithm through minimum spanning tree clustering. In: IEEE 26th International Conference on Tools with Artificial Intelligence, pp. 264–271 (2014)

16. Meila, M.: Comparing clusterings - an information based distance. J. Multivar. Anal. **98**(5), 873–895 (2007)

17. Vinh, N.X., Epps, J., Bailey, J.: Information theoretic measures for clusterings comparison: variants, properties, normalization and correction for chance. J. Mach. Learn. Res. **11**(10), 2837–2854 (2010)

18. Rodriguez, A., Laio, A.: Clustering by fast search and find of density peaks. Science **344**(6191), 1492–1496 (2014)

19. Fayyad, U.M., Irani, K.B.: Multi-interval discretization of continuous-valued attributes for classification learning. In: Proceedings of the IJCAI, pp. 1022–1029 (1993)

20. Yang, H.H., Moody, J.E.: Data visualization and feature selection: new algorithms for nongaussian data. In: Proceedings of the NIPS, pp. 687–693 (1999)

21. Kononenko, I.: Estimating attributes: analysis and extensions of RELIEF. Mach. Learn. **14**, 171–182 (1994)

22. Sotoca, J.M., Pla, F.: Supervised feature selection by clustering using conditional mutual information-based distances. Pattern Recogn. **43**(6), 2068–2081 (2010)

Gene Expression Array Analysis

Joint Sample Expansion and 1D Convolutional Neural Networks for Tumor Classification

Jian Liu, Yuhu Cheng[(✉)], Xuesong Wang, and Yi Kong

School of Information and Control Engineering, China University of Mining and Technology,
Xuzhou 221116, Jiangsu, China
chengyuhu@163.com

Abstract. Since tumors seriously endanger human health, early detection of the tumor is especially critical for the treatment of patients. How to effectively differentiate the tumor samples from normal samples is becoming a notable topic. In this paper, a joint Sample Expansion and 1D Convolutional Neural Network method (SE1DCNN) is proposed for tumor classification. In our method, inspired by the denoising idea, a Sample Expansion (SE) method is proposed. In addition to maintaining the merits of corrupted data, the expanded samples can deal with the problem of insufficiently training samples of gene expression data to a certain extent when using deep learning models. Since CNN models have an excellent performance in classification tasks, the applicability of 1DCNN on gene expression data is analyzed. Finally, we design a 7-layer 1DCNN model to classify the tumor gene expression data by using the expanded samples and raw samples. Experimental studies indicate that SE1DCNN is quite useful in tumor classification task.

Keywords: Tumor classification · Sample expansion · Deep learning · 1DCNN

1 Introduction

Tumors are part of the major malignant diseases that endanger human health in the world. Under routine circumstances, the gene expression data can be obtained from multiple tissue samples. We can get a better insight into the disease pathology by comparing the genes expressed in diseased samples and the ones in normal samples [1]. One of the challenges is that how to effectively differentiate the cancerous gene expression in tumor samples from non-cancerous gene expression in normal samples. To solve this problem, a large number of classification algorithms, such as Support Vector Machine (SVM) [2], have been proposed to classify the tumor data.

Recently, deep learning models draw more and more attention. One advantage of deep models is that they intrinsically learn a high level representation of the data that avoiding laborious work [3]. Another advantage is that compared with shallow models, deep neural networks have exponentially stronger expressive power. According to a review of deep learning in bioinformatics which was proposed by Min et al. [4], deep learning has also been widely used in the bioinformatics domain, including omics, biomedical imaging and biomedical signal processing. However, the only available literature on the application of Stacked Autoencoders (SAE) to tumor classification by using gene expression data was

© Springer International Publishing AG 2017
D.-S. Huang et al. (Eds.): ICIC 2017, Part II, LNCS 10362, pp. 135–141, 2017.
DOI: 10.1007/978-3-319-63312-1_12

proposed by Fakoor et al. [5]. Therefore, we attempt to use other deep learning model to classify the tumor data. In all the deep learning models, the Convolutional Neural Networks (CNN) [6] plays a dominant role in a variety of tasks. In this paper, CNN is considered to be applied to classify the tumor gene expression data.

Generally, the input of CNN is a 2-dimensional image sample that can be utilized to the convolution operation. Since each sample in gene expression data is a 1-dimensional array, the traditional CNN models are not applicable to gene expression data. Fortunately, 1-dimensional CNN (1DCNN), a special CNN model, requires the input is a 1-dimensional vector. Here, we introduced 1DCNN into the tumor classification. In addition, a large number of labeled data are required for training the CNN models, including 1DCNN. However, the number of samples of gene expression data is quite small. To address the problem of insufficient labeled samples, in this paper, a novel Sample Expansion method (SE) is proposed. Stimulated by Denoising Autoencoder (DAE) [7], we simply obtain a large number of samples by randomly cleaning partially corrupted input many times. Then the expanded samples and untreated samples are merged into a matrix as the training samples. In order to benefit from both 1DCNN and SE, we suggest a joint Sample Expansion and 1DCNN (SE1DCNN) method for tumor classification. Firstly, Infinite Feature Selection (Inf-FS) [8], an excellent unsupervised feature selection method, is used as the dimensionality reduction strategy to select genes from gene expression data. In a specific tumor dataset, each feature represents one gene and has its natural meaning. Therefore, feature selection is more convincing than feature extraction when processing gene expression data. Secondly, SE is implemented to expand the number of labeled samples. Finally, the 1DCNN model is utilized to classify the tumor gene expression data by using the expanded samples and raw samples.

The main contributions of our work are explained as follows. Firstly, for the first time, 1DCNN is applied to the tumor gene expression data classification task. Secondly, we propose a novel sample expansion method to address the problem of insufficient training samples when using 1DCNN to implement tumor classification.

2 Methodology

2.1 Infinite Feature Selection

In tumor classification task, the management of high-dimensional gene expression data requires an efficient feature selection method to individuate redundant and/or irrelevant features and avoid overfitting [9]. In [8], Roffo et al. proposed a novel unsupervised feature selection method dubbed Infinite Feature Selection (Inf-FS). The most appealing characteristic of Inf-FS is that the importance of a feature is assessed by considering all the possible subsets of all the features. In addition, the relevance and redundancy of each feature are influenced by all the other ones. Therefore, we adopt Inf-FS as the dimensionality reduction strategy to select genes.

2.2 Sample Expansion Method

Traditional Autoencoders (AE) can produce a useful representation by encoder. But we cannot extract and compose robust features by using AE. Vincent et al. proposed a very different strategy to get a high-level representation: cleaning partially corrupted input, or in short denoising [7]. The denoising idea was successfully used into AE and producing a classical model: Denoising Autoencoders (DAE).

The graphical representation of DAE is described in Fig. 1. Firstly, denote $\mathbf{x}$ as a sample and it is stochastically corrupted to $\tilde{\mathbf{x}}$, where each value filled with black is forced to be 0. Then DAE maps the corrupted data $\tilde{\mathbf{x}}$ to $\mathbf{y}$ via encoder and attempts to reconstruct $\mathbf{x}$ via decoder, generating reconstruction $\mathbf{z}$. Finally, the reconstruction error between $\mathbf{x}$ and $\mathbf{z}$ is calculated with a loss function. Experiments show that the corrupted data is very useful for two reasons: On one hand, comparing with non-corrupted data, the corrupt data can be trained to obtain smaller weight noise; On the other hand, the corrupted data reduces the generation gap between the training and testing data to a certain extent [7].

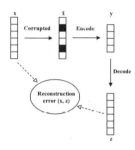

Fig. 1. The Denoising Autoencoders model.

Inspired by the denoising idea, a Sample Expansion method (SE) is proposed to deal with the problem of insufficient training samples. Denote $\mathbf{X} \in \mathbb{R}^{m \times n}$ as a gene expression dataset needs to be trained. For each sample, the values of a $(a \leq m)$ genes are randomly chosen to be set as 0 and the processed sample is saved. Due to the locations of the corrupted genes are required to be non-repeated, this process is repeated $floor(m/a)$ times, where $floor()$ is a function that is rounded down. Therefore, $floor(m/a)$ samples can be obtained in this way. Similarly, we can obtain $n \times floor(m/a)$ expanded samples for all samples. Finally, the expanded samples and the raw samples are combined to be taken as training data.

In order to visualize our idea, the schematic representation of SE is given in Fig. 2. In the gene expression dataset $\mathbf{X} \in \mathbb{R}^{m \times n}$, each row represents a gene and each column represents a sample. Here, a is set to 2. The corrupted gene is filled with black in expanded samples. For Sample 1 (the first sample in $\mathbf{X}$), $floor(m/2)$ expanded samples are obtained. Including Sample 1, we can obtain $floor(m/2) + 1$ samples from one sample. Other samples are processed in the same manner. By combining the expanded samples with n samples in $\mathbf{X}$, $n \times floor(m/2) + n$ samples are stored in a matrix $\mathbf{Y}$ to be taken as the training data. By utilizing SE, we can get a large number of training samples.

The expanded samples not only have the advantages of corrupted data but also address the problem of insufficient training samples of gene expression data to a certain extent.

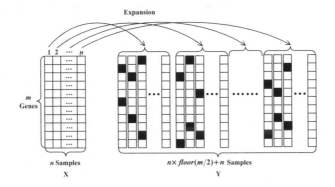

Fig. 2. The schematic representation of Sample Expansion method.

2.3   1DCNN

CNN is a classical deep learning model that exploits spatially local correlation by enforcing a local connectivity pattern between neurons of adjacent layers. A traditional CNN architecture consists of various combinations of convolutional layers, max pooling layers and fully connected layers. Since each sample of gene expression data is a 1-dimensional array, we introduced 1DCNN, which requires the input is a 1-dimensional vector, into the tumor classification.

In Fig. 3, the architecture of 1DCNN is designed with 7 layers: one input layer, two convolutional layers C1 and C2, two max pooling layers M1 and M2, one fully connected layer F and one output layer. In gene expression data, each sample can be taken as the input of 1DCNN which requires the input data to be a vector. In Fig. 3, the input layer is a sample with m_1 genes. Suppose $\mathbf{W}_1$ with size $w_1 \times 1$ and $\mathbf{W}_2$ with size $w_2 \times 1$ are the convolutional kernels of the first convolutional layer C1 and the second convolutional

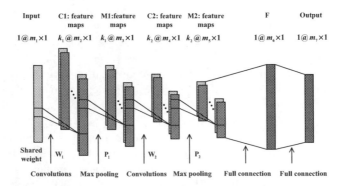

Fig. 3. The 1DCNN architecture consisting of two convolutional layers, two max pooling layers and one fully connected layer.

layer C2, respectively; $\mathbf{P}_1$ with size $p_1 \times 1$ and $\mathbf{P}_2$ with size $p_2 \times 1$ are the filtering kernels of the first max pooling layer M1 and the second max pooling layer M2, respectively; k_1 and k_2 are the number of kernels. After convolving the input layer, C1 contains $k_1 \times m_2 \times 1$ nodes where $m_2 = m_1 - w_1 + 1$. M1 contains $k_1 \times m_3 \times 1$ nodes where $m_3 = m_2 / p_1$. C2 contains $k_2 \times m_4 \times 1$ nodes where $m_4 = m_3 - w_2 + 1$. M2 contains $k_2 \times m_5 \times 1$ nodes where $m_5 = m_4 / p_2$. F and the output layer contain m_6 and m_7 nodes, respectively.

3   Results and Discussion

We test the SE1DCNN method on three tumor datasets: breast cancer [10], leukemia [11] and colon cancer [12]. The descriptions of the three datasets are summarized in Table 1.

Table 1. Descriptions of three tumor datasets.

Tumor dataset	Data matrix	Data labels
Breast cancer	30006×20	1 = non-IBC, 2 = IBC
Leukemia	12600×60	1 = MP, 2 = HDMTX, 3 = HDMTX + MP, 4 = LDMTX + MP
Colon cancer	2000×62	1 = cancer, 2 = normal

For each dataset, Inf-FS is adopted as the dimensionality reduction algorithm to select genes. Generally, the higher the ratio of training samples is, the better the classification results are. To verify the validity of the SE method, we take 20% samples after dimensionality reduction to expand the training samples and the remaining 80% samples as testing samples. In each sample, the corrupted number of genes a is tested. In our experiments, SE1DCNN and 1DCNN are tested. Furthermore, the SAE, SAE with fine tuning and Softmax/SVM (with Gaussian kernel) methods in [5] are used as the competitive algorithms.

In SE1DCNN method, the performance of $a = 1, 2, 3, 4, 5$ are tested. The classification performance on three tumor datasets of SE1DCNN and 1DCNN is given in Table 3. There are a large number of parameters when training SE1DCNN and 1DCNN, the choices of parameters might not be the best but effective for tumor classification. The best performance in Table 2 is indicated by bold. When $a = 3$ on breast cancer and $a = 1$ on colon cancer, SE1DCNN has the same classification accuracies with 1DCNN. In the remaining case, SE1DCNN outperforms 1DCNN which demonstrate SE method is very useful for tumor classification.

In Table 3, the performance comparisons on three tumor datasets of four methods: 1DCNN, SAE, SAE with Fine Tuning and Softmax/SVM. The best performance in Table 2 is indicated by bold. On all the three datasets, 1DCNN has a better performance than the three methods in [5]. It is worth noting that, the three methods in [5] take 80% samples as training samples and 20% samples as testing samples. The results in Tables 2 and 3 indicate SE1DCNN and 1DCNN are applicable and helpful to classify tumor gene expression data.

Table 2. The classification accuracies of SE1DCNN and 1DCNN on three tumor datasets. (%)

	SE1DCNN					1DCNN
	$a = 1$	$a = 2$	$a = 3$	$a = 4$	$a = 5$	
Breast cancer	**100**	**100**	86.67	**100**	**100**	86.67
Leukemia	**59.57**	57.45	57.45	57.45	57.45	55.32
Colon cancer	83.67	**85.71**	**85.71**	**85.71**	**85.71**	83.67

Table 3. Performance comparisons of different methods on three tumor datasets. (%)

	1DCNN	SAE	SAE (Fine tuning)	Softmax/SVM
Breast cancer	**86.67**	63.33	83.33	85.0
Leukemia	**55.32**	33.71	33.71	46.33
Colon cancer	**83.67**	66.67	83.33	83.33

4   Conclusions

In this paper, a joint Sample Expansion and 1D Convolutional Neural Network method is proposed to classify tumor gene expression data. Firstly, an excellent feature selection method Inf-FS is used to reduce the dimensionality of gene expression data. Secondly, inspired by the denoising idea in DAE, a Sample Expansion method is proposed. Finally, since CNN can provide excellent classification effect in many fields, the applicability of 1DCNN on gene expression data is discussed. And a 7-layer 1DCNN model is designed to classify the tumor gene expression data by using the expanded samples and raw samples. The classification results indicate that SE1DCNN has a better performance than the competitive methods.

Acknowledgements. This work was supported by the National Natural Science Foundation of China under Grants 61273143 and 61472424.

References

1. Ge, S.G., Xia, J., Sha, W., Zheng, C.H.: Cancer subtype discovery based on integrative model of multigenomic data. IEEE/ACM Trans. Comput. Biol. Bioinform. **PP**(99), 1 (2016)
2. Guyon, I., Weston, J., Barnhill, S., Vapnik, V.: Gene selection for cancer classification using support vector machines. Mach. Learn. **46**(1–3), 389–422 (2002)
3. Bengio, Y., Courville, A., Vincent, P.: Representation learning: A review and new perspectives. IEEE Trans. Pattern Anal. Mach. Intell. **35**(8), 1798–1828 (2013)
4. Min, S., Lee, B., Yoon, S.: Deep learning in bioinformatics. Briefings Bioinform. (2016, in press)
5. Fakoor, R., Ladhak, F., Nazi, A., Huber, M.: Using deep learning to enhance cancer diagnosis and classification. In: Proceedings of the 30th International Conference on Machine Learning, vol. 28 (2013)
6. Hinton, G.E., Salakhutdinov, R.R.: Reducing the dimensionality of data with neural networks. Science **313**(5786), 504–507 (2006)

7. Vincent, P., larochelle, H., Lajoie, I., Benjio, Y., Manzagol, P.A.: Stacked denoising autoencoders: Learning useful representations in a deep network with a local denoising criterion. J. Mach. Learn. Res. **11**, 3371–3408 (2010)
8. Roffo, G., Melzi, S., Cristani, M.: Infinite feature selection. In: Proceedings of the IEEE International Conference on Computer Vision, pp. 4202–4210 (2015)
9. Guyon, I., Gunn, S., Nikravesh, M., Zadeh, L.A.: Feature Extraction: Foundations and Applications. Studies in Fuzziness and Soft Computing. Springer, New York (2006)
10. Woodward, W.A., Krishnamurthy, S., Yamauchi, H., El-Zein, R., Ogura, D., Kitadai, E., Niwa, S., Cristofanilli, M., Vermeulen, P., Dirix, L.: Genomic and expression analysis of microdissected inflammatory breast cancer. Breast Cancer Res. Treat. **138**(3), 761–772 (2013)
11. Cheok, M.H., Yang, W., Pui, C.-H., Downing, J.R., Cheng, C., Naeve, C.W., Relling, M.V., Evans, W.E.: Treatment-specific changes in gene expression discriminate in vivo drug response in human leukemia cells. Nat. Genet. **34**(1), 85–90 (2003)
12. Alon, U., Barkai, N., Notterman, D.A., Gish, K., Ybarra, S., Mack, D., Levine, A.J.: Broad patterns of gene expression revealed by clustering analysis of tumor and normal colon tissues probed by oligonucleotide arrays. Proc. Natl. Acad. Sci. **96**(12), 6745–6750 (1999)

Systems Biology

2DIs: A SBML Compliant Web Platform for the Design and Modeling of Immune System Interactions

Marzio Pennisi[1], Giulia Russo[2], Giuseppe Sgroi[3], Giuseppe Parasiliti[3],
and Francesco Pappalardo[4(✉)]

[1] Department of Mathematics and Computer Science, University of Catania, Catania, Italy
mpennisi@dmi.unict.it
[2] Department of Biomedical and Biotechnological Sciences, University of Catania, Catania, Italy
giulia.russo@unict.it
[3] University of Catania, Catania, Italy
giuseppesgroi@live.it, giuseppeparasiliti93@gmail.com
[4] Department of Drug Sciences, University of Catania, Catania, Italy
francesco.pappalardo@unict.it

Abstract. We present 2DIs, a web platform that allows the easy design of extracellular models of the immune system function, including the possibility to describe the most important immune system entities and interactions and to produce a validated SBML file of the model. 2DIs permits immunologists to directly describe and share their knowledge with other colleagues and, more importantly, with modelers that can therefore obtain an SBML-compliant modelling template of the immunological process that can be used to develop the computational model. This could introduce a novel way of communicating among immunologists and modelers, reducing the risk of errors and misinterpretations.

Keywords: Immune system modeling · Systems biology markup language · Web applications

1 Introduction

The interaction between immunologists (but this applies, in general, for all researchers involved in the life science field) and computational modelers is based on the idea that each partner can contribute to increase the knowledge of the other one. This interaction can be described by four verbs: to meet, to define, to reflect and to explore.

"To meet" is the first word to describe this interaction because only when these two worlds meet the collaboration can start. The Meeting is the time in which immunologists communicate their problems and needs whereas computational modelers describe their approach and tools. "To define" is the second action to build a strong and productive collaboration. Computational science and immunology are two distinct disciplines that use completely different languages. In order to work proficiently together immunologists and modelers have to share key concepts. This process is surely difficult and time-consuming but a necessary condition to build a robust interaction. "To reflect" is the term to evaluate the interaction. Modelers have to ask themselves how much their models

© Springer International Publishing AG 2017
D.-S. Huang et al. (Eds.): ICIC 2017, Part II, LNCS 10362, pp. 145–154, 2017.
DOI: 10.1007/978-3-319-63312-1_13

reflect the true biological reality. A computational model needs to be able to reproduce the behavior of the biological scenario with its elements and rules. Through careful comparison of the computational model output with bio-models behavior the process of developing a useful model converges.

"To explore" is the final step of the collaboration. Immunologists start to use the mathematical models as tools to explore their biological systems, addressing their studies to evidence new rules and events related to scenario. On the other hand, modelers exploit the validation of their results with in vivo or vitro experiments to refine their models.

The interdisciplinary nature of the computational immunology requires a strong effort of life scientists, which need to go beyond the only data supply, as it is extremely important in defining the biological scenario and ultimately construct a robust and validated mathematical or computational model. Only through an unceasing collaboration of life and computational scientists it is possible to turn software into a valuable tool in life sciences.

In our experience, the first problem we had to face with has been to define a common language among life scientists and modelers. Communication is a fundamental step to start a fruitful collaboration. We initially overcome this issue by using graphical and conceptual languages as those commonly used in computer science to describe complex and dynamical systems. Conceptual models are easily understood by life scientists and allow to transfer qualitative biological knowledge to modelers, which are then able to build mathematical and computational simulators. The parameters of the computational models so constructed are adapted to reproduce the in vivo experimental data available. Also, here the help of life scientists is very much required. They need to provide biological data first and interpretation (checking) of model results after, in order to move from qualitative to quantitative modeling. After validation, the model is ready to predict experimental outcomes, reproducing experimental setups that are either time or money consuming in vivo. Here is where computation turns out to be cheap. Predictions have to be carried by life scientists for in vivo verifications. Model results are finally validated against in vivo or in vitro experiments [12, 13].

Clearly, the involvement of both counterparts is important at every step of the modeling process. The vision must be the same. The interest high on both sides. Without a common effort, it would not be possible to reach any meaningful result.

The establishment of communication standards that improve the dialogue, thus avoiding potential ambiguities, among people with heterogeneous backgrounds may then represent a huge leap ahead. In this field, the systems biology markup language (SBML) [1] has been developed with the initial goal to encode all the possible information required to express a biochemical network model, including its kinetics.

SBML is based on the extended markup language (XML), a widely adopted standard on the Internet and has been continuously developed and improved in the years, becoming a standard that counts nowadays a rather wider community of users, developers and tools.

SBML offers many advantages, such as the interoperability among tools for the design, analysis, simulation and visualization of the biological scenario without the need to rewrite each time the model. Furthermore, the use of a standard language makes easy the sharing and the publishing of models to the research community, thus unlacing the survival of models and the relative intellectual effort from the lifetime of the used software to realize the model.

Presently, there is a broad range of software packages and tools that support SBML. Some are more generic, while others offer more specialized features. There are tools for the simulation of networks [2], for the graphical representation and analysis of networks [3–5], databases of reactions and parameters [6].

Some packages, like Cell Designer [3] and COPASI [2], integrate most of these features in a single software suite. Furthermore, models realized in SBML can be simu lated by using ODE, algebraic equations, stochastic kinetics, and discrete events. One possible drawback of SBML, and probably of the relative tools, is given by the fact that such tools have been mainly oriented for the representation of intracellular and biochemical networks, partially neglecting the representation of extracellular interactions.

We present 2DIs (Dynamic Description of Immune System Interactions), an online graphical tool to easily develop extracellular models of the immune system (IS) function, including the possibility to describe the most important IS entities and interactions and to produce a validated SBML file of the model. In this way, immunologists can directly describe and share their knowledge with other colleagues and, more importantly, with modelers that can therefore obtain an SBML-compliant modelling template of the immunological process that can be used to develop the computational model. This could introduce a novel way of communicating between immunologists and modelers, reducing the risk of errors and misinterpretations. Furthermore, we will show with two examples how in some cases the resulting SBML file can be directly imported into some simulation software that support SBML, and thus simulated just only after the defining of the initial quantities and interaction rates.

2 Materials and Methods

2.1 2DIs Implementation and Description

2DIs was not developed entirely from scratch. The starting point was OrgoShmorgo project (https://github.com/es/orgoShmorgo) developed by Emil Stolarsky. It is an organic molecule visualizer built using the library D3.js. D3.js is a JavaScript library that has been developed for visualizing data in a dynamical and interactive fashion starting from some organized numerical data combining HTML5, Scalable vector Graphics (SVG), and cascading style sheet (CSS) standards.

When used inside an Html page, the D3 library allows the use of pre-built javascript functions to select HTML Document Object Model (DOM) elements, create SVG elements according to predefined graphical styles, and to add transitions and object movements. 2DIs is accessible through RELIEF (ExtendibLe Integrated biomEdicine Framework) web portal (http://vaima.dmi.unict.it). It is freely accessible after registration and its main target is to make possible the sharing of models through the systems biomedicine community and to incorporate several biomedical software. One of this is 2DIs.

When the user launches 2DIs web application it is shown three main panels (Fig. 1). The two at the left and right side (i.e., commands) allow to select the interaction type (left side) and the entity type (right side). The central panel shows in real time the just created immune system interaction. At the beginning, the central panel already contains just a null interaction type.

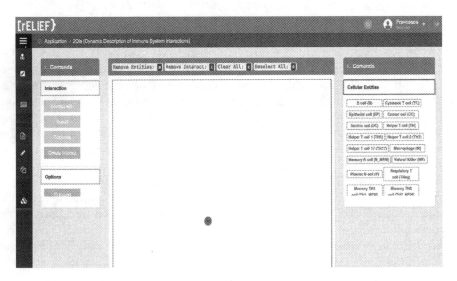

Fig. 1. 2DIs web interface at the first launch.

Suppose that a user would like to describe an immune system interaction between a naive B cell and a soluble antigen. The final product of this interaction, if the B cell has recognized the antigen through its B cell receptor, is the differentiation of the B cell into an antigen presenting B cell that express on its surface the processed antigen complexed with major histocompatibility complex of class II peptides. For doing so, the user just selects from the Entity panel the B and the Ag entities; then connect them with the command "Interact with" with the "int" blue bullet; finally, he/she select the B_APC entity and connect it to the "int" blue bullet with a "Result" type interaction. Figure 2 shows the above described interaction. 2DIs is also able to represent interactions that make use of molecules (i.e., interleukins, chemokines, adjuvants and so on so forth). Suppose, for example, that one would like to represent the differentiation of a Helper T cell into a Helper T cell type 1 (TH1), mediated also through the effect of Interleukin 12 (IL-12) after that a naive Helper T cell has recognized the Antigen complexed with major histocompatibility class II peptides presented on the antigen presenting B cell surface.

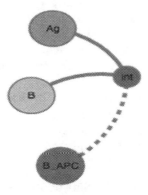

Fig. 2. An example of a well-known immune system interaction. Full lines represent the entities that participate into an interaction, while dashed lines represent the final result (product) of the interaction. (Color figure online)

In this case, the user proceeds as described above i.e., he/she selects the B_APC and TH entities and connect them to the "int" blue bullet through the usage of "Interact with" button. Then he/she add the TH1 entity as a final product, selecting the "Result" button and connecting it to the "int" blue bullet. The final phase is to select the "Result" inter-action. After that a pop-up with a drop-down menu appears, letting the user to select the participating molecules (i.e., IL-12). The final result is depicted in Fig. 3.

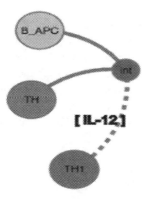

Fig. 3. An interaction that makes use of interleukins. (Color figure online)

In order to enable the "interact with" button the user has to click first on the involved entity, and then on the involved interaction, whereas to enable the "result" button to introduce a new product, it is mandatory first click on the interaction, and then on the entity that represents the product of the interaction. Drag functions that directly allow to create interactions and products by dragging a solid line from an entity to an interaction and a dashed line from an interaction to an entity, respectively, are currently under development.

Through the "export" button it is possible to download a zip file that contains three documents: a screenshot of the developed model, a json file that can be reloaded (using the import button) into 2DIs to modify and/or improve the created model, and a SMBL validated file that can be thus used into any SBML compliant software for further refinements and for the simulations. Finally, the "screenshot" button allows to capture a screenshot of the model, whereas the "clear all" button allows to delete the model in order to start a new one. On the top of the modelling window some important keyboard shortcuts that allow to remove selected entities and interactions, to clear the modelling window, and to deselect all the selected entities and/or interactions are shown.

2.2   2DIs Application Example: Modeling of Melanoma Treatment

In order to show the potential capabilities of the 2DIs graphical tool, we tried to develop an example model to represent the effects of a combined treatment of activated T lymphocytes plus specific antibodies against melanoma in B16-OVA mouse models. This problem has already carried on by both an agent based model [7] and a delayed equation based model [8].

B16-OVA mouse models are mice transduced with the chicken ovalbumin gene that is used as tumor antigen. The goal of the in vivo experiment is to demonstrate the combined action of anti-CD137 monoclonal antibodies and adoptive T cell therapy against B16-OVA melanoma, a poorly immunogenic murine tumor. The treatment includes a single injection of anti-CD137 mAb and OVA-specific TCR-transgenic CD8 cytotoxic T cells (CTLs).

While the ABM model shows an extended vision of the dynamics involved in this problem, the differential equation models tries to briefly reassume the most important features of the biological problem. This latter model uses delayed equations to model the delay that occurs between the injection time of the treatment, and the arrival of the therapy into the skin point where melanoma is developing.

A simplistic vision of the biological background is given as follows. Both antibodies and activated T cells that migrate to the skin compartment work synergistically to improve the infiltration into the tumor mass and the killing of melanoma cancer cells. Killed cancer cells release antigens that are captured by antigen presenting cells and are presented to antigen-specific naive T cells by antigen presenting cells (APC). We note here that APC are not modeled here and in the differential equation based model, so we directly suppose as final result that antigens contribute to stimulate T cell activation, skipping the presentation step. Stimulated naive T cells become active CTLs that will further act against melanoma. More details about the biological background can be found in [7, 8].

The described biological scenario has been modeled using the 2DIs platform, and the resulting model is shown in Fig. 4.

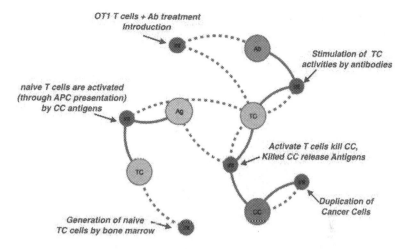

Fig. 4. The 2DIs model for the melanoma treatment. Interactions are represented by small blue bullets, while entities by bigger colored circles. Participation to an interaction is represented by solid lines, while interaction products are represented by dashed lines. (Color figure online)

Once created, we tried to export the 2DIs model into a SBML document. This SBML document can be then directly imported by any modeler into any SBML compliant software for analysis and improvement. We indeed tried to directly use the model for the simulation of the immunological scenario. To this end, we chose to use Snoopy [9], a Petri-net (PN) based software for the modelling and simulation of concurrent systems that is able to interpret and load SBML files. The reason of this choice is twofold. Differently from common SBML compliant software, Snoopy in not exclusively developed to simulate biochemical networks and reactions. Moreover, PN already demonstrated to be a valid alternative for the modelling and simulation of Immune system reactions in respect to classic approaches, as already presented in [10].

Snoopy demonstrated able to load without any problems the validated SBML produced by 2DIs. During the import procedure, we chose to use Continuous Petri Nets, a particular extension of PN that, as in differential equation based models, uses continuous quantities and continuous interaction laws instead of discrete ones for the representation of entities and interactions. Further details about Petri nets can be found in [11]. Once loaded, we only needed to define the initial entity quantities, the interaction coefficients and the arc weights. All the interactions were modeled using the classical mass action law, that is the pre-defined law used by Snoopy when continuous Petri nets are used. Off course, more complex laws to better describe the dynamics of the system can be used, however this goes beyond the scope of the application example. We also needed to modify just only one interaction, that is the one that introduces the combined treatment into the system using a scheduled transition. This transition repeatedly introduces at given times the treatment in order to simulate the continuous migration of the treatment from the injection point on the skin point where melanoma develops. The resulting PN model is represented in Fig. 5.

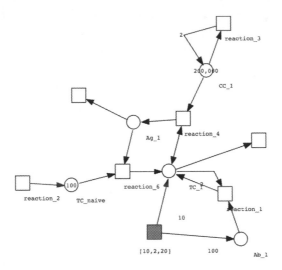

Fig. 5. The Snoopy PN model created by loading the 2DIs SBML model. Circles represent the modelled entities, boxes represent interactions. The gray box represents a scheduled interaction used to model the continuous migration of the treatment (antibodies and activated T cells) from the injection point to the skin point where melanoma develops.

Just after that, the model was ready for the simulation. We run the model for 100 time steps. Results are presented in Fig. 6. The model demonstrated able to qualitatively reproduce the results obtained by both the ABM and the differential equation based models presented in [7, 8], respectively. As expected, the combined treatment demonstrated able to stop the development of the melanoma (see Fig. 6A) by directly attacking the cancer cells and by triggering the activation of naive T cells toward activated T cells, while the lack of treatment (Fig. 6B) leads to the melanoma development. It must be underlined that this simple model is only able to qualitative capture the biological scenario leaving out many details about the mechanisms involved in the immune response, mechanisms that are quantitatively reproduced by the models presented in [7, 8]. Nevertheless, this simple example model shows how the use of the 2DIs platform may represent a strong tool to improve and speed-up the communication between biologists and modelers and the development of immune system models.

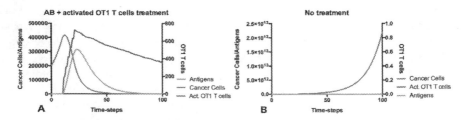

Fig. 6. Simulation of the melanoma treatment model with two different scenarios: (A) administration of the OT1 T cells + Ab combined treatment. (B) simulation of no treatment. Activated OT1 T cells are represented on the right axis to improve readability.

3 Conclusions and Final Remarks

In a multidisciplinary environment one of the most felt problem is represented by the difficulties found in communications among scientists and researchers that come from different disciplines. Communication is a fundamental step to start a fruitful collaboration. One of the solution maybe the usage of graphical and conceptual languages as those commonly used in computer science to describe complex and dynamical systems.

The establishment of communication standards that improve the dialogue, thus avoiding potential ambiguities, among people with heterogeneous backgrounds may then represent a huge leap ahead.

The systems biology markup language has been continuously developed and improved in the years, becoming a standard that counts nowadays a rather wider community of users, developers and tools. It allows the representation and the quantitative description of biological processes in different scales and there are a huge number of tools that provide a graphical front end to facilitate its usage to non-computer science researchers and/or experts.

We presented 2DIs, an online graphical tool to easily develop extracellular models of the immune system function, including the possibility to describe the most important IS entities and interactions and to produce a validated SBML file of the model. 2DIs is still in its first phases of development but it is already able to represent complex immune system models in an impressive easy way. Modelers can then take up the SBML exported file ready to use in a Petri nets modeling framework for further development. In a near future, 2DIs will be extended to include additional features i.e., quantitative data insertion and a fully way to provide the description of mathematical laws that govern interactions. 2DIs is explicitly designed for immune system modeling, but can be used as a starting point to provide the same features in different systems biology applications.

References

1. Hucka, M., Finney, A., Sauro, H.M., Bolouri, J.H., Doyle, C., Kitano, H., Arkin, A.P., Bornstein, B.J., Bray, D., Cornish-Bowden, A.: The systems biology markup language (SBML): a medium for representation and exchange of biochemical network models. Bioinformatics **19**(4), 524–531 (2003)
2. Hoops, S., Sahle, S., Gauges, R., Lee, C., Pahle, J., Simus, N.: COPASI—a COmplex PAthway Simulator. Bioinformatics **22**, 3067–3074 (2006)
3. Funahashi, A., Morohashi, M., Kitano, H., Tanimura, N.: CellDesigner: a process diagram editor for gene-regulatory and biochemical networks. Biosilico **1**(5), 159–162 (2003)
4. Köhler, J., Baumbach, J., Taubert, J., Specht, M., Skusa, A., Rüegg, A., Rawlings, C., Verrier, P., Philippi, S.: Graph-based analysis and visualization of experimental results with ONDEX. Bioinformatics **22**, 1383–1390 (2006)
5. Shannon, P., Markiel, A., Ozier, O., Baliga, N.S., Wang, J.T., Ramage, D., Amin, N., Schwikowski, B., Ideker, T.: Cytoscape: a software environment for integrated models of biomolecular interaction networks. Genome Res. **13**, 2498–2504 (2003)
6. Rojas, I., Golebiewski, M., Kania, R., Krebs, O., Mir, S., Weidemann, A., Wittig, U.: Storing and annotating of kinetic data. In Silico Biol. **7**, 37–44 (2007)

7. Pappalardo, F., Forero, I.M., Pennisi, M., Palazon, A., Melero, I., Motta, S.: SimB16: modeling induced immune system response against B16-melanoma. PLoS ONE **6**(10), e26523 (2011)
8. Pennisi, M.: A mathematical model of immune system-melanoma competition. Comput. Math. Methods Med., Article ID 850754, 13 p. (2012)
9. Heiner, M., Herajy, M., Liu, F., Rohr, C., Schwarick, M.: Snoopy – a unifying Petri Net tool. In: Haddad, S., Pomello, L. (eds.) PETRI NETS 2012. LNCS, vol. 7347, pp. 398–407. Springer, Heidelberg (2012). doi:10.1007/978-3-642-31131-4_22
10. Pennisi, M., Cavalieri, S., Motta, S., Pappalardo, F.: A methodological approach for using High-Level Petri Nets to model the adaptive immune system response. BMC Bioinf. **17**(Suppl 19), 498 (2016). doi:10.1186/s12859-016-1361-6
11. Murata, T.: Petri Nets: properties, analysis and applications. Proc. IEEE **77**(4), 541–580 (1989)
12. Bianca, C., Chiacchio, F., Pappalardo, F., Pennisi, M.: Mathematical modeling of the immune system recognition to mammary carcinoma antigen. BMC Bioinf. **13**(Suppl 17), S21 (2012). doi:10.1186/1471-2105-13-S17-S21
13. Pappalardo, F., Fichera, E., Paparone, N., Lombardo, A., Pennisi, M., Russo, G., Leotta, M., Pappalardo, F., Pedretti, A., De Fiore, F., Motta, S.: A computational model to predict the immune system activation by citrus-derived vaccine adjuvants. Bioinformatics **32**(17), 2672–2680 (2016)

Modeling, Simulation, and Optimization of Biological Systems

Modeling Neuron-Astrocyte Interactions: Towards Understanding Synaptic Plasticity and Learning in the Brain

Riikka Havela[1], Tiina Manninen[1], Ausra Saudargiene[2,3],
and Marja-Leena Linne[1(✉)]

[1] Computational Neuroscience Group, BioMediTech Institute
and Faculty of Biomedical Sciences and Engineering,
Tampere University of Technology, Tampere, Finland
marja-leena.linne@tut.fi
[2] Neuroscience Institute, Lithuanian University of Health Sciences,
Kaunas, Lithuania
[3] Department of Informatics, Vytautas Magnus University, Kaunas, Lithuania

Abstract. Spiking neural networks represent a third generation of artificial neural networks and are inspired by computational principles of neurons and synapses in the brain. In addition to neuronal mechanisms, astrocytic signaling can influence information transmission, plasticity and learning in the brain. In this study, we developed a new computational model to better understand the dynamics of mechanisms that lead to changes in information processing between a postsynaptic neuron and an astrocyte. We used a classical stimulation protocol of long-term plasticity to test the model functionality. The long-term goal of our work is to develop extended synapse models including neuron-astrocyte interactions to address plasticity and learning in cortical synapses. Our modeling studies will advance the development of novel learning algorithms to be used in the extended synapse models and spiking neural networks. The novel algorithms can provide a basis for artificial intelligence systems that can emulate the functionality of mammalian brain.

Keywords: Astrocyte · Neuron · Calcium · Computational model · Synaptic plasticity · Learning

1 Introduction

Understanding the ability of the brain to dynamically reorganize itself and adapt to changing environment is an intriguing topic not only in the field of neuroscience but also in artificial intelligence. In the neuroscience experiments, this ability can be captured as structural, homeostatic, and synaptic forms of plasticity. Structural plasticity includes formation and pruning of axons, dendrites, and synapses. Homeostatic plasticity acts on a slow time scale and rescales all synaptic weights, adjusts the threshold between long-term depression (LTD) and long-term potentiation (LTP), and regulates other neuronal parameters [1]. Synaptic plasticity, on the other hand, is a modification of the synaptic connections between the neurons. Synaptic plasticity is

© Springer International Publishing AG 2017
D.-S. Huang et al. (Eds.): ICIC 2017, Part II, LNCS 10362, pp. 157–168, 2017.
DOI: 10.1007/978-3-319-63312-1_14

believed to underlie the biological basis of learning and memory. In 1949, a Canadian neuropsychologist Donald O. Hebb formulated his famous postulate that can be summarized as "neurons that fire together, wire together" [2]. In other words, when two neurons have a synaptic connection and are active at the same time, the synaptic strength increases, i.e. undergoes LTP. The hypothesis was confirmed by numerous electrophysiological experiments (see, e.g., [3–6]). Hebbian learning rules have been widely applied in artificial learning systems [7–10]. In more recent years, a phenomenon called spike-timing-dependent synaptic plasticity (STDP) was discovered [11–13]: if the presynaptic spike precedes the postsynaptic spike within a short time window, the synapse undergoes LTP, and it exhibits LTD if the temporal order is reversed. Variations on this classic STDP form have been found in hippocampus, cerebral cortex, and basal ganglia, with the activity windows for LTP and LTD depending on the brain area, type of neuron, frequency of input-output spike pairing, the duration of such pairing, spike bursting in the postsynaptic cell, and synaptic location on the dendritic tree [14–20]. Different forms of synaptic plasticity occurring at the molecular, cellular and circuit level coexist.

Astrocytes are a type of glial cell known for their intimate connection with adjacent neurons in a variety of brain areas. They are known to express an overwhelming complexity of molecular and cell-level signaling (see, e.g., recent reviews [21, 22]). They participate in astrocytic and neuronal network formation, act as metabolic support for intensive energy-demanding neuronal activity, and, notably, some lines of evidence suggest that astrocytes actively participate in synaptic signaling by modulating the signals transmitted by neurons (see, e.g., reviews [21–24], and references therein). Based on the experimental literature, the mechanisms involved seem to depend not only on the developmental stage of an animal but also on the brain area, neural circuitry, as well as on the experimental technique used to characterize the phenomena. Moreover, dysfunctioning of astrocytes is a possible contributing factor in pathologies of the central nervous system [23].

Astrocytes connect with synapses using a long, thin extension of their cell body, called endfoot, which tightly wraps around the synapse, allowing the astrocyte to detect substances released by pre- and postsynaptic terminals. Their highly ramified morphology allows them to connect to more than 100,000 synapses of individual neurons [25], and they have been shown to govern their individual, non-overlapping "territories" in neuronal networks. Astrocytes can detect molecular substances released by synaptic terminals and can release substances that may be able to modify properties of synaptic terminals [21]. One well-known mechanism of astrocytic modulation of synaptic transmission is the postsynaptic slow inward current (SIC), induced by glutamate released by an astrocyte which has been stimulated by synaptic activity [26]. The SIC transiently elevates postsynaptic membrane potential, thus inducing a period of more intense postsynaptic firing in response to presynaptic neurotransmitter release. This way the astrocyte can potentially modulate the dynamics of the neuronal network it is connected to.

Typical approach of computational modeling in neuroscience includes different levels of spatial and temporal organization in the brain varying from models of intracellular signaling pathways and associated molecular dynamics, neurons and other cells up to the models of neuronal networks and high-level cognitive systems.

Modeling single neuron dynamics is commonly approached by the Hodgkin-Huxley (HH) model while signaling pathways are traditionally modeled using laws of biochemistry, including the law of mass action and Michaelis-Menten kinetics. Tissue level phenomena are modeled using neural mass theory. Many modeling studies combine different levels and aim to describe the neural system using a multiscale approach, where a certain behavior is explained bottom up starting from molecular kinetics and chemical equations and going up to the microcircuit or system levels.

In our previous studies, we presented the first detailed categorization and evaluation of more than 60 neuron-astrocyte models in a variety of neurophysiological functions [27] and assessed reproducibility and comparability issues of astrocyte and neuron-astrocyte interaction models [27, 28]. We found out that the compared astrocyte models reproduced different responses with the same stimuli or parameter values used and thus cannot be considered to represent exactly the same astrocyte subtype or phenomena [28]. We concluded that only a few models [29, 30] address in a biological, critical way the putative modulatory mechanisms of astrocytes on plasticity and learning. One of the reasons for lack of models is that not all the details of the mechanisms responsible for modulation of plasticity are known experimentally which makes the computational modeling a challenging task. Thus, phenomenological equations are many times used to describe the experimentally unknown parts of the mechanisms.

In this study, we seek to advance our understanding of how several proposed neuronal and astrocytic regulatory molecular mechanisms could modulate information transmission and plasticity in a synapse. We present a new model that takes into account the time courses and dynamics of a variety of postsynaptic neuron and astrocyte properties. We test the model functionality by varying the frequency of stimuli applied on the pre- and postsynaptic neurons. We additionally address the effects of changing the volume of the postsynaptic neuron terminal (i.e. the volume of a spine) on the model functionality. The long-term goal of our work is to develop extended synapse models including neuron-astrocyte interactions to address learning and neuromodulation in the brain. Based on our studies we expect to develop novel algorithms for learning in spiking neural networks and extended synapse models. The novel algorithms can provide a basis for artificial intelligence systems that can emulate the functionality of mammalian brain.

2   Model

We describe the major biophysical and biochemical mechanisms for the two-compartmental postsynaptic and single-compartmental astrocyte terminals (see Fig. 1). The model is complex, with relatively large numbers of equations.

In the present work, we use pairs of pre- and postsynaptic stimulation with varying frequencies to activate the model. We do not model presynaptic neuron explicitly, but instead assume that presynaptic stimulation triggers glutamate release and activates α-amino-3-hydroxy-5-methyl-4-isoxazolepropionic acid (AMPA)-gated channels and metabotropic glutamate receptors (mGluRs) on dendritic membrane of the postsynaptic cell. Postsynaptic stimulation, induced by somatic current injection, triggers a somatic action potential that propagates back into the dendritic compartment and opens

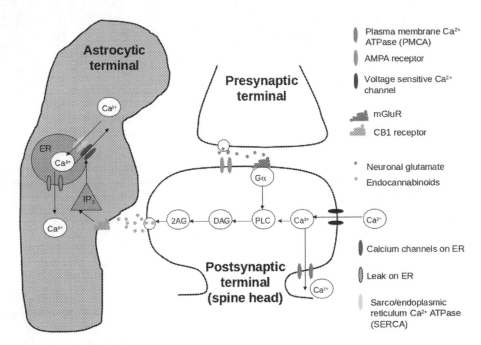

Fig. 1. Illustration of the biophysical and biochemical mechanisms of the model. Glutamate activates postsynaptic AMPA-gated channels and mGluRs. mGluR activation and calcium influx through VSCCs activate PLC and leads to DAG production and 2-AG release. Endocannabinoids (2-AG) bind to the CB_1Rs on the astrocyte membrane, which through IP_3 production activates IP_3Rs and leads to calcium release from the astrocytic ER to the cytosol.

voltage-sensitive calcium channels (VSCCs). Calcium influx into a postsynaptic cell together with the glutamate-activated mGluRs trigger a cascade of biochemical reactions that activates phospholipase C (PLC). This leads to diacylglycerol (DAG) production and 2-arachidonoylglycerol (2-AG) endocannabinoid release. Endocannabinoids bind to the type 1 cannabinoid receptors (CB_1Rs) on the astrocyte membrane and initiate a signaling cascade which activates inositol 1,4,5-trisphosphate (IP_3) receptors (IP_3Rs). IP_3Rs are responsible for calcium release from the endoplasmic reticulum (ER) to the cytosol which leads to the elevated calcium levels in the cytosol of the astrocyte. Our hypothesis is that increased intracellular calcium concentration in astrocytes plays an important role in synaptic plasticity, learning and, ultimately, in memory formation. By developing detailed models we expect to understand the dynamics of events leading to changes in synaptic information transmission, plasticity and learning.

2.1   Postsynaptic Neuron Model

We use a simplified, two-compartment neuron model [31] that describes somatic and dendritic membrane potentials V_s and V_d:

$$C_m \frac{dV_s}{dt} = -I_{Na,s} - I_{Nap,s} - I_{Kdr,s} - I_{L,s} + I_{coupl,s} + I_{ext} \qquad (1)$$

and

$$C_m \frac{dV_d}{dt} = -I_{Na,d} - I_{A,d} - I_{CaL,d} - I_{L,d} - I_{AMPA} + I_{coupl,d}, \qquad (2)$$

where s and d are the somatic and dendritic compartments, respectively. C_m is the membrane capacitance per unit area, I_L are the leak current densities, I_{coupl} are the coupling terms, I_{ext} is the current injected into the soma, and I_{AMPA} is the AMPA synaptic current induced in the dendrite [32]. Initial value for both V_d and V_s is -70 mV.

The soma compartment has sodium current I_{Na}, persistent sodium current I_{Nap}, and delayed rectifier potassium current I_{Kdr} [32]. The dendritic compartment has sodium current I_{Na}, A-type potassium current I_A, and L-type calcium current I_{CaL}. The currents I_{Na}, I_{Nap}, I_{Kdr}, I_A and I_{CaL} are described by the HH formalism [32].

Calcium influx is mediated via VSCCs (L-type calcium channel), and calcium ions are removed from the postsynaptic cell by adenosine triphosphate (ATP)-driven plasma membrane calcium-ATPase (PMCA) pump [33, 34].

AMPA synaptic conductance is modeled by applying an exponential function [35]:

$$g_{AMPA} = \bar{g}_{AMPA} e^{-t/\tau_{AMPA}}, \qquad (3)$$

where $\bar{g}_{AMPA} = 0.2$ mS is the maximal AMPA synaptic conductance, and $\tau_{AMPA} = 5$ ms is the AMPA synaptic conductance decay time constant.

Detailed biochemical mechanisms related to mGluR and endocannabinoid production [36] are included to study the putative interaction of postsynaptic spine terminal with the adjacent astrocyte. The actual reactions can be found in Table 1 in [36]. Glutamate binds to mGluRs and induces dissociation of the G-protein α subunit bound with guanosine-5'-triphosphate (GαGTP) from the molecular complex. Calcium binds to PLC, and, in addition, GαGTP enhances its activity. Active PLC produces IP$_3$ and DAG from phosphatidylinositol 4,5-bisphosphate (PIP$_2$). IP$_3$ is degraded and PIP$_2$ is regenerated by phosphoinositide 3-kinase (PIKin). Calcium binds to DAG lipase (DAGL). This form binds DAG and catalyzes 2-AG synthesis. DAG is inactivated by DAG kinase (DAGK). DAG can also bind to calcium bound form of protein kinase C (PKC) and produce activated PKC.

2.2   Astrocyte Model

Major calcium processes in the astrocyte are described by IP$_3$R channel (J_{IP3R}), leak current into the cytoplasm from the ER (J_{leakER}), and sarco/ER calcium-ATPase (SERCA) pump (J_{SERCA}). We modeled calcium and IP$_3$ concentrations, as well as the active fraction of IP$_3$Rs (h_{IP3R}) following [37–40].

The differential equation for the astrocytic calcium concentration is given as

$$\frac{d[Ca^{2+}]_{astro}}{dt} = J_{IP3R} + J_{leakER} - J_{SERCA}. \tag{4}$$

The differential equation for the astrocytic IP$_3$ concentration is given as

$$\frac{d[IP_3]_{astro}}{dt} = \frac{IP_3^* - [IP_3]_{astro}}{\tau_{IP3}} + r_{IP3}[2AG], \tag{5}$$

where $IP_3^* = 0.16$ μM is the baseline of IP$_3$ in the cytoplasm in the resting state, $\tau_{IP3} = 7000$ ms is the IP$_3$ decay rate, $r_{IP3} = 0.0005$ ms^{-1} is the IP$_3$ production rate, and $[2AG]$ is the concentration of 2-AG released from the postsynaptic neuron.

The differential equation for the active fraction of IP$_3$Rs (h_{IP3R}) is given as

$$\frac{dh_{IP3R}}{dt} = \frac{h_{IP3R\infty} - h_{IP3R}}{\tau_h}, \tag{6}$$

where the equations for $h_{IP3R\infty}$ and τ_h are found as in [38]. Initial values for $[Ca^{2+}]_{astro}$, $[IP_3]_{astro}$, and h_{IP3R} are 0.08 μM, 0.17 μM, and 0.77, respectively.

3 Results

Postsynaptic somatic action potential opens dendritic VSCCs, which leads to the elevated levels of intracellular calcium concentration and PLC activation in the postsynaptic cell. Presynaptically released glutamate opens AMPA-gated channels and binds to mGluRs, which also activates PLC. PLC produces IP$_3$ and DAG. Ultimately, this leads to endocannabinoid synthesis and release. Figure 2 shows postsynaptic calcium concentrations, endocannabinoid production (2-AG), and delayed astrocytic calcium responses induced by pre- and postsynaptic stimulation at 1, 5, and 10 Hz. Postsynaptic neuron activation by somatic current injection preceded presynaptic glutamate release by 10 ms to secure favorable conditions for PLC activation and 2-AG production [36]. The current injected to the soma was a pulse wave with alternating amplitudes of 0 and 25 μA/cm^2 and pulse duration of 5 ms. Concentration of presynaptically released glutamate was a pulse wave with alternating concentrations of 0 and 100 μM and pulse duration of 5 ms. When the stimulation frequency was increased from 1 to 10 Hz, the intracellular calcium concentration reached higher levels from 0.32 up to 0.53 μM in the postsynaptic cell and led to substantial rise in 2-AG production from 0.05 up to 0.35 μM (Fig. 2).

Endocannabinoids released from the postsynaptic cell bind to the CB$_1$Rs on the astrocyte membrane, leading to production of IP$_3$ and activation of astrocytic IP$_3$Rs. IP$_3$Rs release calcium from the astrocytic ER to the cytosol. Endocannabinoid-induced calcium elevations heavily depend on the frequency of pre- and postsynaptic stimulation: for 1 Hz, astrocytic calcium levels reached only 0.24 μM, while for 5 Hz, calcium increased up to 0.69 μM, and for 10 Hz, it increased up to 0.82 μM (Fig. 2).

Postsynaptic calcium and 2-AG and astrocytic calcium

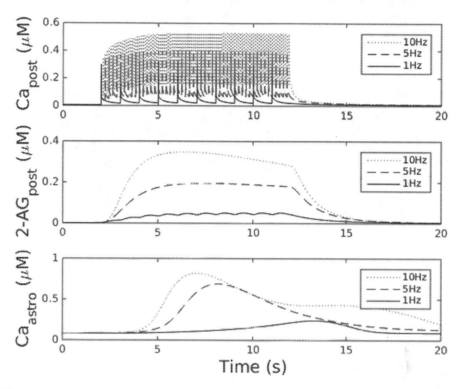

Fig. 2. Simulation of the full model by varying the frequency of the stimuli. 1, 5, and 10 Hz stimulus was given for 10 s between 2 and 12 s. (Top) Postsynaptic calcium level. (Middle) Endocannabinoid (2-AG) production. (Bottom) Astrocytic calcium responses.

The results show that 5 Hz stimulation was required to induce marked changes in the intracellular calcium concentration of the model astrocyte. In addition, astrocytic calcium response is faster with increasing stimulation frequency: the delay between the stimulation onset and the peak value in astrocytic calcium is 11 s for 1 Hz, 6 s for 5 Hz, and 5 s for 10 Hz.

Figure 3 shows the influence of the volume of the postsynaptic terminal (equivalent to a spine) on the astrocytic calcium responses. Increase in the spine radius (increase in the volume of the spine), from 0.5 to 2 μm caused significant changes in the astrocytic calcium responses (see Fig. 3). Specifically, for 1 Hz of pre- and postsynaptic stimulation, calcium levels in a postsynaptic spine decreased from 0.32 to 0.09 μM as the radius of the postsynaptic spine increased from 0.5 to 2 μm. This led to a weaker 2-AG production in the postsynaptic cell and almost negligible calcium elevation in astrocyte (Fig. 3, bottom left panel).

Similar pattern was observed for 5 and 10 Hz pre- and postsynaptic stimulations (Fig. 3, middle column and right column): substantial decrease in the amplitudes of the calcium responses were detected, together with marked delays in the activation of the astrocytic responses, in comparison to original 0.5 μm radius.

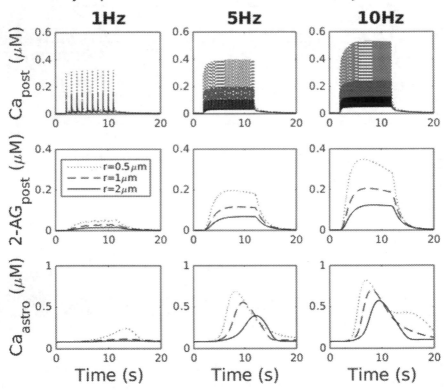

Fig. 3. Simulation of the full model by varying the frequency of the stimuli and the volume of the postsynaptic terminal (equivalent to spine). Stimulus was given for 10 s between 2 and 12 s. Stimulation frequencies used were 1 Hz (left column), 5 Hz (middle column), and 10 Hz (right column). (Top row) Postsynaptic calcium levels with spine radius 0.5, 1, and 2 μm. (Middle row) Endocannabinoid (2-AG) production with spine radius 0.5, 1, and 2 μm. (Bottom row) Astrocytic calcium responses with spine radius 0.5, 1, and 2 μm.

4   Conclusions

In this study, we built a computational model of a synapse that consists of a postsynaptic component and an astrocytic component. We used pairs of pre- and postsynaptic stimulation with varying frequencies to activate the model. We found out that synaptic stimulation induced endocannabinoid production and delayed astrocytic calcium responses. The higher the stimulus frequency, the faster and higher the astrocytic calcium response. In addition, the volume of the postsynaptic terminal significantly affected the astrocytic calcium responses. The larger the volume, the smaller the astrocytic calcium response.

As changes in the calcium levels are considered to be astrocytes' main form of activation, these changes in calcium dynamics can potentially alter the astrocytes' responses to a variety of events in the brain, potentially also synaptic events. Some lines of evidence indicate that calcium elevations in astrocyte can act as the signaling switch regulating release of specific substances, such as glutamate, ATP, and D-serine, from astrocytes (see references, e.g., in [21]). One example is the release of glutamate leading to induction of postsynaptic SICs and altered firing of the postsynaptic neuron [26]. The exact molecular mechanisms behind this type of regulation are still unclear, and require further experimental studies. The data obtained from such studies will be crucial for in silico studies with astrocytes. As demonstrated by our in silico study, morphological changes in the postsynaptic terminal and postsynaptic signaling detected by astrocytes cause increased calcium levels in astrocytes and might also affect the functioning of neuronal networks in the brain (see also [24]).

The long-term goal of our work is to develop detailed models of neuron-astrocyte interactions for different brain areas in order to advance our understanding of how plasticity and learning in spiking neural networks are modulated by astrocytes. The models may help testing hypotheses that are currently impossible to test by wet-lab experiments. They may also clarify various competing theories and controversies in the field of neuro-glioscience (for more discussion, see [21]). Ultimately, the well-validated models can inspire the development of a new generation spiking neural network models with plasticity and learning mechanisms more similar to mammalian brain than the present-day models.

Spiking neural networks represent a third generation of artificial neural networks and are inspired by the computational principles of the brain. Spiking neural networks consist of biophysically realistic spiking neuron models, and information transmission is encoded in precise timing of spikes or sequences of spikes. Synaptic weights in current versions of spiking neural networks are driven by the biologically inspired learning algorithms, such as Hebbian or STDP learning rules [41]. We here provide one of the first steps to extend the synapse models in order to incorporate the functionalities provided by astrocytes, one type of glial cell in the central nervous system.

In the long term, spiking neural networks with mammalian-like plasticity and learning will be increasingly used in real-world engineering and artificial intelligence tasks, such as classification and navigation, speech recognition, and decision making. Emerging application areas are also brain-machine interface systems and neural prostheses. However, simulations of large-scale spiking neural networks are computationally expensive. Therefore, neuromorphic computing platforms are being developed to allow fast and energy-efficient simulations [42]. Such systems, e.g. Spikey, SpiNNaker and TrueNorth, implement physical models of neurons and synapses and highly accelerate network emulations. These systems will evidently include astrocyte regulation of neuronal networks for enhancement of plasticity and learning in the future.

Acknowledgments. The authors wish to thank Tampere University of Technology Graduate School, Emil Aaltonen Foundation, The Finnish Concordia Fund, and Ulla Tuominen Foundation for support for RH.

Funding. This project received funding from the European Union Seventh Framework Programme (FP7) under grant agreement No. 604102 (HBP), European Union's Horizon 2020 research and innovation programme under grant agreement No. 720270, and Academy of Finland (decision No. 297893).

References

1. Turrigiano, G.G., Leslie, K.R., Desai, N.S., Rutherford, L.C., Nelson, S.B.: Activity-dependent scaling of quantal amplitude in neocortical neurons. Nature **391**(6670), 892–896 (1998)
2. Hebb, D.O.: The Organization of Behavior: A Neuropsychological Theory. Wiley, New York (1949)
3. Lømo, T.: Frequency potentiation of excitatory synaptic activity in dentate area of hippocampal formation. Acta Physiol. Scand. **68**(Suppl 277), 128 (1966)
4. Bliss, T.V., Lømo, T.: Plasticity in a monosynaptic cortical pathway. J. Physiol. **207**(2), 61P (1970)
5. Bliss, T.V.P., Lømo, T.: Long-lasting potentiation of synaptic transmission in the dentate area of the anaesthetized rabbit following stimulation of the perforant path. J. Physiol. **232**(2), 331–356 (1973)
6. Douglas, R.M., Goddard, G.V.: Long-term potentiation of the perforant path-granule cell synapse in the rat hippocampus. Brain Res. **86**(2), 205–215 (1975)
7. Steinbuch, K., Jaenicke, W., Reiner, H.: Learning matrix. C.I.P. Office (1965). http://brevets-patents.ic.gc.ca/opic-cipo/cpd/eng/patent/717227/summary.html
8. Willshaw, D.J., Buneman, O.P., Longuet-Higgins, H.C.: Non-holographic associative memory. Nature **222**, 960–962 (1969)
9. Kohonen, T.: Correlation matrix memories. IEEE Trans. Comput. **C-21**(4), 353–359 (1972)
10. Hopfield, J.J.: Neural networks and physical systems with emergent collective computational abilities. Proc. Natl. Acad. Sci. U.S.A. **79**(8), 2554–2558 (1982)
11. Markram, H., Lübke, J., Frotscher, M., Sakmann, B.: Regulation of synaptic efficacy by coincidence of postsynaptic APs and EPSPs. Science **275**(5297), 213–215 (1997)
12. Bi, G.Q., Poo, M.M.: Synaptic modifications in cultured hippocampal neurons: dependence on spike timing, synaptic strength, and postsynaptic cell type. J. Neurosci. **18**(24), 10464–10472 (1998)
13. Bi, G.Q., Poo, M.M.: Synaptic modification by correlated activity: Hebb's postulate revisited. Annu. Rev. Neurosci. **24**(1), 139–166 (2001)
14. Golding, N.L., Staff, N.P., Spruston, N.: Dendritic spikes as a mechanism for cooperative long-term potentiation. Nature **418**(6895), 326–331 (2002)
15. Häusser, M., Mel, B.: Dendrites: bug or feature? Curr. Opin. Neurobiol. **13**(3), 372–383 (2003)
16. Froemke, R.C., Letzkus, J.J., Kampa, B.M., Hang, G.B., Stuart, G.J.: Dendritic synapse location and neocortical spike-timing-dependent plasticity. Front. Syn. Neurosci. **2**, 29 (2010)
17. Letzkus, J.J., Kampa, B.M., Stuart, G.J.: Learning rules for spike timing-dependent plasticity depend on dendritic synapse location. J. Neurosci. **26**(41), 10420–10429 (2006)

18. Sjöström, P.J., Rancz, E.A., Roth, A., Häusser, M.: Dendritic excitability and synaptic plasticity. Physiol. Rev. **88**(2), 769–840 (2008)
19. Wittenberg, G.M., Wang, S.S.H.: Malleability of spike-timing-dependent plasticity at the CA3–CA1 synapse. J. Neurosci. **26**(24), 6610–6617 (2006)
20. Buchanan, K.A., Mellor, J.R.: The activity requirements for spike timing-dependent plasticity in the hippocampus. Front. Syn. Neurosci. **2**, 11 (2010)
21. Volterra, A., Liaudet, N., Savtchouk, I.: Astrocyte Ca^{2+} signalling: an unexpected complexity. Nat. Rev. Neurosci. **15**(5), 327–335 (2014)
22. Magistretti, P.J., Allaman, I.: A cellular perspective on brain energy metabolism and functional imaging. Neuron **86**(4), 883–901 (2015)
23. Dossi, E., Vasile, F., Rouach, N.: Human astrocytes in the diseased brain. Brain Res. Bull. (2017, in Press)
24. De Pittà, M., Brunel, N.: Modulation of synaptic plasticity by glutamatergic gliotransmission: a modeling study. Neural Plast. **2016**, 7607924 (2016)
25. Bushong, E.A., Martone, M.E., Jones, Y.Z., Ellisman, M.H.: Protoplasmic astrocytes in CA1 stratum radiatum occupy separate anatomical domains. J. Neurosci. **22**(1), 183–192 (2002)
26. Pirttimaki, T.M., Hall, S.D., Parri, H.R.: Sustained neuronal activity generated by glial plasticity. J. Neurosci. **31**(21), 7637–7647 (2011)
27. Manninen, T., Havela, R., Linne, M.L.: Computational models of astrocytes and astrocyte-neuron interactions: characterization, reproducibility, and future perspectives. In: De Pittà, M., Berry, H. (eds.) Computational Glioscience. Springer (2017, in Press)
28. Manninen, T., Havela, R., Linne, M.L.: Reproducibility and comparability of computational models for astrocyte calcium excitability. Front. Neuroinform. **11**, 11 (2017)
29. Tewari, S., Majumdar, K.: A mathematical model for astrocytes mediated LTP at single hippocampal synapses. J. Comput. Neurosci. **33**(2), 341–370 (2012)
30. Tewari, S.G., Majumdar, K.K.: A mathematical model of the tripartite synapse: astrocyte-induced synaptic plasticity. J. Biol. Phys. **38**(3), 465–496 (2012)
31. Pinsky, P.F., Rinzel, J.: Intrinsic and network rhythmogenesis in a reduced Traub model for CA3 neurons. J. Comput. Neurosci. **1**(1), 39–60 (1994)
32. Sarid, L., Bruno, R., Sakmann, B., Segev, I., Feldmeyer, D.: Modeling a layer 4-to-layer 2/3 module of a single column in rat neocortex: interweaving in vitro and in vivo experimental observations. Proc. Natl. Acad. Sci. U.S.A. **104**(41), 16353–16358 (2007)
33. Zachariou, M., Alexander, S.P.H., Coombes, S., Christodoulou, C.: A biophysical model of endocannabinoid-mediated short term depression in hippocampal inhibition. PLoS ONE **8**(3), e58296 (2013)
34. Politi, A., Gaspers, L.D., Thomas, A.P., Höfer, T.: Models of IP_3 and Ca^{2+} oscillations: frequency encoding and identification of underlying feedbacks. Biophys. J. **90**(9), 3120–3133 (2006)
35. Destexhe, A., Mainen, Z.F., Sejnowski, T.J.: Kinetic models of synaptic transmission. In: Koch, C., Segev, I. (eds.) Methods in Neuronal Modeling, pp. 1–25. MIT Press, Cambridge (1998)
36. Kim, B., Hawes, S.L., Gillani, F., Wallace, L.J., Blackwell, K.T.: Signaling pathways involved in striatal synaptic plasticity are sensitive to temporal pattern and exhibit spatial specificity. PLoS Comput. Biol. **9**(3), e1002953 (2013)
37. De Young, G.W., Keizer, J.: A single-pool inositol 1,4,5-trisphosphate-receptor-based model for agonist-stimulated oscillations in Ca^{2+} concentration. Proc. Natl. Acad. Sci. U.S.A. **89**(20), 9895–9899 (1992)
38. Li, Y.X., Rinzel, J.: Equations for InsP3 receptor-mediated $[Ca^{2+}]_i$ oscillations derived from a detailed kinetic model: a Hodgkin-Huxley like formalism. J. Theor. Biol. **166**(4), 461–473 (1994)

39. Wade, J., McDaid, L., Harkin, J., Crunelli, V., Kelso, S.: Self-repair in a bidirectionally coupled astrocyte-neuron (AN) system based on retrograde signaling. Front. Comput. Neurosci. **6**, 76 (2012)
40. Nadkarni, S., Jung, P.: Spontaneous oscillations of dressed neurons: a new mechanism for epilepsy? Phys. Rev. Lett. **91**(26), 268101 (2003)
41. Zenke, F., Agnes, E.J., Gerstner, W.: Diverse synaptic plasticity mechanisms orchestrated to form and retrieve memories in spiking neural networks. Nature Commun. **6**, 6922 (2015)
42. Furber, S.: Large-scale neuromorphic computing systems. J. Neural Eng. **13**(5), 051001 (2016)

Modeling PI3K/PDK1/Akt and MAPK Signaling Pathways Using Continuous Petri Nets

Giulia Russo[1], Marzio Pennisi[2], Roberta Boscarino[3], and Francesco Pappalardo[4(✉)]

[1] Department of Biomedical and Biotechnological Sciences, University of Catania, Catania, Italy
giulia.russo@unict.it
[2] Department of Mathematics and Computer Science, University of Catania, Catania, Italy
mpennisi@dmi.unict.it
[3] University of Catania, Catania, Italy
roberta.boscarino@hotmail.it
[4] Department of Drug Sciences, University of Catania, Catania, Italy
francesco.pappalardo@unict.it

Abstract. Malignant melanoma is an invasive skin cancer commonly resistant to conventional therapeutic approaches. Genetic and molecular alterations as mutations of BRAF gene, able to constitutively activate MAPK and PI3K/PDK1/Akt signalling pathways, seem to be responsible of malignant melanocytic transformation and lead to aberrant cellular physiological processes. Specific regulators and modulators of both signaling pathways may represent promising therapeutic targets to investigate drug resistance typical of BRAF-inhibitors such as Dabrafenib. We developed a continuous Petri Net model that simulates both MAPK and PI3K/PDK1/Akt pathways and their interactions in order to analyze the complex kinase cascades in melanoma and to predict new crucial nodes involved in drug resistance like in the Ras arm.

Keywords: Melanoma · Signaling pathways · Drug resistance · Petri nets · Hsa-mir-132

1 Introduction

Cutaneous melanoma is a malignant cancer that originates from melanocytes, cells that make pigment in skin and hair, within the basal layer of epidermis and hair bulbs. While it accounts for the 3–5% of skin cancers, cutaneous melanoma represents the most aggressive form of skin cancer, being also resistant to conventional therapies and entitling a very poor prognosis and a median survival rate of 6 months [1, 2]. The majority of cellular physiological processes, such as proliferation, differentiation and cell survival is orchestrated by fundamental signaling transduction and regulatory networks like PI3K/PDK1/Akt and MAPK signaling pathways. In particular circumstances, these pathways could be activated by different gene alterations [3] that bring towards some detrimental cell behaviors, like an uncontrolled cell proliferation and drug resistance. Malignant melanoma represents a good example to examine the complex mechanisms

© Springer International Publishing AG 2017
D.-S. Huang et al. (Eds.): ICIC 2017, Part II, LNCS 10362, pp. 169–175, 2017.
DOI: 10.1007/978-3-319-63312-1_15

of MAPK and PI3K/PDK1/Akt pathways because of the common mutation of BRAF-V600E (an amino acid substitution at position 600 in BRAF, from a valine to a glutamic acid) that leads to the triggering of the MAPK pathway activation [4]. Furthermore, melanoma seems to be resistant to the common selective BRAF-inhibitors, probably because the constitutive activation of the RAF kinase domain displays itself an increased and transforming kinase activity which leads to a missing response of the negative feedback mechanisms in the MAPK pathway [5].

Computational models are fundamental for the understanding of biological systems and such models can be applied to enhance or predict therapeutic effects at the cellular, tissue and organism level in melanoma disease [6, 7]. Moreover, they are helpful to minutely examine these complex interactions with the purpose to comprehend the resistance dynamics to conventional drugs in the treatment of melanoma [8–11].

In this paper, we present a Petri Nets (PNs) computational model that simulates both PI3K/Akt and MAPK pathways and their interactions in order to analyze the cascade reactions responsible for melanoma development. To this aim, we modeled the behavior of the malignant melanoma A375 cell line, harboring BRAF-V600E mutation, under the treatment of Dabrafenib, a commercial selective BRAF-inhibitor, that has been recently approved in the treatment of patients with BRAF-V600E mutation-positive advanced melanoma [12]. Overall, this model may potentially be used as an in silico lab to study the effects of potential inhibitors that may improve the response to standard treatments.

2 Materials and Methods

2.1 Introduction to Petri Nets

PNs are a diagrammatic modeling tool developed in 1962 by Carl Adam Petri [13, 14] for the description of distributed systems.

In general, a PN can be seen as a set of graphical symbols called places, transitions, arcs and tokens. Places, usually represented by circles or ellipses, represent conditions and/or states of a system. Each place may be named in order to increase the understanding of a Petri net. Bars or boxes are instead used to represent state transitions, i.e. events that may occur and can change the global state of the system. Arcs, represented by arrows, are used to connect places to transitions and vice-versa, but they cannot be used to connect transitions to transitions or places to places. Places usually contain a discrete number of black dots (tokens) which represent the dynamical objects of the system. A Change of State of the net is denoted by a movement of token(s) from place(s) to place(s) passing through a transition. The state (marking) of a Petri net is determined by the distribution of tokens over the places. A transition can have several input and output places. The tokens in the input places are fundamental for the enabling of the transition: a transition is enabled only if each of the input places contain a given required number of tokens. In general, the required number of tokens for each input place depends on the multiplicity of the arcs that connect the input places with the transition. An enabled transition may fire. The firing represents an occurrence of the event or an action taken and corresponds to consuming tokens from the input places and producing tokens for the output places. Again, the number of tokens produced into the output places depends by the multiplicity associated to each arc.

Firing of a transition is an atomic event, in the sense that it represents a single non-interruptible action. Unless an execution policy is defined, when multiple transitions are enabled at the same time, anyone of them may fire (nondeterministic execution).

In continuous Petri Nets, the marking of the net is no more given by the number of tokens in each places, but it is now given by a positive real number, called token value. This token value can be seen as concentration. A transition is now enabled if the token value of all the input places (pre-places) is positive and greater than the minimal quantity required by each involved arc. Different arbitrary firing rates can be defined using mathematical functions, such as mass-action or Henri-Michaelis-Menten laws. The formal definition of a continuous Petri net can be found in [13].

ODE equations can be used to represent the semantic of a continuous Petri net by describing the continuous changes over time on the token values of places. In other words, a continuous Petri net can be seen as a structured description of an ODE-system. This description can sometimes be less error prone and easier to understand than the direct representation of an ODE system.

2.2   The PN Pathway Model

The starting point for the development of our PN model has been represented by the model developed by Brown et al. [15]. We expanded the Brown model by including other missing entities and respective interactions that we considered to be necessary for the modeling infrastructure of PI3K/PDK1/Akt and MAPK signaling pathways. To this end we used Kyoto Encyclopedia of Genes and Genomes (KEGG) PATHWAY Database [16] to gather all the pathway information respectively extracted through KEGG map of MAPK (Kegg reference: ko04010) and the one of PI3K/PDK1/Akt (Kegg reference: hsa04151). We also modeled the Dabrafenib species (represented as a place) in order to reproduce the effects of Dabrafenib inside the MAPK and PI3K/PDK1/Akt pathway and in particular we included: the specific transition rules for representing the physiological drug degradation, according to Dabrafenib half-life of 10 h, as declared on http://www.ema.europa.eu/ema/ in the European public assessment report of the European Medicines Agency (EMA) and the inhibiting effects of Dabrafenib towards bRafMutated species.

Furthermore, we endowed our model with the physiological degradation process of protein receptors (not taken into account in the Brown's model) modeled by an irreversible mass action law that affect both forms of receptor tyrosine kinase (RTK): "free" from the growth factor and "bound" to the growth factor.

Other fundamental kinase cascades triggering ERK and Akt dynamics have respectively been inserted into the model: the activation of the C3G species through bound RTK receptor and the activation of the Rap1 through activated C3G, for what concerns ERK; the activation of mTORC1 pathway to highlight the key role of Akt protein on the activation/deactivation machinery of several proteins.

All these modelling hypotheses led to a model that includes 48 species (represented by places) and 48 biochemical reactions (represented by transitions). For the development of the model we then used continuous Petri Nets and the Snoopy software. All the transition firing rates in our PN model belong to 3 equation kinetics: (*i*) Modified Henri-Michaelis-Menten law for modeling the activation/deactivation reactions that involve both substrate and modifier, important key elements in reactions that activate (and/or deactivate) specific proteins (for example ErkInactive becomes activated (ErkActive) through MekActive); (*ii*) Classical mass action law for the reactions that physiologically inactivate species (i.e. PDK1 deactivation) or represent protein degradation modeled (i.e. Dabrafenib degradation); (*iii*) Irreversible constant flux law for representing the constant production of proteins (for example, the production of free RTK).

To reproduce the dynamics of the BRAF-V600E mutation in the A375 melanoma cell line we (*i*) introduced the species bRafMutated with the same initial concentration of bRafInactive; (*ii*) deleted the Rap1 and Ras induced BRAF activations as bRafMutated is not affected by these signals; (*iii*) we also inhibited the Raf1PPtase induced deactivation of BRAF, since BRAF is no more influenced by Raf1PPtase; (*iv*) we substituted bRafActive with bRafMutated species inside the Mek activation. The complete continuous PN model is presented in Fig. 1.

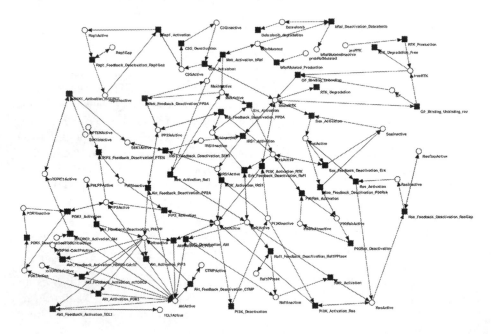

Fig. 1. The continuous Petri net model.

3 Results

We simulated a therapy intervention to analyze the behavior of the A375 cell lines under different concentrations of the Dabrafenib inhibitor. Obtained results are shown in Fig. 2, where the p-ERK and "BRAF mutated" behaviors with and without the

Dabrafenib administration are presented for a time-window of 48 h. Different Dabrafenib dosages have been tested i.e., 0 nM in the firs panel (a), 0.100 nM in the second panel (b) and 0.400 nM in the third panel (c). When no treatment is administered, p-ERK concentration reached 571950 mmol/ml at time 48 h. With different dosages of Dabrafenib, p-ERK concentrations reached 560090 mmol/ml as reported in (b) and 253527 mmol/ml as shown in (c), respectively. From the model results, we can observe that p-ERK levels drop down due to inhibitor activity of Dabrafenib over BRAF-V600E protein.

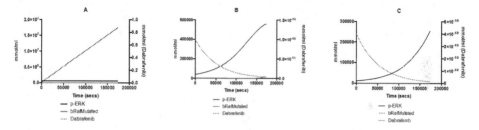

Fig. 2. p-ERK, bRafMutated and Dabrafenib behaviors under different concentrations of Dabrafenib inhibitor. p-ERK and bRafMutated are represented by black and red solid lines respectively. Dabrafenib behavior is represented by a solid green line (right axis). (Color figure online)

The model suggests that Ras could be a leading protein that orchestrates the complex mechanisms of Dabrafenib resistance. To this aim, we simulated a knock-out of Ras arm to investigate how this affects MAPK and PI3K/PDK1/Akt pathways. We report in Fig. 3 the in silico results of ERK activation, respectively with the Ras arm in its physiologically working condition (a), and under Ras knock-out condition (b). In the physiological model, the p-ERK concentration reached 506955 mmol/ml at time of 10 s (a). With Ras knockout, the p-ERK concentrations reached 259762 mmol/ml at time of 10 s, as reported in (b). From these results, we can observe that p-ERK levels undergo a two-fold decrease due to the knock-out of Ras arm, suggesting that the Ras complex protein domain could be potentially involved in the resistance mechanisms of BRAF-inhibitors.

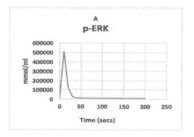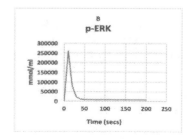

Fig. 3. p-ERK concentrations of the PN model. In each panel is shown: (A) ERK concentrations with the Ras arm and its machinery physiologically working; (B) ERK concentrations under Ras knock-out conditions. These results show a significant p-ERK decrease after Ras arm knock-out in the PN model.

4   Conclusions and Final Remarks

From the PN model results, we observed that p-ERK levels drop down due to the inhibitor activity of Dabrafenib over BRAF-V600E protein. These results are in line with current literature. Moreover, PN model puts the attention on the possible involvement of Ras in BRAF-inhibitors resistance mechanisms. The protein kinase cascades of MAPK (EGF->Sos->Ras->Raf-1->Mek-1->ERK) and the PI3K/PDK1/Akt pathway are connected through the mutual cascade arm of Ras, which has a particular role in the activation of PI3K. PI3K can then phosphorylate Akt through PIP3. According to the findings that come out from the knockout of Ras, the model suggests that Ras could be a leading protein that coordinates the complex set of mechanisms involved into Dabrafenib resistance because, in absence of the Ras machinery, p-ERK is unable to climb itself to the highest concentration and p-ERK levels decrease due to the knock-out of Ras arm. To give strength to this view, Westenskow et al. [13] showed that mir-132 downregulates the expression of p120RasGAP, one of the preeminent modulator of Ras, which strongly influences the two signalling pathways. These mechanisms are still not well understood and there is a lot of interest in knowing more about the involved signaling pathways because they are critical players in melanoma development. They can represent potential therapeutic targets owning many unique features that may differ from individual to individual. In this paper, we are moving towards this direction, and our PN model seems to be a helpful method to investigate the complex scenario of these phenomena. According to our results, the Ras protein and its regulatory machinery are potential targets to overcome metastatic melanoma and BRAF-inhibitors resistance. miR-132, that downregulates the expression levels of p120RasGAP, seems to be an interesting target too, thanks to the possibility to modulate it with the adoption of antisense oligonucleotides technologies. Experimental results dealing with a specific antagomir of hsa-mir-132 are ongoing to modulate mir-132 itself and its target p120RasGAP, involved both in MAPK and PI3K-AKT pathways.

References

1. Russo, A.E., Torrisi, E., Bevelacqua, Y., Perrotta, R., Libra, M., McCubrey, J.A., Spandidos, D.A., Stivala, F., Malaponte, G.: Melanoma: molecular pathogenesis and emerging target therapies (Review). Int. J. Oncol. **34**, 414–420 (2014)
2. Ferlay, J., Soerjomataram, I., Ervik, M., Dikshit, R., Eser, S., Mathers, C., Rebelo, M., Parkin, D.M., Forman, D., Bray, F.: Cancer incidence and mortality worldwide: sources, methods and major patterns in GLOBOCAN 2012. Int. J. Cancer **136**, 359–386 (2013)
3. Manzano, J.L., Layos, L., Bugés, C., de Los Llanos Gil, M., Vila, L., Martínez-Balibrea, E., Martínez-Cardús, A.: Resistant mechanisms to BRAF inhibitors in melanoma. Ann. Transl. Med. **4**, 237 (2016)
4. Libra, M., Malaponte, G., Navolanic, P.M., Gangemi, P., Bevelacqua, V., Proietti, L., Bruni, B., Stivala, F., Mazzarino, M.C., Travali, S., McCubrey, J.A.: Analysis of BRAF mutation in primary and metastatic melanoma. Cell Cycle **4**, 1382–1384 (2005)
5. Garnett, M.J., Marais, R.: Guilty as charged: B-RAF is a human oncogene. Cancer Cell **6**, 313–319 (2004)

6. Brodland, G.W.: How computational models can help unlock biological systems. Semin. Cell Dev. Biol. **47–48**, 62–73 (2015)

7. Pappalardo, F., Fichera, E., Paparone, N., Lombardo, A., Pennisi, M., Russo, G., Leotta, M., Pappalardo, F., Pedretti, A., De Fiore, F., Motta, S.: A computational model to predict the immune system activation by citrus-derived vaccine adjuvants. Bioinformatics **32**, 2672–2680 (2016)

8. Castiglione, F., Pappalardo, F., Bianca, C., Russo, G., Motta, S.: Modeling biology spanning different scales: An open challenge. Biomed. Res. Int. **2014**, 902545 (2014)

9. Gullo, F., van der Garde, M., Russo, G., Pennisi, M., Motta, S., Pappalardo, F., Watt, S.: Computational modeling of the expansion of human cord blood CD133 + hematopoietic stem/progenitor cells with different cytokine combinations. Bioinformatics **31**, 2514–2522 (2015)

10. Pappalardo, F., Palladini, A., Pennisi, M., Castiglione, F., Motta, S.: Mathematical and computational models in tumor immunology. Math. Model. Nat. Phenom. **7**, 186–203 (2012)

11. Pappalardo, F., Russo, G., Candido, S., Pennisi, M., Cavalieri, S., Motta, S., McCubrey, J.A., Nicoletti, F., Libra, M.: Computational modeling of PI3K/AKT and MAPK signaling pathways in melanoma Cancer. PLoS ONE **11**, e0152104 (2016)

12. Vennepureddy, A., Thumallapallya, N., Nehrua, V.M., Atallahb, J.P., Terjanianb, T.: Drugs and Combination Therapies for the Treatment of Metastatic Melanoma. J. Clin. Med. Res. **8**, 63–75 (2016)

13. Murata, T.: Petri nets: properties, analysis and applications. Proc. IEEE **77**, 541–580 (1989)

14. Petri, C.A., Reisig, W.: Petri Net. Scholarpedia **3**, 6477 (2008)

15. Brown, K.S., Hill, C.C., Calero, G.A., Myers, C.R., Lee, K.H., Sethna, J.P., Cerione, R.A.: The statistical mechanics of complex signaling networks: nerve growth factor signaling. Phys. Biol. **1**, 184–195 (2004)

16. Kanchisa, M., Goto, S., Kawashima, S., Okuno, Y., Hattori, M.: The KEGG resource for deciphering the genome. Nucleic Acids Res. **32**, 277–280 (2004)

Keratoconus Diagnosis by Patient-Specific 3D Modelling and Geometric Parameters Analysis

Laurent Bataille[1,2,4,5], Francisco Cavas-Martínez[1(✉)], Daniel G. Fernández-Pacheco[1], Francisco J.F. Cañavate[1], and Jorge L. Alio[3,4,5]

[1] Department of Graphical Expression, Technical University of Cartagena, Cartagena, Spain
{francisco.cavas,daniel.garcia,francisco.canavate}@upct.es
[2] Research and Development Department, Vissum Corporation Alicante, Alicante, Spain
[3] Division of Ophthalmology, Universidad Miguel Hernández, Alicante, Spain
[4] Keratoconus Unit of Vissum Corporation Alicante, Alicante, Spain
[5] Department of Refractive Surgery, Vissum Corporation Alicante, Alicante, Spain
{lbataille,jlalio}@vissum.com

Abstract. The aim of this study is to describe a new technique for diagnosing keratoconus based on Patient-specific 3D modelling. This procedure can diagnose small variations in the morphology of the cornea due to keratoconus disease. The posterior corneal surface was analysed using an optimised computational geometric procedure and raw data provided by a corneal tomographer. A retrospective observational case series study was carried out. A total of 86 eyes from 86 patients were obtained and divided into two groups: one group composed of 43 healthy eyes and the other of 43 eyes diagnosed with keratoconus. The predictive value of each morphogeometric variable was established through a receiver operating characteristic (ROC) analysis. The posterior apex deviation variable showed the best keratoconus diagnosis capability (area: 0.9165, p < 0.000, std. error: 0.035, 95% CI: 0.846-0.986), with a cut-off value of 0.097 mm and an associated sensitivity and specificity of 89% and 88%, respectively. Patient-specific geometric models of the cornea can provide accurate quantitative information about the morphogeometric properties of the cornea on several singular points of the posterior surface and describe changes in the corneal anatomy due to keratoconus disease. This accurate characterisation of the cornea enables new evaluation criteria in the diagnosis of this type of ectasia and demonstrates that a device-independent approach to the diagnosis of keratoconus is feasible.

Keywords: Diagnosis · Geometric modelling · Cornea reconstruction · Scheimpflug · CAD introduction

1 Introduction

The analysis of the corneal geometry is critical for the assessment of vision quality in several clinical applications [1, 2]. Consequently, the detection of changes in morphogeometrical properties of the cornea can improve the diagnosis of corneal pathologies [3, 4]. The Keratoconus (KC) is an ectatic corneal disorder usually bilateral but asymmetric. This corneal pathology is most of the time characterised by a corneal thinning

© Springer International Publishing AG 2017
D.-S. Huang et al. (Eds.): ICIC 2017, Part II, LNCS 10362, pp. 176–187, 2017.
DOI: 10.1007/978-3-319-63312-1_16

which results in corneal protrusion, irregular astigmatism and decreased vision [5, 6]. The frequency of occurrence in the general population is low, between 4/1000 and 6/10001. Its prevalence is higher in areas with important exposure to UV light or by a combination of genetic and environmental factors [7].

A significant corneal steepening is observed in the anterior and posterior corneal surfaces and both curvatures are affected in KC eyes and KC suspect eyes [8, 9]. Recent introduction of Scheimpflug photography for corneal topographic characterisation helps in the study and characterisation of both anterior and posterior corneal surfaces [10].

For the diagnosis of ectasia, numerous quantitative descriptors of the corneal surface provided by corneal tomographers can be found in the scientific literature [1, 12]. Recently, a worldwide group of keratoconus experts established the importance of some singular points located at the posterior surface of the cornea for the diagnosis of ectasia [11]. The point of minimum thickness and the highest point (apex or point of maximum curvature) were both defined as the most significative. Nevertheless, none of the quantitative descriptors described above for the diagnosis of ectasias consider these singular points of the posterior corneal surface, so its quantitative description remains a challenge. These descriptors are based on measures such as central corneal thickness, anterior chamber depth, mean simulated keratometry or mean keratometry, among others. Moreover, the values obtained for these descriptors are different depending on the tomographer used [13–16]. This variability prevents their widespread acceptance and clinical utility.

On the other hand several studies, based on Computer-Aided Design (CAD) and Finite Elements (FE), have developed the concept of patient-specific model [17–30]. These patient-specific models are obtained from the so called "raw data" [31], which are generated by systems based on the projection of a slit of light onto the cornea and on the principle of the Scheimpflug photography [1]. These models provide quantitative results relative to the specific cornea of each patient.

The present study demonstrates a new approach for the diagnosis and detection of keratoconus based on the concept of patient-specific 3D modelling and analysis of singular points of each cornea. A validation study was conducted analysing the statistically significant difference of all these singular points between a group of healthy corneas and a group of corneas diagnosed with keratoconus according to Amsler-Krumeich grading system (AK) [32].

2 Methods

The Sirius system (CSO, Florence, Italy) was used during the study. It is a non-contact digital rotational Scheimpflug tomographer that represents the entire corneal surface as two discrete and finite sets of spatial point representative of both anterior and posterior corneal surfaces, respectively (Fig. 1a). Point cloud coordinates were exported in a CSV table in polar format. Each row represents a circle on the map with a total of 256 points for each circle (radii are incremented in intervals of 0.2 mm), and each column represents a semi-meridian [33, 34]. In this study, an algorithm was implemented with Matlab version R2016a (MathWorks, http://www.mathworks.com) to convert the data of the CSV table from polar format to Cartesian format.

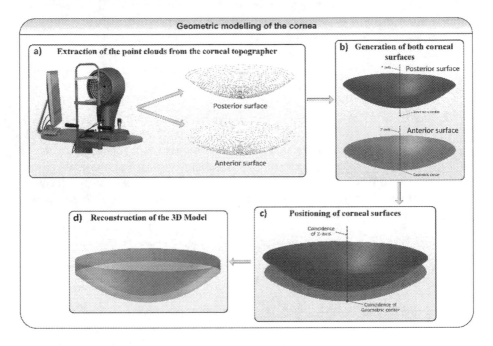

Fig. 1. Geometric modelling process by using DGAO tools.

Subsequently, data were imported into the surface reconstruction CAD Rhinoceros [35] version 5.0 software ((MCNeel & Associates, Seattle, USA). Non-uniform rational B-splines (NURBS) were used to generate surfaces, which are characterised by two parametric directions, u and v, that define the spanning process [36, 37]. In this study, the Rhinoceros' surface from the point grid function was applied to the imported point cloud. This function created a rectangular grid of 21 rows and 256 columns, which was deformed in order to minimise the nominal distance between the spatial points and the grid surface (Fig. 1b). The use of this surface reconstruction function permitted obtaining an average deviation error for the posterior surface of the studied corneas with keratoconus (the most irregular corneal surfaces) of about $4.82 \cdot 10^{-16} \pm 5.09 \cdot 10^{-16}$ mm. By using this procedure, the anterior and posterior corneal surfaces were generated and engaged by their geometrical centre and Z axis (Fig. 1c). These surfaces were then joined with the peripheral one (the bonding surface between both sides in the Z-axis direction) to form a single surface.

The surface reconstructed with Rhinoceros was then exported to the SolidWorks v2016 solid modelling software (Dassault Systèmes, Vélizy-Villacoublay, France), which generates the tri-dimensional reconstruction of a solid representative model of the custom and actual geometry of each cornea (Fig. 1d).

Singular points of the posterior surface were then identified on each solid corneal model (healthy or keratoconic), and a morphogeometrical analysis of discrete landmarks was performed in the local region [38, 39] where curving (caused by progression of keratoconus) manifests gradually.

For this study we analysed the following variables from the posterior surface of the cornea [33, 34]: the Sagittal plane apex area, defined as the area of the cornea within the sagittal plane passing through the Z axis and the highest point (apex, maximum curvature) of the posterior corneal surface (Fig. 2a); Posterior apex deviation (Fig. 2b), defined as the average distance from the Z axis to the highest point (apex, maximum curvature) of the posterior corneal surface; Sagittal plane area at minimum thickness point (Fig. 2c), defined as the area of the cornea within the sagittal plane passing through the Z axis and the minimum thickness point of the posterior corneal surface; Posterior minimum thickness point deviation (Fig. 2d), described as the average distance in the XY plane from the Z axis to the minimum thickness point of the posterior corneal surface. For each cornea, all these variables were measured by the same and unique observer using the SolidWorks calliper function.

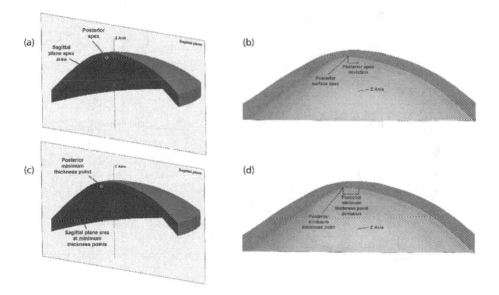

Fig. 2. Geometric variables analysed during the study that achieved the best results: (a) sagittal plane apex area, (b) posterior apex deviation, (c) sagittal plane area at minimum thickness point, (d) posterior minimum thickness point deviation.

2.1 Statistical Analysis

Data distribution was confirmed by means of the Kolmogorov–Smirnov test. According to this analysis, a Student's t-test was performed in order to test the hypothesis according to the aim of the study. A ROC curve analysis was performed in order to obtain the accuracy of the measurements. A ROC curve is a graphical plot that illustrates the performance of a binary classifier system as its discrimination threshold is varied. The curve is created by plotting the true positive rate against the false positive rate at various threshold settings. The accuracy of the test depends on how well the test separates the group being tested into those with and without the disease in question. The area under the ROC curve measures the accuracy. A rough guide for classifying the accuracy of a diagnostic test is the traditional academic point system: excellent if 0.90–1; good if 0.80–0.90, fair if 0.70–0.80, and poor if 0.60–0.70. Statistical analyses were performed using Graphpad Prism version 6 software (GraphPad Software, La Jolla, USA) and SPSS version 17.0 software (SPSS, Chicago, USA).

3   Results

From a total of 86 patients involved in this study, 43 patients with healthy corneas and aged 12–61 years (36.49 ± 14.86) and 43 patients with keratoconic corneas and aged 17–63 years (38.02 ± 14.98) were modelled. In order to obtain an accurate methodology only initial stages of the disease (stage I), according to Amsler-Krumeich grading, were used during this study. The retrospective study adhered to the tenets of the Declaration of Helsinki and was approved by the local Clinical Research Ethics Committee of Vissum Corporation (Alicante, Spain). Patients examined at Vissum Corporation (Alicante, Spain) were retrospectively enrolled. They were selected from a database of candidates for refractive surgery with normal corneas and also a database of cases diagnosed as having keratoconus in both eyes.

All eyes selected (86) underwent a thorough and comprehensive eye and vision examination which included uncorrected distance visual acuity (UDVA), corrected distance visual acuity (CDVA), manifest refraction, Goldmann tonometry, biometry (IOLMaster, Carl Zeiss Meditec AG) and corneal topographic analysis with the Sirius system® (CSO, Florence, Italy), which is a non-invasive system for measuring and characterizing the anterior segment using a rotating Scheimpflug camera. All measurements were performed by the same experienced optometrists, performing three consecutive measurements and taking average values for the posterior analysis.

Table 1. Comparisons within groups (t-tests). Statistical significance (in p-values) between values of different parameters.

Parameter	Healthy group (M ± SD)	Keratoconus group (M ± SD)	t	p
Sagittal plane apex area (mm^2)	4.33 ± 0.27	3.90 ± 0.33	7.398	0.000
Sagittal plane area at minimum thickness point (mm^2)	4.32 ± 0.27	3.88 ± 0.33	7.563	0.000
Posterior apex deviation (mm)	0.08 ± 0.02	0.19 ± 0.09	-11.271	0.000
Posterior minimum thickness point deviation (mm)	0.80 ± 0.24	1.01 ± 0.36	-3.746	0.001

M = mean, SD = standard deviation, $N_{Healthy\ group}$ = 43, $N_{Keratoconus\ group}$ = 43

Table 2. Correlations between parameters.

Parameter	Sagittal plane apex area	Sagittal plane area at minimum thickness point	Posterior apex deviation	Posterior minimum thickness point deviation
Sagittal plane apex area	r = 1	r = 0.998, p < 000	r = −0.374, p < 000	r = −0.285, p < 002
Sagittal plane area at minimum thickness point	r = 0.998, p < 000	r = 1	r = −0.379, p < 000	r = −0.295, p < 001
Posterior apex deviation	r = −0.374, p < 000	r = −0.379, p < 000	r = 1	r = 0.475, p < 000
Posterior minimum thickness point deviation	r = −0.285, p < 002	r = −0.295, p < 001	r = 0.475, p < 000	r = 1

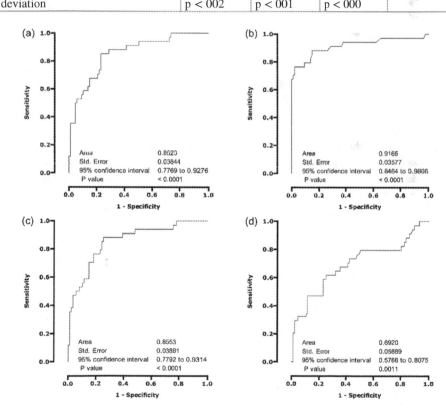

Fig. 3. ROC curve modelling sensitivity versus 1-specificity for the variables that diagnosed the existence of keratoconus disease: (a) sagittal plane apex area, (b) posterior apex deviation, (c) sagittal plane area at minimum thickness point, (d) posterior minimum thickness point deviation.

Regarding area parameters, both sagittal plane areas (posterior apex and posterior minimum thickness point) were statistically higher in the subjects with healthy corneas. As expected, minor deviations for the posterior apex and for the posterior minimum thickness point were also observed in the group of normal corneas (see Table 1). A high correlation ($r = 0.998$; $p < 0.000$) between the Sagittal plane apex area and Sagittal plane area at minimum thickness point was also detected (see Table 2).

The predictive value of each morphogeometric parameter was established by a ROC analysis. Four of these parameters offered an area under the curve (AUC) above 0.69 (Fig. 3). The AUC, independent for each variable, appears to demonstrate an adequate sensitivity and specificity classification between groups. The most accurate variable was the posterior apex deviation (Fig. 3b), followed by the sagittal plane area at minimum thickness point (Fig. 3c), sagittal plane apex area (Fig. 3a) and posterior minimum thickness point deviation (Fig. 3d).

4 Discussion

This study offers an accurate and realistic method of reconstruction for a biological structure: the human cornea. It creates a new understanding of corneal diseases based on data from the posterior surface, using a robust and cost-effective method based on the generation of a patient-specific 3D model. This computational study provides insight into the complex clinical problem of corneal ectatic disease diagnosis.

In scientific literature, some studies report Computer-Aided Design (CAD) and Finite Elements (FE) models based on the concept of patient-specific analysis. In the Finite Elements field, some reports use these data for several purposes: to predict the response to refractive surgeries [23, 26, 28, 30], the response to intrastromal ring segment implantation in corneas with keratoconus [22], to analyse non-surgical corneal modifications, such as applanation tonometry for intraocular pressure measurement [18, 19, 27, 29] or to analyse the behaviour of corneal tissue properties in different scenarios [17, 20, 21, 24, 25]. In all these cases, raw data were provided by the Pentacam [17–26, 30] (Oculus Optikgeräte GmbH, Wetzlar, Germany), the Sirius [27–29] (CSO, Florence, Italy) or the Galilei [20] (Ziemer Ophthalmic Systems AG, Port, Switzerland) devices. However, due to extrinsic errors [1] that occurred during the measurement process, the generated raw data were incomplete. Therefore, the authors of the mentioned studies decided to interpolate data to obtain a complete image of the corneal surfaces, and thus generating an approximated 3D model of the cornea for its posterior use in the Finite Elements analysis.

In the present study, the raw data obtained also presented the previously mentioned extrinsic errors. However, the authors adopted a design protocol based on geometrical and clinical principles in which only real data were used and not any interpolation was performed [33, 34]. Thus, the patient-specific geometric model generated was an authentic and completely personalised cornea model.

In case of keratoconus the deterioration process of the corneal structure is characterised by a significant reduction of the corneal thickness in comparison to healthy eyes. This is triggered by an alteration in corneal collagen fibres causing stromal thinning and breaks in the Bowman's membrane during the different stages of the disease [5, 40]. Furthermore, the presence of corneal irregularities and the influence of the intraocular pressure on these weakened structures will create local steepening and increased radius of curvature which will lead to an increased posterior corneal surface area [5, 6, 40, 41]. In the keratoconus group, the sagittal plane areas were smaller because of their local structural weakening due to fewer collagen fibres in each lamella and the influence of the intraocular pressure [6]. These values are consistent with those published by different studies [41, 42] which have reported that in irregular corneal areas the geometry of posterior corneal surfaces were affected due to the lower number of stromal lamellae and the smaller lamellar interconnection. In other studies where a similar characterisation of the corneas has been described for the differentiation of pathologic eyes, the variables related to corneal thickness and volume, among others, are directly given by the software of the topographers [43–46]. However, in the present study the morpho-geometric variables are calculated from a 3D model generated using only real and non-interpolated raw point cloud data provided by the tomographer.

The average distance from the Z axis to the apex of the posterior corneal surface (deviation of the apex point of posterior corneal surface) differed between groups, with the largest deviations found in the group of eyes with keratoconus. The deviation of the apex on the posterior surface of the cornea was larger in the eyes with keratoconus (0.19 ± 0.09 mm), and there was also a slight deviation in healthy corneas (0.08 ± 0.02 mm) according to the toricity manifested in the subjective refraction [47]. The aforementioned presence of an irregular corneal surface, which created a protrusion in the keratoconic eye, also led to an increased corneal curvature [48] and, therefore, to an increase in the deviation of the point of minimum thickness (maximum curvature) of the posterior corneal surface (average distance from the Z axis to the minimum thickness point of the posterior corneal surface). These deviations were greater in the eyes with keratoconus (1.01 ± 0.36 mm) compared with healthy eyes (0.80 ± 0.24 mm). Some researchers have evaluated certain corneal irregularity ratios and concluded that they were higher for keratoconic corneas [49–51].

The analysis of these variables concluded that the parameter that provides a higher discrimination rate between normal corneas and corneas with keratoconus was the posterior apex deviation (ROC area: 0.9165, p < 0.000, std. error: 0.035, 95% CI: 0.846–0.986), with a cut-off value of 0.097 mm and an associated sensitivity and specificity of 89% and 88%, respectively. This is justified due to the structural instability that keratoconic corneas present in its architecture, being the posterior surface the most susceptible to variations given the forces exerted on the tissue. Several studies have concluded the importance of, and interest in, the posterior corneal surface [50]. However, to the authors' knowledge, this is the first study to describe this phenomenon in disease stage when the degree of corneal protrusion is apparent in the posterior corneal surface, which supports the diagnosis accuracy of the proposed method.

Furthermore, the non-invasive clinical diagnosis method proposed in this study develops the concept of interoperability [52]: a new methodology that uses raw data, which can be shared among the corneal topographers that are based on the projection of a slit of light onto the cornea and on the principle of Scheimpflug photography.

This exchange of information could imply a common benefit for the whole ophthalmic community.

5 Conclusion

In summary, this method provides a feasible and new diagnostic approach of keratoconus, using raw elevation data from any corneal tomographer without the need of proprietary internal algorithms and quantifying the singular points of the posterior corneal surface for diagnosing keratoconus. It is in agreement with the main conclusions drawn by the group of world experts in keratoconus. It also offers a potential comparative tool to analyze data from corneal tomographers from different manufacturers allowing data sharing. This could lead to a better understanding of the ethology and prognosis of this eye disease.

In future studies, our objective will be to increase the number of eyes in each group, to carry out a more advanced statistical analysis and to compare the results with the commercially available multiparametric keratoconus detection indexes.

In future applications, the same method could be used to improve the detection and effects of therapeutic methods used for different grades of keratoconus and other corneal ectatic diseases, such as post-LASIK ectasia.

References

1. Cavas-Martinez, F., De la Cruz Sanchez, E., Nieto Martinez, J., Fernandez Canavate, F.J., Fernandez-Pacheco, D.G.: Corneal topography in keratoconus: state of the art. Eye **3**, 5 (2016). (Lond)
2. Arnalich-Montiel, F., Alio Del Barrio, J.L., Alio, J.L.: Corneal surgery in keratoconus: which type, which technique, which outcomes? Eye Vis. **3**, 2 (2016). (London, England)
3. Montalban, R., Alio, J.L., Javaloy, J., Pinero, D.P.: Correlation of anterior and posterior corneal shape in keratoconus. Cornea **32**, 916–921 (2013)
4. Belin, M.W., Ambrosio, R.: Scheimpflug imaging for keratoconus and ectatic disease. Indian J. Ophthalmol. **61**, 401–406 (2013)
5. Rabinowitz, Y.S.: Keratoconus. Surv. Ophthalmol. **42**, 297–319 (1998)
6. Pinero, D.P., Alio, J.L., Barraquer, R.I., Michael, R., Jimenez, R.: Corneal biomechanics, refraction, and corneal aberrometry in keratoconus: an integrated study. Invest. Ophthalmol. Vis. Sci. **51**, 1948–1955 (2010)
7. Rabinowitz, Y.S., Barbara, A.: Epidemiology of keratoconus. In: Textbook on Keratoconus New Insights (2012)
8. Wilson, S.E., Lin, D.T., Klyce, S.D.: Corneal topography of keratoconus. Cornea **10**, 2–8 (1991)
9. Tomidokoro, A., Oshika, T., Amano, S., Higaki, S., Maeda, N., Miyata, K.: Changes in anterior and posterior corneal curvatures in keratoconus. Ophthalmology **107**, 1328–1332 (2000)
10. Dubbelman, M., Sicam, V.A., Van der Heijde, G.L.: The shape of the anterior and posterior surface of the aging human cornea. Vis. Res. **46**, 993–1001 (2006)

11. Gomes, J.A., Tan, D., Rapuano, C.J., Belin, M.W., Ambrosio Jr., R., Guell, J.L., Malecaze, F., Nishida, K., Sangwan, V.S.: Global consensus on keratoconus and ectatic diseases. Cornea **34**, 359–369 (2015)
12. Belin, M.W., Duncan, J.K.: Keratoconus: the ABCD grading system. Klin. Monatsbl. Augenheilkd. **233**, 701–707 (2016)
13. Anayol, M.A., Guler, E., Yagci, R., Sekeroglu, M.A., Ylmazoglu, M., Trhs, H., Kulak, A.E., Ylmazbas, P.: Comparison of central corneal thickness, thinnest corneal thickness, anterior chamber depth, and simulated keratometry using galilei, Pentacam, and Sirius devices. Cornea **33**, 582–586 (2014)
14. Hernandez-Camarena, J.C., Chirinos-Saldana, P., Navas, A., Ramirez-Miranda, A., de la Mota, A., Jimenez-Corona, A., Graue-Hernindez, E.O.: Repeatability, reproducibility, and agreement between three different Scheimpflug systems in measuring corneal and anterior segment biometry. J. Refract. Surg. **30**, 616–621 (2014). (Thorofare, N.J.1995
15. Savini, G., Carbonelli, M., Sbreglia, A., Barboni, P., Deluigi, G., Hoffer, K.J.: Comparison of anterior segment measurements by 3 Scheimpflug tomographers and 1 Placido corneal topographer. J. Cataract Refract. Surg. **37**, 1679–1685 (2011)
16. Shetty, R., Arora, V., Jayadev, C., Nuijts, R.M., Kumar, M., Puttaiah, N.K., Kummelil, M.K.: Repeatability and agreement of three Scheimpflug-based imaging systems for measuring anterior segment parameters in keratoconus. Invest. Ophthalmol. Vis. Sci. **55**, 5263–5268 (2014)
17. Ariza-Gracia, M.Á., Redondo, S., Llorens, D.P., Calvo, B., Rodriguez Matas, J.F.: A predictive tool for determining patient-specific mechanical properties of human corneal tissue. Comput. Methods Appl. Mech. Eng. **317**, 226–247 (2017)
18. Ariza-Gracia, M.A., Zurita, J., Pinero, D.P., Calvo, B., Rodriguez-Matas, J.F.: Automatized patient-specific methodology for numerical determination of biomechanical corneal response. Ann. Biomed. Eng. **44**, 1753–1772 (2016)
19. Ariza Gracia, M.A., Zurita, J.F., Pinero, D.P., Rodriguez-Matas, J.F., Calvo, B.: Coupled biomechanical response of the cornea assessed by non-contact tonometry. A simulation study. PLoS ONE **10**, e0121486 (2015)
20. Asher, R., Gefen, A., Moisseiev, E., Varssano, D.: An analytical approach to corneal mechanics for determining practical, clinically-meaningful patient-specific tissue mechanical properties in the rehabilitation of vision. Ann. Biomed. Eng. **43**, 274–286 (2015)
21. Dupps Jr., W.J., Seven, I.: A large-scale computational analysis of corneal structural response and ectasia risk in myopic laser refractive surgery. Trans. Am. Ophthalmol. Soc. **114**, T1 (2016)
22. Lago, M.A., Ruperez, M.J., Monserrat, C., Martinez-Martinez, F., Martinez-Sanchis, S., Larra, E., Diez-Ajenjo, M.A., Peris-Martinez, C.: Patient-specific simulation of the intrastromal ring segment implantation in corneas with keratoconus. J. Mech. Behav. Biomed. Mater. **51**, 260–268 (2015)
23. Lanchares, E., Del Buey, M.A., Cristobal, J.A., Calvo, B.: Computational simulation of scleral buckling surgery for rhegmatogenous retinal detachment: on the effect of the band size on the myopization. J. Ophthalmol. **2016**, 3578617 (2016)
24. Roy, A.S., Dupps Jr., W.J.: Patient-specific computational modeling of keratoconus progression and differential responses to collagen cross-linking. Invest. Ophthalmol. Vis. Sci. **52**, 9174–9187 (2011)
25. Seven, I., Roy, A.S., Dupps Jr., W.J.: Patterned corneal collagen crosslinking for astigmatism: computational modeling study. J. Cataract Refract. Surg. **40**, 943–953 (2014)

26. Seven, I., Vahdati, A., De Stefano, V.S., Krueger, R.R., Dupps Jr., W.J.: Comparison of patient-specific computational modeling predictions and clinical outcomes of LASIK for Myopia. Invest. Ophthalmol. Vis. Sci. **57**, 6287–6297 (2016)
27. Simonini, I., Angelillo, M., Pandolfi, A.: Theoretical and numerical analysis of the corneal air puff test. J. Mech. Phys. Solids **93**, 118–134 (2016)
28. Simonini, I., Pandolfi, A.: Customized finite element modelling of the human cornea. PLoS ONE **10**, e0130426 (2015)
29. Simonini, I., Pandolfi, A.: The influence of intraocular pressure and air jet pressure on corneal contactless tonometry tests. J. Mech. Behav. Biomed. Mater. **58**, 75–89 (2016)
30. Vahdati, A., Seven, I., Mysore, N., Randleman, J.B., Dupps Jr., W.J.: Computational biomechanical analysis of asymmetric ectasia risk in unilateral Post-LASIK ectasia. J. Refract. Surg. **32**, 811–820 (2016). (Thorofare, N.J.1995)
31. Ramos-Lopez, D., Martinez-Finkelshtein, A., Castro-Luna, G.M., Pinero, D., Alio, J.L.: Placido-based indices of corneal irregularity. Optom. Vis. Sci. **88**, 1220–1231 (2011)
32. Amsler, M.: Keratocone classique et keratocone fruste, arguments unitaries. Ophthalmologica **111**, 96–101 (1946). Journal international d'ophtalmologie. International journal of ophthalmology. Zeitschrift für Augenheilkunde
33. Cavas-Martinez, F., Fernandez-Pacheco, D.G., De la Cruz-Sanchez, E., Martinez, N.J., Fernandez Canavate, F.J., Vega-Estrada, A., Plaza-Puche, A.B., Alio, J.L.: Geometrical custom modeling of human cornea in vivo and its use for the diagnosis of corneal ectasia. PLoS ONE **9**, e110249 (2014)
34. Cavas-Martínez, F., Fernández-Pacheco, D.G., De La Cruz-Sánchez, E., Martínez, J.N., Cañavate, F.J.F., Alio, J.L.: Virtual biomodelling of a biological structure: the human cornea. Dyna **90**, 647–651 (2015)
35. Browning, J.E., McMann, A.K.: Computational engineering: design, development and applications (2012)
36. Espinosa, J., Mas, D., Pérez, J., Illueca, C.: Optical surface reconstruction technique through combination of zonal and modal fitting. J. Biomed. Opt. **15**, 026022 (2010)
37. Piegl, L.A., Tiller, W.: Approximating surfaces of revolution by nonrational B-splines. IEEE Comput Graph. Appl. **23**, 46–52 (2003)
38. Benítez, H.A., Püschel, T.A.: Modelling shape variance: geometric morphometric applications in evolutionary biology. Int. J. Morphol. **32**, 998–1008 (2014)
39. Klingenberg, C.P.: Analyzing fluctuating asymmetry with geometric morphometrics: concepts, methods, and applications. Symmetry **7**, 843–934 (2015)
40. Parker, J.S., van Dijk, K., Melles, G.R.J.: Treatment options for advanced keratoconus: a review. Surv. Ophthalmol. **60**, 459–480 (2015)
41. Sherwin, T., Brookes, N.H., Loh, I.P., Poole, C.A., Clover, G.M.: Cellular incursion into Bowman's membrane in the peripheral cone of the keratoconic cornea. Exp. Eye Res. **74**, 473–482 (2002)
42. Ozgurhan, E.B., Kara, N., Yildirim, A., Bozkurt, E., Uslu, H., Demirok, A.: Evaluation of corneal microstructure in keratoconus: a confocal microscopy study. Am. J. Ophthalmol. **156**, 885–893 (2013). e882
43. Emre, S., Doganay, S., Yologlu, S.: Evaluation of anterior segment parameters in keratoconic eyes measured with the Pentacam system. J. Cataract Refract. Surg. **33**, 1708–1712 (2007)
44. Ambrosio Jr., R., Alonso, R.S., Luz, A., Velarde, L.G.C.: Corneal-thickness spatial profile and corneal-volume distribution: tomographic indices to detect keratoconus. J. Cataract Refract. Surg. **32**, 1851–1859 (2006)

45. Cervino, A., Gonzalez-Meijome, J.M., Ferrer-Blasco, T., Garcia-Resua, C., Montes-Mico, R., Parafita, M.: Determination of corneal volume from anterior topography and topographic pachymetry: application to healthy and keratoconic eyes. Ophthalmic Physiol. Optics **29**, 652–660 (2009). the journal of the British College of Ophthalmic Opticians (Optometrists)
46. Mannion, L.S., Tromans, C., O'Donnell, C.: Reduction in corneal volume with severity of keratoconus. Curr. Eye Res. **36**, 522–527 (2011)
47. Montalban, R., Pinero, D.P., Javaloy, J., Alio, J.L.: Correlation of the corneal toricity between anterior and posterior corneal surfaces in the normal human eye. Cornea **32**, 791–798 (2013)
48. Ozcura, F., Yildirim, N., Tambova, E., Sahin, A.: Evaluation of Goldmann applanation tonometry, rebound tonometry and dynamic contour tonometry in keratoconus. J. Optom. **10**, 117–122 (2016)
49. Safarzadeh, M., Nasiri, N.: Anterior segment characteristics in normal and keratoconus eyes evaluated with a combined Scheimpflug/Placido corneal imaging device. J. Curr. Ophthalmol. **28**, 106–111 (2016)
50. Schlegel, Z., Hoang-Xuan, T., Gatinel, D.: Comparison of and correlation between anterior and posterior corneal elevation maps in normal eyes and keratoconus-suspect eyes. J. Cataract Refract. Surg. **34**, 789–795 (2008)
51. Tanabe, T., Tomidokoro, A., Samejima, T., Miyata, K., Sato, M., Kaji, Y., Oshika, T.: Corneal regular and irregular astigmatism assessed by Fourier analysis of videokeratography data in normal and pathologic eyes. Ophthalmology **111**, 752–757 (2004)
52. ISO/IEC/IEEE International Standard - Systems and software engineering – Life cycle processes –Requirements engineering. ISO/IEC/IEEE 29148:2011(E), pp. 1–94 (2011)

On Checking Linear Dependence
of Parametric Vectors

Xiaodong Ma[1], Yao Sun[2], Dingkang Wang[3], and Yushan Xue[4(✉)]

[1] College of Science, China Agricultural University, Beijing 100083, China
maxiaodong@cau.edu.cn
[2] SKLOIS, IEE, Chinese Academy of Sciences, Beijing 100093, China
sunyao@iie.ac.cn
[3] KLMM, AMSS, Chinese Academy of Sciences, Beijing 100190, China
dwang@mmrc.iss.ac.cn
[4] School of Mathematics and Statistics,
Central University of Finance and Economics, Beijing 100081, China
cnxueyushan@cufe.edu.cn

Abstract. Checking linear dependence of a finite number of vectors is a basic problem in linear algebra. We aim to extend the theory of linear dependence to parametric vectors where the entries are polynomials. This dependency depends on the specifications of the parameters or values of the variables in the polynomials. We propose a new method to check if parametric vectors are linearly dependent. Furthermore, this new method can also give the maximal linearly independent subset, and by which the remaining vectors are expressed in a linear combination. The new method is based on the computation of comprehensive Gröbner system for a finite set of parametric polynomials.

1 Introduction

One basic problem in linear algebra is to check linear dependence of a finite number of vectors in a vector space over some field [1]. If the entries of the vectors are elements of a field, a classical way to solve this dependency problem is using Gaussian elimination. What if the entries are polynomials? There is a natural question: how to define the linear dependence of a finite number of parametric vectors whose entries are polynomials. The vectors whose entries are polynomials are called polynomial vectors, or parametric vectors without confusion in this paper.

To answer this, we first introduce the definition of **specialization**. Let R be a polynomial ring with variables $u_1, \cdots, u_m$ over the field k, i.e. $R = k[u_1, \cdots, u_m]$. Given a field L, a specialization of R is a homomorphism $\sigma : R \longrightarrow L$. In this paper, we always assume that L is an algebraically closed field containing k, and we only consider the specializations induced by the elements in L^m. That is, for $\bar{a} \in L^m$, the induced specialization $\sigma_{\bar{a}}$ is defined as follows:

$$\sigma_{\bar{a}} : f \longrightarrow f(\bar{a}),$$

© Springer International Publishing AG 2017
D.-S. Huang et al. (Eds.): ICIC 2017, Part II, LNCS 10362, pp. 188–196, 2017.
DOI: 10.1007/978-3-319-63312-1_17

where $f \in R$.

Given a polynomial vector $\mathbf{f} = (f_1, \cdots, f_l) \in R^l$, we can extend a specialization $\sigma_{\bar{a}}$ to R^l:

$$\sigma_{\bar{a}}(\mathbf{f}) = (\sigma_{\bar{a}}(f_1), \cdots, \sigma_{\bar{a}}(f_l)) \in L^l.$$

Note that $\sigma_{\bar{a}}(\mathbf{f})$ does not contain any variable after the above specialization. For a simple example, $f = u_1 u_2 + u_3 \in \mathbb{Q}[u_1, u_2, u_3]$ and $\bar{a} = (1, -1, 2) \in \mathbb{C}^3$, where $\mathbb{Q}$ and $\mathbb{C}$ are the field of rational numbers and the field of complex numbers respectively. Then the specialization $\sigma_{\bar{a}}(f) = f(1, -1, 2) = 1$.

One main goal of this paper is to solve the following interesting problem.

The dependency problem: given a set of parametric vectors $\mathbf{f}_1, \cdots, \mathbf{f}_s \in R^l$, we would like to know for which point $\bar{a} \in L^m$, the vectors $\sigma_{\bar{a}}(\mathbf{f}_1), \cdots, \sigma_{\bar{a}}(\mathbf{f}_s)$ are *linearly independent* over L in the vector space L^l; and for which point $\bar{a} \in L^m$, the vectors $\sigma_{\bar{a}}(\mathbf{f}_1), \cdots, \sigma_{\bar{a}}(\mathbf{f}_s)$ are *linearly dependent* over L.

We will divide the parametric space into finitely many partitions, such that the maximal linearly independent subset are fixed for every specification of the parameters in each partition, and the remaining vectors will be expressed in the linear combinations of the maximal linearly independent subset after specialization.

There is a natural way to solve the dependency problem, and it is based on a generalization of the Gaussian elimination from linear algebra. To some extent it is a generalization of the classical way of checking linear dependence of vectors with entries in a field. The difficulty is that we need to discuss every polynomial entry of the parametric vectors carefully, and the process is very complicated.

In this paper, we propose a new method to solve the dependency problem. It is based on computing a minimal comprehensive Gröbner system for a specific parametric polynomials. The new method is easy to implement, and the advantage is that we can use all existing efficient algorithms for computing comprehensive Grönber systems. In addition, we use a trick to record the computation process in order to find the maximal linearly independent subset, and the remaining vectors will be expressed in the linear combinations of the maximal linearly independent subset after specialization.

The paper is organized as follows. Some preliminaries are given in Sect. 2. In Sect. 3, we propose a new method to solve the dependency problem, which is based on the minimal comprehensive Gröbner system. A complete example is given to illustrate the new method in Sect. 4.

2   Preliminaries

Let k be a field, R be the polynomial ring $k[u]$ in the variables $u = \{u_1, \cdots, u_m\}$, and $R[x]$ be the polynomial ring in the variables $x = \{x_1, \cdots, x_n\}$ over the ring R, where the variables x and u are disjoint.

Lexicographic order and graded reverse lexicographic order are two classic term orders, and are also used in the paper. Let $\succ$ be a term order on x. For a nonzero $f \in k[x]$, the leading monomial, leading term and leading coefficient of f are denoted by $\mathrm{lm}(f)$, $\mathrm{lt}(f)$ and $\mathrm{lc}(f)$ respectively. For a nonzero

$f \in R[x] = k[u][x]$, the leading monomial, leading term and leading coefficient of f w.r.t. $\succ_x$ are denoted by $\text{lm}_x(f)$, $\text{lt}_x(f)$ and $\text{lc}_x(f)$ respectively. Note that $\text{lc}_x(f) \in k[u]$ and $\text{lt}_x(f) = \text{lc}_x(f)\text{lm}_x(f)$.

For a specialization $\sigma : R \longrightarrow L$, it can be extended canonically to a specialization $\sigma : R[x] \longrightarrow L[x]$ by applying σ coefficient-wise. Note that the field L is always assumed to be an algebraically closed field containing k.

Let F be a subset of $R[x]$ and $\bar{a}$ be a given point in L^m. Then we set $\sigma_{\bar{a}}(F) = \{\sigma_{\bar{a}}(f) \mid f \in F\} \subseteq L[x]$, where $\sigma_{\bar{a}}$ is the specialization induced by $\bar{a}$. Let $\langle \sigma_{\bar{a}}(F) \rangle$ be an ideal generated by the set $\sigma_{\bar{a}}(F)$ in $L[x]$.

Let I be an ideal in $k[x]$. The concept of Groebner basis of I w.r.t $\succ$ was proposed by Buchberger, and he also gave an algorithm to compute it. The Groebner basis and minimal Gröbner basis are introduced as follows. For more information, please refer to [2,3].

Definition 1. *Let I be an ideal. A finite set $G = \{g_1, g_2, \cdots, g_t\} \subseteq I$ is called a Gröbner basis for I w.r.t $\succ$, if for any nonzero $f \in I$, $< (f)$ is divisible by $< (g_i)$ for some i.*

Definition 2. *Let G be a Gröbner basis for I. Then G is called a minimal Gröbner basis for I if for any $f \in G$, there does not exist $g \in G \setminus \{f\}$ such that $< (f)$ is divisible by $< (g)$.*

There are many efficient algorithms to compute Gröbner bases and minimal Gröbner bases. If I is an ideal generated by a set of linear polynomials, a minimal basis G for I w.r.t. any given term order $\succ$ can also be obtained by Gaussian elimination of the corresponding coefficient matrix, and all the polynomials in G are linear polynomials.

The concept of a comprehensive Gröbner system (CGS) was introduced by Weispfenning [4]. There are many efficient algorithms to compute CGS, such as [5–11]. It is a powerful tool and widely used in computer science, algebraic geometry, engineering problems, automated geometry theorem proving and automated geometry theorem discovery [12–14]. For a parametric polynomial system $F \subseteq R[x]$, the CGS and minimal CGS are definition below.

Definition 3 (CGS). *For $F \subseteq R[x]$, a finite set $\mathcal{G} = \{(A_1, G_1), \cdots, (A_l, G_t)\}$ is called a comprehensive Gröbner system for F, if for each $1 \leq i \leq t$, $\sigma_{\bar{a}}(G_i)$ is a Gröbner basis for $\langle \sigma_{\bar{a}}(F) \rangle$ in $L[x]$, and for each $g \in G_i$, $\sigma_{\bar{a}}(\text{lc}_x(g)) \neq 0$ for any $\bar{a} \in A_i$, where each A_i is an algebraically constructible set such that $L^m = A_1 \cup \cdots \cup A_t$ and $A_i \cap A_j \neq \emptyset$ for $i \neq j$, and $G_i \subseteq R[x]$.*

A comprehensive Gröbner system $\mathcal{G} = \{(A_1, G_1), \cdots, (A_l, G_l)\}$ for F is said to be **minimal** if for each $1 \leq i \leq t$, $\sigma_{\bar{a}}(G_i)$ is a minimal Gröbner basi s of the ideal $\langle \sigma_{\bar{a}}(F) \rangle \subseteq L[X]$ for $\bar{a} \in A_i$.

For a subset E of $R = k[u]$, the variety definition by E in L^m is the set of all common zeros of the polynomials in E, denoted by $V(E)$. Here, the algebraically constructible set A_i always has the following form:

$$A_i = V(E_i) \setminus V(N_i) = \{\bar{a} \in L^m \mid \bar{a} \in V(E_i), \bar{a} \notin V(N_i)\}.$$

We also call A_i the constraint.

3   Checking Linear Dependence of Parametric Vectors

In this section, we propose a new method to solve the dependency problem, which is based on a minimal comprehensive Gröbner system.

In the following, we will propose a new method to solve the dependency problem for a finite set of parametric vectors $\mathbf{F} = \{\mathbf{f}_1, \cdots, \mathbf{f}_s\} \subseteq R^l$.

Consider the map $\varphi : k^l \to k[z]$, for any for any $\mathbf{f} = (a_1, \cdots, a_l) \in k^l$,

$$\varphi(\mathbf{f}) = a_1 z_1 + \cdots + a_l z_l,$$

where k is a field, and $z = \{z_1, \cdots, z_l\}$ are distinct variables.

In this paper, we consider the parametric vectors. The map can naturally extend to $k[u]$. That is, $\varphi : R^l = (k[u])^l \to k[u][z]$, for any $\mathbf{f} = (a_1, \cdots, a_l) \in R^l$,

$$\varphi(\mathbf{f}) = a_1 z_1 + \cdots + a_l z_l,$$

where $z = \{z_1, \cdots, z_l\}$ are new variables different from $u = \{u_1, \cdots, u_m\}$.

Note that the degree of $\varphi(\mathbf{f})$ w.r.t each variable z_i is one, and $\sigma_{\bar{a}}(\varphi(\mathbf{f}))$ is a linear polynomial in $L[z_1, \cdots, z_l]$ for each $\bar{a} \in L^l$.

Let $\varphi(\mathbf{F}) = \{\varphi(\mathbf{f}_1), \cdots, \varphi(\mathbf{f}_s)\}$. Note that $\varphi(\mathbf{F})$ is a parametric linear system. For the parametric linear systems, Sit W Y has given an algorithm to solve them. For more, please see [15].

Let $\{(A_1, G_1), \cdots, (A_t, G_t)\}$ be a minimal comprehensive Gröbner system for $\varphi(\mathbf{F})$ w.r.t. any given term order on z. It is easy to check that each G_i is a set of liner polynomials in $z_1, \cdots, z_l$ with coefficients in R, and the number of polynomials in G_i is less or equal to s for $1 \leq i \leq t$.

In the following, we will give a result for the minimal comprehensive Gröbner system without proof.

Theorem 1. *Let* $\mathbf{F} = \{\mathbf{f}_1, \cdots, \mathbf{f}_s\}$ *be a subset of* R^l, *and* $\mathcal{G} = \{(A_1, G_1), \cdots, (A_t, G_t)\}$ *be a minimal comprehensive Gröbner system for* $\varphi(\mathbf{F}) \subset k[u][z]$ *w.r.t. any term order in* z, *where* $\varphi(\mathbf{F}) = \{\varphi(\mathbf{f}_1), \cdots, \varphi(\mathbf{f}_s)\}$. *Then for each* $\bar{a} \in A_i$, $\sigma_{\bar{a}}(\mathbf{f}_1), \cdots, \sigma_{\bar{a}}(\mathbf{f}_s)$ *are linearly independent over* L *if and only if the number of polynomials in* G_i *is exactly* s.

Theorem 1 provides a simple way to compute two subsets A and B such that $L^m = A \cup B$, and for each $\bar{a} \in A$, $\sigma_{\bar{a}}(\mathbf{f}_1), \cdots, \sigma_{\bar{a}}(\mathbf{f}_s)$ are linearly independent over L; for each $\bar{a} \in B$, $\sigma_{\bar{a}}(\mathbf{f}_1), \cdots, \sigma_{\bar{a}}(\mathbf{f}_s)$ are linearly dependent over L. If $\mathcal{G} = \{(A_1, G_1), \cdots, (A_t, G_t)\}$ is a minimal comprehensive Groebner system for $\varphi(\mathbf{F}) \subset k[u][z]$, where $\mathcal{G}$ can be obtained by any existing algorithms for computing comprehensive Groebner systems. Then we have

$$A = \bigcup_{|G_i|=s} A_i, \text{ and } B = \bigcup_{|G_i|<s} A_i,$$

where $|G_i|$ is the number of polynomials in G_i.

To solve the dependency problem for $\mathbf{F}$ completely, we also need to compute the maximal linearly independent subset for the dependent part B. We should compute a finite of disjoint subsets $B_1, \cdots, B_t$ and $\mathbf{M}_1, \cdots, \mathbf{M}_t \subseteq \mathbf{F}$ such that $B = B_1 \cup B_2 \cup \cdots \cup B_t$, and for each dependent case B_i, $\mathbf{M}_i$ is the maximal linearly dependent subset for $\mathbf{F}$, and every vector $\mathbf{f}$ in $\mathbf{F} \setminus \mathbf{M}_i$ can be expressed in a linear combination of $\mathbf{M}_i$ after specialization.

For this purpose, we need to compute a minimal Gröbner system for

$$\{\varphi(\mathbf{f}_1) + e_1, \cdots, \varphi(\mathbf{f}_s) + e_s\} \subseteq R[z, e],$$

w.r.t. a block order $\succ := (\succ_z, \succ_e)$, where $e = \{e_1, \cdots, e_s\}$ are new variables different from u and z, $\succ_z$ and $\succ_e$ are two term orders with $z_1 > z_2 > \cdots > z_l$ and $e_1 > e_2 > \cdots > e_s$ respectively.

Here, we say a term order $\succ := (\succ_z, \succ_e)$ is a block order, if $z^{\alpha_1} e^{\beta_1} \succ z^{\alpha_2} e^{\beta_2}$ if and only if $z^{\alpha_1} \succ_z z^{\alpha_2}$, or $z^{\alpha_1} = z^{\alpha_2}$ and $e^{\beta_1} \succ_e e^{\beta_2}$.

The variables $e_1, \cdots, e_s$ are used to record the computation process of the minimal comprehensive Gröbner system. This trick has been used in many ways (such as computing syzygies) in most algebraic computer textbooks, for example [12]. The trick can also help us to compute the maximal linearly independent subset, and the expression of the linear combinations of the maximal linearly independent subset for the remaining vectors after specialization. Thus, we give the following theorem. We use the notation $|G|$ to be the number of the elements in G.

Theorem 2. *Let* $\mathbf{F} = \{\mathbf{f}_1, \cdots, \mathbf{f}_s\}$ *be a subset of* R^l, *and* $\mathcal{G} = \{(A_1, G_1), \cdots, (A_t, G_t)\}$ *be a minimal comprehensive Gröbner system for*

$$\{\varphi(\mathbf{f}_1) + e_1, \cdots, \varphi(\mathbf{f}_s) + e_s\} \subseteq R[z, e]$$

w.r.t. a block order $\succ := (\succ_z, \succ_e)$, *where* $\succ_z$ *and* $\succ_e$ *are two any term orders with* $z_1 > z_2 > \cdots > z_l$ *and* $e_1 > e_2 > \cdots > e_s$ *respectively,* $R = k[u]$, $u = \{u_1, \cdots, u_m\}$, $e = \{e_1, \cdots, e_s\}$ *and* $z = \{z_1, \cdots, z_l\}$.

For each $1 \leq i \leq t$, *suppose that* $G_i' = G_i \cap R[e]$ *and* $G_i'' = G_i \setminus G_i'$.

Then $|G_i| = |G_i'| + |G_i''| = s$, *and if* G_i' *is not empty, the polynomials of* G_i' *have the following form:*

$$g_1 = a_{1i_1} e_{i_1} + \sum_{j=i_1+1}^{s} a_{1j} e_j,$$

$$g_2 = a_{2i_2} e_{i_2} + \sum_{j=i_2+1}^{s} a_{2j} e_j,$$

$$\cdots$$

$$g_r = a_{ri_r} e_{i_r} + \sum_{j=i_r+1}^{s} a_{rj} e_j,$$

with $1 \leq i_1 < i_2 < \cdots < i_r \leq s$, *where* $r = |G_i'|$ *and* $a_{ij} \in R$. *Furthermore,*

(i) if $|G_i''| = s$, then for each $\bar{a} \in A_i$, $\sigma_{\bar{a}}(\mathbf{f}_1), \cdots, \sigma_{\bar{a}}(\mathbf{f}_s)$ are linearly independent over L.

(ii) if $|G_i''| < s$, then for each $\bar{a} \in A_i$, $\sigma_{\bar{a}}(\mathbf{f}_1), \cdots, \sigma_{\bar{a}}(\mathbf{f}_s)$ are linearly dependent over L, and for each $g_k \in G_i'$, $p_1, \cdots, p_s$ are the corresponding coefficients, where p_i is the coefficient of e_i in g_k. Moreover, $\mathbf{M} = \{\mathbf{f}_i \mid \mathbf{f}_i \in \mathbf{F}, i \notin \{i_1, \cdots, i_r\}\}$ the maximal linearly independent subset for $\mathbf{F}$, and every parametric vector of $\mathbf{F} \setminus \mathbf{M}$ can be expressed in the linear combination of $\mathbf{M}$ over L after specialization.

Theorem 2 can be proved by Theorem 1 and the properties of the minimal Groebner system, and here we omit the proof.

4   A Complete Example

In this section, we will use a complete example to show how to apply Theorem 2 to solve the dependency problem.

Example 1. Let $\mathbf{F} = \{\mathbf{f}_1, \mathbf{f}_2, \mathbf{f}_3\} \subset \mathbb{Q}[a, b]^3$, where

$$\mathbf{f}_1 = (a, b, 0), \mathbf{f}_2 = (0, b+1, b), \text{ and } \mathbf{f}_3 = (a+1, 0, -1)$$

are three parametric vectors with the parameters a and b, and $\succ_u$ be the graded reverse lexicographic orders with $a > b$. Here the coefficient field $R = \mathbb{Q}[a, b]$. Next, We use the new method provided by Theorem 2 to solve the dependency problem for $\mathbf{F}$.

To solve the dependency problem for $\mathbf{F}$, we first construct a specific parametric polynomial system using the map φ. Let $f_i = \varphi(\mathbf{f}_i) + e_i$ for $i = 1, 2, 3$. Then we obtain a parametric polynomial system:

$$F = \{f_1 = az_1 + bz_2 + e_1, f_2 = (b+1)z_2 + bz_3 + e_2, f_3 = (a+1)z_1 - z_3 + e_3\} \subseteq \mathbb{Q}[a, b][z, e].$$

Using the algorithm proposed in [6], we can get a minimal comprehensive Gröbner system $\mathcal{G}$ for F w.r.t. a block order $(\succ_z, \succ_e)$, where $\succ_z$ and $\succ_e$ are two lexicographical order with $z_1 > z_2 > z_3$ and $e_1 > e_2 > e_3$ respectively. Here, $\mathcal{G} = \{(A_1, G_1), (A_2, G_2), (A_3, G_3), (A_4, G_4)\}$, where

$$A_1 = L^3 \setminus V((ab^2 - ab + b^2 - a)(b+1)),$$

$$G_1 = \{z_1 + z_2 + (b-1)z_3 - e_1 + e_2 + e_3, (b+1)z_2 + bz_3 + e_2,$$
$$(ab^2 - ab + b^2 - a)z_3 - (ab + a + b + 1)e_1 + (ab + b)e_2 + (ab + a)e_3\},$$

$$A_2 = V(ab^2 - ab + b^2 - a) \setminus V(b+1),$$

$$G_2 = \{z_1 - bz_2 - z_3 - e_1 + e_3, (b+1)z_2 + bz_3 + e_2, (b+1)e_1$$
$$- (ab^2 - ab + b^2 - a + b)e_2 + (ab^2 - ab - a)e_3\},$$

$$A_3 = V(b+1) \setminus V(a+1),$$

$$G_3 = \{z_1 + z_2 - 2z_3 - e_1 + e_2 + e_3, (a+1)z_2 - (a+1)e_1 - ae_2 + ae_3, z_3 - e_2\}.$$

$$A_4 = V(a+1, b+1),$$

$$G_4 = \{z_1 + z_2 - z_3 - e_1 + e_3, z_3 - e_2, e_2 - e_3\},$$

For the minimal comprehensive Gröbner system $\mathcal{G}$, the polynomials in G_i are all linear in z and e with coefficient in $\mathbb{Q}[a, b]$, and the number of polynomials in G_i is exactly 3 for $i = 1, 2, 3$.

By Theorem 2, we get the following results.

(1) For A_1 and A_3, note that

$$G_1' = G_1 \cap R[e] = \emptyset, G_1'' = G_1 \setminus G_1' = G_1,$$

and

$$G_3' = G_3 \cap R[e] = \emptyset, G_3'' = G_1 \setminus G_3' = G_3.$$

We have $|G_1''| = |G_3''| = 3$. Thus, by Theorem 2, $\sigma_{\bar{a}}(\mathbf{f}_1), \sigma_{\bar{a}}(\mathbf{f}_2), \sigma_{\bar{a}}(\mathbf{f}_3)$ are linearly independent over L for each $\bar{a} \in A_1 \cup A_3$.

(2) For case A_2, we have

$$G_2' = G_2 \cap R[e] = \{z_1 - bz_2 - z_3 - e_1 + e_3, (b+1)z_2 + bz_3 + e_2\}$$

and

$$G_2'' = G_2 \setminus G_2' = \{(b+1)e_1 - (ab^2 - ab + b^2 - a + b)e_2 + (ab^2 - ab - a)e_3\}.$$

Note that $|G_2'| < 3$. Then for each $\bar{a} \in A_2$, $\sigma_{\bar{a}}(\mathbf{f}_1), \sigma_{\bar{a}}(\mathbf{f}_2), \sigma_{\bar{a}}(\mathbf{f}_3)$ are linearly dependent over L. Moreover, from the polynomial G_2'', the maximal linearly independent subset is $\mathbf{M}_2 = \{\mathbf{f}_2, \mathbf{f}_3\}$, and we get a linear combination for $\mathbf{f}_1$ after specialization:

$$\sigma_{\bar{a}}(\mathbf{f}_1) = \sigma_{\bar{a}}(b+1)^{-1}(\sigma_{\bar{a}}(ab^2 - ab + b^2 - a + b)\sigma_{\bar{a}}(\mathbf{f}_2) - \sigma_{\bar{a}}(ab^2 - ab - a)\sigma_{\bar{a}}(\mathbf{f}_3)).$$

And $(b+1)$, $-(ab^2 - ab + b^2 - a + b)$ and $(ab^2 - ab - a)$ are the corresponding coefficients for $\mathbf{F}$, such that

$$\sigma_{\bar{a}}((b+1)\mathbf{f}_1 - (ab^2 - ab + b^2 - a + b)\mathbf{f}_2 + (ab^2 - ab - a)\mathbf{f}_3) = (0, 0, 0).$$

(3) For case A_4, we have

$$G'_4 = G_4 \cap R[e] = \{z_1 + z_2 - z_3 - e_1 + e_3, z_3 - e_2\},$$

and $G''_4 = G_4 \setminus G'_4 = \{e_2 - e_3\}$.

Note that $|G'_4| = 2 < 3$. Then for each $\bar{a} \in A_4$, $\sigma_{\bar{a}}(\mathbf{f}_1), \sigma_{\bar{a}}(\mathbf{f}_2), \sigma_{\bar{a}}(\mathbf{f}_3)$ are linearly dependent over L. From G''_4, the maximal linearly independent subset is $\mathbf{M}_4 = \{\mathbf{f}_1, \mathbf{f}_3\}$, and we get a linear combination for $\mathbf{f}_1$ after specialization: $\sigma_{\bar{a}}(\mathbf{f}_2) = \sigma_{\bar{a}}(\mathbf{f}_3)$. And 0, 1 and -1 are corresponding coefficients such that $\sigma_a(\mathbf{f}_2 - \mathbf{f}_3) = (0, 0, 0)$.

Note that $L^3 = A_1 \cup A_2 \cup A_3 \cup A_4$, we have solved the dependency problem for $\mathbf{F}$.

Acknowledgements. This work was supported in part by the Chinese Universities Scientific Fund (Grant No. 2017QC061), and the National Natural Science Foundation of China (Grant No. 11371356).

References

1. Greub, W.H.: Linear Algebra, 3rd edn. Springer, New York (1967)
2. Buchberger, B.: Gröbner-bases: an algorithmic method in polynomial ideal theory. In: Multidimensional Systems Theory - Progress, Directions and Open Problems in Multidimensional Systems, pp. 184–232. Reidel Publishing Company, Dodrecht-Boston-Lancaster 1985
3. Cox, D., Little, J., O'Shea, D.: Using Algebraic Geometry, 2nd edn. Springer, New York (2005)
4. Weispfenning, V.: Comprehensive Groebner bases. J. Symb. Comp. **14**, 1–29 (1992)
5. Kapur, D.: An approach for solving systems of parametric polynomial equations. In: Saraswat, V., Van Hentenryck, P. (eds.) Principles and Practice of Constraint Programming. MIT Press, Cambridge (1995)
6. Kapur, D., Sun, Y., Wang, D.K.: A new algorithm for computing comprehensive Groebner systems. In: Proceedings of ISSAC 2010. ACM Press, New York (2010)
7. Kapur, D., Sun, Y., Wang, D.K.: Computing comprehensive Groebner systems and comprehensive Groebner bases simultaneously. In: Proceedings of ISSAC 2011. ACM Press, New York (2011)
8. Manubens, M., Montes, A.: Minimal canonical comprehensive Groebner system. J. Symb. Comp. **44**, 463–478 (2009)
9. Nabeshima, K.: A speed-up of the algorithm for computing comprehensive Groebner systems. In: Proceedings of ISSAC 2007. ACM Press, New York (2007)
10. Suzuki, A., Sato, Y.: An alternative approach to comprehensive Groebner bases. J. Symb. Comp. **36**, 649–667 (2003)
11. Suzuki, A., Sato, Y.: A simple algorithm to compute comprehensive Groebner bases using Groebner bases. In: Proceedings of ISSAC 2006. ACM Press, New York (2006)
12. Chen, X., Li, P., Lin, L., Wang, D.: Proving geometric theorems by partitioned-parametric Gröbner bases. In: Hong, H., Wang, D. (eds.) ADG 2004. LNCS, vol. 3763, pp. 34–43. Springer, Heidelberg (2006). doi:10.1007/11615798_3

13. Montes, A.: A new algorithm for discussing Groebner basis with parameters. J. Symb. Comp. **33**, 183–208 (2002)
14. Montes, A., Recio, T.: Automatic discovery of geometry theorems using minimal canonical comprehensive Gröbner systems. In: Botana, F., Recio, T. (eds.) ADG 2006. LNCS, vol. 4869, pp. 113–138. Springer, Heidelberg (2007). doi:10.1007/978-3-540-77356-6_8
15. Sit, W.Y.: An algorithm for solving parametric linear systems. J. Symb. Comp. **13**, 353–394 (1992)

Intelligent Computing in Computational Biology

Effective Identification of Hot Spots in PPIs Based on Ensemble Learning

Xiaoli Lin[1,2(✉)], QianQian Huang[2], and Fengli Zhou[1]

[1] Information and Engineering Department of City College, Wuhan University
of Science and Technology, Wuhan 430083, China
aneya@163.com, thinkview@163.com
[2] Hubei Key Laboratory of Intelligent Information Processing
and Real-Time Industrial System, School of Computer Science and Technology,
Wuhan University of Science and Technology, Wuhan 430065, China
1273758784@qq.com

Abstract. The experiment of alanine scanning has shown that most of the binding energies in protein-protein interactions are contributed by a few significant residues at the protein-protein interfaces, and those important residues are called hot spot residues. On the basis of protein-protein interaction, hot spot residues tend to get together to form modules, and those modules are defined as hot regions. So, hot spot residues play an important role in revealing the life activities of organisms. Therefore, how to predict hot spot residues and non-spot residues effectively and accurately is a vital research direction. A new method is proposed combining protein amino acid physicochemical features and structural features to predict the hot spot residues based on the ensemble learning. The experimental results demonstrate that this method of prediction hot spot residues has a good effect.

Keywords: Hot spots · PPI · Classification · Ensemble learning

1 Introduction

As the expression of gene, protein has attracted more and more attention, and the function of protein cannot be separated from the interaction of protein, so the research about protein interaction has become the hot research direction [1–3]. In recent years, the hot spot residues, as a special kind of residues in the protein-protein interface, have attracted the attention of researchers due to their contribution to the free energy. Alanine scan mutation is a very direct method to study the contribution of amino acid residues combined with free energy. The main principle is to change a certain residue into alanine, and to observe the effect of energy change on protein function. The experimental process is to replace the residue of the protein binding surface with alanine, then to measure the change of binding energy of the protein complex before and after replacement. The protein is the basis of biological function, which is a kind of organic compound of amino acid molecules [4, 5]. The protein molecule is a very complex and diverse family. There are two categories amino acids: polar amino acids and non-polar ones. The former

D.-S. Huang et al. (Eds.): ICIC 2017, Part II, LNCS 10362, pp. 199–207, 2017.
DOI: 10.1007/978-3-319-63312-1_18

includes R, N, D, E, Q, H, K, S, T and Y, while the latter includes A, C, G, I, L, M, F, P, W and V. The protein molecules fold to the 3D space conformation which performs their biological function and activities [6]. Most proteins interact with other proteins through the surface area, which usually locate at the specific binding sites. In order to better understand the protein-protein interactions, it is necessary to study the interface residues of protein binding surfaces and their adjacent residues. Based on the experimental data of alanine mutation, the researchers found that the contribution of protein binding surface residues to binding energy is not average distribution, only a small proportion of residues contribute most of the energy in protein-protein interactions, these key residues are called hot spot residues [7].

Furthermore, a large number of experiments show that the hot spot residues are not uniformly distributed on the interface of the protein-protein interaction, but closely clustered in a dense region, and the hot spot residues form a specific conformation between protein and protein, which is called hot region [8, 9]. The distribution characteristics of the hot-spot residues show that the active residues in the same hot region play a role in the stability of the complexes, while the different regions are independent of each other, and the contribution of these independent hot spot residues is cumulative to the stability of protein complexes.

Therefore, the discovery of hot regions helps the understanding of life activities at the micro-level, and has a wide range of biological application value. Such as, the theoretical guidance of biological pharmacy, the drug research based on protein structure [10]. Therefore, how to use computational intelligence method to identify the hot spot residues of protein-protein interaction has become an important applied [11].

With the continuous development and deep research of proteomics, the intelligent calculation method will become more and more important. The research of the prediction method of protein structure and function has important theoretical significance and practical value. Recently many research works have been performed based on the feature-based method [12] and the energy-based method. In molecular biology, the alanine mutation scanning experiment is a prediction method based on energy [13–15].

For hot spots and hot regions in protein-protein interactions, many researches have been carried out [7, 16–18]. Tuncbag proposed a method combining the available surface area and pairing tendency to predict hot spots [19], Cukuroglu showed that central hub protein is found in the correspondence between protein-protein interaction network [20]. Carles [21] predicted the hot region by using various data sets.

The main work in this paper is to study the application of ensemble learning method in the protein-protein interaction. This paper used a hybrid method to predict hot spot residue based on feature.

2 Methods

2.1 Data Sets

Protein structure data come from protein structure database. Alanine mutation data of ASEdb include binding free energy change value $\triangle\triangle G$. In order to analyze the feasibility of our prediction model, this paper will use the same data set in [22–24], which are 16 protein complexes from ASEdb. The interaction data sets are derived from PSAIA (Protein Structure and Interaction Analyzer) [25].

2.2 Feature Selection

Feature selection is an important step in designing classifier, and it can remove redundancy and related features to improve the performance of classifier. There are hundreds of physical and chemical properties of protein complexes, and the classifiers based on such attributes will affect the performance. Li et al. to construct a Profeat computing platform [26], which involves a variety of physical and chemical properties.

Accessible surface area (ASA), relative accessible surface area (RASA) and accessible surface area (ASA) have a strong ability to identify hot spot residues and non-spot residues. Protrusion index (PI) and depth index (DI) also have significant performance characteristics in hot spot residues. In addition to the physical characteristics and structural attributes mentioned above, this paper also extracted another protein related feature, the conservative score of residues, which has been proved to play an important role in protein-protein interaction prediction models. The conservative score of residues is used to describe the parameters of evolution rate. The conservative score of each position in amino acid sequence is not different. Some positions are very conservative, so conservative scores are high. While other places often change, the conservative scores are relatively low. The conservative score reflects the different degree of evolutionary selection on these sites. According to previous studies [27], hot spot residues are more conservative than non-hot residues in the structure, so the conservative score is an important feature to predict hot spot residues.

In theory, the more features are used in prediction model, the ability of identify hot spot residues is stronger. In fact, for a limited data set, when the number of characteristic attributes exceeds a certain value, the accuracy of the prediction model decreases with the increase of characteristic attributes, and it can cause over fitting. Not only that, too many characteristics will consume more training time, so we must select a part of the attributes from these attributes to build the predictive model. Feature selection is the process of selecting a feature subset from the extracted features. The prediction of protein hot spot residues is small sample and high dimensional data, so an effective feature selection algorithm is particularly important.

It is necessary to select the characteristics reasonably and to reduce the dimension appropriately. On the one hand, the redundancy of residual attributes can be eliminated to speed up the calculation speed and to improve the efficiency of classification. On the other hand, the complexity of classifier and the error rate of classification can be reduced. In this paper, minimum-Redundancy-Maximum-Relevance (mRMR) was used to choose a feature subset [28]. The mRMR method is a feature selection method based on heuristic search. The goal of this method is to select a feature subset which can best describe the statistical characteristics of the target classification variable. These selected features are irrelevant and these features are related with the classification target as much as possible.

The mRMR method considers the correlation and redundancy between the features and categories from the perspective of spatial search. Its performance generally outperforms other feature selection method based on sorting, but it still has some shortcomings.

Firstly, mRMR uses only mutual information to measure the correlation between the features, which reduces the scope of the method. As we all know, there are many indicators to evaluate the importance of the characteristics, such as Gini coefficient, fisher score, information gain, and so on. In this paper, the feature correlation redundancy weighting factor α is introduced to refine the measurement of the correlation and redundancy. Through α value is modified, the weighted coefficient of feature relevance and redundancy is adjusted to obtain the best feature selected. The maximum relevant minimum redundancy criteria can be as follows.

$$\max\emptyset(D, R), \qquad \emptyset = \alpha D - (1 - \alpha)R \tag{1}$$

Figure 1 shows the results of the protein properties selected by using the mRMR method. There are the top 10 protein attributes.

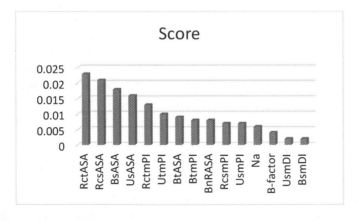

Fig. 1. The results of feature selection

2.3   Classification Based on Improved BoostSVM

As the support vector machine (SVM) has many attractive advantages and promising experimental performance, it receives more and more attention. This technology has become a hot spot in the field of machine learning research, such as text classification, handwritten recognition, image classification, and so on. Many researches have also proved that SVM model can get the better results of hot spot prediction [9, 29, 30].

But support vector machine (SVM) also has its own shortcomings, such as, parameter selection, prediction difficulties for large scale data or unbalanced data. For solving practical problems, the support vector machine has high practical requirements for prediction accuracy. Therefore, many scholars are engaged in researching and improving support vector machine work.

Adaboost is an improved boosting algorithm which can adaptively adjust the error of weak classifier. The AdaBoost algorithm trains data by using weights instead of randomly selected from the training samples. The core idea of AdaBoost algorithm is to select different training sets to train the weak classifier, and to make the weak classifier stronger, which is an iterative algorithm. Each training sample is given the same weight, and select the minimum classification error rate as the best weak classifier of this iteration. For the error sub sample, the algorithm will increase the sample weight in order to make it more attention in the next iteration. The error sample of this iteration will be highlighted in the next iteration. The final strong classifier is combination of several weak classifiers with the weighted sum.

In this paper, we used an improved BoostSVM to predict the hot spot residues and non-spot residues. The steps can be described as follow Table 1.

Table 1. The steps of improved BoostSVM

Algorithm
Training set $S = \{(x_1, y_1), \ldots, (x_n, y_n)\}$

$y_1 = 1$ represents the positive sample and $y_1 = -1$ represents the negative sample

Begin

    Initialize weight for each sample: $D_i = \frac{1}{2}, i = 1, \ldots, n$.

    While ($t = 1, \ldots, T$) {

        Normalized weight: $w_{t,i} = \frac{w_{t,j}}{\sum_{j=1}^{n} w_{t,j}}$

        Select k training data samples

        For each SVM weak classifier h_j, the misclassification rate $\varepsilon_j = \sum_i w_i |h_j(x_i) - y_i|$

        Get a weak classifier with the minimum error rate $h_j: R^d \rightarrow \{-1,1\}$

        Update weight $w_{t+1,i} = w_{t,i} \frac{\varepsilon_t}{1-\varepsilon_t}$

    } **End while**

    Strong classifier

$$H(x) = \begin{cases} 1 & \sum_{t=1}^{T} \alpha_t h_t(x) \geq \frac{1}{2}\sum_{t=1}^{T} \alpha_t \\ 0 & 其他 \end{cases}$$

End

3   Experimental Results and Evaluation

3.1   Measures of Prediction Performance

When the classification model has been constructed, it needs to use evaluation criterion to analyze the performance of prediction model. There are three criterions *Precision*, *Recall*, *F*1 to evaluate the results of prediction. The *Precision* is defined as

$$Precision = \frac{TP}{TP + FP} \tag{2}$$

The *Recall* is the coverage of prediction

$$Recall = \frac{TP}{TP + FN} \tag{3}$$

The *F*1 represents the balance performance between *Precision* and *Recall*

$$F1 = \frac{2 * Precison * Recall}{Precison + Recall} \tag{4}$$

In the above formula, *TP* represents the correct predicted hot spots residue, *FP* indicates that the non-spot residues are predicted to be hot spot residues, *TN* represents the correct predicted non-spot residues and *FN* indicated that the hot-spot residues are predicted to be the non-spot residues.

3.2   Experimental Results

In order to illustrate the efficiency of this improved algorithm, Fig. 2 shows the comparison with previous methods. Here lists respectively the accuracy, the coverage and the

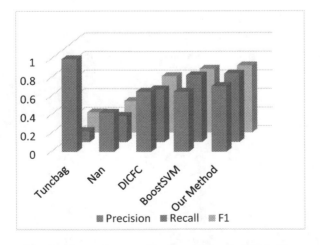

Fig. 2. Comparison results of prediction in ASEdb data

balance performance between accuracy and coverage of five methods (Tuncbag [19], Nan [23], DICFC [24], BoostSVM [31] and our improved method). It can be shown that Tuncbag's precision is quite high, but the balance performance between accuracy and coverage is lower. BoostSVM's precision is consistent with DICFC, but the precision of improved BoostSVM (our method in this paper) is higher than these two methods. Compared with the other methods, our results also show better performance.

4 Conclusion

In order to obtain a good prediction results of hot spot residues and non-spot residues, this paper adopts an improved BoostSVM method combining SVM and AdaBoost. SVM has a perfect theoretical basis and better performance of learning classification. The AdaBoost method, as an ensemble learning, has the advantages of simple, flexible and easy to understand. AdaBoost focuses on error classification sample. Error samples can be learned many times, so that the learning machine is more suitable for these samples. Experimental results show that this predictive method can effectively predict the hot spots residues. How to obtain high reliability and more abundant data sets is an urgent problem in future research. More importantly, in future research, it is necessary to extend the protein-protein interactions between more species.

Acknowledgment.   The authors thank the members of Machine Learning and Artificial Intelligence Laboratory, School of Computer Science and Technology, Wuhan University of Science and Technology, for their helpful discussion within seminars. This work was supported in part by National Natural Science Foundation of China (No. 61502356, 61273225, 61273303).

References

1. Hsu, C.M., Chen, C.Y., Liu, B.J., Huang, C.C.: Identification of hot regions in protein-protein interactions by sequential pattern mining. BMC Bioinform. **8**(Suppl 5), S8 (2007)
2. Keskin, O., Tuncbag, N., Gursoy, A.: Predicting protein-protein interactions from the molecular to the proteome level. Chem. Rev. **116**(8), 4884–4909 (2016)
3. Yu, X., Rangwala, H., Domeniconi, G., Zhang, G.J., Yu, Z.W.: Protein function prediction using multilabel ensemble classification. IEEE/ACM Trans. Comput. Biol. Bioinform. **10**(4), 1045–1057 (2013)
4. Hsu, C.M., Chen, C.Y., Liu, B.J.: MAGIIC-PRO: detecting functional signatures by efficient discovery of long patterns in protein sequences. Nucleic Acids Res. **36**(4), 1400–1406 (2008)
5. Scott, D.E., Bayly, A.R., Abell, C., Skidmore, J.: Small molecules, big targets: drug discovery faces the protein-protein interaction challenge. Nat. Rev. Drug Discov. **15**, 533–550 (2016)
6. Sahu, S.S., Panda, G.: Efficient localization of hot spots in proteins using a novel s-transform based filtering approach. IEEE/ACM Trans. Comput. Biol. Bioinform. **8**(5), 1235–1246 (2011)
7. Tuncbag, N., Gursoy, A., Keskin, O.: Identification of computational hot spots in protein interfaces: combining solvent accessibility and inter-residue potentials improves the accuracy. Bioinformatics **25**(12), 1513–1520 (2009)
8. Reichmann, D., Rahat, O., Albeck, S., Meged, R., Dym, O., Schreiber, G.: The modular architecture of protein-protein binding interfaces. Proc. Natl. Acad. Sci. **102**(1), 57–62 (2005)

9. Ahmad, S., Keskin, O., Sarai, A., Nussinov, R.: Protein–DNA interactions: structural, thermodynamic and clustering patterns of conserved residues in DNA-binding proteins. Nucleic Acids Res. **36**(18), 5922–5932 (2008)
10. Armon, A., Dan, G., Ben-Tal, N.: ConSurf: an algorithmic tool for the identification of functional regions in proteins by surface mapping of phylogenetic information. J. Mol. Biol. **307**(1), 447–463 (2001)
11. Keskin, O., Ma, B.Y., Mol, R.J.: Hot regions in protein-protein interactions: the organization and contribution of structurally conserved hot spot residues. J. Mol. Biol. **345**(5), 1281–1294 (2005)
12. Xu, B., Wei, X.M., Deng, L., Guan, J., Zhou, S.G.: A semi-supervised boosting SVM for predicting hot spots at protein-protein interfaces. BMC Syst. Biol. **2**(2), 1–12 (2012)
13. Morrison, K.L., Weiss, G.A.: Combinatorial alanine-scanning. Curr. Opin. Chem. Biol. **5**(3), 302–307 (2001)
14. Thorn, K.S., Bogan, A.A.: ASEdb: a data base of alanine mutations and their effects on the free energy of binding in protein interactions. Bioinformatics **17**(3), 284–285 (2001)
15. Gonzalez Ruiz, D., Gohlke, H.: Targeting protein-protein interactions with small molecules: challenges and perspectives for computational biding epitope detection and ligand finding. Curr. Med. Chem. **13**(22), 2607–2625 (2006)
16. Ezkurdia, I., Bartoli, L., Fariselli, P., Casadio, R., Valencia, A., Tress, M.L.: Progress and challenges in predicting protein-protein interaction sites. Brief. Bioinform. **10**(10), 233–246 (2009)
17. Lise, S., Buchan, D., Pontil, M., Jones, D.T.: Predictions of hot spot residues at protein-protein interfaces using support vector machines. PLoS One **6**(2), e16774 (2011). doi:10.1371/journal.pone.0016774
18. Lise, S., Archambeau, C., Pontil, M., Jones, D.T.: Prediction of hot spot residues at protein-protein interfaces by combining machine learning and energy-based methods. BMC Bioinform. **10**(1), 365 (2009). doi:10.1186/1471-2105-10-365
19. Tuncbag, N., Keskin, O., Gursoy, A.: HotPoint: hot spot prediction server for protein interfaces. Nucleic Acids Res. **38**, 402–406 (2010)
20. Cukuroglu, E., Gursoy, A., Keskin, O.: Analysis of hot region organization in hub proteins. Ann. Biomed. Eng. **38**(6), 2068–2078 (2010)
21. Carles, P., Fabian, G., Juan, F.: Prediction of protein-binding areas by small world residue networks and application to docking. BMC Bioinform. **12**, 378–388 (2011)
22. Cho, K., Kim, D., Lee, D.: A feature-based approach to modeling protein-protein interaction hot spots. Nucleic Acids Res. **37**(8), 2672–2687 (2009)
23. Nan, D.F., Zhang, X.L.: Prediction of hot regions in protein-protein interactions based on complex network and community detection. In: IEEE International Conference on Bioinformatics and Biomedicine, pp. 17–23 (2013)
24. Hu, J., Zhang, X.L., Liu, X.M., Tang, J.S.: Prediction of hot regions in protein-protein interaction by combining density-based in cremental clustering with feature-based classification. Comput. Biol. Med. **61**, 127–137 (2015)
25. Mihel, J., Sikić, M., Tomić, S., Jeren, B., Vlahovicek, K.: PSAIA-protein structure and interaction analyzer. BMC Struct. Biol. **8**(1), 1–11 (2008)
26. Li, Z.R., Lin, H.H., Han, L.Y., et al.: PROFEAT: a web server for computing structural and physicochemical features of proteins and peptides from amino acid sequence. Nucleic Acids Res. **34**, W32–W37 (2015)
27. Burgoyne, N., Jackson, R.: Predicting protein interaction sites: binding hot-spots in protein-protein and protein-ligand interface. Bioinformatics **22**(11), 1335–1342 (2006)

28. Li, B.Q., Feng, K.Y., Li, C., Huang, T.: Prediction of protein-protein interaction sites by random forest algorithm with mRMR and IFS. PLoS ONE **7**(8), e43927 (2012)

29. Lin, X., Zhang, X.: Identification of hot regions in protein-protein interactions based on detecting local community structure. In: Huang, D.-S., Bevilacqua, V., Premaratne, P. (eds.) ICIC 2016. LNCS, vol. 9771, pp. 432–438. Springer, Cham (2016). doi:10.1007/978-3-319-42291-6_43

30. Yugandhar, K., Gromiha, M.M.: Feature selection and classification of protein-protein complexes based on their binding affinities using machine learning approaches. Proteins Struct. Funct. Bioinform. **82**(9), 2088–2096 (2014)

31. Lin, X., Zhang, X.: Prediction and analysis of hot region in protein-protein interactions. In: BIBM 2016, pp. 1598–1603 (2016)

Fast Significant Matches of Position Weight Matrices Based on Diamond Sampling

Liang-xin Gao[✉], Hong-bo Zhang, and Lin Zhu

Institute of Machine Learning and Systems Biology, College of Electronics and Information Engineering, Tongji University, Caoan Road 4800, Shanghai 201804, China
liangxingao@tongji.edu.cn

Abstract. Position weight matrices are important method for modeling signals or motifs in biological sequences, both in DNA and protein contexts. In this paper, we present techniques for increasing the speed of sequence analysis using position weight matrices. Our techniques also permit the user to specify a p threshold to indicate the desired trade-off between sensitivity and speed for a particular sequence analysis. The resulting increase in speed should allow our algorithm to be used more widely in searching with large-scale sequence and annotation projects.

Keywords: Position weight matrices · Pattern search · Diamond sampling · Lookahead scoring

1 Introduction

Position weight matrices (PWMs), [1–3] are important statistical model for signals, such as transcription factor binding sites, in DNA and in other biological sequences [3]. Nowadays, between weight matrices and sequence, fast search of significant matches is a crucial requirement for biological sequence analysis tools, because of the exponential growth of both protein sequence databases and DNA as well as alignment block databases that the weight matrices can be synthesized (e.g., TRANSFAC [4], PRINTS [22], BLOCKS [5], and JASPAR [6]). Related area need high-performance matrix search tools, such as in genomewide analysis of gene regulation (e.g., 7).

The related algorithms of position weight matrix search should be divided into two direction that substantially differ in their approach, called the index-based algorithms and the online algorithms. The index-based algorithms can be allowed rapid accessing any location of the target sequence through a separately constructed index structure. And utilizing this special index structure, the index may provide a fast search at the cost of possibly large time and space requirement of the index construction. There are many proposed index-based algorithms which usually use suffix trees [8, 9] or suffix arrays [10] as the index. On the contrary, the online algorithms perform the search in a different way which search in one left-to-right scan through the target sequence. The more representative algorithm is lookahead approach [5] which takes advantage of score properties to devise partial thresholds that allow one to terminate the comparison of symbols as

D.-S. Huang et al. (Eds.): ICIC 2017, Part II, LNCS 10362, pp. 208–218, 2017.
DOI: 10.1007/978-3-319-63312-1_19

soon as it is clear that no match will occur. In this paper, we will introduce a promoted search algorithm of the online type for fast search of significant matches.

Actually, the problem, fast searching of significant matches between weight matrices and sequence, can also be transformed into a novel formation. The problem can be abstracted to matrix multiplication which one matrix is unfolded position weight matrix and another one's columns are composed by sequence of sliding window recoding on giving sequences. The original problem has been a maximum match between a vector and a matrix which is a classical k-Nearest Neighbor (KNN) process. With the development of machine learning, nowadays, many variously mature solutions for KNN have been developed. The latest related research is Diamond Sampling for approximate maximum all-pairs dot-product (MAD) searching [11].

Finding similar items is a fundamental problem that underlies numerous problems in data analysis. In real world, many entities are represented as vectors that is in high-dimensional feature space, i.e., $\mathbf{v} \in \mathbb{R}^d$, for some large d. Dot products are involved in many notions of similarity, hence, a usually useful measure of distance between $\mathbf{v}$ and $\mathbf{w}$ is $\mathbf{v} \cdot \mathbf{w}$. When given additional parameter t, more generally definition of those real questions is that find the t index pairs $\left\{ (i_1, j_1), \dots, (i_t, j_t) \right\}$ corresponding to the t largest dot products. Diamond sampling algorithm which is a new randomized approach is creative promoted to solve the MAD problem by applying index sampling methods. Designing a sampling procedure for the MAD problem is the ideal of diamond sampling algorithm. When we do a research in max matching between position weight matrix and sequence, we is also inspired by sampling ideal of diamond sampling algorithm.

Using sampling ideal of diamond sampling algorithm, the order of search sequences can be given permuted ordering instead of standard ordering which match between position weight matrix and sequences in left-to-right order. Permuted ordering is given by sorting the expectation of every column in position weight matrix and related expectation can be calculated through considering the information content of every column, distribution of background in sequence and so on.

We present in Sect. 3 an experimental comparison of the proposed algorithms such as the naive algorithm and the lookahead scoring algorithm.

2   Methods

2.1   The Search Problem

The involved problems, in this paper, is to find some appropriate matches of a position weight matrix in a sequences consisted of symbols in some finite alphabet $\mathfrak{R}$. A position weight matrix can be expressed as such $\mathbf{M} = (\mathbf{M}(j, a))$ and is a real valued $m \times |\mathfrak{R}|$ matrix. In other references, these matrices are also called, e.g., position weighted pattern, profiles, and position-specific scoring matrices. For convenience in this paper, we use matrix or pattern as a short-hand for position weight matrices. The pattern gives a score (weight) for each alphabet symbol a at each position j. We call $\mathfrak{R}$ the alphabet and m the length of the $\mathbf{M}$ (pattern). Table 1 is an example matrix.

Table 1. The pattern was obtained from the count matrix GATA-3 of JASPAR which was transformed into log-odds matrix using background distribution $q_A = 0.278, q_C = 0.312, q_G = 0.212,$ $q_T = 0.198$. A pseudocount q_α was first added to the counts for each alphabet symbol α.

A	0.14	−4.16	1.03	−4.16	0.58	−0.36
C	0.17	−2.31	−4.16	−4.16	−2.31	−1.32
G	−1.06	1.64	−2.32	−0.85	−1.06	1.12
T	0.12	−4.16	−2.64	1.18	0.07	−0.77

Any sequence $\alpha = \alpha_1 \ldots \alpha_m$ of length m in alphabet $\mathfrak{R}$ can be matched by pattern **M**. The degree of match is represented by the match score $W_M(\alpha)$ of α with respect to **M**. The score is defined as

$$W_M(\alpha) = \sum_{j=1}^{m} M(j, \alpha_j).$$

Given an n symbol long sequence $S = s_1 s_2 \ldots s_n$ in alphabet $\mathfrak{R}$. Pattern M given a match score for any m symbol long segment $s_i \ldots s_{i+m-1}$ (called an k-mers) of **S**. The match score of **M** at location i can be denoted as

$$\omega_i = W_M(s_i \ldots s_{i+m-1}).$$

In this paper, the problem studied can be described as follows: Given a real-valued significance threshold k, the weighted pattern search problem with threshold k is to find all location i of sequence S such that $\omega_i \geq k$. In addition to knowing the location i, many applications can be also interesting to knowing the values ω_i.

2.2   Preprocess of Pattern and Significance Thresholding

Under different background sequence distributions, even if the probability distribution of the alphabet symbols is the same, the conservativeness of the corresponding pattern is different. For this reason, in many applications, the weights $M(i, a)$ are in fact log-odds scores of a probabilistic model of a signal to be detected against the background, such as finding putative transcription factors binding sites in DNA. An $m \times |\mathfrak{R}|$ matrix Γ is normally defined as the signal model and $\Gamma(i, \alpha)$ gives the probability of the symbol α to occur in model position j. From the corresponding empirically constructed count matrix [10] possibly with added pseudocounts, these probabilities can be easily obtained.

In actuality, the background is usually involved in the model as an auxiliary model which is an $m \times |\mathfrak{R}|$ matrix Π in which each row have the same probability vector equaling the background probability distribution of the alphabet symbols. For convenience, we use q_α to represent the background probabilities for $\alpha \in \mathfrak{R}$. Hence, $\prod(i, \alpha) = q_\alpha$ for all j.

The log-odds score compares the probability to observe the segment in the background Π and to observe it in the signal model Γ. Thus, the evaluation of the log-odds score decide the match between a sequence and a matrix:

$$\text{Score}(\mathbf{v}) = \log \prod_{i=1}^{m} \frac{\Gamma(i, \mathbf{v}_i)}{\Pi(i, \mathbf{v}_i)} = \sum_{i=1}^{m} \log \frac{\Gamma(i, \mathbf{v}_i)}{\Pi(i, \mathbf{v}_i)}$$

$$= \sum_{i=1}^{m} \log \frac{\Gamma(i, \mathbf{v}_i)}{\mathbf{q}_{\mathbf{v}_i}}$$

If the background Π is fixed, such as estimating searching in the sequence S, that a positionally weight pattern $\mathbf{M}$ can be created from the model Γ:

$$\mathbf{M}(i, \alpha) = \log \frac{\Gamma(i, \alpha)}{\Pi(i, \alpha)}$$

Then, the score obtained as (1) equals the above $Score(\mathbf{v})$.

In a match between a sequence and a matrix, the significance threshold k is usually not easy to obtain and we normally obtain the significance threshold k indirectly by using the standard approach of statistics that adjust the confidence of the findings through a given p-value γ. The corresponding significance threshold k can be obtained from a mapping function $k = k(\gamma)$, called as the quantile function, such that the probability of sequences v such that $W_{\mathbf{M}}(\mathbf{v}) \geq k$ is γ in the background distribution. The mapping function $k = k(\gamma)$ can be implemented by using a well-known dynamic programming algorithm [5, 12], in a pseudopolynomial time.

2.3   Sampled Lookahead Scoring Algorithm

In this section, some well-know search algorithms that will be made an experimental comparison with our fast searching algorithm should be introduced first.

For given S, M and k, the score of every segment ω_i can be easily calculated from (1) and (2) for each $i = 1, 2, \ldots, n - m + 1$. After that, all locations i such that $\omega_i \geq k$ can be reported. This method which solve the weighted pattern search problem be called naïve algorithm (NA). In the implemented process of naïve algorithm, calculating each ω_i from (1) need takes time $O(m)$ such that total matching time of the naïve algorithm need takes time $O(mn)$ where n is the length of the sequence S and m is the length of the matrix.

The naïve algorithm ignore some significant relation between ω_i and significance threshold k. NA algorithm will calculated the value of ω_i, even if it has been previously known that the segment cannot be matched successfully. To utilize this significant relation between ω_i and significance threshold k, the lookahead scoring algorithm is proposed as an improved version of NA from [5]. In order to guarantee each prefix of a candidate segment to have the chance to be a match, the lookahead scoring algorithm precomputes the intermediate score thresholds. For given S, $\mathbf{M}$ and k, the intermediate score thresholds cannot be directly computed because the score of the prefix of a candidate segment is fixed. And yet it can be computed by using evaluating maximal score of the suffix of a candidate segment, defined for $0 \leq t \leq m-1$ as

$$\mathbf{p}_t = \sum_{i=t+1}^{m} \max_{\alpha \in \mathfrak{R}} \mathbf{M}(i, \alpha)$$

For the suffix starting at position $t + 1$ of any given candidate k-mers $\mathbf{s}_i \dots \mathbf{s}_{i+m-1}$, $\mathbf{p}_t$ is the maximum possible score that can be computed. The score of the prefix end at position t of candidate k-mers can be computed as

$$\mathbf{R}_t = W_M(\mathbf{s}_i \dots \mathbf{s}_{i-1+t}) = \sum_{j=1}^{t} \mathbf{M}(j, \mathbf{s}_{i-1+j})$$

If $R_t + P_t$ is below the matching significance threshold k, then we can make sure that the k-mers cannot be a match without considering the remaining symbols because the total score cannot be more than k for any possible k-mers in the given sequence S beyond the position t.

Obviously, the match between sequence and model will be speed up if utilize this relationship. In order to save the calculation of intermediate score thresholds time, the lookahead scoring algorithm first calculate all the intermediate thresholds

$$\mathbf{T}_t = k - \mathbf{P}_t$$

and save it into a array for $t = 0, \dots, m$. The actual process of match is in the form of a sliding window of length m and corresponding position i increment by 1, for example, the current match k-mers is $\mathbf{s}_i \dots \mathbf{s}_{i+m-1}$ and the next match k-mers will be $\mathbf{s}_{i+1} \dots \mathbf{s}_{i+m}$. Then, the strategy of match is that the score is initialized to 0 and the current term $\mathbf{M}(t + 1, \mathbf{S}_{i+t})$ is added to the score only if the current accumulated score is not less than $\mathbf{T}_t$. The search is stopped at the current i and resumed at $i + 1$ if the above-mentioned conditions are not satisfied.

In the first section, one significance concept is mentioned that the problem of matching between sequence and model can be transformed a special case of maximum all-pairs dot-product (MAD) searching problem. Unfortunately, the late research [11] of MAD problem can only give an approximate searching result rather than the precise result that we are looking for. However, the weight sampling concept of diamond sampling algorithm proposed in [11] is worth learning and used in the current searching problem.

In this paper, we propose a further improvement of lookahead scoring algorithm is called the sampling lookahead scoring algorithm (SLS) which utilize the weight relationship, which lookahead scoring algorithm ignored, between the various positions of the model. The order that the algorithm accumulates the scores of the model positions is not necessarily left-to-right but is a given order which is determined specifically for given model to archive faster average speed while the accuracy of the searching results is the same. Thus, unsuccessful k-merss can be detected as early as possible. The matching order is controlled by various factors including the information content of each position in model and the given background distribution. The expected loss is given directly at position i, defined as

$$\mathbf{L}_i = \left| \min_{\alpha \in \mathfrak{R}} \mathbf{M}(i, \alpha) \right| + \max_{\alpha \in \mathfrak{R}} \mathbf{M}(i, \alpha) - \sum_{\alpha \in \mathfrak{R}} \mathbf{q}_\alpha \mathbf{M}(i, \alpha)$$

Where $\mathbf{q}_\alpha$ is the background probability of α. And the matching order can be get by choosing the matrix positions in decreasing order of the expected loss $\mathbf{L}_i$ for $0 \leq i \leq m - 1$.

In theory, this improvement will make the gap of computing score between maximum possible score and the actual score by $\mathbf{L}_i$ more obvious when match at i. The purpose of this treatment is that the score of the prefix of k-mers, when matched in this order, will be the greatest possibility than the intermediate threshold $\mathbf{T}_t$ on average. As a result, the unsuccessful match with k-mers will be end as early as possible and the process of match will take minimal average time. Before matching, it need to be emphasized that the maximum possible score $\mathbf{p}_t$ has to be recalculated which the original left-to-right order should be replaced by the order, determined decreasing expected loss of each model locations.

3   Results

For the experimentation, we implemented the well-known base-line algorithms NA (naïve algorithm) and LS (lookahead scoring algorithm) as well as our SLS (sampling lookahead scoring algorithm). Moreover, using the software implementation by the original authors.

The reported running times are for 3.4 GHz Intel® Core™ i7-2600 processor with 5 gigabytes of main memory, running under Ubuntu 14.04 LTS. Our codes are written in C/C++ and the codes are all single-threaded. The compiler used in the experiments was gcc. We also experimented with 2.6 GHz Intel® Xeon® CPU e5-2650 v2, with essentially similar results (slight differences explained by different cache memory size).

3.1   Datasets

A flood of sequence data is appearing in the nucleotide sequence databases. To analyse these data, mathematical methods and computer algorithms are needed that are simple, logical and sel-consistent. A mathematics that fits these requirements and also connects directly to the physics underlying molecular binding interactions was created by Shannon with the introduction of information theory [13–15]. Information theory has been successfully used to quantify the sequence conservation in nucleotide and protein sequences [16–20]. The total information content (IC) represents the capability of the matrix to make the distinction between a binding site (represented by the matrix) and the background model. Since the information content of the real position weight matrix in gene database has been fixed, it is difficult to find the position weight matrix with a series of specified information content, so we will generate a series of position weight matrix based on the theory of information content. The experiment on real-world and artificial datasets described below.

- *Artificial Sequence S1*: since the background distribution of the sequences in the gene database is usually nonuniform, we try to generate some artificial gene sequences to test the performance of the algorithm. The target sequence was a 50 megabases long amino acid sequence that was randomly created under the condition which the background distribution is similar to $\mathbf{q}_A:\mathbf{q}_C:\mathbf{q}_G:\mathbf{q}_T = 1:1:1:1$. There are only four kinds of alphabets in the sequence, which are A, C, G and T.

- *Artificial Model M1*: Staden (1989) [12] described an efficient method for numerically estimating the probability-generating function for weight-matrix scores based on the theory of information content. If the probability-generating function for the information content of an individual column of an alignment matrix is known, Staden's approach can be directly used to approximate the probability-generating function for a multi-column alignment matrix [21]. According Staden's efficient method, we generate 26 position weight matrix which the length is fixed as 22 and corresponding information content is from 5 to 30. In addition, in order to eliminate the random error caused by the single performance test, we generated 90 sets of the above dataset and averaged as the real performance of the algorithm.

- *Artificial Model M2*: in order to test the effect of the length of the position weight matrix on the performance of the algorithm, and in order to eliminate the interference of the information content to the performance of the algorithm, we fix the information content value at 70% of the maximal information content of the corresponding position weight matrix, and the length of position weight matrix is from 5 to 30. As the same reason, we generated 90 sets of the above dataset and averaged as the real performance of the algorithm.

- *Real-world Sequence S2*: as the target sequence to search the model occurrences we used a 50 megabases long DNA sequence. The sequence contained segments taken from the human and the mouse genome.

- *Real-world model M3*: we collected 368 positionally weight matrices for DNA from the JASPAR database (JASPAR CORE REDUNANT 2016) [6]. The length of these matrices varied from 5 to 21, with average length 13.2. These matrices are divided into 25 groups which the length of matrix contained in each of these groups is from 5 to 21 when some matrices are divided into multiple groups.

3.2 Time and Accuracy Performance

The run time results for searching artificial matrices are summarized in Figs. 1, 2 and for searching Jaspar matrices are summarized in Fig. 3. The matching sequence found by the each algorithm participating in the test is same when we set the same threshold.

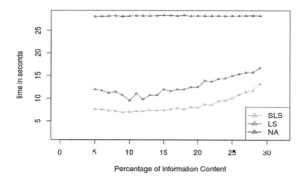

Fig. 1. Average) of different algorithms for model M1 with p-value $\gamma = 0.0001$

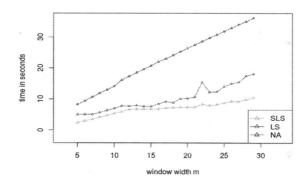

Fig. 2. Average running times (in seconds, preprocessing excluded) of different algorithms for model M2 with p-value $\gamma = 0.0001$

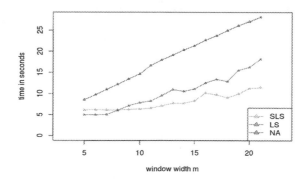

Fig. 3. Average running times (in seconds, preprocessing excluded) of different algorithms for real-world model M3 and sequence S2 with p-value $\gamma = 0.0001$.

The significance threshold k for the experiments was given by a p-value, thus, we experimented to searching a model from JASPAR in sequence M2 with varying p-value and average running times are described by Table 2. The sampled lookahead scoring algorithm always give the best total performance especially for small γ.

Table 2. Average running times (in seconds, preprocessing excluded) of different algorithms for DNA pattern (m = 21) from JASPAR and varying p-values (*Each reported time is an average of 10 runs.*)

γ	10^{-1}	10^{-2}	10^{-3}	10^{-4}	10^{-5}	10^{-6}
NA	34.83	32.53	32.48	32.99	33.85	34.02
LS	30.48	24.06	21.22	19.30	18.08	17.19
SLS	**22.39**	**16.43**	**13.26**	**11.67**	**9.92**	**9.32**

In order to test the behavior of the information content of position weight matrix on the speed of sequence matching, we experimented with model subset M1 of information content from 5 to 30 on the artificial sequence S1 when p-value is given $\gamma = 0.0001$ and the average times are described as Fig. 1. Obviously, the sampled lookahead scoring algorithm is the fastest among contrastive algorithms and the average time as the growing information content slowly increases can be interpreted as that the number of prefix of k-mers attained the requirements will become more, the matching time will increase when the information content is high.

To test the behavior of the length on the sequence match, we run the all algorithms for the searching between model M2 and the artificial sequence S1. The average running times are described as Fig. 2. The sampled lookahead scoring algorithm is still the fastest among contrastive algorithms. Although the average running times increased as the increasing length of position weight matrix, the distinction of average running times between lookahead scoring algorithm and its improved algorithm is more significant.

To test the performance of our algorithm in the real-world databases, we also experimented speed of matching between real-world models M3 and real-world sequence S2. The average running times is described by Fig. 3 and the sampled lookahead scoring algorithm is still the fastest among contrastive algorithms when the length of matrix is more than 8. When the length of matrix is relatively short, the higher average running times can be interpreted as that the number of matching is too short to make the impact of order of matching be non-significant.

4   Conclusion

It should be emphasized that the practical performance of the algorithms studied here quite strongly depends on the implementation details and on the properties of the memory hierarchy of the computer used. However, the sampled lookahead scoring algorithm give a clear speed-up in the case of DNA matrices when it compare to the naïve search (NA) and the lookahead search (LS). Due to the exponential growth of both protein sequence databases and DNA as well as alignment block databases that the weight matrices can be synthesized, the speed-up of sampled lookahead scoring algorithm will be more significant and save more time.

Acknowledgments. This work was supported by the grants of the National Science Foundation of China, Nos. 61672203, 61402334, 61472282, 61520106006, 31571364, U1611265, 61472280, 61532008, 61472173, 61572447, 61373098 and 61672382, China Postdoctoral Science Foundation Grant, Nos. 2016M601646. De-Shuang Huang is the corresponding author of this paper.

References

1. Gribskov, M., Mclachlan, A.D., Eisenberg, D.: Profile analysis: detection of distantly related proteins. Proc. Natl. Acad. Sci. U.S.A. **84**(13), 8 (1987)
2. Zhu, L., Guo, W.L., Deng, S.P., Huang, D.S.: ChIP-PIT: enhancing the analysis of ChIP-Seq data using convex-relaxed pair-wise interaction tensor decomposition. IEEE/ACM Trans. Comput. Biol. Bioinf. **13**(1), 55–63 (2016)
3. Stormo, G.D., Schneider, T.D., Gold, L.M., Ehrenfeucht, A.: Use of the "perceptron" algorithm to distinguish translational initiation sites in E. Coli. Nucleic Acid Res. **10**, 299–3012 (1982)
4. Matys, V., Fricke, E., Geffers, R., Gossling, E., Haubrock, M., Hehl, R., Hornischer, K., Karas, D., Kel, A.E., Kel-margoulils, O.V., Kloos, D.U., Land, S., Lewicki-potapov, B., Michael, H., Munch, R., Reuter, I., Rotert, S., Saxel, H., Scheer, M., Thiele, S., Wingender, E.: TRANSFAC: transcriptional regulation, from patterns to profiles. Nucleic Acids Res. **31**(1), 374–378 (2003)
5. Wu, T.D., Neville-manning, C.G., Brutlag, D.L.: Fast probabilistic analysis of sequence function using scoring matrices. Bioinformatics **16**(3), 233–244 (2000)
6. Sandelin, A., Alkema, W., Engstrom, P., Wasserman, W.W., Lanhard, B.: JASPAR: an open-access database for eukaryotic transcription factor binding profiles. Nucleic Acids Res. **32**, D91–D94 (2004)
7. Hallikas, O., Palin, K., Sinjushina, N., Rautiainen, R., Partanen, J., Ukkonen, E., Taipale, J.: Genome-wide prediction of mammalian enhancers based on analysis of transcription-factor binding affinity. Cell **124**, 47–59 (2006)
8. Dorohonceanu, B., Neville-Manning, C.G.: Accelerating protein classification using suffix trees. In: Proceedings of Eighth International Conference on Intelligent Systems for Molecular Biology (ISMB), pp. 12–133 (2000)
9. Schones, D.E., Smith, A.D., Zhang, M.Q.: Statistical significance of cis-regulatory modules. BMC Bioinform. **8**, 19 (2007)
10. Beckstette, M., Strothmann, D., Homann, R., Giegerich, R., Kurtz, S.: PoSSuMsearch: fast and sensitive matching of position specific scoring matrices using enhanced suffix arrays. In: Proceedings of German Conference on Bioinformatics, pp. 53–64 (2004)
11. Ballard, G., Seshadhri, C.: Diamond sampling for approximate maximum all-pairs dot-product (MAD) search. In: IEEE International Conference on Data Mining, pp. 11–20 (2015)
12. Staden, R.: Methods for calculating the probabilities of finding patterns in sequences. Comput. Appl. Biosci. **5**(2), 89–96 (1989)
13. Shannon, C.E.: A mathematical theory of communication. Bell Syst. Tech. **379–423**, 623–656 (1948)
14. Pierce, J.R.: An Introduction to Information Theory: Symbols, Signals and Noise. Dover Publications, New York (1980)

15. Zhu, L., Deng, S.P., Huang, D.S.: A two stage geometric method for pruning unreliable links in protein-protein networks. IEEE Trans. Nanobiosci. **14**(5), 528–534 (2015)
16. Zhu, L., You, Z.H., Huang, D.S., Wang, B.: t-LSE: a novel robust geometric approach for modeling protein-protein interaction networks. PLoS ONE **8**(4), e58368 (2013). doi:10.1371/journal.pone.0058368
17. Papp, P.P., Chattoraj, D.K., Schneider, T.D.: Information analysis of sequences that bind the replication initiator RepA. J. Mol. Biol. **233**, 219–230 (1993)
18. Schneider, T.D.: Protein patterns as shown by sequence logos. In: Visual Cues-Practical Data Visualization, p. 64. IEEE Press, Piscataway (1993)
19. Pietrokovski, S.: Searching databases of conserved sequence regions by aligning protein multiple-alignments. Nucl. Acids Res. **24**, 3836–3845 (1996)
20. Blom, N., Hansen, J., Blaas, D., Brunak, S.: Cleavage site analysis in picornaviral polyproteins: discovering cellular targets by neural networks. Protein Sci. **5**, 2203–2216 (1996)
21. Hertz, G.Z., Stormo, G.D.: Identifying DNA and protein patterns with statistically significant alignments of multiple sequences. Bioinformatics **15**(7), 56–577 (1999)
22. Attwood, T.K., Beck, M.E.: PRINT—a protein motif finger-print database. Protein Eng. **7**(7), 84–848 (1994)
23. Deng, S.-P., Zhu, L., Huang, D.S.: Mining the bladder cancer-associated genes by an integrated strategy for the construction and analysis of differential co-expression networks. BMC Genom. **16**(Suppl 3), S4 (2015)
24. Deng, S.-P., Huang, D.S.: SFAPS: an R package for structure/function analysis of protein sequences based on informational spectrum method. Methods **69**(3), 207–212 (2014)
25. Huang, D.S., Zhang, L., Han, K., Deng, S., Yang, K., Zhang, H.: Prediction of protein-protein interactions based on protein-protein correlation using least squares regression. Curr. Protein Pept. Sci. **15**(6), 553–560 (2014)
26. Wang, B., Huang, D.S., Jiang, C.: A new strategy for protein interface identification using manifold learning method. IEEE Trans. Nanobiosci. **13**(2), 118–123 (2014)

Combination of EEG Data Time and Frequency Representations in Deep Networks for Sleep Stage Classification

Martí Manzano, Alberto Guillén, Ignacio Rojas, and Luis Javier Herrera[✉]

Department of Computer Architecture and Computer Technology,
University of Granada, Granada, Spain
`jherrera@ugr.es`

Abstract. Almost all of the studies in the literature of sleep stage classification are based on traditional statistical learning techniques from a set of extracted features, which need a relative amount of time and effort. Deep learning offers approaches able to automatically extract patterns and abstractions from different types of data (images, sound, biomedical signals, etc.) to perform classification. However, the application of these techniques in the automatic sleep stage scoring field is less widespread to date. This paper proposes a new approach based on a multi-state deep learning neural network architecture, which we named Asymmetrical Multi-State Neural Network. This new network is able to merge two different neural networks, based on two different architectures receiving different input data: single-channel EEG raw signal in time and the respective spectrum. The proposed Asymmetrical Multi-State Neural Network shows to enhance the separated networks' performance for the given problem on a complete well-known sleep database.

Keywords: Deep learning · Sleep stage classification · Multi-State neural networks

1 Introduction

Sleep is an basic activity in human daily life. During the sleep stages, human body not only performs auto-regulatory activities such as toxins elimination, but for instance several memory consolidation processes enabling learning take place. Cerebral activity during sleep can be measured through a technique called Electroencephalography (EEG), in which a set of electrodes are placed in the patient head and connected to a computer. The capture of EEG together with other signals such as Electrooculography -eye movement- (EOG), Electromyography -muscles movement- (EMG), and breath signals (among others), is called Polysomnography analysis. Diagnosis and monitoring of illnesses such as Alzheimer, epilepsy, sleep disorders and brain dead, can be carried out thanks to this technique.

Criteria for sleep stages classification were standardized for the first time in a manual written by Rechtschaffen and Kales (R&K) in 1968 [1]. The sleep stage set established then included: wakefulness, REM stage (Rapid Eye Movement) and the NREM stage

© Springer International Publishing AG 2017
D.-S. Huang et al. (Eds.): ICIC 2017, Part II, LNCS 10362, pp. 219–229, 2017.
DOI: 10.1007/978-3-319-63312-1_20

(no Rapid Eye Movements stage), which is composed of four differentiated stages (NREM1, NREM2, NREM3 and NREM4). In 2007, a revised manual by R&K was published [2], which reduced the number of NREM stages to three, joining stages NREM3 and NREM4 into a single stage named Slow Wave Sleep (SWS). The reason was the lack of physiological differences among both stages.

The way an hypnogram (series of stages a patient stays along a full sleep period) is obtained from a polysomnogram, is based in the subdivision of the signals in periods of 30 s called epochs, and assigning a single sleep stage to each one of the epochs. In case a transition from one stage to another is observed within an epoch, the most predominant stage in time is assigned. In this manual process, specialists have to deal with difficult decisions in the assignment. Moreover differences in criteria among sleep specialists are normally found, and even the same doctor can perform different hypnogram analysis when given the same polysomnography data in different time moments [3].

Many Machine Learning (ML) techniques have been applied to the problem of sleep stage classification. Most of the published works are based on a preliminary feature extraction from the polysomnogram signals, and later classification [4, 5]. This work moves one step ahead in the automatic sleep stage classification using deep learning as alternative, which avoids the preliminary separated stage of feature extraction. This way, possible loss of information is avoided as the raw biomedical signals are directly processed. This work expands the few deep learning approaches appeared in the literature for this problem [6], in order to gain in understanding and usability of this technique in the application.

Specifically, this work proposes a novel Deep Learning classification model, able to extract a set of features from raw EEG signal from a single channel to perform automatic prediction of sleep stages. It will use separated time and frequency processing in different deep networks (see previous work in [7]) as subnetworks in a joined Multi-state neural network architecture which we named Asymmetrical Multi-State Neural Network. The performance of the proposed deep architecture was assessed over the complete database of 25 patients from the Physionet database [6] using the updated R&K rules.

The rest of the paper is organized as follows: Sect. 2 presents the background and state of the art of the problem. Section 3 introduces the deep learning architecture and the methodology used. Section 4 presents the results obtained in the Physionet database [8]. Section 5 concludes the work.

2 Background

Research work in Automatic Sleep Stage Classification is extensive. As mentioned, most of the revised studies by the authors, follow a classic methodology of feature extraction (followed sometimes by feature selection processes) from the polysomnography and a later design and application of a classification technique. Several alternatives arise in each stage: signal selection (EEG, EMG, EOG, breathing,...), feature extraction (statistical features like mean, variance of the time signals, Hjorth features, wavelets and other signal transformations, etc.), type of classifier (neural networks, fuzzy logic, Linear Discriminant Analysis, etc.). Among the signal selection alternatives, it is to be highlighted that the lowest number of signals needed to perform the classification is in principle desired. This way, data capturing and modeling is simpler, involving cheaper devices, which would for instance promote and facilitate at-home monitoring systems.

With respect to the feature extraction methods, in [4] for instance spectral relative power (RSP) is extracted from the FFT of the EEG signal, from the spectrum previously divided in the relative bands associated to brain activity during sleep (*Alpha, Theta, Delta, Sigma, Beta*). From the RSP a single hidden-layer neural network is designed to perform classification. Other works include combination of different features extraction systems to enhance the classification accuracy [5].

Few works have appeared in the literature applying deep learning techniques to the problem, due to its relative novelty. In [6], a set of experiments using Deep Learning techniques are performed on a dataset of 25 patients. Hand-extracted features are compared with Deep Belief Network extracted features from EEG, EOG and EMG. The deep learning approach attains a 67.4% of accuracy in patient cross-validation, needing a data balance process. However this work did not follow the R&K rules as epochs were 1 s long and transition epochs were eliminated.

Other recently appeared works apply Convolutional Neural Networks on the time domain together with data from previous and following epochs [9] to achieve a 74% of overall accuracy using a selection of 20 patients. In a different work [10], a group of 10 patients is used to train a deep belief network to extract features from the polysomnogram signals (2 EEG channels, EOG and EMG) to attain a 91% of patient CV accuracy. The search for a model using a single EEG channel and the attainment of a robust expected performance measure, using a complete well-known database and patient cross-validation, motivated the presented work.

In a preliminary work [7] the authors compared two deep learning approaches, working in frequency and time domains of single channel EEG epochs. This work extends it and evaluates the combined use of time and frequency domains. A complete database (thus reflecting all possible signal variations among different patients), the Physionet database [6] will be used. Temporal connectivity in the time domain will be operated using a Convolutional Network architecture. Joined use of the two operations over the two domains will be evaluated. In this work only information from the current epoch from a single EEG channel will be used in each epoch classification. Patient cross-validation will be used as the performance measure in the problem.

3 Methodology

3.1 Deep Learning

Deep Learning can be defined as a set of algorithms and statistical models which form part of Statistical Learning and Artificial Intelligence [11]. These algorithms are based on deep neural networks, i.e., neural networks with several layers. They are often trained and used over a large amount of data; due to their inherent complexity, they need more data samples than simpler neural networks models.

One of the most important characteristics of deep learning models is their capability of extracting patterns/properties/data representations thanks to the neural processing layers. This has made this type of classifiers specially popular and useful in image/sound processing. Nevertheless there are several complex deep learning architectures with particular modeling properties. For instance, apart from fully-connected networks, convolutional neural networks, which perform convolution operations over the inputs, and recurrent neural networks, which present neuron connections that can form loops, are other well-known types of deep learning architectures.

In this work we omit many details on Deep Neural Network design and optimization, but readers unfamiliar with Deep Learning operations, can read [11] to get an excellent background on the topic.

3.2 Convolutional Neural Networks

Convolutional Neural Networks (CNNs) are networks composed by a set of connected convolutional layers [7]. The convolution operation applies a filter (or kernel) over a region of the input space. The size of the convolutions is specified during the design phase, but the content of the filter (what is being looked for) is learned during the training from the regional characteristics that the input data presents. Convolutional networks, unlike fully-connected feed forward neural networks, share the synaptic weights. This way, CNNs are suitable to operate with data presenting local connectivity such as images, sound or time series, as their spatial structure can provide pattern and abstraction extraction of different sizes and levels.

The three components of a convolutional neural network are [7, 11]:

- Convolutional layer: It applies a filter over the input space: in a one dimensional problem, along that one; in a 2-D image, along the width and height. These layers represents the main difference with multilayer perceptron. Filters must have lower dimension than the input space size in order to detect the presence or absence of patterns. Filters can be applied several times in a deep set of layers.
- Pooling layer: The pooling layers replace the output of the convolutional layer in a specific point by a combination of its neighborhood. Thus they reduce the spacial dimensions of their inputs. Although loss of information can occur, they are generally convenient so as to reduce over-fitting and diminishing the dimensional complexity of the data flow for the next layers.
- Fully-connected layer: After the successive operation over a set of convolutional and pooling layers, and in order to extract the relevant information from the high-dimensional data, a final higher level processing is carried out through one or several fully-connected layers. These layers will be directly connected to the output layer to perform the classification itself.

3.3 Assessment of the Models

As the dataset available for the problem consists of overnight sleep EEG data from 25 patients [8], assessment of the models has been performed using patient-cross-validation (leave one patient out at a time), as it is expected to provide a better accuracy over any new unseen patient in the extraction of the hypnogram. This technique, as a leave-one-out one, leaves the whole data from one patient out of the model training, and tests it on the left patient; this process is repeated for each of the patients and the mean test performance of the executions is provided. This way, any expectable patient variability in EEG signals recorded or in their manual classification, is reflected in the expected performance of the model, making it a more realistic performance measure. This measure has been used in most of the latest revised works [5, 9, 10].

3.4 Implementation Issues

All experiments have been performed in Python using the *TensorFlow* library [12], which is the innovative deep learning tool provided by Google and widely used in the Deep Learning community. Other libraries such as *NumPy*, *pyEDFlib* (for EDF file reading and processing), *scikit-learn* for quality metrics, *SciPy* (for Short-Time Fourier Transforms with Hanning window) and *Matplotlib* for figures and graphs creation and visualization, have been used.

4 Results and Discussion

4.1 Dataset

Dataset used for the experiments was extracted from the public repository *PhysioNet* [8]. A total of 25 overnight polysomnogram (21 male and 4 female, with ages between 28 and 68-mean age of 50 years old) from the *St. Vicent's University Hospital*, from whom it was suspected that they suffered from any sleeping disturbance, were considered. Recorded polysomnogram signals included: two EEG channels (C3-A2 y C4-A1) at 128 Hz, two EOG -two eyes- at 64 Hz, EMG also at 64 Hz, and other signals related to patient movement, posture and breathing.

Data was provided under EDF format and single EEG channel (C3-A2) was used in this work, which agrees with recommendations in the manual of sleep stage classification of R&K. Before epoch segmentation, preprocessing using a Notch filter at 50 Hz and a high pass filter at 0.3 Hz [6, 13] was performed. Also signal was down-sampled to 64 Hz to reduce the input data size and the temporal correlation among the variables. Each epoch's corresponding sleep stage was tagged by a single expert.

Average sleep data duration from the patients is 6.9 h. In total, 20075 epochs are available from the 25 patients, with a total operation time of 173.2 h. Certain data imbalance was observed as class NREM2 is the most frequent one with a 33.6% of the total epochs classification. The second most frequent stage is the awaken one with a 22.36% of the total, which can be comprehensible as the patients analyzed present symptoms of sleep disorders. SWS stage (union of NREM3 and NREM4) remains with a 12.81% of the total, NREM1 with a 16.37% and REM with a 14.5%.

4.2   Time Domain Classification Through a Convolutional Neural Network

Local connectivity among the variables in the time domain was the objective aimed to take advantage of by using a Convolutional Neural Network. This type of architecture has been previously used with success on different types of signals [14, 15, 16, 17]. Thus taking these as reference, temporal structures from the data were aimed to be extracted automatically using this deep learning technique. Two convolutional layers with pooling were used in order to control the size of the successive input space maps; number of filter and their sizes, strides etc. are shown in Table 1. These hyperparameters were obtained by trial and error. These filter sizes allowed to extract time patterns from lower frequency bands (Delta waves from 0.5–3 Hz) until higher frequency bands (*sleep spindle* or sigma patterns until 16 Hz). Finally a single fully-connected layer to the output was used.

Table 1. Hyperparameters of the designed deep neural networks in the work.

Hyperparameter	Value
Filter size in the first convolutional layer[a,c]	8
Number of filters in the first layer (depth)[a,c]	9
Stride of the first convolutional network[a,c]	2
Stride of the first max-pooling layer[a,c]	2
Filter size in the second convolutional layer[a,c]	16
Number of filters in the second layer (depth) [a,c]	18
Stride of the second convolutional network[a,c]	2
Stride of the second max-pooling network[a,c]	2
Number of neurons in the fully-connected layer[a]	2500
Max. number of iterations in the batch gradient descent[a,b,c]	2000
Batch size[a,b,c]	512
Learning rate[a,b,c]	0.01
Dropout Regularization Keep-rate[a,b,c]	0.5
Number of neurons in the fully-connected layer[b,c]	1600
Number of connections of the concatenation layer[c]	3760
Number of neurons of the fully-conn. bottleneck layer[c]	200

[a]EEG time domain convolutional network
[b]EEG frequency domain fully-connected network
[c]Asymmetrical Multi-State Neural Network

An initial subdivision of training and test from a subset of the data was used for hyperparameter obtaining. Later, patient cross-validation was performed for their final tuning. This holds for the rest of the experiments presented in the paper.

The resulting convolutional neural network architecture designed can be seen in Fig. 1. Most relevant hyper-parameters of the model are detailed in Table 1.

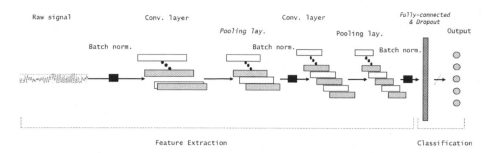

Fig. 1. Neural network architecture operating in the time domain.

This convolutional network when validated using patient-cross-validation, attained a 68.6% of precision and a 54.6% of mean F-measure over the five sleep stages [7].

4.3 Frequency Domain Classification

Spectral analysis over a signal may allow the detection of frequency patterns; visually it can be used to reveal the predominance or absence of characteristic patterns and complexes from the sleep stages.

The use of classifiers based on manually extracted features from the EEG was performed for instance in [4] for sleep stage classification with relatively positive results. However using the raw spectrum of the signal to perform direct classification is a less explored field, which can be tackled by Deep Learning techniques.

A direct neural network working with the raw spectrum obtained by Short-Time Fourier Transform STFT (0.5 Hz to 32 Hz, corresponding to the variation of physiological waves [4]) on 30 s epoch was designed. In order to reduce the width of the limit discontinuities (leakage), a Hanning window was used together with the FFT (which according to our tests provided the best results).

According to [18], spectral phenomena happening in different frequency regions are different, thus weight sharing in the network only have sense on a limited bandwidth. A convolutional network wouldn't therefore be that effective in this problem. So, a fully-connected network with a hidden layer of 1600 neurons using Dropout as regularization to reduce the over-fitting was designed and trained. Most relevant hyper-parameters for this network can be seen in Table 1.

Patient leave-one-out cross-validation led to a 68.9% of accuracy and 57.5% of mean F1 measure. These results are similar to that of the previous convolutional network working in the time domain [7].

4.4 Combined Time-Frequency Domains Classification

The underlying idea after comparing time and frequency domains classifications, is to verify if integrating both sources of information for sleep stage classification may lead to an increase in classification accuracy. Research works such as [17, 19–21] propose classification models based on the two domains after the application of Discrete Wavelet

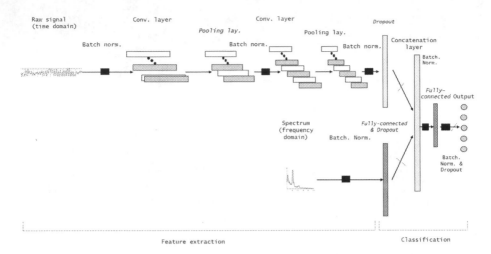

Fig. 2. New network architecture operating on both time and frequency domains. Light gray layers are simple connection-layers (no neurons associated).

Transform performing feature extraction on both domains; this work uses deep learning to perform classification from the raw data.

Networks combining different inputs from the same data source have been explored in the literature, they are usually called multi-state networks. In [22, 23], for instance, convolutional networks for image classification are proposed combining different processings over the same input images. In [24], however, a multi-channel time series classification model is presented which makes use of a convolutional network with several input data channels applying the same convolutional operations over all the channels independently and combining the generated extracted features in a fully-connected set of layers.

In this work, a multi-state deep network combining a convolutional sub-network operating over EEG raw data, and a fully-connected network operating over spectrum was designed. To combine the extracted features from both sub-networks, a concatenation layer was implemented (simply joins the outputs from both sub-networks). This network is able to learn from the data similarly as a simple deep learning architecture would: the concatenation layer allows the back-propagation of the gradients for successive adjustment of the synaptic weights. The successive batch normalization processes allow the processing of the two representations of the input signals on the same scale. This architecture is called *Asymmetrical Multi-State Neural Network* or AMSNN, and it is outlined in Fig. 2.

It can be observed that the designed network is composed by the main parts of the separated designs of the two previous neural networks, but the last fully-connected layer of the time-domain subnetwork which was eliminated. Other modifications such as the concatenation layer, allow combining the features extracted from each specific domain; other include modifications in the placement of the activation functions and Dropout. Moreover, a bottleneck-type fully-connected layer is added to the concatenation layer, whose size is much lower than the previous ones. This way, the network training forces the activations of the mentioned last layer to obtain a compact representation of the mixed time-frequency input signals [25]. Finally activations are passed through Batch Normalization and Dropout processes as ways to normalize and regularize the network. Obtained Hyperparameters of the AMSNN are similar to those of the separated networks (see Table 1), but adding two fully connected layers, the first concatenation layer with 3760 neurons and the bottleneck layer with 200 neurons.

The new AMSNN was trained and validated similarly as for the previous two networks, using patient-cross-validation. Accuracy reached 73.2%, i.e., 4.3 per cent above the frequency domain network and 4.6 over the time domain network. In F1 measure, a 64% was attained, which corresponds to 6.5 and 9 per cent increase respectively with respect to frequency domain and time domain networks (see Table 2).

Table 2. Summary of the results of the three proposed Deep Networks, and comparison with other Deep Learning approaches working with reduced Polysomnography information.

Model	Database	Acc	Std acc	Mean F1
Manzano et al. [7] time domain (EEG 1-channel)	a	68.6%	7.53	54%
Manzano et al. [7] frequency domain (EEG 1-channel)	a	68.9%	7.52	57.5%
Proposed AMSNN double domain (EEG 1-channel)	a	73.2%	6.95	64%
Martin Langkvist et al. [6] (EEG 1-ch, EMG, EOG)	a	67.4%	12.9%	65%
Orestis Tsinales et al. [27] (EEG 1-channel)	b	74%	–	–

Databases: [a]Physionet 25 patients with sleep disorders, [b]Physionet 20 healthy subjects.

In relation to the patient-cross-validation results and the confusion matrix associated, two interesting facts are observed, that also were seen in [6]. In most patients, NREM1 stage was the worst classified, which can be due to the little physiological differences among this one and NREM2 stage [26] as well as the special consideration of the manual classification of the two stages. Second, there is a high variability in the classification accuracies per patient, being the two worst ones 51.3% and 65.3%, and the best ones 85.3% and 83.6%. These differences are expectable due to the different interferences in the polysomnogram recordings, as well as the inter-individual physiological and brain operation differences.

5   Discussion on the Results and Conclusion

As it was shown, the AMSNN network proposed for sleep stage classification, working both in the time frequency domains of a single-channel EEG signal, attains a better performance than the separate networks working in each domain. Due to the complexity of the problem, the 73.2% accuracy obtained in patient-cross-validation is considered very promising taking into account that it uses a single EEG channel with no more information, and it is not making use of other epochs' information to classify the current epoch. Moreover the database used considers patient with sleep disturbances from which larger variations among patients can be expected. Table 2 shows a comparative among different deep learning approaches for sleep stage scoring under a similar framework.

Other works, also working with deep learning techniques include [10] using features extracted using a Deep Belief Network and classifying using an ensemble of classifiers. It attains a very good accuracy (91%, see Sect. 2), although but they used EEG and EOG information, and considered a smaller subset of patients from which inter-patient variability is expected to be lower. Moreover a different work [27] attained greater results (82% of overall accuracy) by considering transition patterns among different sleep stages in their operation, information which was not considered in the present work.

Thus, as future work it is intended to make use of stages transition information in order to enhance the results (for instance Stacked Sequential Learning [5], which demonstrated to highly increase the performance of the sleep stage classification problem, or the approach taken in [27]). Also it is intended to perform a deeper comparison among different databases of healthy and patients with sleep disturbances [28].

References

1. Rechtschaffen, A., Kales, A.: A manual of standardized terminology, techniques and scoring system for sleep stages of human subjects. Public Health Service, U.S. Government Printing Office, Washington, D.C. (1968)
2. American Academy of Sleep Medicine. The AASM Manual for the Scoring of Sleep and Associated Events: Rules, Terminology and Technical Specification (2007)
3. Norman, R., Pal, I., Stewart, C., Walsleben, J., Rappaport, D.: Interobserver agreement among sleep scorers from different centers in a large dataset. Sleep 23, 901–908 (2000)
4. Kerkeni, N., Alexandre, F., Bedoui, M.H., Bougrain, L., Dogui, M.: Automatic classification of sleep stages on a EEG signal by artificial neural networks. In: 5th WSEAS International Conference on SIGNAL, SPEECH and IMAGE, Wisconsin, USA, pp. 1–13 (2005)
5. Herrera, L.J., Fernandes, C.M., Mora, A., Migotina, D., Largo, R., Guillén, A., Rosa, A.: Combination of heterogeneous EEG feature extraction methods and stacked sequential learning for sleep stage classification. Int. J. Neural Syst. 23(3), 1350012 (2013)
6. Längkvist, M., Karlsson, L., Loutfi, A.: Sleep stage classification using unsupervised feature learning. Adv. Artif. Neu. Sys. 5, 5 (2012)
7. Manzano, M., Guillén, A., Rojas, I., Herrera, L.J.: Deep learning using EEG data in time and frequency domains for sleep stage classification. In: Rojas, I., Joya, G., Catala, A. (eds.) IWANN 2017. LNCS, vol. 10305, pp. 132–141. Springer, Cham (2017). doi: 10.1007/978-3-319-59153-7_12
8. Goldberger, A.L., Amaral, L.A.N., Glass, L., Hausdorff, J.M., Ivanov, P.C., Mark, R.G., Mietus, J.E., Moody, G.B., Peng, C.-K., Stanley, H.E.: PhysioBank, PhysioToolkit, and PhysioNet: components of a new research resource for complex physiologic signals. Circulation 101(23), e215–e220, 13 June 2000. doi:10.1161/01.CIR.101.23.e215. Circulation Electronic Pages: http://circ.ahajournals.org/content/101/23/e215.full PMID:1085218
9. Tsinalis, O., Matthews, P.M., Guo, Y., Zafeiriou, S.: Automatic sleep stage scoring with single-channel EEG using convolutional neural networks. Biomed. Eng./Biomed. Tech. arXiv:1610.01683 [stat.ML] (2016)
10. Zhang, J., Wu, Y., Bai, J., Chen, F.: Automatic sleep stage classification based on sparse deep belief net and combination of multiple classifiers. Trans. Inst. Measur. Control 38(4), 435–451 (2015)
11. Goodfellow, I., Bengio, Y., Courville, A.: Deep Learning. MIT Press (2016). http://www.deeplearningbook.org

12. Abadi, M., et al.: TensorFlow: Large-scale machine learning on heterogeneous systems (2015). Software available from tensorflow.org
13. Reddy, A.G., Narava, S.: Artifact removal from eeg signals. Int. J. Comput. Appl. **77**(13), 17–19 (2013)
14. Palaz, D., Magimai-Doss, M., Collobert, R.: Analysis of cnn-based speech recognition system using raw speech as input. In: Proceedings of Interspeech, number Idiap-RR-23-2015, pp. 11–15 (2015)
15. Yang, J.B., Nguyen, M.N., San, P.P., Li, X.L., Krishnaswamy, S.: Deep convolutional neural networks on multichannel time series for human activity recognition. In: Proceedings of the 24th International Conference on Artificial Intelligence, IJCAI 2015, pp. 3995–4001, AAAI Press (2015)
16. Abdel-Hamid, O., Mohamed, A. R., Jiang, H., Penn, G.: Applying convolutional neural networks concepts to hybrid nn-hmm model for speech recognition. In: 2012 IEEE International Conference on Acoustics, Speech and Signal Processing (ICASSP), pp. 4277–4280 (2012)
17. Tóth, L.: Combining time and frequency-domain convolution in convolutional neural network-based phone recognition. In: 2014 IEEE International Conference on Acoustics, Speech and Signal Processing (ICASSP), pp. 190–194 (2014)
18. Abdel-Hamid, D.Y.O., Deng, L.: Exploring convolutional neural network structures and optimization techniques for speech recognition. In: Interspeech (2013)
19. Shufni, S. A., Mashor, M.Y.: Ecg signals classification based on discrete wavelet transform, time domain and frequency domain features. In: 2nd International Conference on Biomedical Engineering (ICoBE), pp. 1–6 (2015)
20. Cecotti, H.: A time–frequency convolutional neural network for the offline classification of steady-state visual evoked potential responses. Pattern Recogn. Lett. **32**(8), 1145–1153 (2011)
21. Tuncer, E., Bolat E.D.: Eeg signal based sleep stage classification using discrete wavelet transform. In: International Conference on Chemistry, Biomedical and Environment Engineering (2014)
22. Sermanet, P., LeCun, Y.: Traffic sign recognition with multi-scale convolutional networks. In: The 2011 International Joint Conference on Neural Networks, pp. 2809–2813 (2011)
23. Bertasius, G., Shi, J., Torresani, L.: Deepedge: a multi-scale bifurcated deep network for top-down contour detection. CoRR, abs/1412.1123 (2014)
24. Zheng, Y., Liu, Q., Chen, E., Ge, Y., Zhao, J.L.: Time series classification using multi-channels deep convolutional neural networks. In: Li, F., Li, Gu., Hwang, S., Yao, B, Zhang, Z. (eds.) WAIM 2014. LNCS, vol. 8485, pp. 298–310. Springer, Cham (2014). doi: 10.1007/978-3-319-08010-9_33
25. Song, Y., McLoughlin, I.V., Dai, L.: Deep bottleneck feature for image classification. In: Proceedings of the 5th ACM on International Conference on Multimedia Retrieval, NY, USA, pp. 491–494 (2015)
26. Bresler, M., Sheffy, K., Pillar, G., Preiszler, M., Herscovici, S.: Differentiating between light and deep sleep stages using an ambulatory device based on peripheral arterial tonometry. Physiol. Meas. **29**(5), 571 (2008)
27. Tsinalis, O., Matthews, P.M., Guo, Y.: Automatic sleep stage scoring using time-frequency analysis and stacked sparse autoencoders. Ann. Biomed. Eng. **44**(5), 1587–1597 (2016)
28. The sleep-edf database [expanded]. https://www.physionet.org/physiobank/database/sleep-edfx/

An Effective Strategy for Trait Combinations in Multiple-Trait Genomic Selection

Zhixu Qiu, Yunjia Tang, and Chuang Ma$^{(\boxtimes)}$

Center of Bioinformatics, College of Life Sciences, Northwest A&F University,
Yangling 712100, Shaanxi, China
chuangma2006@gmail.com

Abstract. Multiple-trait genomic selection (MTGS) is a recently developed method of genomic selection for satisfying the requirements of actual breeding, which usually aims to improve multiple traits simultaneously. Although many efforts have been made to develop MTGS prediction models, how to set the trait combination for the best performance of MTGS prediction models is still under exploration. In this study, we first classified the traits into two groups according to the single-trait genomic selection predictions: traits with a relatively high and low prediction performance. Then, we constructed three trait combinations (High & High, Low & Low, and High & Low) and evaluated their effects on the performance of a state-of-the-art MTGS prediction model using phenotypic and genotypic data from a maize diversity panel. Cross-validation experimental results indicate that single trait predictions could be used as reference for trait combinations in multi-trait genomic selection.

Keywords: Breeding · Cross validation · Genomic selection · Multiple-trait analysis · RR-BLUP · Single-trait analysis

1 Introduction

The key principle of genomic selection (GS) is to build an accurate prediction model based on training populations consisting of individuals with both genotypic and phenotypic data [1]. By using the prediction model, genomic estimated breeding values (GEBVs) are then predicted for genotyped individuals in a breeding population [1]. In contrast to traditional marker-assisted selection, which only focuses on a limited number of markers significantly associated with the traits of interest, GS can increase selection accuracy and reduce breeding cycle time by utilizing the whole genome-wide markers. Therefore, although GS is a relatively new technique, it has already revolutionized the applications of plant breeding [1–4].

Existing GS prediction models can be grouped into two categories based on the number of traits analyzed per time: single-trait GS (STGS) and multi-trait GS (MTGS) [5, 6]. Compared with the STGS, MTGS is more suitable for actual breeding which usually aims to improve multiple traits simultaneously. Moreover, MTGS takes the genetic correlation between traits into account and has been frequently reported to achieve higher prediction accuracy than STGS models [6]. Until now, several MTGS prediction models have been proposed using complex statistical approaches

© Springer International Publishing AG 2017
D.-S. Huang et al. (Eds.): ICIC 2017, Part II, LNCS 10362, pp. 230–239, 2017.
DOI: 10.1007/978-3-319-63312-1_21

(e.g., multi-trait BayesA [7], multi-trait GBLUP [8], and Bayes Cπ [9]). To implement these, plant breeders require a solid background in mathematics and programming skills. In order to facilitate the application of MTGS prediction models in actual breeding programs, a Matlab-based software package MALSAR has recently been used to perform the multi-trait GS analysis [10]. Evaluation results showed that the MTGS prediction model built using the Trace-norm regularized Multitask learning (TNR-MTL) algorithm outperformed the commonly used STGS (ridge regression) on both simulated and real GS data sets [10, 11]. However, a few issues for the application of multi-trait analysis still need attention; one of the most important is the trait combination [12]. Previous researches suggest that two traits with high phenotypic correlation can be used to improve the prediction accuracy of GS in MTGS [7, 9, 10, 13]. Some other researches show that MTGS prediction models can yield improved prediction performance for low-heritability traits when combined with high-heritability traits [14, 15]. Production of a reliable marker-based estimate of heritability is usually affected by several factors, such as the marker density, the marker number, the population size and the linkage disequilibrium [16–18]. Therefore, novel strategies for trait combination in MTGS analysis are still urgently needed.

In this study, we propose a new strategy for trait combination according to the predictions from single-trait analysis. By using the phenotypic and genotypic data from a maize diversity panel, we evaluated the effectiveness of this strategy in the MTGS analysis. The experimental results indicate that this proposed strategy is useful for selecting certain trait combinations to further improve prediction accuracy of GS.

2   Methods and Materials

2.1   Plant Genotyping and Phenotyping

The original genotypic and phenotypic data were obtained from the MaizeGO website (http://www.maizego.org/Resources.html). We filtered out markers with a MAF (minor allele frequency) < 5% by Tassel software (version 5.0; http://www.maizegenetics.net/tassel) [19], and removed individuals with missing genotyping rate > 10%. For the SNP (Single-nucleotide polymorphism) markers, an allele was encoded as 1, 0 or -1, corresponding to the homozygous, heterozygous or other homozygous genotypes, respectively. The final maize GS dataset consisted of 368 individuals and 836,117 SNP markers. Note that markers with missing genotyping data were also filled with 0. The phenotype data of four traits (tassel branch number [TBN], cob diameter [CD], kernel number per row [KNPR], days to anthesis [DTA]) were analyzed in this study.

2.2   Genomic Selection Prediction Models

STGS prediction model

Ridge regression-based best linear unbiased prediction (RR-BLUP) is a commonly used STGS prediction method. The core of RR-BLUP is constructed as a mixed linear model to estimate marker effects of phenotypic traits. It is formulated as

$$Y = Xb + Zg + \varepsilon,$$

where Y is the vector (N $\times$ 1) of phenotypic observations of n individuals, X is the design matrix (N $\times$ r) for fixed planting effects, b is the vector (r $\times$ 1) of planting effects, Z is the design matrix (N $\times$ M) for additive effects of SNP markers, g is the vector (M $\times$ 1) of additive effects of SNP markers, and ε is the vector (N $\times$ 1) of random residuals. It is assumed that $g \sim N\left(0, G\sigma_g^2\right)$, where σ_g^2 is additive variance and G is the realized relationship matrix (N $\times$ N) calculated from the marker matrix Z, and $\varepsilon \sim N\left(0, I\sigma_\varepsilon^2\right)$, where σ_ε^2 is residual variance and I is the M $\times$ M identity matrix. The marker effects, Variance of additive effects and variance of residual effects were simultaneously estimated by solving the mixed model through the restricted maximum likelihood (REML) method implemented in R package 'rrBLUP'. (https://cran.r-project.org/web/packages/rrBLUP/).

MTGS prediction model

The MTGS prediction model was built with the trace-Norm regularization algorithm using the recently developed multi-trait prediction package MALSAR (http://www.public.asu.edu/~jye02/Software/MALSAR/) [20]. This algorithm based MTGS predicted model is defined by

$$y = \mu + \sum_j^M Z_j g_j + e,$$

where y is a matrix (N $\times$ T) of T traits on N individuals, g_j is a vector (1 $\times$ T) for the effects of molecular marker j on all T traits and assumed normally distributed $g \sim N\left(0, \sum g_j\right)$, $\sum g_j$ is the variance–covariance matrix (T $\times$ T) for marker j, e is a matrix (N $\times$ T) for residual error that follows a normal distribution. To estimate marker effects on multiple phenotypic traits for the MT-GS model, the trace-Norm regularization algorithm is used to reduce model complexity and learn features. With the MTGS model, phenotypic values are simultaneously predicted for the analyzed T traits.

2.3   Performance Evaluation Using Five-Fold Cross Validation

Five-fold cross validation was used to evaluate the prediction performance of STGS and MTGS models. The 368 individuals were randomly divided into five groups having an appropriately equal number of individuals (about 73 individuals) in each group; each group's genotyping and phenotyping information was used for testing the performance of STGS and MTGS models, which were trained with data from residual four groups. After testing, the overall predicted results were combined to predict performance evaluation.

Pearson's correlation coefficient (PCC) was used as a measure to globally evaluate the relationship between observed and predicted phenotypic values. PCC analysis in this study was performed using the 'cor.test' function in R programming language. Relative efficiency (*RE*) measurement was used to assess the prediction performance of GS models in selecting individuals with high phenotypic values [21, 22]. For N individuals, predicted scores and observed phenotypic values are formed to be a $N \times 1$ vector of score pairs (X, Y), the *RE* value for selecting top $\alpha\%$ individuals is defined as

$$RE = \frac{\sum_{i=1}^{t} y(i, X) - \mu}{\sum_{i=1}^{t} y(i, Y) - \mu},$$

where $y(i, Y)$ is the ith score of observed phenotypic values Y sorted in an decreasing order, here, $y(1, Y) \geq y(2, Y) \geq \cdots \geq y(N, Y)$. $y(1, X)$ is the corresponding value of Y in the score pairs (X, Y) for the ith value of predicted scores X sorted in an decreasing order, t is the number of extreme $\alpha\%$ individuals, μ is the mean observed phenotypic values of N individuals. *RE* is ranged from -1 to 1. A high *RE* value indicates a high degree that extreme individuals can be predicted by the GS model. The *RE* values were calculated using the R package 'G2P' (https://github.com/cma2015/G2P).

3 Results and Discussions

3.1 Trait Classification Based on STGS-Based Predictions

For each of the analyzed traits, five-fold cross validation was performed to generate STGS -based predictions using the commonly used algorithm RR-BLUP. Figure 1A shows the scatter plots between observed and predicted phenotypic values for the four analyzed traits (DTA, TBN, CD and KNPR). We observed that the PCC values of the RR-BLUP followed the order of: DTA (0.65) > TBN (0.62) > CD (0.41) > KNPR (0.34). The differences in the prediction performance of RR-BLUP are also observed when using the *RE* evaluation measurement for the four analyzed traits (Fig. 1B). The *RE* curve of DTA is higher than that of TBN with $\alpha\%$ increasing from 1% to 34%, and is

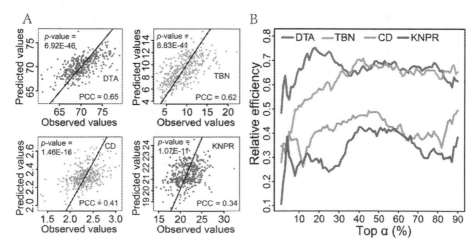

Fig. 1. Prediction performance of four traits for STGT model RR-BLUP. (A) scatter plots of observed and predicted phenotypic values. (B) *RE* curves of four analyzed traits.

close to that of TBN when α% > 34%. Both *RE* curves of DTA and TBN traits are markedly higher than those of CD and KNPR traits when α% varied from 1% to 100% (except α% = 4%). These results indicate that, for the tested STGS model RR-BLUP, CD and KNPR are two traits with a relatively high prediction performance (PCC 0.50; denoted as HPS-Traits), while CD and KNPR are two traits with a relatively low prediction performance (PCC < 0.50; denoted as LPS-Traits).

3.2   Combination of HPS-Traits

The single trait analysis showed good performance for DTA and TBN. PCC analysis showed that the phenotypic correlation between these two traits is 0.36 (*p*-value = 8.07E-13). Therefore, further improvement of the prediction performance can be expected when the phenotypic correlation between the two traits is utilized by using MTGS models. In order to check this expectation, a MTGS prediction model was used to build a multiple-trait GS model with DTA and TBN data. The five-fold cross validation experimental results showed that there is no significant difference when using STGS and MTGS models for the TBN trait. The PCC values of TBN for MTGS and STGS are about 0.62 (Fig. 2A). It is noted that the performance of the MTGS model for DTA, with a PCC value of 0.60, is inferior to the corresponding STGS model with a PCC value of 0.65.

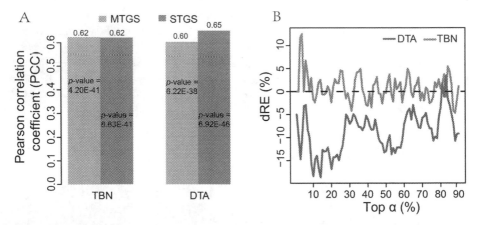

Fig. 2. Prediction performance of HPS-trait combination for MTGS. (A) bar plot of PCC values of MTGS and STGS for two HPS-traits, (B) *dRE* curve of two HPS-traits.

We further compared the prediction performance of STGS and MTGS models in selecting individuals with high TBN values using the *RE* measure. For each possible level of α, we calculated the relative improvement of *RE* value using the formula: $dRE = (RE_{MTGS} - RE_{STGS})/RE_{STGS}$. *dRE* higher than zero indicates that the performance of MTGS is higher than STGS for the tested level of α, whereas *dRE* lower than zero denotes that the performance of MTGS is lower than STGS. For the TBN trait, the *dRE* curve has large fluctuations around zero. In contrast, a *dRE* curve below zero can be seen at almost all possible levels of α (except α% = 83%) for the DTA trait.

Taken together, these results suggest that the combination of two HPS-Traits might not be a good strategy to perform the multi-trait analysis, in spite of the phenotypic correlation between these two traits being 0.36. It might be difficult to improve the performance of the HPS-Traits using the tested MTGS model.

3.3   Combination of LPS-Traits

Because the combination of HPS-traits did not improve prediction performance, we wondered whether the LPS-trait combination shows similar performance. KNPR and CD data were used to construct the MTGS model. The phenotypic correlation between KNPR and CD is -0.031 (p-value $= 0.35$). It is surprising that the MTGS model has better prediction performance than the STGS model RR-BLUP for both KNPR and CD. For the KNPR trait, the PCC values of MTGS and STGS are 0.37 and 0.34, respectively, and those for the CD trait are 0.43 and 0.41 respectively (Fig. 3A). We also observed that the *RE* scores of the two traits have been improved for most α levels (Fig. 3B). Moreover, the *dRE* value of KNPR is significant higher than that of CD, especially when α ranges from 14% to 39% (KNPR *dRE*: 3.78% -53.71%, CD *dRE*: $-19.94\% - 4.03\%$).

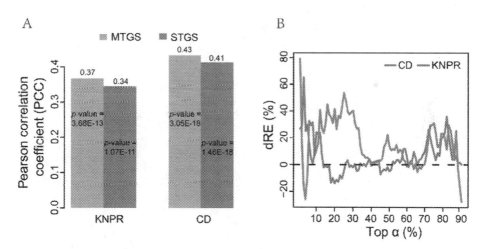

Fig. 3. Prediction performance of LPS-traits using the MTGS model. (A) Bar plot of PCC values of MTGS and STGS for two LPS-traits, (B) *dRE* curves of two LPS-traits.

These results indicate that the prediction performance of LPS-traits could be greatly improved by performing the multi-trait analysis using the tested MTGS model.

3.4   Combination of LPS-Traits and HPS-Traits

Besides combining only HPS-traits or only LPS-traits, another combination strategy for multi-trait analysis is the combination of LPS- and HPS-traits. We tested four combinations among the four traits: KNPR-DTA, KNPR-TBN, CD-DTA, and CD-TBN. Phenotypic correlation analysis indicated that all the paired traits have markedly low PCC values, ranging from -0.049 to 0.053.

As illustrated in Fig. 4, the accuracy of HPS-trait prediction significantly decreased or only slightly increased in the MTGS model compared with the STGS method. This result is different regarding to LPS-traits, which are all predicted with higher accuracy. The performances for KNPR improved even more than for CD using MTGS methods (Fig. 5). Since the single trait predicted performance of CD is better than that of KNPR, this implies that the LPS-trait has a greater promotion potential. For KNPR and CD traits, the predicted accuracies of MTGS models have almost the same improvement no matter which trait it is combined with. This is probably because TBN and DTA have similar performance of STGS models. This suggests that the improvement ability of MTGS for traits that have similar single trait predicted performance is also similar when combined with the same trait. These results have shown MTGS to have higher predicted accuracy than single trait method, and that the LPS-trait prediction can easily be improved in accuracy by using the MTGS prediction model.

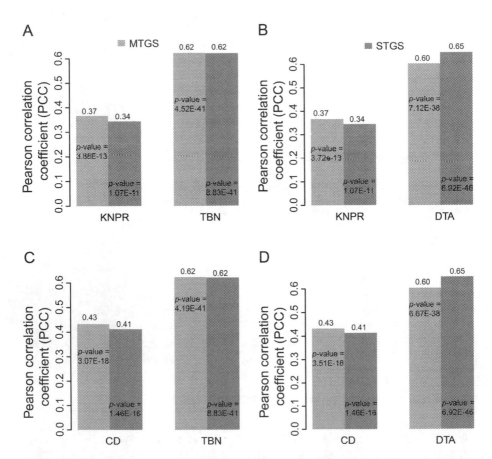

Fig. 4. Prediction performance of the MTGS model for different combinations of LPS-traits and HPS-traits. (A) the trait combination of KNPR and TBN. (B) the trait combination of KNPR and DTA. (C) the trait combination of CD and TBN. (D) the trait combination of CD and DTA.

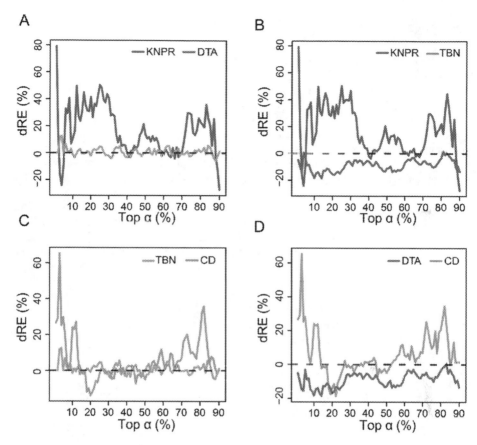

Fig. 5. *dRE* curves of different combinations of LPS-traits and HPS-traits. (A) the trait combination of KNPR and TBN. (B) the trait combination of KNPR and DTA. (C) the trait combination of CD and TBN. (D) the trait combination of CD and DTA.

4 Conclusion

In this study, a new strategy has been proposed to select appropriate trait combinations for building MTGS prediction models. Compared with the STGS method, the MTGS method achieved an improved prediction performance for a LPS-trait combined with another HPS-trait or LPS-trait, even when their phenotypic correlation coefficients were quite low. The experimental results also suggest that integrated multiple trait information exhibits a great advantage of MTGS models in predicting phenotypes from genotypes. An appropriate trait combination may promote the advantage, and results from single trait prediction performance could be used as reference for selecting the trait combination.

Funding. This work was supported by the National Natural Science Foundation of China (31570371), the Agricultural Science and Technology Innovation and Research Project of Shaanxi Province, China (2015NY011), and the Fund of Northwest A & F University.

References

1. Desta, Z.A., Ortiz, R.: Genomic selection: genome-wide prediction in plant improvement. Trends Plant Sci. **19**(9), 592–601 (2014)
2. Jannink, J.L., Lorenz, A.J., Iwata, H.: Genomic selection in plant breeding: from theory to practice. Brief. Funct. Genomics. **9**(2), 166–177 (2010)
3. Hayes, B.J., Bowman, P.J., Chamberlain, A.J., Goddard, M.E.: Invited review: genomic selection in dairy cattle: progress and challenges. J. Dairy Sci. **92**(2), 433–443 (2009)
4. Schmidt, M., Kollers, S., Maasberg-Prelle, A., Grosser, J., Schinkel, B., Tomerius, A., Graner, A., Korzun, V.: Prediction of malting quality traits in barley based on genome-wide marker data to assess the potential of genomic selection. Theor. Appl. Genet. **129**(2), 203–213 (2016)
5. Momen, M., Mehrgardi, A.A., Sheikhy, A., Esmailizadeh, A., Fozi, M.A., Kranis, A., Valente, B.D., Rosa, G.J., Gianola, D.: A predictive assessment of genetic correlations between traits in chickens using markers. Genet. Sel. Evol. **49**(1), 16 (2017)
6. Bao, Y., Kurle, J.E., Anderson, G., Young, N.D.: Association mapping and genomic prediction for resistance to sudden death syndrome in early maturing soybean germplasm. Mol. Breed. **35**(6), 128 (2015)
7. Jia, Y., Jannink, J.L.: Multiple-trait genomic selection methods increase genetic value prediction accuracy. Genetics **192**(4), 1513–1522 (2012)
8. Dos Santos, J.P., Vasconcellos, R.C., Pires, L.P., Balestre, M., Von Pinho, R.G.: Inclusion of dominance effects in the multivariate GBLUP model. PLoS One **11**(4), e0152045 (2016)
9. Calus, M.P., Veerkamp, R.F.: Accuracy of multi-trait genomic selection using different methods. Genet. Sel. Evol. **43**(1), 26 (2011)
10. He, D., Kuhn, D., Parida, L.: Novel applications of multitask learning and multiple output regression to multiple genetic trait prediction. Bioinformatics **32**(12), i37–i43 (2016)
11. Abernethy, J., Bach, F., Evgeniou, T., Vert, J.P.: A new approach to collaborative filtering: operator estimation with spectral regularization. J. Mach. Learn. Res. **10**((Mar)), 803–826 (2009)
12. Schulthess, A.W., Wang, Y., Miedaner, T., Wilde, P., Reif, J.C., Zhao, Y.S.: Multiple-trait- and selection indices-genomic predictions for grain yield and protein content in rye for feeding purposes. Theor. Appl. Genet. **129**(2), 273–287 (2016)
13. Montesinos-Lopez, O.A., Montesinos-Lopez, A., Crossa, J., Toledo, F.H., Perez-Hernandez, O., Eskridge, K.M., Rutkoski, J.: A genomic bayesian multi-trait and multi-environment model. G3 (Bethesda) **6**(9), 2725–2744 (2016)
14. Jiang, J., Zhang, Q., Ma, L., Li, J., Wang, Z., Liu, J.F.: Joint prediction of multiple quantitative traits using a bayesian multivariate antedependence model. Heredity **115**(1), 29–36 (2015)
15. Hayashi, T., Iwata, H.: A bayesian method and its variational approximation for prediction of genomic breeding values in multiple traits. BMC Bioinf. **14**(1), 34 (2013)
16. De los Campos, G., Sorensen, D., Gianola, D.: Genomic Heritability: What Is It? Plos Genetics. **11**(5) (2015)
17. Kruijer, W., Boer, M.P., Malosetti, M., Flood, P.J., Engel, B., Kooke, R., Keurentjes, J.J., van Eeuwijk, F.A.: Marker-based estimation of heritability in immortal populations. Genetics **199**(2), 379–398 (2015)
18. Stanton-Geddes, J., Yoder, J.B., Briskine, R., Young, N.D., Tiffin, P.: Estimating heritability using genomic data. Methods Ecol. Evol. **4**(12), 1151–1158 (2013)

19. Bradbury, P.J., Zhang, Z., Kroon, D.E., Casstevens, T.M., Ramdoss, Y., Buckler, E.S.: TASSEL: software for association mapping of complex traits in diverse samples. Bioinformatics **23**(19), 2633–2635 (2007)
20. Zhou, J., Chen, J., Ye, J.: MALSAR: Multi-task Learning via Structural Regularization (2012). http://www.public.asu.edu/~jye02/Software/MALSAR
21. Qiu, Z., Cheng, Q., Song, J., Tang, Y., Ma, C.: Application of machine learning-based classification to genomic selection and performance improvement. In: Huang, D.-S., Bevilacqua, V., Premaratne, P. (eds.) ICIC 2016. LNCS, vol. 9771, pp. 412–421. Springer, Cham (2016). doi:10.1007/978-3-319-42291-6_41
22. Ornella, L., Perez, P., Tapia, E., Gonzalez-Camacho, J.M., Burgueno, J., Zhang, X., Singh, S., Vicente, F.S., Bonnett, D., Dreisigacker, S., Singh, R., Long, N., Crossa, J.: Genomic-enabled Prediction with classification algorithms. Heredity (Edinb). **112**(6), 616–626 (2014)

A Multiway Semi-supervised Online Sequential Extreme Learning Machine for Facial Expression Recognition with Kinect RGB-D Images

Xibin Jia[1], Xinyuan Chen[1], and Jun Miao[2(✉)]

[1] Faculty of Information Technology,
Beijing University of Technology, Beijing, China
jiaxibin@bjut.edu.cn, xinyuanchen@emails.bjut.edu.cn
[2] School of Computer Science,
Beijing Information Science and Technology University,
Beijing, China
jmiao@bistu.edu.cn

Abstract. This paper aims to develop a facial expression recognition algorithm for a personal digital assistance application. Based on the Kinect RGB-D images, we propose a multiway extreme learning machine (MW-ELM) for facial expression recognition, which reduces the computing complexity significantly by processing the RGB and Depth channels separately at the input layer. Referring to our earlier work on semi-supervised online sequential extreme learning machine (SOS-ELM) that enhances the application to do the fast and incremental learning based on a few labeled samples together with some un-labeled samples of the specific user, we propose to do the parameter training with semi-supervising and on-line sequential methods for the higher hidden layer. The experiment of our proposed multiway semi-supervised online sequential extreme learning machine (MW-SOS-ELM) applying in the facial expression recognition, shows that our proposed approach achieves almost the same recognition accuracy with SOS-ELM, but reduces recognition time significantly, under the same configuration of hidden nodes. Additionally, the experiments show that our semi-supervised learning scheme reduces the requirement of labeled data sharply.

Keywords: Extreme learning machine · Semi-supervising · On-line sequential learning · Multi-way structure · Facial expression recognition

1 Introduction

Facial expression recognition is becoming a focused research in decades, which aims to make the computer/robot comprehend people's emotion. For example, to integrate facial expression recognition into the personal digital assistance application will enhance its intelligence. This expression-sensitive personal digital assistance application facilitates to pushe favorable contents for the user according to understanding the user's current emotion. Considering Kinect has been used widely as a high

© Springer International Publishing AG 2017
D.-S. Huang et al. (Eds.): ICIC 2017, Part II, LNCS 10362, pp. 240–253, 2017.
DOI: 10.1007/978-3-319-63312-1_22

performance RGB-D capture device with low cost [1], we use Kinect for RGB-D based facial expression recognition.

For facial recognition, there exist two problems needs to be considered. Firstly, the manual labeling is a tedious. Secondly, it is not feasible to do the data capturing in one time, for video film consumes time. Additionally the practical expression recognition is normally real time application. Taking into account of these factors, we propose to utilize our former semi-supervising and online sequential extreme learning machine (SOS-ELM) [2] for face recognition. To improve the computing efficient for the multiply channel input, we propose to improve SOS-ELM with multi-way structure. Meanwhile, we design a method with an automatically data capturing strategy. When users watch some built-in emotion stimulus video shot and movies, the application captures training data automatically. Upon some specific moments with strong emotional scenes, the application put the corresponding labels to these face frames accordingly as the labeled data. At other moments without the strong emotional scenes, the application will capture a few unlabeled training data.

In the remainders, Sect. 2 addresses our proposed MW-ELM. Section 3 elaborates the implement of our MW-SOS-ELM based specific user oriented facial expression recognition approach. The experiments and results are given in the Sect. 4. Finally, the conclusion and future work are given at the last section.

2 Proposed MW-ELM

2.1 Structure of Basic ELM

Extreme learning machine (ELM) is a type of machine learning algorithm proposed by Huang [3]. As shown in Fig. 1, ELM is a single hidden layer network with random generated parameters at the first layer and calculating the output vector at the second layer for regression, classification and compression problems, which has been proved having approximate capability. ELM is much faster at learning than most iterative learning based neural network algorithms such as support vector machine SVM [4], due to its one-shot matrix computing scheme.

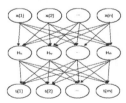

Fig. 1. Structure of ELM network

2.2 Structure of MW-ELM Algorithm

Employing the basic ELM for the multiple input problem, the normal way is to concatenate multiple feature vectors as a single input vector x. Therefore, calculation of each hidden node depends on all feature vectors as shown in Fig. 1. Here, H_{ij} represents the element of translation matric at hidden node j for each input element x_i, a_j represents input layer network weight and b_j is bias for the hidden node j. Note, these parameter a, b are generated randomly. g is an activation function and sigmoid function $g(x) = 1/(1 + e^{-x})$ is used in the paper.

$$H_{ij} = g(a_j x_i + b_j) \tag{1}$$

As we could find, a is a vector with same dimension with input x. When the dimension N of input is expanding, the input node increases. Dimension of weight a_1, ..., $a_{\tilde{N}}$ becomes larger and will occupy lots of memory. Accordingly, the computation of $a_j x_i (i = 1, ..., N\ j = 1, ..., \tilde{N})$ will cost longer time. $\tilde{N}$ is number of hidden node.

For each input of multiple channel is independence, we propose to cope with their feature separately at the first layer to avoid the expansion of input dimension. The structure of our proposed multi-way ELM is shown in Fig. 2. Each channel way calculates its hidden node outputs using its own input vector individually. Then, the hidden output of each channel from the first layer is connected at the second layer.

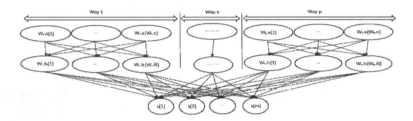

Fig. 2. Structure of MW-ELM network

As shown in Fig. 2, our proposed MW-ELM defines a data structure W for each input channel, denoting as way 1, way 2,..etc. If the machine learning problem has p input vectors, define p channels $W_1, W_2, ..., W_p$. Data structure W contains:

$W_k.n$:   The input vector of way k is a $W_k.n$ dimensional vector. $k = 1, 2, ..., p$.
$W_k.\tilde{N}$:   The number of hidden nodes of way k.
$W_k.x_i$:   The input vector of sample i, way k. It is a $W_k.n$-dimensional vector.
$W_k.a_j$:   The weight vector of hidden node j, way k. It is a $W_k.n$-dimensional vector.
$W_k.b_j$:   The bias of hidden node j, way k.
$W_k.h_i$:   The hidden layer output of sample i, way k. It is a $W_k.\tilde{N}$-dimensional vector. The j-th element of $W_k.h_i$ is calculated by the following equation:

$$W_k.h_i[j] = g(W_k.a_j * W_k.x_i + W_k.b_j) \tag{2}$$

Then we concatenate hidden layer outputs of all ways. The size of matrix H is N rows, $\sum_{k=1}^{p} W_k.\tilde{N}$ columns.

$$
H = \begin{pmatrix}
W_1.h_1^T & W_2.h_1^T & \cdots & W_p.h_1^T \\
W_1.h_2^T & W_2.h_2^T & \cdots & W_p.h_2^T \\
\cdots & \cdots & \cdots & \cdots \\
W_1.h_N^T & W_2.h_N^T & \cdots & W_p.h_N^T
\end{pmatrix} \tag{3}
$$

In basic ELM, output parameter β is calculated as $\beta = H^\dagger T$. The size of matrix β is $\sum_{k=1}^{p} W_k.\tilde{N}$ rows, m columns. $H^\dagger$ is calculated by the Moore-Penrose generalized inverse of matrix H: $H^\dagger = (H^T H)^{-1} H^T$. Where output layer matrix is $T = (t_1^T, \ldots, t_N^T)^T$.

The amount of hidden nodes is adjustable for each channel. It impacts the weight of each feature vector, and impacts the recognition accuracy. To solve the problem of shortage of training data in practical application, we do the parameter calculation based on SOS-ELM. It inherits the properties of semi-supervised extreme learning machine (SS-ELM) [4] and online sequential extreme learning machine (OS-ELM) [5], which realizes the sequential training with gradually capturing data and improves learning performance with unlabeled and labeled data.

2.3   MW-SOS-ELM Algorithm

Given N input training vectors: $[W.x_1, \ldots, W.x_N]^T$ and output vectors $T = [t_1, \ldots, t_N]^T$ as training samples, which includes both labeled and unlabeled data. For labeled sample $W.x_i$, the relative element of the output vector is the label t_i. For unlabeled sample $W.x_j$, the relative element of the output vector t_j is filled with 0. Accordingly, calculate hidden output matrix H as Eq. 3. To implement the online sequential training, the iterative process is given lately with each batch of data including both labeled and unlabeled data. Before that, we address some parameters and their calculation equation first. Define a constant C_0. Then calculate C_i for $i = 1, \ldots, N$, Where class t has N_t samples.

$$
C_i = \begin{cases} 0 & W.x_i \text{ is unlabled data} \\ C_0/N_{t_i} & W.x_i \text{ belongs to class } t_i \end{cases} \tag{4}
$$

Calculate matrix J with the purpose of solving unbalanced training samples:

$$
J_{ij} = \begin{cases} C_i & i = j \\ 0 & i \neq j \end{cases} \tag{5}
$$

Mat_W is calculated as in Eq. 6.

$$
Mat\_W_{ij} = e^{-\sum_{k=1}^{p} \left\| W_k.x_i - W_k.x_j^2 \right\|^2 / 2\sigma^2} \tag{6}
$$

Where σ^2 is the variance of input vectors calculated by following Eq. 7:

$$\bar{x}_k = \frac{\sum_{i=1}^{N} W_k.x_i}{N}, \; \sigma^2 = \frac{\sum_{k=1}^{p} \sum_{i=1}^{N} \| W_k.x_i - \bar{x}_k \|^2}{N} \tag{7}$$

Calculate matrix D as in Eq. 8:

$$D_{ij} = \begin{cases} \sum_{k=1}^{N} Mat\_W_{ik} & i = j \\ 0 & i \neq j \end{cases} \tag{8}$$

Calculate graph Laplacian matrix L as in Eq. 9:

$$L = D - Mat\_W \tag{9}$$

The procedure of MW-SOS-ELM is illustrated as follows:

I. Training stage:

(I). The initial training procedure of MW-SOS-ELM is as follows. Calculate matrix H_0, T_0, J_0, L_0 based on training data with Eqs. 3, 5 and 9. Calculate matrix K_0 as in Eq. 10 and matrix $\beta^{(0)}$ as in Eq. 11.

$$K_0 = I + H_0^T(J_0 + \lambda L_0)H_0 \tag{10}$$

$$\beta^{(0)} = K_0^{-1}H_0^T J_0 T_0 \tag{11}$$

(II). The recursive parameter training procedure of MW-SOS-ELM is:

Given new batch of training data: input data W. x_i and their corresponding output label t_i. $i = 1, ..., N$. Note, if W.x_i is un-labeled data, the corresponding $t_i. = 0$. Calculate matrix H_1, T_1, J_1, L_1 with Eqs. 3, 5 and 9 using the new input vectors and output vectors. Update matrix K as in Eq. 12 and update matrix β as in Eq. 13.

$$K_1 = K_0 + H_1^T(J_1 + \lambda L_1)H_1 \tag{12}$$

$$\beta^{(1)} = \beta^{(0)} + K_1^{-1}H_1^T\left(J_1 T_1 - (J_1 + \lambda L_1)H_1\beta^{(0)}\right) \tag{13}$$

II. The classification/regression stage of MW-SOS-ELM is:

To determine the result of a new input, H is calculated based on this data under-determined as in Eq. 3. Then together with the trained β and the obtained H, the prediction output $\widehat{T}$ is calculated as $\widehat{T} = H\beta$.

2.4   Computing Performance Analysis

In theory, the proposed multi-way ELM uses less memory and runs faster than ELM when they are used to solve feature fusion problems. Some qualitative analysis is made as follows. We set the same amount of hidden nodes for both MW-ELM and ELM, i.e. $\tilde{N} = \sum_{k=1}^{p} W_k.\tilde{N}$. In this case, MW-ELM and ELM have the same size of matrix H and β, so they should have the similar accuracy. The input vector of ELM is concatenation of all feature vectors from each way respectively. So the length of input vector is summation of all feature vector length. $n = \sum_{k=1}^{p} W_k.n$. For high-dimensional input vectors x_i, the ELM hidden layer parameter a_j ($j = 1, ..., \tilde{N}$) occupied lots of memory. The parameter a of MW-ELM occupies less memory than ELM. Other parameters are same size for MW-ELM and ELM. The memory usage of ELM hidden layer parameter a is:

$$\text{size}_{aELM} = \tilde{N} * n * \text{sizeof(double)} = \sum_{j=1}^{p} W_j.\tilde{N} * \sum_{k=1}^{p} W_k.n * \text{sizeof(double)}$$
$$= \sum_{k=1}^{p} W_k.\tilde{N} * (W_1.n + W_2.n + ... + W_k.n + ... + W_p.n) * \text{sizeof(double)} \tag{14}$$

The memory usage of MW-ELM hidden layer parameter a is:

$$size\_a_{MW-ELM} = \sum_{k-1}^{p} (W_k.\tilde{N} * W_k.n) * \text{sizeof(double)} \tag{15}$$

Obviously, $\text{size}\_a_{MW\ ELM} < \text{size}\_a_{ELM}$. For high-dimensional input vectors, MW-ELM occupies less memory than ELM. The calculation of $a_j x_i$ costs longer time.

ELM: N training data, $\tilde{N}$ hidden nodes, n loops each hidden node. Total number of loops is:

$$loop_{ELM} = N * \tilde{N} * n = N * \sum_{k=1}^{p} W_k.\tilde{N} * \sum_{k=1}^{p} W_k.n$$
$$= N * \sum_{k=1}^{p} W_k.\tilde{N} * (W_1.n + W_2.n + ... + W_k.n + ... + W_p.n) \tag{16}$$

MW-ELM: N training data, p ways, $W_k.\tilde{N}$ hidden nodes each way, $W_k.n$ loops each hidden node. Total number of loops is:

$$loop_{MW-ELM} = N * \sum_{k=1}^{p} W_k.\tilde{N} * W_k.n \tag{17}$$

Obviously, $loop_{MW-ELM} < loop_{ELM}$. For feature fusion problems, each hidden node of MW-ELM is only concerned with one feature vector, rather than all feature vectors. For high-dimensional input vectors, MW-ELM saves computing time significantly.

3 MW-SOS-ELM Based Specific User Oriented Facial Expression Recognition

In the paper, we apply our proposed MW-SOS-ELM algorithm in the facial expression recognition based on RGB-D images captured from Kinect. The MW-SOS-ELM is trained for the specific user. To obtain the labeled data for training, we adopt the auto-label method by marking facial images with the synchronizing emotional scenes.

3.1 The Data Capturing and Auto-Label Strategy

The paper aims to implement an approach of facial expression recognition for the future personal digital assistance application. In this application there is none training data at the beginning. To initialize the facial expression recognition module, we provide some build-in emotional stimulating videos and movies. During playing, the relative emotion labels are made on the captured user's data at some certain predefined emotional scenes. In current module, we label the corresponding images with the six basic expression viz. anger, disgust, fear, happiness, sadness, surprise [6, 7], and neutral expression. For doing the later experiments, we develop an interactive labeling interface to capture some lab data with selecting the expression label by the test subject. This is also applied to do the evaluation the reliability of automatic labeled data. Accordingly, the labeled training data are obtained gradually at each watching. We also capture some expression data without labels during play some emotional scenes which might cause the subject's reaction. The amount of unlabeled data is normally fewer than that of labeled data to avoid the decrease of accuracy. To ensure the balance of labeled data among each expression, the application uses same amount of labeled training data for each expression. Under our online sequential learning scheme, the data could be captured gradually. In fact, it is more acceptable for users without requirement of overwhelming movie watching at one time for data capturing.

3.2 Feature Extraction of RGB-D Images

In the paper, Kinect is used to capture RGB and depth images. The Kinect SDK built-in Face Tracking toolkit is utilized for locating 121 interested feature points. Some example of the captured RGB-D images and located points are shown in Fig. 3.

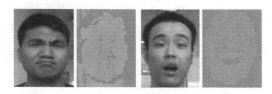

Fig. 3. Sample of RGB and depth images around Facial area

Based on these points, the selected interested regions are partitioned. For Haar-like features [8] have been widely used for texture feature extraction, we extract Haar-like features on each partitioned region of red, green, blue and depth image respectively. Haar-like feature extraction window are created at each feature points. The selected Haar templates shown in Fig. 4 are depicting horizontal, vertical and diagonal texture features of interested regions.

Fig. 4. Used Haar templates

3.3   Parameter Training of MW-SOS-ELM

Based on the calculating the four RGB-D channel's Haar feature on all training data including both labeled and unlabeled data, the 4-way MW-SOS-ELM is trained as the classifier for the expression recognition. The extracted four Haar-like features of RGB-D images are input. The output vector is a 7-dimensional vector indicating seven expressions: neutral, anger, disgust, fear, happiness, sadness and surprise.

Once a few labeled and unlabeled training data are captured and the training procedure is initialized, the facial expression recognition for the specific user is enabled in the personal digital assistance application. Then the application starts to capture more unlabeled training data during daily use, and capture more labeled training data during watching emotional videos and movies. The MW-SOS-ELM parameters for this facial expression recognition module are trained gradually with the online sequential learning scheme of MW-SOS-ELM, which is illustrated in Sect. 2.3.

3.4   Architecture of MW-SOS-ELM Based Facial Expression Recognition

The architecture of MW-SOS-ELM based facial expression recognition approach is shown in Fig. 5. The RGB-D data capture from Kinect is preprocessed using in-build face tracking toolkit in Kinect SDK to obtain interested points. Based on located 121 points, we create 5*5 window at each point, the Haar features are extracted in these windows. We extract 4 Haar feature vectors on each RGB-D way. Then these feature vectors are used as input vectors of a 4-way MW-SOS-ELM to estimate an output vector. Finally the output vector will be converted to an expression.

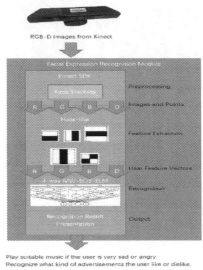

Fig. 5. The architecture of MW-SOS-ELM based facial expression recognition

4 Experiment Results and Analyses

4.1 Dataset

As far as we explore, there is no public dataset of Kinect based facial expression recognition. Therefore, we made the comparison experiments on our self-captured dataset with 8 subjects in this paper. For testing the performance of the algorithm quantitatively, we captured and saved expression data when the user is watching movies, and use these data as our experimental dataset.

Table 1. The illustration of data amount in the experiment

Subject	A	B	C	D	E	F	G	H
Labeled data	28	28	21	49	28	28	28	28
Unlabeled data	14	14	14	26	13	12	13	10
Testing data	385	392	322	798	406	455	392	413

The auto-label strategy together with interactive manual labeling is illustrated in Sect. 3.1 is used during the data capture. The amounts of training data and testing data prepared for each subject are listed in Table 1.

In the comparison experiment, the hidden node is configured as follows. For MW-SOS-ELM, the hidden node numbers of four channels are set at the RHN = 256, GHN = 256, BHN = 256, DHN = 128 separately. For SOS-ELM and ELM, the hidden node number is set at 896. The parameters are determined based on some massive test with good performance. To make the result comparable, the node number is

Table 2. Experiment results

Algorithm	Sub.	Training time	Testing time	Acc. (%)	Sub.	Training time	Testing time	Acc. (%)
MW-SOS-ELM	A	**8779**	**550**	**90.9**	E	9083	**581**	**91.1**
SOS-ELM		8913	2144	90.6		9330	2365	89.9
MW-ELM		8937	562	42.3		**8965**	630	34.8
ELM		9271	2200	37.9		9227	2577	35.7
MW-SOS-ELM	B	**8780**	**553**	**94.8**	F	9111	**626**	**92.5**
SOS-ELM		8925	2231	94.6		9471	2650	91.4
MW-ELM		8954	562	45.4		**9020**	681	38.9
ELM		9211	2269	45.1		9310	2630	39.3
MW-SOS-ELM	C	**9255**	470	87.5	G	9310	613	**88.0**
SOS-ELM		9287	1916	**88.8**		9573	2323	**88.0**
MW-ELM		8962	**463**	27.9		**9081**	**572**	28.0
ELM		9180	1801	22.6		9330	2249	25.5
MW-SOS-ELM	D	9337	**1130**	88.5	H	9320	**601**	**89.3**
SOS-ELM		9604	4692	**87.9**		9551	2460	88.6
MW-ELM		**9237**	1137	32.0		**9088**	641	29.2
ELM		9501	4622	31.3		9322	2545	30.0
Ave.	MW-SOS-ELM	**9122**	**641**	**90.33**	MW-ELM	9031	656	34.81
	SOS-ELM	9332	2598	89.98	ELM	9294	2612	33.43

selected to be similar among all algorithms. Then, the classification accuracy rate is calculated upon each recognition result for every subject and training and test efficient are evaluated for our proposed MW-SOS-ELM and the other three algorithms. The experiment results are given in the Table 2, where the value with minimum training and testing time and maximum accuracy rate are emphasized in bold.

As shown in Table 2, our proposed MW-SOS-ELM outperforms the other three algorithms at the performance of both recognition accuracy and computing efficient. From the aspect of recognition accuracy, our proposed approach achieves the average accuracy of 90.33% which is 0.3% higher than that of SOS-ELM. Meanwhile the MW-ELM and ELM, which only use the labeled data, fail to do the right expression recognition. This proves that semi-supervising method helps to improve the classification performance especially when the labeled data are not enough. The sequential online training also provides gradual increasing with accuracy. From the viewpoint of computing efficient, our MW-SOS-ELM achieves 2.2% faster than SOS-ELM for training on average, and 75.1% faster than SOS-ELM for testing on average. This proves that by separating the multiple channels at the first layer, the computing efficient improves comparing with simple concatenation of multiple channels in one input.

According to the above experiments, we could tell that the accuracy of MW-SOS-ELM is similar with SOS-ELM when total amount of hidden nodes is same. Meanwhile, the recognizing time of MW-SOS-ELM is far less than SOS-ELM. This means MW-SOS-ELM uses less processor time. Due to the semi-supervised online sequential learning scheme, our proposed MW-SOS-ELM uses a few labeled training data to obtain good recognition accuracy. It is also very practical and suitable for the variable data application by gradually updating the performance with incremental learning instead of one time learning. In one word, the high recognition accuracy and efficient computing performance will facilitate the proposed MW-SOS-ELM into practical application.

4.2   Analysis of Impact of Some Parameters for MW-SOS-ELM

To determine the optimization parameters for MW-SOS-ELM, we make some test experiments by adjusting the parameters including the hidden node number of the red channel (RHN), the green channel (GHN), the blue channel (BHN) and the depth channel (DHN), and the penalty coefficient λ in Eq. 5. For comparison, we also did same experiment on SOS-ELM. We adjust the hidden node amount (HN) and the parameter λ of SOS-ELM. To show the corresponding experiments results reasonably, we chose 3 subjects A, B and C, where recognition accuracy for these 3 subjects achieved best, medium and worst results respectively among all 8 subjects. Parts of experiment results showing the corresponding recognition accuracy changing with parameter λ and hidden node numbers are listed in the Table 3.

Table 3. Accuracy of MW-SOS-ELM and SOS-ELM

Subject	λ	MW-SOS-ELM					SOS-ELM	
		RHN	GHN	BHN	DHN	Accuracy	HN	Accuracy
A	0.01	256	256	256	128	89.8%	896	84.9%
	0.001	256	256	256	128	90.9%	896	89.8%
	0.0001	256	256	256	128	90.9%	896	90.3%
	0.00001	256	256	256	128	90.9%	896	90.6%
	0.000001	256	256	256	128	90.6%	896	90.6%
	0.0001	256	256	256	256	90.9%	1024	90.1%
B	0.01	256	256	256	128	91.0%	896	89.2%
	0.001	256	256	256	128	94.3%	896	94.1%
	0.0001	256	256	256	128	94.8%	896	94.6%
	0.00001	256	256	256	128	94.8%	896	94.3%
	0.000001	256	256	256	128	94.8%	896	94.1%
	0.0001	256	256	256	256	94.8%	1024	94.8%
C	0.01	256	256	256	128	77.3%	896	77.9%
	0.001	256	256	256	128	86.3%	896	87.8%
	0.0001	256	256	256	128	87.2%	896	88.8%
	0.00001	256	256	256	128	87.5%	896	88.8%
	0.000001	256	256	256	128	87.5%	896	88.8%
	0.0001	256	256	256	256	87.2%	1024	89.1%

As in Table 3, the accuracies of MW-SOS-ELM and SOS-ELM have similar accuracy at the same λ and same total hidden node number. When the panality parameter λ decreases, both of their recognition accuracy increases until arriving at a certain value. In the experiment λ is set at 0.0001. With this parameter, we have the hidden node number changes, here we only show the changing of depth channel. The experiments show that with increasing of hidden node, the accuracy rate increases.

From the view of computing efficiency, training time and testing time increases with hidden node number increases, as in Table 4. By comparing that between MW-SOS-ELM and SOS-ELM with changing of hidden node number. As shown in Table 4, the training time of MW-SOS-ELM is less than SOS-ELM, and its testing

Table 4. Training time and testing time of MW-SOS-ELM

Subject	MW-SOS-ELM						SOS-ELM		
	RHN	GHN	BHN	DHN	Training time	Testing time	HN	Training time	Testing time
A	256	256	256	128	8779 ms	550 ms	896	8913 ms	2144 ms
	256	256	256	256	13023 ms	625 ms	1024	13204 ms	2375 ms
B	256	256	256	128	8780 ms	553 ms	896	8925 ms	2231 ms
	256	256	256	256	12875 ms	625 ms	1024	13142 ms	2438 ms
C	256	256	256	128	9255 ms	470 ms	896	9287 ms	1916 ms
	256	256	256	256	13519 ms	541 ms	1024	13949 ms	2287 ms

time is far less than SOS-ELM. In this experiment result, we could find that our proposed MW-SOS-ELM achieves much faster recognition speed.

Based on the above experiments, we compared the performance with changing parameters. We found that using 128 hidden nodes for the depth way (DHN = 128) is a good option, because it achieves about the same accuracy with 256 hidden nodes (DHN = 256), but significantly reduces training and testing time.

5 Conclusion

In this paper, we propose a multiway semi-supervised online sequential extreme learning machine (MW-SOS-ELM), and implement a MW-SOS-ELM facial expression recognition algorithm for a personal digital assistance application. This algorithm can achieve a high recognition accuracy with a few training data captured by Kinect. MW-SOS-ELM achieves good recognition results in specific person expression recognition by using a few labeled together with some unlabeled data for training. It achieves the practical result under the support of online sequential learning. To solve the problem of one time data capture problem in video dataset. The high training and recognition computing efficient performance facilitates the proposed MW-SOS-ELM in practical application.

In the future work, the algorithm for evaluation and selection of auto-labeled data will be studied to improve the data capturing strategy and make the labeled data more reliable. Moreover, the transfer learning will be studied to the MW-SOS-ELM based facial expression recognition with trained general algorithm in specific person utilization. Additionally, the further research on larger angel head pose will be done to improve the performance on unconstraint facial expression recognition.

Acknowledgements. This research is partly supported by the National Nature Science Foundation of China (Nos. 91546111, 61672070, 61672071 and 61650201), Beijing Municipal Natural Science Foundation (4152005, 4162058), Key project of Beijing Municipal Education Commission (No. KZ201610005009).

References

1. Cruz, L., Lucio, D., Velho, L.: Kinect and RGBD images: challenges and applications. In: Proceedings: 25th SIBGRAPI - Conference on Graphics, Patterns and Images Tutorials, pp. 36–49. IEEE (2012)
2. Jia, X., Wang, R., Liu, J., Powers, D.M.W.: A semi-supervised online sequential extreme learning machine method. Neurocomputing **174**, 168–178 (2016)
3. Huang, G.-B., Zhu, Q.-Y., Siew, C.-K.: Extreme learning machine: a new learning scheme of feedforward neural networks. In: 2004 IEEE International Joint Conference on Neural Networks Proceedings, pp. 985–990. IEEE (2004)
4. Huang, G., Song, S., Gupta, J.N.D., Wu, C.: Semi-supervised and unsupervised extreme learning machines. IEEE Trans. Cybern. **44**, 2405–2417 (2014)

5. Liang, N.-Y., Huang, G.-B., Saratchandran, P., Sundararajan, N.: A fast and accurate online sequential learning algorithm for feedforward networks. IEEE Trans. Neural Netw. **17**, 1411–1423 (2006)
6. Ekman, P., Friesen, W.V.: Constants across cultures in the face and emotion. J. Pers. Soc. Psychol. **17**, 124–129 (1971)
7. Ekman, P.: Facial expression and emotion. Am. Psychol. **48**, 384–392 (1993)
8. Oualla, M., Sadiq, A., Mbarki, S.: A survey of Haar-Like feature representation. In: International Conference on Multimedia Computing and Systems Proceedings, pp. 1101–1106. IEEE (2014)

Protein Hot Regions Feature Research Based on Evolutionary Conservation

Jing Hu[1,2], Xiaoli Lin[3], and Xiaolong Zhang[1,2(✉)]

[1] School of Computer Science and Technology,
Wuhan University of Science and Technology, Wuhan 430065, Hubei, China
{hujing,xiaolong.zhang}@wust.edu.cn
[2] Hubei Province Key Laboratory of Intelligent Information Processing
and Real-Time Industrial System, Wuhan 430065, Hubei, China
[3] City College, Wuhan University of Science and Technology,
Wuhan 430083, Hubei, China

Abstract. The hot regions of protein interactions refer to the activity scope where hot spots are found to be buried and tightly packing with other residues. The discovery and understanding of hot region is an important way to uncover protein functional activities, such as cell metabolism and signaling pathway, immune recognition and DNA replication, protein synthesis. In this study, machine learning method is used to discover the three aspects features of hot region from sequence conservation, structure conservation and energy conservation, which create conservation scoring algorithm though multiple sequence alignment, module substitute matrix, structural similarity and molecular dynamics simulation. This study has important theoretical and practical significance on promoting hot region research, which also provides a useful way to deeply investigate the functional activities of proteins.

Keywords: Protein interaction · Hot region · Evolution conservation

1 Introduction

Protein functions can be understood by examining protein–protein interactions, which are very useful in understanding the origin of diseases. Hot spots are key binding sites in protein–protein interactions. Instead of being distributed along protein interfaces homogeneously, hot spot residues are usually clustered within tightly packed regions, which are called hot regions [1, 2]. Hot regions of protein–protein interactions play important roles in the functions and stability of protein complexes; they are more important than hot spots in maintaining the stability of protein complexes and exerting the molecular mechanism of biological functions.

In the past, many attempts have been made to research the conservation of hot spots, while there has been little research on the conservation of hot regions. Hot spots in protein-protein interactions are usually more structurally conserved than other surface residues [1, 3, 4], while there is little correlation of conservation between interface residues and other surface residues, either in sequence [5] or in structure [3, 6]. The research

D.-S. Huang et al. (Eds.): ICIC 2017, Part II, LNCS 10362, pp. 254–260, 2017.
DOI: 10.1007/978-3-319-63312-1_23

group [7, 8] in Koc University of Turkey made many contributions to the structure feature of hot regions. Although the proposed research has potential applications for discovering hot region features, it has limitations that evolutionary conservation is not considered.

In this paper, based on evolutionary conservation, machine learning method is used to discover the three aspects features of hot region from sequence conservation, structure conservation and energy conservation, which create conservation scoring algorithm though multiple sequence alignment, module substitute matrix, structural similarity and molecular dynamics simulation.

2   Method

2.1   Evolutionary Conservation on Sequence of Hot Region

Evolutionary conservation on sequence of hot region in different species is illustrated in the following four steps.

(1)   Find isoforms of each gene in the protein complex.

In the first step, isoforms of each gene in protein complexes within hot regions should be identified. Through the Protein data bank [9] (PDB), UniProt ID of each gene of every complex can be obtained according to each chain of every complex. In the UniProt database [10] the sequences of isoforms using UniProt ID are obtained and, if there is more than one isoform in a single gene, the one most closely similar by sequence alignment is chosen as this gene's isoform.

(2)   Find orthologs for each gene in different species using isoforms obtained in the first step.

Orthologs are homologs separated by speciation events. OrthoMCL DB [11, 12] is a database of groups of orthologous protein sequences. Here is a description of the OrthoMCL algorithm:

- All-v-all BLASTP of the proteins
- Compute percent match length. Firstly, Select whichever is shorter, the query or subject sequence. Then call that sequence S.Count all amino acids in S that participate in any HSP. Finally, divide that count by the length of S and multiply by 100.
- Apply thresholds to blast result. Keep matches with E-Value $<1^{e-5}$ percent match length $>=50\%$
- Find potential inparalog, ortholog and co-ortholog pairs using the Orthomcl Pairs program. (These are the pairs that are counted to form the Average Connectivity statistic per group.)
- Use the MCL program to cluster the pairs into groups.

OrthoMCL is a genome-scale algorithm for grouping orthologous protein sequences. It provides not only groups shared by two or more species/genomes, but also groups representing species-specific gene expansion families. So it serves as an important utility for automated eukaryotic genome annotation. OrthoMCL starts with reciprocal best hits within each genome as potential in-paralog/recent paralog pairs and reciprocal best hits across any two genomes as potential ortholog pairs. Related proteins are interlinked in a similarity graph. Then MCL [13] is invoked to split mega-clusters. This process is analogous to the manual review in COG construction. MCL clustering is based on weights between each pair of proteins, so to correct for differences in evolutionary distance the weights are normalized before running MCL.

In this paper, we used isoform sequence to search the OrthoMCL DB to get group id, then we obtained the ortholog distribution status in 150 representative species and all the ortholog sequences in this group by OrthoMCL DB. Among these 150 representative species, there likely is more than a single orthologous sequence in one species, while we just want to select one sequence to represent one species. Therefore we used the isoform sequence as the query sequence and ran an alignment with BioPerl [14] and ClustalW [15] to select the highest scoring ortholog sequence to represent each species.

(3) Perform multiple sequence alignments using orthologs obtained in the second step.

After the above steps, we obtain all the orthologous sequences in ortholog groups that each isoform belongs to, and then used this isoform sequence and all orthologous sequences to do multiple sequence alignments with bioperl and clustalw.

(4) Calculate a conservation score for each hot region using the scoring function and calculate the conservation probability of each hot region compared to other binding sites in different species.

For this paper, we apply BLOSUM62 [16] (BLOcks SUbstitution Matrix) to construct a scoring function. The BLOSUM matrix is a substitution matrix used for sequence alignment of proteins. BLOSUM matrices are used to score alignments between evolutionarily divergent protein sequences [17]. They are based on local alignments and scan the BLOCKS database for very conserved regions of protein families (that do not have gaps in the sequence alignment) and then count the relative frequencies of amino acids and their substitution probabilities. Then, they calculate a log-odds score for each of the 210 possible substitution pairs of the 20 standard amino acids. All BLOSUM matrices are based on observed alignments; they are not extrapolated from comparisons of closely related proteins like the PAM Matrices.

Through BLOSUM, the ith hot spot residue of a hot region in an amino acid sequence changing to another amino acid of the jth ortholog in another species will have a correspond value; we call this value the score$\{i, j\}$. Here i is the position of the hot spot residue in the amino acid sequence, and j is the position of the orthologous gene. We summarize the score$\{i, j\}$ in all different species, and then summarize score$\{i, j\}$ of all hot spot sites within hot region. The conservation score of the whole hot region is described as follows:

$$Hot\ region\_score = \sum_{i=1}^{N} \sum_{j=1}^{K} score\{i, j\} \tag{1}$$

Here K is the number of orthologous genes, and N is the number of hot spot residues in a hot region.

In the next step, we randomly select K sites many times, calculate the conservation scores of these K sites, and then mark M as the number of times there is a higher score than the hot region score. The E-value is the conservation probability of each hot region compared to the other binding sites in different species.

2.2 Evolutionary Conservation on Structure of Hot Region

To study the structural conservation of hot region, the features related to the shape of structure are extracted. After the data normalization, the structure conservation algorithms are constructed from two aspects: the structural similarity and the evolutionary similarity. Structural similarity is the similarity between the shape of the hot region structure and the conservative template in different species, the degree of structural matching reflects the similarity of hot zone structure. Evolutionary similarity is measured by the number of hot spots in the hot region of the protein interaction interface. Polar hotspot residues [18] have strong structural conservation in the evolution. Therefore, the more polar hotspots in hot regions, the more conserved of these hot regions. Specifically, the following three steps can be carried out:

(1) Compare interaction interface structure of hot region with structures in the template interface database.

The protein-protein interaction template interface database [19] stores shape feature of the protein-protein interaction interface, in which there is special docking shape with highly conserved interface as shown in Fig. 1. The above part of Fig. 1 shows a conservative docking shape template. The orange and blue parts represent different proteins, and the black linear regions represent the interaction interface, where have special edge shape. The special edge shape makes this kind of docking template highly conserved. The lower part of Fig. 1 shows visual hot region, where the orange and blue ribbons represent the different protein chains, and the hot spots on the chain are shown as spheres. The protein-protein interaction hot regions like this kind of structure tend to be more conserved in evolution, since docking shape is structurally conserved. For all the protein data to be compared, the first step uses the density-based incremental clustering method to obtain rough hot regions and the second step uses feature-based classification to remove the non-hot spot residues from the clusters obtained by the first step. After extracting the characteristic data of the hot regions, the structural similarity comparison is made between the highly conservative modules in the database of the template interface.

(2) Calculating the number of polar hot spots within hot region.

There are mainly 12 structural physiochemical characters of amino acid, including hydrophobicity, hydrophilicity, polarity, residue accessible surface area, etc. Among these 12 physiochemical characters, polar residues have the most conserved structure, the more polar hot spots in hot region, the more conserved hot region. The number of polar hot spots is an important part in hot region conservation.

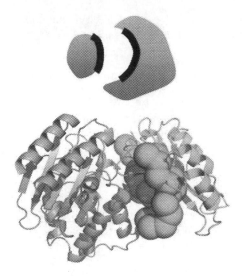

Fig. 1. Docking template shape of hot region (Color figure online)

(3) Calculating structure conservation score of hot region.

The structure conservation $HR_{structure}$ can be calculated by the following formula:

$$HR_{structure} = \alpha f_{evolution} + (1 - \alpha) f_{structure} \tag{2}$$

Here α is the ratio of the relative importance of evolutionary similarity and structural similarity, $f_{evolution}$ is the number of polar hot spots within hot region, $f_{structure}$ is the shape similarity of the hot region and the target template.

2.3　Evolutionary Conservation on Energy of Hot Region

The method to measure the evolutionary conservation on energy of hot region is to calculate the sum of binding free energy of all the hot spots within hot region. Due to the high cost and long cycle, the existing experimental data is only a small amount of protein binding surface resides. For other protein residues that do not have experimental records, it is necessary to use the free energy function to calculate the effect of alanine mutation on the binding free energy of protein complexes. Molecular dynamics simulations are used to estimate changes in binding free energy of protein complexes, as shown in formula (3) and (4).

$$U(x) = \sum_{i,j} \Phi(|X_i - X_j|) = \sum_{i,j} \Phi(r_{i,j}) \tag{3}$$

$$\Phi(r) = 4\varepsilon \left[\left(\frac{\delta}{r} \right)^{12} - \left(\frac{\delta}{r} \right)^{6} \right] \tag{4}$$

Here i and j represent different hot spots within hot region, the formula (3) and (4) use the distance between C^α atom of two residues as variable, which contains two parameters, ε is the depth of potential energy and δ is the distance between two atoms when the potential energy of interaction is zero.

3 Conclusions

The goal of this paper is to illustrate three aspects features of hot region from sequence conservation, structure conservation and energy conservation, which create conservation scoring algorithm though multiple sequence alignment, module substitute matrix, structural similarity and molecular dynamics simulation. In evolutionary conservation of sequence, multiple sequence alignments and score substitution matrix are used to analyze conservative tendency. In evolutionary conservation of structure, evolutionary similarity and structural similarity are used to calculate structure conservation score of hot region. In evolutionary conservation of energy, depth of potential energy and atom distance are used to calculate binding free energy.

We believe the evolutionary conservation of hot regions is very important for analyzing the functions of hot regions and interaction of proteins. In the future, we will continue to collect and publish more experimental data based on wider tests.

Acknowledgment. This work is supported by the National Natural Science Foundation of China (No. 61502356).

References

1. Keskin, O., Ma, B., Nussinov, R.: Hot regions in protein-protein interactions: the organization and contribution of structurally conserved hot spot residues. J. Mol. Biol. **345**, 1281–1294 (2005)
2. Cukuroglu, E., Gursoy, A., Keskin, O.: Analysis of hot region organization in hub proteins. Ann. Biomed. Eng. **38**, 2068–2078 (2010)
3. Bogan, A.A., Thorn, K.S.: Anatomy of hot spots in protein interfaces. J. Mol. Biol. **280**, 1–9 (1998)
4. Ma, B., Elkayam, T., Wolfson, H., Nussinov, R.: Protein-protein interactions: structurally conserved residues distinguish between binding sites and exposed protein surfaces. Proc. Natl. Acad. Sci. U S A **100**, 5772–5777 (2003)
5. Caffrey, D.R., Somaroo, S., Hughes, J.D., Mintseris, J., Huang, E.S.: Are protein-protein interfaces more conserved in sequence than the rest of the protein surface? Pro. Sci. **13**, 190–202 (2004)
6. Aloy, P., Querol, E., Aviles, F.X., Sternberg, M.J.: Automated structure-based prediction of functional sites in proteins: applications to assessing the validity of inheriting protein function from homology in genome annotation and to protein docking. J. Mol. Biol. **311**, 395–408 (2001)
7. Cukuroglu, E., Gursoy, A., Keskin, O.: HotRegion: a database of predicted hot spot clusters. Nucleic Acids Res. **40**, 829–833 (2012)

8. Tuncbag, N., Gursoy, A., Keskin, O.: Identification of computational hot spots in protein interfaces: combining solvent accessibility and inter-residue potentials improves the accuracy. Bioinformatics **25**, 1513–1520 (2009)
9. Berman, H.M., Westbrook, J., Feng, Z., Gilliland, G., Bhat, T.N., Weissig, H., et al.: The protein data bank. Nucleic Acids Res. **28**, 235–242 (2000)
10. Uniport DB. http://www.uniprot.org/uniprot/P01241
11. Fischer, S., Brunk, B.P., Chen, F., Gao, X., Harb, O.S., Iodice, J.B., et al.: Using OrthoMCL to assign proteins to OrthoMCL-DB groups or to cluster proteomes into new ortholog groups. Curr. Protoc. Bioinformatics **12**, 1–19 (2011)
12. Chen, F., Mackey, A.J., Stoeckert Jr., C.J., Roos, D.S.: OrthoMCL-DB: querying a comprehensive multi-species collection of ortholog groups. Nucleic Acids Res. **34**, 363–368 (2006)
13. MCL, Markov Clustering algorithm. http://www.micans.org/mcl
14. BioPerl. http://www.bioperl.org/wiki/Main_Page
15. ClustalW. http://www.ch.embnet.org/software/ClustalW.html
16. BLOSUM62. http://www.uky.edu/Classes/BIO/520/BIO520WWW/blosum62.htm
17. Mount, D.W.: Using BLOSUM in sequence alignments. CSH Protoc. **2008**, 39 (2008)
18. Hu, Z., Ma, B., Wolfson, H., Nussinov, R.: Conservation of polar residues as hot spots at protein interfaces. Proteins Struct. Funct. Bioinf. **39**(4), 331–342 (2000)
19. Protein-Protein Interaction Template Interface Database. http://cosbi.ku.edu.tr/prism

CMFHMDA: Collaborative Matrix Factorization for Human Microbe-Disease Association Prediction

Zhen Shen[✉], Zhichao Jiang, and Wenzheng Bao

School of Electronics and Information Engineering,
Institute of Machine Learning and Systems Biology,
Tongji University, Shanghai 201804, China
zzuliszhen@163.com

Abstract. The research on microorganisms indicates that microbes are abundant in human body, which have closely connection with various human noninfectious diseases. The deep research of microbe-disease associations is not only helpful to timely diagnosis and treatment of human diseases, but also facilitates the development of new drugs. However, the current knowledge in this domain is still limited and far from complete. Here, we proposed the computational model of Collaborative Matrix Factorization for Human Microbe-Disease Association prediction (CMFHMDA) by integrating known microbe-disease associations and Gaussian interaction profile kernel similarity for microbes and diseases. A special matrix factorization algorithm was introduced here to update the correlation matrix about microbes and diseases for inferring the most possible disease-related microbes. Leave-one-out Cross Validation (LOOCV) and k-fold cross Validation were implemented to evaluate the prediction performance of this model. As a result, CMFHMDA obtained AUCs of 0.8858 and 0.8529 based on 5-fold cross validation and Global LOOCV, respectively. It is no doubt that CMFHMDA could be used to identify more potential microbes associated with important noninfectious human diseases.

Keywords: Microbe · Disease · Similarity · Collaborative matrix factorization · Gaussian interaction profile

1 Introduction

The distribution of Microorganism is very wide in oceans, soils, plants, human body and other locations. They include viruses, eukaryotes, bacteria, protozoa, fungi, and archaea [1–3]. For humans, we can find a good amount of microbiome in body, for instance, skin, uterus, lung, saliva, gastrointestinal tracts and so on [4–6]. Existing evidences have shown that microbiome plays extensive and important roles in human physiological activities, including metabolic, intestinal cell proliferation and differentiation, innate and acquired responses to pathogens, and the development of the immune system [7–9]. For example, indigenous microbes can regulate the development of the intestinal villus microvasculature through Paneth cells [4, 10–12]. Colonic microbiotas could boost the metabolism of non-digestible dietary residue, especially non-digestible carbohydrates,

© Springer International Publishing AG 2017
D.-S. Huang et al. (Eds.): ICIC 2017, Part II, LNCS 10362, pp. 261–269, 2017.
DOI: 10.1007/978-3-319-63312-1_24

including large polysaccharides, some oligosaccharides, unabsorbed sugars and alcohols [8, 13–15]. The metabolites of microbiota, like Short chain fatty acids(SCFAs), Tryptophan metabolites, Retinoic acid(RA), has important influence on immune system development and differentiation [16, 17]. The vaginal Lactobacillus species can protect female vaginal health from pathogen invasion by producing hydrogen peroxide and lactic acid [18–20]. The adult intestine contains approximately 10^{14} bacterial cells, which is ten times bigger than the amount of human cell [4, 21, 22]. The microbiome has more than 5 million genes, which is significantly larger than the number of human genes [4, 23–25]. Many of this microbiome can produce enzymes, proteases and glycosidases, and thus they contribute to the host's own biochemical and metabolic capability. Although the researchers have done a great job of microbiome, there still exist some limits about the whole understanding of microbiome.

Existed studies have shown that microbes are closely related with various human noninfectious diseases. Therefore, some researchers have focus on the association between microbes and disease and proposed computational model to identify novel microbe-disease associations. For example, Ma et al. [26] constructed a microbe-disease association network (HMDAD) by using large-scale text mining-based manually curated microbe–disease association data set. The relationships between microbes and disease genes, symptoms, chemical fragments and drugs were also be investigated at the same time. The results show that there are remarkably coherent in microbe-based disease loops. They also outline a theory that microbes with similarity function are often associated with similar diseases. Furthermore, they confirm that this is an effective method to identify potential associations and mechanisms for disease, microbes, genes and drugs by model and analyze the associations between microbe and diseases. In order to predict the association between microbe and diseases, Chen et al. [27] proposed the model of KATZ measure for Human Microbe-Disease Association prediction (KATZHMDA). KATZHMDA could be used to identify potential microbe-diseases associations by integrating known microbe-disease associations and Gaussian interaction profile kernel similarity for microbes and diseases. This model requires only the topology information of known microbe-disease association network as information source, without the use of biological datasets. Therefore, the prediction performance of this model is greatly effect by the reliable and available of known microbe-disease association network.

Considering the flaw of current prediction model and the urgency of drug discovery and disease treatment, we proposed the model of Collaborative Matrix Factorization for Human Microbe-Disease Association prediction (CMFHMDA) to identify novel microbe-disease association by integrating known experimental validated microbe-disease associations and Gaussian interaction profile kernel similarity for microbes and diseases. Based on the known experimental validated microbe-disease associations in HMDAD database, we use two ways: leave-one-out cross validation (LOOCV) and 5-fold cross validation, to evaluate the prediction performance of CMFHMDA. Furthermore, we also compare the prediction performance of CMFHMDA with two other models: Regularized Least Squares [28] and Within and Between Scores [29], which were used to predict novel miRNA-disease associations. As a result, CMFHMDA obtained the AUC of 0.8858 for 5-fold cross validation, and AUC of 0.8529 for LOOCV. Experiment results indicated that

CMFHMDA is a reliable and effective model to predict microbe-disease associations by only using known experimental validated microbe-disease associations.

2 Materials and Methods

2.1 Human Microbe-Disease Associations

We downloaded the microbe-disease associations from the Human Microbe-Disease Association Database (HMDAD, http://www.cuilab.cn/hmdad). The content in HMDAD was mainly collected from the 16s RNA sequencing-based microbiome studies which only give out genus-level information. Including 483 experimentally confirmed human microbe-disease associations about 39 human diseases and 292 microbes. Based on different evidence, we obtained 450 distinct associations by collecting and analyzing the associations between microbe and disease. We defined the microbe-disease association network as adjacency matrix Y. If there is an association between microbe $m(i)$ and disease $d(j)$, the entity $Y(m(i), d(j))$ is 1, otherwise 0. Furthermore, two variables nm and nd were declared to represent the number of microbes and diseases investigated in this paper, respectively.

2.2 Gaussian Interaction Profile Kernel Similarity for Microbes

Under the assumption that microbes with functionally similarity are often related to similar diseases and therefore share the similar interaction and non-interaction patterns with diseases, we cover ways to compute microbe similarity from known microbe-disease association network using Gaussian interaction profile kernel similarity for microbe [30]. This procedure usually consists of two steps. Firstly, the interaction profile of microbe $m(i)$ is declared by a binary vector $AP(m(i))$ for recording whether microbe $m(i)$ has association with each disease or not. In the second step, the kernel similarity between each microbe pair was calculated based on their Gaussian interaction profiles as follows.

$$AKM(m(i), m(j)) = \exp\left(-\gamma_m \|AP(m(i)) - AP(m(j))\|^2\right) \tag{1}$$

$$\gamma_m = \gamma_m' \left/ \left[\frac{1}{nm}\sum_{i=1}^{nm} \|AP(m(i))\|^2\right]\right. \tag{2}$$

where, γ_m regulates the normalized kernel bandwidth based on the new bandwidth parameter γ_m'. Each record $AKM(m(i), m(j))$ represents the Gaussian interaction profile kernel similarity between disease $m(i)$ and $m(j)$.

2.3   Gaussian Interaction Profile Kernel Similarity for Diseases

Based on the assumption that disease which share the functionally similar microbes tend to be similar, the Gaussian interaction profile kernel similarity for disease could be calculated in a similar way as microbes, which is defined as follows:

$$GKD(d(i), d(j)) = \exp\left(-\gamma_d \|GP(d(i)) - GP(d(j))\|^2\right) \tag{3}$$

$$\gamma_d = \gamma_d' \left/ \left[\frac{1}{nd} \sum_{i=1}^{nd} \|GP(d(i))\|^2\right]\right. \tag{4}$$

where, GKD is the Gaussian interaction profile kernel similarity for all investigated diseases and γ_d controls the kernel bandwidth, which could be calculated through normalizing a new bandwidth parameter γ_d' by the average number of associations with microbes per disease.

2.4   CMFHMDA

In this work, we developed the computational model of Collaborative Matrix Factorization for Human Microbe-Disease Association prediction (CMFHMDA) to identify novel microbe-disease associations. Figure 1 shows the flowchart of CMFHMDA.

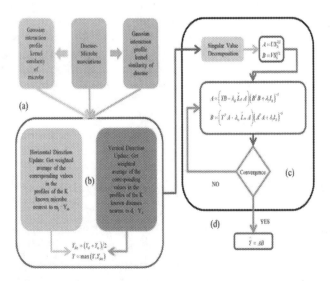

Fig. 1. Flowchart of CMFHMDA based on known microbe-disease association network.

Firstly, microbe similarity matrix *SM* and disease similarity matrix *SD* was obtained by integrating experimentally verified microbe-disease associations and Gaussian interaction profile kernel similarity for microbes and diseases.

Then, WKNKN [31] was used to calculate the association likelihood for these unknown cases based on their known neighbors.

Thirdly, final prediction *Y* could be obtained by the use of Collaborative Matrix Factorization. Three stages were conducted in this step.

(1) For the input matrix *Y*, singular value decomposition was adopted to obtain the initial value of *A* and *B*.

$$[U, S, V] = SVD(Y, k)$$
$$A = US_k^{1/2} \tag{5}$$
$$B = VS_k^{1/2}$$

(2) Here, *L* was used to denote the objection function. a_i and b_j to denotes the ith and j^{th} row vectors of *A* and *B*,respectively. Two alternative update rules (one for updating matrix *A* and one for updating matrix *B*) were derived by setting as formula 6. Based on alternating least squares, these two update rules are run alternatingly until convergence.

$$\frac{\partial L}{\partial A} = 0 \qquad \frac{\partial L}{\partial B} = 0 \tag{6}$$

$$\min_{A,B} \quad \left\| Y - AB^T \right\|_F^2 + \lambda_l \left(\|A\|_F^2 + \|B\|_F^2 \right) + \lambda_m \left\| SM - AA^T \right\|_F^2 + \lambda_d \left\| SD - BB^T \right\|_F^2 \tag{7}$$

$$A = \left(YB + \lambda_d S_d A \right) \left(B^T B + \lambda_l I_k + \lambda_d A^T A \right)^{-1} \tag{8}$$

$$B = \left(Y^T A + \lambda_m S_m B \right) \left(A^T A + \lambda_l I_k + \lambda_m B^T B \right)^{-1} \tag{9}$$

Finally, the predicted matrix for miRNA-disease associations is then obtained by multiplying *A* and *B*.

3   Results

We implemented k-fold cross validation and global LOOCV to evaluate the prediction performance of CMFHMDA and KATZHMDA based on the known microbe-disease associations in the HMDAD database. Furthermore, we also compare the prediction performance of CMFHMDA with two other models: RLSMDA and WBSMDA, which were used to identify novel miRNA-disease associations.

For global LOOCV, all investigated diseases should be taking into account simultaneously. Each known microbe-disease association was left out in turns as a test sample and others for training samples. According to their prediction scores, all unverified microbe-disease associations in HMDAD database would be sorted, including the single test sample. The test sample with a higher rank than the given threshold would be considered as a successful. For k-fold cross validation, all the known microbe-disease association samples were randomly equally divided into k equal parts. And k−1 parts were then used as training samples for model learning while the rest part was used as testing samples for model evaluation. To evaluate the pre-diction performance of CMFHMDA, we obtained the areas under ROC curve (AUC). AUC value of 1 denotes a perfect prediction while the AUC value of 0.5 indicates purely random performance.

As shown in Fig. 2, CMFHMDA obtained a reliable AUC of 0.8858 based on 5-fold cross validation and 0.8529 based on global LOOCV. With the same parameters and datasets, KATZHMDA obtained AUCs of 0.8302 based on 5-fold cross validation and 0.8382 based on global LOOCV. Furthermore, Fig. 3 was used to shown the comparison between CMFHMDA and two other models: RLSMDA and WBSMDA. Under the same circumstance, RLS obtained AUCs of 0.5300 based on 5-fold cross validation and 0.4958 based on global LOOCV, WBS obtained AUCs of 0.3570 based on 5-fold cross validation and 0.2887 based on global LOOCV.

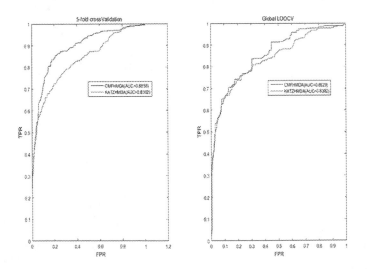

Fig. 2. Performances comparisons between CMFHMDA and KATZHMDA in terms of ROC curve and AUC based on 5-fold cross validation and global LOOCV

Figures 2 and 3 shown the reliable and effective prediction performance of CMFHMDA in the LOOCV and k-fold cross validation. By using the experimentally confirmed microbe-disease associations in HMDAD, we prioritized all the candidate microbes for the diseases recorded in HMDAD database. In future, we expected that we can use experiment to verify the microbe-disease associations with higher ranks.

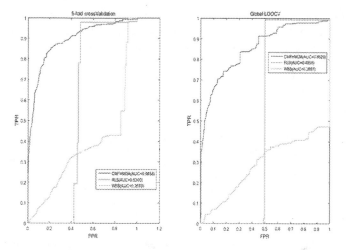

Fig. 3. Prediction performance of different models comparing with CMFHMDA in terms of ROC curve and AUC based on 5-fold cross validation and global LOOCV

4 Discussion and Conclusion

Accumulating evidences indicated that microbes are often associated with various human disease, especially noninfectious human diseases. Modeling and analysis of known microbe-disease associations could improve the discovery of new biomarker and drug. Not only that, it also can help researchers discover new therapies. Therefore, it is no doubt that a reliable and effective computational model is necessary to identify novel microbe-disease associations. In this work, we proposed model CMFHMDA to identify novel microbe-disease associations by integrating known microbe-disease associations and Gaussian interaction profile kernel similarity for microbes and disease. K-fold cross validation and LOOCV were chosen to evaluate the predict performance of CMFHMDA. The results of 5-fold cross validation and LOOCV shows that CMFHMDA has better predict performance than other models. Furthermore, CMFHMDA could be used to identify novel microbes associated with all investigated diseases simultaneously.

Generally speaking, the reliable performance of CMFHMDA could be further attributed to the following reasons. First, the known microbe-disease associations provide by HMDAD is verified by experiments, which can be treated as a reliable biological datasets. Second, the Gaussian interaction profile kernel similarity was used to accurately measure microbe similarity and disease similarity. Finally, Collaborative Matrix Factorization algorithm can fully explore the potential association information in the network for predicting the novel microbe-disease associations.

Of course, CMFHMDA still has some limitations that need to be improved in the future. Firstly, the microbe-disease association obtained from HMDAD is sparse and thus restricted the prediction performance. This problem could be improved by identify and collect more microbe-disease associations in the future. Furthermore, the calculate of Gaussian interaction profile kernel similarity is heavily reliant on the known microbe-disease associations, and therefore would cause inevitable bias towards those well-investigated diseases and microbes. In other words, the diseases with more related microbe records in database would be more possibly predicted to be associated with more potential microbes. And the same goes for the microbes with more related disease records.

Acknowledgement. This work was supported by the grants of the National Science Foundation of China, Nos. 61520106006, 31571364, U1611265, 61672203, 61402334, 61472280, 61472282, 61532008, 61472173, 61572447, 61373098 and 61672382, China Postdoctoral Science Foundation Grant, Nos. 2016M601646.

References

1. Gilbert, J.A., Dupont, C.L.: Microbial metagenomics: beyond the genome. Annu. Rev. Mar. Sci. **3**, 347–371 (2011)
2. Methé, B.A., Nelson, K.E., Pop, M., Creasy, H.H., Giglio, M.G., Huttenhower, C., et al.: A framework for human microbiome research. Nature **486**, 215 (2012)
3. Yu, H.-J., Huang, D.-S.: Normalized feature vectors: a novel alignment-free sequence comparison method based on the numbers of adjacent amino acids. IEEE/ACM Trans. Comput. Biol. Bioinform. (TCBB) **10**, 457–467 (2013)
4. Sommer, F., Bäckhed, F.: The gut microbiota—masters of host development and physiology. Nat. Rev. Microbiol. **11**, 227–238 (2013)
5. Willey, J., Sherwood, L., Woolverton, C.: Prescott's Microbiology. McGraw-Hill Higher Education, New York (2013)
6. Huang, D.-S., Zheng, C.-H.: Independent component analysis-based penalized discriminant method for tumor classification using gene expression data. Bioinformatics **22**, 1855–1862 (2006)
7. Ventura, M., O'Flaherty, S., Claesson, M.J., Turroni, F., Klaenhammer, T.R., van Sinderen, D., et al.: Genome-scale analyses of health-promoting bacteria: probiogenomics. Nat. Rev. Microbiol. **7**, 61–71 (2009)
8. Guarner, F., Malagelada, J.-R.: Gut flora in health and disease. Lancet **361**, 512–519 (2003)
9. Zhu, L., You, Z.-H., Huang, D.-S., Wang, B.: t-LSE: a novel robust geometric approach for modeling protein-protein interaction networks. PLoS ONE **8**, e58368 (2013)
10. Reinhardt, C., Bergentall, M., Greiner, T.U., Schaffner, F., Östergren-Lundén, G., Petersen, L.C., et al.: Tissue factor and PAR1 promote microbiota-induced intestinal vascular remodelling. Nature **483**, 627–631 (2012)
11. Deng, S.-P., Zhu, L., Huang, D.-S.: Predicting hub genes associated with cervical cancer through gene co-expression networks. IEEE/ACM Trans. Comput. Biol. Bioinform. (TCBB) **13**, 27–35 (2016)
12. Huang, D.-S., Du, J.-X.: A constructive hybrid structure optimization methodology for radial basis probabilistic neural networks. IEEE Trans. Neural Netw. **19**, 2099–2115 (2008)
13. Roberfroid, M., Bornet, F., Bouley, C., Cummings, J.: Colonic microflora: nutrition and health. Summary and conclusions of an International Life Sciences Institute (ILSI)[Europe] workshop held in Barcelona, Spain. Nutr. Rev. **53**, 127–130 (1995)
14. Cummings, J.H., Beatty, E.R., Kingman, S.M., Bingham, S.A., Englyst, H.N.: Digestion and physiological properties of resistant starch in the human large bowel. Br. J. Nutr. **75**, 733–747 (1996)

15. Zhu, L., Guo, W.-L., Deng, S.-P., Huang, D.-S.: ChIP-PIT: enhancing the analysis of ChIP-Seq data using convex-relaxed pair-wise interaction tensor decomposition. IEEE/ACM Trans. Comput. Biol. Bioinf. **13**, 55–63 (2016)

16. Levy, M., Thaiss, C.A., Elinav, E.: Metabolites: messengers between the microbiota and the immune system. Genes Dev. **30**, 1589–1597 (2016)

17. Zhu, L., Deng, S.-P., Huang, D.-S.: A two-stage geometric method for pruning unreliable links in protein-protein networks. IEEE Trans. Nanobiosci. **14**, 528–534 (2015)

18. Wang, Z., Yang, Y., Stefka, A., Sun, G., Peng, L.: Review article: fungal microbiota and digestive diseases. Aliment. Pharmacol. Ther. **39**, 751–766 (2014)

19. Petrova, M.I., Lievens, E., Malik, S., Imholz, N., Lebeer, S.: Lactobacillus species as biomarkers and agents that can promote various aspects of vaginal health. Front. Physiol. **6**, 81 (2015)

20. Deng, S.-P., Zhu, L., Huang, D.-S.: Mining the bladder cancer-associated genes by an integrated strategy for the construction and analysis of differential co-expression networks. BMC Genom. **16**, S4 (2015)

21. Xu, J., Gordon, J.I.: Honor thy symbionts. Proc. Natl. Acad. Sci. **100**, 10452–10459 (2003)

22. Deng, S.-P., Huang, D.-S.: SFAPS: an R package for structure/function analysis of protein sequences based on informational spectrum method. Methods **69**, 207–212 (2014)

23. H. M. P. Consortium: Structure, function and diversity of the healthy human microbiome. Nature **486**, 207–214 (2012)

24. Qin, J., Li, R., Raes, J., Arumugam, M., Burgdorf, K.S., Manichanh, C., et al.: A human gut microbial gene catalogue established by metagenomic sequencing. Nature **464**, 59–65 (2010)

25. Huang, D.-S., Zhang, L., Han, K., Deng, S., Yang, K., Zhang, H.: Prediction of protein-protein interactions based on protein-protein correlation using least squares regression. Curr. Protein Pept. Sci. **15**, 553–560 (2014)

26. Ma, W., Zhang, L., Zeng, P., Huang, C., Li, J., Geng, B., et al.: An analysis of human microbe–disease associations. Briefings Bioinform. **18**(1), 85–97 (2016). bbw005

27. Chen, X., Huang, Y.-A., You, Z.-H., Yan, G.-Y., Wang, X.-S.: A novel approach based on KATZ measure to predict associations of human microbiota with non-infectious diseases. Bioinformatics **33**(5), 733–739 (2016). btw715

28. Chen, X., Yan, G.Y.: Semi-supervised learning for potential human microRNA-disease associations inference. Sci. Rep. **4**, 5501 (2014)

29. Chen, X., Yan, C.C., Zhang, X., You, Z.H., Deng, L., Liu, Y., et al.: WBSMDA: Within and between score for MiRNA-disease association prediction. Sci. Rep. **6**, 21106 (2016)

30. Chen, X., Yan, G.Y.: Novel human lncRNA-disease association inference based on lncRNA expression profiles. Bioinformatics **29**, 2617–2624 (2013)

31. Ezzat, A., Zhao, P., Wu, M., Li, X., Kwoh, C.K.: Drug-target interaction prediction with graph regularized matrix factorization. IEEE/ACM Trans. Comput. Biol. Bioinform. **14**(3), 646–656 (2016)

Computational Genomics

Accurately Estimating Tumor Purity of Samples with High Degree of Heterogeneity from Cancer Sequencing Data

Yu Geng[1,4,5], Zhongmeng Zhao[1,3], Ruoyu Liu[2,3], Tian Zheng[2,3],
Jing Xu[1,3], Yi Huang[2,3], Xuanping Zhang[1,3], Xiao Xiao[3,4],
and Jiayin Wang[2,3(✉)]

[1] School of Electronic and Information Engineering,
Xi'an Jiaotong University, Xi'an 710049, China
[2] School of Management, Xi'an Jiaotong University, Xi'an 710049, China
[3] Institute of Data Science and Information Quality,
Shaanxi Engineering Research Center of Medical and Health Big Data,
Xi'an Jiaotong University, Xi'an 710049, China
{zmzhao,wangjiayin}@mail.xjtu.edu.cn
[4] State Key Laboratory of Cancer Biology,
Xijing Hospital of Digestive Diseases, Xi'an 710032, China
[5] Jinzhou Medical University, Jinzhou 121001, China

Abstract. Tumor purity is the proportion of tumor cells in the sampled admixture. Estimating tumor purity is one of the key steps for both understanding the tumor micro-environment and reducing false positives and false negatives in the genomic analysis. However, existing approaches often lose some accuracy when analyzing the samples with high degree of heterogeneity. The patterns of clonal architecture shown in sequencing data interfere with the data signals that the purity estimation algorithms expect. In this article, we propose a computational method, *EMPurity*, which is able to accurately infer the tumor purity of the samples with high degree of heterogeneity. *EMPurity* captures the patterns of both the tumor purity and clonal structure by a probabilistic model. The model parameters are directly calculated from aligned reads, which prevents the errors transferring from the variant calling results. We test *EMPurity* on a series of datasets comparing to three popular approaches, and *EMPurity* outperforms them on different simulation configurations.

Keywords: Cancer genomics · Sequencing data analysis · Tumor purity · Tumor heterogeneity · Probabilistic model

1 Introduction

Benefiting from the great achievements on cancer genomics [1, 2], nowadays, it becomes a routine work to sequence cancer patients in the clinical practice. The somatic mutational events are easily obtained by analyzing the pair-sampled DNA sequencing data, and play an important role in precise diagnosis and treatment. Because

© Springer International Publishing AG 2017
D.-S. Huang et al. (Eds.): ICIC 2017, Part II, LNCS 10362, pp. 273–285, 2017.
DOI: 10.1007/978-3-319-63312-1_25

the popular sequencing technologies require a considerable volume of DNA content, the tumor sample sequenced is actually an admixture that contains non-cancerous cells.

Tumor purity is the proportion of tumor cells in the sampled admixture, which varies widely among samples and cancer types and may reach a quite low degree. A research on pancreatic adenocarcinomas reports that more than 70% of tumor samples are of less than 40% tumor purity. It is also reported that in many samples of early stage breast cancer, the tumor purity is among 55% [3]. Tumor purity not only is a key indicator revealing the tumor micro-environment, but is a major factor that influences the accuracy of the genomic analysis. A recent study reports that poor tumor purity introduces errors in variant calling pipelines. It is estimated that every 2% non-cancerous proportion in tumor sample can raise 166 false positive calls per megabase and 10 false positive calls per megabase in variant calling, even when the known SNP sites are excluded [4]. Unfortunately, many variant calling approaches, which include those designed for cancer genomic data, cannot handle purity issue independently. Some popular tools, such as SomaticSniper [5] and JointSNVMix [6], either assume satisfied tumor purity or require the estimates of the tumor purity to adjust the given sequencing data. Moreover, tumor purity has a strong influence on the data signals representing the ploidy information and the downstream analysis in multiple ways. Thus, accurately inferring tumor purity is urgently needed for analyzing cancer sequencing data.

Tumor purity used to be estimated by visual or image analysis. Recently, several computational approaches are proposed to infer tumor purity from the DNA/RNA microarray data and cancer sequencing data. These state-of-the-art approaches may fall into two categories. One is the array-data based approaches, which include *ABSOLUTE* [7], *ASCAT* [3], *CNAnorm* [8] and *THetA* [9]. A major limitation of these approaches is that an additional microarray test is required for patients. It is not practical for the samples without copy number profiles [10]. The other way of estimating tumor purity is directly from the cancer sequencing data. The basic idea of some popular methods, such as *PurityEst* [11], *AbsCN_seq* [12] and *PurBayes* [13] is to capture the signals of variant allelic frequencies on particular sites and identify the proportion contributed by tumor purity. *PurityEst* sets a statistical model at each selected heterozygous locus with somatic mutations and implements extreme studentized deviate multiple-outlier procedure to obtain the estimations. *PurBayes* introduces a Bayesian framework, which clusters the given somatic mutations into sub-clones. The number of sub-clones is computed by a priori distribution. *AbsCN_seq* infers tumor purity by identifying the patterns of reads mapping to the copy number variations. However, most of the solid tumor samples are also admixtures of clonal populations. The patterns of clonal architecture shown in sequencing data often interfere with the data signals expected by purity estimation algorithms. Thus, for the samples with high degree of heterogeneity, the existing methods significantly lose accuracy.

To overcome this weakness, in this article, we propose a computational approach, *EMPurity*, to accurately estimate tumor purity, for the samples with tumor heterogeneity. *EMPurity* establishes a probabilistic model, where the tumor purity is modeled together with the clonal structure. The model parameters are directly calculated from aligned reads, which prevents the errors transferring from the variant calling results. We compare *EMPurity* to the aforementioned approaches, and it outperforms the others under different simulation configurations, especially when multiple sub-clones exist.

2 Methods

We focus on the problem of inferring the tumor purity when the sample has high-degree of tumor heterogeneity. Assume that we are given a pair-sampled DNA sequencing data. Similar to the existing approaches [11–13], we only consider the heterozygous sites with somatic mutations. For one sample in the pair, the set of possible genotype values at each loci is $G = \{AA, AB, BB\}$. Let N, T and T_M represent the normal sample, virtual pure tumor sample and tumor sample, respectively. Here, the virtual pure tumor sample is actually part of T_M, and designed to facilitate the computation. Then, for the paired samples, the set of possible combined genotype values is a Cartesian product, which is $G \times G = \{(G_N, G_T) : G_N, G_T \in G\}$. Suppose that we are given a total of I sites. For each site, we calculate the numbers of reads supporting the reference allele and mutation, respectively.

2.1 EMPurity Model

Due to tumor heterogeneity, we assume that a tumor sample consists of the sequencing reads from the non-cancerous cells and from multiple clonal populations of tumor cells. Only one founding clone is considered in this model. The observed indicators at each site are the numbers of reads supporting the reference allele and mutation, respectively, while the unknown hidden states include the tumor purity, the proportions of founding clone and sub-clones and the joint genotype. We design a probabilistic model to describe the emission probabilities from the hidden states to the observed indicators and the transition probabilities among the hidden states. As linkage disequilibrium is not considered in somatic mutations, each site is computed independently. This model is solved by an expectation maximization algorithm.

For site i, let $n^i_{N\_ref}$ and $n^i_{T_M\_ref}$ denote the numbers of reads supporting the reference allele in the normal sample and in tumor sample, respectively, each of which follows binomial distribution with parameters μ_N and μ_{T_M}. There are only 9 possible joint genotypes, which follow a polynomial distribution with parameter μ_G. Considering the bias on read depth, we assume that tumor purity follows a normal distribution across all of the given sites, whose parameters are μ_p and λ_p. To facilitate the Bayesian computation, we introduce the conjugate prior distribution of the parameters. And then the posterior distribution is calculated. μ_G, which is the parameter of the polynomial distribution follows a Dirichlet conjugate prior distribution with super-parameter α_G. μ_N and μ_{T_M} follow the Beta conjugate prior distributions, whose super-parameters are $\alpha_{N(G_N)}, \beta_{N(G_N)}$ and $\alpha_{T(G_T)}, \beta_{T(G_T)}$, respectively. μ_p and λ_p follow the normal-gamma conjugate prior distribution, whose super-parameters are μ_0, $\left(\beta_p \lambda_p\right)^{-1}$ and $\left(a_p, b_p\right)$.

Likelihood function. As this model is solved by an expectation maximization algorithm, first we have to establish a likelihood function. A graphical presentation of the model is shown in Fig. 1. For site i, we consider the following values, which are $R^i = \left\{ n^i_{x\_ref}, n^i_{\overline{x\_ref}} \right\}$ and $D^i = \{ n^i_{x\_d} \}$, $x \in \{ N, T, T_M \}$. Let $n^i_{\overline{x\_ref}}$ be the number of reads supporting the mutation in x. Let $n^i_{x\_d}$ represent the read depth in x. For $x \in \{ N, T_M \}$, these values are observed. And then, the estimation of tumor purity $\hat{p} = n^i_{T\_d} / n^i_{T_M\_d}$. Let $\mathcal{G}$ denote the random variable representing the joint genotype $\left\{ G^i_{(G_N, G_T)} \right\}$. Let ϑ represent the set of unknown parameters, which is $\vartheta = \{ \mu_N, \mu_T, \mu_G, \mu_p, \lambda_p^{-1} \}$. Suppose that $\mu_{G(G_N, G_T)}$ is the mixing coefficient of the marginal probability $p \left(G^i_{(G_N, G_T)} \right)$, where $\mu_{G(G_N, G_T)}$ satisfies $0 \le \mu_{G(G_N, G_T)} \le 1$ and $\sum_{G_N \in G} \sum_{G_T \in G} \mu_{G(G_N, G_T)} = 1$.

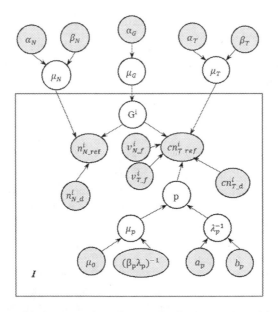

Fig. 1. A graphical presentation of the probabilistic model used in *EMPurity*.

Thus, the likelihood function for the complete set $\{ R, D, \mathcal{G} \}$ is

$$
\begin{aligned}
&L(R, D, \mathcal{G}; \vartheta) \\
&= \prod_{i=1}^{I} \prod_{G_N \in G} \prod_{G_T \in G} \left[\mu_{G(G_N, G_T)} \mathrm{Bin} \left(n^i_{x\_ref} | n^i_{x\_d}, \mu_{x(G_x)} \right) \mathrm{N} \left(p^i_{(G_N, G_T)} | \mu_{p^i_{(G_N, G_T)}}, \lambda^{-1}_{p^i_{(G_N, G_T)}} \right) \right]^{G^i_{(G_N, G_T)}} \quad x \in \{ N, T \}
\end{aligned} \tag{1}
$$

The lower bound of the likelihood function is

$$L(\log p(\vartheta|R,D)) = \log p_{\vartheta}^{(R,D,\mathcal{G})} + \log p(\vartheta) =$$

$$\sum_{i=1}^{I} \left(\sum_{G_N \in G} \sum_{G_T \in G} \mu_{G(G_N,G_T)} \text{Bin}\left(n_{x_{ref}}^{i} | n_{x_d}^{i}, \mu_{x(G_x)}\right) N\left(p_{(G_N,G_T)}^{i} | \mu_{p_{(G_N,G_T)}^{i}}, \lambda_{p_{(G_N,G_T)}^{i}}^{-1}\right) \right)$$

$$+ \sum_{G_N \in G} \left(\alpha_{N(G_N)} - 1\right) \log \mu_{N(G_N)} + \left(\beta_{N(G_N)} - 1\right) \log\left(1 - \mu_{N(G_N)}\right)$$

$$+ \sum_{G_T \in G} \left(\alpha_{T(G_T)} - 1\right) \log \mu_{T(G_T)} + \left(\beta_{T(G_T)} - 1\right) \log\left(1 - \mu_{T(G_T)}\right)$$ (2)

$$+ \sum_{G_N \in G} \sum_{G_T \in G} \left(\delta_{(G_N,G_T)} - 1\right) \log G_{(G_N,G_T)}^{i} + \frac{1}{2}\log \beta_p \lambda_p - \frac{\beta_p \lambda_p \left(\mu_p - \mu_0\right)^2}{2}$$

$$+ \left(a_p - 1\right) \log \lambda_p - b_p \lambda_p + c$$

where c is a constant value.

E-step. The EM algorithm is an iterative process that consists of two steps in each iteration: Expectation step (E-step) and Maximization step (M-step). In the E-step, current parameters ϑ^{old} is used to calculate the posterior distributions of the hidden states and then those are replaced by $p(\mathcal{G}|R^i, D^i, \vartheta^{old})$. The expected value of the likelihood function is then obtained according to the posterior distributions. Thus, the responsibilities of the indicator variables $G_{(G_N,G_T)}^{i}$ are

$$\gamma\left(G_{(G_N,G_T)}^{i}\right) = E\left[G_{(G_N,G_T)}^{i}\right] = \frac{p\left(G_{(G_N,G_T)}^{i} = 1\right) p\left(Y|G_{(G_N,G_T)}^{i} = 1\right)}{\sum_{\mathcal{G}} p(\mathcal{G}=1) p(Y|\mathcal{G}=1)}$$ (3)

And then, the virtual pure tumor sample is computed from the given tumor sample, and $\hat{p}$ is inferred here:

$$n_{T\_ref}^{i} = n_{T_M\_ref}^{i} - p_{(G_N,G_T)}^{i}\left(1 - v_{N\_f}^{i}\right) \times n_{T_M\_d}^{i}$$ (4)

$$n_{T\_d}^{i} = n_{T_M\_d}^{i} - p_{(G_N,G_T)}^{i} \times n_{T_M\_d}^{i}$$ (5)

where $p_{(G_N,G_T)}^{i}$ is the proportion of non-cancerous cells and $v_{N\_f}^{i}$ is the variant allelic frequency in the normal sample.

$$v_{N\_f}^{i} = (n_{N\_d}^{i} - n_{N\_ref}^{i})/n_{N\_d}^{i}$$ (6)

M-step. In the M-step, current responsibilities are used to maximize the objective function, and then a set of new parameters ϑ^{new} is reset accordingly. We have

$$\mathcal{O}(\vartheta, \vartheta^{old}) = \mathrm{E}\big[\log p(R^i, D^i, \mathcal{G}|\vartheta^{old})\big] + \log p\Big(\mu_N, \mu_T, \mu_G, \mu_p, \lambda_p^{-1}\Big) \qquad (7)$$

Each parameter is considered as conditionally independent, and then the priors are:

$$p\Big(\mu_N, \mu_T, \mu_G, \mu_p, \lambda_p^{-1}\Big) = p(\mu_N)p(\mu_T)p(\mu_G)p(\mu_p|\lambda_p)p(\lambda_p)$$

$$= \prod_{G_N \in G} \prod_{G_T \in G} p(\mu_G) \prod_{G_N \in G} p\Big(\mu_{N(G_N)}\Big) \left\{ \prod_{G_T \in G} p\Big(\mu_{T(G_T)}\Big) \left\{ \prod_{G_N \in G} \prod_{G_T \in G} p\Big(\mu_{p(G_N, G_T)}, \lambda_{p(G_N, G_T)}^{-1}\Big) \right\} \right\}$$

$$(8)$$

Once we have the E-step and M-step, the algorithm implements are as follows: First, all of the unknown parameters are initialized, which include the hyper parameters $\alpha_x, \beta_x, a_p, b_p, \beta_p$, the distribution parameters μ_0 and λ_p and the coefficients of the genotypes. Second, the algorithm alters between updating the responsibilities and estimating the parameters. The algorithm in the E-step uses $\gamma\Big(G^i_{(G_N, G_T)}\Big)$ to compute the responsibilities using the current parameters, while the parameters are then estimated based on the current responsibilities in the M-step. The algorithm keeps alternation between E-step and M-step, until the difference between two iterations satisfies the convergence property.

2.2   Initializing and Estimating Model Parameters

For each site, similar to the initializations suggested in [4], we initialize $\mu_N = \mu_T = 0.999$ for wide-type alleles, $\mu_N = \mu_T = 0.6$ for heterozygosis mutations and $\mu_N = \mu_T = 0.001$ for homozygosis mutations. Note that, under ideal conditions, these parameters should be equal to 1 for wide-type alleles, equal to 0.5 for heterozygosis mutations and 0 for homozygosis mutations. However, the sequencing and mapping error may contribute to the bias, and thus the empirical initializations are suggested.

As different genotypes corresponding to different mutation rates, the parameter μ_G of the polynomial distribution is set as follows: (1) If both the normal and tumor samples present a wide-type allele, then we set $\mu_{G(AA,AA)} = 10^6$; (2) If only one of the two samples present a heterozygosis mutation, then we set $\mu_{G(AA,AB)} = \mu_{G(AB,AA)} = 10^2$; (3) If the normal sample presents either a wide-type allele or a homozygosis mutation and the tumor sample only presents a homozygosis mutation, then we set $\mu_{G(AA,BB)} = \mu_{G(AB,BB)} = 10^2$; (4) If both the normal and tumor samples present either a heterozygosis mutation or a homozygosis mutation, then we set $\mu_{G(AB,AB)} = \mu_{G(BB,BB)} = 10^3$; Otherwise, (5) if the normal sample presents a homozygosis mutation but the tumor sample does not, we set $\mu_{G(BB,AA)} = \mu_{G(BB,AB)} = 1$.

To initialize μ_p and λ_p, we sample the sites at which the normal genotypes are wide-type alleles and the tumor genotypes are heterozygosis mutations. The expected value of the variant allelic frequency v_f should be 0.5. Tumor purity and clonal architecture can interfere with the observed variant allelic frequencies. We adopt a binomial-bmm clustering algorithm to cluster the sampled mutations according to the observed variant allelic frequencies. This clustering algorithm is proposed in [14]. To use this clustering algorithm, the number of clusters is set to C, considering cancer type and other prior information. For cluster c, the number of mutations falling into it is set as the weight w_c, which is

$$w_c = n_c \bigg/ \sum_{c \in C} n_c \tag{9}$$

and thus, we initialize

$$\mu_p = 1 - 2 \sum_{c=1}^{C} \sum_{i=1}^{n_c} v_f^i w_c \bigg/ \sum_{c=1}^{C} n_c w_c \tag{10}$$

where n_c represents the number of mutations in cluster c, v_f^i denotes the variant allelic frequency of mutation i. λ_p is set to 10000 for any genotypes.

For the Dirichlet conjugate prior distribution, we set the initial value of α_G equal to μ_G under different joint genotypes. For the Beta conjugate prior distributions, we use the same settings as in [4], which are:

$$\alpha_{x:G_x} = \begin{bmatrix} & AA & AB & BB \\ N & 1000 & 500 & 2 \\ T & 1000 & 500 & 2 \end{bmatrix}, \beta_{x:G_x} = \begin{bmatrix} & AA & AB & BB \\ N & 2 & 500 & 1000 \\ T & 2 & 500 & 1000 \end{bmatrix}$$

For the normal-gamma conjugate prior distribution, we suggest the default values as $a_p = 1000$ and $b_p = 1$. For the initial values of μ_0 and β_p, it is observed that the EM algorithm has a small probability of falling into a local optimal solution. A better selection on the initial values of μ_0 and β_p can greatly reduce the probability. Thus, we suggest the network search strategy. We alter μ_0 ranging from 0.2 to 0.8 by an increment of 0.01 and vary β_p ranging from 0.05 to 1 by an increment of 0.01. And then, a comparative result is suggested as $\mu_0 = 0.6$ and $\beta_p = 0.1$.

3 Experiments and Results

To test the performance of *EMPurity*, we conduct a series of semi-simulation experiments and compare the results to three popular computational approaches, which are *AbsCN_seq*, *PurBayes* and *PurityEST*. In each experiment, we randomly select 10 chromosomes from the reference genome (hg19) and exclude the known self-chain and low complexity regions. Germline variants are randomly planted into this genotype with a mutation rate of 0.1 per hundred bps. Somatic mutations are then randomly planted on the germline genotype to generate the genotype of the founding clone, whose mutation rate is set to 0.01 per hundred bps. For each sub-clone, the mutation rate for planting additional somatic mutations keeps 0.01 per hundred bps. Different clonal structures are designed.

Normal sample is generated by randomly sampling the paired-end reads from the germline genotype. The tumor sample is then generated by randomly sampling the paired-end reads from the germline genotype, founding clone genotype and the genotypes of sub-clones. The probability of a read sampling from germline genotype relies on the preset tumor purity, while the probability of sampling from a particular clonal genotype depends on the proportion of the corresponding clone in the clonal structure. The read length is set to 100 bp. The insert-size is set to 500 bp, which follows a normal distribution with the mean of 500 bp and standard variance of 49 bp. The coverage is set to 200×. Paired-end reads are then mapped to the reference genome by *BWA* [15] with default parameter settings.

3.1   Accuracy Versus Different Levels of Tumor Purity

We first test the accuracy of *EMPurity* by varying the tumor purities from 10% to 90%. Here we select some extreme cases because some cancer sequencing data, such as ctDNA sequencing may reach quite low tumor purity. We preset two clones, where s_1 is founding clone and s_2 is a sub-clone derived from s_1. In the clonal architecture, let the proportion of s_1 be 70% of the whole cancer cell population and let the proportion of s_2 be 30%. For each configuration, we repeat 10 times. The results are shown in Fig. 2. Each mean value of tumor purity estimation is around the true value. And then, we could say that *EMPurity* is able to work in a wide spectrum of tumor purity levels according to these results.

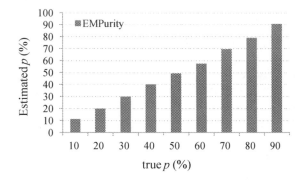

Fig. 2. The estimations of the tumor purity when the true tumor purity levels are preset from 10% to 90%

Note that, the way of generating genotypes could plant some homozygosis somatic mutations on them, especially when high degree of heterogeneity is considered. We use the following formula to balance the results: $p = \lambda p_{G(AA,AB)} + (1 - \lambda)p_{G(AA,BB)}$, where λ is a harmonic parameter. For each site, let $p_{G(AA,AB)}$ denote the proportion of non-cancerous cells in the tumor sample, when the genotypes of normal and tumor samples are wide-type allele and heterozygosis mutation, respectively. Let $p_{G(AA,BB)}$ denote the proportion of the non-cancerous cells in the tumor sample, when the genotypes of normal and tumor samples are wide-type allele and homozygosis muta-tion. λ is estimated by the genotyping results.

We apply the other three approaches on the same datasets. For each configuration, we repeat 10 times. The parameter settings for these approaches are as following: For *AbsCN_seq*, we set the minimum allowed value of tumor purity equal to 0.1, the maximum allowed value of tumor purity equal to 1, tumor ploidy equal to 2, which is the default value. For *PurityEST*, the minimum read depth is set to 6 and the minimum number of somatic mutation is set to 20. For *PurBayes*, we set maximum number of variant populations allowed equal to 5, the number of MCMC sampled for posterior inference equal to 10000. *PurBayes* also requires a prior distribution for tumor purity, and we take one of the suggested settings, which is uniform distribution on the interval 0 to 1. The results of *AbsCN_seq*, *PurBayes* and *PurityEST* are shown in Fig. 3. The estimates from *PurityEST* are always lower than the true values due to sub-clone s_2. If s_2 is removed, then the accuracy of *PurityEST* increases. The estimates from *AbsCN_seq* are often higher than the true values. *PurBayes* can accurately estimate tumor purity when the tumor purity is less than 0.6. According to the comparison results, *EMPurity* outperforms the existing approaches.

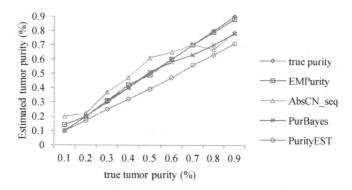

Fig. 3. Tumor purity estimates of four approaches comparing to the preset true values

3.2   Accuracy Versus Different Coverages

We then compare the accuracy when the coverage alters from $100 \times$ to $500 \times$. Other settings remain the same. For each configuration, we repeat 10 times and the mean values are shown in Table 1. It is suggested that *EMPurity* shows stable results when coverage changes.

3.3   Experiments on High Degree of Heterogeneity

Recent studies reported that some tumor cases present high degree of heterogeneity on germline and somatic variants [16, 17]. We first generate two groups of datasets. The first group only considers wide-type allele and heterozygosis mutations on s_1 and

Table 1. The estimates of tumor purity under different coverages

Preset proportion of non-cancerous cells	Coverage (×)	EMPurity	AbsCN_seq	PurBayes	PurityEST
0.2	100	0.219	0.328	0.294	0.343
	200	0.199	0.335	0.333	0.345
	300	0.215	0.340	0.425	0.345
	400	0.199	0.247	0.352	0.357
	500	0.208	0.278	0.366	0.372
0.3	100	0.312	0.313	0.453	0.423
	200	0.300	0.338	0.515	0.437
	300	0.305	0.327	0.443	0.425
	400	0.312	0.347	0.443	0.430
	500	0.308	0.339	0.460	0.434
0.4	100	0.432	0.39	0.430	0.513
	200	0.398	0.300	0.414	0.524
	300	0.396	0.285	0.404	0.528
	400	0.410	0.200	0.382	0.544
	500	0.400	0.305	0.374	0.504

heterozygosis mutations on s_2. The second group considers both heterozygosis somatic mutations and homozygosis somatic mutation on s_1 and s_2. Each group consists of 10 datasets. Table 2 shows the mean values of the estimates given by four approaches, respectively. *AbsCN_seq* still shows limitations on extreme cases. The estimations of *PurityEST* are always higher than the true values. A possible explanation is that the variant allelic frequencies only contributed by s_2 mislead the estimates. *PurBayes* has a better performance but still not as good as *EMPurity*.

Table 2. Estimating tumor impurity in different tumor heterogeneity

Proportion of non-cancerous cells	First group: (AA-AB) + (AB-AB)				Second group: (AA-BB) + (AB-BB)			
	EMPurity	AbsCN_seq	PurBayes	PurityEST	EMPurity	AbsCN_seq	PurBayes	PurityEST
0.1	0.122	0.467	0.299	0.438	0.119	0.454	0.206	0.416
0.2	0.215	0.425	0.378	0.507	0.219	0.394	0.302	0.472
0.3	0.309	0.437	0.455	0.562	0.312	0.441	0.391	0.539
0.4	0.419	0.458	0.533	0.626	0.403	0.443	0.476	0.610
0.5	0.526	0.533	0.608	0.690	0.499	0.562	0.564	0.661
0.6	0.607	0.502	0.688	0.749	0.604	0.599	0.651	0.740
0.7	0.699	0.621	0.761	0.806	0.695	0.700	0.739	0.803
0.8	0.798	0.773	0.843	0.859	0.818	0.798	0.828	0.868
0.9	0.902	0.800	0.921	0.921	0.913	0.800	0.913	0.924

We then generate datasets with more sub-clones: s_3 is a sub-clone derived from s_2. The proportions of s_1, s_2, s_3 are 5:3:2. The proportions of the non-cancerous cells vary from 10% to 90%. We first directly process these dataset by *SciClone*, whose results are shown in Fig. 4. The figures show that when there is few non-cancerous cell contamination (Fig. 4(a) and (b)), *SciClone* is able to provide correct results, which include the identification of the three clones and the proportion of each one. Along with the increasing of the proportion of the non-cancerous cells, the cloning analysis results show more errors (Fig. 4(h) and (i)).

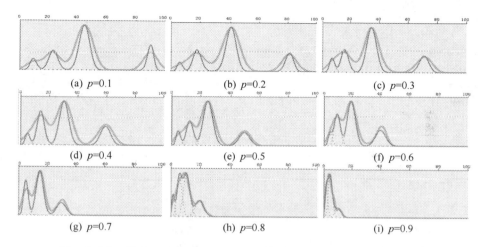

Fig. 4. The cloning analysis results by *SciClone* without purity corrections

We then incorporate *EMPurity* into the processing pipeline. The tumor purity is first inferred by *EMPurity* and then the variant allelic frequencies are corrected by excluding the non-cancerous proportion. The results provided by *SciClone* are shown in Fig. 5. According to the figures, we could conclude that *EMPurity* correctly estimates the tumor purity and successfully removes the bias.

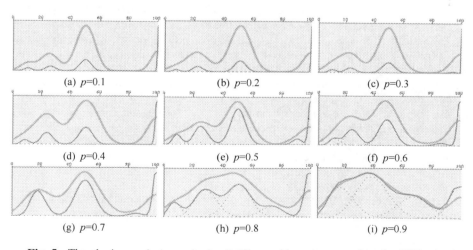

Fig. 5. The cloning analysis results by *SciClone* with purity corrections by *EMPurity*

4 Discussion

In this article, we focus on the computational problem of estimating tumor purity from cancer sequencing data. We develop a novel computational approach, *EMPurity*. *EMPurity* implements on a probabilistic model, which considers tumor purity together with the clonal architecture, and thus it is able to handle the samples with high degree of heterogeneity. An EM algorithm is designed to solve the model. We conduct a series of experiments to test the performance of *EMPurity* under different simulation configurations. *EMPurity* outperforms three popular approaches, especially when the samples have high degree of heterogeneity.

Acknowledgement. This work is supported by the National Science Foundation of China (Grant No: 81400632), Shaanxi Science Plan Project (Grant No: 2014JM8350) and the Fundamental Research Funds for the Central Universities (XJTU).

References

1. The Cancer Genome Atlas Research Network: Comprehensive genomic characterization defines human glioblastoma genes and core pathways. Nature **455**(7216), 1061–1068 (2008)
2. International Cancer Genome Consortium (2016). http://icgc.org
3. Loo, P., Nordgard, S., Lingjærde, O., et al.: Allele-specific copy number analysis of tumors. Proc. Natl. Acad. Sci. U.S.A. **107**(39), 16910–16915 (2010)
4. Cibulskis, K., Lawrence, M., Carter, S., et al.: Sensitive detection of somatic point mutations in impure and heterogeneous cancer samples. Nat. Biotechnol. **31**(3), 213–219 (2013)
5. Larson, D., Harris, C., Chen, K., et al.: SomaticSniper: identification of somatic point mutations in whole genome sequencing data. Bioinformatics **28**(3), 311–317 (2012)
6. Roth, A., Ding, J., Morin, R., et al.: JointSNVMix: a probabilistic model for accurate detection of somatic mutations in normal/tumour paired next-generation sequencing data. Bioinformatics **28**(7), 907–913 (2012)
7. Carter, S., Cibulskis, K., Helman, E., et al.: Absolute quantification of somatic DNA alterations in human cancer. Nat. Biotechnol. **30**(5), 413–421 (2012)
8. Gusnanto, A., Wood, H., Pawitan, Y., et al.: Correcting for cancer genome size and tumour cell content enables better estimation of copy number alterations from next-generation sequence data. Bioinformatics **28**(1), 40–47 (2012)
9. Oesper, L., Mahmoody, A., Raphael, B.: THetA: inferring intra-tumor heterogeneity from high-throughput DNA sequencing data. Genome Biol. **14**(7), R80 (2013)
10. Yoshihara, K., Shahmoradgoli, M., Martínez, E., et al.: Inferring tumour purity and stromal and immune cell admixture from expression data. Nature Commun. **4**(4), 2612 (2013)
11. Su, X., Zhang, L., Zhang, J., et al.: PurityEst: estimating purity of human tumor samples using next-generation sequencing data. Bioinformatics **28**(17), 2265–2266 (2012)
12. Berger, M., Lawrence, M., Demichelis, F., et al.: The genomic complexity of primary human prostatecancer. Nature **470**(7333), 214–220 (2011)
13. Larson, N., Fridley, B.: PurBayes: estimating tumor cellularity and subclonality in next-generation sequencing data. Bioinformatics **29**(15), 1888–1889 (2013)

14. Miller, C., White, B., Dees, N., et al.: SciClone: inferring clonal architecture and tracking the spatial and temporal patterns of tumor evolution. PLoS Comput. Biol. **10**(8), e1003665 (2014)
15. Li, H., Durbin, R.: Fast and accurate short read alignment with Burrows-Wheeler transform. Bioinformatics **25**(14), 1754–1760 (2009)
16. Lu, C., Xie, M., Wendl, M., Wang, J., McLellan, M., Leiserson, M., et al.: Patterns and functional implications of rare germline variants across 12 cancer types. Nature Commun. **6**, 10086 (2015)
17. Xie, M., Lu, C., Wang, J., et al.: Age-related cancer mutations associated with clonal hematopoietic expansion. Nat. Med. **20**(12), 1472–1478 (2014)

Identifying Heterogeneity Patterns of Allelic Imbalance on Germline Variants to Infer Clonal Architecture

Yu Geng[1,3,5], Zhongmeng Zhao[1,3], Jing Xu[1,3], Ruoyu Liu[2,3], Yi Huang[2,3],
Xuanping Zhang[1,3], Xiao Xiao[3,4], Maomao[3], and Jiayin Wang[2,3(✉)]

[1] School of Electronic and Information Engineering,
Xi'an Jiaotong University, Xi'an 710049, China
zmzhao@mail.xjtu.edu.cn
[2] School of Management, Xi'an Jiaotong University, Xi'an 710049, China
wangjiayin@mail.xjtu.edu.cn
[3] Institute of Data Science and Information Quality, Shaanxi Engineering Research Center
of Medical and Health Big Data, Xi'an Jiaotong University, Xi'an 710049, China
[4] State Key Laboratory of Cancer Biology,
Xijing Hospital of Digestive Diseases, Xi'an 710032, China
[5] Jinzhou Medical University, Jinzhou 121001, China

Abstract. It is suggested that the evolution of somatic mutations may be significant impacted by inherited polymorphisms, while the clonal somatic copy-number mutations may contribute to the potential selective advantages of heterozygous germline variants. A fine resolution on clonal architecture of such cooperative germline-somatic dynamics provides insight into tumour heterogeneity and offers clinical implications. Although it is reported that germline allelic imbalance patterns often play important roles, existing approaches for clonal analysis mainly focus on single nucleotide sites. To address this need, we propose a computational method, *GLClone* that identifies and estimates the clonal patterns of the copy-number alterations on germline variants. The core of *GLClone* is a hierarchical probabilistic model. The variant allelic frequencies on germline variants are modeled as observed variables, while the cellular prevalence is designed as hidden states and estimated by Bayesian posteriors. A variational approximation algorithm is proposed to train the model and estimate the unknown variables and model parameters. We examine *GLClone* on several groups of simulation datasets, which are generated by different configurations, and compare to three popular state-of-the-art approaches, and *GLClone* outperforms on accuracy, especially a complex clonal structure exists.

Keywords: Cancer genomics · Germline variant · Variant allelic imbalance · Clonal heterogeneity

1 Introduction

It is thought that the interacting susceptibility germline variants and somatic mutations are involved in more than 3% of cancer cases across multiple cancer types [1]. The "two-hit hypothesis", for example, is one of the popular theories that exemplify such

© Springer International Publishing AG 2017
D.-S. Huang et al. (Eds.): ICIC 2017, Part II, LNCS 10362, pp. 286–297, 2017.
DOI: 10.1007/978-3-319-63312-1_26

interactions. Recent reports associate that germline variants show directly impact on somatic tumor evolution [2], while the somatic copy-number alternations may contribute the potential selective advantages of the germline variants [3]. Existing studies have achieved great success on understanding the clonal expansion [4, 5] and identifying the clonal architecture that implies clinical significance, e.g. some subclones are found to present drug resistance in non-small cell lung cancer cases [6] and chronic lymphocytic leukemia cases [7], a subclone is reported to improve the risk of malignancy development in some multiple myeloma cases [8], et al. However, due to the computational difficulty, it is hard to observe the germline allelic imbalance patterns that interact with the somatic clonal structure.

Along with the cancer sequencing projects, some state-of-the-art approaches have been proposed to explore the clonal architecture based on sequencing data. We roughly categorize them into two classes: The first category mainly rely on the clustering of the read depth and variant allelic frequency at each single nucleotide site, such as *SciClone* [9], *Clomial* [10] and *PyClone* [11]. *SciClone* introduces a Bayesian clustering method on single nucleotide variants (SNVs). *Clomial* adopts a decomposition process. However, both methods limit the SNVs located in copy-number neutral region only. *PyClone* is among the first method that incorporates the variant allelic frequencies (VAFs) with the allele-specific copy numbers, but it is time-consuming to calculate copy-number alternations for each region. The second category pays attention to copy-number alternations, which includes *THetA* [12], *TITAN* [13] and *cloneHD* [14]. *THetA* solves the most likely combination of genomes, but it exposes weaknesses when tumor purity dilutes read depth. And it may present high time-complexity when the number of subclones increases. *TITAN* evaluates the allelic ratios to infer the copy-number alternation segments by comparing the ratio intensity between tumor and normal samples, then a Hidden Markov Model(HMM) is designed for modeling the spatial correlation between two adjacent segments. *cloneHD* is implemented on a factorial HMM, which consists of more comprehensive states that consider both CNVs and SNVs.

The existing approaches are not fit for germline variants because the germline ones exist in all subclones. To address this need, in this paper, we propose a probabilistic model to infer the allelic imbalance patterns on germline variants. This approach, implemented as *GLClone* models the read depth and variant allelic frequency on each variant as observed data, while the copy-number alternation status is coded as hidden variables. An algorithm that computes variational Bayesian distribution is designed to obtain the hidden states and unknown model parameters. We conduct several simulation datasets with different configurations, and *GLClone* robustly outperforms the existing approaches including *TITAN*, *cloneHD* and *PyClone*.

2   The Probabilistic Model and Bayesian-Based Algorithm

Suppose that, for each cancer patient, we are given a pair of sequencing data, one of which is from tumor tissue, while the other one is from adjacent normal tissue or blood (not for leukemia cases). The germline variants and somatic mutations are supposed to be successfully identified by the widely used approaches, such as [15]. We assume that the tumor sample consists of multiple subclones, where a subclone contains subclonal specific copy-number alternations, some of which may harbor germline variants. However, the number of subclones is unknown, and the proposed algorithm, *GLClone* is design to estimate the number of subclones with the absolute copy-number and proportions, and to reconstruct the clonal structure.

To simplify the problem, we only consider the most common case, which we assume that for each single nucleotide site, only one genotype across all subclones is considered; for each copy-number alternation, only the one with the length equal or greater than 10 bp is considered which may harbor small indels. According to literature [3], we limit to the heterozygous germline variants with the maximal ploidy of 5 for each subclone.

Let G represent the genotypes of the normal sample and of all the subclones. The set of subclones represents as $S = \{s_1, \ldots, s_k, \ldots, s_K\}$, where K is the number of subclones. For each germline variant, the observed genotype is a convolution of the genotypes with copy-number alternations, while the hidden state denotes the copy-number alternations. Let M denote the possible de-convolution state, and its number is $|M|$ which equals to $|G| \times K$. Following literatures [11, 15], we also assume that the mutations in the same subclone have similar cellular prevalence. Let ϕ_k represent the cellular prevalence of subclone s_k.

2.1   GLClone Model

Let C_g denote the ploidy of genotype g. If given C_g on s_k, for each single nucleotide site i, the expected copy number is

$$C_i = (1 - \phi_k) \times C_0 + \phi_k \times C_g \tag{1}$$

where C_0 is the copy number of normal sample, which is 2 here. The average read depth is λ when the genotype of s_k is g, then we have

$$\lambda \propto C \times \tau / 2 \tag{2}$$

where τ represents the expected value of the number of reads in the case without CNA. In the normal sample, τ subjects to uniform distribution of which parameter is the sum of reads expectations and assumed to be constant.

It is assumed that read depth follows a Poisson distribution.

$$P(d|\lambda) = \frac{\lambda^d e^{-\lambda}}{d!} \tag{3}$$

where d denotes the reads depth, λ could be obtained by formula (2). For each tumor genotype and subclone type, if cellular prevalence of the subclone is ϕ, and the genotype of s_k is g, the average variant allele frequency is

$$u = \frac{(1 - \phi_k) \times C_0 \times \mu_0 + \phi_k \times C_g \times \mu_g}{(1 - \phi_k) \times C_0 + \phi_k \times C_g} \tag{4}$$

where μ_0 and μ_g are variant allele frequency of normal and tumor genomes, respectively.

It is assumed that the total number of single nucleotide site considered here is N. The reads number of variant alleles $b_{1:N}$ and total reads depth $T_{1:N}$ follow binomial distribution:

$$\text{Bin}(b|T, u) = \binom{T}{b} u^b (1 - u)^{T-b} \tag{5}$$

where u is the expected value of variant allele frequency in the sample.

Finally, the Poisson distribution and binomial distribution are combined to model for $d_{1:N}$, $b_{1:N}$ and $T_{1:N}$, and derive:

$$p(d, T, b|\pi, \lambda, u) = \prod_{i=1}^{N} \sum_{j=1}^{M} \pi_j \text{P}(d_i|\lambda_j) \text{Bin}(b_i|T_i, u_j) \tag{6}$$

where π represents the mixing coefficient which is greater than zero, and the sum of which is 1.

Unobservable random hidden variables can be inferred from the sample of the observable variables. Here, hidden variables $Z = [z_1, \ldots, z_m, \ldots, z_M]$ are introduced, $z_m = 1$ indicates that the single nucleotide site belongs to the jth state, otherwise not. Assuming a given mixing coefficient Π, with an independent indicator variable Z, the condition distribution is:

$$p(Z|\Pi) = \prod_{i=1}^{N} \prod_{j=1}^{M} \pi_j^{z_{ij}} \tag{7}$$

Because the conjugate prior of polynomial distribution is Dirichlet distribution, given the vector set of hidden variables $\{\Pi, \Lambda, U\}$, conditional distribution of $\{d, T, b\}$ and $\{Z\}$ is

$$p(d, T, b|\pi, \lambda, u) = \prod_{i=1}^{N} \prod_{j=1}^{M} [\pi_j \text{P}(d_i|\lambda_j) \text{Bin}(b_i|T_i, u_j)]^{z_{ij}} \tag{8}$$

Set of observable variables is $X = \{d, T, b\}$. Unknown random variable is defined as $\Theta = \{Z, \Pi, \Lambda, U\}$. According to the Bayesian criterion, the joint density function of all random variables is expressed as:

$$p(X, \Theta) = p(d|Z, \lambda)p(b|T, Z, u)p(Z|\pi)p(\pi)$$

$$= \prod_{i=1}^{N} \prod_{j=1}^{M} \left[\frac{\lambda_j^{d_i} e^{-\lambda_j}}{d_i!} \times \frac{T_i!}{b_i!(T_i - b_i)!} u_j^{b_i} (1 - u_j)^{T_i - b_i} \right]^{z_{ij}} \frac{1}{\text{Beta}(\alpha_j)} \prod_{j=1}^{M} \pi_j^{\alpha_j - 1} \tag{9}$$

Thus, *GLClone* is based on a variational Bayesian hierarchical framework. Figure 1 gives a directed graph representation of the model framework.

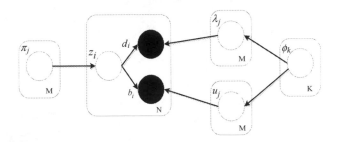

Fig. 1. A graphical representation of *GLClone* model. Each circle denotes a random variable, while the arrows represent the conditional independence between the corresponding variables. The hollow circles represent the model parameters, and the black circles represent the observable variables.

2.2 Variational Approximation

Variational inference [16, 17] is used to estimate the parameters of the model. The main idea of variational study is to find the approximation of the posterior distribution $p(\Theta|X)$ and the true model $p(X)$, X denotes observation. First, the logarithmic marginal probability $\ln p(X)$ could be decomposed into the following form:

$$\ln p(X) = \underbrace{\int Q(\Theta)\ln\frac{p(X,\Theta)}{Q(\Theta)}d\Theta}_{L(Q)} - \underbrace{\int Q(\Theta)\ln\frac{p(\Theta|X)}{Q(\Theta)}d\Theta}_{KL(Q(\Theta)||p(\Theta|X))} \tag{10}$$

In the above formula, $Q(\Theta)$ is the approximation of the true posterior distribution $p(\Theta|X)$.

In the standard variational inference framework, each specific factor $Q_s(\Theta_s)$ is usually expressed as

$$Q_s(\Theta_s) = \frac{\exp\langle\ln p(X,\Theta)\rangle_{i\neq s}}{\int \exp\langle\ln p(X,\Theta)\rangle_{i\neq s}d\Theta} \tag{11}$$

Where s is a specific integer, which is used to differentiate with i. According to formula (11), it could be deduced that the best solution of Z is

$$Q(Z) = \prod_{i=1}^{N} \prod_{j=1}^{M} \gamma_{ij}^{z_{ij}} \tag{12}$$

γ_{ij} is defined as the probability that the ith single nucleotide site belongs to the jth state, E[.] is used to calculate expectations.

$$\gamma_{ij} = \frac{\rho_{ij}}{\sum \rho_{ij}} \tag{13}$$

$$\rho_{ij} \propto \exp\left\{ E\left[\ln \pi_j\right] + d_i E\left[\ln \lambda_j\right] - \overline{\lambda_j} + b_i E\left[\ln u_j\right] + \left(T_i - b_i\right)E[\ln(1 - u_j)] \right\} \tag{14}$$

Here,

$$E[z_{ij}] = \gamma_{ij},\ E\left[\ln \pi_j\right] = \psi\left(\alpha_j\right) - \psi\left(\sum_h \alpha_h\right) \tag{15}$$

$E[\ln \lambda]$, $\overline{\lambda}$, $E[\ln u]$ and $E[\ln(1-u)]$ could be calculated by formula (2) and (4).

2.3   Estimating Lower Bound

The lower bound of the model can usually be used to test the convergence in the process of estimating parameters, and the values will not be reduced in the iterative process. For this variational mixture model, the lower bound is expressed as

$$L = \sum_Z \int q(Z, \pi)\ln\left\{ \frac{p(X, Z, \pi)}{q(Z, \pi)} \right\}d\pi \tag{16}$$

The optimization of the model is similar to the EM algorithm, and the Algorithm 1 gives a complete learning process.

Algorithm 1: *GLClone variational learning*
Step 1: Initialize parameters of Dirichlet distribution.
Step 2: Use K-Means algorithm to initialize the value of γ_{ij}.

Step 3: Variational E step: use the model parameters of the current distribution to calculate the expected value in the formula (2), (4) and (15).

Step 4: Variational M step: update factors according to the formula (11), and use the coordinate descent method to maximize the lower bound and update ϕ..

Step 5: Repeat Step 3~4 until convergence criterion is achieved (for example, the change between current lower bound and the previous lower bound value is less than set threshold value).

---

3   Experiments and Results

In each simulation dataset, we randomly select one chromosome as the reference genome. And then, we use TNSim, a read simulator to sample the reference genome by 100 bp read-length with at least 100 × coverage. 1% single nucleotide loci are randomly planted into the mimic normal genome, while in the tumor genome, some segments are randomly chosen to insert CNA, and the absolute copy number is randomly set as an arbitrary integer value between 0 and 5. Here, 2 represents no copy number variation, while a value less than 2 denotes deletions, and the one greater than 2 denotes amplifi-cations. Because *SciClone* and *Clomial* only consider the copy-number neutral regions, it is unfair to compare with them. Thus, we test the performance and compare to *TITAN* and *cloneHD*. *THetA* is not included because *cloneHD* is reported to outperform *THetA*.

3.1   Estimation of Cellular Prevalence

The generated *FASTQ* file is aligned by *BWA*(version 0.7.5a-r405). All user parameters are set as default. The pipeline also includes *SAMtools* (version 0.1.19-96b5f2294a). And then, *VarScan2* (Version v2.3.9) and *DNAcopy* (version $> = 1.40.0$) are used to produce variant calling and segmentation estimation. The purpose of this experiment is to generate and visualize the specific operational results. We simulate two subpopula-tions, one is 30 × coverage, the other is 70 × coverage, thus ground truth of cellular prevalence are 0.3 and 0.7. Figure 2 shows cellular prevalence of two subclones operated by *GLClone*, *TITAN*, and *PyClone*, respectively. Because *cloneHD* doesn't give all the cell prevalence values of the corresponding single nucleotide sites, it is excluded from the comparison. As the shown in Fig. 2, *TITAN* and *PyClone* estimate only one cellular prevalence value, while the results predicted by *GLClone* are closer to the actual values.

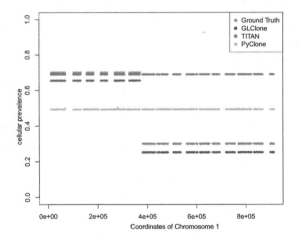

Fig. 2.   Results from *GLClone*, *TITAN* and *PyClone* when 2 subclones exist. Theoretical values of each cellular prevalence are 0.3 and 0.7. The horizontal axis is the location coordinates of chromosome 1, and the vertical axis is the value of cellular prevalence.

3.2   Estimating Absolute Copy Number of Subclones

This dataset is the same as previous section. Figure 3 shows the absolute copy number of subclones estimated by *GLClone* and *TITAN*, when the number of subclones is 2. *PyClone* and *cloneHD* are excluded from the comparison, because both of them do not give the absolute copy number of tumor of each single nucleotide site. *GLClone* demon-strates much more accuracy in estimating absolute copy number of subclones.

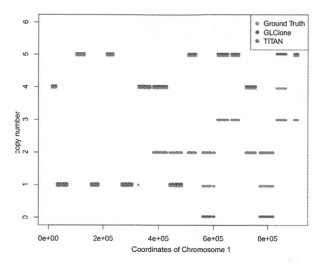

Fig. 3. Results from *GLClone* and *TITAN* when there are two subclones. The horizontal axis is the coordinates of chromosome 1, and the vertical axis is the value of cellular prevalence.

3.3 Comparing to *TITAN*, *CloneHD* and *PyClone*

To verify predictive performance of *GLClone* is superior, results of *GLClone* are compared with *TITAN*, *cloneHD* and *PyClone*.

Table 1. Predicted results of different sequencing coverage

Coverage	30×	40×	50×	60×	80×	100×
GLClone	0.355	0.363	0.349	0.341	0.347	0.342
	0.658	0.667	0.634	0.618	0.641	0.646
TITAN ($K = 1$)	0.749	0.813	0.827	0.945	0.944	0.8873
TITAN ($K = 2$)	0.5	0.473	0.5	0.5	0.5	0.5
	0.749	0.5	0.543	0.658	0.662	0.660
TITAN ($K = 3$)	0.5	0.473	0.507	0.5	0.5	0.5
	0.641	0.473	0.543	0.5	0.5	0.5
	0.749	0.5	0.543	0.657	0.655	0.603
cloneHD	0.268	0.208	0.233	0.248	0.233	0.142
	0.371	0.523	0.654	0.414	0.573	0.831
PyClone	0.349	0.428	0.447	0.430	0.531	0.577
	(0.005)	(0.01)	(0.009)	(0.009)	(0.009)	(0.006)
	0.696	0.766	0.805	0.759	0.422	0.425
	(0.177)	(0.153)	(0.133)	(0.157)	(0.186)	(0.04)
	0.725	0.817	0.809	0.797	0.557	0.470
	(0.161)	(0.108)	(0.131)	(0.123)	(0.214)	(0.084)

Effect of sequencing coverage on the performance of *GLClone*. The cellular prevalence is fixed and set as 0.65 and 0.35, respectively. Sequencing coverage is adjusted to 30×, 40×, 50×, 60×, 80× and 100×. Table 2 shows the running results of *GLClone*, *cloneHD*, *TITAN* and *PyClone*. The numeric in the parentheses of *PyClone* represents the corresponding standard deviation. Because of the large difference between the estimated results by *TITAN* and the true value, the results are computed respectively when the number of subclones is 1 ($K = 1$), 2 ($K = 2$) and 3 ($K = 3$). Results computed by *PyClone* include multiple subclones, among which only a few are showed here. It can be seen from Table 1 that the cellular prevalence of the two tumor subclones predicted by *GLClone* are very close to the actual value, and the influence of the sequencing coverage is minor.

Effects of different clone numbers on estimation. Table 2 shows the cellular prevalence predicted by *GLClone*, *TITAN*, *cloneHD* and *PyClone* with different subclone number. *TITAN* is excluded from the comparison, because of its poor distinction between different subclones. Results predicted by *GLClone* are close to the actual value when the number of subclones is different.

Table 2. Operation results with different subclone number

Subclone number	Ground Truth	GLClone	cloneHD	PyClone
K = 2	0.65	0.617	0.169	0.295(0.006)
	1	0.999	0.972	1(1.03e-6)
K = 3	0.3	0.357	0.013	0.397(0.006)
	0.6	0.605	0.152	0.529(0.022)
	1	0.995	0.987	0.806(0.046)
K = 4	0.2	0.182	0.009	0.342(0.172)
	0.5	0.471	0.076	0.400(0.007)
	0.7	0.675	0.315	0.779(0.100)
	1	0.988	0.987	0.671(0.164)
K = 5	0.2	0.186	0.014	0.312(0.006)
	0.45	0.460	0.042	0.508(0.280)
	0.65	0.632	0.082	0.641(0.197)
	0.85	0.850	0.218	0.687(0.191)
	1	0.977	0.866	0.814(0.021)

Cellular prevalence predicted by *GLClone* is close to the expected value under different conditions (sequencing coverage and subclone numbers), so this model can accurately reconstruct the clonal structure and model tumor data, which could improve the accuracy of detection.

3.4   Detecting Copy Number Variation in Tumor Cells

In order to verify the accuracy in detecting subclone CNA, we perform statistical analysis on the copy number of subclone, which compare predicted by *GLClone* with the actual values, and compare it with that predicted by *TITAN* (*cloneHD* and *PyClone* are excluded from comparison due to they do not have tumor copy number). Accuracy is defined as the proportion of the number of single nucleotide sites which is detected in the subclone region to the total number of single nucleotide loci. The dataset used here is the same as that in Part 3.3. As showed in Figs. 4 and 5. In most cases, *GLClone* shows a relatively high accuracy. There are two groups of data in Fig. 4 showing that the accuracy of *GLClone* is lower than *TITAN*. Because gene signal of tumor sample is affected by tumor genotype and cellular proportion, and either insertion or deletion could lead to a change of gene signal which would be compensated by another. If there are several situations that have close signal value of read depths and variant allele frequency, genotyping is confused to decrease identification accuracy. For example, gene signal of a mixture of 25% tumor genotype (*AAAAB*) and 75% normal genotype (*AB*) could be interpreted as a mixture of 75% tumor genotype (*AAB*) and 25% normal genotype (*AB*).

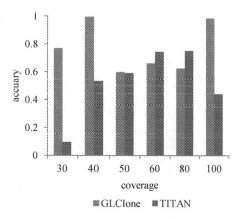

Fig. 4. The accuracy of *GLClone* and *TITAN* in estimating the copy number of tumor cells with different sequencing coverage.

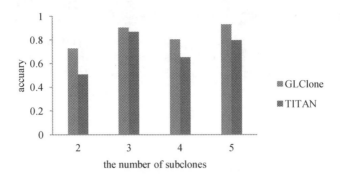

Fig. 5. The accuracy of *GLClone* and *TITAN* in estimating tumor subclone copy number with different subclone numbers.

4 Conclusion

In this paper, we propose a novel method to detect cellular prevalence of subclone and CNA based on paired normal-tumor data. After statistical analysis of the multiple heterogeneous tumor data, we find that *GLClone* demonstrate higher accuracy in detecting cellular prevalence of the tumor subclone and copy number. *GLClone* is superior to *TITAN*, *cloneHD* and *PyClone* in predicting tumor subclone cellular prevalence. Although *TITAN* is better than *GLClone* when predicting copy number of subclone in partial datasets, its performance in differentiating subclones is poor. Referring to *TITIAN*, *GLClone* assumed that each single nucleotide site has one variant genotype, but comprehensive analysis of reads depth and VAF would produce multiple solutions, when only these two parameters are given. In fact, there may be many kinds of variant genotypes, which need to be further explored.

   GLClone is applicable for whole genome sequencing data. In NGS data, the amounts of read counts which reflecting copy number information is large while the amounts of point mutation is small, thus using copy number data is more stable than using allele data to detect subclones. In addition, point mutations include a certain amount of false positive. *GLClone* combines the two kinds of information to improve the predictive ability. Furthermore, the input reads do not need deep sequencing.

   The tumor heterogeneity complicates the understanding of tumor biology, so it is difficult to design an effective treatment plan. Accurately dividing the tumor cells into subclones with different genotype will lead to a great breakthrough in the treatment of tumors. Personalized treatment of cancer can be achieved according to the unique genetic characteristics of each subclone. Accurate estimation of the cellular prevalence of *GLClone* can play an important role in the detection of the tumor heterogeneity, the search for oncogenes and personalized cancer treatment.

Acknowledgement. This work is supported by the National Science Foundation of China (Grant No: 81400632), Shaanxi Science Plan Project (Grant No: 2014JM8350) and the Fundamental Research Funds for the Central Universities (XJTU).

References

1. Rahman, N.: Realizing the promise of cancer predisposition genes. Nature **505**, 302–308 (2014)
2. Carter, H., Marty, R., Hofree, M., et al.: Interaction landscape of inherited polymorphisms with somatic events in cancer. Cancer Discovery (2017) (OnlineFirst). doi: 10.1158/2159-8290.cd-16-1045
3. Lu, C., Xie, M., Wendl, M., Wang, J., McLellan, M., Leiserson, M., et al.: Patterns and functional implications of rare germline variants across 12 cancer types. Nature Commun. **6**, 10086 (2015)
4. Kandoth, C., McLellan, M., Vandin, F., et al.: Mutational landscape and significance across 12 major cancer types. Nature **502**, 333–339 (2013)
5. Xie, M., Lu, C., Wang, J., et al.: Age-related cancer mutations associated with clonal hematopoietic expansion and malignancies. Nat. Med. **20**(12), 1472–1478 (2014)
6. Su, K., Chen, H., Li, K., et al.: Pretreatment epidermal growth factor receptor (EGFR) T790 M mutation predicts shorter EGFR tyrosine kinase inhibitor response duration in patients with non-small-cell lung cancer. J. Clin. Oncol. **30**(4), 433–440 (2012)
7. Landau, D., Carter, S., Stojanov, P., et al.: Evolution and impact of subclonal mutations in chronic lymphocytic leukemia. Cell **152**(4), 714–726 (2013)
8. Magrangeas, F., Avet-Loiseau, H., Gouraud, W., et al.: Minor clone provides a reservoir for relapse in multiple myeloma. Leukemia **27**(2), 473–481 (2013)
9. Miller, C., White, B., Dees, N., et al.: SciClone: inferring clonal architecture and tracking the spatial and temporal patterns of tumor evolution. PLoS Comput. Biol. **10**(8), e1003665 (2014)
10. Zare, H., Wang, J., Hu, A., et al.: Inferring clonal composition from multiple sections of a breast cancer. PLoS Comput. Biol. **10**(7), e1003703 (2014)
11. Roth, A., Khattra, J., Yap, D., et al.: PyClone: statistical inference of clonal population structure in cancer. Nat. Methods **11**(4), 396–398 (2014)
12. Oesper, L., Mahmoody, A., Raphael, B.: Theta: inferring intra-tumor heterogeneity from high-throughput DNA sequencing data. Genome Biol. **14**(7), r80 (2013)
13. Ha, G., Roth, A., Khattra, J., et al.: TITAN: inference of copy number architectures in clonal cell populations from tumor whole-genome sequence data. Genome Res. **24**(11), 1881–1893 (2014)
14. Fischer, A., Vázquez-García, I., Illingworth, C., et al.: High-definition reconstruction of clonal composition in cancer. Cell Rep. **7**(5), 1740–1752 (2014)
15. Xia, H., Li, A., Yu, Z., et al.: A novel framework for analyzing somatic copy number aberrations and tumor subclones for paired heterogeneous tumor samples. Bio-Med. Mater. Eng. **26**(s1), 1845–1853 (2015)
16. Ma, Z., Leijon, A.: Bayesian estimation of beta mixture models with variational inference. IEEE Trans. Pattern Anal. Mach. Intell. **33**(11), 2160–2173 (2011)
17. Fan, W., Bouguila, N.: Variational learning of a Dirichlet process of generalized Dirichlet distributions for simultaneous clustering and feature selection. Pattern Recogn. **46**(10), 2754–2769 (2013)

Computational Proteomics

Predicting Essential Proteins Using a New Method

Xi-wei Tang[1,2(✉)]

[1] National University of Defense Technology, Changsha 410073, China
tangxiwei2010@gmail.com
[2] Hunan First Normal University, Changsha 410205, China

Abstract. Essential proteins are indispensable for the survival of organisms. Computational methods for predicting essential proteins in terms of the global protein-protein interaction (PPI) networks is severely restricted due to the insufficiency of the PPI data, but fortunately the subcellular localization information helps to make up the deficiency. In the study, a new method named CNC is developed to detect essential proteins. First, the subcellular localization information is incorporated into the PPI networks, so each interaction in the networks is weighted. Meanwhile the edge clustering coefficient of each pair interacting proteins is calculated and the second weighted value of each interaction in the networks is gained. The two kinds of weighted values are integrated to build a new weighted PPI networks. The proteins in the new weighted networks are scored by the weighted degree centrality (WDC) and sorted in descending order of their scores. Six methods, i.e., CNC, CIC, DC, NC, PeC and WDC are used to prioritize the proteins in the yeast PPI networks. The results demonstrate that CNC outperforms other state-of-the-art ones. At the same time, the analysis also mean that CNC is an effective technology to identify essential proteins by integrating different biological data.

Keywords: Protein-protein interaction network · Subcellular localization · Edge clustering coefficient

1 Introduction

Essential proteins are indispensable for the growth and development of an organism under a variety of conditions [1–6]. Detecting essential proteins is significant not only for the understanding of the minimal requirements for cellular life, but also for practical implications, e.g., in identifying bacterial drug and vaccine targets.

A variety of centrality metrics such as degree centrality (DC) [7], betweenness centrality (BC) [8], closeness centrality (CC) [9], subgraph centrality (SC) [10], eigenvector centrality (EC) [11], and information centrality (IC) [12] have been proposed to discover essential proteins in the PPI networks in terms of the network topological features. Analysis have indicated that they are far better than pseudorandom selection in detecting essential proteins. Wang et al. develop an essential proteins discovery method by means of edge clustering coefficient called network centrality (NC) [13]. Their experimental results show that NC significantly outperform six measures of gene centrality in detecting essential proteins.

© Springer International Publishing AG 2017
D.-S. Huang et al. (Eds.): ICIC 2017, Part II, LNCS 10362, pp. 301–308, 2017.
DOI: 10.1007/978-3-319-63312-1_27

However, the identification of essential proteins using PPI data alone is insufficient [14–19]. First, protein interaction data produced by high-throughput technologies currently consists of a large amount of false interactions [20]. Since protein interaction measurements descend from a certain range of experimental conditions, only a small number of all possible PPI data is detected. Additionally, the PPI networks include unstable interactions or interactions that take place at different time points, thus the resulting network does not represent the real one but an overlap of many different snapshots [21]. In consideration of the defects from the PPI networks, some researchers begin to weight the PPI networks by other biological data from different sources. Due of the fact that a majority of essential proteins are conservative, Peng et al. build the weighted networks via the integration of PPI data and protein orthology information and develop an algorithm named ION to predict essential proteins [22]. Using logistic regression-based model and function similarity, Li et al. weight the PPI networks and propose an essential protein prediction method [23]. By combination of gene expression profiles and PPI data, two computational approaches are developed [24, 25]. They are called WDC and PeC, respectively. Recently, Peng et al. propose an impressively computational method for predicting essential proteins [26]. They find a novel ways to determine the importance of interactions between proteins with the usage of protein subcellular localization information. And then they propose a compartment importance centrality (CIC) measure to detect essential proteins. The results show that CIC outperforms other weighted methods.

In the present work, based on the integration of protein interaction network topology features and subcellular localization, a new centrality measure Compartment and Network Centrality (CNC), is developed to achieve the reliable prediction of essential proteins. The results show that the integrated approach we present outperforms other state-of-the-art ones.

2   Method

CNC method starts from the new weighted PPI networks which are built by means of the network topology features and protein subcellular localization information. Next, the proteins in the weighted networks are scored based on their weighted degree and sorted in descending of their scores. The top proteins in the list are known as the potentially essential proteins.

Weighting the PPI networks: Our method includes three stages to weight the PPI networks. First, the edge clustering coefficient (ECC) between interacting proteins are calculated and served as the first weighted value of each interaction (edge) in the networks. Second, compartment information is integrated to weight the same networks again. Finally, the two kinds of weighted values are combined to construct a new weighted networks.

(1) Weighting the networks via ECC. Radicchi et al. have developed the edge-clustering coefficient (ECC) according to the usual node-clustering coefficient [27]. ECC of an

edge between nodes i and j is defined as the actual number of triangles $Z_{i,j}^{(3)}$ to which the edge between i and j contributes, divided by the number of possible triangles which are determined by the minimum of the degrees k_i and k_j of the two nodes i and j:

$$ECC(i,j) = \frac{Z_{i,j}^{(3)}}{\min(k_i - 1, k_j - 1)} \tag{1}$$

The value of the edge clustering coefficient is between 0 and 1.

Weighting the networks using the subcellular localization. Study suggests that the importance of interactions which happen in different compartments is different [28]. Therefore, Peng et al. propose a framework to evaluate the interactions in the PPI networks via the subcellular localization data [26]. They research the relationship between the compartments and PPI networks and find that the significance of a compartment is in direct proportion to its protein number. Based on the finding, they quantify the importance of the compartment as the number of proteins in the area, represented by C_X, divided by the number of proteins in the largest size compartment, denoted by C_M. The score of a compartment SC is computed by

$$SC(I) = \frac{C_X(I)}{C_M} \tag{2}$$

where the value of SC ranges from 0 to 1 and $I \in \{1, 2,..., 11\}$. Subsequently, they weight the interactions in terms of the scores of the compartments. $Loc(u)$ represents the set of compartments in which protein u is located. Two interacting proteins, u and v, are likely to be annotated by multiple subcellular localizations, respectively. In other words, the interaction (u, v) between them might be localized at the common compartment. Thus, it can be defined by the shared compartments, i.e., $SLoc(u, v) = Loc(u) \cap Loc(v)$. Moreover, the significance SI of the interaction (u, v) is quantified by

$$SI(u, v) = \begin{cases} \max(SC(I)), \; if \; SLoc(u,v) \neq \emptyset \\ SC(C_N), otherwise \end{cases} \tag{3}$$

The score of the interaction (u, v) is maximized when $SLoc(u, v) \neq \Phi$. In the biological experiments, the compartment information of some proteins might be missing, i.e., $SLoc(u, v) = \Phi$. Under the circumstances, the scores of corresponding interactions are minimized. C_N is the compartment with a minimum of proteins.

Generating new weighted networks. The probability that two proteins interact can be described from the perspective of network topology and is also able to be characterized from the subcellular localization. Therefore, we redefine the importance of the interaction between paired proteins u and v. It is expressed as

$$W(u, v) = ECC(u, v) \times SI(u, v) \tag{4}$$

From the definition, the value of W(u, v) falls within the interval: [0, 1].

(2) Compartment and network centrality: The *CNC* relying on the new weighted PPI networks is computed in terms of the weighted interactions among the protein i and its direct neighbors.

$$\text{CNC}(i) = \sum_{j}^{N_i} W_{i,j} \qquad (5)$$

where N_i stands for the number of neighbors of protein I and $W_{i,j}$ refers to the weighted value of the interaction among proteins i and its neighbor j. We prioritize all proteins in the PPI networks using the CNC method.

3 Results

(1) *Data sources:* Known essential proteins. The known essential proteins of yeast which are downloaded from DEG [29] (Release version DEG 10.0) are used as the benchmark set to estimate the essential protein predictions. Protein-protein interactions. The PPI networks of yeast which are extracted from Biogrid database [30] (Release version BIOGRID-3.2.111) consist of 6, 304 proteins and 81, 614 interactions. Subcellular localization. The subcellular localization data of yeast in the COMPARTMENT database [31] are employed in the experiment. The compartments of yeast are classified into 11 categories, i.e., Nucleus, Cytosol, Cytoskeleton, Peroxisome, Lysosome, Endoplasmic reticulum, Golgi apparatus, Plasma membrane, Endosome, Extracellular space and Mitochondrion.

(2) *Assessment of the proportion of essential proteins in proteins sorted in terms of various prediction methods.* The number of essential proteins in the top proteins is a commonly metric to evaluate the predicting methods [32, 33]. For the 6 identifying approaches, we select the top 100, top 200, etc. of all ranked proteins in the PPI networks and determine how many of these are essential. The results are shown in Fig. 1.

  Figure 1 shows the number of essential proteins in the top proteins ranked by each approach, like CNC, CIC, DC, NC, PeC and WDC. We count the number of essential proteins in the top proteins (top 100, top 200, top 300, top 400, top 500 and top 600). For example, in the top 100 proteins, CNC and other measures (CIC, DC, NC, PeC and WDC) predict 75, 67, 49, 43, 62 and 51 essential proteins, respectively. This means that CNC approach performs dramatically better than other 5 centrality measures. Meanwhile the results also suggest that other proteins except the really essential proteins in the top proteins are likely to be potentially essential proteins. They are valuable references for the biologists.

(3) *ROC curves of various prediction methods.* In essence, identification of essential proteins is a classification problem [34–41], so the receiver operating characteristic (ROC), or ROC curve, can be used to estimate the identifying methods for essential proteins. The ROC curve is generated by plotting the true positive rate (TPR) against the false positive rate (FPR) when its discrimination threshold is varied. In

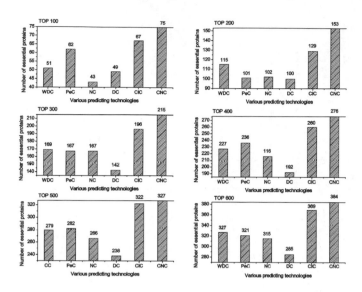

Fig. 1. Number of essential proteins in the top proteins prioritized by each approach. The ordinate and abscissas in the drawing are the number of the essential proteins and different identifying technologies, respectively. Figure 1 gives the number of essential proteins detected by CNC and other detecting approaches, CIC, DC, NC, PeC and WDC, in the top proteins(top 100, top 200, top 300, top 400, top 500 and top 600).

the evaluation of proteins sorted by the prioritizing technologies, the true positive rate, also known as sensitivity, is calculated by the number of essential proteins over a certain discrimination threshold, divided by the number of all essential proteins in the PPI networks. The false positive rate is computed by the number of non-essential proteins over the same threshold, divided by the number of all non-essential proteins. Area under of the curve (AUC) measures the ability of the prioritizing methods to correctly classify the essential proteins and non-essential proteins. The larger AUC of the predicting method suggest the higher discrimination. Figure 2 is plotted by the false positive rate on the horizontal axis and the true positive rate on the vertical. The true positive rate are also known as sensitivity. The false positive rate can be calculated as (1-specificity). AUC represents the area under the curve. From Fig. 2, it can be found that the AUC of the CNC is distinctly bigger than that of each one of other methods including CIC, DC, NC, PeC and WDC. The ROC curves prove that CNC approach is more effective than other approaches in discriminating essential proteins from other proteins in yeast PPI networks.

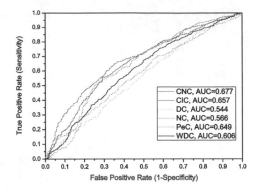

Fig. 2. ROC curves of different methods. Figure 2 shows the ROC curves in terms of the proteins prioritized by sorting technologies.

4   Conclusion

In this research, the subcellular localization information is successfully integrated with the PPI network. Based on CIC and ECC, a new essential protein identifying method named CNC is proposed. The results from the comparison of CNC and other predicting technologies suggest that CNC is more efficient in identifying essential proteins in terms of the measures, i.e., the number of essential proteins in the top proteins and ROC curve analysis. The study suggests that subcellular localization information contributes to identification of essential proteins. The essential proteins have close relationship with causing disease genes, so we will focus on how to merge different data sources to discriminate disease genes from non-disease genes in future.

Acknowledgment.   This work is supported in part by the National Natural Science Foundation of China under Grant Nos.61472133, 61502214, 31560317, 61370172, Hunan Provincial Natural Science Foundation of China Nos. 15JJ2038, 15JJ2037, Research Foundation of Education Bureau of Hunan Province Nos. 14A027, [2015]118, [2013]532, CSC No. 201508430098, Hunan Key Laboratory no. 2015TP1017.

References

1. Judson, N., Mekalanos, J.J.: TnAraOut, a transposon based approach to identify and characterize essential bacterial genes. Nat Biotechnol. **18**(7), 740–745 (2000). Author, F., Author, S.: Title of a proceedings paper. In: Editor, F., Editor, S. (eds.) CONFERENCE 2016. LNCS, vol. 9999, pp. 1–13. Springer, Heidelberg (2016)
2. Li, J.-J., Huang, D.S., Lok, T.-M., Lyu, M.R., Li, Y.-X., Zhu, Y.-P.: Network analysis of the protein chain tertiary structures of heterocomplexes. Protein Peptide Lett. **13**(4), 391–396 (2006)

3. Zhang, G.-Z., Huang, D.S., Quan, Z.H.: Combining a binary input encoding scheme with RBFNN for globulin protein inter-residue contact map prediction. Pattern Recogn. Lett. **26**(10), 1543–1553 (2005)
4. Zhao, X.-M., Huang, D.S., Cheung, Y.-M.: A novel hybrid GA/RBFNN technique for protein classification. Protein Pept. Lett. **12**(4), 383–386 (2005)
5. Zhang, G.-Z., Huang, D.S.: Prediction of inter-residue contacts map based on genetic algorithm optimized radial basis function neural network and binary input encoding scheme. J. Comput. Aided Mol. Des. **18**(12), 797–810 (2004)
6. Zhang, G.-Z., Huang, D.S.: Inter-residue spatial distance prediction by using intergrating GA with RBFNN. Protein Pept. Lett. **11**(6), 571–576 (2004)
7. Vallabhajosyula, R.R., Chakravarti, D., Lutfeali, S., Ray, A., Raval, A.: Identifying hubs in protein interaction networks. PLoS ONE **4**(4), e5344 (2009)
8. Freeman, L.C.: A set of measures of centrality based on betweenness. Sociometry **40**(1), 35–41 (1977)
9. Wuchty, S., Stadler, P.F.: Centers of complex networks. J. Theor. Biol. **223**(1), 45–53 (2003)
10. Estrada, E.: Rodr´ıguez-Vel´azquez, J.A.: Subgraph centrality in complex networks. Phys. Rev. E **71**(5), 056103 (2005)
11. Bonacich, P.: Power and centrality: a family of measures. Am. J. Sociol. **92**(5), 1170–1182 (1987)
12. Stevenson, K., Zelen, M.: Rethinking centrality: methods and examples. Soc. Netw. **11**(1), 1–37 (1989)
13. Wang, H., Li, M., Wang, J., Pan, Y.: A new method for identifying essential proteins based on edge clustering coefficient. In: International Symposium on Bioinformatics Research and Applications, pp. 87–98. Springer, Berlin, Heidelberg, May 2011
14. Shun, P., Huang, D.S.: Cooperative competition clustering for gene selection. J. Cluster Sci. **17**(4), 637–651 (2006)
15. Wang, B., Wong, H.S., Huang, D.S.: Inferring protein-protein interacting sites using residue conservation and evolutionary information. Protein Pept. Lett. **13**(10), 999–1005 (2006)
16. Huang, D.S., Huang, X.: Improved performance in protein secondary structure prediction by combining multiple predictions. Protein Pept. Lett. **13**(10), 985–991 (2006)
17. Huang, D.S., Zhao, X.-M., Huang, G.-B., Cheung, Y.-M.: Classifying protein sequences using hydropathy blocks. Pattern Recogn. **39**(12), 2293–2300 (2006)
18. Huang, D.S., Zheng, C.-H.: Independent component analysis based penalized discriminant method for tumor classification using gene expression data. Bioinformatics **22**(15), 1855–1862 (2006)
19. Li, J.-J., Huang, D.S., Wang, B., Chen, P.: Identifying protein-protein interfacial residues in heterocomplexes using residue conservation scores. Int. J. Biol. Macromol. **38**(3–5), 241–247 (2006)
20. Sprinzak, E., Sattath, S., Margalit, H.: How reliable are experimental protein-protein interaction data? J. Mol. Biol. **327**, 919–923 (2003)
21. Chen, J., Yuan, B.: Detecting functional modules in the yeast protein-protein interaction network. Bioinformatics **22**, 2283–2290 (2006)
22. Peng, W., Wang, J., Wang, W., et al.: Iteration method for predicting essential proteins based on orthology and protein-protein interaction networks. BMC Syst. Biol. **6**(1), 1–17 (2012)
23. Li, M., Wang, J.X., Wang, H., et al.: Identification of essential proteins from weighted protein-protein interaction networks. J. Bioinform. Comput. Biol. **11**(03), 1–19 (2013)
24. Tang, X., Wang, J., Zhong, J., et al.: Predicting essential proteins based on weighted degree centrality. IEEE/ACM Trans. Comput. Biol. Bioinform. **11**(2), 407–418 (2014)

25. Li, M., Zhang, H., Wang, J., et al.: A new essential protein discovery method based on the integration of protein-protein interaction and gene expression data. BMC Syst. Biol. **6**(1), 1–9 (2012)
26. Peng, X., Wang, J., Zhong, J,, et al.: An efficient method to identify essential proteins for different species by integrating protein subcellular localization information. In: IEEE International Conference on Bioinformatics and Biomedicine, vol. 2015, pp. 277–280 (2015)
27. Radicchi, F., Castellano, C., Cecconi, F., Loreto, V., Parisi, D.: Defining and identifying communities in networks. Proc. Natl. Acad. Sci. U.S.A. **101**(9), 2658–2663 (2004)
28. Huh, W.K., Falvo, J.V., et al.: Global analysis of protein localization in budding yeast. Nature **425**(6959), 686–691 (2003)
29. Zhang, R., Lin, Y.: DEG 5.0, a database of essential genes in both prokaryotes and eukaryotes. Nucleic Acids Res. **37**(suppl 1), D455–D458 (2009)
30. Stark, C., Breitkreutz, B.J., et al.: Biogrid: a general repository for interaction datasets. Nucleic Acids Res. **34**(1), D535–D539 (2006)
31. Binder, J.X., Pletscher-Frankild, S., et al.: COMPARTMENTS: unification and visualization of protein subcellular localization evidence. Database **2014**(bau012), 1–9 (2014)
32. Estrada, E.: Virtual identification of essential proteins within the protein interaction network of yeast. Proteomics **6**(1), 35–40 (2006)
33. Ning, K., Ng, H.K., Srihari, S., et al.: Examination of the relationship between essential genes in PPI network and hub proteins in reverse nearest neighbor topology. BMC Bioinform. **11**(1), 1–14 (2010)
34. Deng, S.-P., Zhu, L., Huang, D.S.: Predicting hub genes associated with cervical cancer through gene co-expression networks. IEEE/ACM Trans. Comput. Biol. Bioinform. **13**(1), 27–35 (2016)
35. Zhu, L., Ping, D.-S., Huang, D.S.: A two stage geometric method for pruning unreliable links in protein-protein networks. IEEE Trans. NanoBios-ci. **14**(5), 528–534 (2015)
36. Huang, D.S., Zhang, L., Han, K., Deng, S., Yang, K., Zhang, H.: Prediction of protein-protein interactions based on protein-protein correlation using least squares regression. Curr. Protein Pept. Sci. **15**(6), 553–560 (2014)
37. Zhu, L., You, Z.-H., Huang, D.S., Wang, B.: t-LSE: a novel robust geometric approach for modeling protein-protein interaction networks. PLOS ONE **8**(4), e58368 (2013). doi:10.1371/journal.pone.0058368
38. Huang, D.S., Jiang, W.: A general CPL-AdS methodology for fixing dynamic parameters in dual environments. IEEE Trans. Syst. Man Cybern. Part B **42**(5), 1489–1500 (2012)
39. Xia, J.-F., Zhao, X.-M., Song, J., Huang, D.S.: APIS: accurate prediction of hot spots in protein interfaces by combining protrusion index with solvent accessibility. BMC Bioinform. **11**, 174 (2010)
40. Xia, J.-F., Zhao, X.-M., Huang, D.-S.: Predicting protein-protein interactions from protein sequences using meta predictor. Amino Acids **39**(5), 1595–1599 (2010)
41. Xia, J.-F., Han, K., Huang, D.S.: Sequence-based prediction of protein-protein interactions by means of rotation forest and autocorrelation descriptor. Protein Pept. Lett. **17**(1), 137–145 (2010)

Gene Regulation Modeling and Analysis

Combining Gene Expression and Interactions Data with miRNA Family Information for Identifying miRNA-mRNA Regulatory Modules

Dan Luo[1], Shu-Lin Wang[1(✉)], and Jianwen Fang[2]

[1] College of Computer Science and Electronics Engineering,
Hunan University, Changsha 410082, Hunan, China
smartforesting@gmail.com
[2] Biometric Research Branch, Division of Cancer Treatment and Diagnosis,
National Cancer Institute, Rockville, MD 20850, USA

Abstract. It is well known that microRNAs (miRNAs) play pivotal roles in gene expression, transcriptional regulation and other important biological processes. An impressive body of literature indicates that miRNAs and mRNAs work cooperatively to form an important part of gene regulatory modules which are extensively involved in cancer. However, with the accumulation of available data, it is a great challenge to identify cancer-related miRNA regulatory modules and uncover their precise regulatory mechanism. This paper proposed a novel computational framework by combining gene expression and interaction data with miRNA family information to identify miRNA-mRNA regulatory modules (GIFMRM), which was evaluated on three heterogeneous datasets. Literature survey, biological significance and functional enrichment analysis were used to validate the obtained results. The analysis results show that the modules identified are highly correlated with the biological conditions in their respective datasets, and they enrich in GO biological processes and KEGG pathways.

Keywords: miRNA-mRNA regulation modules · Gene expression · miRNA family

1 Introduction

MicroRNA (miRNA) is a kind of non-protein-coding small RNA which contains approximately 20–24 nucleotides in length. They are involved in a wide variety of biological processes such as cell proliferation, development, and apoptosis [1]. They can bind to complementary sequences on their target mRNAs and cause mRNA degradation, translational inhibition or a combination of them [2]. Differentially regulated miRNAs in various complex diseases have been reported, such as breast cancer, lung cancer, ovarian cancer and a number of neurological disorders including Alzheimer's disease, multiple sclerosis, and so on [3]. To better understand biological process of miRNAs and their roles in the development of disease, it is urgent to uncover the relationship between miRNAs and their targets. However, it is a great

© Springer International Publishing AG 2017
D.-S. Huang et al. (Eds.): ICIC 2017, Part II, LNCS 10362, pp. 311–322, 2017.
DOI: 10.1007/978-3-319-63312-1_28

challenge to discover cancer-related modules for such reasons. (i) Bidirectional regulatory relationship. More than 2000 human mature miRNAs have been reported in miRBase [4]. Each miRNA regulates hundreds of different mRNAs which are estimated by computational predictions of miRNA targets [5] and each mRNA can be regulated by multiple miRNAs [6]. Besides, the regulatory relationship between miRNAs and mRNAs may change in different biological conditions. (ii) The noisy and incompleteness of genomic data could affect experimental results.

Many computational approaches on identifying miRNA-mRNA regulatory modules (MMRMs) (the union of a set of miRNAs and a set of mRNAs) [7] based on gene expression data have been proposed in recent years. Yoon and De Micheli [8] proposed a method that adopt the predicted miRNA-mRNA target information to reconstruct MMRMs. Improved methods based on it were developed by means of coherent expression patterns between miRNAs and mRNAs, or measuring (anti)-correlations between each pair of miRNA and mRNA. Liu, et al. [9] adopted correspondence Latent Dirichlet Allocation (corr-LDA) to identify functional MMRMs only using expression profiles. It does not need any prior target binding information, but there are too many parameters that are sensitive to results and hard to be optimized. Zhang, et al. [7] proposed a method called SNMNMF to predict miRNA regulatory modules. It uses the predicted miRNA-gene interactions, the expression profiles of miRNAs and genes, as well as the gene-gene interaction network, but this method is time-consuming. Furthermore, none of these methods considers the information of miRNA family. miRNA family information has been used to predict the association between miRNAs and diseases. Studies indicate that miRNAs in the same family likely share common sets of targets [10], so they probably have similarity function. And, the miRNAs in the same family target same mRNAs with high probability. Based on the above idea, a novel computational framework by combining gene expression data, miRNA family information and gene interaction data to identify MMRMs (GIFMRM) was proposed.

The framework assigns mRNAs to each miRNA cluster by calculating the ratio of confirmed interactions of miRNA-mRNA in each cluster instead of using clustering method. GIFMRM was tested on three heterogeneous cancer datasets, resulting in 21 modules identified in total. The KEGG pathways and GO biology processes of identified modules are significantly enriched. In addition, a number of miRNAs and mRNAs related to cancers are involved in our modules through literature survey.

2   Materials and Methods

The framework includes three main stages as shown in Fig. 1. In the first stage, the expression profiles of miRNA and mRNA are used to construct the interaction network of miRNA and mRNA S_{miTom} firstly. Then, the family information of miRNA S_{fam} is used to update the weight of miRNA-miRNA similarity network S_{miTomi} generated by Pearson Correlation Coefficient (PCC) based on S_{miTom}. In the second stage, miRNAs are clustered by hierarchical clustering analysis (HCA) [11] based on the updated miRNA-miRNA similarity network S^*. In the third stage, mRNAs are allocated to miRNA clusters according to their membership with each miRNA cluster calculated through miRNA-mRNA interaction matrix S_{con}.

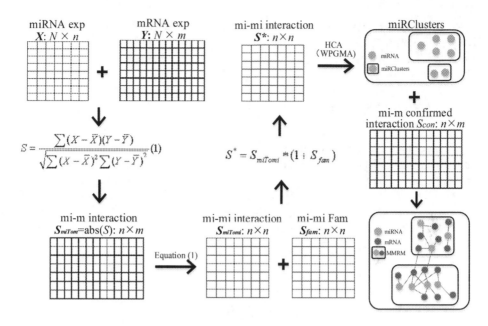

Fig. 1. Workflow of GIFMRM for identifying miRNA-mRNA regulatory modules.

2.1 Data Collection

GIFMRM was tested by three heterogeneous cancer datasets, including NCI-60 data for Epithelia to Mesenchymal Transition (EMT), Multi-class cancer (MCC) data, and 51 human breast cancer cell lines data (BR). They were downloaded from the web site http://europepmc.org/articles/PMC4482624#sec019title [12].

Three Heterogeneous Cancer Datasets. The NCI-60 dataset includes 60 human cancer cell lines that come from National Cancer Institute. 35 probes of miRNAs and 1154 probes of mRNAs were left after differentially analysis. The MCC dataset has 89 samples that contain normal samples and cancer samples. The cancer types include bladder, breast, colon, kidney, lung, pancreas, prostate and uterus. There were 108 miRNAs and 1860 mRNAs left after differentially analysis with p-value < 0.005. Total 725 human miRNAs were measured in the 51 human breast cancer cell lines in the BR dataset. It was divided into 27 luminal samples and 23 basal samples. And there were 92 miRNAs (p-value < 0.2) and 1500 mRNAs (p-value < 0.001) left after differential analysis.

Confirmed Interactions between miRNAs and mRNAs. The interaction datasets were obtained from Masud Karim, et al. [13]. To get more confirmed interactions, MicRNAdb [14], updated as of 2014, was added. Finally, 2345, 6154, and 9455 unique miRNA-mRNA interactions of NCI-60 dataset, BR dataset, and MCC dataset were obtained, respectively.

miRNA Family Information. Family classification of miRNA hairpin sequence was available from miRBase [4]. In our experiments only human miRNAs, including 1025 miRNAs, were selected from this dataset.

2.2   miRNA-miRNA Similarity Network Construction

The similarity between each pair of miRNA is measured by their interaction strength. And the strength interaction of each pair of miRNA is defined as their co-regulatory strength on same mRNAs. So, the strength interaction S_{miTom} between each miRNA and each mRNA is calculated by PCC firstly. X and Y denote the expression profiles of miRNA and mRNA, respectively. Then, the interaction strength S_{miTomi} of each pair of miRNA is calculated based on S_{miTom}. In this step, X denotes the row vector of miRNA i, and Y denotes the row vector of miRNA j. The obtained values are within $[-1,1]$, which represent positive regulation or negative regulation. As our aim is to get the interaction strength, the higher value of PCC indicates the stronger of the interaction, so we use the absolute value of PCC.

$$S = \frac{\sum (X - \bar{X})(Y - \bar{Y})}{\sqrt{\sum (X - \bar{X})^2 \sum (Y - \bar{Y})^2}} \tag{1}$$

$$S_{miTom} = abs(S), S_{miTomi} = S$$

According to that miRNAs in the same family likely share common targets, the miRNA family information is used to improve the interaction strength value of miRNAs that are in the same family. The updated interaction strength value of each pair of miRNA S^* can be defined as follows:

$$S^* = S_{miTomi} * (1 + S_{fam}) \tag{2}$$

S_{fam} denotes the miRNA family information matrix. If the two miRNAs are in the same family, $S_{fam} = 1$, otherwise $S_{fam} = 0$.

2.3   miRNAs Clustering

The miRNAs are clustered based on Hierarchical clustering analysis. It is a method of cluster analysis which seeks to build a hierarchy of clusters. HCA is composed of four basic steps [15]. Firstly, each object is assigned to its own cluster, so there is only one object in each cluster. In this case, the distance between each cluster is equal to the distance between the objects they contain. Secondly, the closest pair of clusters are found and merged into a single cluster. Thirdly, the distance between the new cluster and every old cluster is computed. Finally, the step 2 and 3 are repeated until all object are involved in a single cluster or meet the termination condition.

In our method, the similarity between each miRNA and each miRNA is the evaluation criterion instead of the distance, so miRNAs are clustered on the basis of S^*. Weighted Pair Group Method with Arithmetic Mean (WPGMA) [16] is applied to compute the distance between the new cluster and every old cluster. WPGMA can avoid the interference generated by individual deviation samples and make the distribution of miRNAs better. It constructs a rooted tree that can reflect the structure presented in S^*. The closest clusters i, j are merged into a new cluster $i \cup j$, and the distance $d_{(i \cup j),k}$ between the new cluster $i \cup j$ and every old cluster k is computed as follows:

$$d_{(i \cup j),k} = \frac{d_{i,k} + d_{j,k}}{2} \tag{3}$$

S^* is an initial matrix. The distance matrix is updated after computing the distance between the new cluster and each old cluster. The clusters in which the number of miRNAs is less than 3 are deleted to avoid 'star-shaped' [13].

2.4   mRNAs Assignment of Modules

To enhance the cohesion of modules, miRNA-mRNA interactions S_{con} are took into account. The mRNAs are assigned to each miRNA cluster by calculating the ratio of confirmed interactions of miRNA-mRNA in each cluster instead of using clustering method. The ratio R formula for mRNAs assignment is defined as follows:

$$R_{C_i,j} = \frac{\sum S_{con(k,j)}}{Count(C_i)} \tag{4}$$

where $R_{C_i,j}$ denotes that the ratio of mRNA j is assigned to miRNA cluster C_i. k is miRNA's ID belonging to C_i. $Count(C_i)$ denotes the number of miRNAs in C_i. The greater the R is, the greater the probability of mRNA associated with the miRNA cluster is.

3   Results

GIFMRM was evaluated on three heterogeneous cancer datasets. The number of modules was predefined as 8, 11, and 9 for NCI-60, BR, and MCC datasets, respectively. And the threshold value of mRNA assignment ratio was set as 0.2 after a series of experiments. In the NCI-60 dataset, three modules were deleted as they contained only one miRNA. In the BR dataset, two modules were deleted as they contained only two miRNAs. In the MCC dataset, one module only contained one miRNA and one module did not have any mRNA. Therefore, two modules were deleted. Finally, 5, 9 and 7 modules were left in the NCI-60, BR and MCC datasets, respectively (see Table 1).

$\#M$ denotes the number of modules. $\overline{miRNA}$ is the average number of miRNAs in each module. $\overline{mRNA}$ is the average number of mRNAs in each module. $\overline{mi\_cr}$ is the average number of cancer-related miRNAs that have been confirmed in each module.

Table 1. The experiment results of NCI-60 dataset, BR dataset and MCC dataset.

Dataset	#M	miRNA	mRNA	mi_cr/n/N	Time (s)
NCI-60	5	6	84	3/17/19	13.75
BR	9	9	125	2/21/21	40.49
MCC	7	13	115	8/58/60	70.54

n is the number of cancer-related miRNAs in the identified modules. N is the number of cancer-related miRNAs in the dataset. Time denotes the time consumption.

3.1   Verification of Cancer-Related MiRNAs in Modules

To validate the significance of experiment results, the gold stand cancer-related miRNA association data derived from the Human microRNA Disease Database (HMDD) v2.0 [17] was downloaded. As shown in Table 1, there are 19, 21, and 60 cancer-related miRNAs in the NCI-60, BR, and MCC datasets, respectively. It's nice to find that they are almost involved in the modules identified by GIFMRM. The average probability of all identified cancer-related miRNAs in each dataset is up to 95.38%.

In addition, a number of miRNAs related to cancers are involved in our modules through literature survey. These miRNAs target mRNAs involved in modules to impact on the development of cancer. For example, in moduleBR-3, miR-661 expression in SNAI1-induced epithelial to mesenchymal transition contributes to breast cancer cell invasion by targeting StarD10 messengers [18]. miR-101 expression restoration suppresses multiple malignant phenotypes of hepatocellular carcinoma cells by coordinate repression of a cohort of oncogenes [19] (for instance, STMN1). miR-141 targets ZEB1, which is related to Breast cancer in moduleBR-8. In moduleMCC-5, miR-29 targets ATP1B1, which serves to limit migration and invasion in breast cancer cells. The results show that the modules found are significantly associated with cancer.

3.2   Modularity Analysis of Identified Modules

In the NCI-60 dataset, moduleNCI-5 involves all miR-200 family members (miR-141, miR-200a, miR-200b, miR-200c, and miR-429). miR-18a, miR-18b and miR-106b are involved in moduleNCI-4 and they are related to hepatocellular. miR-7 and miR-7-1* are also involved in moduleNCI-2 which are associated with Breast Cancer.

In the BR dataset, the miRNAs belonging to the same family are in the same modules. For instance, miR-23a and miR-23* related to Atrophy are involved in moduleBR-3. Both miR-221 and miR-222 are involved in moduleBR-9; miR-141, miR-200a and miR-200c are involved in moduleBR-8; miR-361-3p and miR-361-5p are involved in moduleBR-1; miR-500, miR-501-5p and miR-502-5p are involved in moduleBR-4, and so on. What's more, miR-20b*, miR-21*, miR-29c and miR-34a related to Breast cancer are involved in moduleBR-1. In moduleBR-6, 3 of 6 miRNA are related with Breast Neoplasms. Furthermore, 3 of 4 miRNAs are related with Breast Neoplasms in moduleBR-8, and 4 of 7 miRNAs are related with Breast Neoplasms in moduleBR-9. The average probability of cancer-related miRNAs in each module found in the BR dataset is up to 53.87%.

In the MCC dataset, the miRNAs belonging to the same family are also in same module. For example, all members of miR-30 family, let-7 family and miR-200 family are contained in moduleMCC-4, moduleMCC-7 and moduleMCC-8, respectively. Other than that, there are 9 miRNAs relevant to Breast Neoplasms, 4 miRNAs relevant to Carcinoma and 3 of this 4 miRNAs work on Hepatocellular, all of them are involved in moduleMCC-5. In moduleMCC-6, miR-125a and miR-125b worked on Breast Neoplasms, miR-99a and miR-100 related to Esophageal Neoplasms are involved.

Therefore, with that said the miRNAs in the same module have similarity function or belong to the same family, which are proved the modularity, functional of the identified modules and correlation of biological conditions of their respective datasets.

3.3 KEGG Pathways and Go Analysis

Gene Ontology (GO) in Biological Processes (BP) and Kyoto Encyclopedia of Genes and Genomes (KEGG) pathways analysis are applied to assess the biological significance of the modules [20] besides HMDD. As shown in Tables 2, 3 and 4, all modules identified are significantly associated with pathways in cancer. P-values (adj.) have been obtained through hypergeometric analysis corrected by FDR method.

Table 2. KEGG pathways enrichment of moduleNCI-2.

Id	Items	Item_Details	P-value (adj.)	Genes
1	Kegg:04510	Focal adhesion	1.09E-03	ERBB2,FLNA, PRKCA,FN1
2	Kegg:04144	Endocytosis	1.26E-03	RAB11A, RAB11FIP1, PARD6B,RNF41
3	Kegg:05200Kegg:04510	Pathways in cancer Focal adhesion	1.59E-03	ERBB2,PRKCA,FN1
4	Kegg:04010	MAPK signaling pathway	1.70E-03	FGF2,DUSP16, FLNA,PRKCA, MKNK2
5	Kegg:05200	Pathways in cancer	2.22E-03	FGF2,ERBB2,BCR, PRKCA,FN1

For the NCI-60 dataset, FOXA1 (transcriptional factor, which can regulate the expression of E-cadherin) is also involved in the modules beyond CDH1, TWIST1, ZEB1 and ZEB2 (E-cadherin transcriptional repressor, which is usually targeted by miR-200 family).

Besides pathways shown in Table 3, the moudeBR-3 also contains Jak-STAT and MAPK signaling pathways which further indicates cancer pathways enrichment. As we know, Jak-STAT can lead to abnormal growth of many malignancies including breast

Table 3. KEGG pathways enrichment of moduleBR-3.

Id	Items	Item_Details	P-value (adj.)	Genes
1	Kegg:05200	Pathways in cancer	2.54E-12	IL6,MET,COL4A2,TGFBR2, LAMC1,AKT1,EGFR, TGFB1,SMAD3,ITGB1, CEBPA,PDGFB,CBL
2	Kegg:04510	Focal adhesion	1.34E-10	MET,COL4A2,VCL, LAMC1,AKT1,EGFR, ITGB1,COL5A1,PDPK1, PDGFB
3	Kegg:04060 Kegg:05200	Cytokine-cytokine receptor interaction Pathways in cancer	2.64E-10	IL6,MET,TGFBR2,EGFR, TGFB1,PDGFB
4	Kegg:05200 Kegg:04144	Pathways in cancer Endocytosis	2.80E-10	MET,TGFBR2,EGFR, TGFB1,SMAD3,CBL
5	Kegg:05200 Kegg:04510	Pathways in cancer Focal adhesion	3.21E-09	MET,COL4A2,LAMC1, AKT1,EGFR,ITGB1,PDGFB

Table 4. KEGG pathways enrichment of moduleMCC-5.

Id	Items	Item_Details	P-value (adj.)	Genes
1	Kegg:05200	Pathways in cancer	6.72E-08	GSK3B,ITGA6,BIRC5,IGF1R, TGFBR2,AKT3,CDKN1B,LAMC1, HSP90B1,CTNNB1,CDKN2A
2	Kegg:04510	Focal adhesion	9.78E-07	GSK3B,ITGA6,IGF1R,CCND2, AKT3,LAMC1,CTNNB1,PPP1CC
3	Kegg:05200 Kegg:04510	Pathways in cancer Focal adhesion	1.13E-06	GSK3B,ITGA6,IGF1R,AKT3, LAMC1,CTNNB1
4	Kegg:05200 Kegg:05215	Pathways in cancer Prostate cancer	1.15E-06	GSK3B,IGF1R,AKT3,CDKN1B, HSP90B1,CTNNB1
5	Kegg:05200 Kegg:05210	Pathways in cancer Colorectal cancer	5.64E-06	GSK3B,BIRC5,TGFBR2,AKT3, CTNNB1

cancer [21], and MAPK signaling pathway is one of the critical pathways in EMT development.

Focal adhesion is also significant in moduleMCC-5. Previous studies show focal adhesion pathway is involved in multiple tumors. For example, prostate cancer [22], hepatocellular carcinoma tumor [23], and lung cancer [24].

The top 10 GO biological process terms of moduleNCI-2, moduleBR-3 and moduleMCC-5 are all enriched as p-value < 0.05 (see Fig. 2). The most significant biological processes in moduleNCI-2, moduleBR-3 and moduleMCC-5 include wound healing, lung development, positive regulation of peptidyl-serine phosphorylation, axon guidance, negative regulation of apoptotic process and positive regulation of transcription from RNA polymerase II promoter.

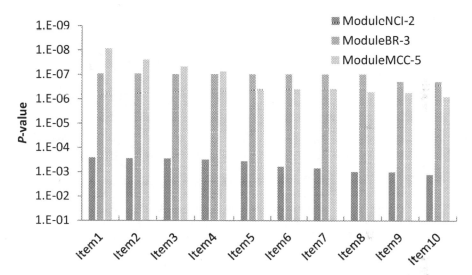

Fig. 2. The top 10 GO (BP) items statistical significance of moduleNCI-2, moduleBR-3 and moduleMCC-5.

3.4 Comparison with Corr-LDA

The NCI-60, BR and MCC datasets are applied to evaluate corr-LDA [9] to compare with GIFMRM. The experimental results of corr-LDA show that the average over-lapping ratio of miRNA in each module is up to 72.78% (This data was obtained by comparing module-01 with other modules identified by corr-LDA and similarly hereinafter) in the NCI-60 dataset. 9 of 10 modules have greater overlapping in the BR dataset, and the average overlapping ratios of miRNAs in these modules are up to 85.19%. In the MCC dataset, 7 of 8 modules have greater overlapping. The average overlapping ratios of these modules are up to 74.29%. Those show a large amount of redundancy of the corr-LDA results. Furthermore, the top 15 KEGG pathways p-values from moduleNCI-2, moduleBR-3 and moduleMCC-5 identified by GIFMRM are more enrichment compared to the top 15 KEGG pathways p-values from moduleNCI-1,

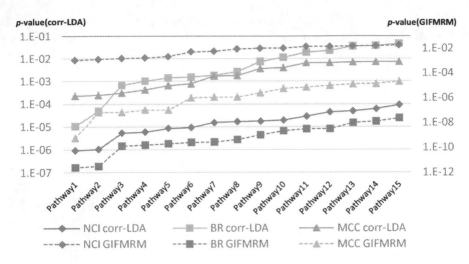

Fig. 3. The top 15 KEGG pathways statistical significance from moduleNCI-2, moduleBR-3 and moduleMCC-5 identified by GIFMRM and moduleNCI-1, moduleBR-1 and moduleMCC-1 identified by corr-LDA.

Table 5. Comparison of running time on NCI-60 dataset, BR dataset and MCC dataset.

Methods	Time of NCI-60	Time of BR	Time of MCC
corr-LDA	127 s	1001 s	1276 s
GIFMRM	13.75 s	40.49 s	70.54 s

moduleBR-1 and moduleMCC-1 identified by corr-LDA (see Fig. 3). We implement GIFMRM and corr-LDA on the same computer. The running time of GIFMRM is much less than corr-LDA (see Table 5).

4   Conclusion and Discussion

The current knowledge indicates that miRNAs play critical roles in many biological processes and differentially expressed miRNAs are involved in various cancers. Therefore, it is a key task to understand the regulatory mechanism of miRNA-mRNA. The existing methods such as corr-LDA [9], SNMNMF [7], Misynery [25] and DICORE [13] do not consider miRNA family information in MMRMs identification.

In this paper, we proposed an effective framework to identify MMRMs, which can directly analyze three kinds of data including gene expression profiles, miRNA family information and miRNA-mRNA interaction data. Our method combines clustering and probability analysis to uncover the regulatory relationship between miRNA and mRNA. And three cancer datasets are applied to evaluate GIFMRM. The KEGG pathways analysis indicates that all identified modules enrich on cancer pathways. And they have nice modularity in which the miRNAs in same module have similarity function or belong to the same family.

Although our framework expands a new visual field of MMRMs identification, the drawback of GIFMRM is that the number of modules and the threshold value of mRNA assignment ratio need to be predefined, which are sensitive to results. Thus future work will focus on designing novel miRNAs clustering methods instead of using HCA (WPGMA).

Acknowledgement. This work was supported by the grants of the National Science Foundation of China (Grant Nos. 61472467, 61672011, and 61471169) and the Collaboration and Innovation Center for Digital Chinese Medicine of 2011 Project of Colleges and Universities in Hunan Province.

References

1. Zhang, W., Zang, J., Jing, X., Sun, Z., Yan, W., Yang, D., Shen, B., Guo, F.: Identification of candidate miRNA biomarkers from mirna regulatory network with application to prostate cancer. J. Transl. Med. **12**(1), 66 (2014)
2. He, L., Hannon, G.J.: MicroRNAs: small RNAs with a big role in gene regulation. Nat. Rev. Genet. **5**(7), 522–531 (2004)
3. Liu, B., Li, J., Cairns, M.J.: Identifying miRNAs, targets and functions. Brief. Bioinform. **15**(1), 1–19 (2014)
4. Kozomara, A., Griffiths-Jones, S.: miRBase: annotating high confidence microRNAs using deep sequencing data. Nucleic Acids Res. **42**(D1), D68–D73 (2014)
5. Friedman, R.C., Farh, K.K.H., Burge, C.B., Bartel, D.P.: Most mammalian mRNAs are conserved targets of micrornas. Genome Res. **19**(1), 92–105 (2009)
6. Lim, L.P., Lau, N.C., Garrett-Engele, P., Grimson, A., Schelter, J.M., Castle, J., Bartel, D.P., Linsley, P.S., Johnson, J.M.: Microarray analysis shows that some microRNAs downregulate large numbers of target mRNAs. Nature **433**(7027), 769–773 (2005)
7. Zhang, S., Li, Q., Liu, J., Zhou, X.J.: A novel computational framework for simultaneous integration of multiple types of genomic data to identify microRNA-gene regulatory modules. Bioinformatics **27**(13), 401–409 (2011)
8. Yoon, S.R., De Micheli, G.: Prediction of regulatory modules comprising microRNAs and target genes. Bioinformatics **21**(2), 93–100 (2005)
9. Liu, B., Liu, L., Tsykin, A., Goodall, G.J., Green, J.E., Zhu, M., Kim, C.H., Li, J.: Identifying functional miRNA-mRNA regulatory modules with correspondence latent dirichlet allocation. Bioinformatics **26**(24), 3105–3111 (2010)
10. Liao, B., Ding, S.M., Chen, H.W., Li, Z.J., Cai, L.J.: Identifying human microRNA-disease associations by a new diffusion-based method. J. Bioinform. Comput. Biol. **13**(4), 1550014 (2015)
11. Johnson, S.C.: Hierarchical clustering schemes. Psychometrika **32**(3), 241–254 (1967)
12. Le, T.D., Zhang, J., Liu, L., Li, J.: Ensemble methods for miRNA target prediction from expression data. PLoS ONE **10**(6), e0131627 (2015)
13. Masud Karim, S.M., Liu, L., Le, T.D., Li, J.: Identification of miRNA-mRNA regulatory modules by exploring collective group relationships. BMC Genom. **17**(Suppl 1), 7 (2016)
14. Wang, D., Gu, J., Wang, T., Ding, Z.: OncomiRDB: a database for the experimentally verified oncogenic and tumor-suppressive microRNAs. Bioinformatics **30**(15), 2237–2238 (2014)
15. Borgatti, S.P.: How to explain hierarchical clustering (1994)

16. Sokal, R.R.: A statistical method for evaluating systematic relationships. Univ. Kansas Sci. Bull. **28**, 1409–1438 (1958)
17. Li, Y., Qiu, C.X., Tu, J., Geng, B., Yang, J.C., Jiang, T.Z., Cui, Q.H.: Hmdd V2.0: a database for experimentally supported human microRNA and disease associations. Nucleic Acids Res. **42**(D1), D1070–D1074 (2014)
18. Vetter, G., Saumet, A., Moes, M., Vallar, L., Béchec, A.L., Laurini, C., Sabbah, M., Arar, K., Theillet, C., Lecellier, C.H.: miR-661 expression in SNAI1-induced epithelial to mesenchymal transition contributes to breast cancer cell invasion by targeting nectin-1 and stard10 messengers. Oncogene **29**(31), 4436 (2010)
19. Wang, L., Zhang, X., Jia, L.T., Hu, S.J., Zhao, J., Yang, J.D., Wen, W.H., Wang, Z., Wang, T., Zhao, J.: C-Myc-mediated epigenetic silencing of MicroRNA-101 contributes to dysregulation of multiple pathways in hepatocellular carcinoma. Hepatology **59**(5), 1850–1863 (2014)
20. Tabas-Madrid, D., Nogales-Cadenas, R., Pascual-Montano, A.: Genecodis3: a non-redundant and modular enrichment analysis tool for functional genomics. Nucleic Acids Res. **40**(w1), W478–W483 (2012)
21. Zhang, Q.: Role of Jak/Stat pathway in the pathogenesis of breast cancer (2010)
22. Zhang, J.S., Gong, A., Gomero, W., Young, C.Y.: ZNF185, a lim-domain protein, is a candidate tumor suppressor in prostate cancer and functions in focal adhesion pathway. Cancer Res. **64**(7), 619–620 (2004)
23. Lee, C., Fan, S., Sit, W., Jor, I.W., Wong, L.L., Man, K., Tan-Un, K., Wan, J.M.: Olive oil enriched diet suppresses hepatocellular carcinoma (Hcc) tumor growth via focal adhesion pathway. Cancer Res. **67**(9 Supplement), LB-60 (2007)
24. Ocak, S., Yamashita, H., Udyavar, A.R., Miller, A.N., Gonzalez, A.L., Zou, Y., Jiang, A., Yi, Y., Shyr, Y., Estrada, L.: DNA copy number aberrations in small-cell lung cancer reveal activation of the focal adhesion pathway. Oncogene **29**(48), 6331–6342 (2010)
25. Li, Y., Liang, C., Wong, K.C., Luo, J.W., Zhang, Z.L.: Mirsynergy: detecting synergistic miRNA regulatory modules by overlapping neighbourhood expansion. Bioinformatics **30**(18), 2627–2635 (2014)

SNPs and Haplotype Analysis

Association Mapping Approach into Type 2 Diabetes Using Biomarkers and Clinical Data

Basma Abdulaimma$^{(\boxtimes)}$, Abir Hussain$^{(\boxtimes)}$, Paul Fergus$^{(\boxtimes)}$,
Dhiya Al-Jumeily$^{(\boxtimes)}$, Casimiro Aday Curbelo Montañez$^{(\boxtimes)}$,
and Jade Hind$^{(\boxtimes)}$

Applied Computing Research Group, Faculty of Engineering and Technology,
Liverpool John Moores University, Byrom Street, Liverpool L3 3AF, UK
{B. T. Abdulaimma, C. A. Curbelomontanez}@2015. ljmu. ac. uk,
{A. Hussain, P. Fergus, D. Aljumeily}@ljmu. ac. uk,
J. Hind@2012. ljmu. ac. uk

Abstract. The global growth in the incidence of Type 2 Diabetes (T2D) has become a major international health concern. As such, understanding the etiology of Type 2 Diabetes is vital. This paper investigates a variety of statistical methodologies at various level of complexity to analyze genotype data and identify biomarkers that show evidence of increased susceptibility to T2D and related traits. A critical overview of several selected statistical methods for population-based association mapping particularly case-control genetic association analysis is presented. A discussion on a dataset accessed in this paper that includes 3435 female subjects for cases and controls with genotype information across 879071 Single Nucleotide Polymorphism (SNPs) is presented. Quality control steps into the dataset through pre-processing phase are performed to remove samples and markers that failed the quality control test. Association analysis is discussed to address which statistical method is appropriate for the dataset. Our genetic association analysis produced promising results and indicated that Allelic association test showed one SNP above the genome-wide significance threshold of 5×10^{-8} which is rs10519107 (Odds Ratio (OR) = 0.7409, P − Value (P) = 1.813×10^{-9}). While there are several SNPs above the suggestive association threshold of 5×10^{-6}, these SNPs should be considered for further investigation. Furthermore, Logistic Regression analysis adjusted for multiple confounder factors indicated that none of the genotyped SNPs had passed genome-wide significance threshold of 5×10^{-8}. Nevertheless, four SNPs (rs10519107, rs4368343, rs6848779, rs11729955) have passed suggestive association threshold.

Keywords: Genetics · Genome-Wide Association Studies (GWAS) · Logistic regression model · P-values · Single Nucleotide Polymorphism (SNP) · Type 2 Diabetes (T2D)

1 Introduction

Currently, the prevalence and the incidence of Type 2 Diabetes (T2D) throughout the world are increasing at an alarming rate. The International Diabetes Federation (IDF) has estimated that the number of diabetic people is expected to rise from 366

© Springer International Publishing AG 2017
D.-S. Huang et al. (Eds.): ICIC 2017, Part II, LNCS 10362, pp. 325–336, 2017.
DOI: 10.1007/978-3-319-63312-1_29

million in 2011 to 552 million by 2030 worldwide [1]. Type 2 Diabetes is a multi-factorial disorder and is the result of the complex interaction between genetic, environment and sedentary lifestyle [2]. However, genetic susceptibility has been established as an essential component of risk [3]. Twins studies have exposed that the concordance rate of T2D in monozygotic twins is approximately 70% compared with 20% to 30% in dizygotic twins [4].

The primary tools for identifying disease susceptibility loci are genetic variations which are termed as Single Nucleotide Polymorphism (SNP). SNP is a single base-pair change in the genetic code, and it is the main cause of human genetic variability [5].

Genome-wide association studies (GWAS) have been widely used and specifically developed for investigating the genetic architecture of human disease in the entire genome [6]. The ultimate aim of GWAS is to identify the genetic risk factors for common complex diseases such as Type 2 Diabetes, Schizophrenia, Epilepsy, Obesity, Cardiovascular Disease, and Hypertension [6]. GWAS is more routinely employed as the increase of less expensive genotyping technologies becomes available [7]. The identification of genetic markers that show evidence of increased susceptibility to T2D and related traits are necessary to advance and facilitate the translation of this genetic information into clinical practice [8]. This advance may help to improve risk prediction [9] of the disease and delay or prevention of disease onset and to mitigate cares expenditures [10]. However, to understand the etiology of such complex diseases, genetic information solely would not be sufficient without considering the non-genetic factors [11].

There are several statistical methods for association mapping including allelic test, genotypic test, dominant test, recessive test, Cochran-Armitage trend test, Fisher exact test, and Logistic regression test [12]. However, it is difficult to specify which association tests to use [13]. It would be ideal to design optimal analyses based on the knowledge about the penetrance patterns of predisposing variants such as additive effect, dominant or recessive effects. Lacking this knowledge forces investigators to use their judgment [13].

This paper considers a case-control study design to conduct several classes of association analysis including; chi-square test based on (Allelic test, Genotypic test, Dominant test, and Recessive test) (results for Genotypic, Dominant, and Recessive tests have not been included in this paper), and Logistic regression. Logistic regression is the preferred approach to performing association analysis as it can readily expand to include covariates such as clinical variables, sociodemographic and environmental factors. Using genetic association analysis would facilitate the investigation of genetic markers that manifest themselves as candidates to increase susceptibility to T2D. These findings provide starting points to researchers and professionals to investigate further and to provide a better understanding to the disease onset and advance the development of medical therapies.

2 Background

Understanding the etiology of complex diseases such as T2D that is caused by the contribution of genetic and non-genetic risk factors is challenging [14]. The development of genetic association mapping has facilitated the discovery of genetic markers predisposing to complex diseases such as T2D. Recently, several GWAS studies accompanied with various statistical methods have been performed in different cohorts and ethnic groups, to measure the association of genetic variants (loci) to disease susceptibility and to test for statistical significant (p-value). A series of publications have addressed various aspects and strategies into T2D genetics studies to be available in the literature for further investigations.

In Cheema's work [15], the authors performed a case-control study to investigate the differences in the association of peroxisome proliferator-activated receptor, gamma, coactivator 1 alpha (PPARGC1A) gene with T2D risk among the population with African origins. The study includes adults aged >30 years old from African Americans (cases = 124, controls = 122) and Haitian Americans (cases = 110, controls = 116). The statistical method used within this study was Chi-squared goodness-fit test that was employed to check genotype counts for each SNP for Hardy-Weinberg Equilibrium. Furthermore, the t-test was used to compare between cases and controls considering demographic (age, sex, BMI, smoking status) and clinical information. Logistic regression approach was also used to calculate adjusted and unadjusted Odds Ratio (OR) with 95% confidence interval (CI). The result indicated that SNP rs7656250 (OR = 0.22, p-value = 0.005) and rs4235308 (OR = 0.42, p-value = 0.026) showed protective association with T2D in Haitian Americans. While in African Americans, SNP rs4235308 (OR = 2.53, p-value = 0.028) showed significant risk association with T2D.

While, in Qiu's study [16] the association analysis was performed in a case-control study to investigate the role of the mutation in KCNJ11 gene (potassium inwardly-rectifying-channel, subfamily-J, member 11) particularly E23K polymorphism (rs5219) in susceptibility to T2D. In this study, 56,349 T2D cases, 81,800 controls, and 483 family trios were collected from 48 published studies. The statistical methods used within the approach included Standard Q-statistic test, subgroup analysis (ethnicity, sample size, BMI, age, and sex) were utilized to explore whether the variation in these studies was due to heterogeneity. Furthermore, the odds ratio with its 95% confidence interval of KCNJ11 E23K polymorphism was calculated to measure the association with T2D. Dominant and Recessive genetic models were applied to examine the association of KCNJ11 E23K polymorphism and T2D risk. The result suggested that KCNJ11 E23K allele of rs5219 (OR = 1.12, $p < 10^{-5}$) was significantly associated with T2D risk. For heterozygous and homozygous allele with (OR = 1.09, $p < 10^{-5}$) and (OR = 1.26, $p < 10^{-5}$) respectively, significant increase of T2D risk was observed. For Dominant and Recessive genetic models, similar results were obtained. This study suggested that a modest but statistically effect of the 23K allele of rs5219 polymorphism in susceptibility to T2D, particularly in East Asians and Caucasians. However, the contribution of these genetic variations to T2D in other ethnic populations (e.g. Indian, African, American, Jews, and Arabian) appears to be relatively low.

Genetic association studies are becoming an important approach for identifying genes mainly SNPs conferring susceptibility to complex diseases. The findings of Disease-SNP associations have been reported consistently using various statistical analysis methods that calculate statistical significant of the SNPs and measure the strength of the association in the study.

3   Materials and Methods

This section provides a description of the dataset that is used in this paper and illustrates the quality control steps taken to pre-process that dataset. It also describes the concept of genetic association analysis and provides in-depth information related to statistical methods that are used in this domain.

3.1   Data Description

The Nurses' Health Study (NHS) cohort data set is used in this paper, and it is provided by the Database of Genotypes and Phenotypes (dbGap) [17]. The NHS was established in 1976. Participants were 121,700 female registered nurses between age 30 to 55 and resided in 11 U.S states. All nurses responded to mailed questionnaire requesting information related to their medical history and lifestyle characteristics. Since then, the Nurses have been invited twice a year to fill the questionnaire and attain an update (for instance information on newly diagnosed illness). Furthermore, all participants were asked to provide blood samples, in which 32,826 members responded. The cases and controls participants were selected from the NHS T2D study. DNA of cases and controls participants were genotyped using the Affymetrix Genome-Wide Human 6.0 array. The final version of the dataset includes 3435 female subjects in which 1581 T2D cases and 1854 controls with genotype information across 879071 SNPs. Participants in this dataset are identified as Hispanic or non-Hispanic, and each belongs to one of four racial categories (White, African-American, Asia or Other). Most participants are White and non-Hispanic representing (97.4%) of the dataset. The NHS dataset also includes corresponding clinical and dietary data, such as age, gender, BMI, alcohol intake, smoking status, physical activity, medical and family history.

3.2   Data Preprocessing

In this paper, the accessed genetic data is in PLINK format. PLINK v1.07 [18] is a whole genome data analysis toolset which is developed for handling SNP data. The files in PLINK format are large and could cause issues with computational performance. As such, we convert these files to binary format using PLINK 1.07 toolset. Transferring to a binary formatted file, resulting in a considerable reduction in file size and significantly enhancing computational efficiency. This step is vital for preparing the dataset for quality control and filtering procedures. We performed data quality control for individuals and genetic data to produce a subset of reliable genetic markers and samples to be used for association analysis phase. Firstly, this study has been restricted to White and non-Hispanic ancestry to reduce potential bias due to population strati-fication. We removed data samples which have been reported with discordant sex information and duplicated or related individuals. Quality control for genetic markers was considered to remove genetic markers (SNPs) with >0.1 missing data and with Minor Allele Frequency (MAF) of <0.05. We further conducted Hardy-Weinberg Equilibrium (HWE) and discarded those SNPs with a p-value <0.001 in control samples. Following the quality control steps, 3255 individuals and 665092 markers remained in the study from the original sample of 3435 and 879071, respectively.

3.3   Association Analysis

An association analysis of a case-control study aims to compare the frequency of alleles or genotypes at genetic marker loci (SNP) between cases and controls from a given population. This analysis will detect if there are any differences in the frequency of alleles between individuals in the study. The testing leads to determine whether the difference in alleles' frequency is statistically significant. In this situation, that alleles (genetic marker) can be recognized as being associated with the phenotype (disease trait) [19]. In other words, association analysis is a series of single-locus statistics tests, exploring each SNP separately for the association to the phenotype.

   In a case-control design study, the association between a single SNP and disease status can be based on standard contingency table tests for independence [13]. Contingency tables are widely used to display genetic markers (SNP) in the format of genotype or allele frequency by disease status (case-control) [19]. Each single SNP consists of minor allele a and major allele A, among case and control groups. These alleles can be represented as a contingency table of the disease status by either genotype count (e.g. AA, Aa, and aa) with a dimension of 2×3 of 2 degrees of freedom (d.f.) or allele count (A and a) with a dimension of 2×2 of 1 d.f. The genetic data can also be analyzed assuming a prespecified genetic model, as contingency tables allow for dif-ferent models of disease penetrance such as dominant model and recessive model. For example, the contingency table of the dominant model of penetrance can be summarized as a 2×2 table with 1 d.f. of genotype count of AA versus Aa or aa, as each copy (one copy) of dominant allele A increases the risk of disease. While to test for a recessive model of penetrance, the contingency table is represented as 2×2 table with 1 d.f. requiring two pairs of recessive allele aa to increase the risk of disease as the genotype count of the recessive model is aa versus the combined count of Aa and AA [20].

   The calculation of degrees of freedom is based on the inheritance models in which representing by genotypic, allelic, recessive and dominant [20]. Therefore, the degrees of freedom of genetic model is calculated based on the (number of rows in the con-tingency table – 1) × (number of columns in the contingency table – 1) [21]. For example, for allelic test where the number of both rows and columns is 2, the degrees of freedom is $(2 - 1) \times (2 - 1) = 1$.

   The contingency table for case and control analyses using various genetic model of penetrance and these summarized in Table 1, where DF represents degrees of freedom.

Table 1. Contingency table for different genetic models

Test	DF	Contingency table representation			
			aa	*Aa*	*AA*
Genotypic test	2	**Cases**	O_{11}	O_{12}	O_{13}
		Controls	O_{21}	O_{22}	O_{23}
			AA	*Aa* or *aa*	
Dominant model	1	**Cases**	O_{11}	O_{12}	
		Controls	O_{21}	O_{22}	
			aa	*Aa* or *AA*	
Recessive model	1	**Cases**	O_{11}	O_{12}	
		Controls	O_{21}	O_{22}	
			a	*A*	
Allelic test	1	**Cases**	O_{11}	O_{12}	
		Controls	O_{21}	O_{22}	

Whereas O_{ij} refers to the observed frequency of individuals in cases and controls, i refers to row number and j to column number. For example, in genotypic model test O_{11} refers to the observed frequency of individuals in cases when genotype aa occurs.

Practically the association test within genetic data of case and control status is to test the null hypothesis of no association between the SNP and phenotype of interest (disease status) in the contingency table. Pearson's chi-squared test (x^2) can be used to test for association. The principle of chi-squared test (x^2) is to compare the distributions of observed and expected values of their contingency tables [22]. Chi-square test summarizes the differences between the observed frequency values and the expected frequency values at a single genetic marker loci (SNP) across cases and controls.

The following equation presents the standard Chi-square test for independence of rows and columns in the contingency table considering a genotypic model for association [20]:

$$x^2 = \sum_i \sum_j \frac{(O_{ij} - E_{ij})^2}{E_{ij}} \tag{1}$$

Where E_{ij} is the expected frequency of allele or genotype in case and control and O_{ij} refers to observe frequency of individuals.

Following the calculation of Chi-Square test, the p-value for Chi-Square is determined based on the degrees of freedom of the test if it has 1 or 2 degrees of freedom. The p-value is a measure of the significance of the Chi-squared test. Formally, the p-value is defined as the probability of seeing a value of test statistic (chi-square statistic test) as equal to or larger than the one that was observed in a given dataset, assuming the null hypothesis (no association) is true [6]. More specifically, the p-value represents the degree of association between the SNP and the phenotype across the entire sample set. It means that lower p-value indicates that it is unlikely for the results to occur under the null hypothesis of (no association) [6].

Logistic regression is defined as a statistical method for predicting binary outcome [19]. Logistic regression model can be used to analyze the contingency table for independence, where disease status accounts as binary traits (0/1) for case and control.

Logistic regression approach can be easily expanded to allow for covariates including further SNPs, sociodemographic and clinical factors. In a case-control study, the strength of an association is measured by the odds ratio (OR) [19]. Odds ratio is the ratio of the odds of disease in the exposed group (cases) compared with a non-exposed group (controls) [19]. For example, based on the variables provided from Table 1, the allelic OR measure the association between disease and allele considering the odds of disease if allele A (major allele) is carried in compared to the odds of disease if allele a (minor allele) is carried. The following formula is used to estimate the allelic OR for allele A [23]. The formula is based on the variables of the contingency table of the allelic test as illustrated in Table 1.

$$OR_A = \frac{(O_{12}/O_{22})}{(O_{11}/O_{21})} = \frac{O_{12}O_{21}}{O_{11}O_{22}} \tag{2}$$

The strength of association of allele A is estimated based on the value of OR. Therefore, OR's value equal to 1 indicates no association, more than 1 indicates an association and less than one indicates a protective association.

3.4   Association Analysis of Geneva NHS Dataset

A case-control association analysis was conducted in an unrelated, white and non-Hispanic racial subpopulation. The analysis was performed to compare the frequency of alleles or genotypes at genetic marker loci (SNP) between cases and controls of Geneva NHS Dataset. The association analyses were performed using PLINK v1.07. We accepted 5×10^{-8} as the threshold level as determined in [24]. Odds ratio with its 95% confidence interval (95% CI) was measured to evaluate the strength of association between SNPs and T2D. To test the null hypothesis of no association, Pearson's chi-squared test (x^2) was used. We conducted Allelic association test to explore the association between single allele of the SNP and the disease trait (Type2 Diabetes). Furthermore, genetic associations were also assessed using an adjusted logistic regression methods. These were performed to calculate adjusted odds ratio with its 95% CI to assess the association of all SNPs in the study with disease status of binary traits (0/1) for case and control. Logistic regression was adjusted for covariate including (age, BMI, smoking status and physical activity) to investigate the additive contribution of the covariates (non-genetic risk factors) to the association test that built from genetic data.

4   Results

Allelic association test's result suggested that there is at least one SNP above the genome-wide significance threshold of 5×10^{-8}. While there are several SNPs above the suggestive association threshold of 5×10^{-6}. Manhattan plot has been used to visualize the results of the association as represented in Fig. 1(a). Allelic test indicated that SNPs rs10519107 $(OR = 0.7409, P = 1.813 \times 10^{-9})$, rs809736 $(OR = 0.7461, P = 7.627 \times 10^{-7})$, rs810517 $(OR = 0.7904, P = 2.682 \times 10^{-6})$, rs12571751 $(OR = 0.7913, P = 2.975 \times 10^{-6})$, rs10181181 $(OR = 0.7738, P = 3.908 \times 10^{-6})$, rs1020731 $(OR = 0.7765, P = 4.882 \times 10^{-6})$ showed protective association with T2D. Significant associations were detected in allelic test with SNPs rs4368343 $(OR = 1.890, P = 9.916 \times 10^{-7})$, rs6848779 $(OR = 1.2760, P = 1.578 \times 10^{-6})$, rs11729955 $(OR = 1.2750, P = 1.812 \times 10^{-6})$, rs11701035 $(OR = 1.3480, P = 2.736 \times 10^{-6})$. Table 2, demonstrated SNPs above suggestive threshold $< 10^{-6}$ with their OR and the p-value.

Table 2. SNPs with the suggestive of association from allelic test

CHR	SNP	P-value	OR	Association type
15	rs10519107	1.813×10^{-9}	0.7409	Protective
15	rs809736	7.627×10^{-7}	0.7461	Protective
2	rs4368343	9.916×10^{-7}	1.2890	Association
4	rs6848779	1.578×10^{-6}	1.2760	Association
4	rs11729955	1.812×10^{-6}	1.2750	Association
10	rs810517	2.682×10^{-6}	0.7904	Protective
21	rs11701035	2.376×10^{-6}	1.3480	Association
10	rs12571751	2.975×10^{-6}	0.7913	Protective
2	rs10181181	3.908×10^{-6}	0.7738	Protective
2	rs1020731	4.882×10^{-6}	0.7765	Protective

Logistic Regression analysis adjusted for multiple confounder factors suggested that none of the genotyped SNPs exceeded genome-wide significance threshold of 5×10^{-8} as represented in Fig. 1(b). Nevertheless, the result also indicated that the SNP rs10519107 $(OR = 0.7599, P = 3.583 \times 10^{-7})$ showed protective association with T2D whereas SNPs rs4368343 $(OR = 1.3210, P = 8.623 \times 10^{-7})$, rs6848779 $(OR = 1.3, P = 2.456 \times 10^{-6})$, rs11729955 $(OR = 1.3, P = 2.521 \times 10^{-6})$ detected significant association with T2D as shown in Table 3.

Table 3. SNPs with the suggestive of association from Logistic regression test

CHR	SNP	P-value	OR	Association type
15	rs10519107	3.583×10^{-7}	0.7599	Protective
2	rs4368343	8.623×10^{-7}	1.3210	Association
4	rs6848779	2.456×10^{-6}	1.3	Association
4	rs11729955	2.521×10^{-6}	1.3	Association

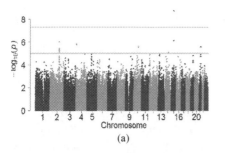
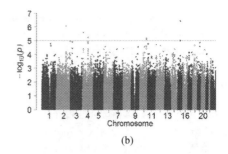

(a)   (b)

Fig. 1. Manhattan plot demonstrated the $-log_{10}(p)$ for association of SNPs in a white racial subpopulation NHS data analysis. (a) Manhattan Plot for Allelic Association Test. (b) Manhattan Plot for Logistic Regression adjusted for confounders including age, BMI, smoking status and physical activity.

We used Q-Q plot as demonstrated in Fig. 2 to visualize the relationship between the expected distribution of p-value (null) and observed distribution of p-value of the association test. The allelic test showed that there is a slight deviation in the upper right tail of the y = x line, this suggests the existence of some form of association in the NHS dataset. Logistic regression adjusted for covariates suggested satisfactory and promising outcomes are observed between the expected p-values and calculated p-values, also showed less possibility of systematic bias (population stratification). As most observed SNPs in the study showed no statistical significance than would be expected, however for some observed SNPs, statistical significance is above the expected, and this indicates a true association between these SNPs and T2D.

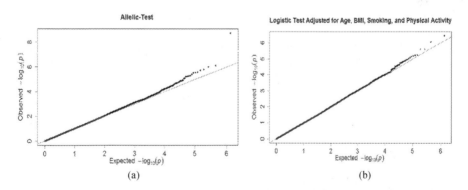

(a)   (b)

Fig. 2. Q-Q plot is showing the expected (null) vs. observed p-value. The red line represents the null hypothesis of no association. While the black dot refers to the observed $-log_{10}(p)$. (a) Q-Q plot for Allelic test. (b) Q-Q plot for logistic test adjusted for confounders (Color figure online)

5   Discussion

In this paper, our discussion of the results presented in Sect. 4 is based on the consideration of the genetic association analysis that is performed to investigate genetic variations that show evidence of increased susceptibility to T2D. These findings may serve as a rigorous ground to advance the improvement of early prediction and prevention of the disease onset. We focused on two widely used association analysis techniques including allelic test and logistic regression model. It is assumed that allelic test has an additive effect and so it is commonly used. However, logistic regression is the preferred approach due to its flexibility to allow for covariates effects including further SNPs, clinical and sociodemographic risk factors.

Of the list of SNPs obtained from allelic association test, only rs10519107 in chromosome 15 passed the genome-wide significance threshold of 5×10^{-8}. However, rs10519107 showed protective association to T2D with a respective odds ratio of 0.7409. The location of rs10519107 is in the Retinoic Acid Receptor-Related Orphan Receptor Alpha (RORα) gene region. RORα gene has known to play a major role in the regulation of lipid and glucose metabolism and insulin expression that are involved in the development of T2D. Researchers in [25] suggested that the genetic variation in RORα gene might be an indication of the individual's susceptibility to T2D. It appears that the effect of rs10519107 to the susceptible to T2D could show risk association if it is investigated in other ethnicity populations. Furthermore, nine SNPs have passed the suggestive association threshold of 5×10^{-6}. However, the risk association of rs11701035 could not reach statistical significance. It is probably due to small sample size effect.

Unlike another study, information obtained from logistic regression model have considered the use of non-genetic risk factors such as age, Body Mass Index (BMI), smoking status, and physical activity. The effects of these factors on the association analysis have shown promising results. However, fewer SNPs have reached the suggestive association threshold, and none of the genotyped SNPs has passed the genome-wide significant threshold. Nevertheless, rs10519107 has shown protective association with statistical significance while the remaining (rs4368343, rs6848779, rs11729955) have shown risk association indicating probably with larger sample size these SNPs could worth further investigation.

Although our analysis generated promising results, other approaches could be considered to model the complexity of non-linearity of genotype-phenotype interactions. Logistic regression has limited power for modeling such interactions. The non-linearity approaches are necessary for discovering the etiology of complex diseases as T2D. Machine learning algorithms have shown considerable promise. Using machine learning techniques will allow researchers to model the relationship between combinations of SNPs, environmental and clinical factors with disease susceptibility and thus to provide an advanced measurement to the etiology of T2D. Moreover, considering the correlations between gene-environment interactions and the effects of epistasis (gene-gene interactions) are fundamental to advance researchers and scientists understanding of disease mechanisms as genetic factors (single SNP) do not act independently to increase disease risk.

6 Conclusions

Association analysis tests have been performed to explore the significant association loci (SNPs) that show evidence of increased susceptibility to T2D. Several genetic models have been chosen for association test, more specifically association under logistic regression adjusted for confounders. Particularly clinical and environmental factors have been examined to measure the strength of association and a significant level of genotype-phenotype information. The analyses revealed satisfactory and promising results with a significance level (p-value) were observed. The preliminary results that have been obtained are encouraging. However, further exploration insights into this dataset remain.

References

1. Whiting, D.R., Guariguata, L., Weil, C., Shaw, J.: IDF Diabetes Atlas: global estimates of the prevalence of diabetes for 2011 and 2030. Diabetes Res. Clin. Pract. **94**, 311–321 (2011)
2. Gulcher, J., Stefansson, K.: Clinical risk factors, DNA variants, and the development of type 2 diabetes. N. Engl. J. Med. **360**, 1360 (2009). Author reply 1361
3. Prasad, R.B., Groop, L.: Genetics of type 2 diabetes—pitfalls and possibilities. Genes (Basel) **6**, 87–123 (2015)
4. Medici, F., Hawa, M., Ianari, A., Pyke, D.A., Leslie, R.D.G.: Concordance rate for type II diabetes mellitus in monozygotic twins: actuarial analysis. Diabetologia **42**, 146–150 (1999)
5. Altshuler, D., Lander, E., Ambrogio, L.: A map of human genome variation from population scale sequencing. Nature **476**, 1061–1073 (2010)
6. Bush, W.S., Moore, J.H.: Chapter 11: genome-wide association studies. PLoS Comput. Biol. **8**, e1002822 (2012)
7. Behjati, S., Tarpey, P.S.: What is next generation sequencing? Arch. Dis. Child. Educ. Pract. Ed. **98**, 236–238 (2013)
8. Lyssenko, V., Laakso, M.: Genetic screening for the risk of type 2 diabetes worthless or valuable? Diabet. Care **36**, S120–S126 (2013)
9. Wang, X., Strizich, G., Hu, Y., Wang, T., Kaplan, R.C., Qi, Q.: Genetic markers of type 2 diabetes: progress in genome-wide association studies and clinical application for risk prediction. J. Diabet. **8**, 24–35 (2016)
10. Hex, N., Bartlett, C., Wright, D., Taylor, M., Varley, D.: Estimating the current and future costs of Type1 and Type2 diabetes in the UK, including direct health costs and indirect societal and productivity costs. Diabet. Med. **29**, 855–862 (2012)
11. Samsom, M., Trivedi, T., Orekoya, O., Vyas, S.: Understanding the importance of gene and environment in the etiology and prevention of type 2 diabetes mellitus in high-risk populations. Oral Heal. Case Rep. **2**, 1–10 (2016)
12. Cortes, A., Medland, S.E., Renteri, M.E.: Using PLINK for Genome-Wide Association Studies (GWAS) and data analysis. In: Gondro, C., van der Werf, J., Hayes, B. (eds.) Genome-Wide Association Studies and Genomic Prediction. Methods in Molecular Biology, vol. 1019, pp. 193–213. Springer Science and Business Media, Heidelberg (2013). doi:10.1007/978-1-62703-447-0_8
13. Balding, D.J.: A tutorial on statistical methods for population association studies. Nat. Rev. Genet. **7**, 781–791 (2006)
14. Tudies, S., Murea, M., Ma, L., Freedman, B.I.: Genetic and environmental factors associated with type 2 diabetes and diabetic vascular complications. Rev. Diabet. Stud. **9**, 6–22 (2012)
15. Cheema, A.K., Li, T., Liuzzi, J.P., Zarini, G.G., Dorak, M.T., Huffman, F.G.: Genetic associations of PPARGC1A with type 2 diabetes: differences among populations with African origins. J. Diabetes Res. **2015**, 921274 (2015)

16. Qiu, L., Na, R., Xu, R., Wang, S., Sheng, H., Wu, W., Qu, Y.: Quantitative assessment of the effect of KCNJ11 gene polymorphism on the risk of type 2 diabetes. PLoS ONE **9**, e93961 (2014)
17. Tryka, K.A., Hao, L., Sturcke, A., Jin, Y., Wang, Z.Y., Ziyabari, L., Lee, M., Popova, N., Sharopova, N., Kimura, M., Feolo, M.: NCBI's database of genotypes and phenotypes: DbGaP. Nucleic Acids Res. **42**, 975–979 (2014)
18. Purcell, S., Neale, B., Todd-Brown, K., Thomas, L., Ferreira, M.A.R., Bender, D., Maller, J., Sklar, P., de Bakker, P.I.W., Daly, M.J., Sham, P.C.: PLINK: a tool set for whole-genome association and population-based linkage analyses. Am. J. Hum. Genet. **81**, 559–575 (2007)
19. Clarke, G.M., Anderson, C.A., Pettersson, F.H., Cardon, L.R., Andrew, P.: Basic statistical analysis in genetic case-control studies. Nat. Am. **6**, 121–133 (2011)
20. Wang, X., Baumgartner, C., Shields, D.C., Deng, H.-W., Beckmann, J.S. (eds.): Application of Clinical Bioinformatics. TB, vol. 11. Springer, Dordrecht (2016). doi:10.1007/978-94-017-7543-4
21. Bland, M.: An Introduction to Medical Statistics. Oxford University Press, Oxford (2015)
22. Chen, Z., Huang, H., Ng, H.K.T.: An improved robust association test for GWAS with multiple diseases. Stat. Probab. Lett. **91**, 153–161 (2014)
23. Li, W.: Three lectures on case-control genetic association analysis. Brief. Bioinform. **9**, 1–13 (2008)
24. Dudbridge, F., Gusnanto, A.: Estimation of significance thresholds for genomewide association scans. Genet. Epidemiol. **32**, 227–234 (2008)
25. Zhang, Y., Liu, Y., Liu, Y., Zhang, Y., Su, Z.: Genetic variants of retinoic acid receptor-related orphan receptor alpha determine susceptibility to type 2 diabetes mellitus in Han Chinese. Genes (Basel) **7**, 54 (2016)

An Ant-Colony Based Approach for Identifying a Minimal Set of Rare Variants Underlying Complex Traits

Xuanping Zhang[1,3], Zhongmeng Zhao[1,3], Yan Chang[1,3],
Aiyuan Yang[1], Yixuan Wang[2,3], Ruoyu Liu[2,3], Maomao[3],
Xiao Xiao[3,4], and Jiayin Wang[2,3(✉)]

[1] School of Electronic and Information Engineering, Xi'an Jiaotong University,
Xi'an 710049, China
zmzhao@mail.xjtu.edu.cn
[2] School of Management, Xi'an Jiaotong University, Xi'an 710049, China
wangjiayin@mail.xjtu.edu.cn
[3] Institute of Data Science and Information Quality,
Shaanxi Engineering Research Center of Medical and Health Big Data,
Xi'an Jiaotong University, Xi'an 710049, China
[4] State Key Laboratory of Cancer Biology,
Xijing Hospital of Digestive Diseases, Xi'an 710032, China

Abstract. Identifying the associations between genetic variants and observed traits is one of the basic problems in genomics. Existing association approaches mainly adopt the collapsing strategy for rare variants. However, these approaches largely rely on the quality of variant selection, and lose statistical power if neutral variants are collapsed together. To overcome the weaknesses, in this article, we propose a novel association approach that aims to obtain a minimal set of candidate variants. This approach incorporates an ant-colony optimization into a collapsing model. Several classes of ants are designed, and each class is assigned to one particular interval in the solution space. An ant prefers to build optimal solution on the region assigned, while it communicates with others and votes for a small number of locally optimal solutions. This framework improves the performance on searching globally optimal solutions. We conduct multiple groups of experiments on semi-simulated datasets with different configurations. The results outperform three popular approaches on both increasing the statistical powers and decreasing the type-I and II errors.

Keywords: Genetic association approach · Rare variants · Ant-colony optimization · Minimal candidate set problem

1 Introduction

The association study on detecting the genetic variants that underlie complex traits is one of the important topics in genetic epidemiology. The "common disease, rare variants" hypothesis mainly focuses on the role of rare variants that contribute to complex traits. While the genome-wide association studies seem falling into a dilemma,

© Springer International Publishing AG 2017
D.-S. Huang et al. (Eds.): ICIC 2017, Part II, LNCS 10362, pp. 337–349, 2017.
DOI: 10.1007/978-3-319-63312-1_30

a series of successful rare variant studies are considered as benefiting supplements to the "missing heritability" issues [1–3]. However, detecting the susceptibility candidates of rare variants is a great computational challenge. Rare variants are those variants whose minor allele frequencies are less than 1% in a given population. Different from the common variants, rare variants may present medium low penetrance, and the total effect of a set of rare variants contributes to the population attributed risk [3]. Thus, an association approach for common variants will encounter a high family error rate when applied on rare variants. The low minor allele frequencies also lead to the loss on statistical power and odd ratio.

Collapsing strategy is proposed for rare variants association studies and then widely used [4], whose basic idea is that merging multiple rare variants to a virtual site, and thus the minor allele frequencies of virtual sites increase. It is reported that collapsing strategy suffers power loss when neutral variants are collapsed to true candidates. To overcome the weaknesses, some approaches adopt weighting strategy, which weights each given variant and reduces the impacts of potential neutral variants on statistical tests. For example, Madsen and Browning [5] propose a weighted-sum method, which calculates the weight by the probability of capturing a pathogenic variant in cases comparing to the probability of capturing it in controls. *RWAS* improves the weighted-sum method by introducing an optimal weighting model [6]. *LRT* replaces the optimal weighting by a likelihood ratio test [7]. The other approaches prefer to select some candidates from the given variants, and only the candidates are considered in statistical tests. For example, *RareCover* adopts a greedy strategy to select the candidates that have positive contributions to a χ^2 test [8]. *RareProb* improves the selection strategy by a hidden Markov random field model [9]. It is further improved by considering regional copy-number alternations [10, 11].

However, the existing approaches are not able to provide a minimal set of pathogenic candidates, which is urgently needed in both research and clinical practices. In this article, we focus on the minimal pathogenic set problem and propose an ant-colony based algorithm to achieve this. Different classes of ants are designed to search the potential associations in parallel. A smart communication mechanism is also designed for better convergence. We conduct several groups of experiments on different simulation configurations. The results demonstrate that the proposed approach obtains better performance than three popular approaches, especially on large-scale datasets.

2   Methods

Let S denote the set of given rare variants across $2N$ genotypes. Let N^+ and N^- represent the numbers of cases and controls, respectively. In the following analysis, we simply assume $N^+ = N^- = N$. If $N^+ \neq N^-$, the following formulas can be used by adding a non-centrality parameter. The aim of an association approach is to identify a subset S' of S, where the elements in S should associate to the phenotype and the size of S' should be minimal. Due to the penetrance, a perfect matching set S' does not exist, and thus, an optimal solution S' should have the highest statistical power. Let $P(S')$ represent the statistic that measures the significance. Thus, the aims are: $\max P(S')$ and $\min \sum_i x_i$, where $x_i = 1$ if variant i is selected in S', 0 otherwise.

We create a virtual site for collapsed rare variants. Let binary rectors $\overrightarrow{P} = \{p_1, p_2, \cdots, p_N\}$ and $\overrightarrow{Q} = \{q_1, q_2, \cdots, q_N\}$ represent the states of the virtual site in cases and controls, respectively. $\overrightarrow{P}$ and $\overrightarrow{Q}$ are divided into K ($K \leq N$) regions. For region i, we have $H_i^P = \left\{ p_{\frac{N}{K}(i-1)+1}, \cdots, p_{\frac{N}{K}i} \right\}$ for $\overrightarrow{P}$ and $H_i^Q = \left\{ q_{\frac{N}{K}(i-1)+1}, \cdots, q_{\frac{N}{K}i} \right\}$ for $\overrightarrow{Q}$. If variant j belongs to region i, let c_{ij}^{case} denote the numbers of genotypes carrying variant j in cases. Let c_{ij}^{ctrl} denote the numbers of genotypes carrying variant j in controls. Then, $\max P(S')$ can be re-written as:

$$Max \sum_{j=1}^{M} c_{ij}^{case} x_j, \quad i = 1, 2, \cdots K \tag{1}$$

and

$$Min \sum_{j=1}^{M} c_{ij}^{ctrl} x_j, \quad i = 1, 2, \cdots K \tag{2}$$

If $K = N$, then it is an ideal solution set.

2.1   Design the Ant Colony Algorithm

If we treat (2) as resource constraints and treat c_{ij}^{ctrl} as cost, the model of association between RV and complex diseases (hereinafter, RV association model) can be simplified to Multidimensional Knapsack Problem (MKP). For a given collection of objects $O = \{O_1, O_2, \cdots O_n\}$ and a given collection of resource constraints $C = \{c_1, c_2, \cdots c_m\}$, MKP can be described as follows.

$$Max \sum_{j=1}^{n} p_j x_j \tag{3}$$

st.

$$\sum_{j=1}^{n} r_{ij} x_j \leq c_i, \quad i = 1, 2, \cdots m \tag{4}$$

where $x_j\{0, 1\}$, and $x_j = 1$ if object O_j is selected to load into knapsack, otherwise $x_j = 0$. p_j is the benefit obtained by selecting object O_j to load into knapsack. r_{ij} represents the cost of resource i required by object O_j.

MKP problem is a typical NP-hard problem [12]. At present, there are two ways to solve it, namely accurate solution and heuristic solution. Accurate solution can only be applied in the small-scale MKP because of its higher time complexity, but heuristic solution, such as genetic algorithm and ant colony algorithm, can be applied to solve big KMP with a good result, because it is easy to obtain the global optimal solution and it has good robustness. For the moment, ant colony algorithm is one of the most effective ways to solve MKP [13–15].

Both RV association model and MKP are looking for the optimal subset subjecting to certain constraints. Because of the similarities between MKP and the RV association model, this paper introduces an ant colony algorithm to detect the association between RV and complex diseases, and takes advantage of weighting and grouping method proposed by [8] to assess a solution set. The basic idea of the method is that each ant takes advantage of the test statistics of weighting and grouping method as heuristic information to assess the solution sets. Assuming that $S_1 = \{s_1, s_3, s_5\}$ is the optimal solution, during each iteration, a few ants may choose some sub-optimal solution sets, such as $\{s_1, s_3, s_6\}$, $\{s_2, s_3, s_5\}$, $\{s_1, s_4, s_5\}$, etc. Though there is no ant finding the optimal solution, after iteration and updating the pheromone, s_1, s_3, s_5 have the higher pheromone concentration, because the numbers of ants choose them is more, which will lead ants to choose them with higher probability during next iteration. Finally, ants will find the optimal solution.

But in the RV association model, there is no clearly resource constrains, because the resource that an object requires cannot be quantified. According to the collapse strategy, we can know that if the distributions of some RV are same, when they are collapsed, they will not increase the resource requirements. The biggest difference between the RV association model and MKP is that in the RV association model, if the distribution difference of two possible RV is small in region H_i^Q, the increases of the cost will be small when they are merged into a group, and if the distribution difference of two possible RV is large in region H_i^P, the increases of the profit will be large when they are merged into a group. We can use this feature to improve ant colony algorithm to detect pathogenic RV.

2.2 An Improved Ant Colony Algorithm

The improved ant colony algorithm is a two-step process. In the first step, the i-th class of ants gives priority to region H_i^P and selects some sub-optimal solutions using greedy strategy. In the second step (hereinafter, post processing), we select the global optimal solution from the sub-optimal solutions. The specific improvement is discussed below.

The probability of selection objects
Using the feature of the RV association model, this paper designs K classes of ants and a_k represents the number of the k-th class of ants. Each class of ants select objects and put them into solution sets during building solutions. The probability $P_j^{k_c}(t)$ of the k-th class of ants selecting object j is determined by

$$P_j^{k_c}(t) = \begin{cases} \dfrac{[\tau_j^k(t)]^\alpha [\eta_j^k(S_c(t))]^\beta}{\sum\limits_{i \in allowed_k(t)} [\tau_i^k(t)]^\alpha [\eta_i^k(S_c(t))]^\beta}, & if\ i \in allowed_{k_c}(t); \\ 0, & otherwise. \end{cases}$$

(5)

where $\tau_j^k(t)$ represents the residual pheromone on object $j.\,j_{\cdot j}^k(S_c(t))$ represents the evaluation function of the k-th class of ants selecting object j, which is called heuristic information.α, β represent the weights that $\tau_j^k(t)$ and $\eta_j^k(S_c(t))$ affect ant to select objects.

The k-th class of ants can only perceive and update the k-th type of corresponding pheromone, namely, if there are k classes of ants, there must be k types of pheromone.

Heuristic information

To define heuristic information of ant colony algorithm, which is suitable for the RV association model, the distance between object i and object j should be given at first. Object i and object j can be either two actual RV sites or a virtual site after collapsing and an actual RV site.

The distance of object i and object j is defined as:

$$L_{ij} = \sum_{k=1}^{K} \left| c_{ki}^{ctrl} - c_{kj}^{ctrl} \right| \tag{6}$$

where c_{ki}^{ctrl} represents the number of mutations of object i in the region H_k^Q. Without considering the region H_k^P, if the mutation number of two objects in region H_k^Q is similar, the closer distribution, the smaller distance and the less to spend after the two collapsing. From the resource consumption point of view, the lower an object cost, the worthier to choose the object. However, besides the cost, the benefit should be taken into account when object is chosen. So, according to the above analysis, the heuristic information of the k-th class of ants selecting object j during the c iteration is defined as:

$$\eta_i^k(S_c(t)) = \frac{P_i}{L_i^k(c,t)} \cdot c_{ki}^{case} \tag{7}$$

where $L_i^k(c,t)$ represents the distance between the k-th class of ants and object i during the c-th iteration; c_{ki}^{case} represents the number of mutations of object i in the region H_k^P. P_i represents the benefit that is got by choosing object i, which can be calculated by Z-score of RV proposed in [8]. It is shown as:

$$P_i = \frac{\hat{p}_i^+ - \hat{p}_i^-}{\sqrt{2/N}\sqrt{\hat{p}_i^\pm(1-\hat{p}_i^\pm)}} \tag{8}$$

where $\hat{p}_i^+$ represents the MAF of the RV observed in case group, $\hat{p}_i^-$ represents the MAF of the RV observed in control group, $\hat{p}_i^+$ represents the MAF of the RV observed in all individuals.

If one ant of the k-th class get a non-empty solution set during the c-th iteration, supposed the virtual site formed by collapsing the solution set is the object j, then

$$L_i^k(c,t) = L_{ij} \tag{9}$$

If the solution set is empty during the c-th iteration, the object j is a zero vector with length 2 N. $L_i^k(c,t)$ is calculated as shown in (9).

Updating pheromone

In the algorithm, the pheromone updating rule is defined as follows. If Ant_{kj}, namely the j-th ant in k-th class, finds a solution set that brings a benefit more than a preset threshold, or Ant_{kj} moves more than the biggest preset steps N_{max}, the ant updates pheromone by (10).

$$\tau_i^k(t+1) = (1 - \rho)\tau_i^k(t) + \sum_{j=1}^{a_k} \Delta\tau_{ij}^k \tag{10}$$

$$\Delta\tau_{ij}^k = \begin{cases} \frac{P(S_{kj})-N_r\overline{P}}{N_r} R^\lambda, & s_i \in S_{kj} \\ 0, & otherwise \end{cases} \tag{11}$$

$$P(S_{kj}) = \frac{\sum \left((\hat{p}_i^+ - \hat{p}_i^-)/\hat{p}_i^\pm\right)}{\sqrt{2/N}\sqrt{\sum \left((1 - \hat{p}_i^\pm)/\hat{p}_i^\pm\right)}} \tag{12}$$

$$R = \frac{\overline{N}}{N_r} \tag{13}$$

In (10)–(13), parameter ρ is the degree of pheromone volatilization, and $\Delta\tau_{ij}^k$ represents the increment of the pheromone that Ant_{kj} leaves on object i. $P(S_{kj})$ is the benefit of the solution set got by Ant_{kj} in this iteration, which is a statistical test proposed in [8] to assess a group of RV and calculated by (13). $\overline{P}$ is a constant representing the average profit of RV. N_r represents the size of solution set that an ant constructs during this iteration, namely the number of RV in solution set. R is a punishment or a compensation. λ represents the weight of compensation. $\overline{N}$ represents the average of the sizes of the solution sets that the ants constructed.

2.3   Association Pipeline

First, the dimension of RV is reduced and the pheromone is initialized. When ant colony algorithm processes large scale problems, because the difference of the information in each path is small in early stages, only after several searches, the amount of information in a better path becomes significantly higher than it in other paths, which can ensure the algorithm eventually converges to the optimal path [16]. For the shortage of ant colony algorithm processing large-scale problems, this paper decreased the scale of problem by reducing the dimension of RV data, and use Z-score of RV to initialize pheromone.

Then, the sub-optimal solutions are built by K classes of ants. In this algorithm, we design K classes of ants corresponding to the K regions $H_i^P (i = 1, 2, \cdots K)$. The k-th class of ants can only perceive and update the pheromone corresponding to k-th region.

So each class of ants can only find the sub-optimal solution for each region. The ants build solutions by iteration. After each iteration, the ants update their corresponding pheromone and save the top-k solutions.

Finally, the global optimal solution is determined by the post processing. When K classes of ants achieved the sub-optimal solutions, they have saved K top-k solutions. In the ideal case, the global solution is the subset of the K top-k solutions. To obtain the global optimal solution, we perform another round of ant colony algorithm, where there are only one class of ants and only one class of pheromone, and there is no regions any more. In this round of ant colony process, the pheromone is initialized using the information of the K top-k solutions, and during each iteration, the ants only save and update one solution which will be final optimal solution.

3   Results and Discussion

We use the same method in [8, 9] to generate the simulation datasets. Rare variants are generated according to MAFs in control group and in case group, which are ρ_s and θ_s, respectively. There are 2,000 individuals with 1,000 cases and 1,000 controls in each RV data. We simulate 100 datasets under different PAR, different numbers of pathogenic RV and different numbers of RV.

Parameter α, β and ρ separately represent the impact weights of pheromone $\tau_j(t)$, the heuristic information $\eta_j(S_k(t))$ on ants choosing objects and the level of pheromone volatilization. According to [14, 15], this paper sets $\alpha = 1, \beta = 5$ and $\rho = 0.2$.

In the first round of ant colony process in the proposed algorithm, the number of regions in case group and control group are divided into $K = N/(2H)$, and the size of each region $H = 100$. The reason why we take 100 for H is that the bigger H is, the lower the algorithm complexity and the algorithm performance are. When $H \rightarrow 1$, the algorithm tends to seek the solution of the optimal virtual sites that is formed after collapsing. But if $H \rightarrow N/2$, there is no evident difference between ant classes. Besides, because MAF of RV is in interval (0.001, 0.01), we take 100 for H, which ensures that there is at least a mutation in each region.

The biggest preset steps of ants moving $N_{max} = 20$, the weight of solution set $\gamma = 1$, the number of iterations $I = 50$, the ant species is 10, the number of each species of ants $a_k = 10$, and the significant threshold δ for constructing each solution is 12.0.

In the second round of ant colony process, $N_{max} = 100$, the ant species is 1, and other parameters are same as the first step. Because it has been found that the numbers of pathogenic RVs for most diseases is less than 50 and the largest one is 151, we set $N_{max} = H = 100$ which can be applicable to most diseases and minimize the time complexity of the algorithm.

The experiments show that the proposed algorithm is sensitive to N_{max} and δ. The smaller N_{max} and δ are, the faster convergence speed the algorithm has, as well as the lower type I error rate and the higher type II error rate are. The algorithm is not sensitive to the number of iterations and the number of ants, and the bigger they are, the higher the algorithm precision is but the higher the computing complexity is. Because the dimension reduction process ensures that the number of candidate RV sites is actually small, the good results can be got by the proposed algorithm with a few iterations.

In order to describe conveniently, we call the method in [6] RareCover, call the method in [8] RWAS, and call the method in [9] LRT. And then, we will compare the proposed method with these three methods under different conditions.

3.1 Performance Comparisons Under Different PARs

We take PAR$\in\{0.2, 0.3, 0.4, 0.5\}$, the number of pathogenic RV$\in\{50, 60, 70, 80, 90\}$, and the total number of RV is 100. The experiments show that the testing powers of the above four methods are 100%. So we omit the result of power index.

Table 1. The comparison of two types of errors under different numbers of PAR and pathogenic RV

PAR	Pathogenic RV	Method in this paper		RareCover	
		Type I	Type II	Type I	Type II
2%	50	8.10%	25.4%	11.90%	45.4%
	60	8.20%	31.5%	4.90%	44.2%
	70	10.40%	32.3%	5.30%	43.9%
	80	13.10%	26.8%	2.40%	52.9%
	90	2.30%	28.7%	2.20%	56.6%
3%	50	4.40%	23.6%	4.80%	39.6%
	60	7.30%	23.2%	3.50%	47.7%
	70	12.80%	23.6%	1.20%	53.9%
	80	14.10%	26.1%	0.70%	59.5%
	90	3.90%	24.7%	0.60%	60.3%
4%	50	1.90%	19.4%	1.10%	49.6%
	60	7.90%	20.8%	0.10%	54.8%
	70	11.60%	17.0%	0.40%	61.3%
	80	4.80%	20.0%	1.10%	64.3%
	90	5.90%	18.2%	0.90%	67.2%
5%	50	4.60%	13.6%	0.90%	53.4%
	60	6.40%	13.5%	0.80%	58.8%
	70	6.90%	16.3%	0.70%	64.9%
	80	4.30%	18.6%	0.60%	70.6%
	90	7.30%	13.8%	0.40%	71.8%

Power is a widely used index that measures the strength of the association between RV. A lot of papers including the comparison methods adopt this index to measure the performance of the algorithm [6–9]. But power cannot fully reflect the performance of the algorithm. Now taking this case into account, some method can always find pathogenic RV and a lot of non-causal sites at the same time, which will reduce the size of power. But, after all, this method finds a subset including pathogenic RV, which makes sense to the further research.

Based on the above analysis, in order to verify the performance of algorithm, Table 4 shows the comparison of two types of errors with different parameters using proposed algorithm and RareCover. Because RWAS and LRT are for a group of RV, we can only get the statistics of association strength between RV and the group's disease, while the real pathogenic site cannot be figured. So we cannot analyze the two errors.

As showed in Table 1, the possibility which first type error happened in RareCover is low, in other words, the possibility of unconcerned RV contained in the candidate pathogenic RV is low. However, the possibility which second type error happened in RareCover is obviously higher than proposed algorithm; that is to say, RareCover is easy to miss the pathogenic RV.

Table 2. The examining conditions under different numbers of PAR and pathogenic rare variants

PAR	Pathogenic RV	Method in this paper		RareCover	
		The total number	The number of right sites	The total number	The number of right sites
2%	50	45.4	37.3	32.3	30.5
	60	49.3	41.1	38.9	34.3
	70	57.8	47.4	41.5	39.7
	80	71.7	58.6	38.2	37.4
	90	66.5	64.2	39.4	39.1
3%	50	42.6	38.2	32.4	30.2
	60	53.4	46.1	33.7	31.6
	70	66.3	53.5	33.2	32.5
	80	73.2	59.1	32.6	32.1
	90	71.7	67.8	36.2	36.1
4%	50	42.2	40.3	26.7	25.0
	60	55.4	47.5	27.5	27.4
	70	69.7	58.1	27.9	27.1
	80	68.8	64	29.8	28.3
	90	79.5	73.6	30.1	29.0
5%	50	47.8	43.2	24.7	23.5
	60	58.3	51.9	25.3	24.6
	70	65.5	58.6	25.8	24.9
	80	69.4	65.1	24.2	23.7
	90	84.9	77.6	25.6	25.1

In order to indicate directly, Table 2 shows the test of two methods. We can see that the larger PAR is, the lower the possibility of second errors in the proposed algorithm is; while RareCover is opposite. PAR indicates the risk of pathogenic RV causing disease. The larger PAR is, the larger risk of pathogenic RV causing disease is. Theoretically speaking, the larger PAR is, the more significant association between pathogenic RV and disease is. So the possibility of finding pathogenic RV is larger. The proposed algorithm matches the expected. However, RareCover chooses the pathogenic RV based on collapsing strategy and greedy strategy, when PAR of pathogenic RV is large, the convergence is fast. Once finding several pathogenic RV, we cannot detect other pathogenic RV easily. This is caused by limitations of collapsing strategy. Besides, we can see that the influence that number of pathogenic RV makes on the proposed algorithm is lower. Take the column which PAR is 3% as example, with the increase of pathogenic RV, the possibility that proposed algorithm has second errors is steady between 23% to 25%, while the possibility of RareCover is between 39% to 60%.We can conclude that fluctuation in RareCover which caused by pathogenic RV more is large. In summary, the proposed algorithm has more advantage than RareCover when detecting small scale RV data.

3.2   Performance Comparisons Under Different Dimensions

Given PAR 5%, Table 3 gives the experiment results of proposed algorithm and contrastive algorithm under different dimensions, where the total number of RV is {200,400,600,800,1000,1500, 2000,3000,4000,5000}. For the power of each algorithm, as shown in Table 3, the proposed algorithm performs better than other three methods when handling high dimension RV data, while RWAS and LRT perform better when handling RV data less than 200, but when the number of RV over 400, the performance decreases significantly, especially when the number of RV over 1000, the power decreases down to 0. RareCover can detect RV data less than 600, but when the number over 800 the power decreases quickly. Besides, we can see that three contrastive methods are influence by the number of pathogenic sites.

Table 3. The comparison between powers

Pathogenic RV	Pathogenic RV	The method in this paper	Rare-Cover	RWAS	LRT
50	200	100%	100%	99.6%	99.9%
	400	100%	100%	85.3%	88.6%
	600	100%	94.6%	54.1%	58.0%
	800	100%	75.4%	33.0%	36.5%
	1000	100%	52.3%	20.7%	22.0%
	2000	100%	20.4%	2.0%	2.0%
	3000	100%	11.2%	0.8%	0.0%
	4000	100%	2.0%	0.4%	0.0%
	5000	100%	0.9%	0.3%	0.0%
20	200	100%	98.7%	32.0%	74.0%
	400	100%	87.5%	12.0%	59.0%
	600	100%	72.1%	1.0%	39.0%
	800	100%	56.2%	1.0%	10.0%
	1000	100%	42.3%	0.0%	1.0%
	2000	100%	3.1%	0.0%	0.0%
	3000	100%	0.0%	0.0%	0.0%
	4000	100%	0.0%	0.0%	0.0%
	5000	100%	0.0%	0.0%	0.0%

Table 4. Two types of errors under different dimensions of rare variants

Pathogenic RV	The number of RV	Method in this paper		RareCover	
		Type I	Type II	Type I	Type II
50	200	2.6%	22.6%	0.1%	53.6%
	400	2.5%	23.0%	0.4%	55.4%
	600	0.7%	7.8%	1.6%	56.8%
	800	0.8%	5.6%	1.5%	52.6%
	1000	0.4%	8.8%	1.6%	55.8%
	2000	0.2%	6.4%	1.5%	53.0%
	3000	0.1%	4.4%	1.2%	55.4%
	4000	0.1%	5.4%	0.8%	54.8%
	5000	0.2%	3.2%	1.1%	47.8%
20	200	0.2%	5.5%	2.3%	10.0%
	400	0.1%	3.0%	2.9%	5.5%
	600	0.0%	2.5%	2.1%	7.0%
	800	0.3%	0.5%	3.1%	1.0%
	1000	0.2%	0.5%	3.1%	4.0%
	2000	0.1%	1.5%	3.1%	1.5%
	3000	0.1%	1.5%	2.0%	4.5%
	4000	0.1%	2.0%	1.6%	3.0%
	5000	0.1%	1.5%	1.3%	6.5%

Table 4 shows two type errors of the proposed algorithm and RareCover. We can see that when the number of pathogenic sites is 50, the proposed algorithm has higher errors if the number of RV is less than 600. This is because the dimension reduction cannot be implemented at this situation. Even so, the type II error happens less than RareCover. When the number reaches 600, with the dimension of RV increases, the proposed algorithm has lower type II error than RareCover, that is to say it has stronger ability to find pathogenic RV. When the number of pathogenic sites is 20, the type I error of the proposed algorithm is lower. That is to say, compared to RareCover, it has lower possibility to mistakenly choose unconcerned RV, especially when the number of pathogenic sites is smaller, it has greater influence for power to mistakenly choose unconcerned sites.

4   Conclusion

For the limitation of the method based on weighted sum groupwise in handling high dimension data, the paper adopts ant colony algorithm. Based on collapsing strategy the paper builds an association model of RV with complex disease. Depending on the feature of models, we design different kinds of ants to build optimal solution. For each kind of ants we give priority to corresponding region. At last, we execute the global ant colony algorithm again as post-processing to improve the ability to find the global solution. For the large solution space, the pheromone to guide the route for ant colony's behavior is weak. To overcome this, the paper proposes a method which can effectively reduce the dimension of RV data. Experimental results show that the proposed algorithm achieves better performance than the existing methods. Especially the proposed algorithm can detect the RV data containing over 5000 sites with better performance, while the existing methods could handle less than 600 sites at most.

Acknowledgement. This work is supported by the National Science Foundation of China (Grant No: 81400632), Shaanxi Science Plan Project (Grant No: 2014JM8350) and the Fundamental Research Funds for the Central Universities (XJTU).

References

1. Ropers, H.H.: New perspectives for the elucidation of genetic disorders. Am. J. Hum. Genet. **81**(2), 199–207 (2007)
2. Manolio, T.A., Collins, F.S., Cox, N.J.: Finding the missing heritability of complex diseases. Nature **461**(7265), 747–753 (2009)
3. Bodmer, W., Bonilla, C.: Common and rare variants in multifactorial susceptibility to common diseases. Nat. Genet. **40**(6), 695–701 (2008)
4. Li, B., Leal, S.M.: Methods for detecting associations with rare variants for common diseases: application to analysis of sequence data. Am. J. Hum. Genet. **83**(3), 311–321 (2008)
5. Madsen, B.E., Browning, S.R.: A groupwise association test for rare mutations using a weighted sum statistic. PLoS Genet. **5**(2), e1000384 (2009)
6. Sul, J.H., Han, B., He, D.: An optimal weighted aggregated association test for identification of rare variants involved in common diseases. Genetics **188**(1), 181–188 (2011)
7. Sul, J.H., Han, B., Eskin, E.: Increasing power of groupwise association test with likelihood ratio test. J. Comput. Biol. **18**(11), 1611–1624 (2011)
8. Bhatia, G., Bansal, V., Harismendy, O.: A covering method for detecting genetic associations between rare variants and common phenotypes. PLoS Comput. Biol. **6**(10), e1000954 (2010)
9. Wang, J., Zhao, Z., Cao, Z., et al.: A probabilistic method for identifying rare variants underlying complex traits. BMC Genom. **14**(Suppl 1), S11 (2013)
10. Geng, Y., Zhao, Z., Zhang, X., et al.: An improved burden-test pipeline for cancer sequencing data. In: Bourgeois, A., Skums, P., Wan, X., Zelikovsky, A. (eds.) Bioinformatics Research & Applications ISBRA 2016. LNCS (LNBI), vol. 9683, pp. 314–315. Springer, Cham (2016)
11. Geng, Y., Zhao, Z., Cui, D., Zheng, T., Zhang, X., Xiao, X., Wang, J.: An Expanded Association Approach for Rare Germline Variants with Copy-Number Alternation. In: Rojas, I., Ortuño, F. (eds.) IWBBIO 2017. LNCS, vol. 10209, pp. 81–94. Springer, Cham (2017). doi:10.1007/978-3-319-56154-7_9
12. Fréville, A.: The multidimensional 0–1 knapsack problem: an overview. Eur. J. Oper. Res. **155**(1), 1–21 (2004)
13. Yu, X.C., Zhang, T.W.: An improved ant algorithm for multidimensional knapsack problem. Chin. J. Comput. **31**(5), 810–819 (2008)

14. Ji, J.Z., Huang, Z., Liu, C.N.: An ant colony optimization algorithm based on mutaion and phromone diffusion for the multidimensional knapsack problems. J. Comput. Res. Dev. **46** (4), 644–654 (2009)
15. Gan, R.W.: Research on ant colony optimization and its application. Sun Yat-sen University, Guangzhou (2009)
16. Bansal, V., Libiger, O., Torkamani, A.: Statistical analysis strategies for association studies involving rare variants. Nat. Rev. Genet. **11**(11), 773–785 (2010)

Evaluation of Phenotype Classification Methods for Obesity Using Direct to Consumer Genetic Data

Casimiro Aday Curbelo Montañez[1(✉)], Paul Fergus[1], Abir Hussain[1], Dhiya Al-Jumeily[1], Mehmet Tevfik Dorak[2], and Rosni Abdullah[3]

[1] Liverpool John Moores University, Liverpool, UK
c.a.curbelomontanez@2015.ljmu.ac.uk,
{p.fergus,a.hussain,d.aljumeily}@ljmu.ac.uk
[2] Liverpool Hope University, Liverpool, UK
dorakm@hope.ac.uk
[3] Universiti Sains Malaysia, George Town, Malaysia
rosni@usm.my

Abstract. Direct-to-Consumer genetic testing services are becoming more ubiquitous. Consumers of such services are sharing their genetic and clinical information with the research community to facilitate the extraction of knowledge about different conditions. In this paper, we build on these services to analyse the genetic data of people with different BMI levels to determine the immediate and long-term risk factors associated with obesity. Using web scraping techniques, a dataset containing publicly available information about 230 participants from the Personal Genome Project is created. Subsequent analysis of the dataset is conducted for the identification of genetic variants associated with high BMI levels via standard quality control and association analysis protocols for Genome Wide Association Analysis. We applied a combination of Random Forest based feature selection algorithm and Support Vector Machine with Radial Basis Function Kernel learning method to the filtered dataset. Using a robust data science methodology our approach identified obesity related genetic variants, to be used as features when predicting individual obesity susceptibility. The results reveal that the subset of features obtained through the Random Forest based algorithm improve the performance of the classifier when compared to the top statistically significant genetic variants identified in logistic regression. Support Vector Machine showed the best results with sensitivity=81%, specificity=83% and area under the curve=92% when the model was trained with the top fifteen features selected by Boruta.

Keywords: Bioinformatics · Data science · Machine learning · Feature selection · Genetics · Obesity · SNPs

© Springer International Publishing AG 2017
D.-S. Huang et al. (Eds.): ICIC 2017, Part II, LNCS 10362, pp. 350–362, 2017.
DOI: 10.1007/978-3-319-63312-1_31

1   Introduction

The global prevalence of obesity has reached epidemic proportions [1]. According to the World Health Organization (WHO)[1], approximately 2.8 million people die each year as a consequence of being overweight or obese [2]. Obesity is a major risk for other chronic diseases which include diabetes, cardiovascular diseases and cancer [3]. The occurrence of obesity is a common problem in high-income countries but, its frequency is also rising in low and middle-income countries [4]. In England, the National Obesity Observatory (NOO) reported that the direct cost to the National Healthcare Service (NHS) for treating overweight, obesity and related morbidities increased from £479.3 million in 1998 to £4.2 billion in 2007[2]. The effects of obesity are so grave that it reduces life expectancy on average by 3 years – in cases of severe obesity this can be between 5 and 13 years [5].

Advances in Human Genomics have provided significant opportunities and research suggests that it might be possible to quantify an individual's susceptibility to obesity from an early age and manage risk as individuals' progress through life [6]. Therefore, combining personalised medicine with genomic information and integrating it into medical care and individualised risk assessments will allow us to mitigate the long-term effects of obesity and its associated co-morbidities. This is being made possible through advances in bioinformatics [7], data science [8] and advanced machine learning algorithms [9].

This paper explores these ideas further and proposes a robust methodology to combine state-of-the-art bioinformatics and data science to investigate genetic profiling and risk factor assessment for obesity. We combined two statistical approaches for Single Nucleotide Polymorphism (SNP) evaluation. Risk-Based approach and Classification-Based Approach. The first approach is applied to identify statistically significant SNPs whilst the second is used to identify a set of SNPs appearing conjointly which can serve to predict obesity. The motivation for this research is to identify strong genetic markers for use in decision support systems. Data science is utilised to automatically build a dataset, using publicly available demographic and genetic information provided by individuals. This dataset and subsequent analysis is intended to provide a starting point for genetic variants data analysis.

2   Background

The decreased costs associated with Deoxyribonucleic acid (DNA) sequencing have made it easier to obtain genomic data. For example, the 100,000 Genomes Project[3], conducted by Genomics England, has sequenced 100,000 genomes from 70,000 NHS patients suffering with rare diseases. The information will be used to create a genomic medicine service for the NHS and enable new scientific discovery and medical insights.

---

[1] http://www.who.int/.

[2] https://www.noo.org.uk/.

[3] http://www.genomicsengland.co.uk/.

In the private sector, genetic screening services are delivered directly to consumers. Individuals provide a saliva sample to a Direct-to-Consumer Genetic Testing (DTCGT) company and obtain genetic information without any health care provider involvement [10]. Many of these DTCGT services use SNP identification to determine ancestry and genetic markers associated with specific diseases with the objective of informing clients about their health and how to change behaviours to improve it [10].

The Personal Genome Project (PGP)[4] is a non-profit organization created to promote the availability and use of personal health information and genome data to help accelerate the understanding of genetic variation in humans. While many object to privacy, confidentiality and anonymity issues, the PGP believes that sharing such data is fundamentally advantageous for the advances in science and society. This is a view endorsed by members of the public who understand the risks and share their personal information. The founding pilot project of the PGP was initiated by the Harvard Personal Genome Project, which now hosts publicly shared genomic and health data from thousands of participants. In 2005 information on 10 fully identified individuals was available; today, more than 4000 US participants have publicly shared their genomic information. There is also evidence that information across initiatives is being shared with genetic data from 23andMe appearing in PGP datasets [11].

Bioinformaticians routinely extract information from websites using web-scrapping techniques to obtain content originally presented for human use [12].

Collecting this data is tedious and time-consuming. Several institutions have invested heavily in data collection, gathering clinical and genetic data within different domains for decades. This has resulted in significant amounts of big data [13] and today organisations, such as the National Institute of Health (NIH), which sponsored the Database of Genomes and Phenotypes (dbGaP), are making this data available to interested parties, subject to specific terms and conditions [14]. However, to access this data, researchers must follow a data request procedure that can be restrictive to general users from other domains that want to make use of genetic data. Consequently, other organisations such as the PGP rely on a different strategy defined by publicly accessible data that anybody from diverse backgrounds can use to get started on genetic data analysis. Having access to such repositories has had a huge positive impact on the scientific community who no longer need to generate their own data for the studies that they conduct.

Approximately 99.5% of the total number of base pairs (nucleotides) in the human genome are identical for any two human individuals [15]. Hence, in genetic association studies, bases where there is variation between humans are commonly considered. Studies utilizing hypothesis-free methodologies such as genome-wide association studies (GWAS) have been used in obesity studies to identify many obesity related loci. GWAS permit the analysis of a large number of genetic variants (whole genome) for association with traits of interest. In statistical association test, logistic regression is often the preferred approach as it has been extensively developed although it is not the only one [16]. Currently, associations of common variants usually should reach threshold levels of $P < 5 \times 10^{-8}$ to be considered significant [17]. Conversely, variants with

---

[4] http://www.personalgenomes.org/.

threshold levels of $P < 10^{-5}$ are termed suggestive SNPs [18] and could be studied further. The importance of GWAS is advancing scientific understanding of disease mechanisms and providing starting points and potential opportunities for researchers to improve the development of medical treatments.

Following an open data initiative, genetic association analysis and predictive modelling strategies are conducted in this study for the analysis of obesity as a binary trait.

3 Materials and Methods

The dataset used in this paper comprises 230 participants from the PGP, which donated genetic data from Direct-to-Consumer genotyping. This data is extracted and analysed by 23andMe using microarray genotyping, which provides an efficient and cost-effective way of evaluating genetic variation in individuals and across populations [19]. In addition to genetic data, clinical information is also provided. Collected contributors are aged between 23 and 79 years of age (average age 46.59) and are all from the United States of America. The average height, body weight and BMI of all participants is 1.74 meters, 78.97 kg, and 25.97 respectively. Of the total population, 150 (65.22%) are males and 80 (34.78%) females. All participants considered in the study reported white as ethnical background.

3.1 Data Collection and Description

During the initial data collection process, 733 observations/participants and 9 variables were scrapped from the PGP website[5]. Table 1 provides a description of the data fields extracted for each participant.

Table 1. Variables selected in the web scraping process.

Variables	Description
Participant_ID	Participant ID
Data link	Genetic data URL
DoB	Date of Birthday
Gender	Gender
Weight	Weight in Kg
Height	Height in meters
T2D	Type 2 Diabetes
Ethnicity	Ethnical background
Blood Type	Blood Type

The Participant_ID is a unique participant identifier assigned in the PGP. The variable Data link provides a Uniform Resource Locator (URL) used to download the genetic profile of each participant. In addition, DoB, Gender, Weight, Height, Ethnicity, and

---

[5] http://www.personalgenomes.org/.

Blood Type contain personal information for each participant. Data about the condition Type 2 diabetes (T2D) was also included, although more features based on the existing variables were subsequently incorporated to the clinical data file.

The resulting dataset contained several empty fields. Only observations with complete values for the variables in Table 1 were retained. Individuals who reported being of ethnicities other than white were excluded to avoid population stratification in our analysis. This reduced the dataset to 235 observations. The data links for five participants were incorrect so these were also discarded from the final dataset, resulting in 230 individuals.

Full genome profiles were downloaded in *txt* format using the variable Data Link identified in Table 1. Only full genome data was included in the analysis i.e., if a participant from the PGP uploaded exon and whole genome data to the PGP website, only the whole genome profiles were considered since full genome provides complete representation of the genome. The genetic profile of each participant contains four variables: *rsid*, *chromosome*, *position* and *genotype*, and several hundred thousand observations that depend on the amount of variants discovered by the genotyping process used by 23andMe [19]. The variables included in the genetic profiles represent genetic variants or SNPs.

Downloaded genetic profiles were converted to binary file format [20]. This type of format allows for a more efficient and convenient way of manipulating SNP data when using open source software for automated GWAS quality control (QC) and analysis, such as PLINK [20]. Subsequently, all 230 genetic profiles were merged into one main binary file (.bed, .bim, .fam). Finally, two main data frames were created – one containing the clinical information and the other containing genetic variants identified by 23andMe for the 230 participants.

Additional features were generated using information from existing columns. These include body mass index (BMI), constructed from the Weight and Height variables and calculated using the metric formula, $BMI = \dfrac{Weight\ (Kg)}{(Height(m))^2}$. A Status feature was also generated from the BMI result. Following the WHO classification for BMI[6], 5 standard weight status categories associated with BMI ranges for adults were derived. Table 2, summarizes the number of participants included in each status category. The category Normal range has the highest representation among the participants (50%) whereas Underweight is the category with the lowest representation (1.74%). The categories Overweight, Obese and Extremely obese, when grouped together, constitute 111 participants. In other words, 48.26% of the participants analysed were included in one of these three status categories. Hence, as shown in Table 2, two closely balanced classes based on the BMI were created, representing the phenotypic variable for risk prediction of obesity. The variables considered in the clinical data frame are: Participant_ID, age, gender, height (m), weight (kg), BMI, Status, T2D, Race and blood type. In the case of the genetic information, the variables considered are: SNP name (rsid), chromosome number, position in the DNA sequence and genotype.

---

[6] http://www.who.int/

Table 2. BMI status among participants included in the study.

Class	Status	Total number
Normal	Underweight	4
51.74%	Normal range	115
Risk	Overweight	65
48.26%	Obese	41
	Extremely obese	5

3.2 Data Pre-processing

Analyses were conducted using PLINK and R software[7]. After the data set construction, and prior to analysis, data QC was performed. Cases and controls in the present study are defined as risk and normal. Following protocols for genetic case-control association studies, QC was performed on individuals and then on markers, to optimise the number of SNPs remaining in the study [21].

In the per-individual QC process, 7 individuals were removed leaving 223 remaining individuals of which, 107 are cases and 116 are controls. Individuals were excluded if they showed abnormal heterozygosity, discordant sex information, were duplicated or related individuals, and individuals of divergent ancestry. Strict values for missing rate were not considered since most samples in the study would be removed. This might be an indicative of poor quality DNA sample [22].

In the per-marker QC process, SNPs with minor allele frequency (MAF<4%), call rate of <98% and deviations from Hardy-Weinberg equilibrium ($p < 1 \times 10^{-3}$) were excluded. A MAF cut point of 4% is commonly applied in small sample settings due to statistical power considerations [23].

3.3 Genetic Association Analysis

For discovery, association analysis on 107 risk cases and 116 controls was performed by testing SNPs and individuals that satisfied quality control. Logistic regression was used to identify SNPs showing a strong association with the trait of interest. However, none of the SNPs reached significance level (P-value $< 5 \times 10^{-8}$) nor were suggestive of association (P-value $< 1 \times 10^{-5}$) as shown in Fig. 1(a), a Manhattan plot of genome genome-wide association analysis results. The figure illustrates, in the y-axis, the level of statistical significance as measured by the negative log of the corresponding P-value, for each SNP. Significant and suggestive levels are represented in red and blue respectively. Each typed SNP is indicated by a dark-blue or orange dot. In the x-axis, SNPs are arranged by chromosomal location.

---

[7] http://www.r-project.org/

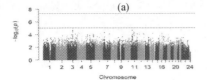 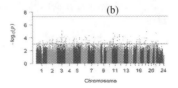

Fig. 1. Manhattan Plot for GWAS: (a) suggestive threshold P-value $< 1\times10^{-5}$, (b) after suggestive threshold modification P-values $< 1\times10^{-3}$.

While no SNPs were identified as significant or suggestive, a subset of SNPs with P-values $< 1\times10^{-3}$ were considered for subsequent analysis as similarly performed in [24]. Consequently, a total of 261 SNPs showing the strongest association with the phenotype (risk or normal) were identified. Extracted features were ordered by statistical significance, being the most important those with lower P-values. In Fig. 1(b), SNPs with P-values lower than 10^{-3} are highlighted in green. The red line represents the significant level while the blue line indicates, this time, the new threshold considered (P-values $< 1\times10^{-3}$). The figure displays a Manhattan plot of SNPs considered after suggestive threshold modification.

4 Feature Selection

After features were extracted, we explored feature selection to determine which features might be the most relevant when discriminating between risk and normal classes.

Some samples had missing genotypes for some individuals so we removed them, resulting in a final number of 185 SNPs considered as features for classification analysis. Additionally, age, gender and T2D were not included in the total set of features i.e., only genetic variants were considered.

Feature selection is performed using Boruta, a random forest (RF) based feature selection method, which provides unbiased and stable selection of important and non-important attributes [25]. Random Forest has been successfully used in genomic data analysis as it is highly data adaptive and accounts for correlation as well as interactions among features [26].

Features selected identified were ranked by importance and divided into three groups. The first group contained the top five most prominent features, the second group the top ten and the third group the top fifteen.

Results were compared against those reported when the top most notable features extracted from association analysis were considered, as we will discuss in the following sections.

5 Results

This section presents the classification results for normal and risk BMI status using data extracted from the PGP website.

After the QC filter process, 722,512 genetic variants and 223 people (145 males and 78 females) remained for the analyses. Subsequent genetic association analysis using logistic regression allowed us to reduce the number of variants to 261. However, missing genotypes in some of the samples caused a further reduction in the number of SNPs (185 SNPs remained).

The top features extracted after QC and association analysis and, those selected by Boruta, are used to model a Support Vector Machine with Radial Basis Function Kernel (SVM) classifier. Support Vector Machine is a well-known machine learning algorithm which provided the best results in previous experiments using similar data [27]. The performance is measured using sensitivity (SE), specificity (SP) and area under the curve (AUC) values. In this study, it is important to predict risk classes, therefore SE are considered higher priority than SP.

K-fold cross validation is used as a prediction metric with 10 folds and 30 repetitions. The average performance obtained from 30 simulations is utilized. This number is considered, by statisticians, to be an adequate number of iterations to obtain an accept-able average. Support Vector Machine was designed and evaluated using appropriate training and testing sets. The selection of hyperparameters to establish an approximately optimal configuration for SVM is addressed using Caret for random search parameter tuning [28]. Tuning parameters, free parameter of the Gaussian radial basis function (sigma) and penalty cost (C), shown in Tables 3 and 5 produced the models with the best receiver operator characteristic (ROC) curve values.

Table 3. Association analysis

Association features	Sensitivity	Specificity	ROC	Best tuning parameters
Top 5 SNPs	0.6322	0.8476	0.7859	$\sigma = 0.0215$ $C = 1.0203$
Top 10 SNPs	0.7856	0.7325	0.8672	$\sigma = 0.0125$ $C = 1.0203$
Top 15 SNPs	0.8399	0.8578	0.9142	$\sigma = 0.0105$ $C = 1.0203$

Sensitivity, Specificity and ROC values for SVM performance in the training data when using extracted features from association analysis.

The performance of SVM when the algorithm is trained and tested with the top features identified in the association analysis ranked by P-value is shown in Tables 3

Table 4. Boruta algorithm

Association features	Sensitivity	Specificity	ROC
Top 5 SNPs	0.7692	0.6897	0.8150
Top 10 SNPs	0.6923	0.9586	0.8622
Top 15 SNPs	0.8462	0.8276	0.9092

Sensitivity, Specificity and AUC values for SVM when predicting the two classes in the test data, using extracted features from association analysis.

and 4 respectively. Conversely, the performance of SVM when fed with the features selected by Boruta are organised in Table 5 for training and Table 6 for testing. Details on the SNPs extracted and selected can be found in Appendix.

Table 5. Association analysis

RF features	Sensitivity	Specificity	ROC	Best tuning parameters
Top 5 SNPs	0.7514	0.7855	0.8318	σ = 0.0183 C = 1.0203
Top 10 SNPs	0.7849	0.7704	0.8482	σ = 0.0106 C = 1.0203
Top 15 SNPs	0.8291	0.8071	0.9011	σ = 0.0120 C = 1.0203

Sensitivity, Specificity and ROC values for SVM performance in the training data when using features selected by Boruta.

Table 6. Boruta algorithm

RF features	Sensitivity	Specificity	ROC
Top 5 SNPs	0.6154	0.7931	0.8176
Top 10 SNPs	0.7692	0.7931	0.8674
Top 15 SNPs	0.8077	0.8276	0.9231

Sensitivity, Specificity and AUC values for SVM when predicting the two classes in the test data, using features selected by Boruta.

To illustrate the performance in binary classification, it is particularly advantageous to use the ROC curve. It is a convenient way of displaying the cut-off values for the false and true positive rates. The ROC curves in Fig. 2 illustrates the SE, SP and AUC values in Tables 4 and 6. The models with ROC curve closer to the top left corner show higher performance as the SE and SP increase. Therefore, the area under the curve increases as the curve moves away from the grey diagonal line towards top left corner of the graph.

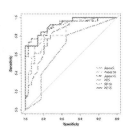

Fig. 2. ROC curves for PGP data when using the subsets of SNPs extracted by association analysis and selected by Boruta.

6   Discussion

The web-scrapping process applied in this study is susceptible to failures in the future if the PGP website structure changes. This is an issue referred to as "medieval torture" [29].

PGP dataset is pre-processed via standard QC and association analysis protocols for GWAS. Although no SNPs were identified as significant or suggestive, we included SNPs with P-values lower than 1×10^{-3} for subsequent analyses as accomplished somewhere else [24]. After QC, 722,512 SNPs were considered for association and lately reduced to 261 SNPs showing certain level of importance among all the variants, using logistic regression. These SNPs are highlighted in Fig. 1(b). Finally, 185 SNPs were considered for classification analysis.

The total 185 features are a subset with the most relevant SNPs obtained after applying QC and logistic analysis to the genetic data binary files. The top most significant features were then organised in three groups to be compared against the most relevant features selected by Boruta.

Using RF-based algorithm as a feature selection technique, three groups of the top features with the highest discriminatory capacity were selected.

Results revealed that using RF-based algorithm ranking of features resulted in an improvement in the performance of SVM when predicting the risk and normal cases. All the ROC values obtained with the three sets of SNPs in Table 7 are higher than those obtained by the set of most statistically significant features listed in Table 5. In most cases, SE were lower than SP, which is not encouraging given that predicting pathological cases is more important than those that are normal. However, when RF-based

Table 7. Ranking of features considered in the study.

Rank	Features extracted in association analysis					Features selected by RF-based algorithm			
	SNP	CHR	BP	Allele	P-Value	SNP	CHR	BP	Allele
1	rs4821758	22	38591190	C	1.980e-05	rs4821758	22	38591190	C
2	rs6768523	3	111962851	C	2.449e-05	rs10790866	11	127063240	G
3	rs9872691	3	111985107	C	2.449e-05	rs7117995	11	19947811	C
4	rs9871650	3	111914624	C	2.449e-05	rs9872691	3	111985107	C
5	rs9288938	3	111921116	A	2.449e-05	rs7574062	2	71716494	T
6	rs441703	11	29822605	T	2.458e-05	rs9871650	3	111914624	C
7	rs4682278	3	111225977	A	3.036e-05	rs9288938	3	111921116	A
8	rs6437989	3	111203691	G	3.036e-05	rs12570718	10	116416959	T
9	rs1553090	3	111180433	A	3.167e-05	rs6768523	3	111962851	C
10	rs10880063	12	41760767	T	3.940e-05	rs2159723	2	230045272	T
11	rs1000147	12	41727678	C	4.659e-05	rs7776422	6	106240162	G
12	rs1451327	11	57991093	A	5.086e-05	rs4483247	9	37045825	G
13	rs10957744	8	75899434	A	6.360e-05	rs5006218	6	133126220	G
14	rs7498886	16	62079202	G	6.526e-05	rs2207900	20	54289508	C
15	rs12579740	12	125177752	T	7.871e-05	rs9493446	6	133125643	T

Ranking of features extracted by association analysis (shown as shaded) and selected by Boruta. The SNPs are listed in order of importance. The top features extracted using logistic regression are ordered by P-value whilst the features selected by RF-based algorithm are ordered by importance. Information about the chromosome number, base pair position and allele is provided. In addition to this information, the P-values for features selected in association analysis are also listed.

method is used, the top fifteen features produced a closely balance SE and SP values of 81% and 83% respectively.

Results in Table 7 indicate that SVM showed the best results with SE=81%, SP=83% and AUC=92% when the model was trained with the top fifteen features selected by Boruta. These features are listed in Appendix section. Reducing the number of features to five did not result in an improvement in the classifier performance.

The ROC curves from Fig. 2 shows how using the top fifteen features selected by Boruta (red ROC curve) allows the highest discrimination between the two classes considered in our study. The lowest performance was achieved with the top five SNPs extracted in association analysis, which is represented in green colour in Fig. 2.

Additionally, the most important feature reported in both approaches was rs4821758 as reported in Appendix.

7 Conclusions

This paper focuses on an approach for selecting informative SNPs from publicly available data collected using web scrapping techniques. The created dataset was built from research-grade data (that is, not for clinical use), and the conductors of the PGP stated that many types of errors are possible. Some of these include errors in the data, failure to report or discover significant genetic issues and ambiguous or false positive findings. This suggests the utilisation of a more reliable data set in future studies, for a solid discovery of genetic risk variants in complex disease prediction.

A small portion of SNPs that have main effects on obesity as binary trait, have been selected after applying QC and association analysis using logistic regression. Subsequent analysis applying a Support Vector Machine with Radial Basis Function Kernel classifier are conducted for the evaluation of the model in two scenarios. First, the algorithm was evaluated using a subset of the most statistically significant genetic variants obtained from GWAS analysis, based on a modified suggestive threshold. Then, results were compared when a subset of features were selected using the Random Forest based algorithm Boruta. Using the selected features improved the performance of SVM although the subset of fifteen SNPs achieved the highest performance.

While the results show specific genetic variants that could serve as good discriminators in the investigation of classification studies, more analysis with a higher representation of samples must be carried out. We propose a set SNPs to be used in future studies as features for the prediction of obesity and other comorbidities such as T2D. The identified genetic variants need to be validated and contrasted with other studies, particularly the SNP rs4821758, which was the most important feature in the association analysis as well as the feature selection process using Boruta. Future work will consider the discriminative capacity of the SNPs identified in this study evaluated in a more complete dataset. A comparison between various feature selection techniques will also be considered.

References

1. James, W.P.T.: WHO recognition of the global obesity epidemic. Int. J. Obes. **32**(Suppl 7), S120–S126 (2008). (Lond)
2. Poloz, Y., Stambolic, V.: Obesity and cancer, a case for insulin signaling. Cell Death Dis. **6**, e2037 (2015)
3. Rao, K.R., Lal, N., Giridharan, N.V.: Genetic & epigenetic approach to human obesity. Indian J. Med. Res. **140**, 589–603 (2015)
4. Li, S., et al.: Physical activity attenuates the genetic predisposition to obesity in 20,000 men and women from EPIC Norfolk prospective population study. PLoS Med. **7**, 1–9 (2010)
5. Bello, A., et al.: Using linked administrative data to study periprocedural mortality in obesity and chronic kidney disease (CKD). Nephrol. Dial. Transpl. **28**, iv57–iv64 (2013)
6. Loos, R.J.F.: Genetic determinants of common obesity and their value in prediction. Best Pract. Res. Clin. Endocrinol. Metab. **26**, 211–226 (2012)
7. Samish, I., Bourne, P.E., Najmanovich, R.J.: Achievements and challenges in structural bioinformatics and computational biophysics. Bioinformatics **31**, 146–150 (2014)
8. Higdon, R., et al.: Unravelling the complexities of life sciences data. Big Data **1**, 17–23 (2012)
9. Tanwani, A.K., Afridi, J., Shafiq, M.Zubair, Farooq, M.: Guidelines to select machine learning scheme for classification of biomedical datasets. In: Pizzuti, C., Ritchie, Marylyn D., Giacobini, M. (eds.) EvoBIO 2009. LNCS, vol. 5483, pp. 128–139. Springer, Heidelberg (2009). doi:10.1007/978-3-642-01184-9_12
10. Su, P.: Direct-to-consumer genetic testing: a comprehensive view. Yale J. Biol. Med. **86**, 59–65 (2013)
11. Ball, M.P., et al.: Harvard personal genome project: lessons from participatory public research. Genome Med. **6**, 10 (2014)
12. Glez-Pena, D., Lourenco, A., Lopez-Fernandez, H., Reboiro-Jato, M., Fdez-Riverola, F.: Web scraping technologies in an API world. Brief. Bioinform. **15**, 788–797 (2014)
13. Marx, V.: Biology: the big challenges of big data. Nature **498**, 255–260 (2013)
14. Tryka, K.A., et al.: NCBI's database of genotypes and phenotypes: dbGaP. Nucleic Acids Res. **42**, D975–D979 (2014)
15. Gonzaga-Jauregui, C., Lupski, J.R., Gibbs, R.A.: Human genome sequencing in health and disease. Annu. Rev. Med. **63**, 35–61 (2012)
16. Bush, W.S., Moore, J.H.: Chapter 11: Genome-wide association studies. PLoS Comput. Biol. **8**, e1002822 (2012). doi:10.1371/journal.pcbi.1002822
17. Fadista, J., Manning, A.K., Florez, J.C., Groop, L.: The (in)famous GWAS P-value threshold revisited and updated for low-frequency variants. Eur. J. Hum. Genet. **24**, 1202–1205 (2016)
18. Zhang, Y.-B., et al.: Genome-wide association study identifies multiple susceptibility loci for craniofacial microsomia. Nat. Commun. **7**, 10605 (2016)
19. Stoeklé, H.-C., Mamzer-Bruneel, M.-F., Vogt, G., Hervé, C.: 23andMe: a new two-sided data-banking market model. BMC Med. Ethics. **17**, 19 (2016)
20. Purcell, S., et al.: PLINK: a tool set for whole-genome association and population-based linkage analyses. Am. J. Hum. Genet. **81**, 559–575 (2007)
21. Anderson, C.A., Pettersson, F.H., Clarke, G.M., Cardon, L.R., Morris, A.P., Zondervan, K.T.: Data quality control in genetic case-control association studies. Nat. Protoc. **5**, 64–73 (2010)
22. Turner, S., et al.: Quality control procedures for genome-wide association studies. Curr. Protoc. Hum. Genet. Chapter 1, Unit1.19 (2011). doi:10.1002/0471142905.hg0119s68
23. Reed, E., Nunez, S., Kulp, D., Qian, J., Reilly, M.P., Foulkes, A.S.: A guide to genome-wide association analysis and post-analytic interrogation. Stat. Med. **34**, 3769–3792 (2015)

24. Gül, H., Aydin Son, Y., Açikel, C.: Discovering missing heritability and early risk prediction for type 2 diabetes: a new perspective for genome-wide association study analysis with the Nurses' Health Study and the Health Professionals' Follow-Up Study. Turkish J. Med. Sci. **44**, 946–954 (2014)
25. Kursa, M.B., Rudnicki, W.R.: Feature Selection with the Boruta package. J. Stat. Softw. **36**, 1–13 (2010)
26. Cordell, H.J.: Detecting gene–gene interactions that underlie human diseases. Nat. Rev. Genet. **10**, 392–404 (2009)
27. Curbelo Montañez, C.A. et al.: Machine learning approaches for the prediction of obesity using publicly available genetic profiles. In: 2017 International Joint Conference on Neural Networks (IJCNN), p. 8, Anchorage, Alaska (2017)
28. Kuhn, M.: Building predictive models in R using the caret package. J. Stat. Softw. **28**, 1–26 (2008)
29. Stein, L.: Creating a bioinformatics nation. Nature **417**, 119–120 (2002)

Protein-Protein Interaction Prediction

Classification of Hub Protein and Analysis of Hot Regions in Protein-Protein Interactions

Xiaoli Lin[1,2], Xiaolong Zhang[1(✉)], and Jing Hu[1]

[1] Hubei Key Laboratory of Intelligent Information Processing and Real-Time Industrial System,
School of Computer Science and Technology, Wuhan University of Science and Technology,
Wuhan 430065, China
aneya@163.com, Xiaolong.Zhang@wust.edu.cn, 1881787@qq.com
[2] Information and Engineering Department of City College,
Wuhan University of Science and Technology, Wuhan 430083, China

Abstract. Proteins are fundamental to most biological processes, which accomplish a vast amount of functions by interacting with other proteins. The research of PPI (protein-protein interaction) and its network has developed into a great importance part in bioinformatics. In the protein-protein interaction networks, most proteins interact with only a few partners, and small number of proteins interact with many partners, these proteins are called hub proteins. The hub proteins can be divided into party hub and date hub. Therefore, in this paper, we do some works about hub proteins. In addition, this paper uses the connectivity and betweenness to classify the hub protein in protein-protein interaction network. On the other hand, the paper studies hub proteins from another perspective (interfaces conformation), which reflects the organization of hot spot residues in hub protein interface.

Keywords: Hot regions · PPI · Hub protein · Machine learning

1 Introduction

Proteins interact with other proteins to complete life activities. The study of protein-protein interaction (PPI) network is helpful to understand the evolutionary process of life. Developing computational methods to predict and analyze protein-protein interaction networks is not only superior to traditional experiment, but also crucial for understanding biological functions. Identification is not complete about the physical interactions of proteins and the functions of many protein are not found.

Protein-protein interaction (PPI) have been researched from multiple perspectives [1]. Each protein is defined as a node in the network, and the interaction between proteins is defined as the linkage. Some proteins have distinctive characteristics and special regions for interacting with other protein.

A large number of research results show that protein-protein interaction (PPI) network is power-law connectivity distribution [2]. It indicates that some proteins are highly connected to other proteins which can be called hub protein, while most proteins interact with only a few proteins [1]. Each hub protein has original conformation for

© Springer International Publishing AG 2017
D.-S. Huang et al. (Eds.): ICIC 2017, Part II, LNCS 10362, pp. 365–374, 2017.
DOI: 10.1007/978-3-319-63312-1_32

interacting in protein-protein interaction (PPI). One of the study perspective in the protein-protein network is how a hub protein can interact with other so many non-hub protein. In some environment, external changes can bring about the new transformation of the space conformation [3], such as PH, temperature, partner concentration and ionic strength. However, since the coverage of the natural protein-protein interaction (PPI) is low, it is still questioned whether the topology structure of the protein-protein interaction (PPI) network can be expressed correctly [4].

In protein-protein interaction (PPI) network, hub protein is a kind of protein with higher number of connections, which plays a key role in driving the evolution of genomes and genetic systems. However, the number of connections does not accurately reflect the role of proteins in protein-protein interaction (PPI) network, because hub proteins with same or similar connect degree are not always equal important role in biological network.

Although the distribution characteristics of protein-protein interaction (PPI) network have not been completely determined, it is obvious that highly connected proteins have certain properties to play an important role [3]. For example, the hub proteins might be crucial in drug design. An understanding of hub protein is necessary for the development of new drugs and drug discovery in modern era. Han [4] indicated that date hub proteins may be responsible for organizing biological modules in the protein-protein interaction (PPI) network.

One of the important characteristic of protein interaction interfaces is the contribution degree of the amino acid to the binding free energy. For many years, a large number of experiments show the binding free energy is not uniformly distributed in the binding of protein-protein interaction. Only a sub-fraction of amino acid residues (these residues are called hot spots) contributes a disproportionately generous to the interaction binding free energy [5]. In addition, some researchers showed that hot spots can be obtained from the alanine mutation energy database [6] to create the hot region models for analyzing the evolutionary mechanism. For the research of hot spot residues in protein-protein interfaces, a lot of computational methods have been proposed [7–9]. Hus [10] and Cukuroglu et al. [11] predicted hot regions in protein-protein interactions from the different perspective. Carles Pons [12] identified and analyzed the protein-protein binding regions conformation.

The PPI depends on distinct space conformation (3D structure) of proteins. Thus, the protein spatial 3D structure can help to analyze the characteristics and the rules of the evolution and functions of life. Our research group made contributions to the prediction of protein spatial 3D structure and hot regions in protein-protein interactions. Lin [13] has proposed the local adjust tabu search algorithm to predict protein spatial folding structure with the 2D off-lattice model and 3D off-lattice model. Zhang [14, 15] used the different heuristic algorithms to predict the protein spatial structure in 3D. Nan [16] and Hu [17] predicted the hot regions in protein-protein interaction by complex network and clustering method.

Tuncbag et al. [18] expressed that hot spot residues are the interface residues with $\triangle\triangle G >= 2.0$ kcal/mol in the protein-protein interaction interface, while nonhot spot residues are the interface residues with $\triangle\triangle G < 0.4$ kcal/mol. In addition, Ozlem Keskin [5], Reichman [19] and Ahmad et al. [20] defined hot regions by different method. This paper defines hot region which have at least three hot spot residues adjacent to each other in the protein-protein interaction interface. Cukuroglu [11] addressed hub proteins yet from interfaces structural, which investigated how hot spots and hot regions can be organized in hub proteins.

This paper consists of the follows sections. Section 2 describes the method of classification of date hub proteins and party hub proteins. Section 3 analyzes the hot region features. Section 4 gives the experiment results and discusses. Section 5 gives conclusion and future directions.

2   Classification of Date Hub and Party Hub

2.1   Classification Based on Average PCC

Han [4] proposed two types of hub proteins by selecting the average PCC: date hubs and party hubs. The former interacts with different proteins at different times or locations, and they are the global connection point between different groups of biological functions. The latter interacts with most of proteins at the same. Therefore, identification and analysis of date hub proteins are critical to the discovery of hidden biological information in protein-protein interaction (PPI) networks. The mRNA expressions are independent and are uncorrelated [21, 22] if hub proteins connected with other proteins by false-positive interactions [23], these proteins could be identified as date hubs. In order to reduce false-positive, Han [4] obtained a high-quality intersecting dataset. For each hub protein, it is necessary to calculated the average PCC of each mRNA expression between the hub proteins and other proteins.

$$\text{Average\_PCC} = \sum \frac{PCC_i}{n} \tag{1}$$

$$PCC_i = \sum (E_x E_y)/N \tag{2}$$

Where E_x, E_y are mRNA expression in different conditions or samples. N is the number of samples. n indicates the number of interaction objects in hub proteins. Party hubs are those with an average PCC higher than the threshold. All other hubs are defined as date hubs.

2.2   Classification Based on Betweenness

According to many researches, the betweenness is also an important topological property beside the connectivity in graph theory. Nodes with higher betweenness values may be the key nodes of the control module in the whole network. The degree of betweenness is the number of the shortest path in a network through a node. Proteins with high betweenness value are the key node to connect multiple important biological pathways in PPI networks. The improved algorithms were proposed to count the betweenness by Girvan and Newman [24, 25]. In this paper, we also use the same definition as Yu [26] that proteins with high betweenness are defined as bottlenecks. To facilitate the comparison and analysis, the proteins with the top 20% betweenness values are selected as bottle-necks, which is consistent with Yu [27]. Then, all proteins in a certain network can be divided into four categories: Hub-Bottleneck Node (HBN, High betweenness and high connectivity), Non-hub-bottleneck node (NHBN, High betweenness and low connectivity), Hub-non-bottleneck node (HNBN, Low betweenness and high connectivity) and Non-hub-non-bottleneck node (NHNBN, Low betweenness and low connectivity).

The connection degree and betweenness of all nodes in the network are calculated, and the betweenness is defined as

$$\text{Betweenness}(v) = \sum_{x,y \in V} \frac{\sigma(x, y | v)}{\sigma(x, y)} \tag{3}$$

Where, $\sigma(x, y)$. represents the shortest path between x and y, $\sigma(x, y | v)$. represents the shortest path between x and y through the node v.

3   Hot Region Features and Classification

Cukuroglu [11] described hub proteins from an interfaces structural point and defined the different types of interfaces. In the study, the interfaces were defined as DD (inter-faces between two date hubs), PP (between two party hubs), and NN (between two non-hub proteins) where D, P, N, and X are for date hub, party hub, non-hub, and any protein, respectively [11].

3.1   Feature Selection

Feature selection technique is a crucial step of classification, which has been widely used in protein-protein interactions [28]. It can contribute to avoid overfitting and enhance the accuracy of classification model, because feature selection can preserve the most primitive and optimal features of amino acid residues. The results of feature selection have a certain effect on the reliability of classification results. Many amino acid residues need to be removed because their biological characteristics are useless. The best feature subset can be obtained by feature selection procedure for improving the performance of the classifier. In this paper, properties of protein are estimated by mRMR (minimum Redundancy Maximum Relevance) algorithm [29].

The basic idea of mRMR algorithm is to meet the minimum redundancy and maximum correlation criterion. That is to say, the redundancy between the residues is analyzed based on the correlation measure, and the correlation function is combined into an objective function. When the objective function is optimal, the correlation between the residues is maximum, and the redundancy of the residues is minimum. Given two random variables X and Y, the mutual information is defined as

$$I(X, Y) = \iint p(x, y) log \frac{p(x, y)}{p(x)p(y)} dx \, dy \tag{4}$$

The maximum correlation criterion is defined as

$$\max D(S, C), D(S, C) = \frac{1}{|S|} \sum_{f_i \in S} I(f_i, C) \tag{5}$$

The minimum redundancy criterion is defined as

$$\min R(S), R = \frac{1}{|S|^2} \sum_{f_i, f_j \in S} I(f_i, f_j) \tag{6}$$

Where S is the characteristic set, and C is the category of targets. $I(f_i, C)$ is the mutual information between feature i and category C. $I(f_i, f_j)$ is the mutual information between feature i and feature j. Combining the above two formulas, the criterion about mRMR can be defined the following.

$$\max \varnothing(D, R), \quad \varnothing = D - R \tag{7}$$

This criterion indicates that a feature subset should be selected with highly correlated and less redundant from the alternative features. Assume that the n feature has been selected as feature subset S_n, then the j feature can be selected from alternative feature set $S - S_t$ and satisfies the following formula.

$$\max_{f_j \in S - S_t} I(f_i, C) - \frac{1}{n} \sum_{f_i \in S_t} I(f_i, f_j) \tag{8}$$

3.2 Classification

Machine learning methods have been widely used in bioinformatics [30–33]. This paper used the support vector machines (SVM) for classification. Support vector machines, developed by Vapnik, are a set of related supervised learning methods which are used for classification and regression. For many years, SVM classifier are more and more widely used in the field of computational biology for predicting protein-protein interaction sites [34–37]. The protein-protein interaction interfaces can be classified as hub and non-hub interfaces according to the different conformation of hot spot residues.

The prediction process of hot regions in protein-protein interaction interfaces adopts 10-fold cross validation. The data set can be separate into ten portion. The training set data has nine portion, and test data is the remained one. The average of the correct rates of the 10 results is used as an estimate of the algorithm's accuracy. The main parameters of the kernel function can be given by the cross validation.

4 Experimental Results and Evaluation

In the experiments, the proportion of various proteins was detected by adjusting the threshold value to classify the date proteins and the party hub protein. According to Fig. 1, under different thresholds, Hub-Bottleneck Node (HBN, High betweenness and high connectivity) class proteins have the highest gene encoding ratio. So, these proteins are most likely to become date hub protein. When the threshold equals 0.1, the proportion of essential gene with Hub-Bottleneck Node (HBN, High betweenness and high connectivity) is the highest. Figure 2 gives the proportion of essential gene with the threshold equals 0.1.

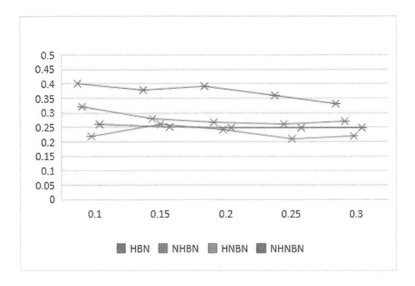

Fig. 1. The proportion of four kinds of proteins with different thresholds

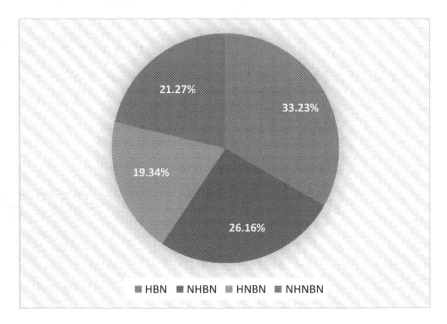

Fig. 2. The proportion with the threshold equals 0.1

It is generally known that polar and hydrophobic interactions play important role in protein-protein interfaces. Therefore, there are two categories amino acids: polar amino acids and non-polar ones. The former includes R, N, D, E, Q, H, K, S, T and Y, while the latter includes A, C, G, I, L, M, F, P, W and V. For the classification, it is necessary to assess the features used for classification. So, Table 1 lists the description of the alternative features, and Table 2 lists their assessment between different types of PPI. These features can be selected to classify if their values are smaller than 0.05.

Table 1. Description of the alternative features

Candidate features	Description
HSO	Hot spot ratio
AHRS	Average hot region size
ANHR	Average number of hot regions
AHR_ΔASA	Average hot region ΔASA to interface ΔASA ratio
PAAFI	Polar amino acid frequencies of interface
PAAFHS	Polar amino acid frequencies of hot spots
PAAFHR	Polar amino acid frequencies of hot regions

Table 2. Assessment of the alternative features

	PP-DD	PP-NN	DD-NN
HSO	<0.05	–	–
AHRS	<0.05	<0.05	–
ANHR	–	<0.05	–
AHR_ΔASA	<0.05	<0.05	–
PAAFI	<0.05	<0.05	–
PAAFHS	<0.05	–	<0.05
PAAFHR	<0.05	–	<0.05

Figure 3 shows that date hub proteins are more inclined to cluster one hot region or more hot regions, which is consistent with Cukuroglu's conclusion. Results also reveal that there are obvious distinguish between different types of interfaces.

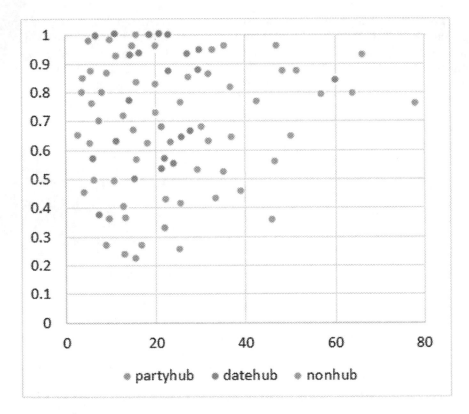

Fig. 3. The distribution of the average fraction of hot spots in the hot regions

5 Conclusion

Hub proteins have high connectivity in protein-protein interaction (PPI) network, which are one of the most significant factors in biological system. Nevertheless, the connectivity cannot entirely illuminate the hub protein's role in protein-protein interaction (PPI) network. The reason is that hub proteins with similar connectivity maybe not equal contribution to the protein-protein interaction (PPI) network. Therefore, the betweenness is considered as the stronger determinant of analyzing and understanding the protein network system. On the other hand, the strong link can be found between hot spot residues or hot regions and hub proteins in the protein-protein interface. There are obviously differences between date hub protein and party hub protein interfaces. One of the future studies is to consider structural properties and energy contribution of different categories of hub proteins.

Acknowledgment. The authors thank the members of Machine Learning and Artificial Intelligence Laboratory, School of Computer Science and Technology, Wuhan University of Science and Technology, for their helpful discussion within seminars. This work was supported in part by National Natural Science Foundation of China (No. 61502356, 61273225, 61273303).

References

1. Jeong, H., Mason, S.P., Barabasi, A.L., Oltvai, Z.N.: Lethality and centrality in protein networks. Nature **411**, 41–42 (2001)
2. Han, J.D., Dupuy, D., Bertin, N., Cusick, M.E., Vidal, M.: Effect of sampling on topology predictions of protein-protein interaction networks. Nat. Biotechnol. **23**, 839–844 (2005)
3. Apic, G., Ignjatovic, T., Boyer, S., Russell, R.B.: Illuminating drug discovery with biological pathways. FEBS Lett. **579**, 1872–1877 (2005)
4. Han, J.D., Bertin, N., Hao, T., Goldberg, D.S., Berriz, F., Zhang, L.V., Dupuy, D., Walhout, A.J., Cusick, M.E., Roth, F.P., Vidal, M.: Evidence for dynamically organized modularity in the yeast protein–protein interaction network. Nature **430**(6995), 88–93 (2004)
5. Keskin, O., Ma, B.Y., Mol, R.J.: Hot regions in protein-protein interactions: the organization and contribution of structurally conserved hot spot residues. J. Mol. Biol. **345**(5), 1281–1294 (2005)
6. Thorn, K.S., Bogan, A.A.: ASEdb: a data base of alanine mutations and their effects on the free energy of binding in protein interactions. Bioinformatics **17**(3), 284–285 (2001)
7. Lise, S., Buchan, D., Pontil, M., Jones, D.T.: Predictions of hot spot residues at protein-protein interfaces using support vector machines. PloS One **6**(2), e16774 (2011)
8. Lise, S., Archambeau, C., Pontil, M., Jones, D.T.: Prediction of hot spot residues at protein-protein interfaces by combining machine learning and energy-based methods. BMC Bioinform. **10**(1), 365 (2009)
9. Tuncbag, N., Keskin, O., Gursoy, A.: HotPoint: hot spot prediction server for protein interfaces. Nucleic Acids Res. **38**, 402–406 (2010)
10. Hsu, C.M., Chen, C.Y., Liu, B.J., Huang, C.C.: Identification of hot regions in protein-protein interactions by sequential pattern mining. BMC Bioinfor. **8**(Suppl 5), S8 (2007)
11. Cukuroglu, E., Gursoy, A., Keskin, O.: Analysis of hot region organization in hub proteins. Ann. Biomed. Eng. **38**(6), 2068–2078 (2010)
12. Carles, P., Fabian, G., Juan, F.: Prediction of protein-binding areas by small world residue networks and application to docking. BMC Bioinform. **12**, 378–388 (2011)
13. Lin, X.L., Zhang, X.L., Zhou, F.L.: Protein structure prediction with local adjust tabu search algorithm. BMC Bioinform. **15**(S15), S1 (2014)
14. Zhang, X.L., Wang, T., Luo, H.P., Yang, J.Y., Deng, Y.P., Tang, J.S., Yang, M.Q.: 3D protein structure prediction with genetic tabu search algorithm. BMC Syst. Biol. **4**(Suppl 1), S6 (2010). doi:10.1186/1752-0509-4-S1-S6
15. Zhang, X.L., Lin, X.L.: Effective 3D protein structure prediction with local adjustment genetic-annealing. Interdisc. Sci. Comput. Life Sci. **2**(3), 256–262 (2010)
16. Nan, D.F., Zhang, X.L.: Prediction of hot regions in protein-protein interactions based on complex network and community detection. In: IEEE International Conference on Bioinformatics and Biomedicine, pp. 17–23 (2013)
17. Hu, J., Zhang, X.L., Liu, X.M., Tang, J.S.: Prediction of hot regions in protein-protein interaction by combining density-based incremental clustering with feature-based classification. Comput. Biol. Med. **61**, 127–137 (2015)
18. Tuncbag, N., Gursoy, A., Keskin, O.: Identification of computational hot spots in protein interfaces: combining solvent accessibility and inter-residue potentials improves the accuracy. Bioinformatics **25**(12), 1513–1520 (2009)
19. Reichmann, D., Rahat, O., Albeck, S., Meged, R., Dym, O., Schreiber, G.: The modular architecture of protein-protein binding interfaces. Proc. Natl. Acad. Sci. **102**(1), 57–62 (2005)

20. Ahmad, S., Keskin, O., Sarai, A., Nussinov, R.: Protein-DNA interactions: structural, thermodynamic and clustering patterns of conserved residues in dna-binding proteins. Nucleic Acids Res. **36**(18), 5922–5932 (2008)
21. Kemmeren, P., et al.: Protein interaction verification and functional annotation by integrated analysis of genome-scale data. Mol. Cell **9**, 1133–1143 (2002)
22. Ge, H., Liu, Z., Church, G.M., Vidal, M.: Correlation between transcriptome and interactome mapping data from Saccharomyces cerevisiae. Nature Genet. **29**, 482–486 (2001)
23. Von Mering, C., et al.: Comparative assessment of large-scale data sets of protein-protein interactions. Nature **417**, 299–403 (2002)
24. Girvan, M., Newman, M.E.: Community structure in social and biological networks. Proc. Natl. Acad. Sci. U.S.A. **99**, 7821–7826 (2002)
25. Newman, M.E.: Scientific collaboration networks. II. Shortest paths, weighted networks, and centrality. Phys. Rev. E **64**, 016132 (2001)
26. Yu, H.Y., Kim, P.M., Sprecher, E., et al.: The importance of bottlenecks in protein net-works: correlation with gene essentiality and expression dynamics. PLoS Comput. Biol. **3**(4), 59 (2007)
27. Yu, H.Y., Greenbaum, D., Xin Lu, H., Zhu, X., Gerstein, M.: Genomic analysis of essentiality within protein networks. Trends Genet. **20**, 227–231 (2004)
28. Yugandhar, K., Gromiha, M.M.: Feature selection and classification of protein-protein complexes based on their binding affinities using machine learning approaches. Proteins: Struct., Funct., Bioinf. **82**(9), 2088–2096 (2014)
29. Li, B.Q., Feng, K.Y., Li, C., Huang, T.: Prediction of protein-protein interaction sites by random forest algorithm with mRMR and IFS. PloS One **7**(8), e43927 (2012)
30. Abnousi, A., Broschat, S.L., Kalyanaraman, A.: An alignment-free approach to cluster proteins using frequency of conserved K-Mers. In: BCB 2015 Proceedings of the 6th ACM Conference on Bioinformatics, Computational Biology and Health Informatics,   pp. 597–606 (2015)
31. Broin, P.Ó., Smith, T.J., Golden, A.A.: Alignment-free clustering of transcription factor binding motifs using a genetic-k-medoids approach. BMC Bioinform. **16**(1), 1–12 (2015)
32. Zheng, C.H., Huang, D.S., Zhang, L., Kong, X.Z.: Tumor clustering using non-negative matrix factorization with gene selection. IEEE Trans. Inf. Technol. Biomed. **13**(4), 599–607 (2009)
33. Zhu, L., Guo, W.L., Deng, S.P., Huang, D.S.: ChIP-PIT: enhancing the analysis of ChIP-Seq data using convex-relaxed pair-wise interaction tensor decomposition. IEEE/ACM Trans. Comput. Biol. Bioinform. **13**(1), 55–63 (2016)
34. Cho, K.I., Kim, D., Lee, D.: A feature-based approach to modeling protein-protein interaction hot spots. Nucleic Acids Res. **37**(8), 2672–2678 (2009)
35. Wong, G.Y., Leung, F.H.F., Ling, S.H.: Predicting protein-ligand binding site using support vector machine with protein properties. IEEE/ACM Trans. Comput. Biol. Bioinform. **10**(6), 1517–1529 (2013)
36. Sriwastava, B.K., Basu, S., Maulik, U.: Predicting protein-protein interaction sites with a novel membership based fuzzy SVM classifier. IEEE/ACM Trans. Comput. Biol. Bioinform. **12**(6), 1394–1404 (2015)
37. Wei, Z.S., Han, K., Yang, J.Y., Shen, H.B., Yu, D.J.: Protein-protein interaction sites prediction by ensembling SVM and sample-weighted random forests. Neurocomputing **193**, 201–212 (2016)

Genome-Wide Identification of Essential Proteins by Integrating RNA-seq, Subcellular Location and Complexes Information

Chunyan Fan and Xiujuan Lei[✉]

School of Computer Science, Shaanxi Normal University, Xi'an 710119, China
xjlei@snnu.edu.cn

Abstract. Essential proteins are significant for understanding the cellular survival and practical purpose, such as the disease diagnosis and drug design. Besides biological experimentally methods, previous computational methods are proposed to predict essential proteins based on topological property of protein-protein interaction (PPI) network. However, these methods ignored the temporal and spatial features of the PPI networks. Moreover, researches show that essentiality is closely tied to the protein complexes to which that protein belongs. Therefore, improving the performance of predicting essential proteins is still a challenging task. In this study, by integrating the RNA-seq data, subcellular location compartments and protein complexes together in the PPI network, we proposed a method called *IUS*. *IUS* is applied to the PPI network of *Saccharomyces cerevisiae*, results based on the multiple evaluate methods show that *IUS* outperform other eight existing methods including DC, BC, EC, IC, SoECC, LAC, PeC and WDC.

Keywords: Essential proteins · RNA-seq · Subcellular location · Protein complexes

1 Introduction

Essential proteins are those proteins indispensable for growth in a rich medium where all required nutrients are available, which result in lethality or infertility when they were missed [1]. Reliable identification of essential proteins can facilitate to several areas, including drug design, disease diagnosis, and medical treatment [2]. Several experimental methods, such as single gene knockouts [3], RNA interference [4] and conditional knockouts [5], have been implemented for the discovery of essential proteins. However, these experimental methods generally require large amounts of resources and are very time-consuming.

Based on the association between the topological features and the essentiality of essential proteins in protein-protein interaction (PPI) network, several studies have been conducted to measure the essentiality of proteins. The useful centrality methods have been proposed to predict essential proteins, such as Degree Centrality (DC) [6], Betweenness Centrality (BC) [7], Eigenvector Centrality (EC) [8], Information Centrality (IC)

© Springer International Publishing AG 2017
D.-S. Huang et al. (Eds.): ICIC 2017, Part II, LNCS 10362, pp. 375–384, 2017.
DOI: 10.1007/978-3-319-63312-1_33

[9], Sum of Edge Clustering Coefficient (SoECC) [10], Local Average of Connectivity (LAC) [11] and so on.

However, these methods have limitations, such as ignoring the biological features and highly depended on the quality of PPI networks. Therefore, heterogeneous source fusion measures were developed to predict essential proteins by integrating topological and biological information. PeC [12] and WDC [13] were proposed, which introduced centrality measures that combined PPI data with gene expression. Besides, PEMC method is proposed based on integrating the modularity and conservatism of proteins [14]. Peng et al. [15]. proposed a method LSED based on localization specificity in the protein subcellular localization interaction network. According to the essential proteins have closely relationship with diseases, Guo et al. [16]. found the enriched tissue-specific transcription factor (TF)-gene regulations are associated with the disease genes, and provide molecular mechanisms of understanding diseases in 13 human tissues. Zhang et al. [17]. identified functional and network biomarkers with an integer programming model on disease state.

In recent years, transcriptome data, genome data in *Saccharomyces cerevisiae* have been released. The rapid accumulation of RNA-seq data provides unprecedented opportunities to study the properties of PPI networks at the transcriptome level. In this study, RNA-seq, protein complexes, subcellular location compartments were integrated with PPI networks to predict the essential proteins, named as Integration of United Scores of complex centrality and subcellular location compartment centrality (*IUS*), which benefit for the foundation of disease proteins.

2   Methods

2.1   Basic Centrality

The PPI network could be considered as an undirected graph $G(V, E)$, where V is the node set and $E = \{e(v_i, v_j)\}$ is the edge set that connecting two proteins v_i and v_j.

Edge clustering coefficient (ECC) has been widely used to describe how close the two protein are [10]. The ECC of $E_{i,j}$ is defined as:

$$ECC(v_i, v_j) = \frac{Z(v_i, v_j)}{\min\{d_i - 1, d_j - 1\}} \tag{1}$$

where $Z(v_i, v_j)$ represent the number of triangles that built on edge(i, j) in the network, d_i (or d_j) denotes the degree of vertex v_i (or v_j).

Pearson correlation coefficient (PCC) was calculated to evaluate how strong two interacting proteins were co-expressed. The PCC is defined as:

$$PCC(e_{i,j}) = \frac{\sum_{i=1}^{T} (x_i - \mu(x))(y_i - \mu(y))}{\sqrt{\sum_{i=1}^{T} (x_i - \mu(x))^2 \cdot \sum_{i=1}^{T} (y_i - \mu(y))^2}} \tag{2}$$

x_i and y_i denote the FPKM gene expression values at the corresponding time point, $\mu(x)$ and $\mu(y)$ are the means of x_i and y_i over all time point.

The definition of in-degree and complex centrality is computed as follows:

$$IN - Degree(v)_i = DC(v)_i \tag{3}$$

$$IDC(v) = \sum_{i \in ComplexSet(v)} IN - Degree(v)_i \tag{4}$$

IN-Degree is the in-degree value of protein v in *ith* protein complexes. *IDC* is identified as the sum of in-degree of node v in all protein complex. $DC(v)_i$ is the degree value of protein v in complex i. *ComplexSet(v)* represent a set of protein complexes include protein v.

2.2   Essential Protein Prediction by Integrating Multiple Data Source

The mainly idea of *IUS* are shown in Fig. 1. To measure the importance of each compartment, we dividing all subcellular location compartments into 11 parts by mapping to the PPI network. Then, importance score (IS) of compartment s was calculated.

$$IS(C_s) = \frac{|C_s|}{|C_{Max}|} \tag{5}$$

$|C_s|$ is the number of proteins in the compartment s, $|C_{Max}|$ denotes the largest protein number among all compartments. So the *IS* value range from 0 to 1.

By integrating PCC and ECC, two proteins may be co-clustered and co-expressed from the perspective of topological view and biological view. Therefore, we calculate *PeC* by integrating *PCC* and *ECC* as follows:

$$PeC(i) = \sum_{j \in N_i} ECC(j, i) \times PCC(j, i) \tag{6}$$

Where N_i denotes the set of neighbors of node i, *PeC* is the sum probabilities of protein i and its neighbors in a same cluster.

A protein may appear in various subcellular location compartments, and different compartment have different centrality scores. So, based on *PeC* and *IS* value, we calculate the United Subcellular location compartment centrality Score (*USS*) as follows:

$$USS(i_s) = 0;$$
$$USS(i_{s+1}) = USS(i_s); \left(PeC(i_s, C_s) \leq USS(i_s) \right)$$
$$USS(i_{s+1}) = USS(i_s) + (PeC(i_s, C_s) - USS(i_s)) * IS(C_s); \left(PeC(i_s, C_s) > USS(i_s) \right) \tag{7}$$

Where $USS(i_s)$ denotes the *USS* of each protein i in the subcellular location compartment s. *PeC(i, Cs)* represent the *PeC(i)* score for each compartment s.

To normalize the *IDC* and *USS* between 0 and 1, the following equations were used:

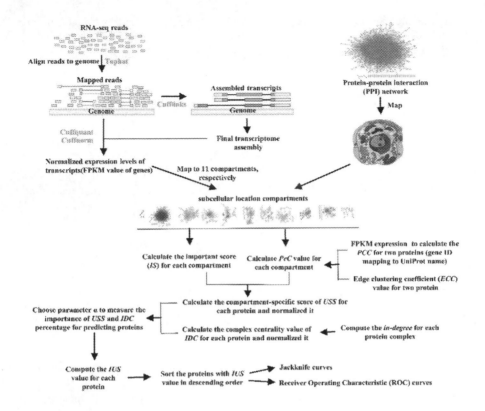

Fig. 1. The schematic of *IUS* method

$$NIDC(i) = \frac{IDC(i) - Min\_IDC(i)}{Max\_IDC(i) - Min\_IDC(i)} \tag{8}$$

$$NUSS(i) = \frac{USS(i) - Min\_USS(i)}{Max\_USS(i) - Min\_IDC(i)} \tag{9}$$

Min_IDC(i) (or *Min_USS(i)*) and *Max_IDC(i)* (or *Max_USS(i)*) denote the minimum value and maximum value of the *IDC* (or *USS*) for all proteins, respectively.

Then, the normalized complex centrality and united subcellular location compartment centrality were integrated by following equations:

$$IUS(i) = \alpha NUSS(i) + (1 - \alpha)NIDC(i) \tag{10}$$

IUS algorithm is introduced as follows:

Input: PPI network; FPKM gene expression value matrix; protein complex set; number of essential proteins *k*; parameter α.

Output: *k* predicted essential proteins

1. PPI network could be regarded as a undirected graph *G(V,E)*, and 11 subcellular location were obtained.

2. *IS(C_s)* was calculated and sorted in descending order in 11 compartment network.
3. For each protein *i* in the compartment *s*, compute the value of *PeC (i_s, C_s)* as the compartment centrality score. And if the protein is absence in the compartment, we set the compartment centrality score of *v* to be 0.
4. The initial value of *USS* was set as 0. When *PeC (i_s, C_s) > USS(i_s)*, calculate the *USS(i_{s+1})*, and finally remain the maximal value of *USS* in the 11 compartments.
5. Calculate the *IDC* value for all protein nodes.
6. Normalized the IDC and USS, and calculate the value of *IUS* and sort proteins in descending order by the *IUS* value.
7. Output the ranked proteins and top *k* proteins as the potential essential proteins.

3   Results

3.1   Experimental Data

The protein-protein interactions of *Saccharomyces cerevisiae* was downloaded from the DIP database (Oct.10, 2010) [18]. There are 24,743 interactions among 5093 proteins in total after the self-interactions and the repeated interactions were filtered.

The subcellular information of proteins were retrieved from knowledge channel of COMPARTMENTS database (April 6, 2017) [19]. There are 5974 proteins and 238620 subcellular locations, which could be classed into 11 categories in yeast.

745 protein complexes was collected from following datasets: CM270 [20], CM425 [21], CYC408 [22] and CYC428 [23].

RNA-seq data were downloaded from NCBI's Gene Expression Omnibus (GEO), which present the progression of the Yeast Metabolic Cycle that concluding 12 time points [24]. Firstly, RNA-seq reads were mapped to the *S. cerevisiae* S288c reference genome sequences by using Tophat2 (version 2.2.1) [25]. Following, transcripts were assembled and the gene expression values were normalized by Fragments Per Kilobase of exon model per Million mapped reads (FPKM) using Cufflinks (version 2.2.1) [26].

The list of essential proteins was integrated from MIPS [20], SGD [27], DEG [2] and SGDP. It includes 1285 essential proteins, which 1167 were shared with 5093 proteins.

3.2   The Effects of Parameter

The ranking scores of proteins were changed with the parameter α, and we choose the value of α range from 0 to 1. When α value take 0, 0.1, 0.2 … 0.9, 1, the prediction accuracy was present to display the performance of *IUS* (Table 1). When α = 0, only in-degree of protein complexes was considered. While when α = 1, the value of NUSS also combined the property of network, gene expression and subcellular location compartments. Therefore, when α values ranging from 0.4 to 0.7, the performance of *IUS* is better. Because the performance of *IUS* have slight difference when predicting the top 15%, 20%, 25% proteins, we consider α is set as 0.7 have the best performance.

Table 1. Influence of parameter α

P	0	0.1	0.2	0.3	0.4	0.5	0.6	0.7	0.8	0.9	1.0
1%	0.75	0.78	0.78	0.80	0.82	0.82	0.82	0.84	0.84	0.86	0.84
5%	0.76	0.77	0.78	0.78	0.79	0.78	0.77	0.78	0.78	0.75	0.74
10%	0.62	0.65	0.65	0.65	0.66	0.68	0.69	0.69	0.68	0.67	0.64
15%	0.57	0.57	0.58	0.59	0.60	0.60	0.60	0.59	0.60	0.59	0.57
20%	0.51	0.52	0.53	0.53	0.54	0.54	0.54	0.54	0.54	0.53	0.51
25%	0.48	0.498	0.502	0.502	0.502	0.502	0.50	0.50	0.49	0.48	0.46

'P' represents the top P percent of ranked proteins according to *IUS* value. Column from 2 to 11 represents prediction accuracy of *IUS* in each percentage of ranked proteins when α was set ranging from 0 to 1.

3.3   Validation with Jackknife Methodology

Jackknife methodology [28] is employed to compare the performance of *IUS* and other previously proposed eight methods including DC, BC, EC, IC, SoECC, LAC, PeC and WDC (Fig. 2). After ranking the score of *IUS* for each protein in descending order, top 100 to 1200 proteins were selected as essential proteins. The area under the curves could be used to compare the prediction performance of *IUS* and other eight methods. Compared with the list of known essential proteins, results show that SoECC, LAC, PeC and WDC combining with multi-information fusion measures have better performance than the topology centrality measures such as DC, BC, EC, IC. While our method show the best performance among them.

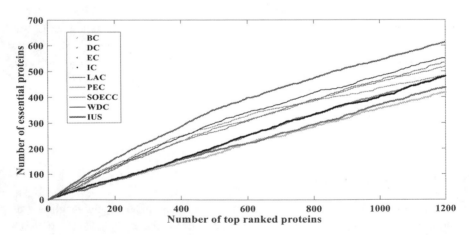

Fig. 2. Jackknife curves of *IUS* and other existing eight methods in DIP. The x-axis from left to right represent the proteins ranked in descending order according to the values with different methods, and the y-axis represents the cumulative number of true essential with respect to the ranked proteins moving left to right.

3.4   Validation with F-Measure and Accuracy

To evaluate the effectiveness of the *IUS* value, several statistical measures were introduced, such as sensitivity (SN), specificity (SP), positive predictive value (PPV), negative predictive value (NPV), F-measure (F), and accuracy (ACC).

$$SN = \frac{TP}{TP + FN} \tag{11}$$

$$SP = \frac{TN}{TN + FP} \tag{12}$$

$$PPV = \frac{TP}{TP + FP} \tag{13}$$

$$NPV = \frac{TN}{TN + FN} \tag{14}$$

$$F = \frac{2 * SN * PPV}{SN + PPV} \tag{15}$$

$$ACC = \frac{TP + TN}{P + N} \tag{16}$$

Where TP, TN, FP, FN are represented as true positives, true negatives, false positives, false negatives, respectively. P is the number of essential proteins and N is the number of non-essential proteins.

The comparison results are shown in Table 2, it is obvious that SN, SP, PPV, NPV, F, ACC, ACC + F of *IUS* are higher than other eight algorithms, which suggest that IUS can identify essential more accurately.

Table 2.   Comparison of the results of SN, SP, PPV, NPV, F, ACC, ACC + F of *IUS* method and other previous eight methods.

Methods	SN	SP	PPV	NPV	F	ACC	ACC + F
DC	0.4319	0.8010	0.3922	0.8258	0.4111	0.7164	1.1275
BC	0.3719	0.7832	0.3377	0.8075	0.3540	0.6889	1.0429
EC	0.4036	0.7926	0.3665	0.8172	0.3842	0.7035	1.0877
IC	0.4336	0.8015	0.3938	0.8264	0.4127	0.7172	1.1299
SoECC	0.4687	0.8120	0.4257	0.8371	0.4462	0.7333	1.1795
LAC	0.4739	0.8135	0.4304	0.8387	0.4511	0.7357	1.1868
PeC	0.4242	0.7987	0.3852	0.8235	0.4038	0.7129	1.1167
WDC	0.4927	0.8191	0.4475	0.8445	0.4690	0.7443	1.2133
IUS	**0.5458**	**0.8349**	**0.4957**	**0.8608**	**0.5196**	**0.7687**	**1.2883**

3.5 Validated by ROC Curves

To further evaluate the prediction of each method, Receiver Operating Characteristic (ROC) curves were plotted, and the Area Under Curve (AUC) are compared to quantitatively assess the performance for the methods (Fig. 3). The larger value of AUC is, the better the method performs. The results show that *IUS* achieve the highest AUC value, which is higher than the AUC value of other eight precious methods including DC, BC, EC, IC, SoECC, LAC, PeC and WDC.

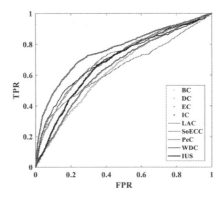

Fig. 3. ROC curves of *IUS* and other eight methods. TPR represent the true positive rate, FPR represent the false positive rate.

4   Conclusion and Discussion

Essential proteins are playing vital roles in the biological activities, and the identification of them becoming a major topic. To overcome the limitation of traditional biological experiment, topological-based and heterogeneous source fusion measures have been widely used to optimize the prediction of essential proteins. In this paper, RNA-seq, protein complexes, PPI network and subcellular location data were integrated, which show better performance than eight other methods including DC, BC, EC, IC, SoECC, LAC, PeC and WDC.

The merits of *IUS* method are shown in following aspects. Firstly, RNA-seq and genome were applied to measure the expression profiles of genes at transcript level, which could increase the predictive accuracy. Secondly, the interactions between proteins were described to tend to co-expressed and co-clustered, which could be measured by the topological property and the correlation coefficient of two proteins. Moreover, subcellular location compartments were divided into 11 parts to specific environment, and the united compartment centrality score was calculated. In addition to, we also take the modular property of essential proteins into account. Although *IUS* has achieved some good results, further research should also be extended. Researches have shown essential proteins are more conserved and PPI network was changed with the different time and conditions, these properties should be investigated further.

Acknowledgements. This paper is supported by the National Natural Science Foundation of China (61672334, 61502290, 61401263), Industrial Research Project of Science and Technology in Shaanxi Province (2015GY016).

References

1. Acencio, M.L., Lemke, N.: Towards the prediction of essential genes by integration of network topology, cellular localization and biological process information. BMC Bioinform. **10**, 290 (2009)
2. Zhang, R., Lin, Y.: DEG 5.0, a database of essential genes in both prokaryotes and eukaryotes. Nucleic Acids Res. **37**, D455–D458 (2009)
3. Giaever, G., Chu, A.M., Ni, L., Connelly, C., Riles, L., Veronneau, S., Dow, S., Lucau-Danila, A., Anderson, K., Andre, B., Arkin, A.P., Astromoff, A., El-Bakkoury, M., Bangham, R., Benito, R., Brachat, S., Campanaro, S., Curtiss, M., Davis, K., Deutschbauer, A., Entian, K.D., Flaherty, P., Foury, F., Garfinkel, D.J., Gerstein, M., Gotte, D., Guldener, U., Hegemann, J.H., Hempel, S., Herman, Z., Jaramillo, D.F., Kelly, D.E., Kelly, S.L., Kotter, P., LaBonte, D., Lamb, D.C., Lan, N., Liang, H., Liao, H., Liu, L., Luo, C., Lussier, M., Mao, R., Menard, P., Ooi, S.L., Revuelta, J.L., Roberts, C.J., Rose, M., Ross-Macdonald, P., Scherens, B., Schimmack, G., Shafer, B., Shoemaker, D.D., Sookhai-Mahadeo, S., Storms, R.K., Strathern, J.N., Valle, G., Voet, M., Volckaert, G., Wang, C.Y., Ward, T.R., Wilhelmy, J., Winzeler, E.A., Yang, Y., Yen, G., Youngman, E., Yu, K., Bussey, H., Boeke, J.D., Snyder, M., Philippsen, P., Davis, R.W., Johnston, M.: Functional profiling of the Saccharomyces cerevisiae genome. Nature **418**, 387–391 (2002)
4. Cullen, L.M., Arndt, G.M.: Genome-wide screening for gene function using RNAi in mammalian cells. Immunol. Cell Biol. **83**, 217–223 (2005)
5. Roemer, T., Jiang, B., Davison, J., Ketela, T., Veillette, K., Breton, A., Tandia, F., Linteau, A., Sillaots, S., Marta, C., Martel, N., Veronneau, S., Lemieux, S., Kauffman, S., Becker, J., Storms, R., Boone, C., Bussey, H.: Large-scale essential gene identification in Candida albicans and applications to antifungal drug discovery. Mol. Microbiol. **50**, 167–181 (2003)
6. Hahn, M.W., Kern, A.D.: Comparative genomics of centrality and essentiality in three eukaryotic protein-interaction networks. Mol. Biol. Evol. **22**, 803–806 (2005)
7. Joy, M.P., Brock, A., Ingber, D.E., Huang, S.: High-betweenness proteins in the yeast protein interaction network. J. Biomed. Biotechnol. **2005**, 96–103 (2005)
8. Bonacich, P.: Power and centrality: a family of measures. Am. J. Sociol. **92**, 1170–1182 (2015)
9. Stephenson, K., Zelen, M.: Rethinking centrality: methods and examples. Soc. Netw. **11**, 1–37 (1989)
10. Wang, J., Li, M., Wang, H., Pan, Y.: Identification of essential proteins based on edge clustering coefficient. IEEE/ACM Trans. Comput. Biolo. Bioinform. **9**, 1070–1080 (2012)
11. Li, M., Wang, J., Chen, X., Wang, H., Pan, Y.: A local average connectivity-based method for identifying essential proteins from the network level. Comput. Biol. Chem. **35**, 143–150 (2011)
12. Li, M., Zhang, H., Wang, J.X., Pan, Y.: A new essential protein discovery method based on the integration of protein-protein interaction and gene expression data. BMC Syst. Biol. **6**, 15 (2012)
13. Tang, X., Wang, J., Zhong, J., Pan, Y.: Predicting essential proteins based on weighted degree centrality. IEEE/ACM Trans. Comput. Biol. Bioinform. **11**, 407–418 (2014)
14. Zhao, B., Wang, J., Li, X., Wu, F.X.: Essential protein discovery based on a combination of modularity and conservatism. Methods **110**, 54–63 (2016). San Diego, California

15. Peng, X., Wang, J., Wang, J., Wu, F.X., Pan, Y.: Rechecking the centrality-lethality rule in the scope of protein subcellular localization interaction networks. PLoS ONE **10**, e0130743 (2015)
16. Guo, W., Zhu, L., Deng, S., Zhao, X., Huang, D.: Understanding tissue-specificity with human tissue-specific regulatory networks. Sci. China Inf. Sci. **59**, 070105 (2016)
17. Zhang, C., Liu, J., Shi, Q., Zeng, T., Chen, L.: Comparative network stratification analysis for identifying functional interpretable network biomarkers. BMC Bioinform. **18**, 48 (2017)
18. Xenarios, I., Salwinski, L., Duan, X.J., Higney, P., Kim, S.M., Eisenberg, D.: DIP, the Database of Interacting Proteins: a research tool for studying cellular networks of protein interactions. Nucleic Acids Res. **30**, 303–305 (2002)
19. Binder, J.X., Pletscher-Frankild, S., Tsafou, K., Stolte, C., O'Donoghue, S.I., Schneider, R., Jensen, L.J.: COMPARTMENTS: unification and visualization of protein subcellular localization evidence. Database: J. Biol. Databases Curation 2014, bau012 (2014)
20. Mewes, H.W., Frishman, D., Mayer, K.F., Munsterkotter, M., Noubibou, O., Pagel, P., Rattei, T., Oesterheld, M., Ruepp, A., Stumpflen, V.: MIPS: analysis and annotation of proteins from whole genomes in 2005. Nucleic Acids Res. **34**, D169–D172 (2006)
21. Friedel, C.C., Krumsiek, J., Zimmer, R.: Bootstrapping the interactome: unsupervised identification of protein complexes in yeast. J. Comput. Biol. J. Comput. Mol. Cell Biol. **16**, 971–987 (2009)
22. Pu, S., Wong, J., Turner, B., Cho, E., Wodak, S.J.: Up-to-date catalogues of yeast protein complexes. Nucleic Acids Res. **37**, 825–831 (2009)
23. Pu, S., Vlasblom, J., Emili, A., Greenblatt, J., Wodak, S.J.: Identifying functional modules in the physical interactome of Saccharomyces cerevisiae. Proteomics **7**, 944–960 (2007)
24. Nocetti, N., Whitehouse, I.: Nucleosome repositioning underlies dynamic gene expression. Genes Dev. **30**, 660–672 (2016)
25. Trapnell, C., Pachter, L., Salzberg, S.L.: TopHat: discovering splice junctions with RNA-Seq. Bioinformatics **25**, 1105–1111 (2009). Oxford, England
26. Trapnell, C., Roberts, A., Goff, L., Pertea, G., Kim, D., Kelley, D.R., Pimentel, H., Salzberg, S.L., Rinn, J.L., Pachter, L.: Differential gene and transcript expression analysis of RNA-seq experiments with TopHat and Cufflinks. Nat. Protoc. **7**, 562–578 (2012)
27. Cherry, J.M., Adler, C., Ball, C., Chervitz, S.A., Dwight, S.S., Hester, E.T., Jia, Y., Juvik, G., Roe, T., Schroeder, M., Weng, S., Botstein, D.: SGD: Saccharomyces Genome Database. Nucleic Acids Res. **26**, 73–79 (1998)
28. Holman, A.G., Davis, P.J., Foster, J.M., Carlow, C.K., Kumar, S.: Computational prediction of essential genes in an unculturable endosymbiotic bacterium, Wolbachia of Brugia malayi. BMC Microbiol. **9**, 243 (2009)

Protein Structure and Function Prediction

Similarity Comparison of 3D Protein Structure Based on Riemannian Manifold

Zhou Fengli[1(✉)] and Lin Xiaoli[1,2]

[1] Faculty of Information Engineering,
City College Wuhan University of Science and Technology, Wuhan 430083, China
thinkvicw@163.com, aneya@163.com
[2] Hubei Key Laboratory of Intelligent Information Processing
and Real-Time Industrial System, School of Computer Science and Technology,
Wuhan University of Science and Technology, Wuhan 430065, China

Abstract. As the representative technology of protein spatial structure exploration, NMR technology provides an unprecedented opportunity for modern life science research. But subsequent large data analysis has become a major problem. It is an important means to study protein structure and functional relationship by known information proteins' three-dimensional structures to predict the unknown spatial structure of proteins. A method for similarity comparison of 3D protein structures based on Riemannian manifold theory is proposed in this paper. By constructing Cα frames and extracting geometric feature of protein, 3D coordinates of proteins are converted into one dimension sequences with rotation and translation invariance. The Riemann distance is used as the three-dimensional structure similarity degree index. Spatial transformation on protein structure is not needed in this method, which avoiding errors when matching two proteins in the traditional method for registration by the least squares fitting. This method is independent of sequence information completely. It has realistic significance for proteins which do not have a similarity between sequences. Three experiments are designed according to 3 sets of data: proteins of different similarity, ten pairs whose protein structures are more difficult to identify proposed by Fischer, 700 proteins in the HOMSTRAD database. Compared with the traditional method, the experiment results show that the matching accuracy of this method has been greatly enhanced.

Keywords: Protein · Cubic spline interpolation · Frame · Riemannian manifold · Structural comparison

1 Introduction

The function of proteins and non-coding RNAs in bio-macromolecule mainly depends on their spatial structure. There have been more than sixty thousand spatial structures of bio-macromolecules to be measured until now, and how to effectively compare the similarity between them has become an important task in life science [1]. Although the determination of protein sequence was completed, directly accessing the 3D protein structures has been a bottleneck, so three-dimensional structures of many proteins that

© Springer International Publishing AG 2017
D.-S. Huang et al. (Eds.): ICIC 2017, Part II, LNCS 10362, pp. 387–397, 2017.
DOI: 10.1007/978-3-319-63312-1_34

sequences are known have not been determined by experiment method. In this case, the main researching methods for protein structure and functional relationship in structural biology are predicting spatial structure of unknown protein by using known proteins' sequence information and spatial structure information. Since obtaining the structural model of protein by experimental method is very difficult and takes a long time, the similarity comparison of 3D protein structures can construct a structural model which required in experiments and propose hypothesis about proteins' function for further experimental work [2]. Therefore, the similarity comparison of protein structures is an important method for protein function analysis, category management and detection.

Now there have a lot of research methods and software tools for protein structure comparison, such as Dali [3], CE [4], VAST [5], STRUCTAL [6], SSM [7], TM-align [8], etc. They have provided a variety of analysis methods for researching protein spatial structure. And protein structure comparison methods were mainly divided into three categories: (1) methods were based on distance matrix between amino acids(Dali, CE); (2) methods were based on protein spatial geometry(STRUCTAL, TM-align); (3) methods were based on protein secondary structure(VAST, SSM).

Traditional protein similarity comparison method was usually depend on the primary sequence of protein structure, but two proteins that have different primary structure and same spatial structure were tend to have same properties. For resolving the shortcomings, this paper has proposed a new method for protein structure similarity comparison based on Riemannian manifolds. Through processing differential manifold, using curvature k and torsion r to replace two dihedral angles ϕ and φ in protein primary structure, thus the analysis of protein structure will convert to a purely mathematical problem.

Louie and Somorjai [9] introduced differential manifold to the research of protein structure in 1982, they took the backbone of protein as a continuous space curve and described it by parameters (b, ω). In this description method, Louie and Somorjai not only gave the index to identify protein structure, but also gave a continuous description for protein backbone structure. On this basis, literatures [10, 11] analyzed the elastic shape of protein structure's three-dimensional curve and extracted the elastic measurement which used in the comparison of protein structure. Literature [12] proposed elastic Riemannian metric further and compared the protein structure in view of manifold. This method had better effect than other structural comparing methods, but it had the shortcoming of computing expensive.

This paper has constructed Cα frames to extract the geometric feature of protein's main chain by curve interpolation and represented the protein structure data by Riemannian metric sequence, thus embedding the relevant information of protein's geometry into the matrix manifold which represented by Riemannian metric sequence. Riemannian geometry can be introduced to extract effective geometric structure through this manifold representation and embedding. This method can reflect protein's structure features more intuitively and provide a new idea to compare the similarity of protein structure.

2 Suggested Method

The data used in this paper are from PDB [13] (Protein Data Bank) database which was established by Brookhaven laboratory in 1971. The database contains crystal structure data of macromolecular proteins and is the unique database that has 3-dimensional structure of biopolymer in Internet. Its contents were obtained from experimental data based on X-ray crystallization and NMR data mainly.

The key of this paper is curve interpolation to chain of protein, and extracts the characteristic with spatial translation and rotation invariance from the 3-dimensional geometry of proteins, then embedding it into the matrix manifold. The specific steps are as follows:

(1) Data preprocessing

Getting proteins' PDB files from the database firstly based on its unique identification PDB ID. If atom H is ignoring, the skeleton of amino acid is a tetrahedron which top is atom Cα and other three tips are amino, carboxyl and R groups respectively. Figure 1 shows the basic atomic framework of two amino acids in protein 1CRN, where R group is CBSG. It can be seen that the skeleton atom of each amino acid is the combination of N-Cα-C. Then extracting the skeleton atom of each amino from PDB file—spatial coordinates of atoms Cα, C and N.

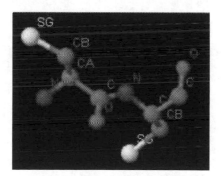

Fig. 1. Amino acid neighboring structure

(2) Constructing Cα coordinate system

Frenet-Serret formula is often used to describe the relationship between tangential, normal and sub-directions of the curve when the particles move on a continuously differentiable curve. Based on the formula, the Cα coordinate system of protein is constructed by the spatial coordinates of atoms Cα, C and N from the data [14].

$$X_i = \frac{C_i - C\alpha_i}{|C_i - C\alpha_i|}, \quad U_i = \frac{N_i - C\alpha_i}{|N_i - C\alpha_i|}, \quad Z_i = \frac{X_i \times U_i}{|X_i \times U_i|}, \quad Y_i = Z_i \times X_i \tag{1}$$

Where $C\alpha$, C and N represent the spatial coordinates of atoms Cα, C and N. X, Y and Z are the column vectors respectively. Therefore Cα coordinate system sequence

$\left[F_1 F_2 \cdots F_i \cdots F_n \right]$. can be composed of 3×3 unit orthogonal matrix, where $F_i = \left(X_i Y_i Z_i \right)$. Each $C\alpha$ coordinate system F_i represents the spatial orientation of its corresponding amino acid skeleton, and each amino acid corresponds to a unit orthogonal matrix, thus the protein amino acid sequence is transformed into a unit orthogonal matrix sequence.

(3) Curve interpolation

Cubic spline interpolation meth has solved the problem of describing the shape of free curve in computational geometry, and maintained the local property at the same time. Therefore, the main cin structure of protein $C\alpha$ is subjected to cubic spline interpolation.

The cubic spline function is defined as follows:

If the division of interval $[a, b]$ is $a = x_0 < x_1 < \cdots < x_n = b$, then

$$S_3(x) = a_0 + a_1 x + \frac{a_2}{2!}x^2 + \frac{a_3}{3!}x^3 + \sum_{j=1}^{n-1} \frac{\beta_j}{3!}\left(x - x_j\right)_+^3 \tag{2}$$

Where $\left(x - x_j\right)_+^3 = \begin{cases} \left(x - x_j\right)^3 & x \geq x_j \\ 0 & x < x_j \end{cases}$, $\quad j = 1, 2, \cdots 3, n-1.$

The cubic spline function is used to interpolate the curve, namely, the interpolation function is cubic spline function. So it is called cubic spline interpolation.

As shown in Fig. 2(b) and (c) are the structural curves of protein 1CRN's cubic spline interpolation before and after, the number is the amino acid sequence number, the position of atom $C\alpha$ is at 'O'. It can be seen that the protein backbone structure after interpolation is transformed into a smooth curve, and each curve between the two atoms $C\alpha$ corresponds to a cic polynomial.

(4) Extracting geometry parameter features

Curvature and torsion respectively reflect the spatial curve's curvature degree and twist degree. They have spatial rotation translation invariance and can fully describe geometric features of the curve. Therefore, this paper chooses each $C\alpha$ atom's curvature, torsion and distance between adjacent $C\alpha$ atoms in the protein structure curve to describe the geometry of protein.

If the parameter equation of spatial curve is known as follow:

$$\begin{cases} x = x(t) \\ y = y(t) \\ z = z(t) \end{cases} \tag{3}$$

Then the calculation formulas of curvature k_t and torsion T_t at parameter t are respectively as follows.

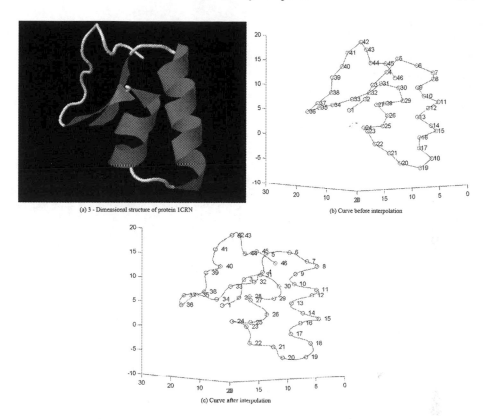

(a) 3 - Dimensional structure of protein 1CRN

(b) Curve before interpolation

(c) Curve after interpolation

Fig. 2. Curve interpolation of protein 1CRN backbone structure

$$
\begin{cases}
k_t = \dfrac{\sqrt{(z''y' - y''z')^2 + (x''z' - z''x')^2 + (y''x' - x''y')^2}}{(x'^2 + y'^2 + z'^2)^{\frac{3}{2}}} \\[2em]
\mathcal{T}_t = \dfrac{x'''(y'z'' - y''z') + y'''(x''z' - x'z'') + z'''(x'y'' - x''y')}{(y'z'' - y''z')^2 + (x''z' - x'z'')^2 + (x'y'' - x''y')^2}
\end{cases}
\tag{4}
$$

Where x', x'' and x''' are x on t's first, second and third derivatives respectively.

According to Eq. (4), curvature k and torsion $\mathcal{T}$ at each Cα point on the protein structure curve can be calculated, then a diagonal matrix can be formed with them and the absolute value of distance d between adjacent Cα atoms together.

$$
\Lambda_i = \begin{bmatrix} |k_i| & & \\ & |\mathcal{T}_i| & \\ & & |d_i| \end{bmatrix}
$$

Where k_i and k_i are the curvature and torsion of i-th atom Cα.

$d_i = \sqrt{\left(x_{i+1} - x_i\right)^2 + \left(y_{i+1} - y_i\right)^2 + \left(z_{i+1} - z_i\right)^2}$, (x_i, y_i, z_i) and $(x_{i+1}, y_{i+1}, z_{i+1})$ are the spatial coordinates of atoms Cα that numbered i and $i + 1$ respectively. In this way, a diagonal matrix sequence which contains the protein's structure curve descriptor is obtained.

(5) Constructing Riemannian metric sequences

Riemannian metric means that the geometry of space should be based on the distance between the infinitely two points $(x_1, x_2, \cdots, x_n)$ and $(x_1 + d_1, x_2 + d_2, \cdots, , x_n + d_n)$. It is a measurement of positive definite quadratic identified with the square of differential arc length, and is also the positive definite symmetric matrices by the functions. In a three-dimensional space, a 3×3 real symmetric matrix can be used to represent the Riemannian metric of a certain point in space.

$$S = \begin{bmatrix} a & b & c \\ d & e & f \\ g & h & i \end{bmatrix} = F \begin{bmatrix} \lambda_1 & & \\ & \lambda_2 & \\ & & \lambda_3 \end{bmatrix} F^T \tag{5}$$

Where $\det(S) > 0$, $\lambda_i > 0$ and the decomposition of positive definite symmetric matrices is reversible.

According to the property of positive definite symmetric matrices in Eq. (2), Riemannian metric sequence can be constructed by using the unit orthogonal matrix and diagonal matrix sequences that obtained in steps (2) and (3).

$$S_i = F_i \Lambda_i F_i^T \tag{6}$$

S_i is the Riemann metric of i-th amino acid.

(6) Calculate Riemann distance between two proteins' Riemannian metric

First dividing the Riemannian metric S into the unit orthogonal matrix F and the diagonal matrix Λ, then calculating the distance $d(F_i, F_{i+1})$ between F_i and F_{i+1}, calculating the distance $d(\Lambda_i, \Lambda_{i+1})$ between Λ_i and Λ_{i+1}, finally the Riemann distance $d(S_i, S_{i+1})$ between S_i and S_{i+1} can be obtained.

$$d^2\left(S_i, S_{i+1}\right) = k\left(\Lambda_i, \Lambda_{i+1}\right) d^2\left(F_i, F_{i+1}\right) + d^2\left(\Lambda_i, \Lambda_{i+1}\right) \tag{7}$$

Where

$$\begin{cases} d\left(F_i, F_{i+1}\right) = \left\| \log\left(F_i^T F_{i+1}\right) \right\| \\ d\left(\Lambda_i, \Lambda_{i+1}\right) = \sqrt{\sum_i \log^2 \left(\dfrac{\lambda_{1,i}}{\lambda_{2,i}}\right)} \end{cases} \tag{8}$$

$\lambda_{1,i}$ and $\lambda_{2,i}$ are the values of i-th row and i-th column in the diagonal matrix Λ_i and Λ_{i+1} respectively. Coefficient k is the weighting factor and its range is between 0 and 1 [16].

$$\begin{cases} \left(H_i, H_{i+1}\right) = \dfrac{1 + \tanh\left(3H_i H_{i+1} - 7\right)}{2} \\ H = log\left(\dfrac{\lambda_{max}}{\lambda_{min}}\right) \end{cases} \tag{9}$$

Where λ_{max} and λ_{min} are the maximum and minimum values in diagonal matrix Λ.

(7) Set threshold, compare and analysis

The Riemann distance sequence $\left[ds_1, ds_2, \cdots, ds_i, \cdots ds_n\right]$. between two proteins can be obtained from step (5). The smaller value has represented that two Riemannian metrics contained proteins' structural geometric information in the Riemannian manifold is closer. So the corresponding proteins' structure is more similar. The threshold of Riemann distance can be taken after experimental validation.

$$k = 0.6\,means(ds) + 0.3std(ds) \tag{10}$$

Where $means(ds)$ and $std(ds)$ respectively represent the mean and variance of Riemann distance sequence. The value which is less than threshold k can be considered that two proteins havehe same structure at the residues i.

3   Experimental Results and Analysis

In order to verify the viability and effectiveness of this method, three groups of experiments have been carried out. First group respectively selects protein data of family, superfamily and non-superfamily from SCOP database to verify whether the results are according with biological significance; Second group experiments on 10 structures with difficult identification that were given by Fischer [17], and it has verified the viability of this method from "same residue ratio"; Third group is based on the HOMSTRAD structure comparison reference library, it has validated whether this method is effective in homologous identification.

3.1   Comparison of Protein Structure with Different Similarity

SCOP (Structural Classification of Proteins) database has provides information about structure and evolutionary relationships between known structural proteins, and the proteins involved include all entries in structural database PDB [18]. SCOP was divided the proteins into all-α-type, all-β-type, α/β-type that mainly composed of parallel folding and $(\alpha + \beta)$-type that mainly composed of antiparallel folding. The experiment has selected protein data that PDB ID are 101 M, 102 M, 1UVY and 1C7Y from SCOP database, they respectively belongs to same family, same superfamily and different superfamily, their similarity has decreased in turn. In general, the closer the relationship between two proteins, the higher the similarity of structure, the smaller the Riemann distance between Riemannian metric sequences.

Table 1 has shown the structural comparison results of protein 101 M and 101 M, 102 M, 1UVY and 1C7Y. Riemann_dis is the mean of Riemann distance between Riemannian metric sequences, we compare it with the evaluation index of traditional protein structure RMSD [19], TM-score [20] and Z-score [21]. RMSD (Root mean square deviation) method compared whether protein skeleton was directly registered, it was the earliest proposed and used most widely. If the RMSD value is smaller, the structures of two proteins are more similar. TM-score and Z-score are on the contrary. The greater the value is, the higher the structure similarity is. From Table 1, it can be seen that if the relationship between proteins becomes farther, four parameters of results show that the similarity of protein structure has been decreased.

Table 1. Structure comparison results of different similarity of protein and protein 101 M

PBD ID	RMSD	TM-score	Z-score	Riemann_dis(mean)	SCOP.superfamily	SCOP.family
101 M	0	1	31.7	0	Globin-like	Globins
102 M	0.15	0.99	30.6	0.0729	Globin-like	Globins
1UXY	3.00	0.55	7.6	0.3525	Globin-like	Truncated hemoglobin
1C7Y	3.7	0.31	0.7	0.9903	C-terminal domain	C-terminal domain

Figure 3 shows the Riemann distance curve of four pairs of proteins. It can be observed that the similarity of protein structure will be decreased and the Riemann distance will be reduced significantly if the relationship between proteins is far away.

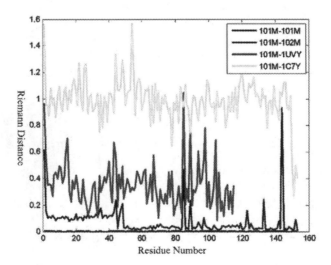

Fig. 3. Curve of Riemann distance between protein 101 M and different similarity of proteins

3.2   Comparison of 10 Protein Structure Pairs with Difficult Identification

Fischer had gave 10 protein pairs which were difficult to compare in literature [23] and we use the method proposed in this paper to compare the structures of these ten protein pairs, then compare its results with three traditional methods named Dali [3], TM-align [8] and SPalign [22]. The results are shown in Table 2. "Equ" represents the number of comparing residues, "ResNum" represents the number of same residues, and "Riemann_dis" represents the average value of Riemann distance in this method. It is generally considered that the larger the same residue ratio (ResNum/Equ), the better the effect is.

As shown in Table 2, the proportion values of method proposed in this paper is about 0.5% higher than Dali method which had the best effect, and is about 2.8% higher than SPalign method which had worse effect. The result proves that the method is not only feasible, but also has better effect than other three methods obviously.

Table 2. Structure comparison average results of 10 more difficult to identify protein structures

Method	Average Equ	Average ResNum	ResNum/Equ
Dali	119.10	14.70	12.34%
TM-align	126.50	14.10	11.15%
Spalign	125.30	13.60	10.85%
Riemann_dis	121.60	15.60	12.83%

3.3   Verification of Reference Library Based on Structure Comparison

HOMSTRAD(Homologous Structure Alignment Database) is a reference library which contains the structure comparison results of homologous proteins. Its results are based on the three procedures MNYFIT, STAMP and COMPARER and through manual adjustment. The current version has 1032 families and each family has members between 2 to 41. The average protein's length in each family is between 17 to 855 residues and the sequence identity is between 0.08 to 0.94. The comparison results of this database have preserved the sequence of structural alignment, the aligned secondary structure, the atomic coordinates after changing and so on. Therefore, we use the alignment structure of HOMSTRAD as a reference to verify the feasibility and accuracy of protein structure comparison method with automatically implemented.

The experiment has used the method proposed in this paper to calculate double-structure instances of 700 proteins from HOMSTRAD. The average matching accuracy is 89%. From the 700 double-structures, 10 pairs of comparison results are selected to compare with Dali, TM-align and SPalign.

Table 3 has shown that the results obtained by these methods are almost identical for proteins with higher similarity. In contrast, the results obtained by these methods are different for proteins with lower similarity. For the comparison with 1C20:A and 1IG6:A, the registration accuracy of this method is 73% while TM-align and Spalign are 53% and 49%, the method proposed by this paper is significantly better than these two methods. And the result obtained by Riemann_dis is also better than others from the average value.

Table 3. Accuracy comparison results of 4 methods for 10 examples in HOMSTRAD

PDB ID1	PDB ID2	Dali	TM-align	Spalign	Riemann_dis
1K6 M:A	IBIF:A	100%	100%	100%	100%
1I2D:A	1G8F:A	95%	95%	95%	96%
1NSD:A	1NHZ:A	80%	76%	82%	84%
1AOR:A	1B25:A	97%	98%	96%	98%
1L21:A	1GC5:A	99%	97%	80%	99%
1IG6:A	1C20:A	71%	53%	49%	73%
1AW5:A	1B4 K:A	94%	98%	95%	98%
1NGI:B	2NGD:A	93%	95%	96%	96%
1FSO:G	1E79:A	88%	88%	80%	90%
1XOL:A	1BGX:A	53%	64%	49%	66%
Average		87%	86%	82%	90%

4   Conclusion

This paper has constructed Cα coordinate system which characterizes the amino acid orientation by using the protein structure data first, and then combined it with three spatial rotation translation invariances of atom Cα in protein structure curve which are curvature, torsion and distance between adjacent Cα to construct the Riemann metric. In this way, the geometrical features of protein structure have been embedded in the matrix manifolds represented by Riemann metric, so Riemann metric has preserved all information of 3D protein structure. Secondly, this paper has used Riemannian metric as a feature descriptor of similarity comparison of 3D protein structure, used Riemannian distance as the similarity measurement, and then calculated the distance between Riemannian metrics which can be used as measurement index of the similarity of protein structure. Finally, this method has been validated from three different experimental data. The results show that the similarity comparison of 3D protein structure is more significant than other methods.

Acknowledgment. The authors thank the members of Machine Learning and Artificial Intelligence Laboratory, School of Computer Science and Technology, Wuhan University of Science and Technology, for their helpful discussion within seminars. This work was supported in part by National Natural Science Foundation of China (No. 61502356).

References

1. Liang, Y.: Structural Biology. Science Press, Ann Arbor (2005)
2. Peng, Q.S., Hu, M.: Approaches for 3D protein structure similarity comparison-a survey. J. Comput. Aided Des. Comput. Graph. **18**(10), 1465–1471 (2006)
3. Holm, L., Sander, C.: Protein structure comparison by alignment of distance matrices. J. Mol. Biol. **233**(1), 123–138 (1993)
4. Shindyalov, I.N., Bourne, P.E.: Protein structure alignment by incremental combinatorial extension(CE) of the optimal path. Protein Eng. **11**(9), 739–747 (1998)

5. Gibrat, J.F., Madej, T., Bryant, S.H.: Surprising similarities in structure comparison. Curr. Opin. Struct. Biol. **6**(3), 377–385 (1996)
6. Levitt, M.: STRUCTAL. A structural alignment program (1994)
7. Krissinel, E., Henrick, K.: Secondary-structure matching (SSM), a new tool for fast protein structure alignment in three dimensions. Acta Crystallogr. Sect. D Biol. Crystallogr. **60**(12), 2256–2268 (2004)
8. Zhang, Y., Skolnick, J.: TM-align: a protein structure alignment algorithm based on the TM-score. Nucleic Acids Res. **33**(7), 2302–2309 (2005)
9. Louie, A.H., Somorjai, R.L.: Differential geometry of proteins: a structural and dynamical representation of patterns. J. Theor. Biol. **98**(2), 189–209 (1982)
10. Joshi, S.H, Klassen, E., Srivastava, A., et al.: A novel representation for Riemannian analysis of elastic curves in Rn. In: IEEE Conference on Computer Vision and Pattern Recognition, Minneapolis, MN, pp. 1–7 (2007)
11. Klassen, E., Srivastava, A., Mio, W., et al.: Analysis of planar shapes using geodesic paths on shape spaces. IEEE Trans. Pattern Anal. Mach. Intell. **26**(3), 372–383 (2004)
12. Liu, W., Srivastava, A., Zhang, J.: A mathematical framework for protein structure comparison. PLoS Comput. Biol. **7**(2), e1001075 (2011)
13. Berman, H.M., Westbrook, J., Feng, Z., et al.: The protein data bank. Nucleic Acids Res. **28**(1), 235–242 (2000)
14. Hanson, A.J., Thakur, S.: Quaternion maps of global protein structure. J. Mol. Graph. Model. **38**, 256–278 (2012)
15. Ji, Y.Q., Xu, Z.C.: Differential Manifold and Riemannian Geometry. Shaanxi Normal University Press, Xi'an (1996)
16. Collard, A., Bonnabel, S., Phillips, C., et al.: An anisotropy preserving metric for DTI processing. arXiv:1210.2826 (2012)
17. Fischer, D., Elofsson, A., Rice, D.W., et al.: Assessing the performance of inverted protein folding methods by means of an extensive benchmark. In: Proceeding of the First Pacific Symposium on Biocomputing, pp. 300–318 (1996)
18. Murzin, A.G., Brenner, S.E., Hubbard, T., et al.: SCOP: a structural classification of proteins database for the investigation of sequences and structures. J. Mol. Biol. **247**(4), 536–540 (1995)
19. Maiorov, V.N., Crippen, G.M.: Significance of root-mean-square deviation in comparing three-dimensional structures of globular proteins. J. Mol. Biol. **235**(2), 625–634 (1994)
20. Zhang, Y., Skolnick, J.: Scoring function for automated assessment of protein structure template quality. Protein Struct. Funct. Bioinf. **57**(4), 702–710 (2004)
21. Shindyalov, I.N., Bourne, P.E.: Protein structure alignment by incremental combinatorial extension(CE) of the optimal path. Protein Eng. **11**(9), 739–747 (1998)
22. Yang, Y., Zhan, J., Zhao, H., et al.: A new size-independent score for pairwise protein structure alignment and its application to structure classification and nucleic-acid binding prediction. Protein Struct. Funct. Bioinf. **80**(8), 2080–2088 (2012)
23. Elofsson, A., Fischer, D., Rice, D.W., et al.: A study of combined structure/sequence profiles. Fold Des. **1**(6), 451–461 (1996)

Protein-Protein Binding Affinity Prediction Based on Wavelet Package Transform and Two-Layer Support Vector Machines

Min Zhu[1], Xiaolai Li[1], Bingyu Sun[1], Jinfu Nie[2], Shujie Wang[2], and Xueling Li[2(✉)]

[1] Hefei Institute of Intelligent Machines, Hefei Institutes of Physical Science, Chinese Academy of Sciences, 350 Shushanhu Road, Hefei 230031, Anhui, People's Republic of China
[2] Center for Medical Physics and Technology, Hefei Institutes of Physical Science, Chinese Academy of Sciences, 350 Shushanhu Road, Hefei 230031, Anhui, People's Republic of China
xlli@cmpt.cas.cn, xuelingli16@foxmail.com

Abstract. Precisely inferring the affinities of protein-protein interaction is essential for evaluating different methods of protein-protein docking and their outputs and also opens a door to inferring real status of cellular protein-protein complex. Accumulation of measured affinities of determined protein complex structures with high resolution facilitate the realization of this ambitious goal. Previous physical model based scoring functions failed to predict the affinities of diverse protein complexes. Therefore, accurate method for binding affinity prediction is still extremely challenging. Machine learning methods are promising to address this problem. However, current machine learning methods are not compatible to this task, which obstructs the effective application of these methods. We propose a Wavelet Package Transform (WPT) combined with two-layer support vector regression (TLSVR-WPT) model to implicitly capture binding contributions that are hard to model explicitly. Wavelet package transform greatly reduced the dimension of input features into machine learning model. The TLSVR circumvents both the descriptor compatibility problem and the need for problematic modeling assumptions. Input features for TLSVR first layer are eight features transformed by Wavelet Transform Package from scores of 2209 interacting atom pairs within each distance bin. The output of the first layer is combined by the next layer to infer the final affinities. A satisfactory result of R = 0.81 and SD = 1.40 was achieved when 2209 features were reduced to eight ones by 3-level Wavelet Package Transform. Results demonstrate that wavelet package transform greatly reduced the dimension of the input features into SVR without reducing the accuracy in predicting the protein binding affinity.

Keywords: Protein-protein binding affinity · Prediction · Wavelet package transform · Two-layer support vector machine

M. Zhu and X. Li—Co-first authors.

© Springer International Publishing AG 2017
D.-S. Huang et al. (Eds.): ICIC 2017, Part II, LNCS 10362, pp. 398–407, 2017.
DOI: 10.1007/978-3-319-63312-1_35

1   Introduction

1.1   Protein-Protein Binding Affinity and the Importance of Its Prediction

Proteins function through binding with their partners [1–3]. The binding of two proteins can be considered as a reversible and rapid process in an equilibrium governed by the law of mass action. The binding affinity is the strength of the interaction between two or more than two molecules that bind reversibly or interact. The interactions are measured at equilibrium. The binding affinity is usually measured by the dissociation constant, K_D, or the association constant, K_A, which is the ratio of association rate, k_a over the dissociation rate, k_d [4–6]. Since the affinity mediates the function and structure, revealing the energetic characteristics of cellular multi-molecular complex is critical to understand the protein function. Although the crystal protein complex structures are not necessary the real cellular structures, since the affinity of protein-protein interaction is usually measured in solution approximating physiological one, predicting protein-protein binding affinity will help to interrogate the real status of protein-protein interaction sub-networks, infer the systems biological behavior with perturbation [7] and design rational protein drugs, etc. Second, protein affinity prediction is also an important part of after protein-docking evaluation. Thus, developing new method to predict protein-protein interaction is very important.

1.2   Current Method of Protein-Protein Binding Affinity Prediction

Four scoring approaches have been developed to infer protein-protein binding affinity (PPBA) [8–10]. Briefly, they are physical-based force fields, empirical scoring functions, knowledge-based statistical potentials, and hybrid scoring functions that combine the components from two or more of the above scoring functions into one function. These energy scoring functions are successful in protein-protein docking evaluation and some are even successful in protein binding affinity prediction. However, one of the challenging problem in existing scoring functions for protein-protein docking is that they usually hold limited capacity to predict the binding affinity of a complex [8–10]. Kastritis and Bonvin assessed the performance of nine commonly used scoring algorithms with a protein-protein binding affinity benchmark consisting of binding constants (Kd's) for 81 complexes. They revealed a poor correlation (R < 0.3) between binding affinity and scores for all the algorithms tested and concluded that accurate prediction of binding affinity remains outside these methods reach. Therefore, improvement in protein-protein interaction affinity is still in great need.

One problem in present methods is that the data set of these computational methods are small, generally containing dozens of protein complexes and usually not generic and heterogeneous enough. If one or a few test data in these small sets are changed, the correlation might be significantly changed. The scoring functions include parameters fitted to experimental or simulation data and their predicted binding affinity, but lack the cross validation process. Actually, in previous work, we developed a distance-independent residue level potential of mean force to predict PPBA on a small data set of PPBA including 80 protein-protein complexes [10]. Performance of this method is comparable with its alternatives at atom level with or without volume correlation on the same small dataset. But we find that when the data set is composed of more generic, diverse and heterogeneous ones, this distance-independent residue level potential of mean force have poor accuracy in PPBA prediction. Therefore, performance on small data sets with high sequence homogenous might be poor. Second, methods such as quantum mechanics, molecular mechanics in physical force field and empirical scoring functions require much prior sophisticate protein physical knowledge and sometimes bear a high computational cost. Moreover, each scoring function assumes a predetermined theory-inspired functional form for the relationship between the variables that characterize the complex. This rigid approach leads to poor predictivity for those complexes that do not conform to the modeling assumptions [8–10].

1.3   Machine Learning Methods for Protein-Protein Binding Affinity Prediction

Machine leaning methods can achieve satisfactory results without assuming any predefined model when used for binding affinity prediction. However, the generalization of methods based on one single classifier is often limited. Furthermore, a great number of features and small data set lead to the curse of dimensionality. Thus, researchers seek for classifier ensembles [9, 11] or multiple instance learning [12] to improve prediction accuracy and generalization and at the same time overcome the high feature dimensional disaster. In fact, a machine learning method, random forest, as an ensemble is recently reported in diverse protein-ligand binding affinity prediction with a higher prediction accuracy [13]. However, due to the use of distance-independent features, the fact that the strength of protein interactions naturally depends on atomic separation was not taken into account. Also, the coarse atom types reduced the generalization ability and interpretability of features in terms of PPBA prediction. The random forest method cannot be directly applied to protein-protein binding affinity (PPBA) prediction. Developing new machine learning approaches and objective estimation of the prediction results is still challenging for generic PPBA prediction.

In our previous study, we used support vector regression models and rough set reduction model to predict TAP-peptide binding affinity and specificity [14, 15]. This method has a high interpretability. In another work, we build a two-layer support vector regression (TLSVR) model with greater generality and prediction accuracy by capturing the non-linear combination effects on affinities of interacting atom pairs within each bin and between bins [8, 9]. Secondly, we construct a new data set by considering that structure diversity will greatly affect the accuracy of PPBA prediction. Finally, we evaluate our method with leave-one out cross validation, which is regarded as a most objective evaluation method.

1.4   The Outline of this Paper

The rest of the article is outlined as follows. In Sect. 2, we describe the data set and methods. Specifically, we collect a generic and heterogeneous data set. We reduce the dimensionality of the representing vectors by Wavelet Package Transform (WPT). We then use TLSVR to account for the non-linear effects on affinities of diverse interacting atoms that cannot be captured by explicit rigid physical model. Finally, leave-one-out cross validation method is used to evaluate the performance of the proposed method and further compare the proposed method with existing state-of-art-methods, DFIRE and Su's method and RF-score. Lastly, in Sect. 3, the results and discussion were present, where results demonstrate that the proposed method is more effective and greatly improves the prediction accuracy.

2   Methods

2.1   Data Set

1056 heterogeneous protein complexes were obtained from PDBbind-CN, 2010 version, which include complexes with a single residue mutation or multiple residue mutations. [16, 17]. The dataset was then filtered with sequence similarity <50% by PDBculled [18] (http://dunbrack.fccc.edu/Guoli/PISCES_InputB.php) with complex entities criteria and other default parameters. We finally integrated 49 proteins from [19] that did not exist in the dataset to get a heterogeneous larger data set. For simplicity, only complexes with two chains are held except 1CHD with four chains (EFG/I). Thus, 180 protein-protein interaction complex were kept in our final data set [8, 9].

2.2   Input Features

Machine learning methods require equal length of vectors. Therefore, how to represent a protein complex structure in an equal-length vector as the input of TLSVR is the first prerequisite. 47 types of heavy atoms were defined as reported by Su, et al. [19]. The occurrence number of 2209 contact atom pairs within each 0.5 Å or 0.2 Å width bin were counted respectively by Eq. 1 at each protein complex interface. Our preliminary experiment shows that TLSVR achieves best prediction results when the cutoff distance of a pair of contact atoms equals 16 Å. Thus, 71 distance bins, i.e., with distance threshold above 1.8 Å were kept. The generated features consist of 2209 elements, i.e., scores of 2209 contact atom pairs obtained with Eq. 2 by multiplying the occurrence number of each contact atom pair with corresponding atom pair potential. The interacting atom pair's potentials were defined as the natural logarithm of the ratio of observed number of interacting atom pairs to those expected. The smoothed and smoothed potentials were respectively generated with Eq. 3 [19]. Thus, 2209 smoothed atom pair scores (features) were finally generated. Specifically, the features

$$f_{ij} = N_{obs}(i, j, r) \times P(i, j, r) \tag{1}$$

where $N_{obs}(i, j, r)$ is the observed number of interacting atom pairs i, j within the distance shell $r(r - \Delta r, r)$ in a given protein-protein binding structure. $\Delta r (\Delta r = 0.2$ Å or $\Delta r = 0.5$ Å) is the bin width for 2.5Å $\leq r_{cut} \leq 16$ Å between two interface chains or proteins at n^{th} distance bin of width 0.2 Å. Here $n = 1, \ldots, 71$ for $\Delta r = 0.2$. Where,

$$P(i,j,r) = -\log(\frac{N_{obs}(i,j,r)}{N_{exp}(i,j,r)})f_{cor}, (i = 1, 2, 3 \ldots 47, j = 1, 2, 3, ..47) \tag{2}$$

The bins $r(r - \Delta r, r)$ ranges from 1.8 Å to 16 Å at 0.2 Å intervals. The interfacial atom pair potential: $N_{exp}(i,j,r) = X_i \times X_j \times N_{total}(r)$ is the expected number of interacting atom pairs of i, j between two interface chains or proteins if there are no preferential interactions between them. X_i is the mole fraction of atom type i and is calculated as N_i/N, where N_i and N are the total number of atom type i and all atoms, respectively, while $N_{total}(r)$ is the total number of interacting atom pairs derived from the reference database. f_{cor} is the correction factor, derived from smoothing $N_{obs}(i,j,r)/N_{exp}(i,j,r)$ ranging from 1.8 Å to 16 Å, by a moving window of 3.5 Å width for bin of width 0.2 Å [19].

2.3 Feature Extraction

High dimensionality greatly increase the SVR computational cost and usually reduces the generality of SVR models. Wavelet Packet Transform(WPT) are a generalization of Discrete Wavelet Transform (DWT) [20]. In WPT signal decomposition, both the approximation and detail coefficients are further decomposed at each level. While in DWT, detail coefficients are transferred down directly, i.e. unchanged to the next level. WPT has a great potential to denoise, which make it especially suitable to the noisy protein complex structure. Many wavelets can be used to extract the feature vector. In this work the Daubechies wavelet function is used to analyze the signal by WPT. In fact, we obtained the minimum standard derivation between the measured affinities and predicted ones using db3 and db6 compared to other types of Daubechies wavelets. Therefore, we use the Daubechies wavelets package decomposition [21] with shannon entropy to reduce the dimensionality of the representing vector of protein complex interface. The three levels decomposition of WPT provides high resolution.

Specifically, each signal (factor) is decomposed into 8 subbands by WPT. The energy of each subband is calculated by (Eq. 3) where $k = 1, 2, \ldots, 8$, $s_k(i)$ are the k-th subband reconstruction coefficients of WPT and N_s is the number of coefficients. Feature extraction was implemented in MATLAB.

$$E_k(s) = \frac{1}{N_s} \sum_{i=1}^{N_s} |s_k(i)| \tag{3}$$

2.4 Two-Layer Support Vector Regression Model

Support Vector Machine (SVM) is introduced by Vapnik [22], They simultaneously minimize the empirical classification error and maximize the geometric margin (reduce structural risk); hence they are also known as maximum margin classifiers. SVM has been extended by a number of other researchers for classification or regression. While SVM classification outputs binary results (binding or nonbinding), Support Vector Regression (SVR) model produces continuous values (affinity absolute value).

Here, we use nonlinear ε-SVR. Suppose we are given training data $\{(x_1, y_1), \ldots, (x_l, y_l)\} \subset \aleph \times \Re$, where $\aleph$ denotes the space of the input patterns (e.g. $\aleph = \Re^d$). In ε-SV regression [22], the goal is to find a function $f(x)$ that has at most ε deviation from the actually obtained targets y_i for all the training data, and at the same time $f(x)$ is as flat as possible. The nonlinear functions $f(x)$, taking the form

$$w = \sum_{i=1}^{l} (\alpha_i - \alpha_i^*)\Phi(x_i) \text{ and } f(x) = \sum_{i=1}^{l} (\alpha_i - \alpha_i^*)k(x_i, x) + b \tag{4}$$

where $k(x_i, x)$ is very often defined as a Gaussian Radial Basis function.

$$K(x_i, x_j) = \exp\left\{-\frac{||x - x_i||^2}{\sigma^2}\right\} \tag{5}$$

Flatness in the case of Eq. 4 means that one seeks the flattest function in feature space, not in input space. w can be described as a linear combination of the training patterns x_i and is not given explicitly. We can write this problem as a convex optimization problem:

$$\begin{aligned} &W(\alpha^{(*)}) = -\frac{1}{2}\sum_{i=1}^{l}(\alpha_i^* - \alpha_i)(\alpha_j^* - \alpha_j)k(x_i, x_j) \\ \text{Maximize}\quad &\\ &-\varepsilon\sum_{i=1}^{l} y_i(\alpha_i^* + \alpha_i) + \sum_{i=1}^{l} y_i(\alpha_i^* - \alpha_i) \end{aligned} \tag{6}$$

$$\text{Subject to } \sum_{i=1}^{l}(\alpha_i - \alpha_i^*) = 0 \text{ and } \alpha_i, \alpha_i^* \in [0, C]$$

Protein-protein binding affinity was predicted by using two-layer SVR (TLSVR) combined with wavelet package transform (TLSVR-WPT). Thus, each input vector at first layer of TLSVR is 2209-dimensional. As depicted above, each real value of a vector represents a score of an atom pair in interface. Namely, input features for each individual SVR at the first layer of TLSVR are 2209 scores of interacting atom pairs within 71 bins, i.e. 1.8 Å:0.2 Å:16 Å of a protein complex. Here the contact atom pairs with distance below 1.8 Å of atom clashes were disregarded. Thus, 71 individual Wavelet Package Transform were performed to generate the input 8-dimensional features for the 71 SVR modes at first layer. As shown in Fig. 1, the predicted values from the individual SVR modes of the first layer were input into the second layer SVR, which is also called the combiner of the SVR ensemble. The output of the combiner was the final predicted affinity. This process was implemented in MATLAB code which is available upon request.

Here, parameters were default in individual SVR models. Finally, the best results were obtained by comparing the RBF, liner, polynomial and sigmoid kernel for SVM. All the computational experiments are carried out with LIBSVM that is available at http://www.csie.ntu.edu.tw/~cjlin/libsvm.

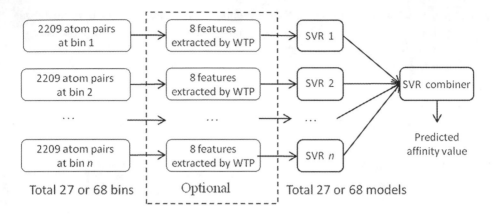

Fig. 1. Scheme of the proposed two-layer SVM prediction system

2.5   Evaluation of TLSVR-WPT Model by Leave-One-Out Validation

The performance of a computational model is often tested by the leave-one-out cross-validation method due to the small size of the data set. The leave-one-out method is the most extreme and accurate type of cross-validation test. In leave-one-out testing, the data set having n complexes is broken in n subsets, each having one example. The classifier was trained on $n - 1$ subset and evaluated on nth subset. The process was repeated n times using each subset as the testing set and the rest of complexes for training set. The results of test subsets are combined to get an overall estimate of training procedure. Finally, the correlation coefficient and standard derivation between the predicted values of all samples and the real ones are calculated.

2.6   Comparison with State-of-Art Methods

Most recently, machine learning method (RF-score) [13] was proposed in protein-ligand binding affinity prediction and achieved very good results. RF-score used distance independent atom pair occurrence counts as features. To implement the RF-score, we generated the 2209 contact atom pairs from 47 atom types. We defined contact atom pairs as those pairs with distance shorter than 12 Å. Other settings were similar to [13] and random forest was implemented in MATLAB.

Knowledge based methods are more compatible with machine learning methods and have more generic application. They have been proved to have equivalent to or better performance than other types of methods in PPBA predictions. Therefore, in this study we only compare our method with two well-known knowledge-based methods, Su's method [19] and DFIRE [23] for protein-protein binding affinity (PPBA) prediction. We generated scores of protein complexes with Su's method as reported in [19], where the scoring function to a protein-protein complex affinity were defined as the summation over all atom pair interactions of the protein-protein complex from all distance bins. Then we fitted the score to the transformed binding free energy. We then generated scores of protein complexes with DFIRE program that was kindly provided by [23], respectively. In the PMF based affinity prediction methods, the non-linear relationships of scores from different bins and the non-linear relationships of the interacting atom pairs within each bin were not taken into account.

3 Results and Discussion

3.1 Feature Extraction by Wavelet Package Transform

The features input into the first layer SVR of TLSVR were 2209 dimensional. To reduce computational cost and overcome the curse of high dimension, we used 3-level decomposition of WPT and extracted eight energy features from the 2209 reconstruction coefficients of the original features of smoothed scores. Table 1 shows that the reduced eight features had similar correlation coefficient (R) to the original 2209 features, and a just slightly lower standard derivation (SD), 1.40 versus 1.32s-1. To our best knowledge, it was the first time that WPT was used for feature reduction in protein affinity prediction problem. Moreover, RBF two-layer SVMs had better performance than linear ones. Polynomial and sigmoid two-layer SVMs failed to predict the PPBA, which is similar to the case in the original features (Table 1).

Table 1. Results of the wavelet package transform combined with radial base kernel TLSVM for affinity prediction

SVM models	RBF		Linear	
	R	SD	R	SD
Small wavelet_db1	0.7944	1.4408	0.6525	1.7819
Small wavelet_db2	0.8009	1.4154	0.5433	2.1575
Small wavelet_db3	0.8072	1.3975	0.7111	1.6402
Small wavelet_db4	0.7987	1.4304	0.638	1.8639
Small wavelet_db5	0.7932	1.4514	0.7094	1.651
Small wavelet_db6	0.8085	1.4011	0.6091	1.8854
Small wavelet_db7	0.8024	1.4214	0.6505	1.7545
Small wavelet_db8	0.801	1.4027	0.6687	1.7345
Without wavelet	0.800	1.3200	–	–

3.2 Comparison with Other State-of Art Methods

To our best knowledge, DFIRE and Su's methods are two best recently proposed PMF-based methods for PPBA prediction. RF-score is recently developed method for protein-ligand affinity prediction. We compare our method with these three representative methods on our collected data set. Comparison results are shown in Table 2. We can see that DFIRE obtained a correlation of 0.12 only. Su's method only obtained a correlation coefficient of 0.18 in predicting the generic and heterogeneous PPBA. TLSVR-WPT model obtained much a higher correlation coefficient of 0.81 between the predicted and experimental affinities than both methods.

Table 2. Comparison of TLSVM-WPT with two state of art methods, DFIRE and Su's method with Jackknife cross-validation on our data set.

On independent test set	R	SD
DFIRE on training set	0.12	2.21
Su's method on training set	0.18	2.18
RF-score	0.18	2.18
TLSVM-WPT	0.81	1.40

Furthermore, we compared DFIRE and Su's methods on benchmark data collected by [24] for protein-docking. TLSVM was implemented with optimized regulation factors C and kernel breadth γ parameters. Results in Table 2 show that TLSVM-WPT has the best prediction. Our method is still not satisfactory, which may be due to the improper collection of this benchmark data for quantitative affinity prediction purpose. We propose leave-one-out cross validation on our collected data may be more objective. Our data set is much more useful in that it includes diverse protein complexes and affinities.

Acknowledgements. This work was supported by the National Science Foundation of China, Nos. 31371340, 61273324 & 31271412, Anhui Provincial Natural Science Foundation Grant 1208085MF96, and the Knowledge Innovation Program of Chinese Academy of Sciences, No. 0823A16121.

References

1. Li, X., Zhao, Y., Tian, B., Jamaluddin, M., Mitra, A., Yang, J., Rowicka, M., Brasier, A.R., Kudlicki, A.: Modulation of gene expression regulated by the transcription factor NF-κB/RelA. J. Biol. Chem. **289**, 11927–11944 (2014)
2. Li, X., Zhu, M., Brasier, A.R., Kudlicki, A.S.: Inferring genome-wide functional modulatory network: a case study on NF-κB/RelA transcription factor. J. Comput. Biol. **22**, 300–312 (2015)
3. Yang, J., Zhao, Y., Kalita, M., Li, X., Jamaluddin, M., Tian, B., Edeh, C.B., Wiktorowicz, J.E., Kudlicki, A., Brasier, A.R.: Systematic determination of human cyclin dependent kinase (CDK)-9 interactome identifies novel functions in RNA splicing mediated by the DEAD box (DDX)-5/17 RNA helicases. Mol. Cell. Proteomics **14**, 2701–2721 (2015)
4. Li, X., Huang, M., Cao, H., Zhao, J., Yang, M.: Study of low molecular weight effectors on the binding between cell membrane receptor IGF-1R and its substrate protein IRS-1 by SPR biosensor. Sens. Actuators B Chem. **124**, 227–236 (2007)
5. Li, X., Fong, C.-C., Huang, M., Cao, H., Zhao, J., Yang, M.: Measurement of binding kinetics between PI3-K and phosphorylated IGF-1R using a surface plasmon resonance biosensor. Microchim. Acta **162**, 253–260 (2008)
6. Li, X.-L., Huang, M.-H., Cao, H.-M., Chen, Y., Zhao, J.-L., Yang, M.-S.: Phosphorylation of receptor-mimic tyrosine peptide on surface plasmon resonance sensor chip and its interaction with downstream proteins. Chin. J. Anal. Chem. **36**, 1327–1332 (2008)
7. Zhang, J.S., Maslov, S., Shakhnovich, E.I.: Constraints imposed by non-functional protein-protein interactions on gene expression and proteome size. Molecular Syst. Biol. **4**, 210 (2008)

8. Li, X., Zhu, M., Li, X., Wang, H.-Q., Wang, S.: Protein-Protein binding affinity prediction based on an SVR ensemble. In: Huang, D.-S., Jiang, C., Bevilacqua, V., Figueroa, J.C. (eds.) ICIC 2012. LNCS, vol. 7389, pp. 145–151. Springer, Heidelberg (2012). doi: 10.1007/978-3-642-31588-6_19

9. Li, X.-L., Zhu, M., Li, X.-L., Wang, H.-Q., Wang, S.: Protein-Protein interaction affinity prediction based on interface descriptors and machine learning. In: Huang, D.-S., Ma, J., Jo, K.-H., Gromiha, M.M. (eds.) ICIC 2012. LNCS, vol. 7390, pp. 205–212. Springer, Heidelberg (2012). doi:10.1007/978-3-642-31576-3_27

10. Li, X.-L., Hou, M.-L., Wang, S.-L.: A residual level potential of mean force based approach to predict protein-protein interaction affinity. In: Huang, D.-S., Zhao, Z., Bevilacqua, V., Figueroa, J.C. (eds.) ICIC 2010. LNCS, vol. 6215, pp. 680–686. Springer, Heidelberg (2010). doi:10.1007/978-3-642-14922-1_85

11. Wolpert, D.H.: Stacked Generalization. Neural Netw. 5, 241–259 (1992)

12. Teramoto, R., Kashima, H.: Prediction of protein-ligand binding affinities using multiple instance learning. J. Molecular Graphics Modelling 29, 492–497 (2010)

13. Ballester, P.J., Mitchell, J.B.O.: A machine learning approach to predicting protein-ligand binding affinity with applications to molecular docking. Bioinformatics 26, 1169–1175 (2010)

14. Li, X.L., Wang, S.L., Hou, M.L.: Specificity of transporter associated with antigen processing protein as revealed by feature selection method. Protein Pept. Lett. 17, 1129–1135 (2010)

15. Li, X.-L., Wang, S.-L.: A comparative study on feature selection in regression for predicting the affinity of TAP binding peptides. In: Huang, D.-S., Zhang, X., Reyes García, C.A., Zhang, L. (eds.) ICIC 2010. LNCS, vol. 6216, pp. 69–75. Springer, Heidelberg (2010). doi: 10.1007/978-3-642-14932-0_9

16. Wang, R.X., Fang, X.L., Lu, Y.P., Yang, C.Y., Wang, S.M.: The PDBbind database: methodologies and updates. J. Med. Chem. 48, 4111–4119 (2005)

17. Wang, R.X., Fang, X.L., Lu, Y.P., Wang, S.M.: The PDBbind database: collection of binding affinities for protein-ligand complexes with known three-dimensional structures. J. Med. Chem. 47, 2977–2980 (2004)

18. Wang, G., Dunbrack Jr., R.L.: PISCES: recent improvements to a PDB sequence culling server. Nucleic Acids Res. 33, W94–W98 (2005)

19. Su, Y., Zhou, A., Xia, X.F., Li, W., Sun, Z.R.: Quantitative prediction of protein-protein binding affinity with a potential of mean force considering volume correction. Protein Sci. 18, 2550–2558 (2009)

20. Zhang, S., Wang, S., Li, X.: Palmprint linear feature extraction and identification based on ridgelet transforms and rough sets. In: Huang, D.-S., Wunsch, D.C., Levine, D.S., Jo, K.-H. (eds.) ICIC 2008. LNCS, vol. 5227, pp. 1101–1108. Springer, Heidelberg (2008). doi: 10.1007/978-3-540-85984-0_132

21. Coifman, R.R., Wickerhauser, M.V.: Entropy-based algorithms for best basis selection. IEEE Trans. Inf. Theory 38, 713–718 (1992)

22. Vapnik, V.N.: Statistical Learning Theory. Springer, New York (1998)

23. Zhang, C., Liu, S., Zhu, Q., Zhou, Y.: A knowledge-based energy function for protein-ligand, protein-protein, and protein-DNA complexes. J. Med. Chem. 48, 2325–2335 (2005)

24. Kastritis, P.L., Bonvin, A.M.J.J.: Are scoring functions in protein-protein docking ready to predict interactomes? Clues from a novel binding affinity benchmark. J. Proteome Res. 9, 2216–2225 (2010)

Prediction of Lysine Pupylation Sites with Machine Learning Methods

Wenzheng Bao[✉] and Zhichao Jiang

Institute of Machine Learning and Systems Biology, Tongji University, Shanghai, China
baowz55555@126.com

Abstract. Post translational modification is a crucial type of protein post-translational modification, which is involved in many important cellular processes and serious diseases. In practice, identification of protein pupylated sites through traditional experiment methods is time-consuming and laborious. Computational methods are not suitable to identify a large number of acetylated sites quickly. Therefore, machine learning methods are still very valuable to accelerate lysine acetylated site finding. Post translational modification of protein is one of the most important biological processions in the field of proteomics and bioinformatics. In this work, the random forest algorithm is employed as the classification model and the PseAAC has been employed as the classification features. Considering the different feature types of PseAAC playing different role in the classification model, the random forest voting method has been proposed in this framework. The results demonstrate that such method will work well in such classification issue.

Keywords: Lysine pupylation · Random forest · Post-translational modification

1 Introduction

Protein post-translational modification (***PTM***) has emerged as a major contributor to variation, localization and control of proteins. In vivo, one of the most efficient biological mechanisms for expanding the genetic code and for regulating cellular physiology is protein post-translational modification [1–3]. Lysine residue in protein can be subjected to many types of PTMs, such as methylation, acetylation, biotinylation, ubiquitination, ubiquitin-like modifications, propionylation and butyrylation, and leading to different complexity of PTM networks. Recently, a new type of PTM, named lysine succinylation, was initially identified by mass spectrometry and protein sequence alignment. Further studies showed that the lysine succinylation responses to different physiological conditions and is evolutionarily conserved [4, 5].

In 2013, Park's team identified 2565 succinylation sites from 779 proteins and they revealed that lysine succinylation have potential impacts on enzymes involved in mitochondrial metabolism including amino acid degradation, tricarboxylic acid cycle (TCA) and fatty acid metabolism [6, 7]. In histones, lysine succinylation is also presented, suggesting that it possibly plays the key role in regulating chromatin structures and functions [8].

D.-S. Huang et al. (Eds.): ICIC 2017, Part II, LNCS 10362, pp. 408–417, 2017.
DOI: 10.1007/978-3-319-63312-1_36

Lysine acetylation is a dynamic and reversible post-translational modification (PTM) that is highly conserved in prokaryotes and eukaryotes [9–11]. Such critical protein function process neuralizes the positive charge on an amino acid and regulates DNA binding, protein-protein interaction, and protein stability [12].

Actually, some efforts have been made in this regard. Since the importance of the topic as well as the urgency of demanding more powerful high-throughput tools in this area, further efforts are definitely needed to enhance the prediction quality. The present study was devoted to developing a more powerful predictor by the pseudo amino acid composition or PseAAC via incorporating a vectorized sequence coupling model into the general form of pseudo amino acid composition (PseAAC) and ensemble random forest approach.

As shown in a series of recent publications in compliance with Chou's five-step rule [13], to establish a really useful sequence-based statistical predictor for a biological system, we should logically follow the four guidelines below and make them crystal clear:

(i) How to construct or select a valid benchmark dataset to train and test the predictor.
(ii) How to formulate the biological sequence samples with an effective mathematical expression that can truly reflect their intrinsic correlation with the target to be predicted.
(iii) How to introduce or develop a powerful algorithm (or engine) to operate the prediction.
(iv) How to properly perform cross-validation tests to objectively evaluate its anticipated accuracy.

In this paper, a novel classification model of two-classification framework has been proposed for lysine acetylation sites of protein. And then random forest is employed as the classifiers in our classification model. Several properties of amino acid residues have been treated as the classification features in such model. The interaction between amino acid side residues and predicted lysine acetylation sites, which have been treated as the center amino acid residues in the protein sequences, has been introduced in this work.

2   Methods and Materials

Several PTMs, such as, propionylation, butyrylation, malonylation, succinylation, and crotonylation, at lysine residues have been discovered in the past few years The biological functions of these novel PTMs are uncertain, and much work is being done to identify their roles in cellular regulation. This review introduces biological roles of lysine acylation in toxic response and discusses recent advances in this active topic in the field of toxicology [14–16].

2.1   Data Collection

The benchmark dataset used in this study was derived from the CPLM, a protein lysine modification database [17]. It contains 2521 lysine succinylation sites and 24,128 non-succinylation sites determined from 896 proteins [18]. All of the corresponding protein sequences were derived from the UniProt [19] database. For facilitating description later, Chou's peptide formulation was adopted. It was used for studying signal peptide cleavage sites [20], HIV protease cleavage sites [21], and protein e protein interaction [22].

2.2   Features

To determine whether pupylation and putative non-pupylation sites have distinct sequence properties, we calculated statistically significant differences in the distribution of amino acid residues in the vicinity of 174 (158 + 16) pupylation peptides and 2207 (1928 + 279) putative non-pupylation segments. The two sample logo of the position-specific residue composition in the vicinity of the pupylation sites and putative non-pupylation sites in a window with length 25 [23–25]. Polar amino acids (G, S, T, Y, C) show as green, acidic (Q, N) purple, basic (K, R, H) blue, positive (D, E) red, and hydrophobic (A, V, L, I, P, W, F, M) amino acids as black in the two sample logo [26, 27].

The two sample logo showed compositional differences between pupylation sites and putative non-pupylation sites. The most distinct feature of pupylation sites is the enrichment of positive amino acid (E) at positions 4 and 7, positive amino acid (D) at positions 5 and 4, hydrophobic amino acids (A, V, and L) at positions 12, 9, 6, 3 and 2, and positively charged amino acid (R) at positions 12, 11, 7 and 12. On the contrary, the depletion of lysine (K) at position 2, polar amino acids (G and S) at positions 7, 4 and 9, positive amino acid (D) at position 12 and hydrophobic amino acid (L) at position -12 are observed around pupylation sites. These statistics show that there are strong correlations between the residues around the pupylation sites, requiring proper encoding methods [28–30].

According to the definition of amino acid composition (AAC), AAC of a protein sequence can be represented by 20 discrete numbers with each denoting the occurrence frequency of one of the 20 native amino acids in the protein. But, if using the 20 dimensional AAC to represent a protein sequence, all its sequence order information would be lost. For instance, AAC cannot catch the strong residue correlations around the pupylation sites as shown in Fig. 1. In view of this, instead of the conventional AAC, we adopt PseAAC to represent a protein sample in a $(20 + \lambda)$ dimensional vector. The first 20 elements in PseAAC reflect the traditional global amino acid composition in the sequence and the later λ elements represent the local correlations among residues [31].

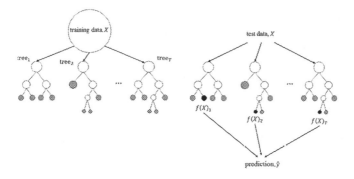

Fig. 1. Random forest model

Since the concept of PseAAC composition was introduced, various PseAAC composition approaches have been developed for enhancing the prediction quality of protein attributes, including protein subcellular localization [32], protein structural class [33], protein oligomer type [34], protein subnuclear localization, protein submitochondria localization, conotoxin superfamily classification, membrane protein type, apoptosis protein subcellular localization, enzyme functional classification, protein fold pattern and signal peptide.

Breiman proposed random forests, which add an additional layer of randomness to bagging. In addition to constructing each tree using a different bootstrap sample of the data, random forests change how the classification or regression trees are constructed. In standard trees, each node is split using the best split among all variables. In a random forest, each node is split using the best among a subset of predictors randomly chosen at that node. This somewhat counterintuitive strategy turns out to perform very well compared to many other classifiers, including discriminant analysis, support vector machines and neural networks, and is robust against overfitting. In addition, it is very user-friendly in the sense that it has only two parameters (the number of variables in the random subset at each node and the number of trees in the forest), and is usually not very sensitive to their values. The detailed procedures and formulation of RF have been very clearly described, and hence there is no need to repeat here [35].

Since most classifiers (including RF) are usually working properly for the benchmark datasets consisting of balanced subsets. For the current skewed dataset, we are to use the asymmetric bootstrap approach to deal with it [36]. The concrete procedures are as follows.

The random forest classifier consists of a combination of tree classifiers where each classifier is generated using a random vector sampled independently from the input vector, and each tree casts a unit vote for the most popular class to classify an input vector [37]. The random forest classifier used for this study consists of using randomly selected features or a combination of features at each node to grow a tree. Bagging, a method to generate a training dataset by randomly drawing with replacement N examples, where N is the size of the original training set, was used for each feature/feature combination selected. Any examples (pixels) are classified by taking the most popular voted class from all the tree predictors in the forest. Design of a decision tree required the choice of an attribute selection measure and a pruning method. There are many approaches to the selection of attributes used for decision tree induction and most approaches assign a quality measure directly to the attribute. The most frequently used attribute selection measures in decision tree induction are the Information Gain Ratio criterion and the Gini Index [38]. The random forest classifier uses the Gini Index as an attribute selection measure, which measures the impurity of an attribute with respect to the classes. For a given training set T, selecting one case (pixel) at random and saying that it belongs to some class Ci, the Gini index can be written as:

Each time a tree is grown to the maximum depth on new training data using a combination of features. These fully grown trees are not pruned. This is one of the major advantages of the random forest classifier over other decision tree methods like the one proposed by Quinlan. The studies suggest that the choice of the pruning methods, and not the attribute selection measures, affect the performance of tree based classifiers [39]. Breiman suggests that as the number of trees increases, the generalization error always converges even without pruning the tree and overfitting is not a problem because of the Strong Law of Large Numbers [40, 41]. The number of features used at each node to generate a tree and the number of trees to be grown are two user-defined parameters required to generate a random forest classifier [14, 42]. At each node, only selected features are searched for the best split. Thus, the random forest classifier consists of N trees, where N is the number of trees to be grown, which can be any value defined by the user [43]. To classify a new dataset, each case of the datasets is passed down to each of the N trees. The forest chooses a class having the most out of N votes, for that case [15].

2.3 Classification Model

In this research, we have introduced random forest as the basic classification model. Considering the specialty of this prediction issue, the lysine acetylation sites prediction seem to a classical binary classification problems. Therefore, the positive samples could easily get the label *1*. At the same time, the negative samples will get the label *0* in this classification framework. Not insignificant, however, is the 2 groups of features hard to get the desired results. So the above mentioned classifier with feature groups will treat as the weak classifiers. Employing an integrated strategy, 2 groups of features run different in two classification model, respectively. The proposed classification model shows in the Fig. 2.

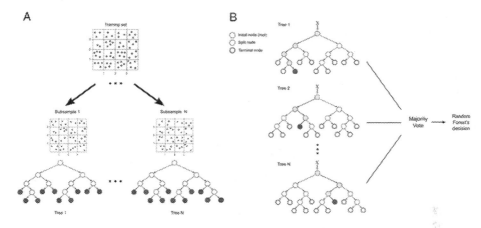

Fig. 2. Random forest voting model

2.4 Prediction Assessment

To test the model of classification, we have employed the Root Mean Square (**RMS**) [20], which has been used as evaluation function of each feature groups. The overall accuracy (**OA**) stands the computing for each dataset. Moreover, the following three measuring performances take advantage of evaluating the prediction accuracy, namely, Sensitivity (**Sens**), Specificity (**Spec**). The **TP** means the true sample has been treat as the positive label. The **TN** means the true sample has been treat as the negative label. The **FN** means the false sample has been treat as the negative label. And then the **FP** means the false sample has been treat as the positive label.

3   Results

The above mentioned features of lysine acetylation sites have been trained by the FNT model in the classification framework, which was proposed in this paper. At the same time, the performances and capabilities will be compared with other existing methods. So the seven features training fitness showed in the Fig. 3. In the Fig. 3, we can easily found out that the training fitness will have a good ability of convergence. Nevertheless, the kind of features can hardly convergent at the same race. The results of each features shows in the Table 1.

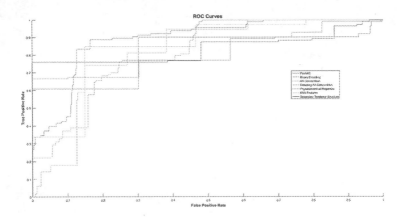

Fig. 3. Training fitness of features

Table 1. Comparison among the proposed & existing features

Features	Sn(%)	Sp(%)	Acc(%)
Binary encoding	43.21	75.53	59.42
AA composition	64.54	52.26	58.13
Grouping AA composition	41.11	76.05	58.08
Physicochemical properties	55.32	63.75	59.64
KNN features	64.64	55.21	60.43
Secondary tendency structure	59.96	57.40	58.68
PseAAC	67.28	75.27	71.28

4   Discussions and Conclusions

From the above research, the proposed method is an attempt to predict the potential pupylation sites with the basic sequence information. With the development of biological experiments, the number of experimentally determined pupylation sites will be growing in the future. Such determined modification data will enrich the current database both in positive and negative samples. Therefore, the current performance of ensemble is helpful to predict some unlabeled pupylation sites from the novel discovered protein. On the one hand, it is necessary to find out some features about the statistical information of the potential modification segments. On the other hand, it is the high time to develop and design some effective algorithms to decrease the rate of the false positive in this classification issue. With these efforts, the precisely identification pupylation modification sites may ultimately lead to development of better drugs and design of better proteins.

Acknowledgments. This work was supported by the grants of the National Science Foundation of China, Nos. U1611265, 61532008, 61672203, 61402334, 61472282, 61520106006, 31571364, 61472280, 61472173, 61572447, 61373098 and 61672382, China Postdoctoral Science Foundation Grant, Nos. 2016M601646.

References

1. Armengaud, J.: Proteogenomics and systems biology: quest for the ultimate missing parts. Expert Rev. Proteomics (2014)
2. Filippakopoulos, P., Knapp, S.: Targeting bromodomains: epigenetic readers of lysine acetylation. Nature Reviews Drug Discovery **13**(5), 337–356 (2014)
3. Scholz, C., Weinert, B., Wagner, S.: Acetylation site specificities of lysine deacetylase inhibitors in human cells. Nat. Biotechnol. **33**(4), 415–423 (2015)
4. Aram, R.Z., Charkari, N.M.: A two-layer classification framework for protein fold recognition. J. Theoret. Biol. **365**, 32–39 (2015)
5. Kouranov, A., et al.: The RCSB PDB information portal for structural genomics. Nucleic Acids Res. **34**(Suppl 1), D302–D305 (2006)
6. Yang, X., Seto, E.: Lysine acetylation: codified crosstalk with other posttranslational modifications. Mol. Cell. **31**(4), 449–461 (2008)
7. Zhao, D., Zou, S., Liu, Y.: Lysine-5 acetylation negatively regulates lactate dehydrogenase a and is decreased in pancreatic cancer. Cancer Cell **23**(4), 464–476 (2013)
8. Wu, X., Oh, M., Schwarz, E.: Lysine acetylation is a widespread protein modification for diverse proteins in arabidopsis. Plant Physiol. **155**(4), 1769–1778 (2011)
9. Sadoul, K., Wang, J., Diagouraga, B.: The tale of protein lysine acetylation in the cytoplasm. BioMed Res. Int. (2010)
10. Hou, T., Zheng, G., Zhang, P.: LAceP: lysine acetylation site prediction using logistic regression classifiers. PLoS ONE **9**(2), e89575–e89575 (2014)
11. Weinert, B., Iesmantavicius, V., Wagner, S.: Acetyl-phosphate is a critical determinant of lysine acetylation in E. coli. Mol. Cell **51**(2), 265–272 (2013)
12. Sol, E., Wagner, S., Weinert, B.: Proteomic investigations of lysine acetylation identify diverse substrates of mitochondrial deacetylase sirt3. PLOS ONE **7**(12), e50545 (2012)
13. Shan, C., Elf, S., Ji, Q.: Lysine acetylation activates 6-phosphogluconate dehydrogenase to promote tumor growth. Mol. Cell **55**(4), 552–565 (2014)
14. Li, Z.R., Lin, H.H., Han, L.Y., Jiang, L., Chen, X., Chen, Y.Z.: PROFEAT: A web server for computing structural and physicochemical features of proteins and peptides from amino acid sequence. Nucleic Acids Res. **34**, W32–W37 (2006)
15. Rao, H.B., Zhu, F., Yang, G.B., Li, Z.R., Chen, Y.Z.: Update of PROFEAT: A web server for computing structural and physicochemical features of proteins and peptides from amino acid sequence. Nucleic Acids Res. **39**, W385–W390 (2011)
16. Bao, W., Chen, Y., Wang, D.: Prediction of protein structure classes with flexible neural tree. Bio-Med. Mater. Eng. **24**, 3797–3806 (2014)
17. Chatterjee, P., Basu, S., Nasipuri, M.: Improving prediction of protein secondary structure using physicochemical properties of amino acids. In: Proceedings of the 2010 International Symposium on Biocomputing (ISB 2010). ACM, New York (2010)
18. Mohri, M., Rostamizadeh, A., Talwalkar, A.: Foundations of Machine Learning. MIT Press, Cambridge (2012)
19. Yang, B., Chen, Y.H., Jiang, M.Y.: Reverse engineering of gene regulatory networks using flexible neural tree models. Neurocomputing **99**, 458–466 (2013)

20. Deng, S.-P., Zhu, L., Huang, D.S.: Mining the bladder cancer-associated genes by an integrated strategy for the construction and analysis of differential co-expression networks. BMC Genomics 16 (Suppl 3), S4 2015

21. Dou, Y., Yao, B., Zhang, C.: PhosphoSVM: prediction of phosphorylation sites by integrating various protein sequence attributes with a support vector machine. Amino Acids **46**(6), 1459–1469 (2014)

22. Minguez, P., Letunic, I., Parca, L., et al.: PTMcode: a database of known and predicted functional associations between post-translational modifications in proteins. Nucleic Acids Res. **41**(D1), D306–D311 (2013)

23. Wang, B., Huang, D.S., Jiang, C.: A new strategy for protein interface identification using manifold learning method. IEEE Trans. Nanobiosci. **13**(2), 118–123 (2014)

24. Xiong, Y., Peng, X., Cheng, Z., et al.: A comprehensive catalog of the lysine-acetylation targets in rice (Oryza sativa) based on proteomic analyses. J. Proteomics **138**, 20–29 (2016)

25. Jia, C., Lin, X., Wang, Z.: Prediction of protein S-nitrosylation sites based on adapted normal distribution bi-profile Bayes and Chou's pseudo amino acid composition. Int. J. Mol. Sci. **15**(6), 10410–10423 (2014)

26. Pougovkina, O., te Brinke, H., Ofman, R., et al.: Mitochondrial protein acetylation is driven by acetyl-CoA from fatty acid oxidation. Hum. Mol. Geneti. **23**(13), 3513–3522 (2014)

27. Zhang, T.L., Ding, Y.S., Chou, K.C.: Prediction protein structural classes with pseudo amino acid composition: approximate entropy and hydrophobicity pattern. J. Theor. Biol. **250**, 186–193 (2008)

28. Yu, H.-J., Huang, D.S.: Graphical representation for DNA sequences via joint diagonalization of matrix pencil. IEEE J. Biomed. Health Inform. **17**(3), 503–511 (2013)

29. Berezovsky, I.N., Kilosanidze, G.T., Tumanyan, V.G., et al.: Amino acid composition of protein termini are biased in different manners. Protein Eng. **12**(1), 23–30 (1999)

30. Andreeva, A., Howorth, D., Chandonia, J.M., Brenner, S.E., Hubbard, T.J.P., Chothia, C., Murzin, A.G.: Data growth and its impact on the SCOP database: new development (2007)

31. Huang, D.S., Zhang, L., Han, K., Deng, S., Yang, K., Zhang, H.: Prediction of protein-protein interactions based on protein-protein correlation using least squares regression. Curr. Protein Pept. Sci. **15**(6), 553–560 (2014)

32. Huang, D.S., Yu, H.-J.: Normalized feature vectors: A novel alignment-free sequence comparison meth-od based on the numbers of adjacent amino acids. IEEE/ACM Trans. Comput. Biol. Bioinform. **10**(2), 457–467 (2013)

33. Ding, C.H.Q., Dubchak, I.: Multi-class protein fold recognition using support vector machines and neural networks. Bioinformatics **17**(4), 349–358 (2001)

34. Chen, K., Kurgan, L.A., Ruan, J.S.: Prediction of protein structural class using novel evolutionary collocation-based sequence representation. J. Comput. Chem. **29**, 1596–1604 (2008)

35. Jones, D.T.: Protein secondary structure prediction based on position-specific scoring matrices. J. Mol. Biol. **292**, 195–202 (1999)

36. Altschul, S.F., Madden, T.L., Schaffer, A.A., Zhang, J., Zhang, Z., Miller, W., Lipman, D.J.: Gapped BLAST and PSI-BLAST: a new generation of protein database search programs. Nucleic Acids Res. **25**, 3389–3402 (1997)

37. Kurgan, L.A., Zhang, T., Zhang, H., Shen, S., Ruan, J.: Secondary structure-based assignment of the protein structural classes. Amino Acids **35**, 551–564 (2008)

38. Kurgan, L., Cios, K., Chen, K.: SCPRED: accurate pre-diction of protein structural class for sequences of twi-light-zone similarity with predicting sequences. BMC Bioinform. **9**, 226 (2008)

39. Liu, T., Jia, C.: A high-accuracy protein structural class prediction algorithm using predicted secondary structural information. J. Theor. Biol. **267**, 272–275 (2010)
40. Lempel, Z.: On the complexity of finite sequences. IEEE Trans. Inf. Theory **22**, 75–81 (1976)
41. Ding, S., Zhang, S., Li, Y., Wang, T.: A novel protein structural classes prediction method based on predicted secondary structure. Biochimie **94**, 1166–1171 (2012)
42. Zheng, C.-H., Huang, D.S., Zhang, L., Kong, X.-Z.: Tumor clustering using non-negative matrix factorization with gene selection. IEEE Trans. Inf. Technol. Biomed. **13**(4), 599–607 (2009)
43. Zhu, L., You, Z.-H., Huang, D.S., Wang, B.: t-LSE: A novel robust geometric approach for modeling protein-protein interaction networks. PLOS ONE 8(4), e58368 (2013). doi:10.1371/journal.pone.0058368,2013

Next-Gen Sequencing and Metagenomics

An Integrative Approach for the Functional Analysis of Metagenomic Studies

Jyotsna Talreja Wassan[1], Haiying Wang[1], Fiona Browne[1], Paul Wash[2],
Brain Kelly[2], Cintia Palu[2,4], Nina Konstantinidou[2], Rainer Roehe[3],
Richard Dewhurst[3], and Huiru Zheng[1(✉)]

[1] School of Computing and Mathematics, Ulster University, Co. Antrim, Northern Ireland, UK
h.zheng@ulster.ac.uk
[2] NSilico Life Science Ltd., Dublin, Ireland
[3] Future Farming Systems, Scotland's Rural College, Edinburgh, Scotland, UK
[4] University College Cork, Cork, Ireland

Abstract. Metagenomics is one of the most prolific "omic" sciences in the context of biological research on environmental microbial communities. The studies related to metagenomics generate high-dimensional, sparse, complex, and biologically rich datasets. In this research, we propose a framework which integrates omics-knowledge to identify suitable-reduced set of microbiome features for gaining insights into functional classification of metagenomic sequences. The proposed approach has been applied to two Use Case studies on: - (1) cattle rumen microbiota samples, differentiating nitrate and vegetable oil treated feed for improving cattle performance and (2) human gut microbiota and classifying them in functionally annotated categories of leanness, obesity, or overweight. A high *accuracy* of 97.5% and *Area Under Curve performance value (AUC)* of 0.972 was achieved for classifying Bos taurus, cattle rumen microbiota using *Logistic Regression (LR)* as classification model as well as feature selector in wrapper based strategy for Use Case 1 and 94.4% *accuracy* with *AUC* of 1.000, for Use Case 2 on human gut microbiota. In general, *LR classifier with wrapper - LR learner* as feature selector, proved to be most robust in our analysis.

Keywords: Metagenomics · OTUs (Operational Taxonomic Units) · Phylogeny · Machine Learning · Classification

1 Introduction

Metagenomics involves the study of gene sequences of microorganisms derived directly from their natural environment such as air, water, human or animal body, and soil etc., following a culture-independent approach [1]. In the past few years, this field has gained prominence due to important projects such as the Human Microbiome Project (http://hmpdacc.org/), Earth Microbiome Project (http://www.earthmicrobiome.org/), and American Gut Project (http://americangut.org/) and due to unprecedented advances in low cost DNA isolation and sequencing strategies such as high throughput Next Generation Sequencing (NGS) over the traditional Sanger approach [2, 3]. Several studies

© Springer International Publishing AG 2017
D.-S. Huang et al. (Eds.): ICIC 2017, Part II, LNCS 10362, pp. 421–427, 2017.
DOI: 10.1007/978-3-319-63312-1_37

have shown relations between microbial composition and host phenotypes. For example, human microbiome is related to various diseases such as diabetes (Type 1 and Type 2), Inflammatory bowel disease (IBD Crohn's Disease or Healthy), Obesity (obese, lean, overweight), and cancer etc. [2, 4]. Belanche et al. [5] studied the impact of supplementing grass hay with Vitamin A on rumen microbiome and its function. Roehe et al. [6] found that host genetics is shaping the rumen microbiome influencing methane production and feed conversion efficiency in cattle.

Metagenomic studies follow a typical metagenomic pipeline consisting of various stages including gene sampling, sequencing, assembly, binning, taxonomic assignment, functional data analysis, and data sharing [7]. The binning of sequences generates Operational Taxonomic Units (OTUs)/taxas. OTU abundance count, relations between OTUs (phylogeny) and sample microbe-microbe interactions contribute effectively in analyzing metagenomic functional roles. Current computational challenges along the metagenomic pipeline concern data management, processing, and analysis of metagenomic datasets. These are due to key characteristics of metagenomic data being massive, high dimensional, sparse, heterogeneous, incomplete, highly-skewed, and noisy [8, 9]. Emergence of NGS has resulted in a gap between the pace of data generation and its analysis [3]. The variance in OTU abundance count also does not follow a normal distribution, and pose statistical challenges [9]. Considering these challenges, we propose an integrative approach, combining omics and data analytics to identify functional roles of metagenomic datasets. This is achieved by identifying a subset of OTU features which offer optimal predictive modelling built upon various Machine Learning (ML) classification algorithms. Selecting a subset of relevant OTU features for ML models is expected to entail improvement in performance.

2   Materials

The study involves analysis over two Use Case datasets; (i) B. *taurus* (cow) rumen microbiome dataset and (ii) human distal gut microbiome dataset. The B. *taurus* microbiome plays an important role in cattle productivity, health, and immunity. To investigate B. *taurus* gut microbiota in the context of these environmental traits, it's community composition was determined in 40 case samples provided by the MetaPlat project[1]. The data consist of 20 samples from an oil based treatment and 20 samples from a nitrate based treatment to reduce methane emissions. Five OTU tables, with different taxonomic levels (Phylum to Genus) of classification, were generated in QIIME by NSilico (http://www.nsilico.com/). The tables consist of 27, 52, 101, 194 and 386 OTU feature vectors for Phylum, Class, Order, Family, and Genus levels respectively. The dataset under consideration for second Use Case was obtained from a study on the human gut microbiome in obese and lean twins conducted by Turn Baugh et al. [4]. To address the factors related to obesity and genotypes, the study considered the microbiome of twin pairs and

---

[1]   MetaPlat (http://www.metaplat.eu) is a 4-year project funded by European Horizon H2020-MSCA-RISE-2015.

their mothers using 16s rRNA genes. The dataset consists of 18 microbiome samples, 756 OTUs at Species level and 3 classes (lean, obese, overweight) for analysis.

3   Methodology

This research was performed using NGS-16S genomic datasets listed in Sect. 2. OTU tables (Biological Observation Matrix), consisting of raw abundance count of taxas, were obtained using the QIIME (http://qiime.org/) or CloVR-metagenomics pipelines (http://data.clovr.org/d/10/obese-lean-twin-gut-metagenome-output) [10]. The samples also associated meta-data describing their relationship with environmental traits. The OTU tables were pre-processed and transposed to fit to ML models. To maximize the performance of our experimental design we followed an integrated workflow (as depicted in Fig. 1), focusing on two major steps: (1) selecting a suitable feature selection method and (2) selecting an appropriate learning classification functional model over selected features in Step (1), by evaluating its performance.

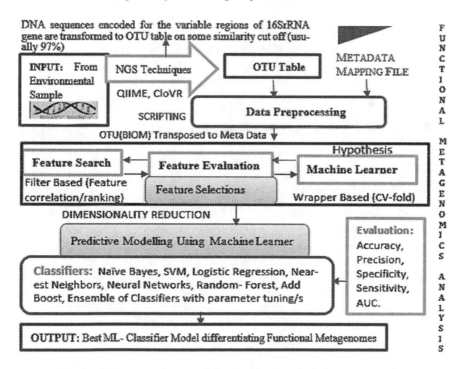

Fig. 1. An integrated approach for functional analysis in metagenomics

3.1   Feature Selection

Feature selection methods remove irrelevant and redundant features. The process primarily consists of two main steps: - (i) feature subset search and (ii) feature subset evaluation. We employed Best First Search (BFS) and Ranker's Method (RM) as feature search strategies [11] to the OTUs at various taxonomical levels. BFS is based on back-tracking the search path for finding OTU subsets till prominent results are attained whereas in RM, OTU features are ranked by their individual evaluations over selection metrics like associated weights, entropy, etc. on a user defined threshold. The features exceeding a threshold defined by user are selected for further analysis. The default value of threshold is set to -1.79769 in our analysis.

The next differentiating factor after subset search is to evaluate the attained subsets for application of ML models. The evaluations are typically inspired from two categories: - (i) filter based techniques (FFS), in which a function evaluates the worth of features by heuristics over general characteristics of OTU data and (ii) wrapper based techniques (WFS), which evaluate the worth by using an embedded ML algorithm over OTUs [11]. The various filter techniques used for evaluating OTU data are:- *Correlation-based feature selection (CFS)*, which select OTU features that are highly correlated with the class but uncorrelated with each other; *Info-gain based feature selection (IFS)* which is driven from probabilistic modelling of nominal valued feature subsets; *Principal Component Analysis (PCA)* that transforms existing features in the subset to new features in lower dimensional feature space; and *Relief based Evaluation (RB)*, which evaluates the worth of OTU features by instance based learning [11]. Wrappers evaluate attribute sets by estimating their accuracy using a learning scheme [11]. The related user defined parameters include the classifier model, the number of associated folds (set to 5), random seed (set to 1) and an associated threshold value (set to 0.01) in our current study. The learning scheme supports a variety of algorithms from Naïve *Bayes, Support Vector Machines, K-Nearest Neighbor, Logistic Regression to Random Forests, Boosting*, etc. The most discriminative OTU feature selection has potential to reduce the complexity and to increase the performance of a ML model.

3.2   Learning Functional Models

A machine learning model works over the knowledge induced from the sample OTU data subsets attained in Sect. 3.1. Applying ML models for categorizing the OTU features, into one of a pre-specified set of functional categories, is the key characteristic step of functional metagenomics. To identify the most suitable model for predicting functional metagenomes, various supervised ML classification algorithms were evaluated for their fitness in the prediction task against the selected OTU feature set. The range of classifiers applied were:- *Naïve Bayes (NB)* with kernel estimator as false; *Neural Network (NN)*, with hidden layers as 01/02/ no = (features + classes)/2, random seed as 1–10, validation threshold as 20, model learning rate as 0.3, momentum as 0.2; *Random Forest (RF)* with maximum tree depth as 0–6; *Support Vector Machine (SVM)* with Poly-Kernel/RBF Kernel and c, seed parameters varying between 0–10; *Logistic Regression (LR)* with ridge estimation; *k-Nearest-Neighbor's classification (K-NN)* with linear nearest neighbor search and number of neighbors as 1–5; *Adaptive Boosting (AdaB)* over decision tree classifier and an *ensemble of classifiers (Zero-R, NN, K-NN, LWL)* [12, 13]. To assess the performance of each prediction model, a 10-folds cross validation procedure was carried out in which OTU data was randomly split into k = 10, mutually exclusive subsets of equal size for overall assessment of classification. The following performance assessment metrics were used for evaluating classification models in our study: - *accuracy (Ac.), precision (Pr.), sensitivity (Se.), and specificity (Sp.), Area under Curve (AUC-ROC)* [13, 14].

4   Experiments and Results

Predictive modelling over the Use Cases supports holistic understanding of input data behavior, and an objective of this study is to identify feature selection methods and classifiers, which are robust and efficient for analysis. The results presented in this section (Fig. 2(a, b)), were obtained after experimenting with the classifiers listed in Sect. 3 and tuning their learning parameters to yield optimum output. The optimal parameters were adjusted by tuning values of batch size, estimator, optimization algorithm, search algorithm, number of iterations, random seed, complexity parameters and weight threshold. The experiments were performed in WEKA 3.8 [11, 13]. Firstly, we applied 8 classification algorithms (*NB, NN, SVM, RF, LR, K-NN, AdaB, Ensemble*) as predictive models, without any feature selection, on both Use Case data sets, for determining the functional classes. The accuracy of functional classification covered range from 25% to 77% over our Use Cases. The four dominant classifiers providing overall good accuracy were: *SVM, LR, NN and RF*. The accuracy of 77.5% achieved by *SVM* at Phylum level of Use Case 1 and accuracy of 50% by *SVM* at Species level of Use Case 2; proved to be best prediction result without feature selection. These results proved useful for further comparative analysis. We thereafter, applied both filter based *(CFS, IFS, PCA, RB)* and wrapper based *(LR, SVM, NN, RF)* feature selections, using BFS and RM search methods. Overall the combination of wrapper based method with *LR* for feature selection and the classification with *LR* model {with parameter settings: - batch size as 100, with ridge estimator for log likelihood as 1.0E-8}, provided highest accuracy in predicting functional classes for metagenomic studies in hand. The highest accuracy of 97.5% was attained for MetaPlat rumen data and accuracy of 94.4% for human microbiota, with the above said combination (Fig. 2(a, b)). The proposed combination, achieved the test average *AUC* of 0.972 with only 12 OTUs, in comparison to *LR*, which has *AUC* of 0.577 with all OTUs (386) in MetaPlat cattle rumen data at genus level of study. Additionally, on human microbiota data, it achieved average *AUC* of 1.000 with only 4 OTU features, serving much better than *LR* model having *AUC* 0.530 over all 756 OTU species. The application of *LR* model substantially depicted higher predictive accuracy in comparison to other state of art conventional ML approaches over feature selections. The OTU abundance count data usually have high variance and is not normally distributed. *LR* proved to be more robust in classification of metagenomic use cases, as it assumes that, the independent OTU features need not to be normally distributed, or have equal variance in each class or functional group.

The findings indicate that feature subset selection provides a drive for comparative very good classification accuracy. *CFS* and wrapper with *LR and RF*, proved to be most effective feature selection methods over metagenomic Use Cases. The bacterial sequences were majorly dominated by *Bacteroidetes, Firmicutes, Actinobacteria, Chloroflexi and Proteobacteria* in all ruminants. The significant biological OTU genera

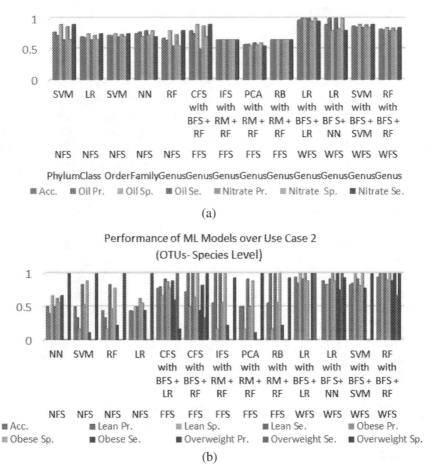

Fig. 2. Performance of predominant classifiers and feature selection methods over a) Use Case 1 and b) Use Case 2 (*Here, NFS: No Feature Selection, FFS: Filter Feature Selection and WFS: Wrapper Feature Selection*)

features selected by these methods in Use Case 1 are *Trichococcus, Tepidimicrobium, Brevibacterium, Methanosphaera, Butyrivibrio, Erwinia and Salana. Bifidobacterium-dentium, Campylobacterconcisus, Helicobacterhepaticus, Mycobacteriummarinum and Borreliaburgdorferi* species proved to be significant in analysis over Use Case 2.

5 Conclusion and Future Work

In this paper, we have presented an integrative approach for characterizing OTU features that are useful in identifying functional roles in metagenomic studies. The results show that feature selection plays an important role in metagenomic analysis. We proposed that the combination of *LR* and wrapper based on *LR* learner with ridge estimation, potentially gives higher validation accuracy for identifying functional roles from human and cattle microbiomes. However, it may be computationally intensive for very large data sets. In this study, we considered features as independent OTUs in classification of metagenomic sequences. In future, we propose to apply new optimization methods with phylogeny-driven classification (i.e. integrating association of OTUs into our framework), for gaining computational efficiency.

Acknowledgement. This work was supported in part by Research Strategy Fund of Ulster University and the MetaPlat project, (http://www.metaplat.eu) funded by H2020-MSCA-RISE-2015.

References

1. Hugenholtz, P., Tyson, G.W.: Metagenomics. Nature **455**, 481–483 (2008)
2. McDonald, D., Amanda, B., Knight, R.: Context and the human microbiome. Microbiome **3**(1), 52 (2015)
3. Schuster, S.C.: Next-generation sequencing transforms today's biology. Nat. Methods **5**, 16–18 (2008)
4. Turnbaugh, P.J., et al.: A core gut microbiome in obese and lean twins. Nature **457**, 480–484 (2009)
5. Belanche, A., et al.: An integrated multi-omics approach reveals the effects of supplementing grass or grass hay with vitamin E on the rumen microbiome and its function. Front. Microbiol. **7**, 1–17 (2016)
6. Roehe, R., et al.: Bovine host genetic variation influences rumen microbial methane production with best selection criterion for low methane emitting and efficiently feed converting hosts based on metagenomic gene abundance. PLoS Genet. **12**(2), e1005846 (2016)
7. Thomas, T., Gilbert, J., Meyer, F.: Metagenomics - a guide from sampling to data analysis. Microb. Inform. Exp. **2**, 3 (2012)
8. Prakash, T., Taylor, D.: Functional assignment of metagenomic data: challenges and applications. Brief. Bioinform. **13**(6), 711–727 (2011)
9. Jonsson, V., Tobias, O., et al.: Statistical evaluation of methods for identification of differentially abundant genes in comparative metagenomics. BMC Genom. **17**(1), 78 (2016)
10. Gonzalez, A., Knight, R.: Advancing analytical algorithms and pipelines for billions of microbial sequences. Curr. Opin. Biotechnol. **23**(1), 64–71 (2012)
11. Mark, H.: Correlation-based feature selection for machine learning. Methodology (1999)
12. Kotsiantis, S.B., Zaharakis, I.D., Pintelas, P.E.: Machine learning: a review of classification and combining techniques. Artif. Intell. Rev. **26**, 159–190 (2006)
13. Mark, H., Frank, E., Holmes, G., Pfahringer, B., Reutemann, P., Witten, I.H.: The WEKA data mining software: an update. ACM SIGKDD Explor. Newsl. **11**(1), 10–18 (2009)
14. Sokolova, M., Lapalme, G.: A systematic analysis of performance measures for classification tasks. Inf. Process. Manag. **45**, 427–437 (2009)

LSLS: A Novel Scaffolding Method Based on Path Extension

Min Li[1(✉)], Li Tang[1], Zhongxiang Liao[1], Junwei Luo[1], Fangxiang Wu[1,2], Yi Pan[1,3], and Jianxin Wang[1(✉)]

[1] School of Information Science and Engineering, Central South University, Changsha, China
{limin,jxwang}@csu.edu.cn
[2] Division of Biomedical Engineering, Department of Mechanical Engineering, University of Saskatchewan, Saskatoon, SK S7N 5A9, Canada
faw341@mail.usask.ca
[3] Department of Computer Science, Georgia State University, Atlanta, GA 30303, USA
pan@cs.gsu.edu

Abstract. While aiming to determine orientations and orders of fragmented contigs, scaffolding is an essential step of assembly pipelines and can make assembly results more complete. Most existing scaffolding tools adopt the scaffold graph approach. However, constructing an accurate scaffold graph is still a challenge task. Removing potential false relationships is a key to achieve a better scaffolding performance, while most scaffolding approaches neglect the impacts of uneven sequencing depth that may cause more sequencing errors, and finally result in many false relationships. In this paper, we present a new scaffolding method LSLS (Loose-Strict-Loose Scaffolding), which is based on path extension. LSLS uses different strategies to extend paths, which can be more adaptive to different sequencing depths. For the problem of multiple paths, we designed a score function, which is based on the distribution of read pairs, to evaluate the reliability of path candidates and extend them with the paths which have the highest score. Besides, LSLS contains a new gap estimation method, which can estimate gap sizes more precisely. The experiment results on the two standard datasets show that LSLS can get better performance.

Keywords: Genome assembly · Scaffolding · Extension of path · Mate-pair · Greedy algorithm

1 Introduction

The next generation sequencing (NGS) technologies facilitate the study of genome research [1]. While de novo assembly is one of the fundamental issues in this field on account of many downstream genome analyses that requires the sequences of organisms [2]. Generally speaking, the process of assemblies mainly consists of two parts, called contig extension and scaffolding. Scaffolding orders contigs, joins adjacent contigs, estimates gap sizes, and fills gaps with 'N' according to the estimated gap sizes. It is an effective way to improve the final performance of assemblies. In order to achieve better

© Springer International Publishing AG 2017
D.-S. Huang et al. (Eds.): ICIC 2017, Part II, LNCS 10362, pp. 428–438, 2017.
DOI: 10.1007/978-3-319-63312-1_38

assembly performance, many methods are proposed. There are mainly two kinds of approaches to scaffolding, the graph-based methods and the greedy-based methods.

The popular graph based methods include GRASS [3], MIP [4], SOPRA [5], Bambus2 [6], SCARPA [7], Opera [8], SGA [9], ABySS [10], ScaffMatch [11], BOSS [12], etc. GRASS and MIP transform the scaffolding problem into a Mixed Integer Programming problem. The difference between GRASS and MIP is that GRASS generates scaffolds in an iterative way while MIP solves the problem by graph partition. SOPRA uses the statistical optimization theory, removing the contradictory vertexes and edges iteratively, until there is no vertex or edge to remove. Bambus 2 removes the contigs which have repeating regions, and reduces the error connection caused by repeated regions, but also makes it hard to excavate many corrects connection at the same time. SCARPA increases the step which can recognize and cut off the wrong contigs at the site, SCARPA adopts to determine the direction of the edge, and then takes the scaffolding connection strategy, and joins the limit function into the scaffolding strategy, to reduce the running time. Operatries to predict the relative position of contigs accurately, but the distance between contigs has not been well calculated. SGA uses a relatively conservative strategy, which makes a very high accuracy rate, but the recall rate is relatively low. AByss restricts the solution space by using gap sizes. ScaffMatch is based on the maximum weight matching of pairs of reverse complement strands representing contigs and further filling the scaffold with skipped short contigs.

The greedy methods include PE [13], Bambus [14], GigAssembler [15], Huson's method [16], and ISEA [17], etc. PE adopts a path scoring function based on contradiction of gap sizes, contig length and orientation. Bambus extends a path with the highest score without the detection of contradiction between the new sequence extended and the old sequence extended. Husonadoptsa strategy that generates short paths firstly, then constructs long paths by using short paths. ISEA corrects reads using its own error correction program, and employs precisely reads mapping algorithms which would enhance the quality of mapping but also lose some potential mapping; during the process of scaffolding, ISEA extends a path with only one contig each time, which promotes the precision of prediction but also miss some potential contigs.

In this paper, we propose a new method for scaffolding, called LSLS (Loose-Strict-Loose Scaffolding) based on path extension. LSLS extends a path by adopting different strategies, and it can be more adaptive to different sequencing depths. To solve the problem of multiple paths, we designed a score function based on the distribution of read pairs which can evaluate the reliability of path candidates and extend them with the path with the highest score. In addition, LSLS contains a new method for the gap estimation, which can estimate gap size more precisely. To test the effectiveness of LSLS, we apply it to two real datasets (staphylococcus aureus and rhodobacter sphaeroides) and the experimental results show that LSLS achieves better results compared with other existing scaffolding methods.

2   Materials and Methods

2.1   Algorithm LSLS

In this paper, we propose a new algorithm LSLS (Loose-Strict-Loose Scaffolding) based on path extension for scaffolding. Firstly, LSLS sorts contigs in a non-increasing order according to their lengths and then it extends contigs one after one. The longer contigs are extended by LSLS earlier as the longer contigs generally get more supports information to make reliable decisions. To produce a scaffold, LSLS extends one contig iteratively and finally forms a path. While forming a path, LSLS constructs a local directed acyclic graph during every iterative extension, and then tries to find a direction to extend the path.

While iteratively extending paths, we need to set some extension strategies to guide whether we should continue an extension or not, and from which direction to extend. The extending decision is mainly based on the distribution of mate-pair supports. Generally speaking, if two contigs are close to each other, there should be more extension mate-pair supports between them. Conversely, if there is more supports between two contigs, the two contigs are more likely to be adjacent. However, due to the impact of uneven sequencing depth, regions with lower sequencing depths have a few supports, often resulting in eliminating true relationships; regions with higher sequencing depths often have many supports, often resulting in many false relationships. The sequencing depths between different datasets may also be different, which gives rise to the same question introduced by uneven sequencing depths. To accommodate uneven sequencing and different depths in different datasets, we employed a relatively flexible extension strategy, known as the "loose - tight - loose" policy. The basic idea of the strategy is to use the loose standards. First, which can adapt to the case that the sequencing depth is low, try the best to find out the adjacency; if the relaxed standards lead to multiple extension paths, it is likely that the sequencing depth is high, and supports information is rich enough, so at this time we use strict extension standards to filter false positive adjacent contigs; if there are still a number of candidate extending paths after filtration, again using relaxed standards, at this time, we use the scoring function of the path to select a path and perform the extension, which allows us not to miss any possible adjacency.

The process of the every time iteration extending algorithm can be described as: LSLS tries to gather the subsequent vertexes of a working vertex in a set, the working vertex refers to the contig located at the end of the path, and compute the size of the gap between each contig in the set and the working vertex. LSLS uses this set to build a local directed acyclic graph, if there is only one extension path in the graph, And then extends the path. If multiple extension paths exist, LSLS thinks that the first loose condition leads the false positive vertexes in the set of subsequent vertexes. Therefore the false vertexes are filtered out, then LSLS gathers the subsequent vertexes in a new set by using more strict criteria. After gathering and building the graph, if there are still a number of extension paths, LSLS uses the scoring function to select a candidate extension path and performs the extension. The flow of algorithm LSLS is described in Fig. 1.

Algorithm LSLS
Input: Contigs Sets, Mate-pair Library
Output: Scaffolds Sets
1: mapping reads to contigs set using bowtie2, the results are stored in sam file
2: resolve sam file, construct pair-supports for every pair of two contigs
3: sort contigs by length in non-increasing order
4: For $i = 1$ to n do // n is number of contigs
5:    If contig i is used, then Continue
6:    Extend(i)
7: report contigs that didn't marked as used

Subroutine:
1: Extend(i)   //Extend contig i
2:    While true
3:       C = CollectCandidate(i) //First collection
4:    If C is empty
5:       If contig i has not been extended for left direction
6:          ReverseAndCompliment(i) //Reverse and compliment
7:          Extend(i) //Extend to another direction
8:       Else break   //Finish the extension
9:    If true == IsOnlyOnePath(C) //Only one path
10:       join contig i and contigs in C properly //join i and this path
11:       refresh contig i and supports information
12:       mark contigs in C as used and continue;
13:    Else
14:       C = CollectCandidate2(i) //Second collection
15:       If C is empty
16:          If contig i has not been extended for left direction
17:             ReverseAndCompliment(i)
18:             Extend(i) //Extend from other side
19:          Else break   //Finish extension
20:       If true == IsOnlyOnePath(C)
21:          join contig i and contigs in C properly
22:          refresh contig i and pair-supports information
23:          mark contigs in C as used and continue;
24:       Else   //Choose path
25:          BuildGraph(C) //Build graph according contig length and gap size
26:          P = ExactPathesByDepthFirst() //Deep first search all paths
27:          ScorePathes (P)
28:          phs = GetPathWithHighestScore(P)
29:          join contig i and path phs properly
30:          refresh contig i and pair-supports information

Fig. 1. Description of algorithm LSLS.

As shown in Fig. 1, LSLS extends contigs from long to short. For one contig, LSLS first extends it from the right direction. If the extension is terminated, LSLS tries to extend the path from the left direction. To save the code, extending from the left direction can be equivalent to extending the complementary sequence from the right direction. Contig extension algorithm can be described as follows: firstly, LSLS tries to collect adjacent contigs in a loose way, and we call it as the first collection of adjacent contigs. If the set of adjacent contigs is empty, the extension is ended. If the extensions on both sides are completed, the extension of the contig is completed. If the set is not empty, a local directed acyclic graph is constructed according to the contig lengths and gap sizes, and then LSLS determines if there is only one path in the local graph. If there is only one path, LSLS extends the path; if there are multiple paths LSLS clear the adjacent contig set and collect adjacent contigs in a strict way, which is also called as the second collection of adjacent contigs in the following sections. If the adjacent contigs set is empty, the extension in this direction is terminated; if the set is not empty, a local directed acyclic graph is constructed according to the contig lengths and gap sizes. If there is only one candidate path in the graph, then the path is extended in the direction of the candidate path; if there are multiple candidate paths, LSLS traverses all the paths, and use the path scoring function to score all the paths, and select the path with highest score as the extending direction.

A critical step in LSLS algorithm is to collect adjacent contigs, which is based on the mate-pair supports information. The characteristics of supports are different because of the variety of sequencing depths, which makes it difficult to find a general principle to collect adjacent contigs. If we use strict and conservative criteria, then when the sequencing depth of the dataset is lower, or the depth of dataset is high but at the end of some contigs sequencing depth are lower, strict criteria may judge the correct associated contigs as weak association, thereby increasing the false negatives. Conversely, if we use loose criteria, when datasets sequencing depth higher, or sequencing depth lower, but at the end of some contigs, sequencing depth are lower, which may introduce false positives, and interfere path selection, making it difficult to choose the path extending direction.

2.2   The First Collection of Adjacent Contigs

When at the first time collecting the set of adjacent contigs, in order to avoid missing the correct adjacent contigs due to the low sequencing depth, we have adopted loose collection rules that traverse all unused contigs. If a contig has the number of supports with the extended paths exceeding the threshold 9, the contig is collected. In the process of collection, the gap sizes of the collected contig between the extended pathsare calculated. The basic principle of the contig gap estimate is based on the mean of insert sizes of the library and the mapping position. Suppose there are M supports between the right contig $C1$ and the left contig $C2$, $d1_i$ is the distance of the left read of i-th supports to the right end of the left contig, $d2_i$ is the distance of the right read of i-th supports to the left end of the right contig. μ is the average insert size of mate-pair reads, the gap size of contig $C1$ and contig $C2$ can be evaluated by Eq. (1).

$$Gap(C_1, C_2) = \mu - \sum_{i=1}^{M} (d1_i + d2_i)/M \tag{1}$$

If two close repeat sequences are the same, the two repeat regions are called tandem repeats. Tandem repeats interfere with the estimated size of the gap. The region a and b are covered by supports on the right end of the left contig, wherein region a is included in the region b, and region b is theoretically mate-side support distribution of left-side contig area. However, since the region a is contained by region b, resulting there are also mate-pair supports in region a. Then, when estimating the size of the gap, taking supports from region a into account, can result in smaller evaluation of the gap size. If tandem repeats region a is in the right side of the region b, it can lead to the large evaluation of the gap size. After getting adjacent contigs and their gap size between the extended paths, LSLS constructs a directed acyclic graph G (V, E)without any distance conflict based on the length of contigs and the gap sizes. Wherein G represents the constructed directed acyclic graph, V is the set of adjacent contigs, E is a directed edge which represent the potential adjacent relationships between contigs, and the start vertex and end vertex of an edge must meet the requirements from the distance conflict. The rule of adding an edge is as follows: for any two vertexes in the adjacent contigs set, an edge is added between them it meets the requirement of the distance conflict free. An edge is distance conflict if its start vertex and end vertex are on the same side on the chromosome, and otherwise it is distance conflict free. Thus, when extending from the vertex which is represented by the end of the path we have extended, distance conflicts between the vertexes of each candidate path do not exist, which can reduce the occurrence of errors.

2.3   The Second Collection of Adjacent Contigs

Since the first loose collection standard introduces the false positives, leading to the failure, the second time we have adopted a stringent collection standard. To extend a sequence, LSLS traverses the contig set which is not extended. If there is a contig having supports between the extended sequence and itself, it is viewed as a potential adjacent contig. Then, LSLS executes a step to verify the potential contig. Considering the problem of tandem repeat regions, the support region is divided into several sets, each set represents a region which is covered by mate-pair support. The division approach is the same as the one at the first time to collect adjacent contig gap sizes. Then according to the left-side support dual reading on extending the sequence alignment position, making the collection which is far from the right end of the extension sequence in front. Next, extract the sorted collections in proper order and do the validation, that is, validate the region corresponding to the collection. If a region correspond to the collection, it is judged as a true adjacent contig, and other collections do not need to be verified.

During the process of adjacent contigs collection, the gap sizes are calculated. According to the gap size and the length of verified contig sequence and extended sequence, the desired extended sequence distribution on the mate-pair support area [*Zoneleft*, *Zoneright*] is estimated. If the mate-pair range of support does not cover the desired area, then the distribution of mate-pair support is viewed as being unreasonable and the validation fails. The *Zoneleft* and *Zoneright* are calculated by Eqs. (2) and (3) as follows.

$$Zoneleft(i, j) \ = \ \mu - gapdis(i, j) - \sigma \tag{2}$$

$$Zoneright(i, j) \ = \ \mu - gapdis(i, j) - j.contigSize \ + \ \sigma \tag{3}$$

Where i represents the extended sequence, j represents the verified contig, μ represents the mean size of the insert, σ represents the variance of the insert size, $gapdis\ (i, j)$ represents an approximate size of the gap between i and j, $j.contigSize$ represents the sequence length verified contig j. The reason that $zoneleft$ needs to subtract the variance and $zoneright$ needs to add variance, is that concerns about mate-pair support insert size is a varying value, and when sequencing uneven, the mate-pair support coverage may not be very good, the expected region size needs to be narrowed.

2.4 The Scoring Function Path

After the second collection of adjacent contigs is completed, to solve the problem of multiple candidate paths selection, we design a scoring function to evaluate the candidate path score, and select the path based on the score. In order to obtain all possible paths, we first build a local directed graph which has the extended sequence and candidate contigs as the vertexes. The path scoring function is shown in Eq. (4), where P represents the extended sequence, Q is a candidate path, $C_i$is the i-th vertex in Q, k is the number of vertexes on the candidate path, i stands for the i-th vertex on the candidate path, $zonelength\ (P, i)$represents the length of the zone between extended sequence P and the vertex i, totalLen refers to the sum of the lengths of expected zones of all the vertexes on the path Q and $P.zoneleft2(P, i)$, $zoneright2(i, j)$, and $zonelength\ (P, i)$ are calculated by Eqs. (5), (6) and (7).

$$ScorePath(P, Q) = \sum_{i=1}^{k} \frac{zonelength(i)}{totalLen} \sum_{j=1}^{h} e^{\frac{-(\mu - d_{j1} - d_{j2} - Gap(P, C_i))^2}{2\sigma^2}} \tag{4}$$

$$zoneleft2(P, i) = \mu - gapdis(P, i) \tag{5}$$

$$zoneright2(i, j) = \mu - gapdis(P, i) - j.contigSize \tag{6}$$

$$zonelength\,(P, i) = zoneleft2(P, i) - zoneright2(P, i) \tag{7}$$

In the Eq. (5), $zoneleft2$ represents a coordinate in the extended sequence, the right side of the extended sequence is 0, if the result of calculation less than 0. However, if $zoneright2$ is or larger than the length of extended sequence, then the varification is required. In the Eq. (6), $zoneright2$ represents a coordinate in the extended sequence, the right side of the extended sequence is 0, and the result of the calculation often is less than 0, rarely larger than the length of the extended sequence, which is yet rare, the varificationis requires. h is the number of supports between the contig i and the expect zone of extended sequence P and i, j represents the j-th support, d_{j1} represents the distance between the mapping site of left mate of the j-th support and the right end of the extended sequence, d_{j2} represents the distance between the mapping site of right mate of the j-th support and the left end of the candidate sequence C_i, $Gap\ (P, C_i)$ indicates the size of the gap between P and C_i, μ represents the average insert size, σ represents insert size variance.

As can be seen from Eq. (4), the more the number of mate-pair support if the candidate path corresponding to the desired coverage area, the higher the score. The total score is the sum of the candidate vertex score on the candidate path, and the score of each vertex is relative to the desired coverage area of mate-pair support and the weight of the vertex. The weight of each vertex is the ratio of the desired coverage area and the total desired coverage area. If the desired coverage area is larger, the statistical significance of the vertex is greater, so the weight of the vertex is greater. $Gap\,(P,\,C_i)$ represents the gap size between P and C_i. And the mate-pair support insert size obeys the distribution of $N(\mu,\,\sigma)$ empirically, we can derive that $\mu - d_{j1} - d_{j2}$ obeys the distribution of $N(Gap\,(P,\,C_i),\,\sigma_2)$, So the score of each support is $e^{-(\mu - d_{j1} - d_{j2} - Gap(P,C_i))^2/(2\sigma^2)}$.

3   Evaluation Metrics

The charactcristics of DNA sequences between eukaryotes and prokaryotes are quite deferent, as shown in Table 1. Not only is the length of the chromosomes of eukaryotes usually much longer than the length of Chromosomes of prokaryotes, but also the genome of eukaryotes often contains more repeat regions, which is one of the major problems in de novo assembly.

Table 1. Details of datasets

Speices	Staphylococcus aureus	Rhodobacter sphaeroides
Number of contigs(M)	3	7
Genome size(Mbp)	2.9	4.6
Library type	Matc pair	Mate pair
Mean value of Insert size	3500	3700
Variance of Insert size	250	300
Read length(bp)	37	101
Number of reads(M)	3.5	2.1
Coverage	~45	~46
Number of contigs	170	577
Potential connection(PC)	167	570

For assessing the quality of scaffolds in a more reasonable way, *Hunt et al. (*2014) proposed a novel evaluation tool [18], which uses four key metrics to evaluate the scaffolding results. The four metrics are: Correct Joins (*CJ*), Incorrect Joins (*IJ*), Skipped Tags (*ST*), and Lost Tags (*LT*).

Due to the different degrees of their importance, we employ the weights of four metrics of *CJ, IJ, LT,* and *ST* as 1, 1, 2, and 0.5, respectively (*Hunt et al.*, 2014). After we get the correct joins (*CJ*) and bad joins (*IJ, ST* and *LT*) of scaffolds, we adopt the *F-score* as a comprehensive metric which is calculated by *TPR*(true positive rate) and *PPV* (*Mandric and Zelikovsky*, 2015).

We compare LSLS with 11 popular scaffolding tools: ABYSS, Bambus2, MIP, Opera, SCARPA, SGA, BESST [19], SOPRA, and SSPACE [20], SOAP2 [21] and ScaffMatch. The results are shown in Tables 2 and 3.

Table 2. Scaffolding results about staphylococcus aureus

Scaffolding tools	CJ	IJ	LT	ST	TPR	PPV	F-score
ABySS	99	2	**0**	13	0.592	0.920	0.721
Bambus2	96	**0**	**0**	27	0.574	0.876	0.694
MIP	2	**0**	**0**	**1**	0.011	0.800	0.023
Opera	112	11	**0**	22	0.670	0.835	0.744
SCARPA	84	8	**0**	9	0.502	0.870	0.637
SGA	83	1	**0**	10	0.497	**0.932**	0.648
BESST	112	11	**0**	21	0.670	0.838	0.745
SOPRA	89	2	**0**	14	0.532	0.908	0.671
SSPACE	110	9	**0**	13	0.658	0.876	0.752
SOAP2	131	12	**0**	13	0.784	0.876	0.827
ScaffMatch	139	14	**0**	23	0.832	0.844	0.838
LSLS	**147**	15	**0**	9	**0.880**	0.882	**0.881**

Table 3. Scaffolding results about rhodobacter sphaeroides

Scaffolding tools	CJ	IJ	LT	ST	TPR	PPV	F-score
ABySS	384	7	0	54	0.673	0.918	0.777
Bambus2	337	3	0	36	0.591	0.941	0.726
MIP	238	11	2	24	0.417	0.904	0.571
Opera	316	**1**	0	23	0.554	0.961	0.703
SCARPA	214	5	0	24	0.375	0.926	0.534
SGA	232	**1**	0	26	0.407	0.943	0.568
BESST	367	2	0	**15**	0.643	**0.974**	0.775
SOPRA	403	3	0	37	0.707	0.949	0.810
SSPACE	357	7	0	49	0.626	0.918	0.744
SOAP2	468	8	0	26	0.821	0.957	0.883
ScaffMatch	482	18	0	40	0.845	0.926	0.884
LSLS	**498**	32	0	34	**0.873**	0.910	**0.891**

LSLS achieves the best *F-score* for the two datasets. In the dataset of rhodobacter sphaeroides, LSLS has the most increase of 56% compared with SGA, and the least increase of 0.7% compared with ScaffMatch. In the dataset of staphylococcus aureus, LSLS also increases at least 5%, which means that LSLS has the best comprehensive evaluation performance. ScaffMatch and SOAP2win the second and third *F-score* for the two datasets, respectively. LSLS always has the best *CJ* and *TPR* value, which means that LSLS can find most potential connections. LSLS has the most correct connections and less *ST*, although it has the most *IJ* for two datasets. The other two better tools, scaffMatch and SOAP2, also have high *IJs*.

LSLS produces 15 IJs while scaffMatch has 14 *IJs*. LSLS just has one more *IJ* than scaffMatch, while LSLS has 14 *STs* less than scaffMatch, and has 8 *CJs* more than scaffMatch. For Rhodobacter sphaeroides, the number of *CJ* of LSLS and scaffMatch are 32 and 18, respectively, LSLS has 14 *IJs* more than scaffMatch, while LSLS has 6 *STs* less than scaffMatch, and 16 *CJs* more than scaffMatch. SOAP2 also has a better *F-score*, and is just behind LSLS and scaffMatch. SOAP2 has less *IJs,* its means that SOAP2 is more conservative and more accurate, but it also lose some potential connection and has a small numbers of *CJs.*

To test the operating efficiency of LSLS, we operated LSLS on a server with 512G memory and 6 cores (Intel Xeon E5-2620 2.00 GHz), the running time of two datasets are 8.6 s and 7.3 s; the peak memory are 51156 kb and 51108 kb, respectively.

4   Conclusion and Discussion

In this paper, we have presented a novel scaffolding method based on path extension. LSLS employs a loose-strict-loose strategy during path extension, which is more adaptive to uneven sequencing depths. Firstly, LSLS uses a relative loose way to extend contigs to avoid losing potential joins. However, the loose extension strategy may lead to false positive adjacent contigs when encountering multiple candidate extension paths. To deal with that problem, LSLS uses strict extension methods to reduce false adjacent contigs. If there are still multiple candidate extension paths after the reduction, LSLS relaxes the extension conditions by evaluating each path and choosing the path with the highest score as the extension direction. Estimating gap sizes is a very important part in scaffolding algorithms and has a great impact on final scaffolding quality. By reducing the negative effects of tandem repeat regions, LSLS improves the accuracy of the valuation of gap sizes. According to the characteristics of the short length of tandem repeats, LSLS filters false adjacent contigs via comparing the pair-end support distribution to the expected distribution. LSLS uses the differences between observed region covered by supports and the expected region that covered by supports to design the path scoring function. The experiment results demonstrate that the scoring function can evaluate the reliability of candidate paths effectively. When comparing with 11 scaffolders (ABySS, Bambus2, MIP, Opera, SCARPA, SGA, BESST, SOPRA, SSPACE, SOAP2, Scaff-Match) on two real datasets, LSLS obtained the best comprehensive performance.

Some ideas of LSLS can also be incorporated into scaffolding algorithms based on graph to improve the accuracy of gap-size estimation and reduce the complexity of graphs. In future, we will try to combine the idea of LSLS with scaffolding based on graph to improve the scaffolding quality.

References

1. Voelkerding, K.V., Dames, S.A., Durtschi, J.D.: Next-generation sequencing: from basic research to diagnostics. Clin. Chem. **55**(4), 641–658 (2009)
2. Luo, J., Wang, J., Zhang, Z., Wu, F.X., Li, M., Pan, Y.: Epga: de novo assembly using the distributions of reads and insert size. Bioinformatics **31**(6), 825–833 (2015)
3. Gritsenko, A.A., Nijkamp, J.F., Reinders, M.J.T., Ridder, D.D.: Grass: a generic algorithm for scaffolding next-generation sequencing assemblies. Bioinformatics **28**(11), 1429 (2012)
4. Salmela, L., Mäkinen, V., Välimäki, N., Ylinen, J., Ukkonen, E.: Fast scaffolding with small independent mixed integer programs. Bioinformatics **27**(23), 3259–3265 (2011)

5. Dayarian, A., Michael, T.P., Sengupta, A.M.: Sopra: scaffolding algorithm for paired reads via statistical optimization. BMC Bioinform. **11**(1), 345 (2010)
6. Koren, S., Treangen, T.J., Pop, M.: Bambus 2: scaffolding metagenomes. Bioinformatics **27**(21), 2964–2971 (2011)
7. Donmez, N., Brudno, M.: Scarpa: scaffolding reads with practical algorithms. Bioinformatics **29**(4), 428 (2013)
8. Gao, S., Nagarajan, N., Sung, W.K.: Opera: reconstructing optimal genomic scaffolds with high-throughput paired-end sequences. J. Comput. Biol. J. Comput. Mol. Cell Biol. **18**(11), 1681–1691 (2011)
9. Simpson, J.T., Durbin, R.: Efficient de novo assembly of large genomes using compressed data structures. Genome Res. **22**(3), 549–556 (2012)
10. Simpson, J.T., Wong, K., Jackman, S.D., et al.: Abyss: a parallel assembler for short read sequence data. Genome Res. **19**(6), 1117 (2009)
11. Mandric, I., Zelikovsky, A.: ScaffMatch: scaffolding algorithm based on maximum weight matching. In: Przytycka, Teresa M. (ed.) RECOMB 2015. LNCS, vol. 9029, pp. 222–223. Springer, Cham (2015). doi:10.1007/978-3-319-16706-0_22
12. Luo, J., Wang, J., Zhen, Z., Min, L., Wu, F.X.: Boss: a novel scaffolding algorithm based on an optimized scaffold graph. Bioinformatics **33**, 169–176 (2016). btw597
13. Ariyaratne, P.N., Sung, W.K.: Pe-assembler: de novo assembler using short paired-end reads. Bioinformatics **27**(2), 167 (2011)
14. Pop, M., Kosack, D.S., Salzberg, S.L.: Hierarchical scaffolding with bambus. Genome Res. **14**(1), 149–159 (2004)
15. Kent, W.J., Haussler, D.: Assembly of the working draft of the human genome with gigassembler. Genome Res. **11**(9), 1541–1548 (2001)
16. Huson, D.H., Reinert, K., Myers, E.W.: The greedy path-merging algorithm for contig scaffolding. J. ACM **49**(5), 603–615 (2002)
17. Min, L., Liao, Z., He, Y., Wang, J., Luo, J., Yi, P.: Isea: iterative seed-extension algorithm for de novo assembly using paired-end information and insert size distribution. IEEE/ACM Trans. Comput. Biol. Bioinform. **PP**(99), 1 (2016)
18. Hunt, M., et al.: A comprehensive evaluation of assembly scaffolding tools. Genome Biol. **15**(3), 1–15 (2014)
19. Sahlin, K., Vezzi, F., Nystedt, B., Lundeberg, J., Arvestad, L.: Besst - efficient scaffolding of large fragmented assemblies. BMC Bioinform. **15**(1), 281 (2014)
20. Boetzer, M., Henkel, C.V., Jansen, H.J., Butler, D., Pirovano, W.: Scaffolding pre-assembled contigs using sspace. Bioinformatics **27**(4), 578–579 (2011)
21. Li, R., Yu, C., Li, Y., Lam, T.W., Yiu, S.M., Kristiansen, K., et al.: Soap2: an improved ultrafast tool for short read alignment. Bioinformatics **25**(15), 1966–1967 (2009)

Structure Prediction and Folding

The Hasp Motif: A New Type of RNA Tertiary Interactions

Ying Shen and Lin Zhang[(⊠)]

School of Software Engineering, Tongji University, Shanghai, China
{yingshen, cslinzhang}@tongji.edu.cn

Abstract. RNA structural motifs are recurrent structural elements occurring in RNA molecules. They play essential roles in consolidating RNA tertiary structures and in binding proteins. Recently, we discovered a new type of RNA structural motif, namely the hasp motif, from 27 RNA molecules. The hasp motif comprises three nucleotides which form a structure similar to a hasp. Two consecutive nucleotides in the motif come from a double helix and the third one comes from a remote stand. The hasp motif makes two helices approach each other, which leads to RNA structure folding. All the identified hasp motifs reveal a consensus structural pattern although their sequences are not conserved. Hasp motifs are observed to reside both inside and on the surface of RNA molecules. Those inside RNA molecules help consolidate RNA tertiary structures while the others locating on the surface are evidenced to interact with proteins. The wide existence of hasp motifs indicates that hasp motifs are quite essential in both keeping RNA structures' stableness and helping RNA perform their functions in biological processes.

Keywords: Hasp motif · RNA structural motif

1 Introduction

RNA structural motifs are recurrent substructures occurring in RNA tertiary structures. They play an important role in various biological processes. Some of them are involved in stabilizing 3D structures of ribosomes by binding with ribosomal proteins [1]. While others can help RNAs performing their functions through binding with other specific proteins in biological processes [2–6]. On the other hand, RNA structural motifs are also essential in consolidating RNA tertiary structures. For example, two types of structural motifs, D-loop and T-loop, are essential in maintaining the 3D structures of tRNAs [7]. In addition, many other motifs, e.g. ribose zipper (Cate et al. 1996), tetraloop-tetraloop receptor [8], etc., help connect remote parts of RNAs together and contribute to the formation of stable and functional RNA structures. Under researchers' endeavor, abundant RNA structural motifs have been discovered recently, such as tetraloop [9], A-minor motif [10], kink-turn [11], C-loop [12], etc.

In the last decade, more new motifs have proliferated, for instance, UA_handle [13], G-ribo motif [14], Adenosine-wedge [15], GA-minor motif [16], Right angle motif [17], λ-turn [18], to name a few. These new motifs show distinct sequence and structural patters and greatly enrich researchers' understanding in RNA tertiary

© Springer International Publishing AG 2017
D.-S. Huang et al. (Eds.): ICIC 2017, Part II, LNCS 10362, pp. 441–453, 2017.
DOI: 10.1007/978-3-319-63312-1_39

structures and the relationships with their functions. However, despite of abundant identified motifs, we believe that many new RNA structural motifs are awaiting to be discovered, the task of which is quite significant and helpful for the research in RNA structure prediction and RNA folding.

Recently, we discovered a new type of RNA structural motif, namely the hasp motif. We successfully identified 89 hasp motifs from 27 RNA molecules ranging from ribosomal RNAs, transcription factors to riboswitches. The hasp motif comprises three nucleotides which form a structure similar to a hasp. Its structure is conserved across different RNA molecules, although its sequence does not show strong consensus patterns. Based on our observation, two structural characteristics of the hasp motif are summarized as follows. Firstly, the hasp motif is composed by two consecutive nucleotides and a remote one. Secondly, the remote nucleotide interacts with the two consecutive nucleotides simultaneously. Hasp motifs locate both inside and on the surface of RNA molecules, which implies that they both contribute to consolidating RNA tertiary structures and helping bind proteins. The wide existence of hasp motifs indicates that they play an important role in various RNA molecules.

2   Results

2.1   Definition of the Hasp Motif

The hasp motif is a tertiary interaction composed of three nucleotides. In a standard hasp motif, the first two nucleotides (NT1 and NT2 shown in Fig. 1(A)) come from a strand which is part of a helix. The third one (NT3) interacts with NT1 and NT2. Specifically, the base of NT3 interacts with the 2'-OH of NT1. The 2'-OH of NT3 interacts with the base and the sugar of NT2. In Fig. 1, we present 3D structures of seven hasp motifs identified in HM 50S (PDB ID: 1S72). The hasp motif helps two strands approach each other in a way that looks like a hasp. We also aligned all the 89 instances in order to demonstrate the consensus structural pattern of hasp motifs. The structure alignment has been shown in Fig. 1(H). From the figure, we can see that the positions of three nucleotides from different instances do not vary much.

2.2   Identification of Hasp Motifs

We searched 550 non-redundant RNA molecules sorted by Leontis and Zirbel [19]. The non-redundant RNA list can be downloaded from http://rna.bgsu.edu/rna3dhub/ nrlist/ (release ID 0.8). Through our experiment, we identified 89 instances of the hasp motif from 27 RNA molecules, the classes of which include ribosomal RNA, ribonuclease P RNA, transcription factor, tRNA-splicing endonuclease, and riboswitch. The diversity of RNA classes indicates the importance of the hasp motif in various RNA structures.

The sequences of all 89 hasp motifs are listed in Table 1. The sequences shown in Table 1 are presented in the form of: type of each nucleotide (A/U/G/C) + its sequence id + chain ids of all nucleotides. For example, sequence "A1500 C1501 A1394 AAA" contains three nucleotides whose ids in the sequence are 1500, 1501, and 1394,

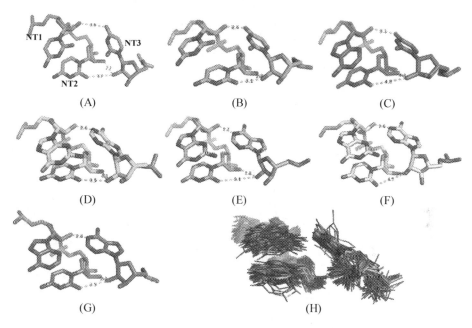

Fig. 1. The structures of hasp motifs in 1S72. (A) 0:C2318 C2319 U2322; (B) 0:C1403 C1404 U1408; (C) 0:A106 U107 U68; (D) 0:G2365 C2366 A2370; (E) 0:G1849 U1850 A1941; (F) 0: G1832 U1833 A874; (G) 0:A1242 C1243 A1247; (H) structure alignment of all hasp motifs in 27 RNA molecules.

respectively. In addition, the types of three nucleotides are adenine, cytosine, and adenine, respectively. Moreover, all of them are located on chain A. As the consequence, there are three As affixed at the end of the sequence.

Table 1. The sequences of 89 hasp motifs identified in 27 RNA molecules.

PDB	Class	Hasp motif	PDB	Class	Hasp motif
1FJG	Ribosome	A1500 C1501 A1394 AAA G1295 C1296 U1302 AAA G867 C868 A873 AAA C1209 C1210 C1214 AAA	3BBN	Ribosome	G375 U376 A446 AAA A1449 C1450 A1343 AAA
1NBS	Ribonuclease P RNA	G182 C183 U101 BBA	3IZ7	Ribosome	U1094 C1095 A1098 AAA

(continued)

Table 1. (*continued*)

PDB	Class	Hasp motif	PDB	Class	Hasp motif
1PNU	Ribosome	C1784 A1785 A1883 000	3IZ9	Ribosome	C2656 C2657 A2660 AAA G2703 U2704 A2708 AAA C1481 C1482 A1487 AAA
1PNX	Ribosome	A1500 C1501 A1394 AAA G867 C868 A873 AAA G1295 C1296 U1302 AAA	3IZF	Ribosome	C2653 C2654 A2657 AAA C1478 U1479 U1484 AAA
1S1H	Ribosome	C1209 C1210 C1214 AAA G867 C868 A873 AAA G1295 C1296 U1302 AAA	3JYV	Ribosome	U1086 C1087 A1090 AAA
1S1I	Ribosome	C2318 C2319 U 2322 333 A106 U107 U68 333 C1403 C1404 U1408 333 G1849 U1850 A1941 333 G2365 C2366 A2370 333 G1832 U1833 A874 333 A1242 C1243 A1247 333 G68 U69 A11 444	3JYX	Ribosome	C1478 U1479 G1483 555
1S72	Ribosome	C2318 C2319 U2322 000 C1403 C1404 U1408 000 A106 U107 U68 000	3SD3	Tetrahydrofolate riboswitch	G65 C66 A56 AAA

(*continued*)

Table 1. (*continued*)

PDB	Class	Hasp motif	PDB	Class	Hasp motif
		G2365 C2366			
		A2370 000			
		G1849 U1850			
		A1941 000			
		G1832 U1833			
		A874 000			
		A1242 C1243			
		A1247 000			
1UN6	Transcription factor IIIA	G66 C67 A11 FFF	3SUX	Riboswitch	G74 C75 A61 XXX
2AW7	Ribosome	A1500 C1501 A1394 AAA C1209 C1210 C1214 AAA G867 C868 A873 AAA G220 C221 A197 AAA	3U5F	Ribosome	C1441 U1442 C1447 666 A1763 C1764 A1631 666 U1089 C1090 A1093 666
2GJW	tRNA-splicing endonuclease	G9 A10 A15 EEH	3U5H	Ribosome	C2653 C2654 A2657 555 C2151 A2152 A2243 555 G2700 U2701 A2705 555 C1478 U1479 U1484 555 G2134 U2135 A913 555 A97 U98 U60 888 G67 C68 A11 777
2J37	Ribosome	C67 C68 U72 ZZZ	3UXR	Ribosome/ Antibiotic	C1297 C1298 A1302 AAA C2284 C2285 A2288 AAA A111 U112 U72 AAA C1793 U1794 A1900 AAA G2331 U2332 A2336 AAA

(*continued*)

Table 1. (*continued*)

PDB	Class	Hasp motif	PDB	Class	Hasp motif
					A1269 C1270 G1325 AAA G768 G769 A1379 AAA G1776 U1777 A781 AAA G1139 C1140 A1143 AAA C1604 C1605 A1610 AAA G69 C70 A13 BBB
2QBG	Ribosome	C1297 C1298 A1302 BBB A2284 C2285 A2288 BBB C1793 A1794 A1900 BBB A111 U112 U72 BBB G2331 C2332 A2336 BBB G1139 C1140 A1143 BBB A1269 C1270 U1325 BBB G1776 U1777 A781 BBB	4A1B	Ribosome	C2642 C2643 A2646 111 G2689 U2690 A2694 111 C2147 A2148 A2238 111 C1504 C1505 U1510 111 G1172 A1173 U1358 111 G2130 U2131 A938 111
2ZJR	Ribosome	C2263 C2264 A2267 XXX C1784 A1785 A1883 XXX C1310 C1311 A1315 XXX A109 U110 A71 XXX G1767 U1768 A794 XXX	4A1C	Ribosome	G66 C67 A11 333 A98 U99 U63 222
3B31	RNA	G6192 U6193 A6209 AAB			

In the searching process, we used a small cutoff value to guarantee that all the searched results are true positive instances. Therefore, we believe that there will be more hasp motifs in the searched RNAs. In addition, with more non-redundant RNAs being released, more instances of hasp motif will be recognized.

2.3   Consensus Pattern of Hasp Motifs

We aligned the sequences of 89 instances and drew a consensus logo for the hasp motif (see Fig. 2(A)). It can be seen that all four types of nucleotides (A/U/G/C) occur in the first position (NT1) and the frequency of each type of nucleotide does not vary much except uracil ($freq_A$ = 0.191; $freq_U$ = 0.034; $freq_G$ = 0.438; $freq_C$ = 0.3371). In the second position, only A, U, and C occur and their frequencies are 0.079, 0.315, and 0.596, respectively. In the third position, although all four types of nucleotide occur in this position, adenine and uracil occur more frequently, which can be seen from their occurring frequencies ($freq_A$ = 0.708; $freq_U$ = 0.225; $freq_G$ = 0.003; $freq_C$ = 0.045). It can be seen that the nucleotides in the second and the third positions (NT2 and NT3) is more conserved than the first position. It is interesting to reach to this conclusion. We will explain why different position has different conserveness in the following paragraphs.

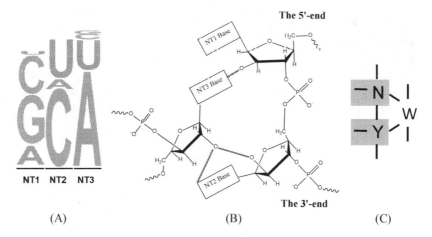

Fig. 2. Consensus pattern of the hasp motif. (A) consensus logo of the motif; (B) secondary structure of the hasp motif; (C) interaction diagram of the motif. (Color figure online)

There are three bonds (shown as red lines in Fig. 2(B)) connecting NT3 and NT1/NT2. The first bond connects the base of NT3 and the sugar of NT1. The other two bonds connect the sugar of NT3 and the base/sugar of NT2, respectively. Because only the sugar of NT1 involves in the connection, the base of NT1 can be substituted freely to any nucleotides. However, both bases of NT2 and NT3 participate in the interaction, only certain types of nucleotides can occur in these two positions. Therefore, NT1 is not so conserved compared with NT2 and NT3.

According to occurring frequencies of nucleotides in each position, we infer the consensus sequence pattern of the hasp motif, which has been shown in Fig. 2(C). Based on the previous analysis, NT1 is not as conserved as NT2 or NT3. Therefore, letter N is used for NT1 to represent any nucleotide occurring in this position. Similarly, because the frequencies of uracil and cytosine are much higher than that of adenine, letter Y is used in position of NT2 to represent a pyrimidine. In the end, letter W represents an adenine or a uracil which occur more frequently in position of NT3.

The structure alignment of all identified hasp motifs has been shown in Fig. 1(H). It reveals distinct characteristics of hasp motif's structure.

2.4   The Hasp Motif vs. G-Ribo Motif

G-ribo motif is a large motif that comprises two double helices which are arranged side by side (see Fig. 3(A)). Two helices in the G-ribo motif approach each other and form three layers of bases. The stability of G-ribo motif relies on the existence of a hasp motif (shown in red in Fig. 3(A)). A nucleotide from the right helix extrudes out and interacts with the backbones of two consecutive nucleotides from the left helix. These three nucleotides form a hasp motif, which brings two helices together closely. In view of this, the hasp motif is an indispensable substructure of the G-ribo motif.

However, the hasp motif does not merely exist as a substructure of G-ribo motif. It also exists independently in different places in RNA tertiary structures. For example, in a ribosomal RNA 1S1I, a hasp motif G1849 U1850 A1941 333 (i.e. red nucleotides shown in Fig. 3(B)) draws two double helices together. Unlike the hasp motif shown in Fig. 3(A), there is no G-ribo motif here, which means that hasp motifs can occur independently. From this example, we can also infer that one important function of the hasp motif is to consolidate RNA tertiary structure, in a way that is similar to a ribose zipper. Using this mechanism, different parts, especially helices, in RNA tertiary structures are assembled together.

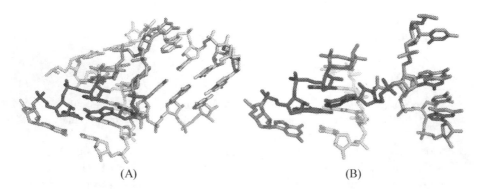

(A)   (B)

Fig. 3. G-ribo motif versus independent hasp motif (A) a G-ribo motif observed in 2QBE. The sequence number is B:G1296-C1305/B:G1623-G1627/B:C1639-C1644. The red nucleotides form a hasp motif; (B) a hasp motif from 1S1I that does not belong to any G-ribo motif. The sequence number of the whole structure is 3:G1848-G1851/3:C1880-U1883/3:U1939-G1944. (Color figure online)

2.5   Variations in Hasp Motifs

Typically, three bonds that connect two pairs of nucleotides (NT3-NT1 and NT3-NT2) help to form a hasp motif. However, sometimes one of the bonds connecting NT3 and NT2 may be missing. Under this circumstance, other part, such as phosphate group, of NT3 or even another nucleotide tends to be involved. Three examples are shown in Fig. 4. In Fig. 4(A), the phosphate group of a neighboring nucleotide (the lower right nucleotide in orange) involves in the connection, which helps to stabilize the structure of the hasp motif. In Fig. 4(B), the sugar of NT2 of the hasp motif directly interacts with the phosphate group of a neighboring nucleotide of NT3 (the lower nucleotide in orange), instead of the sugar of NT3.

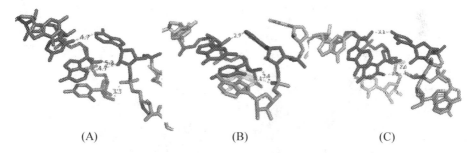

(A)   (B)   (C)

Fig. 4. Variance in the hasp motif's structure. (A) the sequence number of the first example is 0: G1783-C1786/G1882-A1884 (PDB ID: 1PNU); (B) the sequence number of the second example is A:A374-G377/U445-A447 (PDB ID: 3BBN); (C) the sequence number of the third example is 0:G105-U108/A67-A69 (PDB ID: 1S72) (Color figure online)

2.6   Locations of Hasp Motifs and Protein Recognition

Hasp motifs locate both inside and on the surface of RNA molecules. Take hasp motifs occurring in chain A of 3UXR as an example. 3UXR is a ribosome containing a 23S rRNA and a 5S rRNA as well as 28 ribosomal proteins. The structure of 3UXR can better reveal interactions between RNAs and proteins. There are ten hasp motifs located on chain A of 3UXR (23S rRNA) and they have been shown in Table 2. Each of them is given an identifier for the convenience of reference. Five of ten hasp motifs locate on the surface of 23S rRNA, which include hm2, hm3, hm4, hm5, and hm9. While the other motifs reside inside 23S rRNA. The positions of hm2, hm3, hm4, hm5, and hm9 have been shown in Fig. 5. Among these five hasp motifs, four of them (hm3, hm4, hm5, and hm9) interact with ribosomal proteins in 3UXR (see Fig. 6).

From the example, we can see that 50% hasp motifs in 23S RNA locate inside and the other 50% locate on the surface. In addition, four out of five motifs on the surface interact with ribosomal proteins. All these evidences imply that hasp motifs are essential in both consolidating RNA tertiary structures and binding with proteins.

Table 2. Hasp motifs occurring in chain A of 3UXR

ID	Sequence
hm1	C1297 C1298 A1302 AAA
hm2	C2284 C2285 A2288 AAA
hm3	A111 U112 U72 AAA
hm4	C1793 U1794 A1900 AAA
hm5	G2331 U2332 A2336 AAA
hm6	A1269 C1270 G1325 AAA
hm7	G768 G769 A1379 AAA
hm8	G1776 U1777 A781 AAA
hm9	G1139 C1140 A1143 AAA
hm10	C1604 C1605 A1610 AAA

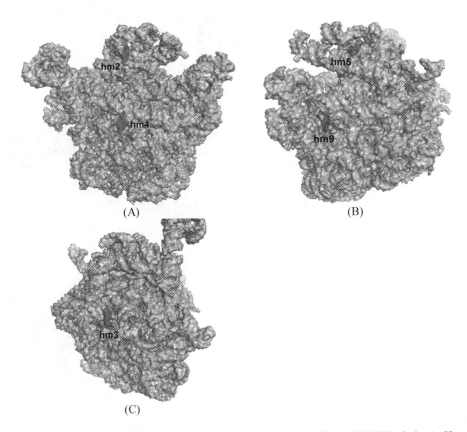

Fig. 5. Location of hm2, hm3, hm4, hm5, and hm9 on the surface of 3URX, chain A. Hasp motifs are colored in red. (Color figure online)

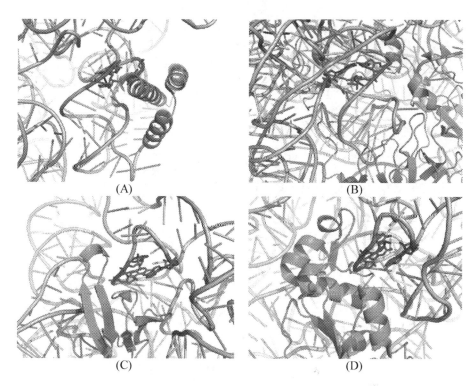

Fig. 6. Interactions between hasp motifs and ribosomal proteins. Hasp motifs are colored in red and proteins are colored in green. (A) Motif hm3 interacts with ribosomal protein L29 (chain 2); (B) hm4 interacts with ribosomal protein L2 (chain D); (C) hm5 interacts with ribosomal protein L27 (chain 0); (D) hm9 interacts with ribosomal protein L13 (chain N). (Color figure online)

3 Method

We used the method described in [20] and discovered many new RNA structural motifs, which include four hasp motifs coming from HM 50S (PDB ID 1S72). Then we used four hasp motifs in 1S72 as templates and searched 550 non-redundant RNA molecules using the method LS-RSMR [21] to find other instances. The cutoff parameter used in LS-RSMR was set 6 which is a bit smaller than the suggested one because that a small cutoff value could guarantee that all the output sequences are true hasp motifs. As the consequence, we obtained 89 hasp motifs coming from 27 RNA molecules ranging from ribosomal RNAs, transcription factors to riboswitches.

4 Conclusion

In this paper, we describe a new type RNA structural motif, namely the hasp motif, which wide exist in different types of RNAs. A hasp motif comprises three nucleotides, two of which come from a strand of a helix and one of which comes from a remote strand. It brings two helices approaching each other and helps to stabilize RNA tertiary structures. The structures of hasp motifs have a consensus structural pattern, although their sequences do not reveal strong patterns. Sometimes the connections between nucleotides in a hasp motif can be strengthened by involving the interactions with the phosphate group of a neighboring nucleotide. A hasp motif can exist as a substructure of another large RNA structural motif, G-ribo, or occur independently.

Hasp motifs are observed to locate both inside and on the surface of RNA molecules. Those located inside RNA molecules help to stabilize RNA tertiary structures, while the others located on the surface tend to bind with proteins. All these evidences imply that hasp motifs are essential in both consolidating RNA tertiary structures and helping RNAs performing their functions in biological processes.

Acknowledgement. The work described in this paper is supported partially by the Natural Science Foundation of China under grants no. 61303112 and no. 61672380.

References

1. Becker, M.M., Lapouge, K., Segnitz, B., Wild, K., Sinning, I.: Structures of human SRP72 complexes provide insights into SRP RNA remodeling and ribosome interaction. Nucleic Acids Res. **45**, 470–481 (2017)
2. Cate, J.H., Gooding, A.R., Podell, E., Zhou, K., Golden, B.L., Szewczak, A.A., Kundrot, C.E., Cech, T.R., Doudna, J.A.: RNA tertiary structure mediation by adenosine platforms. Science **273**, 1696–1699 (1996)
3. Ciriello, G., Gallina, C., Guerra, C.: Analysis of interactions between ribosomal proteins and RNA structural motifs. BMC Bioinform. **11**(Suppl 1), S41 (2010)
4. Correll, C.C., Swinger, K.: Common and distinctive features of GNRA tetraloops based on a GUAA tetraloop structure at 1.4 Å resolution. RNA **9**, 355–363 (2003)
5. François, B., Russell, R.J., Murray, J.B., Aboul-ela, F., Masquida, B., Vicens, Q., Westhof, E.: Crystal structures of complexes between aminoglycosides and decoding a site oligonucleotides: role of the number of rings and positive charges in the specific binding leading to miscoding. Nucleic Acids Res. **33**, 5677–5690 (2005)
6. Gagnon, M.G., Steinberg, S.V.: The adenosine wedge: A new structural motif in ribosomal RNA. RNA **16**, 375–381 (1020)
7. Grabow, W.W., Zhuang, Z., Shea, J.-E., Jaeger, L.: The GA-minor submotif as a case study of RNA modularity, prediction, and design. Wiley Interdiscip Rev. RNA **4**, 181–203 (2013)
8. Grabow, W.W., Zhuang, Z., Swank, Z.N., Shea, J.-E., Jaeger, L.: The right angle (RA) motif: a prevalent ribosomal RNA structural pattern found in group I introns. J. Mol. Biol. **424**, 1–2 (2012)
9. Hendrix, D.K., Brenner, S.E., Holbrook, S.R.: RNA structural motifs: building blocks of a modular biomolecule. Q. Rev. Biophys. **38**, 221–243 (2005)
10. Huang, L., Wang, J., Lilley, D.M.: A critical base pair in k-turns determines the conformational class adopted, and correlates with biological function. Nucleic Acids Res. **44**, 5390–5398 (2016)
11. Jaeger, L., Verzemnieks, E.J., Geary, C.: The UA_handle: a versatile submotif in stable RNA architectures. Nucleic Acids Res. **37**, 215–230 (2009)

12. Kim, S.H., Suddath, F.L., Quigley, G.J., McPherson, A., Sussman, J.L., Wang, A.H.J., Seeman, N.C., Rich, A.: Three-dimensional tertiary structure of yeast phenylalanine transfer RNA. Science **185**, 435–440 (1974)
13. Klein, D.J., Schmeing, T.M., Moore, P.B., Steitz, T.A.: The kink-turn: a new RNA secondary structure motif. EMBO J. **20**, 4214–4221 (2001)
14. Leontis, N.B., Westhof, E.: Analysis of RNA motifs. Curr. Opin. Struct. Biol. **16**, 300–308 (2003)
15. Leontis, N.B., Lescoute, A., Westhof, E.: The building blocks and motifs of RNA architecture. Curr. Opin. Struct. Biol. **16**, 279–287 (2006)
16. Leontis, N.B., Zirbel, C.L.: Nonredundant 3D structure datasets for RNA knowledge extraction and benchmarking. In: Leontis, N., Westhof, E. (eds.) RNA 3D Structure Analysis and Prediction. Nucleic Acids and Molecular Biology, vol. 27, pp. 281–298. Springer, Heidelberg (2012)
17. Moore, P.B.: Structural motifs in RNA. Annu. Rev. Biochem. **68**, 287–300 (1999)
18. Nissen, P., Ippolito, J.A., Ban, N., Moore, P.B., Steitz, T.A.: RNA tertiary interactions in the large ribosomal subunit: the a-minor motif. Proc. Natl. Acad. Sci. U.S.A. **98**, 4899–4903 (2001)
19. Shen, Y., Wong, H.-S., Zhang, S., Zhang, L.: RNA structural motif recognition based on least-squares distance. RNA **19**, 1183–1191 (2013)
20. Ren, H., Shen, Y., Zhang, L.: The λ-Turn: a new structural motif in ribosomal RNA. In: Huang, D.-S., Jo, K.-H., Hussain, A. (eds.) ICIC 2015. LNCS, vol. 9226, pp. 456–466. Springer, Cham (2015). doi:10.1007/978-3-319-22186-1_45
21. Steinberg, S.V., Boutorine, Y.I.: G-ribo: a new structural motif in ribosomal RNA. RNA **13**, 549–554 (2007)

Biomarker Discovery

SPYSMDA: SPY Strategy-Based MiRNA-Disease Association Prediction

Zhi-Chao Jiang[✉], Zhen Shen, and Wenzheng Bao

School of Electronics and Information Engineering, Tongji University Shanghai,
Shanghai 201804, China
1531680@tongji.edu.cn

Abstract. Developing computational models to identify potential miRNA-disease associations in large scale, which could provide better understanding of disease pathology and further boost disease diagnostic and prognostic, has attracted more and more attention. Considering various disadvantages of previous computational models, we proposed the model of SPY Strategy-based MiRNA-Disease Association (SPYSMDA) prediction to infer potential miRNA-disease associations by integrating known miRNA-disease associations, disease semantic similarity network and miRNA functional similarity network. Due to the large amount of 'missing' associations in the unlabeled miRNA-disease pairs, simply regarding unlabeled instances as negative training samples would lead to high false negative rates of predicted associations. In this paper, we introduced the concept of 'spy instances' to identify reliable negatives for model performance improvement. As a result, SPYSMDA achieved excellent AUCs of 0.8827, 0.8416, and 0.8802 in global leave-one-out cross validation, local leave-one-out cross validation and 5-fold cross validation, respectively. Furthermore, Esophageal Neoplasms was taken as a case study, where 47 out of top 50 predicted miRNAs were successfully confirmed by recent biological experimental literatures.

Keywords: microRNA · Disease · Association prediction · Spy strategy · Regularized least squares classifier

1 Introduction

MiRNAs are increasingly recognized as key regulatory players in a range of human fundamental physiologic processes, including cell proliferation, differentiation, development, metabolism, aging, apoptosis and so on. Not surprisingly, functional dysregulation of miRNAs has great impact on various complex human diseases such as cardiovascular diseases, diabetes mellitus and cancers. For example, expression of miR-199a, miR-21, miR-24, miR-125b, miR-195 and miR-214 were found up-regulated in patients with cardiac hypertrophy, while expression of miR-150, miR-181b, miR-93 and miR-29c were found down-regulated [1]. MiR-10b* was identified as a master inhibitor of cell cycle in breast cancer cell line, down-regulation of which would increase expression level of corresponding target messenger RNAs BUB1, CCNA2 and PLK1 and further reduce the survival rates of breast cancer patients [2]. What's more, miR-141,

© Springer International Publishing AG 2017
D.-S. Huang et al. (Eds.): ICIC 2017, Part II, LNCS 10362, pp. 457–466, 2017.
DOI: 10.1007/978-3-319-63312-1_40

miR-375 and miR-378* were discovered significantly over-expressed in serum samples collected from prostate cancer patients [3]. These three miRNAs could serve as important prognostic biomarkers for prostate cancer progression. In addition, miR-103 and miR-107 are responsible for decrease of insulin sensitivity and destruction of glucose homeostasis and could be regarded as novel targets for treatment of type 2 diabetes mellitus [4]. As mentioned above, known associations between miRNAs and diseases could provide valuable clues for disease prevention, diagnosis, treatment and prognosis [5, 6]. However, current knowledge of miRNA-disease associations can hardly satisfy the demand of medical research. Identifying potential miRNA-disease associations for better understanding of disease molecular mechanisms has become a research focus in recent years. Although more and more high-throughput sequencing technologies (e.g. Perturb-seq) have been developed to perform a systematic analysis of functional perturbations in the genome, validate every miRNA-disease pair through these methods is completely unfeasible for the vast demand of time and expense. Nowadays, more and more miRNA-related biological datasets were generated, including miR2Disease [7], dbDEMC [8] and HMDD [9]. Powerful computational models for novel miRNA-disease association prediction could be designed based on the association records collected in these databases.

Based on the assumption that miRNAs with associations with similar diseases are functionally more related and diseases with associations with common miRNAs are phenotypically more similar, Jiang et al. [10] designed a hypergeometric distribution-based scoring system to prioritize the entire human microRNAome for diseases of interest by constructing human phenome-microRNAome network and functionally related microRNA network. However, only miRNA neighbor information was taken into consideration, which resulted in poor prediction performance of proposed model. Mørk et al. [11] coupled known miRNA-protein interactions with protein-disease associations and developed a protein-driven model called miRNA-Protein-Disease (miRPD). MiRPD is the first biological resource to link miRNAs and diseases via proteins that are likely to mediate the associations, which inspires us to take more biological datasets into account for novel association discovery. Shi et al. [12] further presented a computational framework for miRNA-disease association prediction by integrating known miRNA-target associations, disease-gene associations and protein-protein interaction (PPI) network. In this study, they mapped miRNA targets and disease genes on PPI network and then prioritized candidate miRNA-disease pairs by random walk analysis. Prediction performance of above three models is far from satisfaction because of the high false-positive and false-negative rates of predicted miRNA-target interactions. To improve prediction accuracy, Xuan et al. [13] calculated miRNA functional similarity by incorporating known miRNA-disease associations, disease phenotype similarity and disease semantic similarity obtained from information content of diseases terms. The model novelty lied in that members from the same miRNA family or cluster were set with higher weights, which remarkably enhanced the stability of miRNA functional similarity network. Then, they implemented weighted K most similar neighbors as core classifier and a reliable miRNA-disease association prediction model named HDMP was presented. However, the number of known associated miRNAs for different diseases was significantly different, thus how to select the value of parameter

K properly remained unsolved. Chen *et al.* [14] enhanced miRNA and disease similarity measures by integrating Gaussian interaction profile similarity for miRNAs and diseases. After that, the model of Within and Between Score for MiRNA-Disease Association prediction (WBSMDA) was proposed based on the integrated miRNA similarity network and integrated disease similarity network. Although HDMP and WBSMDA do not rely on predicted miRNA-target interactions, prediction performance of these network-similarity-based methods is still not satisfactory.

In this paper, we comprehensively analyzed the limitations of existing model and proposed a novel computational model of SPY Strategy-based MiRNA-Disease Association prediction (SPYSMDA) to infer potential associations between miRNAs and investigated diseases. Since unlabeled miRNA-disease pairs contain a large number of 'missing' miRNA-disease associations, simply regarding them as negative instances would lead to high false negative rates of predicted associations. In this study, we introduce a semi-supervised strategy named Spy to identify reliable negatives (RN). Computational results have confirmed that Spy strategy could significantly improve prediction performance of proposed model. We implemented global leave-one-out cross validation (global LOOCV), local leave-one-out cross validation (local LOOCV), and 5-fold cross validation (5-fold CV) to evaluate the performance of SPYSMDA. As a result, SPYSMDA achieved AUCs of 0.8827, 0.8416, and 0.8802 in global LOOCV, local LOOCV, and 5 fold CV, respectively, which showed superior performance to previous classical prediction models. In the case study of Esophageal Neoplasms, 47 out of top 50 predicted miRNAs were confirmed by experimental evidences collected in miR2Disease database [7] and dbDEMC database [8].

2   Materials and Methods

The related works on cloud platform which use OSGi framework as the architectural foundation [1, 9, 10] has researched its viability and proved it scientific. Based on the present study on the cloud platform faults, we divided various faults that might appear into several fault diagnosis models.

2.1   MiRNA-Disease Associations

The Human microRNA Disease Database (HMDD, http://www.cuilab.cn/hmdd) [9] collected 10368 high-quality miRNA-disease entries from 3511 publications. We removed duplicated association records and obtained 5430 distinct miRNA-disease associations, including 383 diseases and 495 miRNA genes. Adjacency matrix A was then adopted to quantify the relationship between miRNAs and diseases, where binary element $A(i,j)$ denotes the presence or absence of association between disease $d(i)$ and miRNA $m(j)$ ('0' represents absence while '1' represents presence). Furthermore, to present the number of diseases and miRNAs investigated in this article, variables nd and nm were respectively defined.

2.2   MiRNA Functional Similarity Network

In previous study, Wang *et al.* [15] developed miRNA functional similarity calculation based on the assumption that miRNAs with similar functions are always associated with diseases with similar phenotypes. We derived miRNA functional similarity from http://www.cuilab.cn/files/images/cuilab/misim.zip and constructed miRNA functional similarity matrix *SM*, where functional similarity score between miRNA $m(i)$ and $m(j)$ is stored in entity $SM(i,j)$.

2.3   Disease Semantic Similarity Network

Disease semantic similarity between disease i and disease j was calculated based on previous work [15]. After that, disease semantic similarity matrix *SD* was constructed.

2.4   SPYSMDA

The main process of SPYSMDA could be divided into two steps (See Fig. 1).

Since the unlabeled miRNA-disease pairs contain a large number of 'missing' associations, simply regarding them as negative training samples would lead to high false negative rates of predicted associations. Thus, we first utilized a semi-supervised strategy, Spy [16], to identify the reliable negatives from unlabeled instances. Here, known miRNA-disease associations were denoted as P (positive instances) and all the unlabeled miRNA-disease pairs were represented as U (unlabeled instances). We randomly selected a small set of positive instances, S ('spy' instances), from P and injected them into U. The remaining positive instances were denoted as P' and the union set of U and S were represented as U'. The reliable negatives could be identified from U by investigating the behavior of S by following steps: (1) instances in P' and U' were labeled as positive training samples and negative training samples respectively to build a classifier; (2) each sample in U' was assigned with a confidence score to be a positive by the classifier trained above; (3) the minimum of confidence scores of 'spy' instances was selected as threshold, and instances in U with confidence scores less than the threshold were determined as RN (reliable negatives); (4) a new classifier could be trained based on P and RN to predict potential miRNA-disease associations (See Step 1, Fig. 1). To be clear, we randomly selected 10% of P as 'spy' instances every time and repeated the strategy for 10 times to reduce the potential deviations caused by sample selection. The final threshold used for RN identification was obtained by averaging the 10 thresholds.

Based on the underlying assumption that miRNAs with similar functions always show similar relation or non-relation patterns with similar diseases and vice versa, we here developed the model of SPYSMDA to predict potential associated miRNAs for diseases of interest. The core classifier utilized in this study is Regularized Least Squares (RLS) classifier [17] for its elegant prediction ability at new test samples. Aforementioned assumption could be formulated into two RLS classifiers in disease space and miRNA space, respectively (See Step 2, Fig. 1). RLS classifier is designed to construct a continuous classification function which can meet the following two criterions: (1) it should comply with known miRNA-disease associations as accurately as possible; (2)

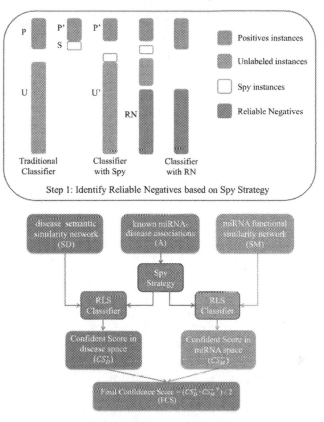

Fig. 1. Flowchart of potential miRNA-disease association prediction based on the computational model of SPYSMDA.

it should be smooth over the disease space and miRNA space, i.e. for a given miRNA (disease), similar diseases (miRNAs) would obtain similar confidence scores, which reflects the basic assumption of our model. In miRNA space, the optimal classification function could be obtained by solving the following optimization problem [17]:

$$\min_{CS_M}[||A^T - CS_M||_F^2 + \lambda_M * ||CS_M * SM * CS_M^T||_F^2] \tag{7}$$

where $|| \cdot ||_F$ is the Frobenius norm and λ_M is the trade-off parameter. By calculating the derivative of this objective function, the solution of this optimization problem could be obtained as follows:

$$CS_M^* = SM * (SM + \lambda_M * I_M)^{-1} * A^T \tag{8}$$

where I_M is an identity matrix with the same size as miRNA functional similarity matrix SM.

Optimal classification function in disease space could be obtained in the similar way as follows:

$$\min_{CS_D}[||A - CS_D||_F^2 + \lambda_D * ||CS_D * SD * CS_D^T||_F^2] \tag{9}$$

$$CS_D^* = SD * (SD + \lambda_D * I_D)^{-1} * A \tag{10}$$

where λ_D is also the trade-off parameter. Here, we set $\lambda_M = 1$ and $\lambda_D = 1$ according to previous studies [18].

Finally, we combined the optimal classification function in disease space and miRNA space by a single average operation and obtained final confidence score FCS, where element $FCS(i,j)$ in row i column j measures the association probability between disease $d(i)$ and miRNA $m(j)$.

3   Results

3.1   Performance Evaluation

In this study, SPYSMDA was compared with two state-of-art miRNA-disease association prediction models, namely RLSMDA [19] and WBSMDA [14]. As a result, SPYSMDA, RLSMDA, and WBSMDA obtained AUCs of 0.8827, 0.8653, and 0.8030 in global LOOCV framework, respectively. Furthermore, SPYSMDA yielded AUC of 0.8416 in local LOOCV framework, which had 0.0197, 0.0370 increase compared with

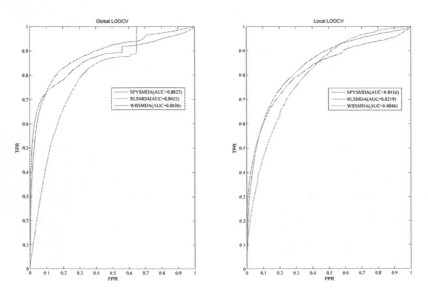

Fig. 2. Performance comparisons between SPYSMDA and two state-of-art miRNA-disease association prediction models (RLSMDA and WBSMDA) in terms of ROC curve and AUC. As a result, SPYSMDA achieved AUCs of 0.8827 and 0.8416 based on global and local LOOCV, significantly outperforming previous classification models.

RLSMDA and WBSMDA (See Fig. 2). When 5-fold CV was implemented, SPYSMDA achieved a reliable performance with AUC of 0.8802 ± 0.0040. In conclusion, SPYSMDA has a reliable performance in the framework of cross validations.

3.2 Case Study

SPYSMDA was applied to prioritize candidate miRNAs for all the diseases investigated in this study. For further prediction ability evaluation, Esophageal Neoplasms was taken as independent case study. Based on recent published experimental literatures, predicted miRNAs ranked in top 50 were validated, respectively. It should be noted that only miRNA-disease pairs without known association evidences in HMDD would be class into validation datasets, which guarantee the complete independence between validation candidates and known associations used for model training.

Table 1. Case study on esophageal neoplasms.

Top 1–25		Top 26–50	
miRNA	Evidence	miRNA	Evidence
hsa-mir-200b	dbDEMC	hsa-mir-132	dbDEMC
hsa-mir-125a	dbDEMC	hsa-mir-9	dbDEMC
hsa-mir-18a	dbDEMC	hsa-mir-222	dbDEMC
has-let-7e	dbDEMC	hsa-mir-30c	dbDEMC
hsa-mir-17	dbDEMC	hsa-mir-191	dbDEMC
hsa-mir-429	dbDEMC	hsa-let-7 g	unconfirmed
has-let-7f	unconfirmed	hsa-mir-15b	dbDEMC
hsa-mir-10b	dbDEMC	hsa-mir-93	dbDEMC
hsa-mir-16	dbDEMC	hsa-mir-107	dbDEMC;miR2Disease
hsa-mir-1	dbDEMC	hsa-mir-302c	dbDEMC
hsa-mir-18b	dbDEMC	hsa-mir-24	dbDEMC
hsa-mir-142	dbDEMC	hsa-mir-29b	dbDEMC
hsa-mir-106b	dbDEMC	hsa-mir-106a	dbDEMC
hsa-mir-199b	dbDEMC	hsa-mir-7	dbDEMC
hsa-mir-125b	dbDEMC	hsa-mir-302b	dbDEMC
hsa-mir-218	unconfirmed	hsa-mir-30a	dbDEMC
hsa-mir-19b	dbDEMC	hsa-mir-182	dbDEMC
hsa-let-7d	dbDEMC	hsa-mir-181b	dbDEMC
hsa-mir-29a	dbDEMC	hsa-mir-135a	dbDEMC
hsa-mir-146b	dbDEMC	hsa-mir-133b	dbDEMC
hsa-mir-127	dbDEMC	hsa-mir-20b	dbDEMC
hsa-let-7i	dbDEMC	hsa-mir-373	dbDEMC;miR2Disease
hsa-mir-221	dbDEMC	hsa-mir-194	dbDEMC;miR2Disease
hsa-mir-181a	dbDEMC	hsa-mir-224	dbDEMC
hsa-mir-195	dbDEMC	hsa-mir-497	dbDEMC

Esophageal Neoplasms (EN) rapidly becomes a serious threat to people from all over the world, annual deaths of with climbed from 345,000 in 1990 to 400,000 in 2012 [20]. Because of late accurate diagnosis, outcomes of EN are generally fairly poor, with five-year survival rates of 15–18% [21]. For now, curative surgery of early-stage lesions to remove all or part of the esophagus is the most effective treatment to EN patients [22]. Recent studies demonstrated that miRNAs could serve as sensitive biomarkers for early stage esophageal cancer detection. For example, expression level of miR-101, miR-143, miR-31, miR-200a, miR-196b, miR-210 and miR-27a was discovered more than doubled in esophageal cancer tissues compared with normal esophageal tissues, while expression of miR-126, miR-223, miR-454, miR-486, miR-574 was found down-regulated more than half [23]. We implemented SPYSMDA to identify potential miRNAs associated with EN. As a result, 9 out of top 10, 18 out of top 20 and 47 out of top 50 predicted miRNAs were successfully verified by recent experimental literatures (See Table 1). Typically, down-regulation of miR-200b (1st in the prediction list) in esophageal squamous cell carcinoma (ESCC) cells was significantly related with shortened survival, lymph node metastasis and advanced clinical stage in ESCC patients [24]. Down-regulation of miR-125a (2nd in the prediction list) as well as up-regulation of miR-18a, let-7e and miR-17 (3rd, 4th and 5th in the prediction list, respectively) was also discovered in EN tissues [25–28].

For further biological and clinical experiment validation, we prioritized and publicly released the prediction of all the unknown miRNA-disease pairs. It is anticipated that the candidate miRNA-disease pairs with higher ranks could offer valuable clues and would be confirmed by experimental observation in the near future

4   Discussion

With the rapid development of high-throughput sequencing techniques, increasing literatures have demonstrated that functional disorders of miRNAs play critical roles in a variety of complex human diseases. Identifying potential miRNA-disease associations for better understanding of disease pathology and novel discovery of drugs has attracted lots of attention in recent years. Based on the assumption that miRNAs with similar functions tend to be associated with diseases with similar phenotypes, more and more computational models have been designed to reduce experimental time and cost that traditional biological experiments suffer. In this study, we presented a novel computational model of SPYSMDA to prioritize candidate miRNA-disease pairs for further biological experiment validation. SPYSMDA is a multiple network-based prediction model taking advantages of miRNA functional similarity network, disease semantic similarity network and known miRNA-disease associations derived from HMDD database. SPYSMDA achieved reliable prediction performance with AUCs of 0.8827, 0.8416 and average AUC of 0.8802 in global LOOCV, local LOOCV and 5-fold CV, respectively.

In conclusion, the following factors drove the excellent prediction performance of SPYSMDA. First of all, three kinds of heterogeneous biological network (i.e. known miRNA-disease association network, disease semantic similarity network and miRNA functional similarity network) were integrated into SPYSMDA, which effectively improved the data completeness and further reduce model prediction bias. Furthermore, traditional supervised prediction models simply regards unlabeled miRNA-disease pairs as negative instances for model training, which would lead to high false negative rates of predicted associations. To cope with it, we introduced a semi-supervised strategy, Spy, to identify reliable negatives (RN) from unlabeled instances. Spy strategy enabled model training on positives and reliable negative and significantly improved prediction performance of proposed model. In addition, SPYSMDA could be applied to new diseases without any known related miRNAs as well as new miRNAs without any known related diseases. Finally, SPYSMDA is a global ranking model and could prioritize all candidate miRNA-disease pairs for all investigated disease in large-scale. It is anticipated that SPYSMDA would be an important biological resource for experimental guidance.

Acknowledgments. This work was supported by the grants of the National Science Foundation of China, Nos. 61402334, 61472282, 61520106006, 31571364, U1611265, 61672203, 61472280, 61532008, 61472173, 61572447, 61373098 and 61672382, China Postdoctoral Science Foundation Grant, Nos. 2016M601646. De-Shuang Huang is the corresponding author of this paper.

References

1. van Rooij, E., Sutherland, L.B., Liu, N., Williams, A.H., McAnally, J., Gerard, R.D., et al.: A signature pattern of stress-responsive microRNAs that can evoke cardiac hypertrophy and heart failure. Proc. Natl. Acad. Sci. U.S.A. **103**, 18255–18260 (2006)
2. Biagioni, F., Ben-Moshe, N.B., Fontemaggi, G., Canu, V., Mori, F., Antoniani, B., et al.: miR-10b*, a master inhibitor of the cell cycle, is down-regulated in human breast tumours. EMBO Mol. Med. **4**, 1214–1229 (2012)
3. Nguyen, H.C., Xie, W., Yang, M., Hsieh, C.L., Drouin, S., Lee, G.S., et al.: Expression differences of circulating microRNAs in metastatic castration resistant prostate cancer and low-risk, localized prostate cancer. Prostate **73**, 346–354 (2013)
4. Trajkovski, M., Hausser, J., Soutschek, J., Bhat, B., Akin, A., Zavolan, M., et al.: MicroRNAs 103 and 107 regulate insulin sensitivity. Nature **474**, 649–653 (2011)
5. Yanaihara, N., Caplen, N., Bowman, E., Seike, M., Kumamoto, K., Yi, M., et al.: Unique microRNA molecular profiles in lung cancer diagnosis and prognosis. Cancer Cell **9**, 189–198 (2006)
6. Weinberg, M.S., Wood, M.J.: Short non-coding RNA biology and neurodegenerative disorders: novel disease targets and therapeutics. Hum. Mol. Genet. **18**, R27–R39 (2009)
7. Jiang, Q., Wang, Y., Hao, Y., Juan, L., Teng, M., Zhang, X., et al.: miR2Disease: a manually curated database for microRNA deregulation in human disease. Nucleic Acids Res. **37**, D98–104 (2009)
8. Yang, Z., Ren, F., Liu, C., He, S., Sun, G., Gao, Q., et al.: dbDEMC: a database of differentially expressed miRNAs in human cancers. BMC Genom. **11**(Suppl 4), S5 (2010)
9. Li, Y., Qiu, C., Tu, J., Geng, B., Yang, J., Jiang, T., et al.: HMDD v2.0: a database for experimentally supported human microRNA and disease associations. Nucleic Acids Res. **42**, D1070–D1074 (2014)
10. Jiang, Q., Hao, Y., Wang, G., Juan, L., Zhang, T., Teng, M., et al.: Prioritization of disease microRNAs through a human phenome-microRNAome network. BMC Syst. Biol. **4**(Suppl 1), S2 (2010)

11. Mork, S., Pletscher-Frankild, S., Palleja Caro, A., Gorodkin, J., Jensen, L.J.: Protein-driven inference of miRNA-disease associations. Bioinformatics **30**, 392–397 (2014)
12. Huang, D.S., Zhang, L., Han, K., Deng, S., Yang, K., Zhang, H.: Prediction of protein-protein interactions based on protein-protein correlation using least squares regression. Curr. Protein Pept. Sci. **15**(6), 553–560 (2014)
13. Xuan, P., Han, K., Guo, M., Guo, Y., Li, J., Ding, J., et al.: Prediction of microRNAs associated with human diseases based on weighted k most similar neighbors. PLoS ONE **8**, e70204 (2013)
14. Chen, X., Yan, C.C., Zhang, X., You, Z.H., Deng, L., Liu, Y., et al.: WBSMDA: within and between score for MiRNA-disease association prediction. Sci. Rep. **6**, 21106 (2016)
15. Wang, D., Wang, J., Lu, M., Song, F., Cui, Q.: Inferring the human microRNA functional similarity and functional network based on microRNA-associated diseases. Bioinformatics **26**, 1644–1650 (2010)
16. Liu, B., Lee, W.S., Yu, P.S., Li, X.: Partially supervised classification of text documents. In: Nineteenth International Conference on Machine Learning, pp. 387–394 (2003)
17. Rifkin, R., Klautau, A.: In defense of one-vs-all classification. J. Mach. Learn. Res. **5**, 101–141 (2004)
18. van Laarhoven, T., Nabuurs, S.B., Marchiori, E.: Gaussian interaction profile kernels for predicting drug-target interaction. Bioinformatics **27**, 3036–3043 (2011)
19. Chen, X., Yan, G.Y.: Semi-supervised learning for potential human microRNA-disease associations inference. Sci. Rep. **4**, 5501 (2014)
20. Lozano, R., Naghavi, M., Foreman, K., Lim, S., Shibuya, K., Aboyans, V., et al.: Global and regional mortality from 235 causes of death for 20 age groups in 1990 and 2010: a systematic analysis for the Global Burden of Disease Study 2010. Lancet **380**, 2095–2128 (2012)
21. Deng, S.-P., Zhu, L., Huang, D.S.: Mining the bladder cancer-associated genes by an integrated strategy for the construction and analysis of differential co-expression networks. BMC Genom. **16**(Suppl 3), S4 (2015)
22. Zheng, C.-H., Zhang, L., Ng, V.T.Y., Shiu, S.C.-K., Huang, D.S.: Molecular pattern discovery based on penalized matrix decomposition. IEEE/ACM Trans. Comput. Biol. Bioinform. **8**(6), 1592–1603 (2011)
23. Huang, D.S., Yu, H.-J.: Normalized feature vectors: a novel alignment-free sequence comparison method based on the numbers of adjacent amino acids. IEEE/ACM Trans. Comput. Biol. Bioinform. **10**(2), 457–467 (2013)
24. Deng, S.-P., Zhu, L., Huang, D.S.: Predicting hub genes associated with cervical cancer through gene co-expression networks. IEEE/ACM Trans. Comput. Biol. Bioinform. **13**(1), 27–35 (2016)
25. Fassan, M., Pizzi, M., Realdon, S., Balistreri, M., Guzzardo, V., Zagonel, V., et al.: The HER2-miR125a5p/miR125b loop in gastric and esophageal carcinogenesis. Hum. Pathol. **44**, 1804–1810 (2013)
26. Zhang, W., Lei, C., Fan, J., Wang, J.: miR-18a promotes cell proliferation of esophageal squamous cell carcinoma cells by increasing cylin D1 via regulating PTEN-PI3K-AKT-mTOR signaling axis. Biochem. Biophys. Res. Commun. **477**, 144–149 (2016)
27. Deng, S.P., Huang, D.S.: SFAPS: an R package for structure/function analysis of protein sequences based on informational spectrum method. Methods **69**(3), 207–212 (2014)
28. Ogawa, R., Ishiguro, H., Kuwabara, Y., Kimura, M., Mitsui, A., Katada, T., et al.: Expression profiling of micro-RNAs in human esophageal squamous cell carcinoma using RT-PCR. Med. Mol. Morphol. **42**, 102–109 (2009)

Applications of Machine Learning Techniques to Computational Proteomics, Genomics, and Biological Sequence Analysis

SOFM-Top: Protein Remote Homology Detection and Fold Recognition Based on Sequence-Order Frequency Matrix

Junjie Chen[1], Mingyue Guo[1], Xiaolong Wang[1,2], and Bin Liu[1,2(✉)]

[1] School of Computer Science and Technology, Harbin Institute of Technology
Shenzhen Graduate School, Shenzhen 518055, Guangdong, China
junjie.chen.hit@gmail.com, gmy362717784@icloud.com,
wangxl@insun.hit.edu.cn, bliu@hit.edu.cn
[2] Key Laboratory of Network Oriented Intelligent Computation,
Harbin Institute of Technology Shenzhen Graduate School,
Shenzhen 518055, Guangdong, China

Abstract. Protein remote homology detection and fold recognition are critical for the studies of protein structure and function. Currently, the profile-based methods showed the state-of-the-art performance in this field, which are based on widely used sequence profiles, such as Position-Specific Frequency Matrix (PSFM) and Position-Specific Scoring Matrix (PSSM). However, these approaches ignore the sequence-order effects along protein sequence. In this study, we proposed a novel profile, called Sequence-Order Frequency Matrix (SOFM), which can incorporate the sequence-order information and extract the evolutionary information from Multiple Sequence Alignment (MSA). Statistical tests and experimental results demonstrated its effects. Combined with a previously proposed approach Top-n grams, the SOFM was then applied to remote homology detection and fold recognition, and a computational predictor called **SOFM-Top** was proposed. Evaluated on four benchmark datasets, it outperformed other state-of-the-art methods in this filed, indicating that **SOFM-Top** would be a more useful tool, and SOFM is a richer representation than PSFM and PSSM. SOFM will have many potential applications since profiles have been widely used for constructing computational predictors in the studies of protein structure and function.

Keywords: Sequence-Order Frequency Matrix · Profile representation · Protein remote homology detection · Protein fold recognition · Top-n-grams

1 Introduction

Protein remote homology detection and fold recognition are two fundamental problems for inferring the structure of a newly sequenced protein with low sequence similarities [1–4]. Some computational methods have been proposed, and among them,

Electronic supplementary material The online version of this chapter (doi:10.1007/978-3-319-63312-1_41) contains supplementary material, which is available to authorized users.

© Springer International Publishing AG 2017
D.-S. Huang et al. (Eds.): ICIC 2017, Part II, LNCS 10362, pp. 469–480, 2017.
DOI: 10.1007/978-3-319-63312-1_41

profile-based methods showed the state-of-the-art performance [5]. A profile is calculated based on the Multiple Sequence Alignments (MSAs) obtained by searching against a non-redundant database [6] in an unsupervised manner. Each protein sequence in a MSA has statistically significant sequence identity with the query. The comprehensive review and comparison of different computational methods for protein remote homology detection can be found in [5].

Position-Specific Frequency Matrix (PSFM) and Position-Specific Scoring Matrix (PSSM) are two most widely used profiles. However, they assume that each amino acid residue in a MSA is independent, and the matrix scores are calculated based on the residues appearing in the given column in a MSA. As a result, the sequence-order information in MSA is totally lost. However, some previous studies showed that sequence-order effect is important for representing protein sequences, and some methods have been proposed to extract the sequence-order information from protein sequences, such as kmer [7], distance pair [8, 9], and mismatch [10].

In order to overcome the disadvantages of the aforementioned profiles, in this study, we are to propose a novel profile, called Sequence-Order Frequency Matrix (SOFM) to incorporate the sequence-order effects and the evolutionary information from MSAs. To our best knowledge, SOFM is the first profile considering the sequence-order effects in MSAs. It is then applied to protein remote homology detection and fold recognition, and a predictor called **SOFM-Top** is proposed. Evaluated on four benchmark datasets, it outperforms other state-of-the-art methods in this filed, indicating that SOFM-Top would be a more useful tool, and SOFM is a richer representation than PSFM and PSSM.

2   Materials and Methods

2.1   Benchmark Datasets

Remote homology detection and fold recognition are simulated as classification problems at superfamily level and fold level respectively. Two commonly used superfamily benchmark datasets and two commonly used fold benchmark datasets are used to evaluate the performance, which have been used by many studies of remote homology detection methods and fold recognition [11–15].

The first superfamily benchmark dataset [14] was constructed based on SCOP version 1.53. It contains 4352 proteins from 54 families. The second one [12] was constructed based on SCOP version 1.67, containing 4019 sequences from 102 families.

Fold recognition is a more challenging task, because it is simulated at fold level rather than superfamily level, and the sequence identities at fold level are much lower than that at superfamily level. The first fold benchmark [13] was constructed from SCOP version 1.53, containing 4352 sequences from 23 superfamilies. The second fold benchmark [12] was constructed from SCOP version 1.67 with 3840 sequences from 86 superfamilies.

The sequence identities of any pair of proteins in all these benchmark datasets are no more than 95%.

2.2   Constructing SOFM Based on Multiple Sequence Alignment

There are two main steps for constructing the SOFM: (1) extracting the substrings with sequence-order information; (2) computing the scores of substrings in SOFM. Here, we will introduce how to calculate the SOFM by extracting the adjacent amino acids substrings and counting the frequency of substrings as SOFM scores.

Suppose a protein **P** with L amino acids residues:

$$\mathbf{P} = \quad A_1 \quad A_2 \quad A_3 \quad A_4 \quad \cdots \quad A_{L-1} \quad A_L \tag{1}$$

where A_i ($i = 1, 2, \ldots, L$) represents the i-th amino acid residues in the protein **P**. Its corresponding MSA can be represented as:

$$
\begin{aligned}
\mathbf{P}_1 &: \quad A_{1,1} \quad A_{1,2} \quad A_{1,3} \quad A_{1,4} \quad \cdots \quad A_{1,L-1} \quad A_{1,L} \\
\mathbf{P}_2 &: \quad A_{2,1} \quad A_{2,2} \quad A_{2,3} \quad A_{2,4} \quad \cdots \quad A_{2,L-1} \quad A_{2,L} \\
\mathbf{P}_3 &: \quad A_{3,1} \quad A_{3,2} \quad A_{3,3} \quad A_{3,4} \quad \cdots \quad A_{3,L-1} \quad A_{3,L} \\
&\cdots \\
\mathbf{P}_N &: \quad A_{N,1} \quad A_{N,2} \quad A_{N,3} \quad A_{N,4} \quad \cdots \quad A_{N,L-1} \quad A_{N,L}
\end{aligned}
\tag{2}
$$

where the $\mathbf{P}_i$ ($i = 1, 2, \ldots, N$) represents the i-th sequence in the MSA and $A_{i,j}$ ($i = 1, 2, \ldots, N; j = 1, 2, \ldots, L$) represents an amino acid or a gap in the j-th position in sequence $\mathbf{P}_i$. In this study, the MSA for each protein is generated by PSI-BLAST version 2.3.0 searching against a non-redundant protein NCBI's nrdb90 database [16]. The parameters of PSI-BLAST were set as default except that the number of iterative was set as 5.

A subsequence $s_{i,j}$ with k amino acids at position j in sequence $\mathbf{P}_i$ can be represented as:

$$s_{i,j} = A_{i,j} A_{i,j+1} A_{i,j+2} \cdots A_{i,j+k-1} \tag{3}$$

where k ($k = 1, 2, \ldots, L$) represents the length of subsequence $s_{i,j}$. Let $\mathbb{S}_j$ to represent the collection that contains all the subsequences with k amino acids at position j.

$$\mathbb{S}_j = \left\{ A_{i,j} A_{i,j+1} A_{i,j+2} \cdots A_{i,j+k-1} | \forall i \in [1, 2, \cdots, N] \right\} \tag{4}$$

where the elements in $\mathbb{S}_j$ are repeatable and the size of $\mathbb{S}_j$ equals to the number of protein sequences in MSA.

The sequence-order effects can be incorporated if the subsequence appearing frequency in the given column of a MSA is used to generate the profile. In this regard, the SOFM scores were calculated based on frequency of subsequence $s_{i,j}$ appearing the $\mathbb{S}_j$.

A SOFM can be represented as a matrix:

$$
\mathbf{M} = \begin{bmatrix}
m_{1,1} & m_{1,2} & \cdots & m_{1,L-k+1} \\
m_{2,1} & m_{2,2} & \cdots & m_{2,L-k+1} \\
\vdots & \vdots & \vdots & \vdots \\
m_{20^k,1} & m_{20^k,2} & \cdots & m_{20^k,L-k+1}
\end{bmatrix}
\tag{5}
$$

where L is the length of the protein; 20 represents the total number of standard amino acids, and 20^k is the total number of all the possible subsequences φ_i ($i = 1, 2, \ldots, 20^k$) of length k ($k = 1, 2, \ldots, L$). The $m_{i,j}$ ($0 \le m_{i,j} \le 1$) is the occurring probability of subsequence φ_i ($i = 1, 2, \ldots, 20^k$) occurring in the position j ($j = 1, 2, \ldots, L - k + 1$) during evolutionary processes, which is defined as:

$$m_{i,j} = \frac{\mathsf{F}_{i,j}(\varphi_i, \mathbb{S}_j)}{\sum\limits_{i=1}^{20^k} \mathsf{F}_{i,j}(\varphi_i, \mathbb{S}_j)} \tag{6}$$

$$\mathsf{F}_{i,j}(\varphi_i, \mathbb{S}_j) = \begin{cases} f(\varphi_i | \mathbb{S}_j), & \varphi_i \in \mathbb{S}_j \\ 0, & \varphi_i \notin \mathbb{S}_j \end{cases} \tag{7}$$

where $f(\varphi_i | \mathbb{S}_j)$ is the occurring frequency of subsequence φ_i in collection $\mathbb{S}_j$. $\mathsf{F}_{i,j}(\varphi_i, \mathbb{S}_j)$ equals to $f(\varphi_i | \mathbb{S}_j)$ if the subsequence φ_i appears in the j-th position of MSA, and otherwise is 0. $\sum\limits_{i=1}^{20^k} \mathsf{F}_{i,j}(\varphi_i, \mathbb{S}_j)$ represents the total occurring frequency of all subsequences in position j. For each column of $\mathbf{M}$, the scores add up to one. Higher score $m_{i,j}$ indicates more likely appearing of subsequence φ_i at j-th position of protein sequence $\mathbf{P}$. In this study, SOFM with $k = 1, 2$ and 3 were investigated, and their dimensions are $20 \times L$, $400 \times (L - 1)$ and $8000 \times (L - 2)$, respectively.

SOFM variants can be easily constructed based on different sequence-order extraction methods (such as distance pair substrings and mismatch substrings) or based on different weighted schemes (such as considering the background frequency of substrings and log-transformation).

2.3 Converting SOFM into Top-n-grams

The SOFM is a matrix that contains local sequence order information extracted from the MSA, but it can't be directly incorporated into the predictors. To deal with this problem, Top-n-gram approach [11] is employed to extract the building blocks from SOFM, and then the building blocks can be used to generate a fix-length vector for training the classifier **SOFM-Top**.

In previous studies, Top-n-gram extracts the most frequent subsequence in each column of a PSFM. The building blocks extracted by Top-n-gram [11] have shown promising performance, because the subsequence with higher frequency is more conserved. In this study, SOFM was combined with Top-n-gram for protein remote homology detection and fold recognition.

Given a SOFM, 20^k subsequences in each column are sorted in descending order by their frequencies, and then the most frequent one is selected.

$$m_j^{\text{max}} = \max\{m_{1,j}, m_{2,j}, m_{3,j}, \cdots, m_{20^k,j}\} \tag{8}$$

where $m_{1,j}, m_{2,j}, m_{3,j}, \cdots, m_{20^k,j}$ are the probability scores of 20^k subsequences at position j of protein $\mathbf{P}$. m_j^{max} represents the highest frequency, and the φ_j^{max} is the corresponding most frequent subsequence, which is called the Top-1-gram at position j. For a protein $\mathbf{P}$ with L amino acids residues, the dimension of its SOFM is $20^k \times (L - k + 1)$. Thus, the number of different Top-1-grams is $L - k + 1$, which can be represented as $\varphi_1^{max}, \varphi_2^{max}, \varphi_3^{max}, \cdots, \varphi_{L-k+1}^{max}$.

A fixed-length feature vector $\mathbf{V}(\mathbf{P})$ with 20^k dimensions can be generated by counting the occurrence frequency of each Top-1-gram:

$$\mathbf{V}(P) = [f(\varphi_1) \quad f(\varphi_2) \quad f(\varphi_3) \quad \cdots \quad f(\varphi_{20^k})] \tag{9}$$

where the $f(\varphi_i)$ represents the appearing frequencies of φ_i. In this study, **SOFM-Top** with $k = 1$, 2 and 3 were investigated, and the dimensions of their resulting feature vectors are 20, 400 and 8000, respectively. The main process is shown in Fig. 1. For more information of Top-n-grams, please refer to [11].

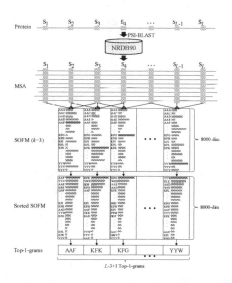

Fig. 1. The flow chart of generating Top-1-grams based on SOFM ($k = 3$). The query protein is first input into PSI-BLAST to generate the MSA. Then, the SOFM is constructed by counting the frequency of all kinds of subsequences with size k. For a given protein $\mathbf{P}$ with length of L, the dimension of SOFM ($k = 3$) is 8000 $\times$ ($L - 2$). The histograms represent the occurring frequencies of all possible subsequences in each column, where the most frequent subsequence is colored with red. Each column of SOFM is sorted in descending order to extract the Top-1-grams. Finally, there are $L - k + 1$ Top-1-grams for protein $\mathbf{P}$ with length of L, and the feature vectors of $\mathbf{P}$ can be generated based on the extracted Top-1-grams by using Eq. (9). (Color figure online)

2.4   Construction of SVM Classifiers

Support Vector Machine (SVM) was employed to construct the proposed **SOFM-Top** predictor, which has exhibited excellent performance in practice and has a strong theoretical foundation of statistical learning. In this study, the publicly available Gist SVM package (http://www.chibi.ubc.ca/gist/) was employed as the implementation. All parameters in Gist SVM were set as default.

2.5   Evaluation Methodology

The receiver operating characteristics (ROC) score is used to evaluate the performance of all kinds of methods. Because the positive and negative samples are not evenly distributed, ROC score is a better way to evaluate the trade-off between the specificity and sensitivity. An ROC score is the normalized area under a curve that plots true positives against false positives for different classification thresholds. ROC50 score is the area under the ROC curve up to the first 50 false positives. The score of 1 means perfect separation of positive samples from negative ones, whereas a score of 0.5 means that the results are randomly predicted.

3   Result and Analysis

3.1   The Amino Acid Residues in MSAs are Dependent

The Pearson's chi-square test was performed to calculate the correlations between two adjacent amino acids according to their observation frequencies in a large non-redundant protein database [17, 18].

The first residue is represented by variable x, and the second residue is represented by variable y. The Chi-square test χ^2 of two adjacent amino acids x and y is defined as:

$$\chi^2 = \sum_{(x,y)}^{(20,20)} \frac{\left(O_{x,y} - E_{x,y}\right)^2}{E_{x,y}} \tag{10}$$

where $O_{x,y}$ is the observation frequency of two amino acid x and y co-occurring; $E_{x,y}$ is the expected frequency of $O_{x,y}$, which can be calculated by the following equation:

$$E_{x,y} = \frac{n_x \times n_y}{N} \tag{11}$$

where n_x is the sum of observation frequencies that the first positions are amino acid x; the n_y is the sum of observation frequencies that the second positions are amino acid y; N is the total number of all samples.

In this study, the observation frequencies of two adjacent amino acids were counted according to NCBI's nrdb90 which contains 534,936 proteins with the identities of any pairs of proteins no more than 90%. The observation frequencies were shown in **Supplementary S1**. The chi-square value is 2117407.25 computed by using Eq. (10), and the critical value is 406.30 under the degree of freedom 361. Thus, the chi-square value χ^2 is much larger than the critical value, indicating that the two adjacent amino acids are obviously correlated.

The widely-used profiles PSFM and PSSM assume that the amino acids are independent from each other. As a result, all the sequence-order information is totally lost. In contract, SOFM is able to incorporate the sequence-order effects and overcomes the disadvantages of PSFM and PSSM. In the following sections, we will evaluate its performance for protein homology detection and fold recognition.

3.2 The Influence of K Value on the Performance of SOFM-Top Predictor

As introduced above, there is a parameter k in **SOFM-Top**. In this section, we will investigate its impact on the predictive performance.

Table 1 shows the performance of **SOFM-Top** with different k values on the four benchmark datasets. For remote homology detection, in terms of ROC score, **SOFM-Top** performs well for $k = 1$ and $k = 2$, and its performance slightly decreases for $k = 3$. For fold recognition, in terms of ROC score, **SOFM-Top** ($k = 3$) achieves the best performance. These results show that smaller k values are preferred for remote homology detection, while larger values are suggested for fold recognition. It is interesting to explore if these three kinds of **SOFM-Top** ($k = 1, 2, 3$) are complementary or not. In this regard, pairwise performance comparisons among the three predictors are conducted, and the results are shown in **Supplementary S2** and **S3**. If the predictors have totally same performance, the identical results of the two methods will fall on the diagonal line. From the results, we can see that the three predictors are complementary because only a few points fall on the diagonal line. Therefore, these three predictors are combined to further improve the performance. As shown in Table 2, the results are obviously improved.

Table 1. Performance of **SOFM-Top** with different k values on the four benchmark datasets.

Parameters	SCOP 1.53				SCOP 1.67			
	Superfamily		Fold		Superfamily		Fold	
	ROC	ROC50	ROC	ROC50	ROC	ROC50	ROC	ROC50
$k = 1$	0.9014	0.7306	0.7763	0.3407	0.9062	0.7670	0.7908	0.6204
$k = 2$	**0.9249**	**0.7695**	0.7889	0.2927	**0.9252**	**0.8026**	0.8289	**0.6531**
$k = 3$	0.8827	0.7397	**0.8308**	**0.3702**	0.9086	0.7872	**0.8332**	0.6464

Table 2. The performance of **SOFM-Top** with different k value combinations on the four benchmark datasets.

Parameters	SCOP 1.53				SCOP 1.67			
	Superfamily		Fold		Superfamily		Fold	
	ROC	ROC50	ROC	ROC50	ROC	ROC50	ROC	ROC50
$k = 1$ & $k = 2$	0.9365	0.7637	0.8280	0.3682	0.9397	0.8066	0.8487	0.6697
$k = 2$ & $k = 3$	0.9340	**0.7861**	0.8342	**0.3912**	0.9338	0.8094	0.8469	**0.6874**
$k = 1$ & $k = 2$ & $k = 3$	**0.9414**	0.7644	**0.8610**	0.3877	**0.9443**	**0.8243**	**0.8635**	0.6844

3.3   Performance Comparison with Other Related Methods

Tables 3 and 4 summarize the performance of **SOFM-Top** and some state-of-the-art methods on superfamily and fold benchmark datasets, including SVM-Bprofile [19], SVM-Top-n-gram [11], SVM-Ngram [20], SVM-pattern [20], SVM-motif [20], PSI-BLAST [21], SVM-pairwise [22], GPkernel [23], LSTM [23], SVM-LA [22].

From these two tables, we can see that **SOFM-Top** outperforms other methods on both superfamily and fold benchmark datasets, indicating that the SOFM profile is efficient.

Table 3. Performance comparison on SCOP 1.53 superfamily benchmark dataset.

Methods	ROC	ROC50	Source
SOFM-Top	0.941	0.764	This study
SVM-Bprofile ($Ph = 0.13$)	0.903	0.681	[19]
SVM-Top-n-gram	0.933	0.763	[11]
SVM-Ngram	0.791	0.584	[20]
SVM-Pattern	0.835	0.589	[20]
SVM-Motif	0.814	0.616	[20]
PSI-BLAST	0.675	0.330	[20]
SVM-Pairwise	0.896	0.464	[22]
GPkernel	0.899	–	[23]
Mismatch	0.872	0.400	[10]
SVM-LA	0.925	0.649	[22]
LSTM	0.932	0.652	[23]

From Table 4 we can see that **SOFM-Top** outperforms SVM-Top-n-gram by 5% in terms of ROC score. These two approaches are similar. Their only difference is that they are based on different profiles, i.e. SOFM for **SOFM-Top,** PSFM for SVM-Top-n-gram. These results indicate that the SOFM is a richer representation of MSA than PSFM, and it can improve the performance for both remote homology detection and fold recognition.

For a protein with length L, the shape of its PSFM is $20 \times L$, and the shape of SOFM (k) is $20^k \times (L - k + 1)$. Although the size of SOFM is exponential growth as the k increases, the time complexities of both PSFM and SOFM(k) are $O(L)$, because the scores of SOFM(k) are computed according the frequency of substrings in MSA.

And the input feature vector of **SOFM-Top(k)** can be generated with $O(L)$ time complexity based on **SOFM(k)**. Thus, the time complexities of both PSFM and SOFM (k) are the same. Furthermore, the feature vector length of **SOFM-Top(k)** is 20^k where $L - k + 1$ scores are non-zeros. If we set the k too large, the generated feature vector will be very sparse.

Table 4. Performance comparison on SCOP 1.67 fold benchmark dataset.

Methods	ROC	ROC50	Source
SOFM-Top	0.864	0.684	This study
SVM-Bprofile ($Ph = 0.11$)	0.804	0.644	[19]
SVM-Top-n-gram	0.818	0.677	[11]
SVM-Motif	0.698	0.308	[20]
PSI-BLAST	0.501	0.010	[23]
Mismatch	0.814	0.467	[10]
SVM-Pairwise	0.724	0.359	[22]
SVM-LA	0.834	0.504	[22]
GPkernel	0.844	0.514	[23]
LSTM	0.821	0.571	[23]

3.4   More Information Content in SOFM Than PSFM

In this section, we will explore the reason why SOFM is a richer profile representation than PSFM.

The information content is an important quantitative indicator, which measures the conservation at a given position in profiles [24–26]. The positions with high information content are highly conserved and have a low tolerance for mutations, while positions with the low information content have a high tolerance for mutations. The information content in SOFM is calculated by:

$$I = H - H' \tag{12}$$

$$H = -\sum_{i=1}^{20^k} \frac{1}{20^k} \log_2 \frac{1}{20^k} \tag{13}$$

$$H' = -\sum_{i=1}^{20^k} p(\varphi_i) \log_2 p(\varphi_i) \tag{14}$$

where 20^k is the number of all possible subsequences with length of k; $p(\varphi_i)$ is the present probability of subsequence φ_i in Eq. (6) in one specific position. The information content is measured in bits. In the case of SOFM, a position in SOFM that all amino acids occur with equal probability has an information content of 0 bit, while a position that only single amino acid can occur has an information content of I bits. H' is information entropy, which measures the uncertainty of occurring amino acids at one position. The maximum information content of a position in SOFM($k = 1$) is $\log_2 1/20$ bits, and in SOFM($k = 2$), the maximum is $2 \times \log_2 1/20$ bits. The maximum in SOFM ($k = 3$) is $3 \times \log_2 1/20$ bits.

We take the *Phage T4 lysozyme* protein as an example which has 163 amino acid residues. Its SCOP ID is 'd119l_' belonging to *Phage lysozyme* family and *Lysozyme-like* superfamily. The information content at each position in the protein is computed with the subsequence length $k = 1$, 2 and 3. As shown in Fig. 2, the three

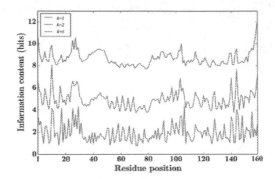

Fig. 2. The comparison of information content per position. We take *Phage T4 lysozyme* protein as an example, and computing the information content per position in SOFM with $k = 1$, 2 and 3. The three lines have the same trends, and the information content is richer as the increasement of k value.

lines have the same trends, and SOFM ($k = 3$) has the highest information content, while SOFM ($k = 1$) has the lowest information content. The reason is that the longer subsequence contains richer sequence-order information, and more sequence-order effects from MSAs are incorporated into SOFM. Therefore, SOFM is a richer representation than PSFM.

4   Conclusion and Discussion

In this study, we presented a novel profile called SOFM, which was constructed based on the appearing local subsequence frequency in MSA. Thus, it could contain the sequence-order information and evolutionary information from MSAs. SOFM was then combined with a feature extraction approach Top-n-gram, and a predictor called **SOFM-Top** was proposed for protein remote homology detection and fold recognition. Experimental results on four widely used benchmark datasets showed that **SOFM-Top** achieved promising predictive performance. Various statistical tests showed that SOFM is a richer representation than the widely-used PSFM and PSSM.

Profiles have been widely used as a key feature to represent the proteins, and therefore, it is anticipant that SOFM will have many potential applications.

Authors' Contribution. Bin Liu and Junjie Chen conceived of this study. Junjie Chen and Mingyue Gou carried out the study, coding the experiments, performing the statistical analysis, and drafting the manuscript. Bin Liu and Xiaolong Wang participated in drafting and modifying the manuscript. Informed consent was obtained from all individual participants included in the study.

Acknowledgements. This work was supported by the National Natural Science Foundation of China (No. 61672184, and 61573118), the Natural Science Foundation of Guangdong Province (2014A030313695), Guangdong Natural Science Funds for Distinguished Young Scholars (2016A030306008), and Scientific Research Foundation in Shenzhen (Grant No. JCYJ2015 0626110425228).

References

1. Blake, J.D., Cohen, F.E.: Pairwise sequence alignment below the twilight zone. J. Mol. Biol. **307**(2), 721–735 (2001)
2. Rost, B.: Twilight zone of protein sequence alignments. Protein Eng. **12**(2), 85–94 (1999)
3. Ananthalakshmi, P., Kumar, C.K., Jeyasimhan, M., Sumathi, K., Sekar, K.: Fragment finder: a web-based software to identify similar three-dimensional structural motif. Nucl. Acids Res. **33**(suppl 2), W85–W88 (2005)
4. Nagarajan, R., Siva Balan, S., Sabarinathan, R., Kirti Vaishnavi, M., Sekar, K.: Fragment finder 2.0: a computing server to identify structurally similar fragments. J. Appl. Crystallogr. **45**(2), 332–334 (2012)
5. Chen, J., Guo, M., Wang, X., Liu, B.: A comprehensive review and comparison of different computational methods for protein remote homology detection. Brief. Bioinform. bbw108 (2016). doi:10.1093/bib/bbw108
6. Gribskov, M., McLachlan, A.D., Eisenberg, D.: Profile analysis: detection of distantly related proteins. Proc. Natl. Acad. Sci. **84**(13), 4355–4358 (1987)
7. Leslie, C.S., Eskin, E., Noble, W.S.: The spectrum kernel: a string kernel for SVM protein classification. In: Pacific Symposium on Biocomputing, pp. 566–575 (2002)
8. Liu, B., Chen, J., Wang, X.: Protein remote homology detection by combining Chou's distance-pair pseudo amino acid composition and principal component analysis. Mol. Genet. Genomics **290**(5), 1919–1931 (2015)
9. Lingner, T., Meinicke, P.: Remote homology detection based on oligomer distances. Bioinformatics **22**(18), 2224–2231 (2006)
10. Eskin, E., Weston, J., Noble, W.S., Leslie, C.S.: Mismatch string kernels for SVM protein classification. In: Advances in Neural Information Processing Systems, pp. 1417–1424 (2002)
11. Liu, B., Wang, X., Lin, L., Dong, Q., Wang, X.: A discriminative method for protein remote homology detection and fold recognition combining Top-n-grams and latent semantic analysis. BMC Bioinform. **9**, 510 (2008)
12. Håndstad, T., Hestnes, A.J., Sætrom, P.: Motif kernel generated by genetic programming improves remote homology and fold detection. BMC Bioinform. **8**(1), 1 (2007)
13. Rangwala, H., Karypis, G.: Profile-based direct kernels for remote homology detection and fold recognition. Bioinformatics **21**(23), 4239–4247 (2005)
14. Liao, L., Noble, W.S.: Combining pairwise sequence similarity and support vector machines for detecting remote protein evolutionary and structural relationships. J. Comput. Biol. **10**(6), 857–868 (2003)
15. Liu, X., Zhao, L., Dong, Q.: Protein remote homology detection based on auto-cross covariance transformation. Comput. Biol. Med. **41**(8), 640–647 (2011)
16. Holm, L., Sander, C.: Removing near-neighbour redundancy from large protein sequence collections. Bioinformatics **14**(5), 423–429 (1998)
17. Tomovic, A., Oakeley, E.J.: Position dependencies in transcription factor binding sites. Bioinformatics **23**(8), 933–941 (2007)
18. McHugh, M.L.: The chi-square test of independence. Biochemia medica **23**(2), 143–149 (2013)
19. Dong, Q., Lin, L., Wang, X.: Protein remote homology detection based on binary profiles. In: Hochreiter, S., Wagner, R. (eds.) BIRD 2007. LNCS, vol. 4414, pp. 212–223. Springer, Heidelberg (2007). doi:10.1007/978-3-540-71233-6_17
20. Dong, Q.-W., Wang, X.-L., Lin, L.: Application of latent semantic analysis to protein remote homology detection. Bioinformatics **22**(3), 285–290 (2006)

21. Altschul, S.F., Madden, T.L., Schäffer, A.A., Zhang, J., Zhang, Z., Miller, W., Lipman, D.J.: Gapped BLAST and PSI-BLAST: a new generation of protein database search programs. Nucl. Acids Res. **25**(17), 3389–3402 (1997)
22. Saigo, H., Vert, J.P., Ueda, N., Akutsu, T.: Protein homology detection using string alignment kernels. Bioinformatics **20**(11), 1682–1689 (2004)
23. Hochreiter, S., Heusel, M., Obermayer, K.: Fast model-based protein homology detection without alignment. Bioinformatics **23**(14), 1728–1736 (2007)
24. Zhang, S.-B., Tang, Q.-R.: Predicting protein subcellular localization based on information content of gene ontology terms. Comput. Biol. Chem. **65**, 1–7 (2016)
25. Schneider, T.D., Stormo, G.D., Gold, L., Ehrenfeucht, A.: Information content of binding sites on nucleotide sequences. J. Mol. Biol. **188**(3), 415–431 (1986)
26. Crooks, G.E., Hon, G., Chandonia, J.-M., Brenner, S.E.: WebLogo: a sequence logo generator. Genome Res. **14**(6), 1188–1190 (2004)

Biomedical Image Analysis

A Supervised Breast Lesion Images Classification from Tomosynthesis Technique

Vitoantonio Bevilacqua[1(✉)], Daniele Altini[1], Martino Bruni[1], Marco Riezzo[1],
Antonio Brunetti[1], Claudio Loconsole[1], Andrea Guerriero[1],
Gianpaolo Francesco Trotta[2], Rocco Fasano[3], Marica Di Pirchio[4], Cristina Tartaglia[4],
Elena Ventrella[4], Michele Telegrafo[5], and Marco Moschetta[6]

[1] Department of Electrical and Information Engineering (DEI), Polytechnic University of Bari,
Via Orabona 4, 70126 Bari, Italy
vitoantonio.bevilacqua@poliba.it
[2] Department of Mechanics, Mathematics and Management (DMMM),
Polytechnic University of Bari, Via Orabona 4, 70126 Bari, Italy
[3] SARIS, Policlinico Hospital of Bari, Bari, Italy
[4] D.I.M., University of Bari Medical School, Bari, Italy
[5] Department of Radiology, Miulli Hospital, Acquaviva delle Fonti, BA, Italy
[6] DETO, University of Bari Medical School, Bari, Italy

Abstract. In this paper, we propose a deep learning approach for breast lesions classification, by processing breast images obtained using an innovative acquisition system, the Tomosynthesis, a medical instrument able to acquire high-resolution images using a lower radiographic dose than normal Computed Tomography (CT). The acquired images were processed to obtain Regions Of Interest (ROIs) containing lesions of different categories. Subsequently, several pre-trained Convolutional Neural Network (CNN) models were evaluated as feature extractors and coupled with non-neural classifiers for discriminate among the different categories of lesions. Results showed that the use of CNNs as feature extractor and the subsequent classification using a non-neural classifier reaches high values of Accuracy, Sensitivity and Specificity.

Keywords: Breast cancer · Tomosynthesis · Image processing · Segmentation · Deep learning · Convolutional · Neural network

1 Introduction

The increased life expectancy and the higher incidence of breast cancer in the population require an accurate assessment of the breast glands with the imaging techniques, and mammography still represents the gold standard imaging tool in this field [1]. Digital breast tomosynthesis (DBT) has been recently introduced for breast imaging and consists in a new radiological technique used as a valuable addition to breast cancer screening and cancer detection. This technique reduces the effect of breast tissue superimposition found with conventional planar digital mammography and improves the

© Springer International Publishing AG 2017
D.-S. Huang et al. (Eds.): ICIC 2017, Part II, LNCS 10362, pp. 483–489, 2017.
DOI: 10.1007/978-3-319-63312-1_42

visualization of masses and architectural distortions. Lesion edges are better defined with an improvement in breast cancer visibility and observer performance [2].

Computer Aided Diagnosis (CAD) systems, based on supervised classification approaches and evolutionary strategies, are useful tools to support physicians in the diagnosis of different pathologies, such as breast cancer or compatibility for kidney transplant [3, 4]. Deep Learning and Convolutional Neural Networks (CNNs) have been used in many applications for images segmentation and classification, including medical field [5].

In this work, a Deep Learning approach based on Convolutional Neural Networks for breast lesions classification using images from tomosynthesis is proposed; in particular, different pre-trained CNN models are evaluated as feature extractors for the subsequent classification by non-neural classifiers. The following sections are organized as follows: in Sect. 2, both patients and image acquisition system are introduced. In Sect. 3, the image processing steps preceding the classification procedure are described; in Sect. 4, the different approaches followed for classification are described, whereas the results are reported and discussed in Sect. 5. In Sect. 6, the conclusions are presented.

2 Materials

2.1 Patients

From January to November 2016, 16 patients underwent breast tomosynthesis examination. All the patients were women aged between 35 years and 65 years (mean 49.8 ± 9.2 years). During this time period, some women underwent more than a single tomosynthesis examination, thus reaching a dataset composed of a total of 39 exams.

2.2 Images Description and Acquisition Protocol

In DBT, several images of the compressed breast are obtained projecting X-ray from different emission sources [6]. Starting from these images, a reconstruction algorithm is performed to produce a tridimensional view of the breast tissue, slice by slice, mutually parallel and suitably spaced. In this work, the acquisition system was the 2nd generation DBT Giotto Tomo (http://www.tomosynthesis-giotto.com/), whose emission sources are included in a wide scanning angle (40°) and the reconstruction is made through an innovative iterative algorithm. Despite the wide acquisition range, the number of exposures is limited to 13 exposures, and each one is characterized by an optimized dose.

3 Methods

Several pre-trained CNN models were used as feature extractors. For this aim, the classification layer and the related full-connected layer were removed, so that the output of the last considered layer consisted in a set of features describing the input image. Before the classification phase, images had to be processed to extract only significant Regions of Interest (ROIs) for the following computation phases.

3.1   Image Preprocessing

The first step of the proposed approach was the preprocessing of the acquired images. Initially, the image contrast was adjusted performing a Windowing operation [7]. After this preliminary step, correction of the contrast, median filtering and border removal were performed to enhance the ROIs to be processed (Fig. 1). The correction of the contrast was applied on the image resulting from the previous windowing operation (Fig. 1a) to map the intensity value of the pixels to new values, such that 1% of the data was saturated in both low and high intensities (Fig. 1b). Then, a median filter was applied to remove noise from the image [8], considering a 3×3 squared kernel (Fig. 1c). Finally, the removal of the breast external border, which could interfere in the subsequent segmentation phase, was performed; this last operation was executed by expanding the background (the black pixels) using a disk with 16-pixel radius (Fig. 1d).

Fig. 1. The result of each image preprocessing step: (a) windowing; (b) contrast correction; (c) median filtering; (d) external border removal.

3.2   Segmentation

After the preprocessing phase, the segmentation of the images was performed to automatically detect interesting areas. As preprocessing, even the segmentation phase was composed of several steps leading to the extraction of ROIs containing suspicious areas. The first step of segmentation was the binarization of the image based on grey levels; adaptive thresholding was used to perform this task. This approach is used to segment ROIs both in similar contexts, e.g. detection of masses in mammography [9], and in different contexts, such as the retinal vessel extraction [10]. The adaptivity of threshold avoided both an overly selective segmentation, that could remove interesting regions, and an excessively tolerant segmentation, that could include non-interesting areas. The threshold was empirically determined considering the grey levels of the considered areas. Subsequently, ROIs with area below a threshold were removed, and an hole filling procedure was executed. Since the remaining regions included some undesired vessels, the aspect ratio was considered, thus removing areas with aspect ratio not near to 1. At the end of the segmentation phase, a dilatation of each segmented ROI was performed and the mean values of gray levels in both the initial and the extended ROIs were evaluated; if the difference between these two values was higher than a further threshold (the grey levels of the two areas are different) the segmented ROI was left as it was; otherwise, an active contour procedure was performed [11, 12], setting the border coming from the binarization process as the initial curve.

3.3   Images Extraction

The previous segmentation step returned the ROIs to be classified. Before the classification step, ROIs were cropped from the starting image and resized to 227×227 pixels; each image was labelled according to the considered 4 classes: **None**, segmented ROI not containing any kind of lesion (Fig. 2a); **Ori**, segmented ROI containing an **irregular opacity** (Fig. 2b); **Oro**, segmented ROI containing a **regular opacity** (Fig. 2c); **Ost**, segmented ROI containing a **stellar opacity** (Fig. 2d).

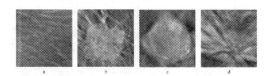

Fig. 2. Images extracted after the segmentation phase: (a) ROI with no lesions; (b) ROI with irregular opacity; (c) ROI with regular opacity; (d) ROI with stellar opacity.

4   Classification

Classification was performed coupling CNNs with different non-neural classifiers. Following this approach, the CNNs were used as feature extractor, and the non-neural classifiers, in cascade to the network, were used to discriminate among the classes of ROIs.

4.1   Convolutional Neural Networks

Since CNNs work on images, a preliminary step for the training of CNNs consists in the augmentation of the images [13], meaning that all the images were replicated with specific transformations, such as rotation, translation, skewing and distortions. In this work, since the Positive samples (Oro, Ori, Ost) were less than the Negative ones, a combination of flips and rotations (0°, 90°, 180° and 270°) was used; in this way, 8 additional images were obtained for each Positive image sample. To evaluate the results variability and the mean performance of the classifiers, tests were conducted training and testing the classifier 50 times, randomly shuffling the dataset each iteration.

Several pre-trained models were considered for features extraction, which were: GoogLeNet, ResNet, AlexNet, VGG-verydeep-19, VGG-F, VGG-M and VGG-S [14–17]. The output of the CNN models was used to train a new classifier; to do this, several learners were evaluated: Linear Support Vector Machine (Linear SVM), K-Nearest Neighbor (KNN), Naive Bayes, Decision Tree, Linear Discriminant Analysis (LDA) [18–22].

5   Results

In this section, all the results obtained in the classification sessions are reported, considering all the CNN pre-trained models combined with different learners. In particular, some preliminary tests were firstly performed to find the optimal working set to improve the classifiers performances.

5.1   Activations Normalization and Images Augmentation

In the first approach, all the CNN models were evaluated comparing the classification performance considering the activations (input to the learners) without and then with normalization. Tests conducted considering the two approaches showed that normalization does not significantly affect the accuracy, but a slight improvement can be observed in almost all cases.

Subsequently, all the CNN models were evaluated comparing the classification performances considering the dataset with and without the augmented images. Unlike the activations normalization, the augmentation of the images improves the classifiers performance. The accuracy improvement is approximately 10% for VGG nets (very-deep, F, M and S) and AlexNet while it is greater (about 15%) for GoogLeNet and ResNet.

5.2   Learners Comparison

The CNN models performing better were used with different learners to classify the different kinds of lesions. Except for GoogLeNet, all the CNN models reached accuracies near to 90%. For this reason, final tests could be performed on a subset of the available CNNs; in particular, VGG-F, VGG-S and VGG-S were considered since they showed the higher mean accuracies and the lowest processing time. Results are reported in Table 1.

Table 1.   Classification accuracy for all the considered learners using VGG-F, M, S models.

	KNN		LDA		Linear SVM		Naïve bayes		Decision tree	
	Acc (%)	std (%)	Acc (%)	std (%)	Acc (%)	std (%)	Acc (%)	std (%)	Acc (%)	std (%)
VGG-F	91,63	0,41	64,57	0,66	67,29	2,02	43,82	0,59	59,68	1,07
VGG-M	90,74	0,48	66,25	0,60	69,50	2,16	42,85	0,57	57,03	0,98
VGG-S	92,02	0,48	65,24	0,80	68,84	1,89	44,89	0,60	56,16	0,93

As reported in Table 1, naïve Bayes (NBA) classifier was not recommended in this classification; decision Trees (TREE) allowed to improve the mean accuracy in comparison to naïve Bayes, but it was far to be considered as a reliable classifier. The linear classifiers (SVM and LDA) further increased the performance. Despite of these results, mean accuracies were considered too low compared to the expectations. Different outcomes were obtained from KNN classifiers. In this case, mean performances reached very good levels of accuracy, specificity and sensitivity with low variability. Furthermore, to substantiate the high level of performance in terms of accuracy, it is worth to mention that sensitivity and specificity for positive samples were higher than 95%, as reported in Table 2 were the results were calculated using a *1-vs-all* approach.

Table 2. Sensitivity and Specificity for the lesions evaluated through 1-vs-all approach.

	Ori vs all		Oro vs all		Ost vs all	
	Sensitivity	Specificity	Sensitivity	Specificity	Sensitivity	Specificity
VGG-F	98,67%	97,07%	96,01%	95,98%	97,24%	96,93%
VGG-M	98,14%	96,64%	95,00%	95,76%	97,18%	96,62%
VGG-S	98,36%	97,13%	95,61%	96,33%	96,67%	97,25%

6   Conclusions

In this work, Convolution Neural Networks have been used for breast lesion classification starting from tomosynthesis images; several CNN models were tested as feature extractors using this output as input for different non-neural learners. Regarding the training of CNNs coupled with learners, it was found that the activation normalization is useful for a slight performance improvement, while the images augmentation is necessary to get good accuracy levels. At the end, VGG-F, VGG-M and VGG-S were considered for final tests and final results in terms of Accuracy, Sensitivity and Specificity are higher than 90%. The reported results confirm that using CNNs as features extractors is a very promising approach for this kind of classification. In the future, these performances will be compared with a classical approach for features extraction and Artificial Neural Networks for classification.

References

1. Vestito, A., Mangieri, F.F., Gatta, G., Moschetta, M., Turi, B., Ancona, A.: Breast carcinoma in elderly women. Our experience. Il giornale di chirurgia **32**, 411–416 (2011)
2. Korhonen, K.E., Weinstein, S.P., McDonald, E.S., Conant, E.F.: Strategies to increase cancer detection: review of true-positive and false-negative results at digital breast tomosynthesis screening. RadioGraphics **36**, 1954–1965 (2016)
3. Bevilacqua, V., Brunetti, A., Triggiani, M., Magaletti, D., Telegrafo, M., Moschetta, M.: An optimized feed-forward artificial neural network topology to support radiologists in breast lesions classification. In: Proceedings of the 2016 on Genetic and Evolutionary Computation Conference Companion, pp. 1385–1392. ACM (2016)
4. Bevilacqua, V., Pietroleonardo, N., Triggiani, V., Brunetti, A., Di Palma, A.M., Rossini, M., Gesualdo, L.: An innovative neural network framework to classify blood vessels and tubules based on Haralick features evaluated in histological images of kidney biopsy. Neurocomputing **228**, 143–153 (2017)
5. Cha, K.H., Hadjiiski, L., Samala, R.K., Chan, H.P., Caoili, E.M., Cohan, R.H.: Urinary bladder segmentation in CT urography using deep-learning convolutional neural network and level sets. Med. Phys. **43**, 1882–1896 (2016)
6. Niklason, L.T., Christian, B.T., Niklason, L.E., Kopans, D.B., Castleberry, D.E., Opsahl-Ong, B., Landberg, C.E., Slanetz, P.J., Giardino, A.A., Moore, R.: Digital tomosynthesis in breast imaging. Radiology **205**, 399–406 (1997)

7. Lehmann, T.M., Gonner, C., Spitzer, K.: Survey: interpolation methods in medical image processing. IEEE Trans. Med. Imaging **18**, 1049–1075 (1999)
8. Lim, J.S.: Two-Dimensional Signal and Image Processing, 710 p. Prentice Hall, Englewood Cliffs (1990)
9. Kom, G., Tiedeu, A., Kom, M.: Automated detection of masses in mammograms by local adaptive thresholding. Comput. Biol. Med. **37**, 37–48 (2007)
10. Carnimeo, L., Bevilacqua, V., Cariello, L., Mastronardi, G.: Retinal vessel extraction by a combined neural network–wavelet enhancement method. In: Huang, D.-S., Jo, K.-H., Lee, H.-H., Kang, H.-J., Bevilacqua, V. (eds.) ICIC 2009. LNCS, vol. 5755, pp. 1106–1116. Springer, Heidelberg (2009). doi:10.1007/978-3-642-04020-7_118
11. Kass, M., Witkin, A., Terzopoulos, D.: Snakes: active contour models. Int. J. Comput. Vis. **1**, 321–331 (1988)
12. Bevilacqua, V., Mastronardi, G., Piazzolla, A.: An evolutionary method for model-based automatic segmentation of lower abdomen CT images for radiotherapy planning. Appl. Evol. Comput. **6024**, 320–327 (2010)
13. Simard, P.Y., Steinkraus, D., Platt, J.C.: Best practices for convolutional neural networks applied to visual document analysis. In: ICDAR, pp. 958–962. Citeseer (2003)
14. Szegedy, C., Liu, W., Jia, Y., Sermanet, P., Reed, S., Anguelov, D., Erhan, D., Vanhoucke, V., Rabinovich, A.: Going deeper with convolutions. In: Proceedings of the IEEE Conference on Computer Vision and Pattern Recognition, pp. 1–9. IEEE (2015)
15. He, K., Zhang, X., Ren, S., Sun, J.: Deep residual learning for image recognition. In: Proceedings of the IEEE Conference on Computer Vision and Pattern Recognition, pp. 770–778. IEEE (2016)
16. Krizhevsky, A., Sutskever, I., Hinton, G.E.: Imagenet classification with deep convolutional neural networks. In: Advances in Neural Information Processing Systems, pp. 1097–1105. NIPS (2012)
17. Chatfield, K., Simonyan, K., Vedaldi, A., Zisserman, A.: Return of the devil in the details: delving deep into convolutional nets, pp. 1–11 (2014). arXiv preprint arXiv:1405.3531
18. Burges, C.J.: A tutorial on support vector machines for pattern recognition. Data Min. Knowl. Disc. **2**, 121–167 (1998)
19. Beyer, K., Goldstein, J., Ramakrishnan, R., Shaft, U.: When is "nearest neighbor" meaningful? In: Beeri, C., Buneman, P. (eds.) ICDT 1999. LNCS, vol. 1540, pp. 217–235. Springer, Heidelberg (1999). doi:10.1007/3-540-49257-7_15
20. Domingos, P., Pazzani, M.: On the optimality of the simple Bayesian classifier under zero-one loss. Mach. Learn. **29**, 103–130 (1997)
21. Rokach, L., Maimon, O.: Data Mining with Decision Trees: Theory and Applications. World scientific, River Edge (2014)
22. Fisher, R.A.: The use of multiple measurements in taxonomic problems. Ann. Eugen. **7**, 179–188 (1936)

Human-Machine Interaction: Shaping Tools Which Will Shape Us

Computer Vision and EMG-Based Handwriting Analysis for Classification in Parkinson's Disease

Claudio Loconsole[1(✉)], Gianpaolo Francesco Trotta[2], Antonio Brunetti[1], Joseph Trotta[1], Angelo Schiavone[1], Sabina Ilaria Tatò[3], Giacomo Losavio[3], and Vitoantonio Bevilacqua[1]

[1] Department of Electrical and Information Engineering (DEI), Polytechnic University of Bari, Via Orabona 4, 70126 Bari, Italy
claudio.loconsole@poliba.it
[2] Department of Mechanics, Mathematics and Management (DMMM), Polytechnic University of Bari, Viale Japigia 182, 70126 Bari, Italy
[3] Medica Sud s.r.l., Viale della Resistenza n. 82, Bari, BA, Italy

Abstract. Handwriting analysis represents an important research area in different fields. From forensic science to graphology, the automatic dynamic and static analyses of handwriting tasks allow researchers to attribute the paternity of a signature to a specific person or to infer medical and psychological patients' conditions. An emerging research field for exploiting handwriting analysis results is the one related to Neurodegenerative Diseases (NDs). Patients suffering from a ND are characterized by an abnormal handwriting activity since they have difficulties in motor coordination and a decline in cognition.

In this paper, we propose an approach for differentiating Parkinson's disease patients from healthy subjects using a handwriting analysis tool based on a limited number of features extracted by means of both computer vision and Electro-MyoGraphy (EMG) signal-processing techniques and processed using an Artificial Neural Network-based classifier.

Finally, we report and discuss the results of an experimental test conducted with both healthy and Parkinson's Disease patients using the proposed approach.

Keywords: Handwriting analysis · Parkinson's Disease · EMG · Computer vision

1 Introduction

Subjects suffering from neurodegenerative diseases can have severe difficulties in motor coordination.

As an example, Parkinson's Disease (PD) patients tend to perform sequential movements in a more segmented fashion spending more time for each movement component (also recurring to pauses between them) than healthy people [5]. For these reasons, several research works studied the effects of neurodegenerative diseases on handwriting tasks.

© Springer International Publishing AG 2017
D.-S. Huang et al. (Eds.): ICIC 2017, Part II, LNCS 10362, pp. 493–503, 2017.
DOI: 10.1007/978-3-319-63312-1_43

Handwriting is a highly overlearned fine and complex manual skill, as well as one of the most common activities performed by people of all ages in a variety of leisure and professional settings involving an intricate blend of cognitive, sensory and perceptual-motor components [7].

Therefore, it is not surprising that abnormal handwriting is a well-recognized manifestation of a neurodegenerative disease.

In detail, in PD, handwriting tasks are mainly characterized by:

- impairment of the amplitude of the movement control, i.e., decreased letter size (micrographia) and failing in maintaining stroke size of the characters as writing pregresses [8, 11, 17–19, 26, 31, 32];
- changes in kinematic aspects including increased movement time, decreased velocities and accelerations, and increased number of changes in velocity and acceleration (in particular irregular and bradykinetic movements) [9, 12, 20–22, 28].

Indeed, the automatic processing of handwriting movements in PD is consistently disturbed, thus showing a shift from an "open-loop" (that is, automatic) toward a "closed-loop" (that is, controlled) mode of movement generation [25].

The kinematics of a "closed-loop" performance of handwriting movements is characterized by flattened velocity profiles with an increased number of inversions in velocity per single stroke [10, 17]. Moreover, because of a loss of automatic performance, handwriting movements of patients with PD are slower compared to normal control subjects [10, 27].

It follows that, if patients with PD are less able to increase speed in handwriting, movement amplitude should have a linear relationship to movement duration. Thus, in contrast to the peak acceleration in handwriting of age matched controls, which is governed by the isochrony principle, peak acceleration in parkinsonian handwriting could be unrelated to writing size [30].

Regarding the study and the development of computer-aided handwriting analysis tools for differentiation of PD patients from healthy subjects, in the literature, it is possible to find some significant examples.

In [13], Helsper et al. proposed an approach for detecting differences between the handwriting of preclinical PD patients and healthy controls. The approach was based on a five-processing step method aiming to extract 10 features for each segment composing the written text. Then, the authors used all the obtained numerous segment features for calculating a set of overall features based on the average, the standard deviation and the frequencies of occurrence. In the experiments, subjects were asked to write a text and, subsequently, two lines of handwriting which could be scanned most easily were selected. The authors statistically proved the validity of their approach confirming that preclinical PD handwriting may demonstrate unique features many years prior to their actual diagnosis.

In [16], Longstaff et al. investigated the ability of PD patients to scale the size and speed of drawing movements and the impact that this has on movement variability. For the experiment they used a stylus on a graphics tablet and geometrical writing pattern (circles and spirals with different sizes). Through the analysis of the geometrical extracted features the authors showed that in PD patients there is a substantial divergence in the quality of movements with respect to the control group.

In [29], Ünlü *et al.* introduced an electronic pen able to write on normal paper sheets and to record both the exerted pressure (3-axis) and the inclination relative to the x-y plane during writing. Their approach was based on the extraction of a total of 8 features evaluated through a Receiver Operating Characteristic (ROC) analysis to characterize the diagnostic possibilities not only by their sensitivity, but also by the specificity of the diagnostic approach. The obtained Area Under the Curve (AUC) in the best case resulted to be equal to 0.963.

In [24], instead, Rosenblum *et al.* used a tablet for acquiring X-Y position and an electronic pen for displacement, pressure, and pen-tip angle. In the experiments, the participants were requested to perform some writing tasks according to two writing patterns (his/her name and a fixed address). Mean pressure and mean velocity were measured for the entire task and the spatial and temporal characteristics were measured for each stroke. The high number of considered features allowed them to differentiate PD patients from healthy subjects with a sensitivity of 95%.

In [9], Drotàr *et al.* proposed a methodology based on 11 features for building the predictive model of PD from kinematic handwriting features obtained from different handwriting tasks. The resulting accuracy they obtained was 79.4% with similar values for specificity and sensitivity.

In this paper, we aim at proposing an approach for differentiating PD patients from healthy subjects using a reduced set of features (4 features) by exploiting computer vision techniques applied on the scan of common paper sheets (no tablet use) and ElectroMyoG-raphy (EMG) signal processing already used in medical applications [14, 15].

An important aspect regarding the handwriting task is the type of the surface on which the person is writing on, since it can deeply affect the task itself.

On smooth and slippery surfaces, such as that of a tablet or of the back of a credit card, people can have difficulty in writing and/or signing [34]. The sensation of sliding over a slippery surface suggests that the fine motor control required for adjusting pen movements can be disturbed [1].

For this reason, to not introduce further variables in the proposed approach, we decided to extract all the features by processing both EMG signals acquired at the arm level (time feature) and scans of normal paper sheets (vision-based features). Further-more, the introduction of EMG signal processing plays a significant role for future works dealing with the extension of our approach to PD early diagnosis, since further dynamic biometric features can be extracted and their performance can be investigated.

The remainder of the paper is organized as follows. Section 2 presents the proposed approach in terms of feature extraction and Artificial Neural Network-based classification algorithms. Section 3 describes and reports the results of the experimental tests conducted with both healthy and PD patients, whereas Sect. 4 is dedicated to the conclusion and future works.

2   The Proposed Approach

2.1   Handwriting Feature Extraction

To reduce the complexity of the entire handwriting analysis process, we selected a limited number of features to be extracted from biometric signals.

However, to have a significant overview of the biometric signals, we selected two dynamic and two static features for a total of four features: (i) *execution time*, (ii) *execution average linear speed*, (iii) *density ratio*, and (iv) *height ratio*.

In next subsections, we will focus on each selected feature.

I. Execution time feature: Execution time is a dynamic feature and is the time interval (in s) during which the subject performs the writing pattern. The feature extraction from the EMG signal can be performed off-line (i.e., after the completion of the task) and is based on an adaptive threshold to detect the starting and the end instants of time of the writing task. Since it is assumed that just before and after the writing task the user is resting at least for x s, the adaptive threshold is fixed to the highest peak value of the EMG signal over a time interval corresponding to the sum of the first and the last x s, thus resulting in a time interval of $2 \times x$ s. Among all EMG signal values exceeding the adaptive threshold, the starting and the end instants of the execution time interval are those corresponding to first and the last exceedances of EMG signal value, respectively.

II. Execution average linear speed feature: Execution average linear speed is a dynamic feature and is the ratio between the sum of the width of each written word and the execution time (as reported in [23]). It is measured in cm/s and the spatial measurement of the word width is performed through a 1-cm visual marker on the paper sheet, thus resulting in a feature extraction process independent from the scanning device and its resolution.

III. Density ratio feature: Density ratio is a static feature. The variation of the pixel density as defined in [35] (dimensionless value) can represent a potential assessment index of micrography. To extract this feature, firstly the entire written sentence is subdivided in an arbitrary number of cells characterized by having the same width and the height determined by the text upper and lower bounds. Secondly, black pixels contained in each cell are counted and the result is divided by the area of the belonging cell, thus obtaining the cell density value. In case of micrography, since the size of the written text is progressively reduced from left to right, the expected result is that the ratio of the density of the first cells and that of the last cells (density ratio) is greater than 1. Specifically, since it is possible to suppose that the first cell contains a capital letter (thus resulting in a larger cell area), for the density ratio feature extraction, we use the ratio between the second cell and the last cell.

IV. Height ratio feature: Height ratio is a static feature corresponding to a measure of the user's ability to maintain the writing task fixed in size (dimensionless value). The feature extraction process is similar to the density ratio, especially for cell subdivision. However, the feature value corresponds to the ratio between the height of the second cell and that of the last cell. Also in this case, micrography should result in a height ratio value greater than 1.

2.2 Classification Algorithms and Techniques

Several approaches for finding the best topology for an ANN classifier [4] by exploiting Genetic Algorithms are described in literature; in [3], the authors searched for the optimal topology of ANN using a Multi-Objective Genetic Algorithm (MOGA). In this work, we found the topology of the ANN using a mono-objective Genetic Algorithm (GA) searching for the best topology by maximizing the average test accuracy on a certain number of training, validation and test iterations for each ANN topology using different permutations of the dataset [2].

For each input dataset, the GA was executed to find the optimal topology in terms of: (i) number of hidden layers (ranging from 1 to 3), (ii) number of neurons per layer (ranging from 1 to 256 for the first hidden layer, and from 0 to 255 for other hidden layers), and (iii) activation functions in *log-sigmoid* (logsig), the *hyperpolic tangent sigmoid* (tansig), the *pure linear* (purelin) and the *symmetric saturating linear* (satlins), for all the neurons per-single layer.

The performance of the classifiers were evaluated in terms of accuracy (Eq. 1) specificity (Eq. 2) and sensitivity (Eq. 3), using the confusion matrix as follows (Table 1):

$$Accuracy = \frac{TP + TN}{TP + TN + FP + FN} \tag{1}$$

$$Specificity = \frac{TN}{TN + FP} \tag{2}$$

$$Accuracy = \frac{TP}{TP + FN} \tag{3}$$

Table 1. The configuration of the confusion matrix

		True condition	
		Positive	*Negative*
Predicted condition	*Positive*	*TP*	*FP*
	Negative	*FN*	*TN*

3 Experiments and Results

3.1 Participants

Eleven participants (all males, age: 48 ± 25 years old) took part to the experimental tests. In detail, the control group was composed of 7 healthy subjects (age: 31 ± 11 years old), whereas the PD group was composed of 4 subjects (age: 77 ± 3).

All subjects signed informed consent forms.

3.2 Experimental Description

To validate the proposed approach, we selected three writing patterns corresponding to as many writing tasks. They were properly differentiated according to a writing size constrained/unconstrained point of view:

1. a fixed sentence in Italian composed of 8 words all containing equal sized letters (e.g., no d,f,g,h,l,p,q,t) to be written in italic;
2. a sequence of 8 "l" with a size of 2 cm (2 cm visual marker reference on the left of the paper sheet);
3. a sequence of 8 "l" with a size of 5 cm (5 cm visual marker reference on the left of the paper sheet).

In writing tasks no. 2 and 3, since they consisted in writing only one word composed of 8-"l", for density and height ratio feature extraction, each written text was subdivided in 8 cells.

Furthermore, while in writing task no. 1, the height ratio feature was calculated as previously described (i.e., the ratio between the height of the second and of the last cells), in tasks no. 2 and 3 the height ratio was calculated as the ratio between the average height of the 8 "l" and the expected letter height, that is 2 cm and 5 cm, respectively. Moreover, it is worth to mention that while writing task no.1 can be considered not constrained in size (the letter size is up to the patient), tasks no. 2 and 3 are size constrained, thus forcing the subject to write using a predefined size.

To familiarize with the exercise, each subject was asked to perform all three writing tasks once. Then, each subject was asked to perform 4 repetitions of all 3 writing tasks, thus resulting in a total of 12 writing samples per subject, 132 (12 × 11 subjects) total writing samples and 528 (132 × 4 features) total features to be processed and analyzed.

According to the proposed acquisition protocol, both before and after each writing task, the subject was asked to rest for at least 3 s allowing the algorithm to identify the adaptive threshold for execution time feature extraction.

3.3 System Setup and Description

The setup (left) and the paper sheet template (right) used for the experimental tests are illustrated in Fig. 1.

In detail, in Fig. 1 left, it is possible to see the MyoTM Gesture Control Armband (www.myo.com) for acquiring EMG signals from 8 different points of the arm. It needs a simple and a brief calibration to start the acquisition.

In Fig. 1 right, instead, on the printed template of the paper sheet, it is possible to identify two (vertical) visual marker size references of 2-cm and 5-cm, respectively, for writing tasks no. 2 and 3 and three (horizontal) 1-cm markers used for spatial mapping needed to extract the execution average linear speed feature for all three writing tasks.

Regarding feature extraction, execution time feature was extracted according to the method described in Sect. 2.1 by processing the average signal of the 8 EMG signals acquired from the 8 electrodes of the Myo Armband to let the entire process be independent from the actual position of the armband on the subject's arm. Finally, for extracting the other three features, we use classical computer vision techniques (e.g., morphology operators, image segmentation process, etc.) to process the scanned paper sheet compiled by the subject's during the tests (Fig. 2).

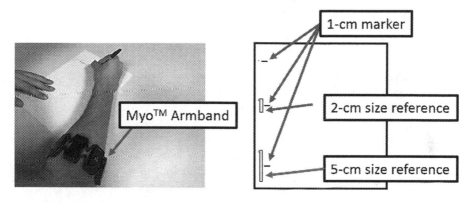

Fig. 1. The system setup used for the experimental tests to validate the proposed approach (left). The paper sheet template with two (vertical) visual marker size references of 2-cm and 5-cm, respectively, for writing tasks no. 2 and 3 and three (horizontal) 1-cm markers used for spatial mapping needed to extract the execution average linear speed feature for all three writing tasks.

3.4 Results and Discussion

Two samples of a repetition of all three writing tasks respectively performed by a healthy subject and a PD subject are reported in Fig. 2.

Fig. 2. Two samples of a repetition of all three writing tasks respectively performed by a healthy subject (left) and by a PD subject (right).

Regarding the classification algorithms and techniques, the ANN optimal topologies specified by the Genetic Algorithm in 2 cases of feature groups are:

Case 1. Dataset with all 4 features (2 dynamic and 2 static). ANN with: 2 layers, 42 neurons for the hidden layer, and 2 neurons for the output layer. The activation functions found by the GA were *hyperbolic tangent sigmoid tranfer (tansig)* function for the hidden layer, *softmax* function for the neurons of the output layer.

Case 2. Dataset with only the 2 dynamic features since their performance are better ranked by the Information Gain Ranking Filter [33]. ANN with 4 layers: 124 neurons for the first hidden layer, 131 for the second, 16 for the third and 2 neurons for the output layer. The activation functions found by the GA were *linear transfer function* for the first hidden layer, *log-sigmoid transfer* function for the second and the third hidden layers, *softmax* function for the output layer.

The ANN training, validation, and test sets were obtained from the input dataset with 60% of samples for the training, 20% for the validation, and 20% for the test. Specifically, at each iteration, the above sets were obtained through a random permutation of the input dataset, keeping constant the ratio between the two classes. Moreover, the classification thresholds were determined using Receiver Operating Characteristic (ROC) curves [6], by evaluating the True Positive Rate (TPR) against the False Positive Rate (FPR) at various threshold settings, in order to find the value able to achieve the best discrimination performance between the two classes.

Results are expressed in terms of mean values, considering 200 iterations, for accuracy, specificity, and sensitivity. In the first case, accuracy was 95.81% (std = 0.0335), specificity was 0.9488 (std = 0.0569), and sensitivity was 0.9680 (std = 0.0393). In the second case accuracy was 95.52% (std = 0.0337), specificity was 0.9343 (std = 0.0620), and sensitivity was 0.9737 (std = 0.0385).

Due to the small number of samples, the performance in all cases were quite similar; in particular, the maximum value for the accuracy was the same in the two cases and equal to 100%. The confusion matrix is shown in Table 2.

Table 2. Confusion matrix for the best case

		True condition	
		Positive	*Negative*
Predicted condition	*Positive*	*10*	*1*
	Negative	*1*	*17*

It is worth to observe that Information Gain Ranking Filter ranked better the dynamic features with respect to the static ones, thus suggesting to privilege dynamic features for further research works in handwriting analysis for differentiating PD from healthy subjects.

The obtained results demonstrated also that a reduced set of features (2 features instead of 4) did not significantly degrade the achieved performance.

In fact, the difference between the performance of the two-feature group for accuracy, specificity and sensitivity were 0.29%, 0.0145 and −0.0057, respectively.

4   Conclusion and Future Works

The results of our study on the proposed approach showed that PD subjects can be differentiated with a good accuracy from healthy subjects using a limited number of features. The performance obtained considering two feature groups were analyzed. The first feature group contained all 4 features (2 dynamic and 2 static), whereas the second group contained only the 2 dynamic features, since the Information Gain Ranking Filter significantly ranked better the execution time and the execution average linear speed features. Although the number of cases we could analyzed was relatively small, the obtained accuracy, specificity and sensitivity for both feature groups were higher than 95%, thus preliminarily validating the proposed approach.

Future works will deal with investigating more significant biometric dynamic features (based also on EMG processing) and increasing the number of subjects under test both for achieving better performances in differentiating PD patients from healthy people and for conducting studies on early diagnosis for PD in patients.

Acknowledgments. This work has been funded from the FutureInResearch program of the Regione Puglia - project n. JTFWZV0 ABIOSAN - Advanced BIOmetric analysiS Against Neuromuscular disease.

References

1. Alamargot, D., Morin, M.-F.: Does handwriting on a tablet screen a effect students graphomotor execution? a comparison between grades two and nine. Hum. Mov. Sci. **44**, 32–41 (2015)
2. Bevilacqua, V., Brunetti, A., Triggiani, M., Magaletti, D., Telegrafo, M., Moschetta, M.: An optimized feed-forward artificial neural network topology to support radiologists in breast lesions classification. In: Proceedings of the 2016 on Genetic and Evolutionary Computation Conference Companion, pp. 1385–1392. ACM (2016)
3. Bevilacqua, V., Mastronardi, G., Menolascina, F., Pannarale, P., Pedone, A.: A novel multi-objective genetic algorithm approach to artificial neural network topology optimisation: the breast cancer classification problem. In: 2006 International Joint Conference on Neural Networks IJCNN 2006, pp. 1958–1965. IEEE (2006)
4. Bevilacqua, V.: Three-dimensional virtual colonoscopy for automatic polyps detection by artificial neural network approach: New tests on an enlarged cohort of polyps. Neurocomputing **116**, 62–75 (2013)
5. Bidet-Ildei, C., Pollak, P., Kandel, S., Fraix, V., Orliaguet, J.-P.: Handwriting in patients with parkinson disease: Effect of l-dopa and stimulation of the subthalamic nucleus on motor anticipation. Hum. Mov. Sci. **30**(4), 783–791 (2011)
6. Bradley, A.P.: The use of the area under the roc curve in the evaluation of machine learning algorithms. Pattern Recogn. **30**(7), 1145–1159 (1997)
7. Carmeli, E., Patish, H., Coleman, R.: The aging hand. J. Gerontol. Ser. A: Biol. Sci. Med. Sci. **58**(2), M146–M152 (2003)
8. Contreras-Vidal, J.L., Teulings, H.-L., Stelmach, G.E.: Micrographia in parkinson's disease. NeuroReport **6**(15), 2089–2092 (1995)

9. Drotar, P., Mekyska, J., Smekal, Z., Rektorova, I., Masarova, L., Faundez-Zanuy, M.: Prediction potential of different handwriting tasks for diagnosis of parkinson's. In: 2013 E-Health and Bioengineering Conference (EHB), pp. 1–4. IEEE (2013)
10. Eichhorn, T., Gasser, T., Mai, N., Marquardt, C., Arnold, G., Schwarz, J., Oertel, W.: Computational analysis of open loop handwriting movements in parkinson's disease: a rapid method to detect dopamimetic effects. Mov. Disord. **11**(3), 289–297 (1996)
11. Flash, T., Inzelberg, R., Schechtman, E., Korczyn, A.D.: Kinematic analysis of upper limb trajectories in parkinson's disease. Exp. Neurol. **118**(2), 215–226 (1992)
12. Gordon, A.M.: Task-dependent deficits during object release in parkinson's disease. Exp. Neurol. **153**(2), 287–298 (1998)
13. Helsper, E., Teulings, H.-L., Karamat, E., Stelmach, G.E.: Preclinical Parkinson features in optically scanned handwriting. In: Handwriting and Drawing Research: Basic and Applied Issues, pp. 241–250. IOS Press, Amsterdam (1996)
14. Leonardis, D., Barsotti, M., Loconsole, C., Solazzi, M., Troncossi, M., Mazzotti, C., Castelli, V.P., Procopio, C., Lamola, G., Chisari, C., Bergamasco, M., Frisoli, A.: An EMG-controlled robotic hand exoskeleton for bilateral rehabilitation. IEEE Trans. Haptics **8**(2), 140–151 (2015)
15. Loconsole, C., Leonardis, D., Barsotti, M., Solazzi, M., Frisoli, A., Bergamasco, M., Troncossi, M., Foumashi, M.M., Mazzotti, C., Castelli, V.P.: An EMG-based robotic hand exoskeleton for bilateral training of grasp. In: 2013 World Haptics Conference (WHC), pp. 537–542. IEEE (2013)
16. Longstaff, M.G., Mahant, P.R., Stacy, M.A., Van Gemmert, A.W., Leis, B.C., Stelmach, G.E.: Discrete and dynamic scaling of the size of continuous graphic movements of parkinsonian patients and elderly controls. J. Neurol. Neurosurg. Psychiatry **74**(3), 299–304 (2003)
17. Margolin, D.I., Wing, A.M.: Agraphia and micrographia: Clinical manifestations of motor programming and performance disorders. Acta Physiol. (Oxf) **54**(1), 263–283 (1983)
18. McLennan, J., Nakano, K., Tyler, H., Schwab, R.: Micrographia in parkinson's disease. J. Neurol. Sci. **15**(2), 141–152 (1972)
19. Müller, F., Stelmach, G.: Prehension movements in parkinson's disease. Adv. Psychol. **87**, 307–319 (1992)
20. Nutt, J.G., Lea, E.S., Van Houten, L., Schuff, R.A., Sexton, G.J.: Determinants of tapping speed in normal control subjects and subjects with parkinson's disease: differing effects of brief and continued practice. Mov. Disord. **15**(5), 843–849 (2000)
21. Nutt, J.G., Wooten, G.F.: Diagnosis and initial management of parkinson's disease. N. Engl. J. Med. **353**(10), 1021–1027 (2005)
22. Rand, M.K., Stelmach, G.E., Bloedel, J.R.: Movement accuracy constraints in parkinsons disease patients. Neuropsychologia **38**(2), 203–212 (2000)
23. Raudmann, M., Taba, P., Medijainen, K.: Handwriting speed and size in individuals with parkinsons disease compared to healthy controls: the possible effect of cueing. Acta Kinesiologiae Universitatis Tartuensis **20**, 40–47 (2014)
24. Rosenblum, S., Samuel, M., Zlotnik, S., Erikh, I., Schlesinger, I.: Handwriting as an objective tool for parkinsons disease diagnosis. J. Neurol. **260**(9), 2357–2361 (2013)
25. Siebner, H.R., Ceballos-Baumann, A., Standhardt, H., Auer, C., Conrad, B., Alesch, F.: Changes in handwriting resulting from bilateral high-frequency stimulation of the subthalamic nucleus in parkinson's disease. Mov. Disord. **14**(6), 964–971 (1999)
26. Teulings, H., Contreras-Vidal, J.L., Stelmach, G., Adler, C.H.: Adaptation of handwriting size under distorted visual feedback in patients with parkinson's disease and elderly and young controls. J. Neurol. Neurosurg. Psychiatry **72**(3), 315–324 (2002)

27. Teulings, H.-L., Stelmach, G.E.: Control of stroke size, peak acceleration, and stroke duration in parkinsonian handwriting. Hum. Mov. Sci. **10**(2), 315–334 (1991)
28. Tresilian, J.R., Stelmach, G.E., Adler, C.H.: Stability of reach-to-grasp movement patterns in parkinson's disease. Brain **120**(11), 2093–2111 (1997)
29. Ünlü, A., Brause, R., Krakow, K.: Handwriting analysis for diagnosis and prognosis of parkinson's disease. In: Maglaveras, N., Chouvarda, I., Koutkias, V., Brause, R. (eds.) ISBMDA 2006. LNCS, vol. 4345, pp. 441–450. Springer, Heidelberg (2006). doi: 10.1007/11946465_40
30. Van Gemmert, A., Adler, C.H., Stelmach, G.: Parkinsons disease patients undershoot target size in handwriting and similar tasks. J. Neurol. Neurosurg. Psychiatry **74**(11), 1502–1508 (2003)
31. Van Gemmert, A., Teulings, H.-L., Contreras-Vidal, J.L., Stelmach, G.: Parkinsons disease and the control of size and speed in handwriting. Neuropsychologia **37**(6), 685–694 (1999)
32. Van Gemmert, A.W., Teulings, H.-L., Stelmach, G.E.: Parkinsonian patients reduce their stroke size with increased processing demands. Brain Cogn. **47**(3), 504–512 (2001)
33. Wang, G., Lochovsky, F.H.: Feature selection with conditional mutual information maximin in text categorization. In: Proceedings of the Thirteenth ACM International Conference on Information and Knowledge Management, pp. 342–349. ACM (2004)
34. Wann, J., Nimmo-Smith, I.: The control of pen pressure in handwriting: A subtle point. Hum. Mov. Sci. **10**(2), 223–246 (1991)
35. Zhi, N., Jaeger, B., Gouldstone, A., Sipahi, R., Frank, S.: Toward monitoring parkinsons through analysis of static handwriting samples: A quantitative analytical framework. IEEE J. Biomed. Health Inform. **21**(2), 488–495 (2016)

A Novel Approach in Combination of 3D Gait Analysis Data for Aiding Clinical Decision-Making in Patients with Parkinson's Disease

Ilaria Bortone[1], Gianpaolo Francesco Trotta[2], Antonio Brunetti[3], Giacomo Donato Cascarano[3], Claudio Loconsole[3], Nadia Agnello[4], Alberto Argentiero[4], Giuseppe Nicolardi[5], Antonio Frisoli[1], and Vitoantonio Bevilacqua[3(✉)]

[1] PERCRO Laboratory, TeCIP Institute Scuola Superiore Sant'Anna, Pisa, Italy
[2] Department of Mechanics, Mathematics and Management Engineering, Polytechnic University of Bari, Bari, Italy
[3] Department of Electrical and Information Engineering, Polytechnic University of Bari, Bari, Italy
vitoantonio.bevilacqua@poliba.it
[4] ISBEM S.C.p.A, Brindisi, Italy
[5] Laboratory of Human Anatomy and Neuroscience, Department of Biological and Environmental Technologies and Sciences, University of Salento, Lecce, Italy
http://www.kisshealth.it

Abstract. The most common methods used by neurologist to evaluate Parkinson's Disease (PD) patients are rating scales, that are affected by subjective and non-repeatable observations. Since several research studies have revealed that walking is a sensitive indicator

for the progression of PD. In this paper, we propose an innovative set of features derived from three-dimensional Gait Analysis in order to classify motor signs of motor impairment in PD and differentiate PD patients from healthy subjects or patients suffering from other neurological diseases. We consider kinematic data from Gait Analysis as Gait Variables Score (GVS), Gait Profile Score (GPS) and spatio-temporal data for all enrolled patients. We then carry out experiments evaluating the extracted features using an Artificial Neural Network (ANN) classifier. The obtained results are promising with the best classifier score accuracy equal to 95.05%.

Keywords: Parkinson's disease · Gait analysis · Artificial neural network · Classification

1 Introduction

Parkinson's Disease (PD) is the most common neurodegenerative cause of Parkinsonism, a clinical syndrome characterized by lesions in the basal ganglia, predominantly in the substantia nigra. PD alone represents approximately the 80% of cases of Parkinsonism, which is a general term referring to a group of neurological disorders causing

© Springer International Publishing AG 2017
D.-S. Huang et al. (Eds.): ICIC 2017, Part II, LNCS 10362, pp. 504–514, 2017.
DOI: 10.1007/978-3-319-63312-1_44

movement impairments similar to those provoked by the Parkinson's disease, e.g. tremors, slow movement and stiffness [1] which can lead to dangerous conditions such as subject's fall [2].

The pathological changes of PD may appear as early as three decades before the appearance of clinical signs. Early in the disease process, it is often hard to know whether a person has idiopathic PD or a syndrome mimicking it. The symptoms caused by PD include an ongoing loss of motor control (resting tremors, stiffness, slow movement, postural instability) as well as a wide range of non-motor symptoms (such as depression, loss of sense of smell, gastric problems, cognitive changes, and many others). There is no definitive test to detect PD or Parkinsonism. For the diagnosis, doctors take a thorough medical history and may request a number of motor tests. Because of the observational nature of the diagnosis, Parkinson's can sometimes be confused with Parkinsonism, and the diagnosis may need to be revised over time based on the speed of disease progression, the response to medications and other factors [3].

The Movement Disorder Society - Unified Parkinson Disease Rating Scale (MDS - UPDRS) - Part III is the most commonly used scale to rate motor symptoms in PD [4], and is widely accepted test to determine the efficacy of intervention in clinical studies. As the disease progresses, gait impairment increases and motor symptoms, mainly subjectively recorded in patient diaries, start to fluctuate. Thus, subjective information and rating of motor signs are the basis for the daily clinician's diagnostic workup and guide them toward decisions. With this evidence, the main disadvantage is the subjective measurements, particularly concerning accuracy, and precision, which have a negative effect on the diagnosis, follow-up, and treatment of the pathologics.

In contrast to this background, progress in new technologies has given rise to devices and techniques which allow clinicians to perform an objective evaluation of different gait parameters, resulting in more efficient measurement and providing specialists with a large amount of reliable information on patients' gaits. This also reduces the error margin caused by subjective techniques. In particular, Human Motion Analysis has been recognized as gold standard method for gait assessment and walking has been considered as a sensitive indicator for the progression of PD [5], as individuals present an altered gait pattern with increased cadence and reduced stride lengths [6] and reduced velocity [7, 8].

Biomechanical studies have addressed spatio-temporal gait parameters in PD [9, 10], but few have focused on angular parameters. A reduction in the angular excursion of lower limb joints has been noted in parkinsonian syndromes with the primary gait deficit in PD having been described as an inability to generate sufficient range of motion [11, 12].

In the present work, we proposed to establish and validate an objective Gait Analysis system that used classification algorithms to discriminate PD with respect to normal and other neurological diseases through a set of kinematic features. The paper is organized as follows: Sect. 2 describes the enrolled patients, the procedures for data collection with Human Motion Analysis Laboratory and the Artificial Neural Network for feature classification; Sect. 3 illustrates the achieved results and discusses general comments to the adopted methods. Finally in Sect. 4, we presents the conclusions of our work.

2 Materials and Methods

2.1 Participants

The GAPAR (Gait Analysis in PARkinson's disease) protocol was an observational study where a convenient sample of patients attending the Neurological Department of Local Hospital of Brindisi were recruited. The study focused on the implementation of a co-creative model of diagnosis and follow up between Neurological and Rehabilitative Department of the Hospital and the Human Motion Analysis Laboratory (KISS-Health). With this evidence, each patient would be followed by a specialized team of neurologist, physiatrist, physiotherapists, and bioengineer in order to ensure the best outcome and at the same time reduce the whole costs for the Healthcare System.

After giving their written consent, an experienced neurologist clinically examined the eligible patients. The inclusion criteria were an ability to independently walk bare-foot without a gait-assistance device; absence of any other neurologic disorder or dementia and clinical signs of PD. There were no particular exclusion criteria since the objective was to test the reliability of Gait Analysis as a clinical support decision tool.

Twelve patients (4 male and 8 females, mean age 75.83 years ±8.32) were enrolled and descriptive and demographic characteristics at baseline for all the participants are reported in Table 1.

Table 1. Characteristics of study population.

ID	GENDER	AGE	BMI [m/sec]	UPDRS Gait Score	Diagnosis	Groups
1	Male	78	25.86	3	CCD[a]	1
2	Female	71	24.49	1	MS[b]	1
3	Male	58	20.94	0	PD[c]	2
4	Male	87	27.32	2	PD	2
5	Male	79	34.82	1	(A)PAR[d]	1
6	Female	87	24.77	2	PD	2
7	Male	79	37.46	3	(A)PAR	1
8	Female	81	37.31	2	PAR[d]	1
9	Female	75	34.14	3	PD	2
10	Female	75	30.11	2	PAR	1
11	Female	65	25.96	2	PAR	1
12	Female	75	22.03	1	(A)PAR	1

[a] CCD: Chronic Cerebrovascul Disease
[b] MS: Myelodysplastic Syndrome
[c] PD: Parkinson's Disease
[d] (A)PAR: (Akinetic) PARkinsonism

According to the neurologist's diagnosis, we divided the patients into two groups, one including all the patients with no diagnosis of PD - identified as 1 - and the other group identified as 2 including all the patients with a diagnosis of PD.

2.2   Gait Analysis

The participants were informed regarding the data acquisition procedures, familiarized with the place at which data would be collected and trained so that gait would be as normal as possible. The participants did not use any gait-assistance devices and absolute silence in the laboratory was requested during data acquisition so that no noises interfered with the participant's attention during the tasks.

All measurements have been performed in the KISS-Health Human Motion Analysis Laboratory, located in Mesagne (Brindisi, Italy). Spatio-temporal, kinematic, and kinetic parameters of gait were acquired using an optoelectronic system equipped with eight SMART-DX 7000 infrared cameras (BTS Bioengineering, Italy) set at a sampling rate of 500 Hz and two force platforms (BTS P6000, BTS Bioengineering, Italy) embedded in a 6-m long walkway. Nineteen retro-reflective passive markers of a diameter of 14 mm, were attached to the participant' skin at specific landmarks (Fig. 1), according to the simplified version of the protocol described by Davis et al. [13].

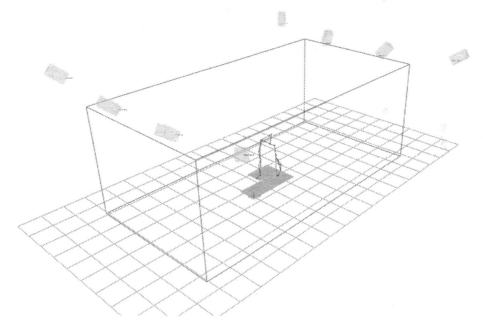

Fig. 1. 3D Tracking during Gait Analysis Processing in BTS Smart Clinic environment: infrared cameras (orange), force platforms (magenta), biomechanical model (red segments), calibrated volume (green parallelepiped). (Color figure online)

Before marker placement, the motion analysis team took the relevant anthropometric measures required by the protocol. Then, the participants were asked to walk barefoot at a comfortable and self-selected speed in the most natural possible manner along the walkway, and several walking trials were recorded. The raw data were then processed using a dedicated software Smart Analyzer (BTS Bioengineering, Italy) in order to compute:

- six spatio-temporal parameters (stride length, stance, swing and double support phase duration, gait speed and cadence);
- nine kinematic parameters, namely foot progression, ankle dorsi-plantar-flexion, knee flexion-extension, hip flexion-extension, adduction-abduction and rotation, pelvic tilt, rotation and obliquity;
- nine kinetic parameters, namely ground reaction forces (medio-lateral, antero-posterior and vertical components), flexion-extension and power for ankle, knee and hip;
- dynamic range of motion (ROM) for hip, knee and ankle joints on the sagittal plane calculated as the difference between maximum and minimum value of the flexion-extension and dorsi-plantar-flexion angle during the whole gait cycle extracted from each trial.

Smart Analyzer software works with the TDF files, acquired by the SMART Motion Capture System and has a series of files to store protocols, archives and normative band files. In this work to read some information about the output of acquisition the MDX files are used. The MDX file is used to store all data contained in an acquisition that is both data read from TDF file and data create by protocol operators. The kinematic data, read from MDX archive file, were summarized using concise measures of gait quality, Gait Variable Score (GVS) and Gait Profile Score (GPS) [14]. The computations for GPS were performed on data written in TDF files by Smart Analyzer software containing spatial positions of all body markers during a Gait session and MDX files containing parameters of the skeletal model such as angles, velocities and orientations. As reference data, we took in consideration the healthy individuals extracted from BTS Smart Analyzer Software (50 healthy people over 65 years old).

Gait Profile Score. Instrumented three-dimensional gait analysis generates kinematic measurements of a wide range of variables across the gait cycle. Clinical decisions are generally based on the interpretation of the complex information contained in those data. It could be useful, then, to have a single measure that can quantify the overall severity of a condition affecting walking, progress monitoring, or evaluation of the outcome of an intervention prescribed to improve the gait pattern.

Although other measures have been proposed, the only one to have a widespread clinical acceptance is the Gillette Gait Index (GGI [15]), which quantifies the difference between data from one gait cycle for a particular individual and the average of a reference dataset from people exhibiting no gait pathology. The GGI, however, has several shortcomings.

The Gait Profile Score (GPS) is based on various GVSs, each of which represents the RMS difference between a specific time-normalized gait variable and the mean data from a reference population calculated across the gait cycle. Thus, if $x_{i,t}$ is the value of a gait variable, GPS is calculated at a specific point in the gait cycle t, and it is the mean value of that variable at the same point in the gait cycle for the reference population, then the i_{th} gait variable score is given by Eq. 1, where T is the number of instants into which the gait cycle has been divided.

$$GVS_i = \frac{1}{T} \sum_{t=1}^{T} \left(x_{i,t} - \dot{x}_{i,t}^{ref} \right)^2 \qquad (1)$$

$$GPS = \frac{1}{N} \sum_{i=1}^{N} GVS_i^2 \tag{2}$$

The GPS is thus the RMS average of the GVS variables, as showed in Eq. 2.

The overall GPS proposed in the original paper [16] is based upon 15 clinically important kinematic variables (pelvic tilt, obliquity and rotation of the left side and hip flexion, abduction, internal rotation, knee flexion, dorsiflexion and foot progression for left and right sides). In the present work, we calculated a GPS score for each side based on all nine GVS for that side, as reported in [17].

2.3   Classification Algorithm

As stated in Sect. 2.1, we divided patients in two groups, in order to perform a binary classification. The first group contains subjects with a diagnosis of PD; conversely, patients with no diagnosis or with other neurological syndromes are grouped in the second set. In this section, we describe the Artificial Neural Network (ANN) [18] used to differentiate PD patients from patients suffering from other neurological diseases by using the information available from the two groups mentioned above based on Gait Analysis parameters.

Data Analysis. The resulting dataset was composed of **40** entries in the first group (Parkinsonism and Others) and **64** samples in the second one (Parkinson's Disease). We also considered different number of features when testing the proposed classifier obtaining the following three feature classes:

- Class 1 containing 25 features, composed of all GVS for both side, the GPS and six temporal features (left and right step cadence, left and right velocity, and left and right stride length);
- Class 2 containing 19 features composed of all nine GVS for each side and GPS;
- Class 3 containing only the spatio-temporal features.

Classifiers. Several approaches have been described in literature in order to find the best topology for an ANN classifier with Genetic Algorithms; in [19], the authors identified the optimal topology of ANN using a Multi-Objective Genetic Algorithm (MOGA) [20]. In the present work, we found the topology of the ANN using a mono-objective Genetic Algorithm (GA): the algorithm looked for the best topology maximizing the average test accuracy on a certain number of training, validation, and test iterations. It investigated each topology of the ANN using different permutations of the dataset [22].

For each input dataset, the GA was executed to find the optimal topology in each case. Specifically, the GA found the topology of the ANN in terms of number of hidden layers (ranging from 1 to 3), number of neurons per layer (ranging from 1 to 256 for the first hidden layer, and from 0 to 255 for the other hidden layers), and activation functions in *log-sigmoid* (logsig), the *hyperbolic tangent sigmoid* (tansig), the *pure linear* (purelin) and the *symmetric saturating linear* (satlins), for all the neurons per-single layer.

The performances of the classifiers were evaluated in terms of Accuracy (Eq. 3), Specificity (Eq. 4) and Sensitivity (Eq. 5), using the confusion matrix as follows (Table 2):

Table 2. The configuration of confusion matrix.

	True	
Predicted	True	False
Positive	TP	FP
Negative	TN	FN

$$Accuracy = \frac{TP + TN}{TP + TN + FP + FN} \tag{3}$$

$$Specificity = \frac{TN}{TN + FP} \tag{4}$$

$$Sensitivity = \frac{TN}{TP + FN} \tag{5}$$

3   Results and Discussion

All the participants were able to complete both clinical and instrumented evaluations. The distribution of the study subjects by age, gender and main clinical and demographic characteristics did not show significant difference between groups. The mean and the standard deviation values of the spatio-temporal parameters for both groups are summarized in Table 3.

Table 3. Mean Value (Standard Deviation) of spatio-temporal parameters and GPS in both groups (Parkinsonism and Others (1) and Parkinson's Disease (2)).

Parameters	Group 1	Group 2
Right cadence [step/min]	54.0 (17.9)	73.0 (11.0)
Right velocity [m/sec]	0.314 (0.128)	0.648 (0.233)
Right stride length [m]	0.581 (0.134)	0.863 (0.235)
Left cadence [step/min]	51.8 (12.7)	75.2 (25.4)
Left velocity [m/sec]	0.317 (0.135)	0.631 (0.221)
Left stride length [m]	0.593 (0.144)	0.842 (0.232)
GPS	11.613 (3.461)	10.645 (2.976)

Regarding the classification algorithms and techniques, the optimal topologies for ANNs specified by the Genetic Algorithm in the 3 feature classes were (Fig. 2):

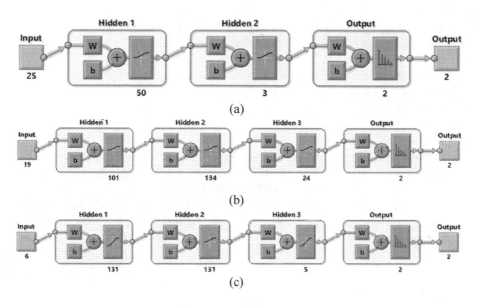

Fig. 2. Optimal topologies for ANNs specified by the Genetic Algorithm: (a) Case 1: Dataset with all 25 Features, (b) Case 2: Dataset with only GVS, (c) Case 3: Dataset with only spatio-temporal features.

Case 1. Dataset (64 positive sample and 40 negative samples) with all the 25 features
ANN with: 2 layers, 50 neurons for the first hidden layer, 3 for the second and 2 neurons for the output layer. The activation functions found by the GA were *log-sigmoid* for the first and the second layer, while the *soft max* transfer function was set for the output layer (Fig. 2(a)).

Case 2. Dataset (64 positive sample and 40 negative samples) with only GVS features
ANN with 3 layers: 101 neurons for the first hidden layer, 134 for the second, 24 for the third layer and 2 neurons for the output layer. The activation functions found by the GA were *log-sigmoid* for all the hidden layers, while the *soft max* transfer function was set for the output layer (Fig. 2(b)).

Case 3. Dataset (64 positive sample and 40 negative samples) with only the spatio-temporal features
ANN with 3 layers: 131 neurons for the first hidden layer, 131 for the second, 5 for the third layer and 2 neurons for the output layer. The activation functions found by the GA were *tansig* for the first and the third layer, *log-sigmoid* for the second layer, while the *soft max* transfer function was set for the output layer (Fig. 2(c)).

The ANN training, validation, and test sets were obtained from the input dataset with 60% of samples for training, 20% for validation, and 20% for test set. Specifically, at each iteration, the above sets were obtained through a random permutation of the dataset, keeping constant the ratio between the two classes. Moreover, the classification thresholds were determined using Receiver Operating Characteristic (ROC) curves [21], by evaluating the True Positive Rate (TPR) against the False Positive Rate (FPR) at various thresholds setting, in order to find the value which could achieve the best discrimination between the two classes.

Results are expressed in terms of mean values, considering 200 iterations, for Accuracy, Specificity, and Sensitivity, as expressed in the Eqs. 3, 4 and 5. In the first case, Accuracy was 95.05% (std = 0.0435), Specificity was 0.9731 (std = 0.0404), and Sensitivity was 0.9268 (std = 0.0877).

In the second case, Accuracy was 94.07% (std = 0.0566), Specificity was 0.9704 (std = 0.0498), and Sensitivity was 0.9080 (std = 0.0981).

In the third case, Accuracy was 86.11% (std = 0.0664), Specificity was 0.9207 (std = 0.0693), and Sensitivity was 0.8044 (std = 0.1203).

Due to the small number of samples, the performances in all the cases are quite similar; in particular, the maximum value for the accuracy was the same in the three cases and equal to 100%. The confusion matrix for the three set of features is shown in Table 4.

Table 4. Confusion Matrix for the three set of features

	True	
Predicted	True	False
Positive	14	0
Negative	0	8

4   Conclusions

Nowadays, Parkinson's disease influences a large part of worldwide population. About 1% of the population over 55 years of age is affected by this disease. Most of the current methods used for evaluating PD heavily rely on human expertise.

In this work, we proposed an innovative set of features derived from three-dimensional gait analysis in order to classify motor signs of motor impairment in PD and differentiate PD patients from healthy subjects or patients suffering from other neurological diseases. Implementations were carried out on the PD dataset to diagnose Parkinson's disease in a fully automatic and computer-aided.

Various evaluation schemes were employed for calculating the performance score of the classifiers. The ANN classifier with all 25 the features yielded the best score. The experimental results gained 95.05% classification accuracy for neural networks.

Acknowledgments. This work was partially supported by the Italian Ministry of Education University and Research under the Framework "Social Innovation" (DD 84 Ric, March 2nd 2012) with the Grant PON04a3_00097.

References

1. Twelves, D., Perkins, K.S., Counsell, C.: Systematic review of incidence studies of Parkinson's disease. Mov. Disord. **18**(1), 19–31 (2003)
2. Bevilacqua, V., Nuzzolese, N., Barone, D., Pantaleo, M., Suma, M., D'Ambruoso, D., Volpe, A., Loconsole, C., Stroppa, F. Fall detection in indoor environment with kinect sensor. In: INISTA 2014 – Proceedings of the IEEE International Symposium on Innovations in Intelligent Systems and Applications, pp. 319–324 (2014). doi:10.1109/INISTA. 2014.6873638
3. Magdalinou, N., Morris, Huw R.: Clinical features and differential diagnosis of parkinson's disease. In: Falup-Pecurariu, C., Ferreira, J., Martinez-Martin, P., Chaudhuri, K.R. (eds.) Movement Disorders Curricula, pp. 103–115. Springer, Vienna (2017). doi: 10.1007/978-3-7091-1628-9_11
4. Song, J., Fisher, B.E., Petzinger, G., Wu, A., Gordon, J., Salem, G.J.: The relationships between the unified Parkinson's disease rating scale and lower extremity functional performance in persons with early-stage Parkinson's disease. Neurorehabilit. Neural Repair **23**(7), 657–661 (2009)
5. Patel, S., Chen, B.R., Mancinelli, C., Paganoni, S., Shih, L., Welsh, M., Dy, J., Bonato, P.: Longitudinal monitoring of patients with Parkinson's disease via wearable sensor technology in the home setting. In: Engineering in Medicine and Biology Society, EMBC, 2011 Annual International Conference of the IEEE, pp. 1552–1555. IEEE (2011)
6. Esser, P., Dawes, H., Collett, J., Feltham, M.G., Howells, K.: Assessment of spatio–temporal gait parameters using inertial measurement units in neurological populations. Gait Posture **34**(4), 558–560 (2011)
7. Morris, M.E., Huxham, F., McGinley, J., Dodd, K., Iansek, R.: The biomechanics and motor control of gait in Parkinson disease. Clin. Biomech. **16**(6), 459–470 (2001)
8. Blin, O., Ferrandez, A.M., Serratrice, G.: Quantitative analysis of gait in Parkinson patients: increased variability of stride length. J. Neurol. Sci. **98**(1), 91–97 (1990)
9. Lewis, G.N., Byblow, W.D., Walt, S.E.: Stride length regulation in Parkinson's disease: the use of extrinsic, visual cues. Brain **123**(10), 2077–2090 (2000)
10. Bloem, B.R., Valkenburg, V.V., Slabbekoorn, M., Willemsen, M.D.: The Multiple Tasks Test: development and normal strategies. Gait Posture **14**(3), 191202 (2001)
11. Morris, M., Iansek, R., McGinley, J., Matyas, T., Huxham, F.: Threedimensional gait biomechanics in Parkinson's disease: Evidence for a centrally mediated amplitude regulation disorder. Mov. Disord. **20**(1), 40–50 (2005)
12. Delval, A., Salleron, J., Bourriez, J.L., Bleuse, S., Moreau, C., Krystkowiak, P., Defebvre, L., Devos, P., Duhamel, A.: Kinematic angular parameters in PD: reliability of joint angle curves and comparison with healthy subjects. Gait Posture **28**(3), 495501 (2008)
13. Davis, R.B., Ounpuu, S., Tyburski, D., Gage, J.R.: A gait analysis data collection and reduction technique. Hum. Mov. Sci. **10**(5), 575–587 (1991)
14. Baker, R., McGinley, J.L., Schwartz, M.H., Beynon, S., Rozumalski, A., Graham, H.K., Tirosh, O.: The gait profile score and movement analysis profile. Gait Posture **30**(3), 265–269 (2009)
15. Schutte, L.M., Narayanan, U., Stout, J.L., Selber, P., Gage, J.R., Schwartz, M.H.: An index for quantifying deviations from normal gait. Gait Posture **11**(1), 25–31 (2000)
16. Baker, R., McGinley, J.L., Schwartz, M., Thomason, P., Rodda, J., Graham, H.K.: The minimal clinically important difference for the Gait Profile Score. Gait Posture **35**(4), 612–615 (2012)

17. Mazurowski, M.A., Habas, P.A., Zurada, J.M., Lo, J.Y., Baker, J.A., Tourassi, G.D.: Training neural network classifiers for medical decision making: The effects of imbalanced datasets on classification performance. Neural Netw. **21**(2), 427–436 (2008)

18. Bevilacqua, V., Mastronardi, G., Menolascina, F., Pannarale, P., Pedone, A.: A novel multi-objective genetic algorithm approach to artificial neural network topology optimization: the breast cancer classification problem. In: International Joint Conference on Neural Networks, IJCNN 2006, pp. 1958–1965. IEEE, July 2006

19. Bevilacqua, V., Pacelli, V., Saladino, S.: A novel multi objective genetic algorithm for the portfolio optimization. In: Huang, D.-S., Gan, Y., Bevilacqua, V., Figueroa, J.C. (eds.) ICIC 2011. LNCS, vol. 6838, pp. 186–193. Springer, Heidelberg (2011). doi:10.1007/978-3-642-24728-6_25

20. Bevilacqua, V., Tattoli, G., Buongiorno, D., Loconsole, C., Leonardis, D., Barsotti, M., Frisoli A., Bergamasco, M.: A novel BCI-SSVEP based approach for control of walking in virtual environment using a convolutional neural network. In: 2014 International Joint Conference on Neural Networks (IJCNN), pp. 4121–4128. IEEE, July 2014

21. Bevilacqua, V., Brunetti, A., Triggiani, M., Magaletti, D., Telegrafo, M., Moschetta, M.: An optimized feed-forward artificial neural network topology to support radiologists in breast lesions classification. In: Proceedings of the 2016 on Genetic and Evolutionary Computation Conference Companion, pp. 1385–1392. ACM, July 2016

22. Bradley, A.P.: The use of the area under the ROC curve in the evaluation of machine learning algorithms. Pattern Recogn. **30**(7), 1145–1159 (1997)

Protein and Gene Bioinformatics:
Analysis, Algorithms and Applications

Identification of Candidate Drugs for Heart Failure Using Tensor Decomposition-Based Unsupervised Feature Extraction Applied to Integrated Analysis of Gene Expression Between Heart Failure and DrugMatrix Datasets

Y-h. Taguchi$^{(\boxtimes)}$

Department of Physics, Chuo University, Tokyo 112-8551, Japan
tag@granular.com

Abstract. Identifying drug target genes in gene expression profiles is not straightforward. Because a drug targets not mRNAs but proteins, mRNA expression of drug target genes is not always altered. In addition, the interaction between a drug and protein can be context dependent; this means that simple drug incubation experiments on cell lines do not always reflect the real situation during active disease. In this paper, I apply tensor decomposition-based unsupervised feature extraction to the integrated analysis of gene expression between heart failure and the DrugMatrix dataset where comprehensive data on gene expression during various drug treatments of rats were reported. I found that this strategy, in a fully unsupervised manner, enables us to identify a combined set of genes and compounds, for which various associations with heart failure were reported.

Keywords: Tensor decomposition · Drug discovery · Heart diseases

1 Introduction

In silico drug discovery is an important task because experimental identification/ verification of therapeutic compounds is a time-consuming and expensive process. There are two major trends of *in silico* drug discovery: the ligand-based approach [1] and structure-based [2] approach. The former is very straightforward; new drug candidates are identified based upon the similarity with known drugs no matter how the similarity is evaluated. Although it is a powerful method, there are some drawbacks; if there are no known drugs for target proteins, then there is no way to find new drug candidates. Even if there are many known drugs for the target protein, it is hopeless to find compounds that are effective but without any similarity with known drugs. The second, structure-based approach, can address these weaknesses. It can identify new therapeutic compounds even without the information about known drugs. Of course, there are some drawbacks in this strategy, too. If the target protein structure is not known, it must be predicted prior to the drug discovery process. Even if the target protein's structure is known, because we need

© Springer International Publishing AG 2017
D.-S. Huang et al. (Eds.): ICIC 2017, Part II, LNCS 10362, pp. 517–528, 2017.
DOI: 10.1007/978-3-319-63312-1_45

to numerically verify the binding affinity between the ligand compound and target protein, which also requires a large amount of computational resources, structure-based *in silico* drug discovery is still far from easy to perform. In addition, prediction accuracy of protein structure and of ligand-binding structure is not very high at all. Thus, it would be helpful to have an additional/alternative strategy for *in silico* drug discovery.

Recently, an alternative approach was proposed that is aimed at finding drug candidates computationally using gene expression profiles of cell lines treated with compounds [3, 4]. This third approach is not straightforward at all. First of all, because compounds target not mRNAs but proteins, mRNA expression of drug target proteins is not always affected. Therefore, direct identification of a drug target protein in gene expression data cannot be done. Second, gene expression alteration caused by treatment with a compound may be context dependent; in other words, in a cell line, the gene expression difference caused by incubation with a compound may differ from that in diseases. To compensate these difficulties, the gene expression signature strategy was developed. In this approach, gene expression alteration profiles caused by treatment of a cell line with various drug candidates are compared with those of known drugs. If the profiles are similar, then new drug candidates are expected to function similarly to known drugs. Although this third strategy is a useful one, if there are no known drugs for the target disease, this approach cannot function at all as in the case of ligand-based approaches.

Some examples aiming to identify drug-target interaction from gene expression based upon previous knowledges are as follows. Wang et al. [5] tried to identify on and off-targets of drugs based upon similarity between drug-induced *in vitro* genomic expression changes. Iwata et al. [6] explored potential target proteins with cell-specific transcriptional similarity using chemical–protein interactome. Lee et al. [7] tried drug repositioning for cancer therapy based on large-scale drug-induced transcriptional signatures. Although these are only a few examples, they need pre-knowledge about drug-target interactions. Alternatively, instead of drug-target interactions, drug-disease interaction is investigated. For example, Cheng et al. [8] tried to measure the connectivity between disease gene expression signatures and compound-induced gene expression profiles. Sirota et al. [9] also integrated gene expression measurements from 100 diseases and gene expression measurements on 164 drug compounds, yielding predicted therapeutic potentials for these drugs. Iorio et al. [10] investigated compound-targeted biological pathways based upon gene expression similarity. They are unsupervised approaches to some extent, but target genes cannot be exploited.

In this paper, I propose a strategy that can infer drug candidates from drug treatment-associated gene expression profiles without the information about known compounds for diseases. In this strategy, I employ the tensor decomposition (TD)-based unsupervised feature extraction (FE) approach, which is an extension of the recently proposed principal component analysis (PCA)-based unsupervised FE; PCA-based unsupervised FE successfully solved various bioinformatic problems [11–28]. In this TD-based strategy, tensors were generated using a mathematical product of a gene expression profile of drug-treated cell lines and of a gene expression profile of a disease. Then, pairs of compounds and genes are identified whose mRNA expression alteration is associated with drug-treated cell lines and is coincident with such alteration during disease progression. Biological evaluation of the identified genes and compounds based upon past studies turned out to be promising.

2 Materials and Methods

2.1 Mathematics of TD

In this subsection, I briefly discuss what the TD is and how I apply TD to the present problem. Suppose an m-mode tensor $x_{j_1 \ldots j_{m-1} i}$ represents gene expression of the ith gene under the j_k ($k = 1, \ldots, m-1, j_k = 1, \ldots, N_k$) conditions, examples of which are diseases, patients, tissues, and time points. Then, TD is defined as

$$x_{j_1 \ldots j_{m-1} i} = \sum_{l_1 \ldots l_m}^{N_1 \ldots N_m} G(l_1 \ldots l_m) x_{l_m i} \prod_{k=1}^{m-1} x_{l_k j_k} \tag{1}$$

where $G(l_1 \ldots l_m)$ is a core tensor and $x_{l_m i}$ and $x_{l_k j_k}$ are singular value matrices that are supposed to be orthogonal to one another. Because $G(l_1 \ldots l_m)$ is assumed to be as large as $x_{j_1 \ldots j_{m-1} i}$, it is obviously an overcomplete problem; thus, there are no unique solutions. To solve TD uniquely, I specifically employed the higher-order singular value decomposition [29] (HOSVD) algorithm that tries to attain TD such that smaller number of core tensors and singular value vectors can represent $x_{j_1 \ldots j_{m-1} i}$ as much as possible.

2.2 Tensor Generation for Integrated Analysis

It is quite common when there is a set of gene expression profiles of human cell lines or model animals treated with various compounds at multiple dose densities. For example, Drug Matrix[1] and LINCS [30] are good examples, although the former comprises only temporal gene expression after drug treatments. Nonetheless, it is not easy to infer a drug's action on diseases by means of only these gene expression profiles; some kind of integrated analysis with disease gene expression profiles is required, but it is not so straightforward. Candidate drugs should satisfy these conditions:

- Gene expression in these profiles must significantly decrease or increase with the increasing dose density of compounds.
- Gene expression alteration caused by drug treatment must be significantly coincident with that associated with disease progression.

How these two independent significance values can be evaluated is unclear. For example, we can have two sets of significant gene expression alterations of the ith gene, $\{\Delta x_i\}$, caused by drug treatment and those of the i'th gene, $\{\Delta x'_{i'}\}$, during disease progression, respectively. First, we need to test whether the two sets of genes are

---

[1] https://ntp.niehs.nih.gov/drugmatrix/index.html.

significantly overlapping. Next, when there is a significant overlap, we have to determine whether these two gene expression alteration profiles correlate significantly. Furthermore, because the analysis is usually conducted among multiple compounds, all the significance evaluation must be corrected based upon a multiple comparison criterion. It is obviously a complicated and not a promising strategy.

Nevertheless, if we can have gene expression profiles expressed via a tensor, $x_{j_1 \ldots j_{m-1} i}$, where j_k ($k = 1, \ldots, m-1$) corresponds to drug candidates, dose density, and disease progression, we can easily evaluate a candidate drug using TD, Eq. (1). If there are $x_{l_k j_k}$ values that represent significant dependence upon dose densities and disease progression, genes' and compounds' singular value vectors that share core tensor G with larger absolute values with these $x_{l_k j_k}$ s can be used for the selection of genes as well as compounds as follows.

Suppose $\{l_k\}$ is a set of indices of genes' or compounds' singular value vectors that are associated with significant dose density dependence as well as disease progression dependence. Genes and compounds can be identified as being associated with significant singular value vector components. For this purpose, P-values are attributed to each ith gene/j_k th compound assuming a χ^2 distribution,

$$P_i = P_{\chi^2}[> \sum_{\{l_m\}} \left(\frac{x_{l_m i}}{\sigma_{l_m}} \right)^2] \text{ or } P_{j_k} = P_{\chi^2}[> \sum_{\{l_k\}} \left(\frac{x_{l_k j_k}}{\sigma_{l_k}} \right)^2] \tag{2}$$

where $P_{\chi^2}[> x]$ is the cumulative probability that the argument is greater than x assuming the χ^2 distribution and σ_{l_m} and σ_{l_k} are standard deviation. After adjusting P-values using the Benjamini–Hochberg (BH) criterion [31], genes and compounds that have significant P-values, e.g., less than 0.01, are selected as those contributing to the specified singular value vectors. Nevertheless, because such a tensor can be obtained only when drug treatment is performed on patients, this strategy is useless; if we can test drug efficiency directly on patients, then there is no need for *in silico* drug discovery. To overcome this discrepancy, I replace $x_{j_1 \ldots j_{m-1} i}$ with a product, $x_{j_1 \ldots j_{m'} i} x_{j_1 \ldots j_{m''} i}$, where $x_{j_1 \ldots j_{m'-1} i}$ is gene expression for the drug treatment of cell lines/model animals, while $x_{j_1 \ldots j_{m''-1} i}$ is gene expression for the patients ($m - 1 = m' + m''$). Because these two can be obtained independently, we can test any kind of combinations of drug treatments and diseases even after all measurements were performed.

The process performed in the present study following the above procedures is illustrated in Fig. 1.

2.3   Gene Expression Profiles

Gene expression profiles for drug treatments of rats were retrieved from Drug Matrix under the gene expression omnibus (GEO) ID GSE59905, while heart failure human gene expression was taken from GEO ID 57345. For both datasets, expression files of genes, GSE57345-GPL11532_series_matrix.txt.gz, GSE59905-GPL5426_series_matrix.txt.gz, and GSE59905-GPL5425_series_matrix.txt.gz were directly downloaded from the series matrix.

2.4   Various Servers for Enrichment Analysis

To Enrichr [32] and TargetMine [33], 274 gene symbols were uploaded. For TargetMine, human was assumed as an organism under study, and the BH criterion was used for P-value correction.

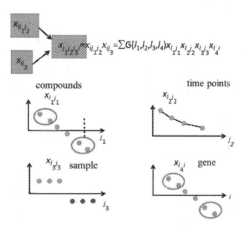

Fig. 1. Schematic that illustrates gene-drug pairs identification processes performed in the present study. Gene expression from DrugMatrix, $x_{j,j_2,i}$, and heart failure, $x_{j,i}$, are multiplied and tensor, x_{j,j_2j_3i}, is generated. It is decomposed into core matrix, G, and compound singular value matrix x_{l,j_1}, time point singular value matrix, $x_{l_2j_2}$, human sample singular value matrix, $x_{l_3j_3}$, and gene singular value matrix, x_{l_4i}, are generated. After the temporal dependent $x_{l_2j_2}$ and disease dependent $x_{l_3j_3}$ are identified, associated genes and compounds $x_{l_4j_4}$ and $x_{l_1j_1}$ are selected. Based upon them, genes and compounds as outliers are selected by assuming a χ^2 distribution.

2.5   Statistical Analysis

All the statistical analyses were performed within the R software. HOSVD was carried out using the `hosvd` function in the rTensor package.

3   Results

3.1   TD-Based Unsupervised FE Was Applied to a Combined Tensor

From gene expression profiles of the rat left ventricle (LV) treated with 218 drugs, we selected four time points (1/4, 1, 3, and 5 days after treatment). Although these do not directly represent drug dose dependence, time course observations can be replaced with dose dependence, because drug dose density is expected to monotonically decrease with time. On the other hand, human heart gene expression profiles are composed of 82 idiopathic dilated cardiomyopathy patients, 95 ischemic patients, and 136 healthy controls, respectively. Among them, 3937 genes sharing gene symbols between human and rat were considered. Then, the generated tensor is

$$x_{j_1 j_2 j_3 i} = x_{j_1 j_2 i} x_{j_3 i}, 1 \leq j_1 \leq 218, 1 \leq j_2 \leq 4, 1 \leq j_3 \leq 313, 1 \leq i \leq 3937 \qquad (3)$$

which represents the products of gene expression of the ith gene of LV treated with j_1 compound at the j_2 th time point after the drug treatment and gene expression of the j_3 th human heart, respectively. HOSVD was applied to $x_{j_1 j_2 j_3 i}$ and core tensor $G(l_1 l_2 l_3 l_4), 1 \leq l_1 \leq 218, 1 \leq l_2 \leq 4, 1 \leq l_3 \leq 313, 1 \leq l_4 \leq 3937$, compound singular value matrix $x_{l_1 j_1}$, time point singular value matrix, $x_{l_2 j_2}$, human sample singular value matrix, $x_{l_3 j_3}$, and gene singular value matrix, $x_{l_4 i}$, were obtained. Prior to selection of genes and compounds, we need to know which time points singular value vector represents time dependence and which human sample singular value vector represents the distinction between patients and healthy controls (Fig. 2). As for time point singular value vectors, I decided to use the 2nd time point singular value vector, $x_{l_2 j_2}, l_2 = 2$, because it has the strongest correlation with days. It also represents reasonable time development. After drug treatment, gene expression gradually increases because it takes a while for a drug treatment to have an effect. Then, after it has a peak on day 1, a monotonic decrease follows. On the other hand, for human sample singular value vectors, the 2nd and 3rd ones were selected because they have a clear distinction between patients and healthy controls.

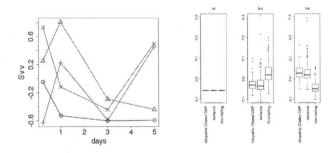

Fig. 2. Left: Time points' singular value vectors. Black circle: 1st, red triangle: 2nd, green cross: 3rd, and blue cross: 4th singular value vectors, respectively. Pearson's correlation coefficients toward days are $-0.72, -0.82, 0.51$, and -0.09, respectively. **Right**: A box plot of human sample singular value vectors. From left to right, the $1^{st}, 2^{nd}$, and 3rd singular value vectors are shown. (Color figure online)

Next, I tried to identify gene singular value vectors and compound singular value vectors associated with core tensor $G(l_1 l_2 l_3 l_4), l_2 = 2, 2 \leq l_3 \leq 3$ that have larger absolute values (Table 1). One can see that the 2nd singular value vector of compounds is always associated with top 20 core tensors. The selection of gene singular value vectors is not so trivial. First of all, generally low-ranked gene singular value vectors are listed. This means that gene expression associated with disease progression is not a majority. This is a common situation because the disease usually affects only a limited number of genes. Then, tentatively, I decided to select top 10 gene singular value vectors, 21st, 25th, 27th, 28th, 33rd, 36th, 37th, 38th, 41st, and 42nd singular value vectors of genes. Using these singular value vectors, P values were attributed to genes and compounds. The attributed P values were adjusted by the BH criterion. Then, 281 probes and 0 compounds associated with adjusted P values less than 0.01 were selected. Because no compounds pass our criteria, I sought another way to select compounds. Figure 3 shows the histogram of the 2nd singular value vectors of compounds. There are obviously some outliers. Then, tentatively, I selected 43 compounds having the absolute 2nd singular value vector components larger than 0.1.

3.2   Biological Evaluation of the Selected Compounds and Genes

To see if we can successfully identify biologically relevant compounds and genes, we evaluated these selected genes and compounds. At first, a literature search was performed on the 43 drugs. Then, some heart failure-related studies were identified for most of the 43 drugs (Table 2). This means that biologically relevant drugs were likely to be identified successfully. As for the genes identified, 274 genes associated with the identified 281 probes are shown in Table 3.

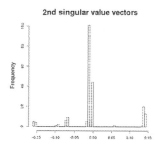

Fig. 3. A histogram of 2nd singular value vectors of compounds.

To evaluate biological reliability of these 274 genes, they were uploaded to various enrichment servers. When they were uploaded to TargetMine, top five tissue enrichment results were related to the heart (Table 4). Top four significant disease enrichment results represent heart failure (Table 5). When they were uploaded to Enrichr, top three OMIM disease enrichment results were related to heart failure (Table 6). Two out of top three MGI Mammalian Phenotype Level 3 enrichment results were also related to heart failure (Table 7). Thus, our identification of genes was also successful.

Table 1. $G(l_1l_2l_3l_4), l_2 = 2, 2 \leq l_3 \leq 3$, in the order of larger absolute values of G

l_1	l_3	l_4	G	l_1	l_3	l_4	G
2	2	27	66.2	2	3	40	−25.5
2	3	38	−43.7	2	2	29	25.2
2	2	33	40.6	2	2	31	−22.6
2	2	28	−40.2	2	3	39	21.8
2	3	41	38.2	2	2	32	20.7
2	3	37	−31.6	2	3	33	−19.7
2	2	21	28.5	2	2	26	−19.5
2	3	36	−26.8	2	2	11	−18.2
2	3	42	−26.2	2	2	18	−17.3
2	2	25	−26.2	2	3	31	15.4

Table 2. Literature search performed on 43 drugs identified by TD-based unsupervised FE. Numbers are Pubmed IDs (https://www.ncbi.nlm.nih.gov/pubmed/) that report the relation to heart failure.

Amitriptyline [27994924]	Atropine [24279866]	Baclofen [27682809]	Bezafibrate [26957517]	Caffeine [25944789]
Calcitriol [27209698]	Chlorambucil [8164221]	Cimetidine [18656805]	Citalopram [25326372]	Clemastine [16288909]
Clonazepam [15699937]	Cyclophosphamide [24467219]	D-Tubocurarine Chloride [14839919]	Dexamethasone [25923220]	Dexchlorpheniramine [—]
Digitoxin [27082032]	Diphenhydramine [22158278]	Doxazosin [26515144]	Ebastine [21410688]	Fenofibrate [26497978]
Fluphenazine [23395964]	Gabapentin [19195912]	Ifosfamide [14586140]	Iproniazid [13688979]	Lacidipine [23911888]
Loratadine [21880544]	Nevirapine [15526045]	Nimodipine [17191657]	Nitrendipine [22750214]	Ofloxacin [21559378]
Oxymetazoline [22855901]	Paroxetine [26216863]	Phenacemide [—]	Phenytoin [24172819]	Rosiglitazone [21666037]
Sparteine [4408029]	Stavudine [—]	Valsartan [26992459]	Vecuronium Bromide [—]	Venlafaxine [23301719]
Vinblastine [25537132]	Vincristine [—]	Zidovudine [25838291]		

Enrichr also outputted many epigenetic feature enrichment results. Top most significant ENCODE TF-ChiP-seq 2015 is POLR2A_heart_mm9; POLR2A is a transcription factor (TF) reported to be a stable reference gene for gene expression alteration in gene expression studies on rodent and human heart failure [34]. This finding suggested that *POLR2A* is constantly expressive in heart failure, which is coincident with our analysis. Top most significant TF-LOF Expression result from

Table 3. The 274 genes associated with 281 probes identified by TD-based unsupervised FE.

Atp6v1 h Smad4 Tfam Ramp2 Vdac2 Sfrp4 Accn3 Pdxk Ccnl1 Kcnk3 Pdk4Nfe2l2 Nexn Ccl2
Lphn3 P2rx3 Odz2 Mpp3 Kcnt1 Gapdh Ncoa2 Pacsin2 Slpi Tnfaip6 Prelp Ppp2r2d Sharpin
Slc38a2 Col5a1 Steap3 Ppp1r14a Bves Nsf Sox18 Ndfip1 Yme1l1 Gosr1 Nf1 Fndc5 Pold4
Wbp4 Immt Sdhd Dlc1 Itga6 Eif2s2 Bmpr1a Abcb10 Mknk2 Kpna1 Bag3 F8 Lrp1 Vezt Aqp4
Pdcl3 Schip1 Gbe1 Synj1 Map2k4 Laptm4b Psmd12 Mtus1 Ddit4 l Mlycd Ppm1b Mterf Ing4
Vsnl1 Rhoa Ltbp4 Dhrs1 Txndc12 Tnfrsf12a Itm2c Samm50 B4galt7 Fbl Chchd4 Pdrg1 Pycr2
Rplp1 Rps20 Bzw1 Fos Cybb Sccpdh Smpd1 Kcmf1 Gna12 Nedd4 l Bpgm Akap1 Actr1b
Msn Dnajc5 Lcp1 Agpat1 Tarbp2 Git2 Usp14 Nfatc4 Rxrg Uqcrc2 Actn1 Ndufs2 Rps18
Slc40a1 Chdh Rela Ciapin1 Fbxo22 mrpl9 Ppp1r14c Btbd9 Obscn Cmklr1 Fyttd1 Sirt5 Flt1
Grwd1 Hrc Trpc4ap Dcps Idh3a Tmem30a Fut8 Pi4k2a Cdh23 Eif4a1 C1qa Gpx3 Slc25a4
Fgf9 Psmc1 Rbm10 Nr0b2 Acsl1 A2 m Alas1 Suclg1 Acads Atp5a1 Ccnd2 Csnk2b Psmb4
Canx Cd36 Pggt1b Pde4b Npr3 Hspa5 Nr3c1 Apob Got2 Actg2 Nr3c2 Egfr Ldha Adcy3 Cryab
Man2c1 Il6r Slc6a1 Adra1b Ednra Tnfrsf1a Atf3 Mapk6 Agrn Rab15 Ywhae Arf4 Pdia4 Ppara
Il6st Adrb2 Egr1 Got1 Myc Myl2 Mme Spin2b Stat3 Slc2a4 Apod Dpp4 Mapk10 Azgp1
Ephx1 Htr4 Mgp Spp1 Adora3 Eef2 k Hmgb1 Nes Ptgds Slc5a1 Ywhah Cd74 Aoc3 Atp1b1
Itpr3 Ak3 Lcat Pccb Ppm1a Ppp2ca Sod1 Glul Ghr Kcnj8 Areg Cd63 Ctf1 Tnni3 Rps6
Serpinh1 Uchl1 Btg2 Mapk9 Tpm1 Vtn Hapln1 Mgat3 Ca3 Tpsab1 Anxa2 Ccrl Junb Gnb3
Stx7 Gnb2l1 Il1rl1 Fstl1 Gatm Pdk2 Ces1 Fabp5 Csda Txnip Lss Acvr1c Scn2b Mfn2 Mxd3
Ptger2 Mvd Gucy1a3 Ppif Mapk14 Gnb1 Ttn Acta1 Gstp1 Hmbs C3 Vim Cebpg Amhr2
Idh3 g Csrp3 Acox3 Cyb5b Cast

Table 4. Top five significant tissue enrichment results of TargetMine.

Tissue	p-value	Matches
left ventricular apex samples	1.880e–37	101
heart atrium	1.357c–36	155
heart	8.637e–36	153
ventricular myocardium	4.306e–35	95
atrial myocardium	1.716e–34	98

Table 5. Top four disease enrichment results of TargetMine.

Disease	p-Value	Matches
Myocardial Ischemia	1.546e–7	22
Infarction, Middle Cerebral Artery	6.072e–6	9
Reperfusion Injury	9.567e–5	13
Cardiomyopathies	1.448e–4	13

Table 6. Top three significant OMIM Disease enrichment results of Enrichr.

Name	Overlap	P-value	Adjusted p-value
cardiomyopathy, hypertrophic	5/17	2.516E–05	9.814E–05
cardiomyopathy	5/42	2.615E–04	5.099E–03
cardiomyopathy, dilated	4/33	1.031E–03	1.341E–02

Table 7. Top three significant MGI Mammalian Phenotype Level 3 enrichment results of Enrichr.

Term	Overlap	P-value	Adjusted P-value
MP0001544_abnormal_cardiovascular_system_physiology	54/1130	5.133E−016	3.233E−014
MP0002106_abnormal_muscle_physiology	35/671	1.171E−011	1.844E−010
MP0002127_abnormal_cardiovascular_system_morphology	50/1223	2.818E−012	5.919E−011

GEO is yy1_227711985_skeletal_muscle_lof_mouse_gpl8321_gse39009_up. YY1 is a TF reported to play critical roles in cardiac morphogenesis [35]. The top most significant ENCODE Histone Modifications 2015 result is H3K36me3_myocyte_mm9; H3K36me3 was reported to play a crucial role in cardiomyocyte differentiation [36]. These TFs as well as histone modifications identified by our strategy can be possible drug targets.

4 Conclusions

In this paper, I introduced a new strategy that integrates disease (heart failure) gene expression profiles with drug treatment-related tissue gene expression profiles. The identified genes as well as compounds have been widely reported to be related to heart failure. Thus, this strategy turned out to be useful for *in silico* drug discovery.

References

1. Favia, A.D.: Theoretical and computational approaches to ligand-based drug discovery. Front. Biosci. (Landmark Ed.) **16**, 1276–1290 (2011)
2. Lionta, E., Spyrou, G., Vassilatis, D., Cournia, Z.: Structure-based virtual screening for drug discovery: principles, applications and recent advances. Curr. Top. Med. Chem. **14**, 1923–1938 (2014)
3. Liu, C., Su, J., Yang, F., Wei, K., Ma, J., Zhou, X.: Compound signature detection on LINCS L1000 big data. Mol. BioSyst. **11**, 714–722 (2015)
4. Hizukuri, Y., Sawada, R., Yamanishi, Y.: Predicting target proteins for drug candidate compounds based on drug-induced gene expression data in a chemical structure-independent manner. BMC Med. Genomics **8**, 82 (2015)
5. Wang, K., Sun, J., Zhou, S., Wan, C., Qin, S., Li, C., He, L., Yang, L.: Prediction of drug-target interactions for drug repositioning only based on genomic expression similarity. PLoS Comput. Biol. **9**, e1003315 (2013)
6. Iwata, M., Sawada, R., Iwata, H., Kotera, M., Yamanishi, Y.: Elucidating the modes of action for bioactive compounds in a cell-specific manner by large-scale chemically-induced transcriptomics. Sci. Rep. **7**, 40164 (2017)

7. Lee, H., Kang, S., Kim, W., Fedorov, O., Filippakopoulos, P., Hunt, J.: Drug repositioning for cancer therapy based on large-scale drug-induced transcriptional signatures. PLoS One **11**, e0150460 (2016)

8. Cheng, J., Yang, L., Kumar, V., Agarwal, P.: Systematic evaluation of connectivity map for disease indications. Genome Med. **6**, 95 (2014)

9. Sirota, M., Dudley, J.T., Kim, J., Chiang, A.P., Morgan, A.A., Sweet-Cordero, A., Sage, J., Butte, A.J.: Discovery and preclinical validation of drug indications using compendia of public gene expression data. Sci. Transl. Med. **3** (2011)

10. Iorio, F., Bosotti, R., Scacheri, E., Belcastro, V., Mithbaokar, P., Ferriero, R., Murino, I.., Tagliaferri, R., Brunetti-Pierri, N., Isacchi, A., di Bernardo, D.: Discovery of drug mode of action and drug repositioning from transcriptional responses. Proc. Natl. Acad. Sci. U.S.A. **107**, 14621–14626 (2010)

11. Kinoshita, R., Iwadate, M., Umeyama, H., Taguchi, Y.H.: Genes associated with genotype-specific DNA methylation in squamous cell carcinoma as candidate drug targets. BMC Syst. Biol. **8**(Suppl 1), S4 (2014)

12. Taguchi, Y., Iwadate, M., Umeyama, H., Murakami, Y., Okamoto, A.: Heuristic principal component analysis-aased unsupervised feature extraction and its application to bioinformatics. In: Wang, B., Li, R., Perrizo, W. (eds.) Big Data Analytics in Bioinformatics and Healthcare, pp. 138–162 (2015)

13. Murakami, Y., Kubo, S., Tamori, A., Itami, S., Kawamura, E., Iwaisako, K., Ikeda, K., Kawada, N., Ochiya, T., Taguchi, Y.H.: Comprehensive analysis of transcriptome and metabolome analysis in intrahepatic cholangiocarcinoma and hepatocellular carcinoma. Sci Rep. **5**, 16294 (2015)

14. Taguchi, Y.-H., Iwadate, M., Umeyama, H.: Heuristic principal component analysis-based unsupervised feature extraction and its application to gene expression analysis of amyotrophic lateral sclerosis data sets. In: 2015 IEEE Conference Computational Intelligence Bioinformatics Computing Biology (2015)

15. Umeyama, H., Iwadate, M., Taguchi, Y.: TINAGL1 and B3GALNT1 are potential therapy target genes to suppress metastasis in non-small cell lung cancer. BMC Genom. **15**, S2 (2014)

16. Taguchi, Y., Murakami, Y.: Principal component analysis based feature extraction approach to identify circulating microRNA biomarkers. PLoS One (2013)

17. Taguchi, Y.-H., Murakami, Y.: Universal disease biomarker: can a fixed set of blood microRNAs diagnose multiple diseases? BMC Res. Notes. **7**, 581 (2014)

18. Murakami, Y., Tanahashi, T., Okada, R., Toyoda, H., Kumada, T., Enomoto, M., Tamori, A., Kawada, N., Taguchi, Y.H., Azuma, T.: Comparison of hepatocellular carcinoma miRNA expression profiling as evaluated by next generation sequencing and microarray. PLoS One **9** (2014)

19. Taguchi, Y.-h., Iwadate, M., Umeyama, H.: Principal component analysis-based unsupervised feature extraction applied to in silico drug discovery for posttraumatic stress disorder-mediated heart disease. BMC Bioinform. **16**, 139 (2015)

20. Taguchi, Y-h.: Identification of more feasible MicroRNA–mRNA interactions within multiple cancers using principal component analysis based unsupervised feature extraction. Int. J. Mol. Sci. **17**, E696 (2016)

21. Taguchi, Y.-H., Iwadate, M., Umeyama, H.: Heuristic principal component analysis-based unsupervised feature extraction and its application to gene expression analysis of amyotrophic lateral sclerosis data sets. In: 2015 IEEE Conference on Computational Intelligence in Bioinformatics and Computational Biology (CIBCB), pp. 1–10 (2015)

22. Taguchi, Y.-H.: Principal component analysis based unsupervised feature extraction applied to publicly available gene expression profiles provides new insights into the mechanisms of action of histone deacetylase inhibitors. NEPIG (2016)

23. Taguchi, Y.-H., Iwadate, M., Umeyama, H.: SFRP1 is a possible candidate for epigenetic therapy in non-small cell lung cancer. BMC Med. Genomics. **9** (2016)

24. Taguchi, Y.-H.: microRNA-mRNA interaction identification in Wilms tumor using principal component analysis based unsupervised feature extraction. In: 2016 IEEE 16th International Conference on Bioinformatics and Bioengineering (BIBE), pp. 71–78 (2016)

25. Taguchi, Y.-H.: Principal component analysis based unsupervised feature extraction applied to budding yeast temporally periodic gene expression. BioData Min. **9**, 22 (2016)

26. Murakami, Y., Toyoda, H., Tanahashi, T., et al.: Comprehensive miRNA expression analysis in peripheral blood can diagnose liver disease. PLoS ONE **7**, e48366 (2012)

27. Ishida, S., Umeyama, H., Iwadate, M., Taguchi, Y.H.: Bioinformatic Screening of Autoimmune Disease Genes and Protein Structure prediction with FAMS for drug discovery. Protein Pept. Lett. **21**, 828–839 (2014)

28. Taguchi, Y.: Principal components analysis based unsupervised feature extraction applied to gene expression analysis of blood from dengue haemorrhagic fever patients. Sci. Rep. **7**, 44016 (2017)

29. De Lathauwer, L., De Moor, B., Vandewalle, J.: a multilinear singular value decomposition. SIAM J. Matrix Anal. Appl. **21**, 1253–1278 (2000)

30. Duan, Q., Reid, S.P., Clark, N.R., Wang, Z., Fernandez, N.F., Rouillard, A.D., Readhead, B., Tritsch, S.R., Hodos, R., Hafner, M., Niepel, M., Sorger, P.K., Dudley, J.T., Bavari, S., Panchal, R.G., Ma'ayan, A.: L1000CDS2: LINCS L1000 characteristic direction signatures search engine. npj Syst. Biol. Appl. **0** (2016). 16015

31. Benjamini, Y., Hochberg, Y.: Controlling the false discovery rate: a practical and powerful approach to multiple testing. J. R. Stat. Soc. **B57**, 289–300 (1995)

32. Kuleshov, M.V., Jones, M.R., Rouillard, A.D., Fernandez, N.F., Duan, Q., Wang, Z., Koplev, S., Jenkins, S.L., Jagodnik, K.M., Lachmann, A., McDermott, M.G., Monteiro, C. D., Gundersen, G.W., Ma'ayan, A.: Enrichr: a comprehensive gene set enrichment analysis web server 2016 update. Nucleic Acids Res. **44**, W90–W97 (2016)

33. Chen, Y.-A., Tripathi, L.P., Mizuguchi, K.: TargetMine, an integrated data warehouse for candidate gene prioritisation and target discovery. PLoS ONE **6**, e17844 (2011)

34. Brattelid, T., Winer, L.H., Levy, F.O., Liestøl, K., Sejersted, O.M., Andersson, K.B.: Reference gene alternatives to Gapdh in rodent and human heart failure gene expression studies. BMC Mol. Biol. **11**, 22 (2010)

35. Beketaev, I., Zhang, Y., Kim, E.Y., Yu, W., Qian, L., Wang, J.: Critical role of YY1 in cardiac morphogenesis. Dev. Dyn. **244**, 669–680 (2015)

36. Cattaneo, P., Kunderfranco, P., Greco, C., Guffanti, A., Stirparo, G.G., Rusconi, F., Rizzi, R., Di Pasquale, E., Locatelli, S.L., Latronico, M.V.G., Bearzi, C., Papait, R., Condorelli, G.: DOT1L-mediated H3K79me2 modification critically regulates gene expression during cardiomyocyte differentiation. Cell Death Differ. **23**, 555–564 (2016)

Calculating Kolmogorov Complexity from the Transcriptome Data

Panpaki Seekaki[1] and Norichika Ogata[1,2(✉)]

[1] Nihon BioData Corporation, 3-2-1 Sakado, Takatsu-Ku,
Kawasaki, Kanagawa, Japan
norichik@nbiodata.com
[2] Medical Mechanica Incorporated, 3-2-1 Sakado, Takatsu-Ku,
Kawasaki, Kanagawa, Japan

Abstract. Information entropy is used to summarize transcriptome data, but ignoring zero count data contained them. Ignoring zero count data causes loss of information and sometimes it was difficult to distinguish between multiple transcriptomes. Here, we estimate Kolmogorov complexity of transcriptome treating zero count data and distinguish similar transcriptome data.

Keywords: Kolmogorov complexity · Transcriptome · Information entropy

1 Introduction

When environmental conditions change abruptly, living cells must coordinate adjustments in their genome expression to accommodate the changing environment [1]. It is possible that the degree of change in the environment affects the degree of change in the gene expression pattern. However, it is difficult to completely understand the amount of change that occurs in the transcriptome, given that this would involve thousands of gene expression measurements. A previous study defined transcriptome diversity as the Shannon entropy of its frequency distribution, and made it possible to express the transcriptome as a single value. Dimensionality Reduction methods e.g. Principle component analysis and t-SNE had been used to transcriptome analyses [2, 3], but biological meanings of value of results of these methods was unclear. Transcriptome diversity also reduces the dimensions of transcriptome data and the biological meaning of transcriptome diversity is clear. The first research on transcriptome diversity was performed with human tissues [1], and later research compared cancer cells with normal cells [4]. In plant research, transcriptome diversity has been used to compare several wounded leaves [5]. Transcriptome diversity also used for measurement of cellular dedifferentiation [6]. Our Previous study compared transcriptome diversity between cells cultured in vitro in media supplemented with several concentrations of phenobarbital, to investigate how the amount of environmental change (in terms of drug

D.-S. Huang et al. (Eds.): ICIC 2017, Part II, LNCS 10362, pp. 529–540, 2017.
DOI: 10.1007/978-3 319-63312-1_46

concentration) affects the amount of transcriptome change and multi-stability of the genome expression system was indicated by an observation of hysteretic phenomenon of transcriptome diversities [7].

Information entropy of transcriptome were described in a previous study [1], the transcriptomes of each tissue as a set of relative frequencies, P_{ij}, for the ith gene (i = 1, 2, ..., g) in the jth tissue (j = 1, 2, ..., t); and then quantified transcriptome diversity using an adaptation of Shannon's entropy formula (1):

$$H_{ij} = -\sum_{i=1}^{g} P_{ij} \log_2(P_{ij}) \tag{1}$$

Transcriptome data obtained using RNA-seq contains zero count data. In Shannon entropy, these zero count data were ignored. There is possibility that this founder mental character of information entropy makes it difficult to distinguish similar transcriptomes, ignoring zero count data causes loss of information. We hypothesized that developing a transcriptome-summarizing method treating zero count makes distinguish similar transcriptomes possible.

The elementary theories of Shannon information and Kolmogorov complexity have a common purpose [8]. Kolmogorov complexity is the minimum number of bits from which a particular message or file can effectively be reconstructed. Since zero count data contained transcriptome data have a message, Kolmogorov complexity treat zero count data.

Here, we developed a Kolmogorov complexity calculating method from transcriptome data containing zero count data. Similar transcriptomes that were not distinguished using Information entropy were distinguished using Kolmogorov complexity. Monte Carlo simulation indicated that Kolmogorov complexity does not detect genes order in transcriptome data files.

2 Materials and Methods

2.1 Transcriptome Data

Transcriptome sequence data from two cellular culture conditions were obtained DDBJ SRA. Short-read data have been deposited in the DNA Data Bank of Japan (DDBJ)'s Short Read Archive, under project ID DRA002853. We choose DRX025343 and DRX025346 as similar transcriptomes that were not well distinguished using Information entropy. These two data sets contain three replicates, respectively. Short-read sequences were mapped to an annotated silkworm transcript sequence as previously described [7]. I show the first four lines of the file as an example (Table 1).

Table 1. An example of transcriptome data.

rownames(data)	DRX3_1	DRX3_2	DRX3_3	DRX6_1	DRX6_2	DRX6_3
BGIBMGA000001	6	9	7	6	6	5
BGIBMGA000002	39	18	25	22	36	23
BGIBMGA000003	1	0	0	0	0	0

2.2 Calculation of Kolmogorov Complexity

Kolmogorov complexity in a universal computer u having a data string expressed as a finite length character string x is defined by the following Eq. (2):

$$K_u(x) = \min_{p:u(p)=x} l(p) \qquad (2)$$

Here, p is a program for computer u, and u (p) is a character string outputted when it is executed. l (p) represents the length of the program. However, the program does not have an input, and always returns a fixed output. Here, the universal computer means a computer having the same capability as a universal Turing machine. Kolmogorov complexity is not a computable function and we cannot compute Kolmogorov complexity of a given individual file, but we can estimate Kolmogorov complexity by the compression of given individual files.

Calculation of Kolmogorov Complexity was performed using UNIX and R 3.0.2 [9]. At first in R, we converted reads count data to RPM data and converted RPM data to rounded RPM data. Reads count data was named "for_R_count.txt".

```
data<-read.table("for_R_count.txt",header=T,row.names=1)
out_f<-"for_R_RPM.txt"
param1 <- 1000000
norm_factor <- param1/colSums(data)
out <- sweep(data, 2, norm_factor, "*")
tmp <- cbind(rownames(data), out)
write.table(tmp,out_f,sep="\t",append=F, quote=F,
row.names=F, col.names=T)
data<-read.table("for_R_RPM.txt",header=T,row.names=1)
out_f <- "temp1"
write.table(round(data), out_f, sep="\t", append=F,
quote=F, row.names=T, col.names=F)
```

Then we made rounded RPM data in UNIX.

```
head -n 1 for_R_RPM.txt > temp2
cat temp2 temp1 > for_R_RPMround.txt
```

We converted rounded RPM data from decimal number to 22-digit binary number For example, "33" was converted "0000000000000000100001".

```
more temp2 | tr "\t" "\n" | awk '{print "more temp1 | awk
@{print $"NR"}@ > "$1}' | grep -v rowname | tr "@" "'" >
do_single.sh
sh do_single.sh
more temp2 | tr "\t" "\n" | grep -v rowname | awk '{print
"more "$1" | awk @{print \"echo \\\"obase=2;ibase=10;
\"$1\"\\\" | bc | awk V{printf(\\\"%022d\\\\n\\\",$1)}V
>> 2shin_"$1".text\"}@ | tr \"V\" \"@\" >>
do_conv_10_2.sh"}' | tr "@" "'" > make_do_conv_10_2.sh
sh make_do_conv_10_2.sh
sh do_conv_10_2.sh
```

We compressed rounded RPM data (22-digit binary number) using zip command.

```
more temp2 | tr "\t" "\n" | grep -v rownames | awk
'{print "zip 2shin_"$1".zip 2shin_"$1".text"}' >
do_zip.sh
sh do_zip.sh
```

We compared file sizes between rounded RPM data (22-digit binary number) and zip compressed rounded RPM data (22-digit binary number).

```
ls -al | grep 2shin |  ls -al | grep 2shin | grep zip |
awk '{print $9"\t"$5}' > temp3
ls -al | grep 2shin |  ls -al | grep 2shin | grep text |
awk '{print $9"\t"$5}' > temp4
paste temp3 temp4 | tr "\." "\t" | sed -e 's/2shin_//g' |
awk '{print $1"\t"($3/$6*1.00)"\t"$3"\t"$6}' > temp5
echo "sampleID@kc@zip_filesize@raw_filesize" | tr "@"
"\t" > temp6
cat temp6 temp5 >
for_R_sampleID_kc_ZIPfilesize_RAWfilesize.txt
```

These codes were written in a file named "cal_KC.sh". I show the first four lines of the output file as an example (Table 2).

Table 2. An example of the Output file.

sampleID	kc	zip_filesize	raw_filesize
DRX3_1	0.0692269	23283	336329
DRX3_2	0.0692209	23281	336329
DRX3_3	0.0697502	23459	336329

2.3 Monte Carlo Simulation

We choose DRX025343 and DRX025346 for MC simulation since these transcriptomes were similar and were not clearly distinguished using information entropy. We extracted data of these two samples in UNIX.

```
more for_R_RPMround.txt | cut -f 1,5,6,7,20,21,22 >
for_R_RPMround_ovs+.txt
more for_R_RPMround_ovs+.txt | grep -v row | cut -f 2- >
temp01.txt
```

We obtained random numbers for each gene in R. We performed MC simulation 13 times.

```
a<-runif(14623,0,1)
[...]
n<-runif(14623,0,1)
x<- data.frame(a,b,c,d,e,f,g,h,i,j,k,l,m,n)
write.table(x,"runif.txt")
```

We gave random numbers to each gene and sorted genes in the order of given random numbers.

```
more runif.txt | tr -d "\"" | tr " " "\t" | cut -f 2- |
tail -n 14623 > unif_table.txt
paste temp01.txt unif_table.txt | awk '{print $7"\t"$0}'
| sort | cut -f 2,3,4,5,6,7 > sorted_7.txt
[...]
paste temp01.txt unif_table.txt | awk '{print $19"\t"$0}'
| sort | cut -f 2,3,4,5,6,7 > sorted_19.txt
```

To calculate Kolmogorov complexity of 13 simulations using modified "cal_KC. sh". The following "cal_KC_mod.sh" is displayed.

```
more temp2 | tr "\t" "\n" | awk '{print "more temp1 | awk
@{print $"NR"}@ > "$1}' | grep -v rowname | tr "@" "'" >
do_single.sh
sh do_single.sh
more temp2 | tr "\t" "\n" | grep -v rowname | awk '{print
"more "$1" | awk @{print \"echo \\\"obase=2;ibase=10;
\"$1\"\\\" | bc | awk V{printf(\\\"%022d\\\n\\\",$1)}V
>> 2shin_"$1".text\"}@ | tr \"V\" \"@\" >>
do_conv_10_2.sh"}' | tr "@" "'" > make_do_conv_10_2.sh
sh make_do_conv_10_2.sh
sh do_conv_10_2.sh
more temp2 | tr "\t" "\n" | grep -v rownames | awk
'{print "zip 2shin_"$1".zip 2shin_"$1".text"}' >
do_zip.sh
sh do_zip.sh
ls -al | grep 2shin |  ls -al | grep 2shin | grep zip |
awk '{print $9"\t"$5}' > temp3
ls -al | grep 2shin |  ls -al | grep 2shin | grep text |
awk '{print $9"\t"$5}' > temp4
paste temp3 temp4 | tr "\." "\t" | sed -e 's/2shin_//g' |
awk '{print $1"\t"($3/$6*1.00)"\t"$3"\t"$6}' > temp5
echo "sampleID@kc@zip_filesize@raw_filesize" | tr "@"
"\t" > temp6
cat temp6 temp5 >
for_R_sampleID_kc_ZIPfilesize_RAWfilesize.txt
head -n 1 for_R_RPM.txt | tr "\t" "\n" | grep -v rownames
| awk '{print "rm "$1}' > clean_single_lanes.sh
```

2.4 Additional Example Using Mammalian Cells

Validate relationships information entropy and Kolmogorov complexity; we cultured adherent mammalian cells in several concentrations of drug and calculated information entropy and Kolmogorov complexity of transcriptomes. Experiments were performed as previously described [7]. We compared the magnitude of the correlation between drug concentrations and information entropy and the correlation between drug concentrations and Kolmogorov complexity in R.

```
data<-
read.table("temp04_condtion_sampleID_kc_ie.txt",header=T)
attach(data)
model<-lm(drug_c~kc+ie)
summary(model)
```

3   Results and Discussion

3.1   Comparison of Information Entropy and Kolmogorov Complexity

We calculated Kolmogorov complexity from the transcriptome data deposited in the DNA Data Bank of Japan (DDBJ)'s Short Read Archive, under project ID DRA002853. Previous study presented a comparison between the amount of environmental change and the amount of transcriptome change. In that research, the amount of environmental change was defined by drug concentration and the amount of transcriptome change was defined by information entropy (transcriptome diversity) of transcriptomes. Here, we change definition of the amount of transcriptome change, to Kolmogorov complexity from information entropy. We compared the relationship between Kolmogorov complexity of transcriptome and drug concentration and the relationship between information entropy of transcriptome and drug concentration (see Table 3, Fig. 1).

Table 3. Drug concentration and information entropy and Kolmogorov complexity.

DDBJ_ID	pb	hyst	i.e.	kc
DRR027746	0	F	9.958201	0.067109883
DRR027747	0	F	10.39145	0.068986023
DRR027748	0	F	10.55041	0.069116847
DRR027752	0.25	F	10.17702	0.068641122
DRR027753	0.25	F	10.3602	0.068572737
DRR027754	0.25	F	10.43436	0.069113874
DRR027758	0	T	10.12837	0.067062311
DRR027759	0	T	9.751655	0.065798667
DRR027760	0	T	10.26499	0.067508303
DRR027761	0.25	T	9.349972	0.063761971
DRR027762	0.25	T	10.22082	0.067332879
DRR027763	0.25	T	9.110797	0.063574655
DRR027755	1	F	8.171864	0.060137544
DRR027756	1	F	7.569893	0.059911575
DRR027757	1	F	8.002486	0.055995766
DRR027764	2.5	F	7.681749	0.058249512
DRR027765	2.5	F	7.334465	0.057211837
DRR027766	2.5	F	8.059749	0.055249473
DRR027767	12.5	F	8.004586	0.058974992
DRR027768	12.5	F	7.519157	0.056935322
DRR027769	12.5	F	7.931476	0.058570626

(A)   **(B)**

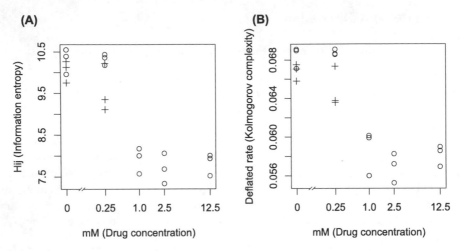

Fig. 1. Scatter plots of the amount of environmental change vs the amount of transcriptome change. Transcriptome data deposited in the DNA Data Bank of Japan (DDBJ)'s Short Read Archive, under project ID DRA002853 were used. (A) Scatter plot of drug concentration vs information entropy of transcriptome. (B) Scatter plot of drug concentration vs Kolmogorov complexity of transcriptome.

Information entropy and Kolmogorov complexity of cells cultured in media containing 0.25 mM phenobarbital after previous cultivation (cultivation for 80 h in MGM-450 insect medium without phenobarbital, followed by cultivation for 10 h in 1.0 mM phenobarbital-supplemented MGM-450 insect medium) (see Fig. 1 plot "+") were different from cells cultured in media containing 0.25 mM phenobarbital after previous cultivation (cultivation for 80 h in MGM-450 insect medium without phenobarbital) (see Fig. 1 plot "O"). The hysteretic phenomenon of transcriptome was reproduced using Kolmogorov complexity. In plot (A), using information entropy, a range of plot "+" and a range plot "o" overlapped. In plot (B), using Kolmogorov complexity, a range of plot "+" and a range plot "o" did not overlap. These results indicated that the similar transcriptomes that were not well distinguished using Information entropy were well distinguished using Kolmogorov complexity.

3.2 Kolmogorov Complexity and the Order of Genes of Transcriptomes

Generally, genes expression data are saved as count tables (see Table 1). The order of those tables is changed easily since that is determined by the short-read sequences mapping process and the annotation process. Generally, genes expression data are saved as count tables (See Table 1). The order of those tables is changed easily since that is determined by the short-read sequences mapping process and the annotation process. The order of genes also has any message as well as zero count data and have a possibility to affect Kolmogorov complexity. Monte Carlo simulations were performed to know the amount of effect of the order of genes to Kolmogorov complexity. We simulated 13 time, we gave random numbers for each gene and randomized the order of genes of transcriptomes. As a result, the order of genes in transcriptome does not correlate Kolmogorov complexity (see Table 4, Fig. 2).

Table 4. Kolmogorov complexity of 13 MC simulations.

MC	DRX3_1	DRX3_2	DRX3_3	DRX6_1	DRX6_2	DRX6_3
1	0.0692269	0.0692209	0.0697502	0.0643685	0.0681773	0.0639493
2	0.0691614	0.0693428	0.0694617	0.0643983	0.0681654	0.0641782
3	0.0692952	0.0692477	0.0696729	0.0645529	0.0681862	0.0641842
4	0.0694439	0.0692952	0.0696283	0.0642317	0.068097	0.063872
5	0.0691496	0.0693815	0.0697026	0.0642704	0.0680792	0.0643864
6	0.0694707	0.0694588	0.0695093	0.0643864	0.0679335	0.0641426
7	0.0695063	0.0693577	0.0695034	0.0645261	0.0679989	0.064095
8	0.0694112	0.0691942	0.0695837	0.0640861	0.0681505	0.0640504
9	0.0692209	0.0694201	0.0696699	0.0641842	0.0679602	0.0641634
10	0.069432	0.0695093	0.0697383	0.0643477	0.0679573	0.0638868
11	0.0693666	0.0693042	0.0699256	0.0645023	0.0678978	0.0636579
12	0.0694796	0.069545	0.069322	0.0643328	0.0681713	0.0642763
13	0.0694766	0.0693636	0.069542	0.0643091	0.0680286	0.0641574

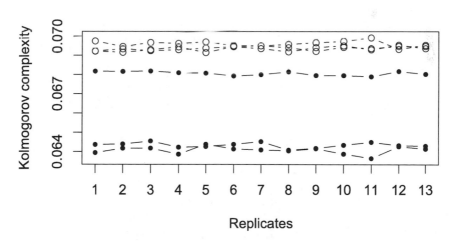

Fig. 2. Kolmogorov complexity of genes order randomized transcriptome. Gene orders of 6 transcriptomes were randomized by Monte Carlo simulation.

3.3   Additional Example Using Mammalian Cells

To validate relationships information entropy and Kolmogorov complexity, we cultured adherent mammalian cells in several concentrations of drug and calculated information entropy and Kolmogorov complexity of transcriptomes (Table 5). We named Table 5, "temp04_condtion_sampleID_kc_i.e.txt".

Table 5. Relative drug concentration and information entropy and Kolmogorov complexity.

drug_c	i.e	kc
0	11.79063	0.0607232
0	11.71702	0.0609433
0	11.8423	0.0610302
0.1	11.92365	0.0610032
0.1	11.82417	0.0609453
0.1	11.82852	0.0610109
1	11.51067	0.0597326
1	11.84796	0.0612445
1	11.98672	0.0611866
10	11.79985	0.0610824
10	11.79642	0.0611094
10	11.84944	0.0613025

We compared the correlation between information entropies and drug concentrations and the correlation between Kolmogorov complexities and drug concentrations using Fitting Linear Models in R.

```
Residuals:
     Min        1Q     Median        3Q        Max
-0.57247  -0.27188  -0.03609    0.25045    0.54018

Coefficients:
             Estimate  Std. Error  t value  Pr(>|t|)
(Intercept)   -28.083      18.967   -1.481    0.1728
kc           1080.540     558.570    1.934    0.0851 .
ie             -3.175       1.980   -1.603    0.1434
---
Signif. codes: 0.001 '**' 0.01 '*' 0.05 '.' 0.1 '' 1

Residual standard error: 0.4065 on 9 degrees of freedom
Multiple R-squared:  0.2939, Adjusted R-squared:   0.137
F-statistic: 1.873 on 2 and 9 DF, p-value: 0.2089
```

The correlation between Kolmogorov complexities and drug concentrations was stronger than that of information entropies (see Fig. 3).

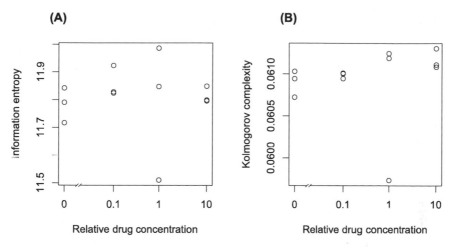

Fig. 3. Scatter plots of the amount of environmental change vs the amount of transcriptome change in mammalian cells. (A) Scatter plot of relative drug concentration vs information entropy of transcriptome. (B) Scatter plot of relative drug concentration vs Kolmogorov complexity of transcriptome.

4   Conclusions

Transcriptome measurement technologies and applications of information theories to transcriptome data allow studying the cellular systems. Especially, Shannon's theory of information explained several phenomena [10]. In this study, to expand the use of information theories, we improved weakness "Ignoring Zero Counts" of Shannon's information entropy in transcriptome analyses using Kolmogorov complexity. Kolmogorov complexity treating zero count data in transcriptome data and distinguished similar transcriptomes that were not well distinguished using information entropy.

References

1. Martinez, O., Reyes-Valdes, M.H.: Defining diversity specialization and gene specificity in transcriptomes through information theory. Proc. Natl. Acad. Sci. U.S.A. **105**, 9709–9714 (2008)
2. Buettner, F., Natarajan, K.N., Casale, F.P., Proserpio, V., Scialdone, A., Theis, F.J.: Computational analysis of cell-to-cell heterogeneity in single-cell RNA-sequencing data reveals hidden subpopulations of cells. Nat. Biotechnol. **33**, 155–160 (2015)
3. Saadatpour, A., Guo, G., Orkin, S.H., Yuan, G.C.: Characterizing heterogeneity in leukemic cells using single-cell gene expression analysis. Genome Biol. **15**, 525 (2014)
4. Martinez, O., Reyes-Valdes, M.H., Herrera-Estrella, L.: Cancer reduces transcriptome specialization. PLoS ONE **5**, e10398 (2010)
5. Heil, M., Ibarra-Laclette, E., Adame-Alvarez, R.M., Martinez, O., Ramirez-Chavez, E., Molina-Torres, J.: How plants sense wounds: damaged-self recognition is based on plant-derived elicitors and induces octadecanoid signaling. PLoS ONE **7**, e30537 (2012)

6. Ogata, N., Yokoyama, T., Iwabuchi, K.: Transcriptome responses of insect fat body cells to tissue culture environment. PLoS ONE **7**, e34940 (2012)
7. Ogata, N., Kozaki, T., Yokoyama, T., Hata, T., Iwabuchi, K.: Comparison between the amount of environmental change and the amount of transcriptome change. PLoS ONE **10** (12), e0144822 (2015)
8. Peter, G., Shannon, P.: Information and kolmogorov complexity (2017). http://homepages. cwi.nl/ ∼ paulv/papers/info.pdf
9. R.: a language and environment for statistical computing (2017). http://www.R-project.org/
10. Philippe, K.: The natural defense system and the normative self model. F1000Res. **5**, 797 (2016)

Influence of Amino Acid Properties for Characterizing Amyloid Peptides in Human Proteome

R. Prabakaran[1], Rahul Nikam[1], Sandeep Kumar[2], and M. Michael Gromiha[1(✉)]

[1] Department of Biotechnology, Bhupat and Jyoti Metha School of Biosciences, Indian Institute of Technology Madras, Chennai 600036, Tamilnadu, India
gromiha@iitm.ac.in
[2] Biotherapeutics Pharmaceutical Sciences, Pfizer Inc., 700 Chesterfield Parkway West, Chesterfield, MO 63017, USA

Abstract. Amyloidosis denotes the medical disorders associated with deposition of insoluble protein fibrillar aggregates and it is associated with various human diseases. Presence of aggregation prone regions plays an important role in determining the aggregation propensity of a protein, hence understanding the characteristics of these regions is of keen interest in academia and industry. In this work, we have identified 465 aggregation prone regions with 353 unique peptides in human proteome. Evaluation of the performance of available methods for identifying these 353 peptides showed a sensitivity in the range of 15% to 90%. Further, we identified the amino acid properties enthalpy, entropy, free energy and hydrophobicity are important for promoting aggregation. Utilizing these properties, we have developed a model for distinguishing between amyloid forming and non-amyloid peptides, which showed an accuracy of 71% with a balance between sensitivity and specificity. We suggest that the results obtained in this work could be effectively used to improve the prediction performance of existing methods.

Keywords: Protein aggregation · Amyloid fibrils · Physicochemical · Prediction · Aggregation propensity · Proteome

1 Introduction

Understanding the mechanism of protein aggregation has become one of the most important areas of research, due to its association with human maladies and prospective applications in biomaterials. Many human diseases are been linked to protein misfolding and aggregation. Though aggregation is a complex process controlled by many intrinsic and extrinsic factors, presence of aggregation prone regions within a protein sequence is a key trait in determining the aggregation propensity of the protein [1, 2]. APRs are 5–15 residues-long, hydrophobic-rich segments of low charge which are capable of forming extended beta-steric zipper motifs, which forms the core of the amyloid fibrils [3, 4]. Over years, hundreds of such amyloidogenic peptides have been identified experimentally, yet the mechanism of aggregation remains elusive. Exploring the characteristics of these amyloidogenic peptides would give us deep insights about the complex

© Springer International Publishing AG 2017
D.-S. Huang et al. (Eds.): ICIC 2017, Part II, LNCS 10362, pp. 541–548, 2017.
DOI: 10.1007/978-3-319-63312-1_47

formation of protein aggregates and influence our understanding of human protein conformation disorders.

Several studies have been performed in the past to understand the characteristics of aggregation prone regions (APRs) in human proteins [3, 5, 6]. These studies highlighted that APRs are mostly buried and flanked by charged and proline residues to prevent intermolecular hydrophobic contacts, which lead to nucleate protein aggregation. However, major drawbacks of such studies were lack of sufficiently large experimentally validated amyloidogenic peptides and use of prediction algorithms to identify aggregation prone regions.

In this work, we have systematically analyzed the characteristics of experimentally validated amyloidogenic peptide sequences in human proteome. We have analyzed various physicochemical, energetic and conformational properties for these peptides against a control dataset of non-amyloidogenic peptides, to identify the amino acid properties influencing the amyloidogenicity. We have also compared the sensitivity of existing prediction algorithms in detecting these amyloidogenic peptides. From the analysis, we were able to handpick amino acid properties essential to discriminate amyloidogenic peptides from non-amyloidogenic peptides. This work elucidate the importance of fundamental physiochemical properties in determining the complex aggregation propensity of peptides.

2   Materials and Methods

2.1   Datasets and Amino Acid Properties

We have collected experimentally validated 700 amyloidogenic and 1049 non-amyloidogenic peptide sequences of minimum length 6 from three different databases CPAD, AmyLoad and WaltzDB [7–9]. Out of the 700, 353 amyloidogenic peptides [4] (pep353) were found to occur at least once in 20135 human protein sequences which were collected from UniProt [10]. Length of these 353 peptide sequences ranges from 6 to 93 residues. 53 dataset was used for this study along with pep1049 as control dataset. We have utilized a set of 50 amino acid residue properties to identify the ones which influence the amyloidogenicity of pep353. These involve 49 physicochemical, energetic and conformational properties [11–13] and a residue aggregation propensity from AGGRESCAN [14].

2.2   Sequence Analysis

We have calculated the amino acid composition using the equation: $comp(i) = \Sigma n(i)/N$, where, $n(i)$ is the number of residues of each amino acid residue and i varies from 1 to 20. N is the total number of residues in a protein. The average property value of each peptide is computed using equation: $P = \sum p(i).n(i)/N$, where, $p(i)$ and $n(i)$ are, respectively, the property value of the ith amino acid residue and the number of amino acids of ith type in a protein. N is the total number of residues in a protein.

2.3   Prediction of Aggregation Prone Regions

To compare the sensitivity of aggregation prone region prediction tools to detect the 353 amyloidogenic peptides, we used 11 prediction algorithms, namely AGGRESCAN [14], Amyloidogenic pattern [15], Average Packing Density [16], Beta-strand contiguity [17], Hexapeptide Conformational Energy [18], NetCSSP [19], Pafig [20], Secstr [21], TANGO [22], WALTZ [23] and GAP [24]. Except GAP, results from other algorithms were obtained through AmylPred2 [25], a binary prediction, metaserver.

3   Results and Discussion

3.1   Occurrence of Amyloidogenic Peptides in Human Proteome

20135 reviewed human protein sequences were collected from UniProt [10] and scanned for exact match of known 700 amyloidogenic peptides (Amyloid700). We found that 353 (50.4%) unique peptides are present in 465 occurrences. As expected more than 66% of the hexapeptides in Amyloid700 were also found in human proteome. Specifically, aggregating peptides with lengths 6, 13, 14, 23 and 27 occur frequently (more than 60%) in human proteome. The only one peptide with a length of 27 residues in experimentally known 700 peptides is also present in human proteome.

3.2   Sensitivity of Prediction Tools to Amyloidogenic Peptides

We studied the sensitivity of the existing prediction tools to detect pep353. Using AmylPred2 and GAP. The results clearly indicates that individual methods except GAP are not sufficient to detect all pep353 peptides. Each of these prediction algorithms rely

Table 1. Sensitivity of various prediction tools

Prediction tools	Overall sensitivity	Sensitivity of peptides of length				
		6	7–10	11–20	21–30	>30
Number of peptides	353	200	53	71	18	11
AGGRESCAN	61.8	46.0	67.9	88.7	100.0	81.8
Amyloidogenic Pattern	39.7	39.5	28.3	29.6	88.9	81.8
Average Packing Density	34.8	21.0	34.0	57.7	72.2	81.8
Beta-strand contiguity	21.8	0.5	28.3	54.9	77.8	72.7
Hexapeptide conformational energy	65.7	54.5	67.9	84.5	94.4	90.9
NetCSSP	34.3	11.5	32.1	77.5	88.9	90.9
Pafig	58.1	34.5	86.8	87.3	94.4	100.0
SecStr	15.0	1.5	13.2	35.2	55.6	72.7
TANGO	25.8	5.0	45.3	54.9	61.1	63.6
WALTZ	74.2	69.0	71.7	85.9	77.8	100.0
GAP	90.1	92.5	86.8	95.8	88.9	90.9

on selective features of aggregation but do not sufficiently cover all the factors which determine aggregation. Consensus prediction tools such as AmylPred and MetAmyl are built to overcome such shortcomings. On the other hand, GAP which is based on pair-preference of position specific interactions show remarkable sensitivity. These results highlight the need to study the relationship between various physiochemical properties and aggregation propensity to understand the process of protein aggregation. Further the sensitivity of the prediction tools was calculated for pep353 by grouping the pep353 peptides based on the sequence lengths. The results summarized in Table 1 shows the variability associated with the sensitivity of the prediction tools. GAP [24] and WALTZ [23] among other tools, show the most consistent results.

3.3 Amino Acid Properties

We have computed the average values for each of the pep353 peptides based on the 50 amino acid residue properties as described in methods section. Correlation matrix between the 50 amino acid residue properties which include the aggregation propensity from Aggrescan (Agg) was computed to study the redundancy in the properties. Figure 1 shows the correlation matrix highlights that many of these properties are inter-dependent. For example Buriedness (B_r), Solvent accessible reduction ratio (R_a),

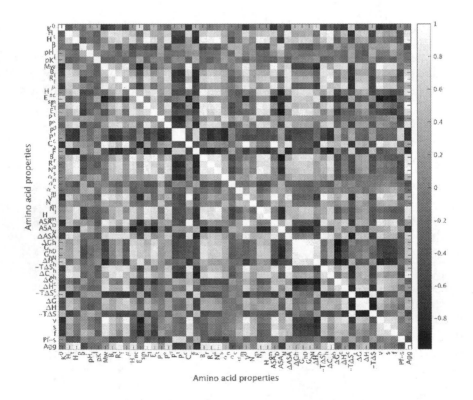

Fig. 1. Correlation between the 50 amino acid properties

Average number of surrounding residues (N_s), Solvent accessible surface of protein unfolding (ΔASA), Power to be at N-terminal of alpha-helix (α_N), Gibbs free energy change of hydration for unfolding (ΔG_h), Gibbs free energy change of hydration for denatured protein (G_{hD}), Gibbs free energy change of hydration for native protein (G_{hN}) and unfolding enthalpy change of hydration (ΔH_h) are highly correlated with each other.

Earlier analysis showed that hydrophobicity and secondary structure tendency such as β-strand propensity and α-helical propensity are important features for amyloidogenic peptides because of the amyloid fibrillar structure formation. Most prediction tools have incorporated these features directly or indirectly in the algorithm. The correlation values of α-helical tendency (P_α), β-strand tendency (P_β) and surrounding hydrophobicity (H_p) against Aggregation propensity (Agg) are 0.19, 0.84 and 0.83, respectively.

3.4   Discrimination of Amyloidogenic Peptides Using Amino Acid Properties

The average value for each of the 50 amino acid properties in pep353 and control dataset, pep1049, showed that most properties except Isoelectric point (pH_i), Equilibrium constant with reference to the ionization property of COOH group (pK'), Molecular weight (Mw), Volume (v), Shape (s) and Flexibility (f), had statistically significant difference (P-value < 0.01) between them (Fig. 2).

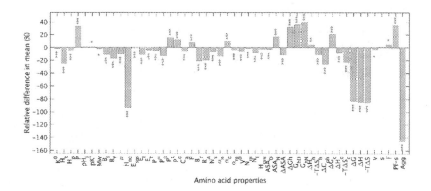

Fig. 2. Relative difference between amyloidogenic and non-amyloidogenic peptides in mean of 50 amino acid properties. The significance of the difference, as tested by Wilcoxon ranksum test is denoted over the bar: * , ** and *** denotes significant (P-value $<= 0.05$), very significant (P-value $<= 0.01$) and extremely significant (P-value $<= 0.001$) respectively.

Further to identify properties which significantly differentiate the pep353 from pep1049, we calculated the ratio of the mean to determine the magnitude of the difference in property value. Interestingly Normalized consensus hydrophobicity (H_{nc}), unfolding Gibbs free energy change (ΔG), unfolding enthalpy change (ΔH), unfolding entropy change ($-T\Delta S$) and AGGRESCAN-Aggregation propensity (Agg) showed significant difference in magnitude (|ratio| > 2).

Figure 3 shows the difference in distribution of average H_{nc}, ΔG, ΔH, $-T\Delta S$ and Agg values of pep353 and pep1049. To confirm whether these amino acid properties can discriminate amyloidogenic peptides (pep353) from non-amyloidogenic peptides (pep1049), we built a logistic regression model using H_{nc}, ΔG, ΔH and $-T\Delta S$. Aggregation propensity (Agg) was excluded from the model since similar analysis using the property have been done elsewhere [14]. The coefficients for H_{nc}, ΔG, ΔH and $-T\Delta S$ were determined as -2.5, 109.2, -109.6 and -109.5, respectively, with 1.63 as constant. We obtained reasonable accuracy of 71% with these four identified amino acid properties. We showed that the inclusion of these properties could improve the prediction performance of existing methods. The work on combining GAP with these properties is in progress.

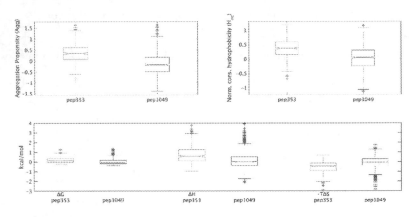

Fig. 3. Distribution of H_{nc}, ΔG, ΔH, $-T\Delta S$ and aggregation propensity (Agg)

4   Conclusions

We have investigated the similarities and variations of various sequences, structural and energetic properties of amyloidogenic and non-amyloidogenic peptides. There exist clear difference in amino acid composition of these peptides which is directly observed in the average amino acid property values. Based on the analysis we have identified amino acid properties such as H_{nc}, ΔG, ΔH, $-T\Delta S$ and Agg, with significant difference between amyloidogenic and non-amyloidogenic peptides. These properties can in turn be used to discriminate amyloidogenic peptides as elucidated by the logistic regression model.

Acknowledgements. We thank Indian Institute of Technology Madras for the computational facilities.

References

1. Wang, X., Das, T.K., Singh, S.K., Kumar, S.: Potential aggregation prone regions in biotherapeutics: a survey of commercial monoclonal antibodies. MAbs **1**, 254–267 (2009)
2. Mittag, T., Marzahn, M.R.: Short aggregation-prone peptide detectives: finding proteins and truths about aggregation. J. Mol. Biol. **427**, 221–224 (2015)
3. Reumers, J., Maurer-Stroh, S., Schymkowitz, J., Rousseau, F.: Protein sequences encode safeguards against aggregation. Hum. Mutat. **30**, 431–437 (2009)
4. Prabakaran, R., Goel, D., Kumar, S., Gromiha, M.M.: Aggregation prone regions in human proteome: insights from large-scale data analyses. Proteins Struct. Funct. Bioinforma. **85**, 1099–1118 (2017)
5. Tzotzos, S., Doig, A.J.: Amyloidogenic sequences in native protein structures. Protein Sci. **19**, 327–348 (2010)
6. Monsellier, E., Ramazzotti, M., Taddei, N., Chiti, F.: Aggregation propensity of the human proteome. PLoS Comput. Biol. **4**, e1000199 (2008)
7. Thangakani, A.M., Nagarajan, R., Kumar, S., Sakthivel, R., Velmurugan, D., Gromiha, M.M.: CPAD, curated protein aggregation database: a repository of manually curated experimental data on protein and peptide aggregation. PLoS ONE **11**, e0152949 (2016)
8. Wozniak, P.P., Kotulska, M.: AmyLoad: website dedicated to amyloidogenic protein fragments. Bioinformatics **31**, 3395–3397 (2015)
9. Beerten, J., Van Durme, J., Gallardo, R., Capriotti, E., Serpell, L., Rousseau, F., Schymkowitz, J.: WALTZ-DB: a benchmark database of amyloidogenic hexapeptides. Bioinformatics **31**, 1698–1700 (2014)
10. Wasmuth, E.V., Lima, C.D.: UniProt: the universal protein knowledgebase. Nucleic Acids Res. **45**, 1–12 (2016)
11. Gromiha, M.M., Oobatake, M., Sarai, A.: Important amino acid properties for enhanced thermostability from mesophilic to thermophilic proteins. Biophys. Chem. **82**, 51–67 (1999)
12. Gromiha, M.M., Oobatake, M., Kono, H., Uedaira, H., Sarai, A.: Importance of mutant position in Ramachandran plot for predicting protein stability of surface mutations. Biopolymers **64**, 210–220 (2002)
13. Gromiha, M.M.: Importance of native-state topology for determining the folding rate of two-state proteins. J. Chem. Inf. Comput. Sci. **43**, 1481–1485 (2003)
14. Conchillo-Solé, O., de Groot, N.S., Avilés, F.X., Vendrell, J., Daura, X., Ventura, S.: AGGRESCAN: a server for the prediction and evaluation of "hot spots" of aggregation in polypeptides. BMC Bioinform. **8**, 65 (2007)
15. Ventura, S., Zurdo, J., Narayanan, S., Parreño, M., Mangues, R., Reif, B., Chiti, F., Giannoni, E., Dobson, C.M., Aviles, F.X., Serrano, L.: Short amino acid stretches can mediate amyloid formation in globular proteins: the Src homology 3 (SH3) case. Proc. Natl. Acad. Sci. U.S.A. **101**, 7258–7263 (2004)
16. Galzitskaya, O.V., Garbuzynskiy, S.O., Lobanov, M.Y.: Prediction of amyloidogenic and disordered regions in protein chains. PLoS Comput. Biol. **2**, 1639–1648 (2006)
17. Zibaee, S., Makin, O.S., Goedert, M., Serpell, L.C.: A simple algorithm locates β-strands in the amyloid fibril core of α-synuclein, Aβ, and tau using the amino acid sequence alone. Protein Sci. **16**(5), 906–918 (2007)
18. Zhang, Z., Chen, H., Lai, L.: Identification of amyloid fibril-forming segments based on structure and residue-based statistical potential. Bioinformatics **23**, 2218–2225 (2007)
19. Kim, C., Choi, J., Lee, S.J., Welsh, W.J., Yoon, S.: NetCSSP: web application for predicting chameleon sequences and amyloid fibril formation. Nucleic Acids Res. **37**, 469–473 (2009)

20. Tian, J., Wu, N., Guo, J., Fan, Y.: Prediction of amyloid fibril-forming segments based on a support vector machine. BMC Bioinform. **10**(Suppl 1), 1–8 (2009)
21. Hamodrakas, S.J., Liappa, C., Iconomidou, V.A.: Consensus prediction of amyloidogenic determinants in amyloid fibril-forming proteins. Int. J. Biol. Macromol. **41**, 295–300 (2007)
22. Fernandez-Escamilla, A.-M., Rousseau, F., Schymkowitz, J., Serrano, L.: Prediction of sequence-dependent and mutational effects on the aggregation of peptides and proteins. Nat. Biotechnol. **22**, 1302–1306 (2004)
23. Maurer-Stroh, S., Debulpaep, M., Kuemmerer, N., Lopez de la Paz, M., Martins, I.C., Reumers, J., Morris, K.L., Copland, A., Serpell, L.C., Serrano, L., Schymkowitz, J.W.H., Rousseau, F.: Exploring the sequence determinants of amyloid structure using position-specific scoring matrices. Nat. Methods **7**, 237–242 (2010)
24. Thangakani, A.M., Kumar, S., Nagarajan, R., Velmurugan, D., Gromiha, M.M.: GAP: towards almost 100 percent prediction for β-strand-mediated aggregating peptides with distinct morphologies. Bioinformatics **30**, 1983–1990 (2014)
25. Tsolis, A.C., Papandreou, N.C., Iconomidou, V.A., Hamodrakas, S.J.: A consensus method for the prediction of "aggregation-prone" peptides in globular proteins. PLoS ONE **8**, e54175 (2013)

Link Mining for Kernel-Based Compound-Protein Interaction Predictions Using a Chemogenomics Approach

Masahito Ohue[1,2,3,4(✉)], Takuro Yamazaki[3], Tomohiro Ban[4],
and Yutaka Akiyama[1,2,3,4(✉)]

[1] Department of Computer Science, School of Computing,
Tokyo Institute of Technology, Tokyo, Japan
{ohue,akiyama}@c.titech.ac.jp
[2] Advanced Computational Drug Discovery Unit, Institute of Innovative Research,
Tokyo Institute of Technology, Tokyo, Japan
[3] Department of Computer Science, Faculty of Engineering,
Tokyo Institute of Technology, Tokyo, Japan
[4] Department of Computer Science, Graduate School of Information Science and Engineering,
Tokyo Institute of Technology, Tokyo, Japan

Abstract. Virtual screening (VS) is widely used during computational drug discovery to reduce costs. Chemogenomics-based virtual screening (CGBVS) can be used to predict new compound-protein interactions (CPIs) from known CPI network data using several methods, including machine learning and data mining. Although CGBVS facilitates highly efficient and accurate CPI prediction, it has poor performance for prediction of new compounds for which CPIs are unknown. The pairwise kernel method (PKM) is a state-of-the-art CGBVS method and shows high accuracy for prediction of new compounds. In this study, on the basis of link mining, we improved the PKM by combining link indicator kernel (LIK) and chemical similarity and evaluated the accuracy of these methods. The proposed method obtained an average area under the precision-recall curve (AUPR) value of 0.562, which was higher than that achieved by the conventional Gaussian interaction profile (GIP) method (0.425), and the calculation time was only increased by a few percent.

Keywords: Virtual screening · Compound-protein interactions (CPIs) · Pairwise kernel · Link mining · Link indicator kernels (LIKs)

1 Introduction

Virtual screening (VS), in which drug candidate compounds are selected by a computational method, is one of the main processes in the early stages of drug discovery. There are three main approaches to VS: ligand-based VS (LBVS) [1] using known activity information for the target protein of the drug; structure-based VS (SBVS) [2] using structural information for the target protein of the drug; and chemogenomics-based VS (CGBVS) [3] based on known interaction information for multiple proteins and multiple compounds (also called drug-target interaction prediction). Both LBVS and CGBVS do

© Springer International Publishing AG 2017
D.-S. Huang et al. (Eds.): ICIC 2017, Part II, LNCS 10362, pp. 549–558, 2017.
DOI: 10.1007/978-3-319-63312-1_48

not require a protein tertiary structure, and both depend on statistical machine learning using known experimental activity data. However, CGBVS yields more robust prediction results by handling multiple types of proteins. CGBVS has been well studied in recent years [4–9], and a review of CGBVS was recently published by Ding *et al.* [3].

In CGBVS, computations are mainly performed using a similarity matrix of proteins, a similarity matrix of compounds, and an interaction profile matrix composed of binary values with and without interactions (Fig. 1).

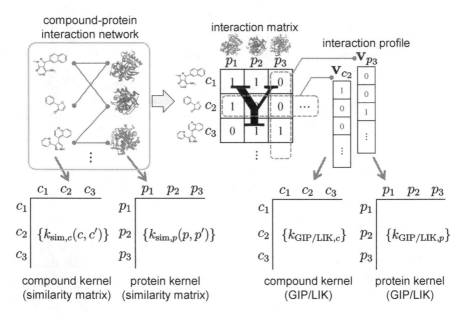

Fig. 1. Schematic diagram of information used in CGBVS. Information on interaction matrices and interaction profiles can be obtained from CPI data. Kernel matrices were obtained from the relationship between the compounds and the proteins.

The kernel method is often applied for prediction [5–7]. Conventionally, in CGBVS, an interaction profile matrix is used only as labeled training data; however, an increasing number of frameworks have recently been described that utilize interaction profile matrices as their main features. The Gaussian interaction profile (GIP) is one of these frameworks [5]. In addition to information regarding similarity matrices, GIP uses similarities between vectors when an interaction profile matrix is viewed as vertical and horizontal vectors. The GIP kernel functions of proteins and compounds are represented as follows (details are described in Sect. 2):

$$k_{\text{GIP},c}(c, c') = \exp\left(-\gamma_c \|\mathbf{v}_c - \mathbf{v}_{c'}\|^2\right), \quad k_{\text{GIP},p}(p, p') = \exp\left(-\gamma_p \|\mathbf{v}_p - \mathbf{v}_{p'}\|^2\right). \quad (1)$$

The problem with GIP kernels is that '0' bit (interaction is unknown) is taken into account, similarly to '1' bit (interaction). Thus, k_{GIP} shows a maximum value (which is equivalent to two compounds with common interaction partners) for two novel compounds when the interactions with all proteins are unknown. Since '0' potentially includes both no interactions and unknown interactions in general CGBVS problems and benchmark datasets, information for '1' should be considered more reliable.

In this way, as a framework that mainly considers '1' rather than '0', link mining has emerged for calculation of links within a network. Link mining is a framework applied to analyze networks such as social networks and the World Wide Web. Nodes and edges (links) of a network are used in the calculation. If nodes of the network are proteins/compounds, and edges are drawn in the interacting compound-protein pair, analysis using the framework of link mining becomes possible. Some reports have also applied the method of link mining directly to the problem of CGBVS [8, 9]. However, these methods do not use the framework of the kernel method.

Therefore, in this study, we propose to use the link indicator kernel (LIK), based on link indicators used in link mining, as the kernel with an interaction profile matrix such as that formed from GIP kernels. According to a review by Ding et al., the pairwise kernel method (PKM) [7] using a support vector machine (SVM) as a kernel learning scheme is superior in learning performance to CGBVS [3]. Thus, we integrated GIP and LIK kernels to the PKM and showed that LIK kernels could capture the effects of interaction profiles.

2   Materials and Methods

2.1   Preliminary

An overview of the compound-protein interaction prediction problem is shown in Fig. 1. Similarities were defined between compounds and between proteins, with the Tanimoto coefficient of fingerprints (e.g. ECFP [10], SIMCOMP [11]) or Euclid distance of physicochemical properties for compounds, and the Euclid distance of k-mer amino acid sequence profiles or Smith-Waterman alignment scores [12] for proteins.

The interaction $y(c, p)$ between a compound c and a protein p is defined as binary $\{0, 1\}$, where '1' represents an interaction (e.g., c is the active compound for the protein p) and '0' represents no interaction (often including unknowns). For learning, the $n_c \times n_p$ matrix $\mathbf{Y} = \{y(c, p)\}_{c, p}$ (called the interaction matrix) was used as training data, where n_c is the number of target compounds and n_p is the number of target proteins. Interactions were predicted for pairs of compounds and proteins using the learned model. When looking at each row and each column of the interaction matrix as a vector, the vector was called the interaction profile. The interaction profile $\mathbf{v}_c$ of compound c was $\mathbf{v}_c = (y(c, p_1), y(c, p_2), \dots, y(c, p_{np}))^{\mathrm{T}}$, and the interaction profile $\mathbf{v}_p$ of protein p was $\mathbf{v}_p = (y(c_1, p), y(c_2, p), \dots, y(c_{nc}, p))^{\mathrm{T}}$.

2.2   PKM

The pairwise kernel method (PKM) [7] developed by Jacob et al. is based on pairwise kernels and tensor product representation for compound and protein vectors. Normally, a map $\Phi(c, p)$ for a pair of compounds and proteins (c, p) is required for a learning scheme. In the PKM, the learning scheme involves the tensor product of the map of compound $\Phi_c(c)$ and the map of protein $\Phi_p(p)$. Therefore, $\Phi(c, p)$ is represented as follows:

$$\Phi(c,p) = \Phi_c(c) \otimes \Phi_p(p), \tag{2}$$

where $\otimes$ is the tensor product operator. Pairwise kernel k is defined between two pairs of proteins and compounds (c, p) and (c', p') as follows:

$$\begin{aligned} k((c,p),(c',p')) &\equiv \Phi(c,p)^T \Phi(c',p') \\ &= (\Phi_c(c) \otimes \Phi_p(p))^T (\Phi_c(c') \otimes \Phi_p(p')) \\ &= \Phi_c(c)^T \Phi_c(c') \times \Phi_p(p)^T \Phi_p(p') \\ &\equiv k_c(c,c') \times k_p(p,p'), \end{aligned} \tag{3}$$

where k_c is a compound kernel between two compounds, and k_p is a protein kernel between two proteins. Thus, it is possible to find the kernel k between two compound-protein pairs with the scalar product of k_c and k_p. If both k_c and k_p are positive definite kernels, k is also a positive definite kernel. The similarity value mentioned in Sect. 2.1 is often used for k_c and k_p [4].

2.3   GIP

The Gaussian interaction profile (GIP) [5] was developed to incorporate interaction profiles into kernel learning by van Laarhoven *et al.* The GIP kernel k_{GIP} based on the radial basis function shown in Eq. (1) is used for the compound kernel k_c and protein kernel k_p in Eq. (3). Here,

$$\gamma_c = \left(\frac{1}{n_c} \sum_{i=1}^{n_c} \left\| \mathbf{v}_{c_i} \right\|^2 \right)^{-1}, \quad \gamma_p = \left(\frac{1}{n_p} \sum_{i=1}^{n_p} \left\| \mathbf{v}_{p_i} \right\|^2 \right)^{-1}. \tag{4}$$

Importantly, k_{GIP} is never used alone for k_c and k_p, but is used as a multiple kernel (simple weighted average) in combination with similarity-based kernels:

$$\begin{aligned} k_c(c,c') &= k_{\mathrm{sim},c}(c,c') + w_{k,c} k_{\mathrm{GIP},c}(c,c') \\ k_p(p,p') &= k_{\mathrm{sim},p}(p,p') + w_{k,p} k_{\mathrm{GIP},p}(p,p'), \end{aligned} \tag{5}$$

where $w_{k,c}$ and $w_{k,p}$ are weighted parameters for multiple kernels.

2.4   LIK

The link indicator is an index used for network structural analysis, such as analysis of the hyperlink structure of the World Wide Web and friend relationships in social network services. In this study, we proposed link indicator kernels (LIKs) based on the link indicators for compound-protein interaction networks to incorporate interaction profiles into kernel learning. We selected three link indicators:

$$Jaccard\ index \quad k_{\text{LIK - Jac}}(\mathbf{v}, \mathbf{v}') = \frac{\mathbf{v}^{\mathrm{T}}\mathbf{v}'}{\|\mathbf{v}\|^2 + \|\mathbf{v}'\|^2 - \mathbf{v}^{\mathrm{T}}\mathbf{v}'} \tag{6}$$

$$Cosine\ similarity \quad k_{\text{LIK}-\cos}(\mathbf{v}, \mathbf{v}') = \frac{\mathbf{v}^{\mathrm{T}}\mathbf{v}'}{\|\mathbf{v}\|\|\mathbf{v}'\|} \tag{7}$$

$$LHN \quad k_{\text{LIK - LHN}}(\mathbf{v}, \mathbf{v}') = \frac{\mathbf{v}^{\mathrm{T}}\mathbf{v}'}{\|\mathbf{v}\|^2\|\mathbf{v}'\|^2}. \tag{8}$$

These link indicators become positive definite kernels when used as kernels. Cosine similarity and LHN are positive definite kernels because of the properties of the kernel function[1] and the positive definite of the inner product between the two vectors, and the Jaccard index was previously proven to be positive definite by Bouchard et al. [13]. There are other link indicators, such as the Adamic-Adar index and graph distance; however, because these are not positive definite kernels, as required for kernel methods, they were not used in this study.

For integration of LIK and PKM (similarity kernels), the same method applied for GIP was adopted. That is, considering multiple kernels, the kernels were defined as:

$$
\begin{aligned}
k_c(c, c') &= k_{\text{sim},c}(c, c') + w_{k,c}k_{\text{LIK},c}(c, c') \\
k_p(p, p') &= k_{\text{sim},p}(p, p') + w_{k,p}k_{\text{LIK},p}(p, p'),
\end{aligned}
\tag{9}
$$

where $k_{\text{LIK},c}$ and $k_{\text{LIK},p}$ are LIKs for two compounds and two proteins, respectively.

2.5 Implementation

In this study, we used scikit-learn [14], a Python library for machine learning, to implement PKM, GIP, and LIK. As a kernel learning method, SVM can be used for scikit-learn based on LIBSVM [15]. For the link indicator calculation of LIK, we used the python library networkx [16].

2.6 Dataset and Performance Evaluation

We used the benchmark dataset of CPI predictions published by Yamanishi et al. [4] according to the review of Ding et al. [3]. It is a well-known and well-used benchmark dataset in the field. The dataset consisted of four target protein groups ("Nuclear Receptor", "GPCR", "Ion Channel", and "Enzyme"). The SIMCOMP score [11] was used for compound similarity, and the normalized Smith-Waterman score [12] was used

---

[1] Let $k{:}X \times X \to \mathbb{R}$ be a positive definite kernel and $f{:}X \to \mathbb{R}$ be an arbitrary function. Then, the kernel $k'(\mathbf{x}, \mathbf{y}) = f(\mathbf{x})k(\mathbf{x}, \mathbf{y})f(\mathbf{y})$ $(\mathbf{x}, \mathbf{y} \in X)$ is also positive definite.

for protein similarity, as calculated by Yamanishi *et al*. [4]. Information on the interaction matrix was also provided by Yamanishi *et al*. [4]. Evaluation was performed by cross validation (CV). Three types of CVs were tested: randomly selected from all compound-protein pairs (pairwise CV), randomly selected compounds (compound CV), and randomly selected proteins (protein CV). The outlines of these CVs are shown in Fig. 2. According to Ding *et al*. [3], the area under the receiver operating characteristic curve (AUROC) and the area under the precision-recall curve (AUPR) were calculated for the evaluation value of 10-fold CVs. Each accuracy value was averaged five times for 10-fold CVs with different random seeds. Note that the cost parameter C of SVM was optimized from $\{0.1, 1, 10, 100\}$ in 3-fold CVs according to Ding *et al*. [3]. The multiple kernel weights w_k of Eqs. (5) and (9) have the same values for proteins and compounds, and we evaluated $\{0.1, 0.3, 0.5, 1\}$.

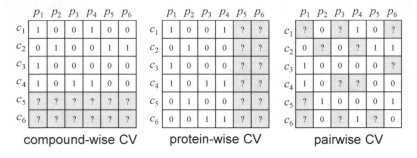

Fig. 2. Conceptual diagram of three types of cross-validations (CVs): compound-wise CV, protein-wise CV, and pairwise CV. A case with a 3-fold CV is shown as an example. The "?" indicates samples to be used for the test set.

3   Results and Discussion

3.1   Performance of the Proposed Method for Cross-Validation Benchmarking

Figure 3 shows the results for the prediction accuracy of the average values of three types of CVs in the four Yamanishi datasets (i.e., average values of 12 prediction accuracy values). We tested multiple kernel weights w_k in four patterns, and LIK with cosine similarity was the most accurate for both AUPR and AUROC (AUPR: 0.562 and AUROC: 0.906). In the case of cosine similarity, the weight $w_k = 0.5$ showed the best performance. Compared with GIP, the prediction accuracy of LIK showed higher accuracy overall.

Figure 4 shows the mean value of the prediction accuracy for each CV, including compound-wise, protein-wise, and pairwise CVs. Division of the dataset in each CV was randomly tried five times, and the values were averaged. The multiple kernel weight w_k was set to the best value in the cross-validation results (shown in Fig. 3). In the evaluation of AUROC, GIP showed accuracy comparable to that of the three LIKs; however, similar results were not observed for AUPR. In particular, we found that LIK showed a high value in the AUPR evaluation for compound-wise and protein-wise CVs,

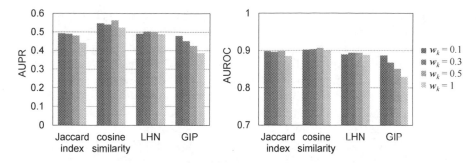

Fig. 3. Overall prediction accuracy for each CPI prediction method in 10-fold CV tests. The AUPR and AUROC values are averaged values of three types of CVs and four types of datasets (total average for 12 AUPR/AUROC values). For 10-fold CVs, calculations were performed five times with different random seeds, and the accuracy values were then averaged.

which could be evaluated for predictive performance for novel compounds and proteins, respectively.

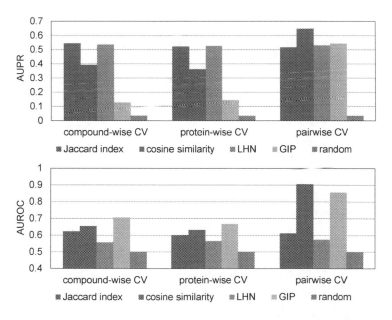

Fig. 4. Prediction accuracy of protein-wise, compound-wise, and pairwise CVs. In random prediction cases, an AUROC value of 0.5 and an AUPR value of 0.035 were obtained (averaged values depending on the ratio of positive samples on the dataset).

3.2 Observed Distribution of Link Indicator Frequency

Distribution of values of four protein interaction profile similarities ($k_{\text{LIK}}(p, p')$) were calculated using each link indicator to determine why LIKs with cosine similarity showed better results. The histograms of similarity values are shown in Fig. 5. From this result, the Jaccard index and LHN were found to have relatively similar distributions of similarity values between 0 and 1 (i.e., the intermediate value was low). Additionally, the number of pairs whose similarities ranged from 0.95 to 1 had the highest cosine similarity. This may be related to the AUROC value of cosine similarity, which tended to be higher. Conversely, for LHN, which showed the lowest similarity from 0.95 to 1, precision may be higher, and AUPR may tend to be higher. GIP also consisted of a few intermediate values similarly to LHN. Overall, cosine similarity showed the best performance in this study. A gentle distribution using an intermediate value may be more effective as information for the compound-protein network structure.

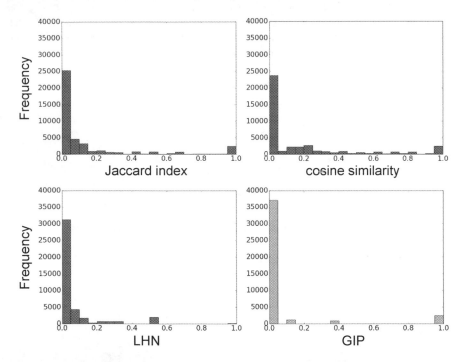

Fig. 5. Distribution of values of LIKs (Jaccard index, cosine similarity, and LHN) and GIP given all protein interaction profiles in the Yamanishi dataset.

3.3 Computational Complexity

The computational complexity for constructing the prediction model for the PKM is $O(n_c^3 n_p^3)$. However, the computational complexity for calculating the link indicators used in this study was $O(n_c n_p (n_c + n_p))$. Thus, the computational complexity of our proposed method was $O(n_c^3 n_p^3 + n_c n_p (n_c + n_p))$. Here, n_c and n_p were greater than 1 in general, and thus, $n_c^3 n_p^3$ was greater than $n_c n_p (n_c + n_p)$. Therefore, the computational complexity was $O(n_c^3 n_p^3)$, which was the same as those of PKM and GIP. Our method could predict CPIs without a major increase in calculation time. The execution time of one run of 10 runs of 10-fold CVs is shown in Table 1. The results in the table are shown for computations running on an ordinary personal computer with an Intel Core i5 CPU. Thus, our proposed method showed a slight increase in the execution time by several percentage points as compared with that of PKM ("Nuclear Receptor" had the highest rate of increase due to the small dataset).

Table 1. Comparison of calculation times for the PKM and proposed method in each dataset. The time taken to calculate one time out of 10 calculations of 10-fold CVs is shown.

	Nuclear receptor	GPCR	Ion channel	Enzyme
Conventional (PKM) [sec]	0.0680	4.86	24.1	232
Proposed (PKM plus LIK) [sec]	0.0850	5.17	24.8	239
Increase rate (%)	25	6.4	2.9	3.3

3.4 Limitations and Challenges

The proposed method can be directly applied to prediction based on the GIP (e.g., WNNGIP [6], KBMF2 K [17], and KronRLS-MKL [18]), and improvement of prediction accuracy is expected. For example, WNNGIP can provide robust predictions for compounds and proteins with less interaction information by complementing the interaction matrix with the weighted nearest neighbor method in advance. However, kernel-based methods, including the proposed method, are restricted to the framework using kernel functions. For example, it is not possible to simply combine LIK or GIP with the method based on matrix factorization (e.g., NRLMF [19]). Further mathematical ideas and computational experiments are needed to develop integrated methods.

4   Conclusions

In this study, we proposed a kernel method using link indicators from the viewpoint of link mining to utilize the information of the CPI network for machine learning. We attempted to utilize three link indicators (Jaccard index, cosine similarity, and LHN) for construction of positive definite kernels and compared them with the GIP method when combined with the SVM-based PKM method. As a result, learning by multiple kernels using LIK with cosine similarity and setting of the kernel weight w_k to 0.5 showed the best prediction accuracy (averaged AUPR = 0.562). In both AUROC and AUPR, the improvement of LIK accuracy was confirmed compared with that of GIP.

Acknowledgments. This work was partially supported by the Japan Society for the Promotion of Science (JSPS) KAKENHI (grant numbers 24240044 and 15K16081), and Core Research for Evolutional Science and Technology (CREST) "Extreme Big Data" (grant number JPMJCR1303) from the Japan Science and Technology Agency (JST).

References

1. Lavecchia, A.: Machine-learning approaches in drug discovery: methods and applications. Drug Discov. Today. **20**, 318–331 (2015)
2. Drwal, M.N., Griffith, R.: Combination of ligand- and structure-based methods in virtual screening. Drug Discov. Today Technol. **10**, e395–e401 (2013)
3. Ding, H., Takigawa, I., Mamitsuka, H., Zhu, S.: Similarity-based machine learning methods for predicting drug-target interactions: a brief review. Brief. Bioinform. **15**, 734–747 (2014)
4. Yamanishi, Y., Araki, M., Gutteridge, A., Honda, W., Kanehisa, M.: Prediction of drug-target interaction networks from the integration of chemical and genomic spaces. Bioinformatics **24**, i232–i240 (2008)
5. van Laarhoven, T., Nabuurs, S.B., Marchiori, E.: Gaussian interaction profile kernels for predicting drug-target interaction. Bioinformatics **27**, 3036–3043 (2011)
6. van Laarhoven, T., Marchiori, E.: Predicting drug-target interactions for new drug compounds using a weighted nearest neighbor profile. PLoS ONE **8**, e66952 (2013)
7. Jacob, L., Vert, J.P.: Protein-ligand interaction prediction: an improved chemogenomics approach. Bioinformatics **24**, 2149–2156 (2008)
8. Daminelli, S., Thomas, J.M., Duran, C., Cannistraci, C.V.: Common neighbours and the local-community-paradigm for topological link prediction in bipartite net-works. New J. Phys. **17**, 113037 (2015)
9. Duran, C., Daminelli, S., Thomas, J.M., Haupt, V.J., Schroeder, M., Cannistraci, C.V.: Pioneering topological methods for network-based drug-target prediction by exploiting a brain-network self-organization theory. Brief Bioinform. (2017) [Epub ahead of print]
10. Rogers, D., Hahn, M.: Extended-connectivity fingerprints. J. Chem. Inf. Model. **50**, 742–754 (2010)
11. Smith, T.F., Waterman, M.S.: Identification of common molecular subsequences. J. Mol. Biol. **147**, 195–197 (1981)
12. Hattori, M., Okuno, Y., Goto, S., Kanehisa, M.: Development of a chemical structure comparison method for integrated analysis of chemical and genomic information in the metabolic pathways. J. Am. Chem. Soc. **125**, 11853–11865 (2003)
13. Bouchard, M., Jousselme, A.-L., Doré, P.-E.: A proof for the positive definiteness of the Jaccard index matrix. Int. J. Approx. Reason. **54**, 615–626 (2013)
14. scikit-learn: machine learning in Python. http://scikit-learn.org/stable/. Accessed 27 March 2017
15. Chang, C.-C., Lin, C.: LIBSVM: A library for support vector machines. ACM Trans. Intell. Syst. Technol. **2**, 1–27 (2011)
16. NetworkX-High-productivity software for complex networks. https://networkx.github.io/. Accessed 27 March 2017
17. Gonen, M.: Predicting drug-target interactions from chemical and genomic kernels using Bayesian matrix factorization. Bioinformatics **28**, 2304–2310 (2012)
18. Nascimento, A.C.A., Prudêncio, R.B.C., Costa, I.G.: A multiple kernel learning algorithm for drug-target interaction prediction. BMC Bioinformatics **17**, 46 (2016)
19. Liu, Y., Wu, M., Miao, C., Zhao, P., Li, X.L.: Neighborhood regularized logistic matrix factorization for drug-target interaction prediction. PLoS Comput. Biol. **12**, e1004760 (2016)

Investigating Alzheimer's Disease Candidate Genes Based on Combined Network Using Subnetwork Extraction Algorithms

Xiaojuan Wang[1], Hua Yan[2], Di Zhang[1], Le Zhao[1], Yannan Bin[1(✉)], and Junfeng Xia[1]

[1] Institute of Health Sciences, Anhui University, Hefei, Anhui, China
ynbin@ahu.edu.cn
[2] School of Life Sciences, Anhui University, Hefei, Anhui, China

Abstract. There is increasing need for accurate Alzheimer's disease (AD) related genes prediction to inform study design, but available genes estimates are limited. In this study, the subnetwork extraction algorithms were applied to extract subnetworks and mine candidate genes based on a combined network, which was constructed by integrating the information of protein-protein interactions and gene-gene co-expression network. We obtained seven candidate genes with high possibility during AD progression. The application of subnetwork extraction algorithms based on combined network would provide a new insight into predicting the AD-related genes.

Keywords: Alzheimer's disease · Protein-protein interactions · Gene-gene co-expression · Subnetwork extraction algorithm · Candidate genes

1 Introduction

At present, there are about 49 million people have dementia, and this number is projected increase to more than 131 million by 2050 [1]. Consequently, prevention and treatment of Alzheimer's disease (AD), the most common cause of dementia, have become important. It is estimated that genetic factors cause about 60–80% of AD risk [2]. There is increasing need for accurate AD-related genes prediction to inform study design with the onset of AD prevention and treatment, but available genes estimates are limited.

The computational approaches to investigate gene function with the known functional properties of genes are more cost-effective than experiment method. Protein-protein interactions (PPIs) network is the most useful molecular network constructed experimentally [3], and gene-gene co-expression network (GGCEN) is usually used in identifying disease related genes with topological differences [4]. Subnetworks extracted from the reference network may disclose gene relationships in the whole network. There have been several web-based tools that implement various subnetwork extraction algorithms, *e.g.* GenRev, NeAT and Genes2Networks [5]. Compared to NeAT and Genes2Networks limiting by specific purposes for the query or metabolic networks, GenRev is uniquely designed for a general purpose approach [5]. As a

© Springer International Publishing AG 2017
D.-S. Huang et al. (Eds.): ICIC 2017, Part II, LNCS 10362, pp. 559–565, 2017.
DOI: 10.1007/978-3-319-63312-1_49

package, GenRev is applied to investigate the functional relevance of genes from high-throughput biological data, including Klein-Ravi algorithm, limited kWalks algorithm and heuristic local search algorithm.

In this study, we developed a computational method to reveal AD candidate genes by known AD-related genes and a combined network containing both PPIs and GGCEN information. The subnetwork extraction algorithms in GenRev were used to obtain the subnetworks containing AD-related genes (seed genes, also called terminal genes) and AD candidate genes (linker genes). The application of combined network and subnetwork extraction algorithms would provide a new insight into predicting the AD-related genes and investigating new methods for AD prevention and treatment.

2   Materials and Methods

2.1   Data Collection

We obtained the expression arrays of human cortex samples from GSE15222 in GEO (https://www.ncbi.nlm.nih.gov/geo/), including 24,350 probes, 186 normal controls and 176 AD samples [6]. T-test analysis was performed in the genes expression between AD groups and normal groups from GSE15222, and p-value <0.01 was considered statistically significant difference.

Genes in a PPI are more likely have common functions, consequently, the genes connecting to known AD-related genes mined by PPIs may play key roles in AD development. In this study, the PPIs data were retrieved from STRING (http://string-db.org/) [7], including 8,548,002 PPIs among 19,247 human proteins (9606.protein.links.v10.txt.gz).

The genes identified to be associated with AD were collected from the following data sources: (1) 41 AD-related genes in OMIM (http://omim.org/) [8]; (2) top ten most strongly associated genes in AlzGene (http://www.alzgene.org/) [9]; (3) 27 genes in AlzBase (http://alz.big.ac.cn/alzBase/) with frequency distribution of dysregulation >18 times [10]. After combining the above genes, we obtained 73 non-redundant AD-related genes (as seed genes).

2.2   GGCEN and PPIs Network

In GSE15222, probe entries were converted to protein-coding genes based on GPL2700 and "Retrieve/ID mapping" tool from the UniProt database (http://www.uniprot.org/uploadlists/) [11]. If multiple probe IDs mapping to the same gene, the median values were collected, and the unmapping probe IDs were abandoned. Using the function adjacency of WGCNA in R (threshold is 0.01), we constructed a GGCEN based on the AD brain transcriptome datasets by calculating Pearson's correlations between each pair of genes [12]. In file "9606.protein.links.v10.txt.gz", ensemble IDs in PPIs represent proteins and were converted into protein-coding genes by "Retrieve/ID mapping" tool, and the unconverted IDs were discarded.

2.3   Subnetwork Extraction Algorithms

The seed genes and combined network were imported into GenRev to extract subnetworks by Klein-Ravi algorithm, limited kWalks algorithm and heuristic local search algorithm, respectively. The Pearson's correlation coefficients of each gene pair were used as edge weights, and node scores were calculated by GenRank [5]. The subnetworks were visualized by Cytoscape software (version 3.4.0). The framework included three stages as shown in Figure 1.

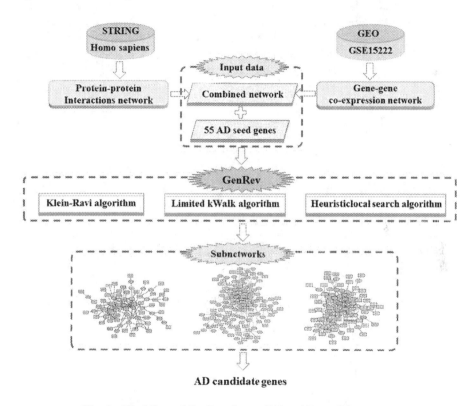

Fig. 1. Workflow of GenRev for predicting AD candidate genes.

Klein-Ravi algorithm is proposed to solve the node-weighted Steiner tree problem and finds a subnetwork with a minimum score that could connect all the seed genes. The subnetwork score is the sum of its nodes score, and the nodes are scored proportional to properties of interest in most biological studies. Limited kWalks algorithm is a universal method based on random walks and Markov chain. The relevance of an edge and a node in relation to the seed genes is evaluated by the expected times random walk passed starting from one seed to any of the others. The maximum walk length L was set to 5. Heuristic local search algorithm based on a previously published method uses a local expansion approach to find the highest scored nodes connected to the seeds.

Shortest path is used to link the node to network. The local search radius d was set to 2 and the score increment rate r was 0.1.

3   Results and Discussion

3.1   Combined Network and AD Seed Genes

There are certain relationships between gene-gene co-expression information and disease genes, and GGCEN provides more novel gene-gene interactions information than PPIs network [4]. The combined network was constructed by integrating the gene pairs information of PPIs and the co-expression data of GGCEN, including 10,200 nodes and 169,004 edges. We obtained 55 genes in AD-related genes of Sect. 2.1 and also in the combined network.

3.2   Subnetworks and Candidate Genes

The combined network and seed genes were imported into GenRev for exploring the function relevance genes. The three algorithms in GenRev compared with each other, and each of them has advantages and disadvantages. The limitid kWalk algorithm finds the most relevant nodes in the information flow, while the Klein-Ravi algorithm explores the minimum number of nodes to connect terminals [5]. Different from the above two algorithms, heuristic local search algorithm looks for paths and genes connecting the seed genes. The numbers of linker genes, terminal genes and edges in three subnetworks were exhibited in Table 1.

Table 1. Numbers of linker, terminal and edges in subnetworks

Algorithm	Linker	Terminal	Edges
Klein-Ravi algorithm	18	55	131
Limited kWalk algorithm	79	55	272
Heuristic local search algorithm	37	55	214

In the three subnetworks, the terminal genes were AD seed genes, and the linker genes emerged in more than one subnetwork were considered as the AD candidate genes. We obtained twelve overlapping genes in two subnetworks. Among these candidate genes, the p-values of *SLIT3*, *JAG1*, *RGS1*, *LDHB* and *B2M* larger than 0.01 and the five genes were cut out. The remaining seven genes might have a strong relationship with AD, and the details of these candidate genes emerged in two subnetworks were displayed in Table 2.

Table 2. Description and p-value of seven candidate genes ($p < 0.01$)

Algorithm	Gene	Description	p-value
Klein-Ravi and limited kWalk algorithms	DBI	Diazepam binding inhibitor	5.89E-7
	SNCB	β-synuclein	7.16E-6
Klein-Ravi and heuristic local search algorithms	BAX	Apoptosis regulator	9.86E-10
	ITGB1	Integrin subunit	7.86E-9
	NCOR1	Nuclear receptor corepressor	6.71E-5
	CSF1R	Colony-stimulating factor receptor	2.37E-4
Limited kWalk and heuristic local search algorithms	CACNG3	Calcium voltage-gated chanel auxiliary subunit	4.49E-20

3.3 AD Candidate Genes Annotation

We searched the function of these candidate genes in published literature to test whether the genes were useful for AD researches. Among these genes, five genes (*DBI*, *SNCB*, *BAX*, *NCOR1* and *CSF1R*) were related to AD, the other two genes (*CACNG3* and *ITGB1*) might have potential roles in AD condition. These genes were discussed one by one in the following.

The *DBI* gene encodes diazepam-binding inhibitor, which is identified as an upregulation genes in AD. The loading of cholesterol into the mitochondrial inner membrane promoting by *DBI* is associate with mitochondrial hypothesis of AD [13]. The *SNCB* gene encodes β-synuclein protein, might decrease formation of Lewy bodies (a common cause of dementia) by preventing abnormal α-synuclein aggregation [14]. The *BAX* gene encodes protein in hippocampal and promotes apoptosis. Growing evidence illustrates that the apoptosis protein activation interferes humanin antiapoptotic activity by specifically binding to humanin [15]. The *NCOR1* gene encodes nuclear receptor corepressor protein, and the *NCOR1* transcript is reduced by the overexpression of tau protein. It has been suggested that *NCOR1* is modestly upregulated when $A\beta_{1-42}$ treatment combined with P301L tau expression [16]. Colony-stimulating factor 1 receptor is encoded by *CSF1R* gene, and its activation regulates the proliferative activity of microglial cells. The biological process of this receptor regulation is a hallmark of AD conditions and provides a target for preventing the progression of neurodegenerative conditions [17].

Based on the existing literature, it is discovered that the two candidate genes (*CACNG3* and *ITGB1*) are associated with the other neurodegenerative diseases or cancer development. The *CACNG3* gene encodes the γ subunit of L-typer voltage-dependent calcium channels, and it has been identified that *CACNG3* is the candidate gene of age-related macular degeneration and childhood absence epilepsy [18]. As a cancer associated gene, *ITGB1* encodes integrin subunit and overexpresses in prostate cancer clinical specimens. Knockdown of *ITGB1* significantly inhibits cancer cell migration and invasion in prostate cancer cells by regulating downstream signaling [19]. The two genes might have a potential role with AD, and it needs further researches to reveal their roles in AD.

4   Conclusion

In this study, we used subnetwork extraction algorithms based on the combined network and seed genes to identify AD candidate genes. The combined network was constructed by integrating the PPIs network and GGCEN. Klein-Ravi algorithm, limited kWalks algorithm and heuristic local search algorithm in GenRev were applied to extract subnetworks and mine candidate genes. As a result, we obtained seven candidate genes with high possibility for AD, and among these genes, five genes (*DBI, SNCB, BAX, NCOR1* and *CSF1R*) were related to AD, the other two genes (*CACNG3* and *ITGB1*) might have potential roles in AD condition. This work provides a global overview of using subnetwork extraction algorithms to mine the targets or biomarkers of AD based on GGCEN and PPIs information, and further researches of these genes would provide more information for AD therapeutics.

Acknowledgments. This work was supported by National Natural Science Foundation of China (61672037 and 21601001), the Initial Foundation of Doctoral Scientific Research in Anhui University (J01001319), and Anhui Provincial Outstanding Young Talent Support Plan (No. gxyqZD2017005).

References

1. Alzheimer's Disease International, World Alzheimer Report 2016. London: Alzheimer's Disease International (2016)
2. Nicolas, G., Charbonnier, C., Campion, D.: From common to rare variants: the genetic component of Alzheimer disease. Hum. Hered. **81**(3), 129–141 (2016). doi:10.1159/000452256
3. Vidal, M., Cusick Michael, E., Barabási, A.-L.: Interactome networks and human disease. Cell **144**(6), 986–998 (2011). doi:10.1016/j.cell.2011.02.016
4. Sun, G.P., Jiang, T., Xie, P.F., et al.: Identification of the disease-associated genes in periodontitis using the co-expression network. Mol. Biol. (Mosk) **50**(1), 143–150 (2016). doi: 10.7868/s0026898416010195
5. Zheng, S., Zhao, Z.: GenRev: exploring functional relevance of genes in molecular networks. Genomics **99**(3), 183-188. doi:10.1016/j.ygeno.2011.12.005
6. Webster, J.A., Gibbs, J.R., Clarke, J., et al.: Genetic control of human brain transcript expression in Alzheimer disease. Am. J. Hum. Genet. **84**(4), 445–458 (2009). doi:10.1016/j.ajhg.2009.03.011
7. Franceschini, A., Szklarczyk, D., Frankild, S., et al.: STRING v9.1: protein-protein interaction networks, with increased coverage and integration. Nucleic Acids Res. **41**(Database issue), D808–D815 (2013). doi:10.1093/nar/gks1094
8. Amberger, J.S., Bocchini, C.A., Schiettecatte, F., et al.: OMIM.org: Online Mendelian Inheritance in Man (OMIM(R)), an online catalog of human genes and genetic disorders. Nucleic Acids Res. **43**(Database issue), D789–D798 (2015). doi:10.1093/nar/gku1205
9. Bertram, L., McQueen, M.B., Mullin, K., et al.: Systematic meta-analyses of Alzheimer disease genetic association studies: the AlzGene database. Nat. Genet. **39**(1), 17–23 (2007). doi:10.1038/ng1934
10. Bai, Z., Han, G., Xie, B., et al.: AlzBase: an integrative database for gene dysregulation in Alzheimer's disease. Mol. Neurobiol. **53**(1), 310–319 (2016). doi:10.1007/s12035-014-9011-3

11. Consortium TU, UniProt: the universal protein knowledgebase. Nucleic Acids Res. **45**(D1), D158–D169 (2017). doi:10.1093/nar/gkw1099

12. Langfelder PHorvath S, WGCNA: an R package for weighted correlation network analysis. BMC Bioinform. **9**, 559 (2008). doi:10.1186/1471-2105-9-559

13. Mills, J.D., Nalpathamkalam, T., Jacobs, H.I., et al.: RNA-Seq analysis of the parietal cortex in Alzheimer's disease reveals alternatively spliced isoforms related to lipid metabolism. Neurosci. Lett. **536**, 90–95 (2013). doi:10.1016/j.neulet.2012.12.042

14. Windisch, M., Hutter-Paier, B., Rockenstein, E., et al.: Development of a new treatment for Alzheimer's disease and Parkinson's disease using anti-aggregatory beta-synuclein-derived peptides. J. Mol. Neurosci. **19**(1–2), 63–69 (2002). doi:10.1007/s12031-002-0012-8

15. Obulesu, M., Lakshmi, M.J.: Apoptosis in Alzheimer's disease: an understanding of the physiology, pathology and therapeutic avenues. Neurochem Res. **39**(12), 2301–12 (2014). doi:10.1007/s11064-014-1454-4

16. Hoerndli, F.J., Pelech, S., Papassotiropoulos, A., et al.: Aβ treatment and P301L tau expression in an Alzheimer's disease tissue culture model act synergistically to promote aberrant cell cycle re-entry. Eur. J. Neurosci. **26**(1), 60–72 (2007). doi:10.1111/j.1460-9568.2007.05618.x

17. Olmos-Alonso, A., Schetters, S.T., Sri, S., et al.: Pharmacological targeting of CSF1R inhibits microglial proliferation and prevents the progression of Alzheimer's-like pathology. Brain **139**(Pt 3), 891–907 (2016). doi:10.1093/brain/awv379

18. Spencer, K.L., Olson, L.M., Schnetz-Boutaud, N., et al.: Dissection of chromosome 16p12 linkage peak suggests a possible role for CACNG3 variants in age-related macular degeneration susceptibility. Invest. Ophthalmol. Vis. Sci. **52**(3), 1748–1754 (2011). doi:10.1167/iovs.09-5112

19. Kurozumi, A., Goto, Y., Matsushita, R., et al.: Tumor-suppressive microRNA-223 inhibits cancer cell migration and invasion by targeting ITGA3/ITGB1 signaling in prostate cancer. Cancer Sci. **107**(1), 84–94 (2016). doi:10.1111/cas.12842

Special Session on Computer Vision based Navigation

A Comparative Analysis Among Dual Tree Complex Wavelet and Other Wavelet Transforms Based on Image Compression

Inas Jawad Kadhim[✉], Prashan Premaratne[✉], Peter James Vial[✉], and Brendan Halloran[✉]

School of Electrical and Computer and Telecommunications Engineering, University of Wollongong, North Wollongong, NSW 2522, Australia
{ijk720,bh294}@uowmail.edu.au, {prashan,Peter_Vial}@uow.edu.au

Abstract. Recently, the demand for efficient image compression algorithms have peeked due to storing and transmitting image requirements over long distance communication purposes. Image applications are now highly prominent in multimedia production, medical imaging, law enforcement forensics and defense industries. Hence, effective image compression offers the ability to record, store, transmit and analyze images for these applications in a very efficient manner. This paper offers a comparative analysis between the Dual Tree Complex Wavelet Transform (DTCWT) and other wavelet transforms such as Embedded Zerotree Wavelet (EZW), Spatial orientation Transform Wavelet (STW) and Lifting Wavelet Transform (LWT) for compressing gray scale images. The performances of these transforms will be compared by using objective measures such as peak signal to noise ratio (PSNR), mean squared error (MSE), compression ratio (CR), bit per pixel (BPP) and computational time (CT). The experimental results show that DTCWT provides better performance in term of PSNR and MSE and better reconstruction of image than other methods.

Keywords: Image compression · Wavelet transformer · DTCWT · EZW · STW · LWT

1 Introduction

Image compression is fundamental to applications including transmission and storing capacity needs in databases. Its goal is to reduce the bit rate for transmission or stor-age capacity while retaining an adequate image quality. This goal is usually realized as the pixel neighborhood contains higher correlation and images carry redundant data. For the most part, the aim is to discover the representation of image in which the image pixels are decorrelated [1]. Compression systems can be fundamentally separated into two parts: the spatial domain and frequency domain compression techniques. In frequency domain techniques, there are many techniques like Fourier, Cosine, and Wavelet Transform etc. Based on the quality of resulting images, the image compression can also be categorized into lossless and lossy techniques [2]. Lossless image compression does not lose any information in the process of compression. This means that the

© Springer International Publishing AG 2017
D.-S. Huang et al. (Eds.): ICIC 2017, Part II, LNCS 10362, pp. 569–580, 2017.
DOI: 10.1007/978-3-319-63312-1_50

lossless compression has the ability to be decompressed perfectly (i.e. original image is recovered). Lossy image compression on the other hand, loses information during compression. Wavelet Transform (WT) is an apparatus that allows multi resolution analysis of an image. It can extract important information from an image and can adjust to cater for human visual characteristics. WT decomposes an image into a different set of resolution or scale sub-images. These sub-images correspond to different frequency groups and provide a multi-resolution representation of the image with localization in both spatial and frequency domains. Popular techniques in wavelet family are Embedded Zerotree Wavelet (EZW) [3], Spatial orientation Tree Wavelet (STW) [4], and Lifting Wavelet Transform (LWT) [5]. EZW and STW widely use progressive significant coefficients methods to compress images into a lower bit rate code which further can be used for storing and transmission purposes. These methods can improve the compression ratio of images using methods like discrete cosine transform. LWT has gained increasing interest in image compression as it has low computational complexity due to utilization of an ordinary wavelet filter into lifting steps [5]. Despite their efficiencies, they have three main limitations in the implementation; one being the failure to have a variance in a shift is a major drawback in the process. A slight shift in the wavelength is not proportionally reflected in the process. In fact, a shift in the projection is one of the ways through which the processed image under the algorithm is distorted [6]. Another disadvantage is associated with the inefficiency in directional selectivity. In multiple dimensions, orientations are difficult to be distinguished, which are important in image and signal processing. The m-dimensional transform coefficients of them reveal only a few feature orientations in the spatial domain. Third disadvantage is the absence of phase information that leaves a huge gap in the usefulness of the information in many image processing applications. The Dual Tree Complex Wavelet Transform (DTCWT) overcomes these restrictions of wavelet transform mentioned above; it is oriented in 2D and is approximately shift-invariant. The 2D yields six subbands at each scale, each of which is powerfully oriented at distinctive angles [7].

We will introduce in this paper DTCWT which is highly computationally effective in dealing with shift invariance and gives much better directional selectivity during processing of images. The comparative analysis is introduced to determine the potentials and disadvantages of using these techniques in order to assist in choosing the suitability of them in a certain application in terms of PSNR, MSE, CR, CT and BPP. The experimental results show that the method of DTCWT shows better performance when compared with EZW, STW and LWT in the aspects of images compression fidelity criteria.

The paper is organized as follows. Section 2 discusses the related work. Section 3 describes the wavelet based image decomposition technique followed by discussion on DTCWT, EZW, STW and LWT methods. Experimental results and discussions are discussed in Sect. 4 for various test images. Finally, the concluding remarks are given in Sect. 5.

2   Related Work

Image compression is an area that has received a lot of attention from the research community in the recent past in ways to represent an image in a more compact way, so that one can store more images in a given memory or transmit images faster. Many image compression techniques still suffer from large redundant data due to limitations in the implementation. Dual tree complex wavelet transform [8] was proposed for compressing image based on arithmetic encoding algorithm. This method brings coefficient of wavelet close to zero by using the thresholding to produce more zeros in order to achieve higher compression ratios. The experimental result was implemented on (Lena image 256 × 256). The result has improved performance in terms of compression ratio compared with DCT combined with arithmetic and Huffman coding respectively. Reddy [9] proposed combination method between 2DDTCWT and Huffman encoder in order to enhance the quality of image recovery with higher CR. The algorithm was tested on (cameraman image) and using different values of thresholding. The performance was compared with EZW combination with Huffman coding. The result showed that 2DDTCWT was better than EZW. In [10, 11] a new image compression was proposed based on DTCWT and simple Set Partitioning in Hierarchical Tree (SPIHT). The advantage of DTCWT is approximate shift invariance and good direction selectivity to improve the compression ratio. This method has exploited the relationship between target bit rate and initial threshold in order to effectively decrease the space of storage and computation time. The experimental result carried on three images namely (Barbara, boat and Lena) and the result was compared with DWT in terms of CR and PSNR. In [12] a comparative analysis of famous different wavelet transform was proposed for still image compression. These algorithms are: wavelet difference reduction (WDR), STW, EZW and modified SPIHT. The result show that modified SPIHT method work better than other algorithms. While in [13] the result show that WDR outperform in term of CR. In [14] a comparative study for true and virtual image coder was performed using improved EZW. For fully embedded code, special design of bit in the bit stream was used in order to achieve the requirement. The result of proposed method compared with different wavelet filters such as Biorthogonal, Coiflets, Daubechies, Symlets and Reverse Biorthogonal Filters. The experimental result carried on true Rice and Human Spine image and the result was compared in term of MSE and PSNR. The results show that improvement on EZW algorithm work better than other wavelet filters. In later works [15–17] lifting wavelet has gained increasing interest in image compression as it has low computational complexity due to utilization of an ordinary wavelet filter into lifting steps. LWT was also employed for transformation and SPIHT for coded which give the best result in term of compression ratio under the image quality after tested on medical and natural images.

To sum up these works also have the limitation of quality image, CR and computational time. Therefore, the comparative analysis is introduced to determine the potential disadvantages of using these techniques in order to assist in choosing the suitability of them in a certain application in terms of PSNR, MSE, CR, CT and BPP.

3 Wavelet Based Compression Algorithms

In majority of the images, the image contains redundant information and their neighboring pixels are correlated. By compressing an image, we ought to locate a less correlated representation of the image. Image compression depends on reduction of redundancy and irrelevancy. When redundant information and irrelevant information are removed from the image, the quality of the image does not change due to the way the Human Visual System (HVS) operates. The redundancies in an image can be distinguished as spatial redundancy, psycho visual redundancy and coding redundancy. In Image compression, numbers of bits to represent data are reduced by expelling redundant bits from the image [18]. The basic steps of an image compression system can be divided into two stages: encoder and decoder. The encoder involves mapper, quantizer and symbol coder while the decoder is the inverse procedure of the encoder. In compression, an image is mapped into frequency domain function of the lower and higher spectrum of coefficient [18]. After generating the coefficients of frequencies, these coefficients are encoded by an appropriate method available to encode the image into lesser code of array. Now for reconstruction of the same image, a decoder of the same technique which is used to encode the signal is used to decode it into a function of frequency. After applying final step of using inverse transform, a reconstructed image is obtained back which is almost similar to the input image. After getting the reconstructed image compression parameters are analyzed to measure the performance index of the particular coding technique. These parameters are PSNR, MSE, BPP, CT etc. Discrete wavelet transform (DWT) is considered as one of the best methods for enhancing the performance of image in term of PSNR [19]. It produces both time and frequency representation of a signal with great efficiency [20]. To transform the image into frequency spectrum coefficients, wavelet transform is used to transform a signal into the frequency domain. With the help of low pass and high pass filter banks, wavelet transform is the best way to achieve multi-resolution analysis [21, 22]. After applying the wavelet transform, coefficients of lower and higher frequency are analyzed to compress the data.

All of the above steps of compression system are invertible, consequently lossless, with the exception of the quantization step. Quantization results in a reduction of the accuracy of the floating point values (FPV) of the wavelet transforms, which are commonly either 32-bit or 64-bit floating point numbers [18]. To utilize less bits in the compressed transform which is vital if compression of 8 bpp or 12 bpp images are to be accomplished. These transforms values must be expressed with less bits for each value leading to rounding errors. These approximate, quantized, wavelet transform coefficients will create approximations to the image creating lossy compression.

3.1 Dual Tree Complex Transform Wavelet

In 1999 Kingsbury proposed the dual tree complex wavelet has the following properties [23, 24]:

- Near shift invariance;
- Good directional selectivity in 2-dimensions (2-D) with Gabor-like filters (additionally valid for higher dimensionality, m-D);
- Perfect reconstruction (PR) (filters having low linear phase).
- Scales are independent and minimal redundancy, E.g. 2: 1 for 1-D (2 m: 1for m-D)
- Efficient order-N computation only twice the simple DWT for 1-D

The dual-tree wavelet transform has been applied to solve a myriad of image processing issues such as the compressed data, the removal of noise as well as the approximation of movement [6].

The DTCWT consists of two parallel wavelet filter bank trees, a and b. The parallel filters are designed with different delays in order to minimize the aliasing effect caused by down sampling. The subband signals of (Tree a) corresponds to the real part of a complex wavelet transform, whereas the sub-band signals of (Tree b) corresponds to the imaginary part of the complex wavelet transform. If the filters are de-signed under those conditions, the DT-CWT is nearly shift-invariant. In order to de-sign the correct filters, the total delay difference of a given level must be the same as the sum of the delay differences of all previous levels. As can be seen in Fig. 1, all of the filters are difference to each other and DTCWT consists of two normal inverses discrete wavelet transforms, and they are merged when reconstructing the image. Hence there is no requirement to calculate of complex arithmetic.

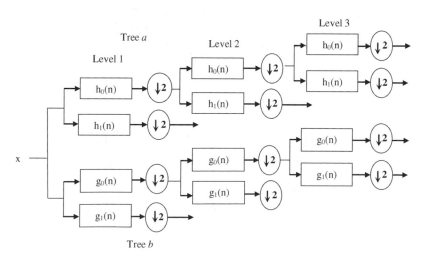

Fig. 1. Two trees of real filter for DTCWT

3.2 Embedded Zerotree Wavelet

In 1993 Shapiro introduced the embedded coding using zero-tree coefficients of wavelets [25]. It is considered a powerful progressive algorithm for the compression system by combining bit stream coding with zero-tree data. The strategy of quantization and coding based on discrete wavelet is intended to accomplish an ideal representation of coding positions utilizing huge coefficients in one frequency band to foresee the value of significant and location coefficients in other bands [26].

The initial step is to set up an initial threshold in this method. If the coefficient value is larger than the threshold, then any coefficient in the wavelet may be significant. Lower bands are spatially related to every coefficient in a hierarchical subband system. Bands having higher bands coefficients are called 'descendants'. It is coded a 'positive significant' (ps) when coefficient is positive and significant. Similarly, it will be coded a 'negative significant' (ns) when a coefficient is negative and significant. It will be coded a 'zero tree root' (ztr) when a coefficient is insignificant and all its descendants are insignificant as well. It will be coded an 'insignificant zero' (iz) when a coefficient is insignificant and all its descendants are insignificant.

3.3 Spatial Orientation Tree Wavelet

The spatial orientation tree is basically for the SPIHT algorithm. The energy of the image is accumulated on the low frequency components, so the variance reductions from the high levels to the low levels of the subband pyramid. Moreover the coefficients are getting better when going down on the pyramid with the same spatial orientation. This spatial relationship is naturally represented as a tree. As we can see in Fig. 3, the spatial orientation tree is clearly a pyramid that is constructed with four sub-band splitting. In Fig. 2, Each node of the tree can be considered a pixel and its child node is the pixels of the same spatial orientation in the next level of the pyramid.

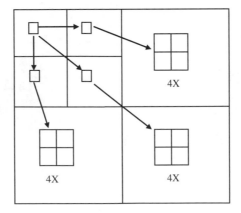

Fig. 2. The spatial orientation tree

In lowest frequency subband, a node is the root of a tree and it has three child nodes, respectively. The nodes in the other frequency subband have four child nodes, which can be considered as the leaves of the tree. The node index is same as the index of the pixel. In ordering algorithm, the subset information is stored in an ordered list and evaluated the next list order. If a subset is considered to be significant, it is removed from the ordered list and partitioned. This ordered list is called dominant list [4].

3.4 Lifting Wavelet Transform

In 1998 Sweldens proposed Lifting wavelet transform [5]. LWT is nonlinear wavelet transform without up and down sampling. LWT is an alternative solution for DWT, with less computation and memory requirements and more efficient in real time applications, resulting in better than the traditional wavelet transforms. LWT which is derived from a poly-phase matrix representation that is able to distinguish between even and odd samples. In order to derive the lifting wavelet transform, the original filter is divided into a series of shorter filters using the filter factoring algorithm. Those filters are designed as low-pass and high-pass filters with lifting steps. There are three stages in the LWT: split, prediction and update. In the split stage the input image data is divided into two sets: odd and even samples. The prediction stage is used to estimate the odd from even samples which can be seen in an error image. Then the update operator is used to detail image and combining the result to even sample, the resultant can be regarded as smooth part of the original image. To design prediction and update operator can be used for interpolating subdivision. To reconstruct LWT, the inverse operation can be applied by reversing the predication and update step. This reversing can be done by changing each positive value into negative value and vice versa.

4 Experimental Result and Discussion

Our experiments are implemented on MATLAB and we used twenty gray images of size 256×256. In order to evaluate the quality of the decompressed images, the fidelity criteria is applied to measure the amount of the information required in the decompressed image, and can be divided into two types: (1) objective fidelity criteria and (2) subjective fidelity criteria. Commonly used objective measures like MSE, PSNR, CR, BPP and CT. The MSE is the cumulative squared error between the compressed and the original image, whereas PSNR is a measure of the peak error and BPP the number of bits required to store one pixel of the image.

The mathematical formulae are:

$$\text{MSE} = \frac{1}{MN} \sum\nolimits_{y=1}^{M} \sum\nolimits_{x=1}^{N} \left[I(x,y) - \left(I'(x,y) \right) \right]^2 \tag{1}$$

$$\text{PSNR} = 20 \times \log_{10} \left(\frac{MAX_1}{\sqrt{\text{MSE}}} \right) \tag{2}$$

$$CR = \frac{original\ image\ size\ in\ bits}{compressed\ image\ size\ in\ bits} \qquad (3)$$

Where $I(x, y)$ is the original image and $I'(x, y)$ is the decompressed image (which is actually the approximated original) and M, N are the dimensions of the image. Here, MAX_1 is the maximum possible pixel value of the image. The value of MAX_1 is equal to 255 when the pixels are represented by 8 bits per sample. A lower value for MSE means lesser error, and as seen from the inverse relation between the MSE and PSNR, this translates to a high value of PSNR. Logically, a higher value of PSNR is good because it means that the ratio of signal to noise is higher. In this comparative analysis, the images are compressed by DTCWT, EZW, STW and LWT with respect to the five performance parameter i.e. PSNR, Computation Time, MSE, BPP and Compression Ratio. Wavelet decomposition is evaluated up to three levels. Number of passes of coding was set on five. All results of methods are obtained by graphs and the observation table. The original

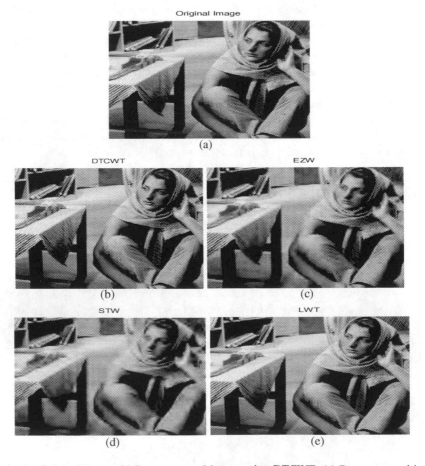

Fig. 3. (a) Original image (b) Decompressed image using DTCWT, (c) Decompressed image using EZW, (d) Decompressed image using STW, (e) Decompressed image using LWT

Barbara image is shown in Fig. 3 (a) and reconstructed (decompressed) images are shown in Fig. 3 (b)–(e) by using DTCWT, EZW, STW, LWT methods respectively. It can clearly show that the reconstructed image using DTCWT has better visual results compared to those obtained from different wavelet methods. Also, it can be seen that the result from EZW and STW have blurred and resulting in visual distortion. In Fig. 4 graphs for Barbara image illustrates the comparison performance of DTCWT and different wavelet methods in term of PSNR, ESM, CR, CT and BPP. It is easy to see that DTCWT has better performance in term of PSNR and ESM than other methods. The graph in Fig. 5 illustrates that the average of matrix performance on twenty images for different compression wavelet methods. It can be seen that DTCWT has better PSNR and MSE performance than others methods but not better than LWT in terms of computation time and BPP. The results of six from twenty images obtained from the different compression wavelet methods can be compared in Table 1. Table 2 shows that the average performance for four methods is implemented on twenty images. The results in Tables 1 and 2 indicate that PSNR of DTCWT higher than other performance methods and ESM lower than other wavelet methods when decomposition level is three. However, the EZW has higher performance in terms of CR with deteriorating image quality.

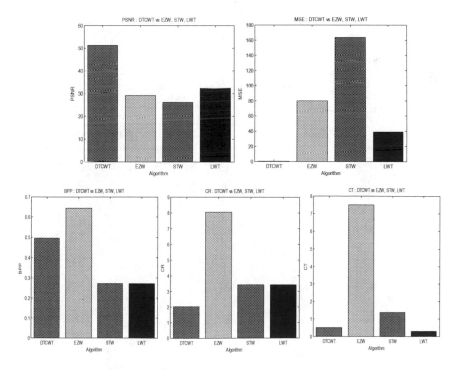

Fig. 4. Comparison graph for one image (Barbara)

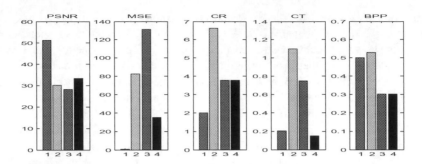

Fig. 5. Average performance on 20, 256 × 256 images graph, 1:DTCWT, 2:EZW, 3:STW, 4:LWT

Table 1. Performance of different compression algorithms on six 256 × 256 images

Method	Image	PSNR	ESM	CR	CT	BPP
DTCWT	1	51.13346	0.50088	2.05841	0.35007	0.48581
	2	51.15536	0.49836	2.00570	0.34384	0.49858
	3	51.14353	0.49972	2.01189	0.51402	0.49704
	4	51.15362	0.49856	2.11454	0.35073	0.47292
	5	51.13836	0.50032	1.94597	0.29692	0.51388
	6	51.17494	0.49612	1.95605	0.32948	0.51123
EZW	1	23.47448	292.16776	3.63770	1.69240	0.29102
	2	29.04085	81.09521	8.69598	6.73205	0.69568
	3	29.11256	79.76727	8.04596	7.50653	0.64368
	4	25.64557	177.22443	3.57971	1.68535	0.28638
	5	29.05183	80.89039	9.11255	6.59273	0.72900
	6	30.92292	52.57613	8.31757	2.29989	0.66541
STW	1	23.51794	289.25839	3.61481	1.36002	0.28918
	2	25.64829	177.11338	3.46985	2.11664	0.27759
	3	25.99407	163.55848	3.39813	1.37277	0.27185
	4	25.71044	174.59695	3.59192	1.32909	0.28735
	5	25.37110	188.78641	3.60870	2.65554	0.28870
	6	27.01637	129.25373	3.94592	1.34344	0.31567
LWT	1	31.60376	44.94740	3.61481	0.31103	0.28918
	2	32.78014	34.28204	3.46985	0.24246	0.27759
	3	32.26018	38.64225	3.39813	0.28377	0.27185
	4	31.65510	44.41915	3.59192	0.22000	0.28735
	5	32.28703	38.40405	3.60870	0.25403	0.28870
	6	31.90724	41.91369	3.94592	0.23230	0.31567

Table 2. Average performance on twenty images (The best performance value is highlighted)

Method	PSNR	MSE	CR	CT	BPP
DTCWT	**51.12629**	**0.50173**	2.00087	0.19729	0.50033
EZW	29.43424	85.49816	**6.36192**	1.06421	0.50895
STW	27.65598	134.00296	3.67867	0.73622	**0.29429**
LWT	34.15140	25.00000	3.67867	**0.15331**	**0.29429**

5 Conclusion

The objective of the paper is to offer a comparative analysis between different wavelet transforms based on image compression algorithms such as DTCWT, EZW, SWT and LWT in the aspects of PSNR, MSE, CR, CT and BPP. These algorithms are successfully tested on twenty gray images. The experimental results show that, the DTCWT method gives the higher PSNR and less MSE than other methods because it is oriented in 2D and it is approximately shift-invariant. However, in the aspects of compression ratio, EZW achieves compression but with degradation of the image quality. While CR and BPP value for STW and LWT retain the same. Moreover, for the calculation time and BPP, DTCWT is not better than LTW. The improving CR and reducing CT are an important issue for future research.

References

1. Rabbani, M., Jones, P.W.: Digital Image Compression Techniques. SPIE Press, Bellingham (1991)
2. Wallace, G.K.: The JPEG still picture compression standard. IEEE Trans. Consum. Electron. **38**(1), xviii–xxxiv (1992)
3. Creusere, C.D.: A new method of robust image compression based on the embedded zerotree wavelet algorithm. IEEE Trans. Image Process. **6**(10), 1436–1442 (1997). doi:10.1109/83.624967
4. Said, A., Pearlman, W.A.: A new, fast, and efficient image codec based on set partitioning in hierarchical trees. IEEE Trans. Circ. Syst. Video Technol. **6**(3), 243–250 (1996)
5. Daubechies, I., Sweldens, W.: Factoring wavelet transforms into lifting steps. J. Fourier Anal. Appl. **4**(3), 247–269 (1998)
6. Neumann, J., Steidl, G.: Dual-tree complex wavelet transform in the frequency domain and an application to signal classification. Int. J. Wavelets Multiresolut. Inf. Process. **3**(1), 43–65 (2005)
7. Selesnick, I.W., Baraniuk, R.G., Kingsbury, N.C.: The dual-tree complex wavelet transform. IEEE Sig. Process. Mag. **22**(6), 123–151 (2005). doi:10.1109/MSP.2005.1550194
8. Indiradevi, K., Shanmugalakshmi, R.: Dual tree complex wavelet transform based image compression using thresholding. ARPN J. Eng. Appl. Sci. **10**(8), 3772–3776 (2015)
9. Reddy, D.S., Varadarajan, S., Giriprasad, M.N.: 2D dual-tree complex wavelet transform based image analysis. Contemp. Eng. Sci. **5**(3), 127–136 (2012)
10. Fang, L.H., Feng, M.G., Jie, X.H.: Images compression using dual tree complex wavelet transform. In: International Conference of Information Science and Management Engineering, pp. 559–562. IEEE (2010). doi:10.1109/ISME.2010.213

11. Wagh, S.A.: Performance evaluation of DWT and DT-CWT with SPIHT progressive image coding for natural image compression. Int. J. Adv. Res. Electr. Electron. Instrum. Eng. **1**(4), 245–251 (2012)

12. Kourav, A., Sharma, A.: Comparative analysis of wavelet transform algorithms for image compression. In: International Conference on Communications and Signal Processing, pp. 414–418. IEEE (2014). doi:10.1109/ICCSP.2014.6949874

13. Taujuddin, M., Afifi, N.S., Ibrahim, R.: A comparative analysis on the wavelet-based image compression techniques. J. Comput. Sci. Eng. **21**(1), 1–6 (2013)

14. Singh, A.P., Singh, B.P.: A comparative study of improved Embedded Zerotree Wavelet image coder for true and virtual images. In: Students Conference on Engineering and Systems, pp. 1–5. IEEE (2012). doi:10.1109/SCES.2012.6199064

15. Kabir, M.A., Khan, M.M., Islam, M.T., Hossain, M.L., Mitul, A.F.: Image compression using lifting based wavelet transform coupled with SPIHT algorithm. In: International Conference on Informatics, Electronics & Vision, pp. 1–4. IEEE (2013). doi:10.1109/ICIEV. 2013.6572638

16. Fan, W., Chen, J., Zhen, J.: SPIHT algorithm based on fast lifting wavelet transform in image compression. In: Hao, Y., Liu, J., Wang, Y.-P., Cheung, Y.-M., Yin, H., Jiao, L., Ma, J., Jiao, Y.-C. (eds.) CIS 2005. LNCS, vol. 3802, pp. 838–844. Springer, Heidelberg (2005). doi: 10.1007/11596981_122

17. Nautiyal, A., Tyagi, I., Pathela, M.: PSNR comparison of lifting wavelet decomposed modified SPIHT coded image with normal SPIHT coding. Int. J. Comput. Appl. **102**(15), 16–21 (2014)

18. Bhaskaran, V., Konstantinides, K.: Image and Video Compression Standards: Algorithms and Architectures. Springer Science and Business Media, New York (1997)

19. Grgic, S., Grgic, M., Zovko-Cihlar, B.: Performance analysis of image compression using wavelets. IEEE Trans. Industr. Electron. **48**(3), 682–695 (2001). doi:10.1109/41.925596

20. Zettler, W. R., Huffman, J. C., Linden, D. C.: Application of compactly supported wavelets to image compression. In: International Society for Optics and Photonics, Electronic Imaging 1990, pp. 150–160. Santa Clara (1990)

21. Du, K., Peng, L.: New algorithms for preserving edges in low-bit-rate wavelet-based image compression. IEEJ Trans. Electr. Electron. Eng. **7**(6), 539–545 (2012)

22. Kekre, H.B., Sarode, T.K., Vig, R.: A new multi-resolution hybrid wavelet for analysis and image compression. Int. J. Electron. **102**(12), 2108–2126 (2015). doi:10.1080/00207217.2015.1020882

23. Kingsbury, N.: Complex wavelets for shift invariant analysis and filtering of signals. Appl. Comput. Harmonic Anal. **10**(3), 234–253 (2001)

24. Kingsbury, N.G.: The dual-tree complex wavelet transform: a new technique for shift invariance and directional filters. In: Proceedings 8th IEEE DSP Workshop, vol. **8**, p. 86. Utah (1998)

25. Shapiro, J.M.: Embedded image coding using zerotrees of wavelet coefficients. IEEE Trans. Sig. Process. **41**(12), 3445–3462 (1993)

26. Liu, Y., Liu, Z.: An improved image compression algorithm based on embedded zerotree wavelets transform. Int. J. Future Comput. Commun. **1**(4), 1097–1102 (2012)

Distributed One Dimensional Calibration and Localisation of a Camera Sensor Network

Brendan Halloran$^{(\boxtimes)}$, Prashan Premaratne, Peter Vial,
and Inas Kadhim

University of Wollongong, Wollongong, NSW 2522, Australia
bh294@uowmail.edu.au

Abstract. Metric calibration and localisation are crucial requirements for many higher-level robotic vision tasks, such as visual navigation and tracking. Furthermore, distributed algorithms are being increasingly used to create scalable camera sensor networks (CSN) which are resistant to node failure. We present a distributed algorithm for the calibration and localisation of a CSN. Our method involves a robust local calibration at each node using a 1D calibration object, consisting of collinear points moving about a single fixed point. Next, each node builds a vision graph and performs cluster-based bundle adjustment, utilising the structure of calibration object to produce pose estimates for its cluster. Finally, these estimates are brought to global consensus through Gaussian belief propagation. Experimental results validate our algorithm, showing that it has comparable performance to centralised algorithms, despite being distributed in nature.

Keywords: Camera calibration · Localisation · Distributed algorithms · Gaussian belief propagation

1 Introduction

Calibration and localisation of a CSN is an important first step for many higher-level computer vision tasks, allowing 3D information to be derived from 2D images. Extensive work has gone into studying calibration, however most multiview solutions require a centralised processor with access to data from all camera nodes, whilst distributed algorithms are generally slower and don't recover the scale of the scene. Distributed algorithms are becoming increasingly important in CSNs where centralised processing is not resistant to node failure and communication with a central node can be expensive, such as in battery powered applications. This promotes the need for a simpler distributed algorithm that can accurately calibrate and localise a CSN to known scale.

The most common calibration method is '2D' calibration, where each camera observes a planar pattern at several arbitrary orientations [1]. This method is popular due to its simplicity; however, it's not suitable for large CSNs due to the pattern's self-occlusion at wider angles. Another popular method is 'self-calibration', which does not use a calibration object [2]. Here, each camera matches feature points from a static scene and solves the structure from motion (SfM) problem to calibrate and localise each node. This method is more suited to large CSNs than 2D calibration as there is no

© Springer International Publishing AG 2017
D.-S. Huang et al. (Eds.): ICIC 2017, Part II, LNCS 10362, pp. 581–593, 2017.
DOI: 10.1007/978-3-319-63312-1_51

self-occlusion, however it is more computationally complex and only calibrates and localises the system up to an unknown scale, which is insufficient for many applications.

The calibration method being utilised in this paper is known as '1D' calibration, which involves observing three or more collinear 3D points moving about one of which is fixed [3]. This method also doesn't have self-occlusion problems and determines the scale of the scene. The other major component of our algorithm is Gaussian belief propagation (GaBP) [4, 5]. Here, each node on a probabilistic graph sends its neighbours messages based on initial potentials and current belief, which iteratively form new beliefs until the convergence. GaBP models the variables as Gaussians, such that each message is simply two scalars representing the mean and variance of the density [6].

1.1   Related Work

First proposed by Zhang, 1D calibration has received many improvements to greatly improve its accuracy [3]. Hammarstendt et al. simplified the closed-form solution and analyse critical configurations [7], while other researchers have relaxed the fixed-point constraint to allow for planar motion of the calibration object [8–10]. De Franca et al. improved accuracy by normalising the image points [11] and Shi et al. improved accuracy with a weighted equation based on orientation of the 1D object [12]. Recently Wang et al. demonstrated that a linear matrix inequality (LMI) relaxation algorithm could replace the non-linear optimisation, speeding up the algorithm [13]. There have also been improvements that allow for general motion of the 1D object in multiview systems, however the results are not as accurate as fixed point methods [14–16].

There has also been work on distributed calibration algorithms, generally based on self-calibration. Devarajan et al. proposed a method that operates on a vision graph, which models the system as a Markov Random Field (MRF) where edges represent overlap in the fields of view [17, 18]. Each node builds a cluster with its neighbours and performs SfM on these clusters. They later brought these estimates to global consensus through belief propagation [19, 20]. Tron and Vidal proposed an algorithm whereby pairwise pose estimates of a CSN were brought to alignment using average consensus [21], and later extended their algorithm to operate on Riemannian manifolds [22].

1.2   Paper Contributions

Our proposed algorithm consists of modelling the network as an MRF. Firstly, we perform a robust local 1D calibration to estimate the intrinsic parameters at each node and 3D points of the calibration object. We then utilise the structure of the calibration object to construct a vision graph and perform a cluster-based bundle adjustment such that each node obtains estimates of their cluster's extrinsic parameters. Finally, we use a Gaussian belief propagation algorithm to align the estimates into global consensus.

The paper is organised as follows. Section 2 reviews the preliminaries for the graph and camera models. Section 3 describes the local 1D calibration algorithm. Section 4 explores the vision graph and the initialisation of pose estimates. Section 5, describes the application of Gaussian belief propagation. Section 6 presents our experimental results and analyses the performance, and finally, Sect. 7 concludes the paper.

2 Preliminaries

2.1 Camera Sensor Networks as Markov Random Fields

Two undirected graphs describe a distributed CSN – the communication graph, $\mathcal{G}_c = \{\mathcal{V}, \mathcal{E}_c\}$, and the vision graph, $\mathcal{G}_v = \{\mathcal{V}, \mathcal{E}_v\}$. In these graphs, $\mathcal{V} = \{1, \ldots, M\}$ is the set of camera nodes. In the communication graph, $\mathcal{E}_c \subseteq \mathcal{V} \times \mathcal{V}$ is the set of edges containing $(i, j) \in \mathcal{E}_c$ pairs of nodes in direct communication. For the vision graph, $\mathcal{E}_v \subseteq \mathcal{V} \times \mathcal{V}$ is the set of edges containing $(i, j) \in \mathcal{E}_v$ pairs of nodes with overlapping fields of view.

In either graph, we denote the direct neighbours of node i as $\mathcal{N}_i = \{j \in \mathcal{V} : (i, j) \in \mathcal{E}\}$, with degree $d_i = |\mathcal{N}_i|$. In this paper, we operate on the vision graph, however the communication graph is necessary for implementation on a physical system. The communication and vision graphs can be significantly different, as can be seen in Fig. 1.

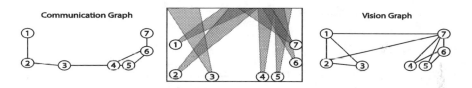

Fig. 1. An example camera sensor network, with fields of view shown. A possible communication graph is shown to the left and the vision graph is shown to the right.

2.2 Camera Model

Considering the 2D point on the image plane $\boldsymbol{m} = [x, y]^T$ as a projection of the 3D world point $\boldsymbol{M} = [X, Y, Z]^T$, we represent these in homogeneous form as $\tilde{\boldsymbol{m}} = [x, y, 1]^T$ and $\tilde{\boldsymbol{M}} = [X, Y, Z, 1]^T$. The relationship between the 2D and 3D point is given by

$$s\tilde{\boldsymbol{m}} = \boldsymbol{K}[\boldsymbol{R}|\boldsymbol{t}]\tilde{\boldsymbol{M}}, \quad where\, \boldsymbol{K} = \begin{bmatrix} \alpha & \gamma & u_0 \\ 0 & \beta & v_0 \\ 0 & 0 & 1 \end{bmatrix}.$$

Here, s is projective depth whilst extrinsic parameters $\boldsymbol{R}$ and $\boldsymbol{t}$ are the rotation and translation of each camera in the world coordinate system. In the intrinsic matrix $\boldsymbol{K}$, α and β are scale factors along the $x-$ and $y-$axes of the image, (u_0, v_0) is the principal point, and γ is the skew.

3   One-Dimensional Camera Calibration

Firstly, we performed local calibration at each node to determine the intrinsic parameters and 3D world points. We used a 1D calibration object consisting of three collinear points, one of which is fixed. The input to our algorithm is the image points from the 1D calibration object at N different orientations, rotated arbitrarily about its fixed point. These images are taken from M camera nodes, as shown in Fig. 2.

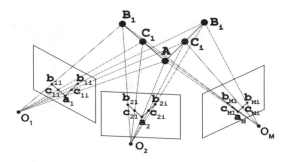

Fig. 2. The 1D calibration object and its appearance to various camera nodes.

3.1   Zhang's Algorithm

Consider three collinear 3D world points A, B and C, where A is a fixed point and C lies between A and B. The known length from A to B is given by

$$L = \|B - A\|. \tag{1}$$

The point C is found using known ratio of λ_A and λ_B,

$$C = \lambda_A A + \lambda_B B. \tag{2}$$

We define the 3D points in the camera coordinate system, such that $[R|t] = [I|0]$. Therefore, the image points $\tilde{a}$, $\tilde{b}$ and $\tilde{c}$ are simply related to their respective 3D points by the intrinsic matrix and unknown projective depths:

$$A = z_A K^{-1} \tilde{a} \tag{3}$$

$$B = z_B K^{-1} \tilde{b} \tag{4}$$

$$C = z_C K^{-1} \tilde{c} \tag{5}$$

Substituting these expressions into Eq. 2 and rearranging allows us to relate the projective depths

$$z_B = -z_A \cdot \frac{\lambda_A(\tilde{\boldsymbol{a}} \times \tilde{\boldsymbol{c}}) \cdot (\tilde{\boldsymbol{b}} \times \tilde{\boldsymbol{c}})}{\lambda_B(\tilde{\boldsymbol{b}} \times \tilde{\boldsymbol{c}}) \cdot (\tilde{\boldsymbol{b}} \times \tilde{\boldsymbol{c}})}. \tag{6}$$

This can be substituted into Eq. 1 to get a new expression for length.

$$z_A^2 \boldsymbol{h}^T \boldsymbol{K}^{-T} \boldsymbol{K}^{-1} \boldsymbol{h} = L^2$$

$$where \quad \boldsymbol{h} = \tilde{\boldsymbol{a}} + \frac{\lambda_A(\tilde{\boldsymbol{a}} \times \tilde{\boldsymbol{c}}) \cdot (\tilde{\boldsymbol{b}} \times \tilde{\boldsymbol{c}})}{\lambda_B(\tilde{\boldsymbol{b}} \times \tilde{\boldsymbol{c}}) \cdot (\tilde{\boldsymbol{b}} \times \tilde{\boldsymbol{c}})} \tilde{\boldsymbol{b}}$$

Now we consider the vector $\boldsymbol{\omega} = [\Omega_0 \ \Omega_1 \ \Omega_2 \ \Omega_3 \ \Omega_4 \ \Omega_5]^T$ which is taken from

$$\boldsymbol{\Omega} = \boldsymbol{K}^{-T}\boldsymbol{K}^{-1} = \begin{bmatrix} \Omega_0 & \Omega_1 & \Omega_3 \\ \Omega_1 & \Omega_2 & \Omega_4 \\ \Omega_3 & \Omega_4 & \Omega_5 \end{bmatrix}.$$

If we define $\boldsymbol{h}_i = \begin{bmatrix} h_{i,0} & h_{i,1} & h_{i,2} \end{bmatrix}^T$ for each image i, and $\boldsymbol{x} = z_A^2 \boldsymbol{\omega}$, then we can rewrite Eq. 4 as

$$\boldsymbol{v}_i^T \boldsymbol{x} = L^2 \boldsymbol{1} \tag{7}$$

$$where \ \boldsymbol{v}_i = \begin{bmatrix} h_{i,0}^2 & 2h_{i,0}h_{i,1} & h_{i,1}^2 & 2h_{i,0}h_{i,2} & 2h_{i,1}h_{i,2} & h_{i,2}^2 \end{bmatrix}^T.$$

For N images, we stack each Eq. 7 to get the $N \times 6$ matrix $\boldsymbol{V} = [\ \boldsymbol{v}_1 \ \boldsymbol{v}_2 \ \cdots \ \boldsymbol{v}_N\ ]^T$, which is related to $\boldsymbol{x}$ by

$$\boldsymbol{V}\boldsymbol{x} = L^2 \boldsymbol{1}. \tag{8}$$

This equation is solved for $\boldsymbol{x}$ using linear least squares, with intrinsic parameters and z_A obtained by Cholesky decomposition of $\boldsymbol{x}$. We can determine z_B using Eq. 6, and the points $\boldsymbol{A}$, $\boldsymbol{B}$ and $\boldsymbol{C}$ for each image using Eqs. 2–4. Furthermore, we can refine these parameters by performing non-linear optimisation on the cost function:

$$E = \sum_{i=1}^{N} \left(\|\boldsymbol{a}_i - \pi(\boldsymbol{K},\boldsymbol{A})\|^2 + \|\boldsymbol{b}_i - \pi(\boldsymbol{K},\boldsymbol{B}_i)\|^2 + \|\boldsymbol{c}_i - \pi(\boldsymbol{K},\boldsymbol{C}_i)\|^2 \right)$$

Here, $\pi(\boldsymbol{K},\boldsymbol{M})$ is a function projecting the point $\boldsymbol{M}$ onto the image plane to get $\boldsymbol{m}$. This optimisation refines $8 + 2N$ parameters with five parameters for the intrinsic matrix, three parameters for the fixed point $\boldsymbol{A}$, and $2N$ parameters encoding points $\boldsymbol{B}_i$ and $\boldsymbol{C}_i$ at each orientation, using spherical coordinates, θ and ϕ, and the relationship

$$\boldsymbol{B}_i = \boldsymbol{A} + L \begin{bmatrix} sin\theta_i cos\phi_i \\ sin\theta_i sin\phi_i \\ cos\theta_i \end{bmatrix}. \tag{9}$$

3.2   Improvements to Zhang's Method

Two major improvements that have been utilised in our algorithm are based on work by de Franca et al. and Shi et al. [11, 12]. To improve the numerical conditioning of the calibration, image data was normalised using a transform T based on the image size, with the inverse transformation recovering the unnormalized intrinsic parameters.

$$\hat{m} = Tm$$

$$K = T^{-1}\hat{K}$$

$$where \quad T = \begin{bmatrix} 2/width & 0 & -1 \\ 0 & 2/height & -1 \\ 0 & 0 & 1 \end{bmatrix}$$

The weighting system used in our implementation is based on the relative depths from Eq. 6 and the distance between image point a and b.

$$\beta_i = \frac{\lambda_A (\tilde{a}_l \times \tilde{c}_l) \cdot (\tilde{b}_l \times \tilde{c}_l)}{\lambda_B (\tilde{b}_l \times \tilde{c}_l) \cdot (\tilde{b}_l \times \tilde{c}_l)}$$

$$w_i = \frac{\|a_i - b_i\|}{\beta_i^2}$$

For N images we define the weighting matrix as $W = diag(w_1, w_2, \ldots, w_N)$. With this, we replace Eq. 8 with the following and solve for x:

$$WVx = W(L^2 \cdot \mathbf{1})$$

4   Initialising the Distributed Network

4.1   Connecting the Vision Graph

Now that the intrinsic parameters and 3D points were estimated, we initialised the vision graph for the system to determine each node's neighbours. The vision graph is an undirected graph with cameras as nodes and edges representing a overlap in fields of view. Since all cameras are viewing the same object, we could assume a fully connected vision graph, however an overly connected graph can prevent the belief propagation from converging. We limited these connections to the four nearest neighbours based on spherical coordinates. Each node j calculates its 'principal spherical coordinates' by averaging the spherical coordinates of each orientation in Eq. 9, and uses the difference to find the closed four neighbours in rotation-wise distance.

$$\begin{bmatrix} \Theta_j \\ \Phi_j \end{bmatrix} = \frac{1}{N} \sum_{i=1}^{N} \begin{bmatrix} \theta_{ij} \\ \phi_{ij} \end{bmatrix}$$

4.2 Estimating Extrinsic Parameters

A node then considers its neighbours as a cluster centred about itself, and calculates the poses of its cluster, relative to itself, by comparing world points. Firstly, it calculates the centre of gravity of each node's world points,

$$G_j = \frac{1}{3N} \sum_{i=1}^{N} (A_j + B_{ij} + C_{ij}).$$

To find the rotation between node j and node k, we find the covariance between the points of the two nodes and apply singular value decomposition.

$$E_{j \to k} = \sum_{i=1}^{3N} (M_{ij} - G_j)(M_{ik} - G_k)^T = USV^T, \, for\, M \in \{A, B, C\}$$

$$R_{j \to k} = VU^T$$

$$t_{j \to k} = -R_{j \to k} G_j + G_k.$$

4.3 Cluster-Based Bundle Adjustment

Each node j has estimates for the poses of all nodes $k \in \mathcal{N}_j$ relative to itself, which we now refine using bundle adjustment. Like the local calibration stage, the number of parameters to be optimised is minimised by utilising the structure of the calibration object. In addition to the three parameters for the fixed point A, we represent B_i and C_i using the two parameters θ_i and ϕ_i as shown in Eq. 9. Also, the poses require three translation variables and three Euclidean angles. Therefore, the number of parameters to be optimised for each cluster is $3 + 2N + 6d_j$, where d_j is the number of neighbours for node j. The cost function is given by the following:

$$E = \sum_{j=1}^{M} \sum_{i=1}^{N} \left(\left\| a_{ij} - \pi_A (K_j, R_j, t_j, A) \right\|^2 + \left\| b_{ij} - \pi_B (K_j, R_j, t_j, \theta_i, \phi_i) \right\|^2 \right.$$

$$\left. + \left\| c_{ij} - \pi_C (K_j, R_j, t_j, \theta_i, \phi_i) \right\|^2 \right)$$

In this equation, π_A, π_B and π_C are the reprojection functions for a, b and c respectively. During this stage, we also estimated the covariance that will be required by the Gaussian belief propagation. As we had covariance estimates for our image point data, V_A, we propagated this through the bundle adjustment Jacobian J with mean-squared error E_{MS}, using

$$V_B = (E_{MS}(J^T V_A J))^{-1}.$$

5 Gaussian Belief Propagation

We now model the true state of each node j as a random vector Y_j, containing the translation variables and Euclidean angles for itself and all neighbours $k \in \mathcal{N}_j$, producing a $6(d_j + 1)$ vector. Z_j is the noisy observation of Y_j obtained from the bundle adjustment. We wish to estimate Y_j from Z_j by marginalising the joint density

$$p(Y_j|Z_1, \ldots, Z_M) = \int_{(Y_k, k \neq j)} p(Y_1, \ldots, Y_M | Z_1, \ldots, Z_M) dY_k.$$

The joint density can be factorised into node potentials ϕ_j and edge potentials ψ_{jk},

$$p(Y_1, \ldots, Y_M) \propto \prod_{j \in \mathcal{V}} \phi_j(Y_j) \prod_{(j,k) \in \mathcal{E}} \psi_{jk}(Y_j, Y_k).$$

Belief propagation is the name given to Pearl's message-passing algorithm which can perform inference on graphs [5]. This algorithm provides exact inference on trees, although it still performs accurately on loopy graphs [4]. It is an iterative algorithm where messages $m^t_{j \to k}(Y_k)$ are sent from node j to k across edges at every iteration t.

$$m^t_{j \to k}(Y_k) \propto \int_{Y_j} \phi_j(Y_j) \psi_{jk}(Y_j, Y_k) \prod_{l \in \mathcal{N}_j \setminus k} m^{t-1}_{l \to j}(Y_j) dY_j \qquad (10)$$

Here, $\mathcal{N}_j \setminus k$ means all neighbours of j except k. The messages are the product of incoming messages from the last iteration, combined with the potentials and marginalised over variables not common to both nodes. Figure 3 shows how all messages received at the sending node, except from the target node, are incorporated into a new message.

The belief at iteration t is simply the product of the node potential and all messages from that iteration.

$$b^t_j(Y_j) \propto \phi_j(Y_j) \prod_{k \in \mathcal{N}_j} m^t_{k \to j}(Y_j)$$

In Gaussian belief propagation, our random variables are assumed to be Gaussian densities, with node potentials, messages and beliefs modelled as two scalars – mean and variance. The node potential mean, μ_{jj}, is our noisy measurement from the bundle adjustment while the node potential variance, V_{jj}, is the diagonal of the covariance. In our algorithm, the edge potentials are selector functions which only select the common variables between the two nodes. The messages from Eq. 10 now have the form

$$V^t_{j \to k} = \left(V^{-1}_{jj} + \sum_{l \in \mathcal{N}_j \setminus k} (V^{t-1}_{l \to j})^{-1} \right)^{-1}$$

$$\mu^t_{j \to k} = V^t_{j \to k} \left(\mu_{jj} V^{-1}_{jj} + \sum_{l \in \mathcal{N}_j \setminus k} \left(\mu^{t-1}_{l \to j} (V^{t-1}_{l \to j})^{-1} \right) \right),$$

and the belief has the form

$$V_j^t = \left(V_{jj}^{-1} + \sum_{k \in \mathcal{N}_j} (V_{k \to j}^t)^{-1}\right)^{-1}$$

$$\mu_j^t = V_j^t \left(\mu_{jj} V_{jj}^{-1} + \sum_{k \in \mathcal{N}_j} \left(\mu_{k \to j}^t (V_{k \to j}^t)^{-1}\right)\right).$$

Once the belief means at each node converge, we take this as the estimate for all Y_j and thus the global consensus for the localisation.

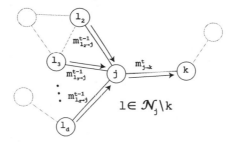

Fig. 3. Messages from node i to j include all messages received at node i, excluding node j.

5.1   Frame Alignment

The nodes initially have the pose estimates for their cluster in their own coordinate frames, and therefore cannot converge to a meaningful value. As such, we need all pose measurements to be aligned. To do this we give each node an arbitrary unique identifier. The lowest numbered node is taken as the basis node, b, and all pose measurements of its neighbours, $j \in \mathcal{N}_b$, are aligned to its frame, using node b's estimate of j's pose. As each node becomes aligned to the basis, this is propagated to its unaligned neighbours.

6   Experiments and Results

6.1   Evaluating the One-Dimensional Calibration

We first simulated a camera with scale factors $\alpha = \beta = 655px$, zero skew and principal point at the centre of the $1024 \times 768px$ image. The calibration object was simulated to be 20 cm in length with the fixed point located at $A = [0 \quad 0 \quad 75]^T$. Image points for the object were then generated at 100 different orientations by sampling θ in $\left[0, \frac{\pi}{4}\right]$ and ϕ in $[0, 2\pi]$ with uniform distribution. We calculated the average error across 200 independent trials for each level of Gaussian noise of zero mean and σ between 0.1 to 2.0 pixels. We measured the relative error of the all parameters with respect to α, giving $|\Delta\alpha/\alpha|$, $|\Delta\beta/\alpha|$, $|\Delta\gamma/\alpha|$, $|\Delta u_0/\alpha|$ and $|\Delta v_0/\alpha|$, as proposed by Triggs [23]. The results are shown in Fig. 4, with errors staying below 5% at 2 pixels of noise.

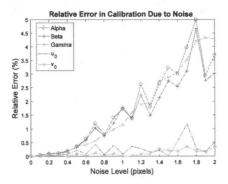

Fig. 4. Relative error for one-dimensional calibration parameters across different noise levels.

6.2   Evaluating the Distributed Localisation

Next, we performed a simulation of the full algorithm. In this experiment, we used the same intrinsic parameters as previous and similarly used a 20 cm calibration object uniformly sampled 100 times for each trial from θ in $\left[0, \frac{\pi}{4}\right]$ and ϕ in $[0, 2\pi]$. Our cameras were located in a mostly circular fashion on the xy-plane around the fixed point and spaced $2\pi/M$ apart at distances uniformly sampled from $[75, 95]$ in centimetres. For our experiment, we used the number of camera nodes as $M = 10$. The cameras were initially directed towards the fixed point but had their yy-axis rotations perturbed by a uniform sampling of $[-5°, 5°]$. Once again, we took the average error of 200 trials at each noise level from 0.1 to 2.0 pixels. An example ground truth layout from our simulation is shown in Fig. 5, with the cameras shown located about the world points. In these simulations, our method of constructing the vision graph resulted in each node being connected to its two neighbours either side.

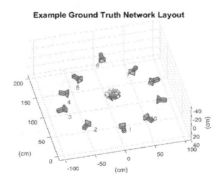

Fig. 5. The ground truth layout for one trial of our simulation, using ten cameras nodes.

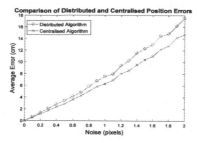

Fig. 6. The position error for noise levels using the distributed and centralized algorithms.

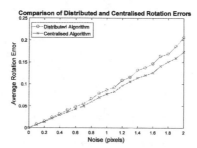

Fig. 7. The rotation error for noise level using the distributed and centralized algorithms.

In our simulation, we first performed the local intrinsic calibration and used the estimates of the calibration object to construct our vision graph. We then performed cluster-based bundle adjustment followed by Gaussian belief propagation until convergence. Our criteria for convergence was that the proportional change in belief $\left| b_j^t - b_j^{t-1} \right| / b_j^t$ was less than 0.0001 for all nodes. We found that convergence occurred in between 12 and 40 iterations for all trials.

The error metric we used for the localisation was that proposed by Devarajan [19], using the error in position as $\|\Delta C\|$ and the error in orientation as $2\sqrt{1 - \cos(\Delta\theta)}$, where $\Delta\theta$ is the relative rotation between the ground truth and estimated value. We averaged these errors across all nodes and trials for each noise level and compared the values to the errors of a centralised global bundle adjustment. The position errors in the distributed and centralised algorithms can be seen in Fig. 6, while the equivalent rotations errors can be seen in Fig. 7.

As can be seen from Figs. 6 and 7, the centralised algorithm only performed slightly better than the distributed algorithm across all noise levels. This shows that the benefits of the distributed algorithm, such as being resistant to single node failure and being scalable without causing communication bottlenecks, are achievable without substantial degradation to the quality of the localisation.

7 Conclusion

In this paper, we have presented a distributed algorithm for camera sensor network calibration and localisation with known scale. Our method models the network as a Markov Random Field based on a vision graph. Following local one-dimensional calibration, a vision graph is constructed and extrinsic parameters are estimated from cluster-based bundle adjustment. These estimates are brought to a globally accurate consensus through Gaussian belief propagation. Our method was shown to be comparable to centralised calibration and bundle adjustment whilst having the advantages of distributed algorithms, namely resistance to node failure and scalability.

Acknowledgements. This research has been conducted with the support of the Australian Government Research Training Program Scholarship.

References

1. Zhang, Z.: A flexible new technique for camera calibration. IEEE Trans. Pattern Anal. Mach. Intell. **22**(11), 1330–1334 (2000)
2. Luong, Q.T., Faugeras, O.D.: Self-calibration of a moving camera from point correspondences and fundamental matrices. Int. J. Comput. Vision **22**(3), 261–289 (1997)
3. Zhang, Z.: Camera calibration with one-dimensional objects. IEEE Trans. Pattern Anal. Mach. Intell. **26**(7), 892–899 (2004)
4. Bickson, D.: Gaussian belief propagation: Theory and application. arXiv preprint arXiv: 0811.2518 (2008)
5. Pearl, J.: Probabilistic Reasoning in Intelligent Systems: Networks of Plausible Inference. Morgan Kaufmann, Burlington (2014)
6. Weiss, Y., Freeman, W.T.: Correctness of belief propagation in Gaussian graphical models of arbitrary topology. Neural Comput. **13**(10), 2173–2200 (2001)
7. Hammarstedt, P., Sturm, P., Heyden, A.: Degenerate cases and closed-form solutions for camera calibration with one-dimensional objects. In: Tenth IEEE International Conference on Computer Vision, ICCV 2005, vol. 1, pp. 317–324. IEEE (2005)
8. Wu, F., Hu, Z., Zhu, H.: Camera calibration with moving one-dimensional objects. Pattern Recogn. **38**(5), 755–765 (2005)
9. Qi, F., Li, Q., Luo, Y., Hu, D.: Camera calibration with one-dimensional objects moving under gravity. Pattern Recogn. **40**(1), 343–345 (2007)
10. Qi, F., Li, Q., Luo, Y., Hu, D.: Constraints on general motions for camera calibration with one-dimensional objects. Pattern Recogn. **40**(6), 1785–1792 (2007)
11. De Franca, J.A., Stemmer, M.R., de M. Franca, M.B., Alves, E.G.: Revisiting zhang's 1D calibration algorithm. Pattern Recogn. **43**(3), 1180–1187 (2010)
12. Shi, K., Dong, Q., Wu, F.: Weighted similarity-invariant linear algorithm for camera calibration with rotating 1D objects. IEEE Trans. Image Process. **21**(8), 3806–3812 (2012)
13. Wang, L., Wang, W., Shen, C., Duan, F.: A convex relaxation optimization algorithm for multi-camera calibration with 1D objects. Neurocomputing **215**, 82–89 (2016)
14. Kojima, Y., Fujii, T., Tanimoto, M.: New multiple-camera calibration method for a large number of cameras. In: Electronic Imaging 2005, pp. 156–163. International Society for Optics and Photonics (2005)

15. Wang, L., Wu, F., Hu, Z.: Multi-camera calibration with one-dimensional object under general motions. In: IEEE 11th International Conference on Computer Vision, ICCV 2007, pp. 1–7. IEEE (2007)
16. De Franca, J.A., Stemmer, M.R., Franca, M.B.d.M, Piai, J.C.: A new robust algorithmic for multi-camera calibration with a 1D object under general motions without prior knowledge of any camera intrinsic parameter. Pattern Recogn. 45(10), 3636–3647 (2012)
17. Devarajan, D., Radke, R.J.: Distributed metric calibration of large camera networks. In: Proceedings of the 1st Workshop on Broadband Advanced Sensor Networks, vol. 3(4), pp. 5–24 (2004)
18. Devarajan, D., Radke, R.J., Chung, H.: Distributed metric calibration of ad hoc camera networks. ACM Trans. Sens. Netw. (TOSN) 2(3), 380–403 (2006)
19. Devarajan, D., Radke, R.J.: Calibrating distributed camera networks using belief propagation. EURASIP J. Appl. Sig. Process. 2007(1), 221–222 (2007)
20. Devarajan, D., Cheng, Z., Radke, R.J.: Calibrating distributed camera networks. Proc. IEEE 96(10), 1625–1639 (2008)
21. Tron, R., Vidal, R.: Distributed image-based 3D localization of camera sensor networks. In: Proceedings of the 48th IEEE Conference on Decision and Control, Held Jointly with the 2009 28th Chinese Control Conference. CDC/CCC 2009, pp. 901–908. IEEE (2009)
22. Tron, R., Vidal, R.: Distributed 3D localization of camera sensor networks from 2D image measurements. IEEE Trans. Autom. Control 59(12), 3325–3340 (2014)
23. Triggs, B.: Autocalibration from planar scenes. In: Burkhardt, H., Neumann, B. (eds.) ECCV 1998. LNCS, vol. 1406, pp. 89–105. Springer, Heidelberg (1998). doi:10.1007/BFb0055661

Neural Networks: Theory
and Application

CAPTCHA Recognition Based on Faster R-CNN

Feng-Lin Du[1], Jia-Xing Li[1], Zhi Yang[1], Peng Chen[2], Bing Wang[3,4], and Jun Zhang[1(✉)]

[1] School of Electrical Engineering and Automation, Anhui University,
Hefei 230601, Anhui, China
wwwzhangjun@163.com
[2] Institute of Health Sciences, Anhui University, Hefei 230601, Anhui, China
[3] School of Electrical and Information Engineering, Anhui University of Technology,
Ma'anshan 243002, China
[4] Key Laboratory of Metallurgical Emission Reduction & Resources Recycling,
Ministry of Education, Ma'anshan 243002, China

Abstract. In this paper, Faster R-CNN was employed to recognize the CAPTCHA (Completely Automated Public Turing test to tell Computers and Humans Apart). Unlike traditional method, the proposed method is based on deep learning object detection framework. By inputting the database into the network and training the Faster R-CNN, the feature map can be obtained through the convolutional layers. The proposed method can recognize the character and it is location. Experiments show that Faster R-CNN can be used in CAPTCHA recognition with promising speed and accuracy. The experimental results also show that the mAP (mean average precision) value will improve with the depth of the network increasing.

Keywords: Faster R-CNN · Deep learning · CAPTCHA recognition

1 Introduction

A CAPTCHA (Completely Automated Public Turing test to tell Computers and Humans Apart) is a type of challenge-response test which used to determine whether a user providing the response is human or not. Most CAPTCHAs on the internet use the combination of characters and numbers to form the picture. Traditionally, CAPTCHA recognition tasks are always divided into four steps: pretreatment (the noise reduction), image segmentation (the character segmentation), feature extraction and character recognition. This makes problems between the composite of multiple steps. However, these multiple steps coordinate not very well, each step of system has different optimization method. The real CAPTCHA picture usually are very complicated, such as the characters and noise of different images may be very similar, or the conjoint characters is not easy distinguished, these reasons make the traditional methods can not achieve the best recognition performance.

Contrasting to traditional methods, deep learning architecture has its advantage. CNN (Convolutional Neural network) [1] of the most common deep learning architecture, which can extract the effective features from the original images. In recent year, object detection framework based on deep-learning makes big progress. R-CNN

D.-S. Huang et al. (Eds.): ICIC 2017, Part II, LNCS 10362, pp. 597–605, 2017.
DOI: 10.1007/978-3-319-63312-1_52

(Regions with Convolutional Neural Network Features) [2], SPP-NET (Spatial Pyramid Pooling in Deep Convolutional Networks) [3], Fast R-CNN (Fast Region-based Convolutional Network) [4] and Faster R-CNN [5]. Faster R-CNN is end-to-end object framework, which can be much faster than other methods. Usually, the framework is divided into four parts: region proposal, feature extraction, classification and location refinement. Faster R-CNN is an end-to-end network which integrates these four parts into a framework, can achieve the object detection and regression the location of object. Faster R-CNN has been used in many object recognition such as pedestrian [6] and very small object-birds in the sky [7]. In this paper, Faster R-CNN was used to recognize the CAPTCHA.

2   Faster R-CNN

2.1   CNN (Convolutional Neural Networks)

CNN [1] is a widely used common deep learning architecture, which can be used for image classification. CNN usually consists three different neural network layers: convolutional layer, pooling layer and activation function layer. The convolutional layer learned to extract the features from the input images. The pooling layer used to reduce the connection from the convolutional layers and this was good for reducing the computation complexity. Activation function layer was used to join the nonlinear factors, because the linear model expression ability was not enough. CNN achieves better classification performance than traditional neural networks because of the local receptive field and the weights of shared. Each neuron only needs localized awareness, and then in the higher level, the information of local comprehensive up to get the global information. Each neuron connects with next layer to use the same parameters. These will help CNN to reduce the parameters, and then we can add more convolution kernels, so that we can extract features fully.

2.2   The Development of Faster R-CNN

The initial implementation of object detection framework based on deep-learning is R-CNN. R-CNN uses selective search [8] to generate 2000 region proposals from original picture. Since the input image of CNN must be the same size, so all these region proposals must wrapped into fixed dimension (227×227) and input them into CNN. In other words, each picture needs generate 2000 region proposals, each region proposal needs to extract features, it is time-consuming. In the output layer, the feature vectors will be output, and then input these feature vectors into SVM (support vector machine) classifier. R-CNN training needs to be divided into multiple stages: we need to fine-tune the net, use selective search to generate region proposals and use SVM to classify. To solve these problems, the SPP-NET and the Fast R-CNN are developed.

SPP-NET doesn't use selective search to generate region proposal. For each picture, one feature map is formed and all the region proposals based on the feature map of the original picture are generated. But the problem is: region proposals have different size. To solve this question, spatial pyramid pooling was designed. The same size feature maps must be input in the CNN's connection layer, using spatial pyramid pooling can make these different size maps into same size. SPP-NET has a faster speed in the rate of object detection than R-CNN. Considering recognition precision, SPP-NET has poor performance. Fast R-CNN is a special case of SPP-NET, which is the integration of R-CNN and SPP-NET. Fast R-CNN uses selective search to generate region proposals, and then puts these region proposals into convolutional layers to get feature vectors. In the next layer, Fast R-CNN uses a ROI (region of interest) pooling layer, which is a simplified version of SPP-NET, it uses a pyramid mapping to each region proposal. So we can get the same dimension vector to the full connection layer. Instead of SVM (Support Vector Machine), Fast R-CNN uses softmax to classify, it also uses multi-tasking loss function frame regression to join the network, which makes the training process is an end-to-end process. Fast R-CNN improves the precision in object detection, but the speed is slow due to the selective search. Then the Faster R-CNN is designed to overcome this problem.

2.3 The Faster R-CNN

Faster R-CNN is an end-to-end object detection framework based on deep-learning and has no repetitive computation, these merits make Faster R-CNN works in much faster speed and can achieve higher object recognition accuracy. Faster R-CNN network can be trained for the object detection and obtain the location of object using regression. When in the feature extraction process, Faster R-CNN uses the classified network of ImageNet [9] as the front network. Faster R-CNN provides three different network models with different level convolutional layers. By Inputting an image to Faster R-CNN network, going through the multiple convolutional layers of front network, we can get the feature maps. Faster R-CNN designs a RPN (Region Proposal Network) layer to replace the fully connected layer of classified network. RPN network is a fully convolutional network [10] which trains the data end-to-end with back propagation and SGD (stochastic gradient descent) [11]. The construction of RPN network is shown in Fig. 1.

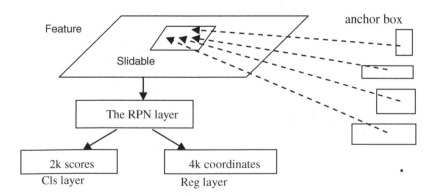

Fig. 1. The construction of RPN network

The RPN network is connected behind the last convolutional layer, and the final shareable convolutional layer output is the convolutional feature map. A slidable network will be designed on the feature map. Each sliding window maps to a lower–dimensional vector. This vector exports to two fully-connected layers which contain a box-regression layer (reg layer) and a box-classification layer (cls layer). The k anchor boxes mapped to each sliding window are predicted. The box-regression layer creates 4 k coordinates of the k anchor boxes and the box-classification layer creates 2 k scores that estimate probability of object/not object for the k anchor boxes. These anchor boxes can correspond to the object with nine different size and scale, which make the region proposal boxes more suitable to actual situation.

The loss function of Faster R-CNN is followed the minimized loss functions with the multi-task loss of Fast R-CNN. For an image, the loss function is defined as:

$$L(p_i, t_i) = \frac{1}{N_{cls}} \sum_i L_{cls}(p_i, p_i^*) + \lambda \frac{1}{N_{reg}} \sum_i p_i^* L_{reg}(t_i, t_i^*) \tag{1}$$

The P_i is the predicted probability of the object. We obtain the IOU (Intersection-over-Union) by the region proposal boxes and the ground truth boxes. If the IOU is greater than 0.7, the $P_i^* = 1$ and the corresponding anchor is positive. Otherwise, $P_i^* = 0$ and the corresponding anchor is negative. The $t_i = (t_x, t_y, t_w, t_h)$ is a vector which indicates the four coordinates of the region proposal boxes. The t_i^* is the coordinate of the ground truth box which corresponds with the positive anchor. $\lambda = 10$, $N_{cls} = 25$, $N_{reg} = 2,400$. The loss function L_{cls} which belongs to the object or not is defined as:

$$L_{cls}(p_i, p_i^*) = -\log[p_i^* p_i + (1 - p_i^*)(1 - p_i)] \tag{2}$$

The loss function L_{reg} is defined as:

$$L_{reg}(t_i, t_i^*) = R(t_i - t_i^*) \tag{3}$$

The loss function R is defined as:

$$R(x) = \begin{cases} 0.5x^2, |x| < 1 \\ |x| - 0.5, |x| \geq 1 \end{cases} \tag{4}$$

3   Experiments and Discussion

3.1   Dataset Preparation

In this paper, we train a network to recognize characters on our datasets of CAPTCHA which include 62 classes of upper and lower case letters and numbers. These CAPTCHAs are divided into three different types and each type of them consists five thousand images. These images have different backgrounds and noises, the letters and numbers in the images are also carried on deformation and rotation. These images are crawled from three large CAPTCHA web sites, so all these data are very effective and real. In all experiments, three datasets are used to train the network. Some of the images of the CAPTCHA data are shown in Fig. 2.

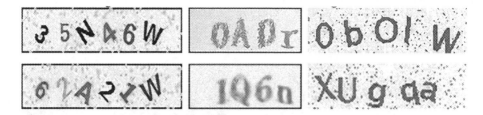

Fig. 2. The images of the CAPTCHA data

All these pictures are process manually according to the PASCALVOC2007 [12] data format. We use a MATLAB script to create a TXT file which includes all of the file names of these pictures. Then, we use a GUI program to circle characters in a box from each picture. Some of the labels are shown in Fig. 3.

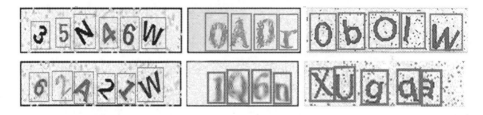

Fig. 3. Data labels making

The GUI program can save every box information which includes Xmin, Ymin, Xmax, Ymax. The information will be saved with the corresponding file name. After doing this work, the XML file format is converted from original txt file. Finally, we divide each dataset into training, validation and test data. The training and validation dataset consists of nine-tenth whole data, the test dataset has only one-tenth data, it uses to test when the model parameters were determined and test the performance model.

3.2   Train

We download three existing models (ZF [13], VGG_CNN_M_1024, VGG16 [14]) which have different shareable convolutional layers to fine-tune with our dataset. The ZF is a minitype network, and the VGG_CNN_M_1024 is a medium network, the VGG16 is a large network. They have different convolutional layers. The architecture and training process of Faster R-CNN is shown in Fig. 4. Then we use the processed data to train the system. Training the system needs two stages. In the first stage we train the RPN layer and adjust the convolutional layers, calculate the RPN results in the train/ test dataset. Then use the RPN region proposals to train the Fast R-CNN and adjust the convolutional layers, this stage can assess the mAP(mean Average Precision). In the second stage, we train the RPN layer and fix the convolutional layers, also calculate the RPN results in the train/test dataset. Then use the RPN region proposals to train the Fast R-CNN and fix the convolutional layers, assess the mAP. The Faster R-CNN training process is shown in Fig. 4.

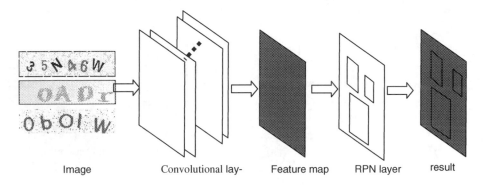

Image   Convolutional lay-   Feature map   RPN layer   result

Fig. 4. Faster R-CNN training process

3.3 Parameter Setting

The network has some parameters to set: Stage1_fast_R-CNN_train.pt: base_lr: 0.001, lr_policy: "step", gamma: 0.1, stepsize: 30000, display: 20, average_loss: 100, momentum: 0.9, weight_decay: 0.0005. Stage1_rpn_train.pt: base_lr: 0.001, lr_policy: "step", gamma: 0.1, stepsize: 60000, display: 20, average_loss: 100, momentum: 0.9, weight_decay: 0.0005. Stage2_fast_R-CNN_train.pt: base_lr: 0.001, lr_policy: "step", gamma: 0.1, stepsize: 30000, display: 20, average_loss: 100, momentum: 0.9, weight_decay: 0.0005. Stage2_rpn_train.pt: base_lr: 0.001, lr_policy: "step", gamma: 0.1, stepsize: 60000, display: 20, average_loss: 100, momentum: 0.9, weight_decay: 0.0005. The partial parameters of the network are set as follows: data_param_str_num_classes: 63, cls_score_num_output: 63, bbox_pred_num_output: 252.

3.4 Experimental Result

Three different depth of the networks are used to train and test the data, they are ZF [13], VGG_CNN_M_1024 and VGG16 [14]. We use mAP (mean Average Precision) to evaluate the experimental results. For a single character, it has a great result. Such as the characters of A and B, the accuracy reached 0.9995 and 0.9996. But for the characters of O and 6, the accuracy reached 0.9098 and 0.9091, which are lower comparing with other characters. To the whole CAPTCHA pictures, we train three kinds of CAPTCHA individually, then train all of the CAPTCHAs. The training results of the three kinds of networks are shown in Table 1.

Table 1. The training results of the three kinds of networks

MODEL	CAPTCHA1	CAPTCHA2	CAPTCHA3	All CAPTCHA
ZF	0.974	0.965	0.966	0.965
VGG_CNN_M_1024	0.982	0.973	0.972	0.974
VGG16	0.985	0.978	0.975	0.976

From Table 1 we can see that when the depth of the network increased, the mAP value of the training results will improve. The VGG16 network has the highest accuracy, because the deeper convolutional layer can learn more information from the image, which is more powerful to classify and regresses the location of the object. And we also obtain the figure of the loss function, which used the VGG16 network in Fig. 5. In coordinates, the x-axis is the iterations, and the y-axis is the value of loss. Stage1 and stage2 are two stages to train the Faster R-CNN network. We can see that the loss has a decrease trend with the iteration increase which means that the experimental results have become better.

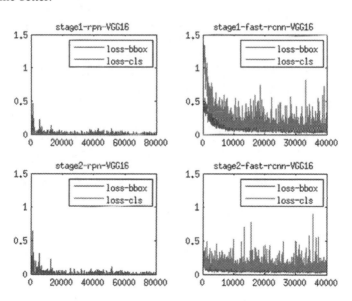

Fig. 5. The loss changed with iterations

Some example detections using Faster R-CNN on CAPTCHA are shown in Fig. 6. We see that these CAPTCHAs can be recognized well. We set the threshold as 0.8, when the predicted object bounding box probability is greater than this threshold, the predicted object bounding box will be consider as true recognition. The predicted object bounding box will be more accurate by this way, because the threshold can exclude the low probability boxes.

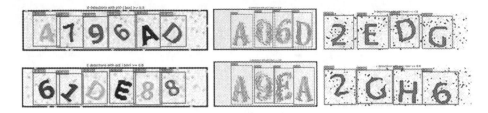

Fig. 6. The detection results of CAPTCHA

4 Conclusion

In this paper, Faster R-CNN was used to recognize CAPTCHA. The experimental results show that this method is feasible and also can get satisfactory results. CAPTCHA recognition method based on Faster R-CNN can recognize CAPTCHA with a high accuracy. This method is different with traditional recognition method in several steps, this makes our method efficient and accurate. But this approach has limitation, we can only recognize these three CAPTCHAs. For the other CAPTCHAs, you need to add more data about them to train the Faster R-CNN network. And we have reason to believe that if we have greater amount of data which include a wider variety of CAPTCHAs, this method will recognize more CAPTCHAs with a promising result.

Acknowledgements. This work was supported by National Natural Science Foundation of China under grant Nos. 61271098, 61472282, 61300058, 61672035 and 61032007, Provincial Natural Science Research Program of Higher Education Institutions of Anhui Province under grant No. KJ2012A005, and Natural Science Foundation of Anhui Province under Grant No. 1508085MF129. This work was also supported in part by the Key Laboratory of Metallurgical Emission Reduction & Resources Recycling (Anhui University of Technology), Ministry of Education, under Grant No. KF 17-02.

References

1. Krizhevsky, A., Sutskever, I., Hinton, G.E.: ImageNet classification with deep convolutional neural networks. In: International Conference on Neural Information Processing Systems, pp. 1097–1105. Curran Associates Inc. (2012)
2. Girshick, R., Donahue, J., Darrell, T., et al.: Rich feature hierarchies for accurate object detection and semantic segmentation, pp. 580–587 (2014)
3. He, K., Zhang, X., Ren, S., et al.: Spatial pyramid pooling in deep convolutional networks for visual recognition. IEEE Trans. Pattern Anal. Mach. Intell. **37**(9), 1904–1916 (2015)
4. Girshick, R.: Fast R-CNN. In: Computer Science (2015)
5. Ren, S., He, K., Girshick, R., et al.: Faster R-CNN: towards real-time object detection with region proposal networks. In: International Conference on Neural Information Processing Systems. MIT Press, pp. 91–99 (2015)
6. Zhang, L., Lin, L., Liang, X., et al.: Is faster R-CNN doing well for pedestrian detection? (2016)

7. Takeki, A., Tu, T.T., Yoshihashi, R., et al.: Combining deep features for object detection at various scales: finding small birds in landscape images. IPSJ Trans. Comput. Vis. Appl. **8**(1), 5 (2016)
8. Uijlings, R., van de Sande, A., Gevers, T., et al.: Selective search for object recognition. Int. J. Comput. Vis. **104**(2), 154–171 (2013)
9. Russakovsky, O., Deng, J., Su, H., et al.: ImageNet large scale visual recognition challenge. Int. J. Comput. Vis. **115**(3), 211–252 (2015)
10. Long, J., Shelhamer, E., Darrell, T.: Fully convolutional networks for semantic segmentation. In: IEEE Conference on Computer Vision and Pattern Recognition, pp. 3431–3440. IEEE (2015)
11. Lecun, Y., Boser, B., Denker, J.S., et al.: Backpropagation applied to handwritten zip code recognition. Neural Comput. **1**(4), 541–551 (2014)
12. Everingham, M., Van Gool, L., Williams, C.K., Winn, J., Zisserman, A.: The pascal visual object classes (voc) challenge. Int. J. Comput. Vis. **88**(2), 303–338 (2010)
13. Zeiler, M.D., Fergus, R.: Visualizing and understanding convolutional networks, pp. 818–833 (2013)
14. Simonyan, K., Zisserman, A.: Very deep convolutional networks for large-scale image recognition. In: Computer Science (2014)

Prediction of Subcellular Localization of Multi-site Virus Proteins Based on Convolutional Neural Networks

Lei Wang[1,2], Dong Wang[1,2(✉)], Yaou Zhao[1,2], and Yuehui Chen[1,2(✉)]

[1] School of Information Science and Engineering, University of Jinan, Jinan 250022, China
3027630499@qq.com, {ise_wangd,ise_zhaoyo,yhchen}@ujn.edu.cn
[2] Shandong Provincial Key Laboratory of Network Based Intelligent Computing, Jinan 250022, China

Abstract. Prediction of subcellular localization is critical for the analysis of mechanism and functions of proteins and biological research. A series of efficient methods have been proposed to identify subcellular localization, but challenges still exist. In this paper, a novel feature extraction method, denoted as F-Dipe, is proposed to identify subcellular localization. F-Dipe, which is based on dipeptide pseudo amino acid composition method, improves the performance of multi-site prediction by increasing the focus information of proteins. Besides, convolution neural networks, denoted as CNN, is utilized to predict the subcellular localization of multi-site virus proteins. The multi-label k-nearest neighbor algorithm, denoted as MLKNN, is a base classifier to verify the performance of F-Dipe and CNN. The best overall accuracy of F-Dipe on dataset S from the predictor of MLKNN is 59.92%, higher than the accuracy of pseudo amino acid based features method, denoted as PseAAC, 57.14% and the best overall accuracy of F-Dipe on database S from the predictor of CNN is 62.3%, better than from the predictor of MLKNN 59.92%.

Keywords: Subcellular localization · Multi-site prediction · Feature extraction · F-Dipe · CNN

1 Introduction

The virus is closely related to various kinds of organisms and diseases of human. A virus is a small particle that infects cells in various kinds of organisms. As acellular organisms and obligate intracellular parasites, viruses can reproduce themselves only by invading and taking over other cells as they lack the cellular machinery for self-reproduction. Although viruses are acellular organisms, virus proteins are required to reside in different cellular compartments of the host cell or virus infected cell to perform their functions [1].

Viruses use the host synthetic machinery to replicate. They have evolved mechanisms to exploit the host nucleic acid replication and protein translation apparatus and have also developed strategies to evade humoral immune surveillance. Virus proteins require targeting to the appropriate subcellular compartments of the host cell to fulfill their roles [2]. In terms of the importance of viruses, knowledge of the subcellular localization of virus proteins is very helpful in studying the function of virus proteins and

© Springer International Publishing AG 2017
D.-S. Huang et al. (Eds.): ICIC 2017, Part II, LNCS 10362, pp. 606–615, 2017.
DOI: 10.1007/978-3-319-63312-1_53

designing antiviral drugs because it is closely related to their destructive tendencies and consequences [3].

With the avalanche of protein sequences generated in the post-genomic age, many computational methods are established for timely identifying their subcellular localization according to the sequence information alone [4]. They can be roughly classified into two series [5]. One is the "PLoc" series and the other is "iLoc" series. The "PLoc" series contains the six web-servers [6, 7] to deal with eukaryotic, virus, human, plant, Gram positive and Gram negative proteins, while the "iLoc" series contains the seven web-servers [8, 9] to deal with eukaryotic, virus, plant, human, Gram positive, Gram negative, and animal proteins, respectively. Studies have shown that improvement of prediction is achieved by developing feature extraction methods and classifiers.

In recent years, a great diversity of feature extraction methods have existed to improve the performance of prediction. (1) amino acid composition, denoted as AAC [10]; (2) homology-based [11]; (3) sorting-signals based [12]; (4) pseudo amino acid based features method [13]. All these methods have indicated excellent performance, but still can be improved. AAC is lack of location information of protein sequences; Homology-based is more suitable for high homology rather than low homology of protein sequences; PseAAC can reflect order information of sequences, but lack the ability of extraction of focus information. Therefore, F-Dipe is proposed to improve the ability of extraction of focus information. Besides, the present studies are devoted to develop a new and more powerful predictor for identifying subcellular localization of virus proteins.

In this study, a novel feature extraction method, F-Dipe, is proposed to improve the performance of multi-site prediction by increasing the focus information of proteins. Based on F-Dipe, the convolution neural networks is employed to predict subcellular localization of multi-site virus proteins.

2 Dataset

Dataset S in establishing Virus-mPloc is the benchmark dataset for the study [4]. The selection reasons of dataset S are as follows. (1) The dataset is established specialized for viral proteins. (2) None of proteins included in S has ≥25% pairwise sequence identity to any other in a same location. (3) They also contain proteins with more than one location and therefore could be used to do with subcellular localization of multi-site virus proeins [4].

The dataset S is classified into six locations and contains 207 different proteins, of which 165 belong to one subcellular location, 39 to two locations, 3 to three locations, and none to four or more locations. $S1$ is the subset for the subcellular location of "host viral capsid", $S2$ for "host cell membrane", and so forth (Table 1). The dataset can be formulated as Eq. 1.

$$S = S1 \cup S2 \cup S3 \cup S4 \cup S5 \cup S6 \tag{1}$$

Table 1. The viral protein benchmark dataset S taken from Virus-mPloc [4]

Subset	Subcellular location	Number of proteins
S1	Host viral capsid	8
S2	Host cell membrane	33
S3	Host endoplasmic reticulum	20
S4	Host cytoplasm	87
S5	Host nucleus	84
S6	Secreted	20
Total number of locative proteins		252
Total number of different proteins		207

Where $\cup$ represents the symbol for "union" in set theory.

Here, locative protein sequences and different protein sequences [4] can be formulated as Eq. 2.

$$N(loc) = N(dif) + \sum_{m=1}^{M} (m-1)N(m) \tag{2}$$

Where $N(loc)$ is the number of total locative proteins. $N(dif)$ is the number of total different proteins. The m is the number of locations where the specific protein is identified, $N(m)$ is the number of proteins which are identified in m mocaltions.

3   Methods

3.1   F-Dipe

F-Dipe increases the focus information by moving windows to extract features.

Firstly, the number of amino acid residues of 20 types in the given protein can be calculated by Eq. 3. Then, the number of residues is in standard conversion formulated as Eq. 4.

$$R = [r_1, r_2, r_3, \ldots \ldots, r_i, \ldots \ldots, r_{20}] \tag{3}$$

Where, r_i is the number of $i\text{-}th$ residue of every protein sequence of all the proteins.

$$r_i^* = \frac{r_i - \mu}{\sigma} \tag{4}$$

Where, r_i^* is the standard conversion of r_i. μ denotes the mean of r_i and σ represents the standard deviation of r_i.

Then, the spacing between two windows and window size can be set by users. The Window size is smaller than the minimum length of all protein sequences. Two better protein subsequences are combined to create a new database according to prediction results. The new database has repetitive information that is the focus information. Focus information of the new database improves the performance of prediction of subcellular localization.

Lastly, dipeptide pseudo amino acid composition method, denoted as Dipe, is used for the new database. Dipe will generate 400 components, i.e. AA, AC, ..., YW, YY. The number of 400 components needs to be calculated and then be in standard conversion.

3.2   CNN

CNN has the ability of learning a hierarchical representation of the input data without requiring any effort to design handcrafted features. Besides, different layers of the network are capable of different levels of abstraction and capture different amount of structure from the patterns present in the image [14]. Just for these reasons, CNN has demonstrated remarkable performance in image classification and attracted a lot of attention in recent years [15–17].

In this paper, CNN is proposed to identify subcellular localization of virus proteins.

The structure and relevant parameters of CNN are described as a profile. Input dimension, output dimension, file path of training dataset and testing dataset, sample number of training dataset and testing dataset, and so forth are contained in the first part of profile. Then, Parameters of learning rate, learning sample number, and layers structure are set in the second part of profile.

Firstly, the sequences of virus proteins are converted to 400-dimensional feature vector by F-Dipe. The 400-dimensional feature vector is the result of initial feature extraction. Then, it is sent into the predictor of CNN. The structure of CNN is described as Fig. 1.

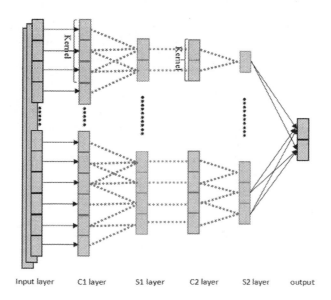

Input layer   C1 layer   S1 layer   C2 layer   S2 layer   output

Fig. 1. The structure of CNN

Where, Fig. 1 shows the structure of CNN. The number of layers is six. The layers are input layer, convolutional layer (C1 layer), pooling layer (S1 layer), convolutional layer (C2 layer), pooling layer (S2 layer), output layer, respectively. The 400-dimensional feature vectors of training dataset are sent to C1 layer. The convolution kernel is 3 and the feature map is 5 of C1 layer. The convolution kernel is 2 and the feature map is 8 of C2 layer. Thus, the number of neurons can be calculated by Eq. 5.

$$N = (I - C_k + 1) * F_m \tag{5}$$

Where, N represents the number of neurons. I is the input dimension of former layer. C_k is the convolution kernel. F_m is the feature map.

Therefore, the number of neurons of C1 layer is 1990 and the number of neurons of C2 layer is 3176. The S1 layer and S2 layer are designed to maintain the integrity of structure of CNN. They don't play an important role in performance of prediction of subcellular localization. CNN is utilized for binary classification. Thus, the number of output neurons is two.

The input layer is the first layer and it has a tanh activation function. The output layer is the last layer and it has a softmax activation function for binary classification. C1 layer and C2 layer is the second and the third feature extraction. The activation functions of convolutional layers are rectified linear units. The activation functions of pooling layers are none. The activation function of output layer is softmax.

The log-likelihood cost function of CNN can be described as Eq. 6.

$$C = - \sum_k y_k \log a_k \tag{6}$$

Where, a_k represents the output of the k-th neuron in CNN algorithm. y_k is the real value of the k-th neuron. The value of y_k is one or zero.

3.3 MLKNN

MLKNN is a base classifier to verify the performance of F-Dipe and CNN. MLKNN is a simple non-parametric multi-label classifier, which uses the k-nearest neighbor algorithm to collect the category tag information of neighbor samples, and exploits the principle of maximum posterior probability to inference the no example of label set [18]. MLKNN can be formulated as Eqs. 7 and 8.

$$T_j = \sum_{(x,Y) \in N(X)} \{y_j \in Y\} \tag{7}$$

Where, T_j denotes the number of neighbors of x part of the $N(x)$ class. y_j is a category label and Y is the category label set.

$$h(x) = \left\{ y_j \left| \frac{P(H_j|T_j)}{P(\neg H_j|T_j)} \right| > 0.5, 1 \leq j \leq q \right\}$$ (8)

Where H_j represents this event about x containing the category of y_j. $P(H_j|y_j)$ represents the posterior probability established H_j that when $N(x)$ includes the number samples of T_j with a category label y_j.

4 Evaluation Functions

A new scale, the so-called "absolute true" overall accuracy, to provide a more intuitive and easier-to-understand measurement and reflect the accuracy of a predictor, can be formulated as Eq. 9 [19].

$$\Lambda = \frac{\sum_{i=1}^{N} \Delta(i)}{N}$$ (9)

Where Λ is the absolute true rate, N denotes the number of entire proteins investigated, and $\Delta(i) = 1$ or $\Delta(i) = 0$.

Where, all the subcellular locations of the $i\text{-}th$ protein will be tested. Every subcellular location of the $i\text{-}th$ protein is correctly predicted, $\Delta(i) = 1$, otherwise, $\Lambda(i) = 0$.

5 Experiments and Results

In this study, the spacing between two windows and window size are set to one and thirty respectively by a series of experiments. Thirty window size generates 24 groups, and thus, the database is divided into 24 groups. The starting position of the first group is the zero position of the original database. The starting position of the second group is the first position of the original database, and so forth. The number of residues of every group can be calculated by Eqs. 3 and 4. The overall accuracy of every group can be shown in Table 2.

The best overall accuracy of groups is 57.54%, better than of original dataset 55.16%. Group 5 and group 7 demonstrate the best overall accuracy of groups and they are combined to create a new database. For example, group 5 {EFGHI} and group 7 {GHIKL} are converted to a new database {EFGHIGHIKL}. Thus, the extraction of focus information is {GHI}.

Table 2. Overall accuracy of 24 groups

Group number	Overall accuracy of every group
1	50.00%
2	45.63%
3	50.00%
4	54.76%
5	**57.54%**
6	53.57%
7	**57.54%**
8	47.22%
9	55.56%
10	46.43%
11	44.05%
12	41.27%
13	49.21%
14	50.00%
15	43.25%
16	50.79%
17	50.79%
18	51.98%
19	55.56%
20	56.35%
21	46.83%
22	46.03%
23	46.03%
24	44.05%

As shown in Table 3, AAC and Dipe of the new database are better than two methods of original database. Methods of the original database cannot show a better performance because it contains the redundant information. The new database increases the focus information of subsequences that is equivalent to increasing the weight of key residues.

Table 3. AAC and Dipe of original database and the new database

AAC of original database	F-AAC	Dipe of original database	F-Dipe
55.16%	**58.33%**	54.76%	**59.92%**

Then, F-Dipe is compared with PseAAC [20]. Two methods, F-Dipe and PseAAC, are utilized to identify the subcellular localization of multi-site virus proteins on the new database by MLKNN.

The result of PseAAC is obtained by a web-server called PseAAC at http://www.csbio.sjtu.edu.cn/bioinf/PseAAC/#. The ω is 0.05. The λ is 40. F-Dipe achieves 59.92% accuracy and PseAAC of the web-server shows 57.14% accuracy for MLKNN algorithm. Therefore, F-Dipe demonstrates a better performance of multi-site prediction.

As shown in Table 4, the number of correct prediction of every location of virus proteins is calculated by Eq. 9. The best overall accuracy of F-Dipe on database S from the predictor of CNN is 62.3%, better than from the predictor of MLKNN 59.92% in the majority of locations.

Table 4. Overall accuracy of MLKNN and CNN based on F-Dipe

Subcellular location	Overall accuracy	
	MLKNN	CNN
Viral capsid	7/8 = 87.5%	6/8 = 75%
Host cell membrane	12/33 = 36.36%	15/33 = 45.45%
Host endoplasmic reticulum	11/20 = 55%	7/20 = 35%
Host cytoplasm	49/87 = 56.32%	57/87 = 65.52%
Host nucleus	59/84 = 70.24%	59/84 = 70.24%
Secreted	13/20 = 65%	13/20 = 65%
Overall accuracy	151/252 = 59.92%	157/252 = 62.30%

6   Conclusion and Discussion

In this study, a novel feature extraction method, F-Dipe, is utilized to improve the performance of prediction of subcellular localization of virus proteins. F-Dipe method demonstrates a superior performance than PseAAC method of web-server for the predictor of MLKNN. Thus, the approach of increasing focus information of proteins is effective on the prediction of subcellular localization. Besides, CNN also obtains a higher success rate based on F-Dipe than MLKNN based on F-Dipe in the majority of locations. Thus, multiple feature extraction of CNN is critical for the performance of multi-site prediction [21–24, 29].

F-Dipe has demonstrated good performance of subcellular localization, but still can be improved. For example, the size of appropriate window, the spacing between two windows can be considered. The size of them is always set by experiments and experience. Besides, due to the index growth of complexity, CNN is employed in two classification. If CNN can be converted into the predictor of multi-site of multi-classification and has relative low complexity, the performance of multi-site prediction may be more excellent [25–28].

Acknowledgment.   This research was supported by the National Key Research And Development Program of China (No. 2016YFC0106000), National Natural Science Foundation of China (Grant No. 61302128, 61573166, 61572230, 61671220, 61640218), the Youth Science and Technology Star Program of Jinan City (201406003), the Natural Science Foundation of Shandong Province (ZR2013FL002), the Shandong Distinguished Middle-aged and Young Scientist Encourage and Reward Foundation, China (Grant No. ZR2016FB14), the Project of Shandong Province Higher Educational Science and Technology Program, China (Grant No. J16LN07), the Shandong Province Key Research and Development Program, China (Grant No. 2016GGX101022), Research Fund for the Doctoral Program of University of Jinan (No. XBS1604).

References

1. Wang, X., Li, G.Z., Lu, W.C.: Virus-ECC-mPLoc: a multi-label predictor for predicting the subcellular localization of virus proteins with both single and multiple sites based on a general form of Chou's pseudo amino acid composition. Protein Pept. Lett. **20**, 309–317 (2013)
2. Scott, M.S., Oomen, R., Thomas, D.Y., Hallett, M.T.: Predicting the subcellular localization of viral proteins within a mammalian host cell. Virol. J. **3**, 24 (2006)
3. Accquaah-Mensah, G.K., Leach, S.M., Guda, C.: Predicting the subcellular localization of human proteins using machine learning and exploratory data analysis. Genomics Proteomics Bioinform. **4**(2), 120–133 (2006)
4. Xiao, X., Wu, Z.C., et al.: iLoc-Virus: a multi-label learning classifier for identifying the subcellular localization of virus proteins with both single and multiple sites. J. Theor. Biol. **284**(1), 42–51 (2011)
5. Chou, K.C.: Impacts of bioinformatics to medicinal chemistry. Med. Chem. **11**(3), 218–234 (2015)
6. Ji, Z., Wu, D., Zhao, W., et al.: Systemic modeling myeloma-osteoclast interactions under normoxic/hypoxic condition using a novel computational approach. Sci. Rep. **5**, 13291 (2015)
7. Wang, B., Zhang, J., Chen, P., et al.: Prediction of peptide drift time in ion mobility mass spectrometry from sequence-based features. BMC Bioinform. **14**(8), S9 (2013)
8. Shen, H.B., Chou, K.C.: A top-down approach to enhance the power of predicting human protein subcellular localization: Hum-mPLoc 2.0. Anal. Biochem. **394**(2), 269–274 (2009)
9. Chou, K.C., Shen, H.B.: A new method for predicting the subcellular localization of eukaryotic proteins with both single and multiple sites: Euk-mPLoc 2.0. PLoS ONE **5**(4), e9931 (2010)
10. Chou, K.C., Wu, Z.C., Xiao, X.: iLoc-Hum: using the accumulation-label scale to predict subcellular locations of human proteins with both single and multiple sites. Mol. BioSyst. **8**(2), 629–641 (2012)
11. Huang, D.S.: Systematic Theory of Neural Networks for Pattern Recognition (in Chinese). Publishing House of Electronic Industry of China, Beijing (1996)
12. Wu, Z.C., Xiao, X., Chou, K.C.: iLoc-Gpos: a multi-layer classifier for predicting the subcellular localization of singleplex and multiplex Gram-positive bacterial proteins. Protein Pept. Lett. **19**(1), 4–14 (2012)
13. You, Z.-H., Lei, Y.-K., Huang, D.S., Zhou, X.: Using manifold embedding for assessing and predicting protein interactions from high-throughput experimental data. Bioinformatics **26**(21), 2744–2751 (2010)
14. Deng, Y., Luo, Y.L., et al.: Effect of different drying methods on the myosin structure, amino acid composition, protein digestibility and volatile profile of squid fillets. Food Chem. **171**(15), 168–176 (2015)
15. Dehzangi, A., Heffernan, R., et al.: Gram-positive and gram-negative protein subcellular localization by incorporating evolutionary-based descriptors into Chou's general PseAAC. Theor. Biol. **364**, 284–294 (2015)
16. Emanuelsson, O., Nielsen, H., et al.: Predicting subcellular localization of proteins based on their N-terminal amino acid sequence. Mol. Biol. **300**(4), 1005–1016 (2000)
17. Shen, H.B., Chou, K.C.: PseAAC: a flexible web server for generating various kinds of protein pseudo amino acid composition. Analyt. Biochem. **373**(2), 386–388 (2007)
18. Milletari, F., Ahmadi, S.A., Kroll, C., et al.: Hough-CNN: deep learning for segmentation of deep brain regions in MRI and ultrasound. Comput. Vis. Image Underst. (2017). doi:10.1016/j.cviu.2017.04.002

19. Huang, D.S., Yu, H.-J.: Normalized feature vectors: a novel alignment-free sequence comparison method based on the numbers of adjacent amino acids. IEEE/ACM Trans. Comput. Biol. Bioinf. **10**(2), 457–467 (2013)
20. Ji, Z., Wu, G., Hu, M.: Feature selection based on adaptive genetic algorithm and SVM. Comput. Eng. **14**, 072 (2009)
21. Yu, S.Q., Jia, S., Xu, C.Y.: Convolutional neural networks for hyperspectral image classification. Neurocomputing **219**(5), 88–98 (2017)
22. Han, S.Y., Chen, Y.H., Tang, G.Y.: Sensor fault and delay tolerant control for networked control systems subject to external disturbances. Sensors **17**(4), 700 (2017)
23. Han, S.Y., Zhang, C.H., Tang, G.Y.: Approximation optimal vibration for networked nonlinear vehicle active suspension with actuator time delay. Asian J. Control (2017). doi: 10.1002/asjc.1419
24. Zhang, M.L., Zhou, Z.H.: ML-KNN: a lazy learning approach to multi-label learning. Pattern Recogn. **40**(7), 2038–2048 (2007)
25. Xiao, X., Wu, Z.C., et al.: A multi-label classifier for predicting the subcellular localization of gram-negative bacterial proteins with both single and multiple sites. PLoS ONE **6**, e20592 (2011)
26. Bao, W., Chen, Y., Wang, D.: Prediction of protein structure classes with flexible neural tree. Bio-Med. Mater. Eng. **24**(6), 3797–3806 (2014)
27. Ji, Z., Wang, B., Deng, S.P., et al.: Predicting dynamic deformation of retaining structure by LSSVR-based time series method. Neurocomputing **137**, 165–172 (2014)
28. Chou, K.C.: Pseudo amino acid composition and its applications in bioinformatics, proteomics and system biology. Curr. Proteomics **6**(4), 262–274 (2009)
29. Han, S.Y., Chen, Y.H., Tang, G.Y.: Fault diagnosis and fault-tolerant tracking control for discrete-time systems with faults and delays in actuator and measurement. J. Franklin Inst. **354**(12), 4719–4738 (2017)

Improved Convolutional Neural Networks for Identifying Subcellular Localization of Gram-Negative Bacterial Proteins

Lei Wang[1,2], Dong Wang[1,2(✉)], Yaou Zhao[1,2], and Yuehui Chen[1,2(✉)]

[1] School of Information Science and Engineering,
University of Jinan, Jinan 250022, China
3027630499@qq.com,
{ise_wangd,ise_zhaoyo,yhchen}@ujn.edu.cn
[2] Shandong Provincial Key Laboratory of Network
Based Intelligent Computing, Jinan 250022, China

Abstract. Prediction of subcellular localization of Gram-negative bacterial proteins plays a vital role in the development of antibacterial drugs. Computational approaches have made remarkable progress in bacterial protein subcellular localization, but disadvantages still exist. Recently, deep learning has received significant attention in bioinformatics and one of the key steps in prediction of subcellular localization is developing a powerful predictor. Therefore, improved convolutional neural networks (ICNN) is used to improve the performance of multi-site prediction. First of all, Amphiphilic pseudo amino acid based features (Ampseaac) is used to extract features. Then, compared to the multi-label k-nearest neighbor algorithm (MLKNN), ICNN is developed to identify the subcellular localization of Gram-negative bacterial proteins. The best overall accuracy of Ampseaac from ICNN predictor is 65.25%, better than MLKNN predictor 58.58%.

Keywords: Subcellular localization · Bacterial proteins · Ampseaac · ICNN

1 Introduction

Identifying the location where a protein residues within a cell plays a crucial role in cell biology, as it contributes to our knowledge about the molecular function of a protein, and the biological pathway in which it is involved, [1, 2] which will in turn provide insights into drug targets identification and drug design [3].

Bacteria have a highly organized internal architecture at the cellular level. Identifying the subcellular localization of bacterial proteins is critical to studying their functions and design antibacterial drugs. [4] The shapes of bacteria are in varied forms, such as spirals, rods and spheres. With the Gram-staining technique, bacteria are often divided into Gram-negative bacteria and Gram-positive bacteria. Bacteria can form complex relationships with other organisms, including parasitism, mutualism and

© Springer International Publishing AG 2017
D.-S. Huang et al. (Eds.): ICIC 2017, Part II, LNCS 10362, pp. 616–625, 2017.
DOI: 10.1007/978-3-319-63312-1_54

commensalism. [4] In terms of functions of bacteria, pathogenic bacteria can lead to a series of human diseases, such as foodborne illness, tuberculosis, leprosy, tetanus and typhoid fever [5], but over thousand types of bacteria in the normal human gut flora contribute to vitamins synthesis, gut immunity and sugars-to-lactic acid conversion. [6] Thus, studies of bacteria are of paramount significance for antibacterial drug design and basic research.

With the explosion of protein sequences generated in the post-genomic age, it is highly desired to develop computational methods for fast and accurately identifying subcellular localization of newly discovered bactcrial proteins based on their sequence information alone because this kind of knowledge will be very useful for screening candidates in drug design, or selecting proteins for a special target. [7] Although a series of predictors are developed for prediction of subcellular localization in various organisms, those that are specialized for handling Gram-negative proteins are only a few. [7] Besides, machine learning methods cannot achieve the effect of multiple feature extraction. Its learning degree is not sufficient and its structure is not complex. However, deep learning has advantage of network structure and multiple feature extraction. Therefore, ICNN is selected as our prcdictor in this paper.

In this study, amphiphilic pseudo amino acid based features (Ampseaac) [8] is used to extract features. Then, 60 dimensional feature vector is formed and sent into MLKNN and ICNN. MLKNN is used to identify the location of Gram-negative bacterial proteins while ICNN is utilized to implement multiple feature extraction and then identify the location.

2 Dataset

Dataset S in establishing Gneg-mPLoc contains 1392 different protein sequences and 1456 locative proteins. The reason why we select dataset S as the benchmark dataset is that the dataset is constructed specialized for bacterial proteins. In addition, none of proteins included in S has $\geq 25\%$ pairwise sequence identity to any other in a same location. [9] They also contain proteins with more than one location and hence can be utilized to deal with prediction of subcellular localization of Gram-negative bacterial proteins.

The dataset S contains 1392 different proteins, of which 1328 belong to one subcellular location, 64 to two locations, and none to three or more locations. The dataset is classified into 8 subcellular locations, and can be represented as Eq. 1.

$$S = S_1 \cup S_2 \cup S_3 \cup S_4 \cup S_5 \cup S_6 \cup S_7 \cup S_8 \tag{1}$$

Where S_1 represents the subset for the subcellular location of "cell inner membrane", S_2 for "cell outer membrane", S_3 for "cytoplasm", S_4 for "extracellular", S_5 for "fimbrium", S_6 for "flagellum", S_7 for "nucleoid", S_8 for "periplasm" (Table 1); while $\cup$ denotes the symbol for "union" in set theory.

The 1392 different proteins actually correspond to 1456 locative proteins. Brief description of them can be formulated as Eq. 2.

$$N(Loc) = N(Dif) + \sum_{m}^{M} (m-1)N(m) \tag{2}$$

Where $N(Loc)$ is the number of all locative proteins. $N(Dif)$ is the number of all different proteins. The m is the number of locations where the specific protein is predicted, $N(m)$ is the number of proteins which are identified in m mocaltions.

Table 1. The benchmark dataset S [7]

Subset	Subcellular location	Number of proteins
S_1	Cell inner membrane	557
S_2	Cell outer membrane	124
S_3	Cytoplasm	410
S_4	Extracellular	133
S_5	Fimbrium	32
S_6	Flagellum	12
S_7	Nucleoid	8
S_8	Periplasm	180
Total number of different proteins		1392
Total number of locative proteins		1456

3   Methods

3.1   Ampseaac

Different from pseudo amino acid composition (Pseaac), Ampseaac offers another sequence order information. Details of sequence order information can be described as Eq. 3.

$$\begin{cases} \psi_{2\theta-1} = \sum_{i=1}^{L-\theta} \frac{C^1(R_i, R_{i+1})}{L-\theta} \\ \psi_{2\theta} = \sum_{i=1}^{L-\theta} \frac{C^2(R_i, R_{i+1})}{L-\theta} \end{cases}, \theta = 1, 2, \ldots, \lambda, \, and \, \lambda < L \tag{3}$$

Where, θ denotes different ranks of sequence order correlation factors. Ψ_θ denotes the θ_{th} correlation factor and L represents the length of bacterial protein. $C^1(R_i, R_{i+1})$ and $C^2(R_i, R_{i+1})$ are formulated as Eq. 4.

$$\begin{cases} C^1(R_i, R_j) = C^1(R_i) * C^1(R_j) \\ C^2(R_i, R_j) = C^2(R_i) * C^2(R_j) \end{cases} \tag{4}$$

Where $C^1(R_i)$ and $C^2(R_i)$ are the values of hydrophobicity and hydrophilicity respectively. These values are utilized after a standard conversion, constructed by Eq. 6. Supposing a vector T, which can be described as Eq. 5.

$$T = t_1 t_2 t_3 t_4 \ldots \ldots t_n \tag{5}$$

Where, t_i is the i-th value of vector T and n denotes the total dimension of T.

$$t_i^* = \frac{t_i - \mu}{\sigma} \tag{6}$$

Where, t_i^* is the standard conversion of t_i. μ denotes the mean of v_i. σ represents the standard deviation of t_i.

3.2 ICNN

Deep learning has been widely utilized in bioinformatics, such as prediction of protein tertiary structural classes, sequence analysis, [10–12] and so on. Recently, CNN has achieved the state-of-the-art performance in image classification task. With the rebirth of CNN, various kinds of feature extractors stacked from low-level to high-level can be automatically learned from training data in an end-to-end manner. A great number of works have indicated that these learned feature extractors can be successfully transferred to other tasks. [13] In this paper, ICNN is utilized to identify subcellular localization of Gram-negative bacterial proteins.

The structure and related factors of ICNN [14–17] are set by experiments. Experiments follow a series of principles. For example, the size of convolution kernel in the lower layer is smaller than in the upper layer. This design strategy contributes to multiple feature extraction. Then, the feature map in the lower layer is bigger than in the upper layer. Feature maps is useful for feature extraction either. Besides, activation function is key to the speed and accuracy of training. Last but not least, data processing, related parameters and number of layers are also key to the prediction of subcellular localization.

First of all, Ampseaac is utilized to extract feature initially. The sequences of Gram-negative bacterial proteins are converted to 60-dimensional feature vector by Ampseaac. Then, 60-dimensional feature vector is sent into the predictor of ICNN. Details of structure in terms of principles from what has been said above are as follows in Fig. 1.

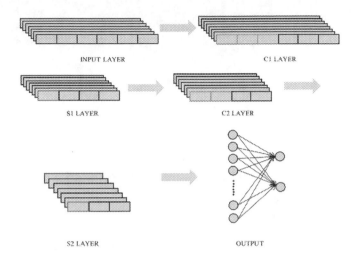

Fig. 1. Structure of improved convolutional neural networks

Where, Fig. 1. shows the general structure of ICNN. Fully connected layer, convolutional layer and pooling layer are concluded in the structure of ICNN.

The activation function of output layer is softmax and the log-likelihood cost function can be formulated as Eq. 7.

$$L = -\sum_{k} y_k \log a_k \qquad (7)$$

Where, L represents log-likelihood cost function. y_k is the real value of the k-th neuron and a_k denotes the result of predictor of ICNN. y_k is a binary value of one or zero.

Besides, some improvements of ICNN are designed. The micron is set to regulate learning rate by hessian matrix. Adaptive learning rate is set to learn better and speed up convergence. Adjustment of learning rate can be described as Eqs. 8 to 11.

First of all, second derivative of every layer needs to be computed. Second derivative of output layer to regulate weight can be formulated as Eq. 8.

$$\Delta \omega = \Delta \omega * (a_k - y_k)^2 \qquad (8)$$

Where, $\Delta \omega$ is the variation of weight. y_k is the real value of the k-th neuron and a_k denotes the result of predictor of ICNN. $(a_k - y_{k)}^2$ is the second derivative of output layer.

Second derivative of other layers to regulate weight can be described as Eq. 9.

$$\Delta \omega = \Delta \omega * (1 - y_k * y_k)^2 \qquad (9)$$

Where, $(1 - y_k * y_k)^2$ is the second derivative of other layers.

Then, the average of sum of the variation of weight can be showed as Eq. 10.

$$\overline{\Delta\omega_{sum}} = \frac{\Delta\omega * y_k * y_k}{HSN} \tag{10}$$

Where, $\overline{\Delta\omega_{sum}}$ is the average of sum of the variation of weight. $\Delta\omega$ is the variation of weight. HSN represents the sample number of Hessian matrix.

Lastly, the learning rate can be regulated by Eq. 11.

$$\eta^{'} = \frac{\eta}{\Delta\omega_{sum} + micron} \tag{11}$$

Where, $\eta^{'}$ is the learning rate after regulation and η is the last learning rate. Micron is a small positive number to regulate learning rate by hessian matrix. Typically, micron is between zero and one.

3.3   MLKNN

MLKNN is a multi-label classifier, which uses the k-nearest neighbor algorithm to collect the category tag information of neighbor samples, and exploits the principle of maximum posterior probability to inference the no example of label set. [18] MLKNN can be described as Eqs. 12 and 13.

$$M_i - \sum_{(x,Y)\in N(x)} \{y_i \in Y\} \tag{12}$$

Where M_i is the number of neighbors of x belonging to the $N(x)$ class.

$$h(x) = \left\{ y_i \left| \frac{P(H_i|M_i)}{P(\neg H_i|M_i)} > 0.5, 1 \le i \le q \right. \right\} \tag{13}$$

Where H_i is this event about x containing the category of y_i. $P(H_i|y_i)$ denotes the posterior probability constructed H_i that when $N(x)$ includes the number samples of M_i with a category label y_i.

The principle of MLKNN from what has been said above can be shown in Fig. 2 approximately.

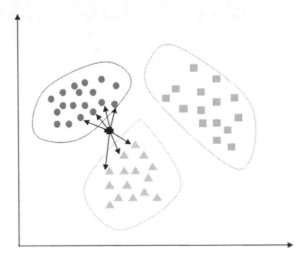

Fig. 2. The principle of MLKNN

4   Evaluation Functions

A new scale, the so-called "absolute rate" overall accuracy is set by Xiao and Wu to reflect the accuracy of a predictor. [19] The so-called "absolute rate" can be formulated as Eq. 14.

$$ATR = \frac{\sum\limits_{i=1}^{N} \Delta(i)}{N} \tag{14}$$

Where, *ATR* represents the absolute true rate and N is the number of proteins investigated. $\Delta(i)$ is a binary value of zero or one.

5   Experiments and Results

In this study, a profile of ICNN is designed for convenience of establishment of structure. In the profile, file path of training dataset and testing dataset is concluded. Learning rate and the number of learning sample are one. Micron is set to regulate learning rate by hessian matrix and its value is 0.3. Layer number is six, concluding forward layer, convolutional layer, pooling layer, convolutional layer, pooling layer and output layer, respectively. The activation function of input layer, convolutional layer are tanh and relu, respectively. The size of convolution kernel in second and forth layer are three and two. Feature maps in second and forth layer are five and eight by experiments.

The improvement performance of ICNN can be shown in Table 2.

Table 2. The iterations and corresponding likelihood

Iterations	likelihood
155001	0.0115846
156001	0.0115775
157001	0.0115765
158001	0.0117854
159001	0.0123622
160001	0.0112785
161001	0.0113042
162001	0.0111569
163001	0.0115152
164001	0.011236
165001	0.0110264
166001	0.0109135
167001	0.0108608
168001	0.0108244
169001	0.010754
170001	0.0108499
171001	0.010715
172001	0.0107471
173001	0.0108102
174001	0.010756

Where, Table 2 shows the effect of adaptive learning rate by mean squared error (MSE). Learning rate is regulated by micron and MSE shows a steady downward trend.

The overall accuracy of feature extraction method of Ampseaac in the predictor of ICNN and in the predictor of MLKNN is shown in Table 3.

Table 3. Overall accuracy of MLKNN and ICNN based on Ampseaac

Subcellular location	Overall accuracy	
	MLKNN	ICNN
Cell inner membrane	470/557 = 84.38%	444/557 = 79.71%
Cell outer membrane	57/124 = 45.97%	65/124 = 52.42%
Cytoplasm	247/410 = 60.24%	287/410 = 70%
Extracellular	60/133 = 45.11%	54/133 = 40.6%
Fimbrium	0/32 = 0%	13/32 = 40.63%
Flagellum	12/12 = 100%	12/12 = 100%
Nucleoid	7/8 = 87.5%	7/8 = 87.5%
Periplasm	0/180 = 0%	68/180 = 37.78%
Overall accuracy	853/1456 = 58.59%	950/1456 = 65.25%

6 Conclusion and Discussion

In this study, different from Pseaac, Ampseaac offers another sequence order information to extract feature initially. The predictor of ICNN shows better performance than of MLKNN in the majority of locations of Gram-negative bacterial proteins. There are a lot of reasons of this performance. First of all, multiple feature extraction of ICNN learn better than MLKNN. Then, structure of ICNN is more complicated than MLKNN. Besides, we do some improvement of ICNN to speed up convergence by micron. Adaptive learning rate contributes to the steady downward trend. All these reasons demonstrate the performance of ICNN [20–23, 26].

ICNN shows better performance of subcellular localization than MLKNN in the majority of locations of Gram-negative bacterial proteins, but still can be improved. If ICNN can be a predictor of multi-site of multi-classification, the performance may be more excellent [24, 25]. However, the structure of ICNN will be more complicated and the complexity will show index explosion. Our future will focus on how to reduce complexity of ICNN to make improvement.

Acknowledgment. This research was supported by the National Key Research And Development Program of China (No. 2016YFC0106000), National Natural Science Foundation of China (Grant No. 61302128, 61573166, 61572230, 61671220, 61640218), the Youth Science and Technology Star Program of Jinan City (201406003), the Natural Science Foundation of Shandong Province (ZR2013FL002), the Shandong Distinguished Middle-aged and Young Scientist Encourage and Reward Foundation, China (Grant No. ZR2016FB14), the Project of Shandong Province Higher Educational Science and Technology Program, China (Grant No. J16LN07), the Shandong Province Key Research and Development Program, China (Grant No. 2016GGX101022), Research Fund for the Doctoral Program of University of Jinan (No.XBS1604).

References

1. Cocco, L., Manzoli, L., Barnabei, O., et al.: Significance of subnuclear localization of key players of inositol lipid cycle. Adv. Enzyme Regul. **44**(1), 51–60 (2004)
2. Park, T.J., Gray, R.S., Sato, A., et al.: Subcellular localization and signaling properties of dishevelled in developing vertebrate embryos. Curr. Biol. **15**(11), 1039–1044 (2005)
3. Cai, Y.D., Chou, K.C.: Predicting protein localization in budding yeast. Bioinformatics **21** (7), 944–950 (2005)
4. Wan, S., Mak, M.W., et al.: Gram-LocEN: Interpretable prediction of subcellular multi-localization of Gram-positive and Gram-negative bacterial proteins. Chemometr. Intell. Lab. Syst. **162**(15), 1–9 (2016)
5. Harvey, R.A., Nau, C.C., Fisher, B.D.: Microbiology (3rd edn.), vol. 32, pp. 332–353. Lippincott Williams & Wilkins/Wolters Kluwer (2013)
6. O'Hara, A.M., Shanahan, F.: The gut flora as a forgotten organ. EMBO **7**(7), 688–693 (2006)
7. Shen, H.B., Chou, K.C.: Gneg-mPLoc: A top-down strategy to enhance the quality of predicting subcellular localization of Gram-negative bacterial proteins. J. Theoret. Biol. **264**, 326–333 (2010)

8. Huang, C., Yuan, J.Q.: Predicting protein subchloroplast locations with both single and multiple sites via three different modes of Chou's pseudo amino acid compositions. J. Theoret. Biol. **335**, 205–212 (2013)

9. Ji, Z., Wu, D., Zhao, W., et al.: Systemic modeling myeloma-osteoclast interactions under normoxic/hypoxic condition using a novel computational approach. Scientific reports, vol. 5 (2015)

10. Ji, Z., Wang, B., Deng, S.P., et al.: Predicting dynamic deformation of retaining structure by LSSVR-based time series method. Neurocomputing **137**, 165–172 (2014)

11. Shao, H., Peng, T., Ji, Z., et al.: Systematically studying kinase inhibitor induced signaling network signatures by integrating both therapeutic and side effects. PLoS ONE **8**(12), e80832 (2013)

12. Xiao, X., Wu, Z.C., et al.: A multi-label classifier for predicting the subcellular localization of gram-negative bacterial proteins with both single and multiple sites. PLoS ONE **6**, e20592 (2012)

13. Wang, B., Zhang, J., Chen, P., et al.: Prediction of peptide drift time in ion mobility mass spectrometry from sequence-based features. BMC Bioinformatics **14**(8), S9 (2013)

14. Ji, L., Pu, X.R., Qu, H., Liu, G.: One-dimensional pairwise CNN for the global alignment of two DNA sequences. Neurocomputing. **149**, 505–514 (2015)

15. Miki, Y., Muramatsu, C., et al.: Classification of teeth in cone-beam CT using deep convolutional neural network. Comput. Biol. Med. **80**(1), 24–29 (2017)

16. Perlin, H.A., Lopes, H.S.: Extracting human attributes using a convolutional neural network approach. Pattern Recogn. Lett. **68**, 250–259 (2015)

17. Yu, W., Yang, K., Yao, H., Sun, X., et al.: Exploiting the complementary strengths of multi-layer CNN features for image retrieval. Neurocomputing **237**, 235–241 (2016)

18. Bai, S.: Growing random forest on deep convolutional neural networks for scene categorization. Expert Syst. Appl. **71**, 279–287 (2017)

19. Pang, S., Yu, Z., et al.: A novel end-to-end classifier using domain transferred deep convolutional neural networks for biomedical images. Comput. Methods Programs Biomed. **140**, 283–293 (2017)

20. Han, S.Y., Chen, Y.H., Tang, G.Y.: Sensor fault and delay tolerant control for networked control systems subject to external disturbances. Sensors **17**(4), 700 (2017)

21. Han, S.Y., Zhang, C.H., Tang, G.Y.: Approximation optimal vibration for networked nonlinear vehicle active suspension with actuator time delay. Asian J. Control. (2017). doi:10.1002/asjc.1419

22. Zhang, M.L., Zhou, Z.H.: ML-KNN: A lazy learning approach to multi-label learning. Pattern Recogn. **40**(7), 2038–2048 (2007)

23. Bao, W., Chen, Y., Wang, D.: Prediction of protein structure classes with flexible neural tree. Bio-Med. Mater. Eng. **24**(6), 3797–3806 (2014)

24. You, Z.-H., Lei, Y.-K., Huang, D.S., Zhou, X.: Using manifold embedding for assessing and predicting protein interactions from high-throughput experimental data. Bioinformatics **26**(21), 2744–2751 (2010)

25. Xiao, X., Wu, Z.C., et al.: A multi-label classifier for predicting the subcellular localization of gram-negative bacterial proteins with both single and multiple sites. PLoS ONE **6**, e20592 (2011)

26. Han, S.Y., Chen, Y.H., Tang, G.Y.: Fault diagnosis and fault-tolerant tracking control for discrete-time systems with faults and delays in actuator and measurement. J. Franklin Inst.**354**(12), 4719–4738 (2017)

Emotion Recognition from Noisy Mandarin Speech Preprocessed by Compressed Sensing

Xiaoqing Jiang[1,2(✉)], Dapeng He[3], Xinghai Yang[1],
and Lingyin Wang[1]

[1] School of Information Science and Engineering, University of Jinan,
Jinan 250022, People's Republic of China
ise_jiangxq@ujn.edu.cn
[2] School of Electronics and Information Engineering,
Hebei University of Technology, Tianjin 300401, People's Republic of China
[3] Engineering Training Center, School of Mechanical Engineering,
University of Jinan, Jinan 250022, People's Republic of China

Abstract. Noisy speech emotion recognition is significant in Artificial Intelligence (AI) and Human-Computer Interaction (HCI). In this paper, Compressed Sensing (CS) theory is adopted in preprocessing procedure to remove the added noise on the samples in a mandarin emotional speech corpus. A novel binary tree structure is utilized in the designing of the multi-class classifier. Acoustic features are selected to build feature subset with better emotional recognizability. The recognition accuracies and corresponding confusion matrices of the original, noisy and reconstructed speech samples are compared. The recognition performance of the reconstructed samples is better than the samples contaminated by noise and similar as the performance of original samples. The experimental results show that Compressed Sensing is feasible and effective in noisy speech emotion recognition as a preprocess method.

Keywords: Speech emotion recognition · Compressed sensing · Feature selection · Support vector machine

1 Introduction

Speech, including both semantic and emotional information, is the most convenient and direct mode of communication. The emotional information usually implies the real meaning of the speech signals. So effective processing of emotional information influences the naturalness of Human-Computer Interaction (HCI) [1–3]. Nowadays, one of the problems to be solved in Speech Emotion Recognition (SER) is the adaptability and robustness of models built on the clean samples in the real world. In most research of SER, the emotional speech samples are often recorded or captured in quiet environment and pronounced clearly, but speech signals in the real applications usually carry various noises, which degrades the performance of SER system. Thus, the study about noisy SER is necessary in the real realization of Artificial Intelligence (AI) and HCI. The research of this paper mainly focuses on the elimination of the effects of noise in preprocessing procedure and the designing of the classifier structure in the recognition phase.

© Springer International Publishing AG 2017
D.-S. Huang et al. (Eds.): ICIC 2017, Part II, LNCS 10362, pp. 626–636, 2017.
DOI: 10.1007/978-3-319-63312-1_55

Noise influences the extraction of features and the recognition accuracy. Traditionally noise is reduced by various speech enhancement algorithms. The research about noisy SER mainly started from 2006. Schuller et al. selected feature subset from a 4 k feature set to recognize emotions from noisy speech samples [4]. You et al. proposed enhanced Lipschitz embedding to reduce the influence of noise [5]. Donoho et al. proposed Compress Sensing theory, which provides a promising method to handle the noise in sparse signals [6, 7]. Speech signals can be compressible under some proper orthogonal basis [8]. So Compressed Sensing is feasible as alternative preprocessing method before emotion recognition with the advantages of reduction of data rate and elimination of noise effects.

SER is a pattern recognition problem that can be handled by various classifiers. SVM is one of the most efficient binary classifier by mapping to the feature space [9]. In the design of multi-class classifier, the structure should be fully considered. Typically, there are one-versus-one, on-versus-all and hierarchy structures with different performance. So the structure design is important issue to achieve lower error accumulation as well as higher accuracies.

In this paper, emotional samples of a mandarin speech corpus including five basic emotions are studied from the aspect of speaker-dependent SER to test the performance of CS in the preprocessing of noisy speech signals. A novel binary tree structure is adopted to design the multi-class classifier. And the recognition results of the clean samples, the noisy samples and the corresponding reconstructed speech are compared and the confusion matrices are analyzed to demonstrate the effectiveness and robustness of the proposed method.

The rest of this paper are arranged as the followings: Sect. 2 introduces the basic conception of CS and analyze the compressible characteristics of the speech signals; Sect. 3 gives a brief description of the acoustic features and feature selection; Sect. 4 reviews concepts of SVM and the structure design of the multi-class classifier; Sect. 5 is about experimental results of the classifier structure analysis and noisy SER; Sect. 6 gives the conclusions.

2 Analysis About the Compressibility of Speech Signals

Compressed Sensing (CS), also named Compressed Sampling, can reconstruct sparse signals exactly under the key conception of sparsity and Restricted Isometry Property (RIP).

In N-dimensional space R^N and N is the number of samples, $x = [x(1), x(2), \ldots, x(N)]^T$ can be represented by the orthogonal basis vector $\{\psi_i | i = 1, 2, \ldots, N\}$:

$$x = \sum_{i=1}^{N} \alpha_i \psi_i = \Psi\alpha \tag{1}$$

where Ψ is the representation matrix, When the signal x only has k non-zero α_i coefficients and $k \ll N$, x can be considered k-sparse with sparse representation of Eq. (1).

Then the sensing process is:

$$y = \Phi x \tag{2}$$

where Φ is the measurement matrix and $y \in R^M$ is the measurement vector in M-dimensional space R^M ($M \ll N$). With Eq. (2), Eq. (1) can be rewritten in terms of α as:

$$y = \Theta \alpha \tag{3}$$

where $\Theta = \Phi \Psi$ is the reconstruction matrix, and α is k-sparse vector representing the projection coefficients of x in Ψ domain. When the signal is sparse and Θ satisfies the Restricted Isometry Property (RIP) condition, a sparse approximation solution to Eq. (3) can be obtained by minimizing the l_1-norm. The solving process of Eq. (3) is a convex optimization problem and α is the optimization variable in:

$$\begin{aligned} &\min \|\alpha\|_1 \\ &\text{subject to } y = \Phi \Psi \alpha \end{aligned} \tag{4}$$

The l_1 minimization problem in Eq. (4) can be considered as Basis Pursuit that can be efficiently solved by optimization techniques [10, 11]. In the above theory, if $x = [x(1), x(2), \ldots, x(N)]^T$ are speech signals, the basic conception of sparsity (or k-sparsity) and RIP must be satisfied under proper orthogonal basis.

According to the production principle of the speech signals, the most energy is carried by the voiced speech. The Discrete Cosine Transformation (DCT) is proper used as the orthogonal basis to achieve the sparse representations of speech signals because of its characteristics in the spectrum. Figure 1(a) (b) (c) and (d) give the waveforms and the sparsity of the voiced and unvoiced speech in DCT domain. The voiced speech has more obvious sparsity than unvoiced speech and has better reconstructed quality, which is illustrated in Fig. 1(e) and (f).

From the research of Jarvis Haupt et al., the compressible signals can be accurately recovered from random projection contaminated with noise [12]. Thus noisy voiced segments of speech signals can be reconstructed with high quality. At the same time, because of the noisy similarity in the production of unvoiced segments, the reconstruction of unvoiced speech of worse than the voiced ones, which makes a much clearer boundary between voiced and unvoiced speech in the reconstruction samples. In the extraction of features, the differentiation of voiced and unvoiced speech is significant to the precision of the feature values and most of the acoustic features are calculated based on the voiced speech. Thus the high reconstructed quality in voiced speech and the depressed reconstruction in unvoiced speech are beneficial to improve the emotional recognizability of the feature set in SER.

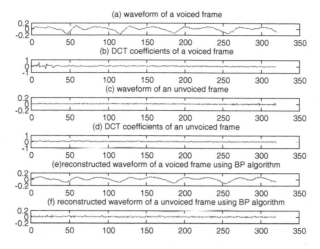

Fig. 1. The sparsity of speech signals in DCT domain

3   Acoustic Features and Feature Selection

3.1   Acoustic Features

In this paper, we extract pitch, energy, duration, formant, and MFCC and their statistics parameters, and total dimension the feature vector is 45. Table 1 lists the acoustic features in the following research.

Table 1. Acoustic features

Type	Feature	Statistic Parameters
Prosody Feature	Pitch	Maximum, Minimum, Range, Mean, Std, First Quartile, Median, Third Quartile, Inter-quartile range
	Energy	Maximum, Minimum, Range, Mean, Std, First Quartile, Median, Third Quartile, Inter-quartile range
	Duration	Total frames, Voiced frames, Unvoiced frames, Ratio of voiced frames versus unvoiced frames, Ratio of voiced frames versus total frames, Ratio of unvoiced frames versus total frames
Voice Quality Feature	Formant	F_1: Mean, Std, Median
		F_2: Mean, Std, Median
		F_3: Mean, Std, Median
Spectral Feature	MFCC	12 MFCC

3.2   Feature Selection

Feature selection is necessary to select the most optimal feature subset for SER. In the feature selection methods, the criteria based on information theory are common and effective.

In this paper, Double Input Symmetrical Relevance (DISR) method is adopted using the variable SR to measure the inherent properties between two inputted features [13]. For random variables X and Y, The Symmetric Relevance (SR) is defined as:

$$SR(X;Y) = \frac{I(X,Y)}{H(X,Y)} \tag{5}$$

where $I(X,Y)$ is the mutual information between random variable X and Y, and $H(X,Y)$ is their joint conditional entropy. The criterion of DISR algorithm is:

$$F_{\text{DISR}} = \arg \max_{f_q \in F_{-s}} \left\{ \sum_{f_p \in F_s} SR(\{f_p, f_q\}; C) \right\} \tag{6}$$

where F_s and F_{-s} are the selected subset and unselected subset and C stands for the classes of the emotion. SR takes use of the complementarity information between the double features f_p and f_q to select more effective feature subset with less attribution dimensions. The main advantage of DISR criterion is the selected complementary variable has much higher probability relevance on all of the double inputs in the subset.

4   SVM and Structure Design of the Classifier

The Support Vector Machine (SVM) is a discriminative classifier proposed for binary classification problem and based on the theory of structural risk minimization. In this section basic concepts and the correlative equations of SVM are introduced briefly.

Given l training patterns $\{(x_i, y_i)\}^l$ where x_i is the input vector and y_i is the class label of x_i. Then in the feature space induced by mapping function ϕ, we can find a hyperplane with the maximum margin, which can classify two classes with discriminant function:

$$f(x) = \langle w, \phi(x) \rangle + b \tag{7}$$

where w and the b are weight vector and offset that can be computed by solving a quadratic optimization problem:

$$\min_{w,b} \frac{1}{2} w^T w \tag{8}$$
$$\text{subject to } y_i(w^T \phi(x_i) + b) \geq 1, \ i = 1, 2, \ldots, l$$

In Eq. (8), the constraint is hard and in many cases such separation is impossible. To make the method more flexible and robust, a hyperplane can be constructed by relaxing constrains in Eq. (8), which leads to the following soft margin formulation with the introduction of slack variables ξ_i to account for misclassifications. The objective function and constraints can be formulated as:

$$\min_{w,b} \frac{1}{2} w^T w + C \sum_{i=1}^{l} \xi_i \tag{9}$$

$$\text{subject to } y_i(w^T \phi(x_i) + b) \geq 1 - \xi_i, \xi_i \geq 0. \, i = 1, 2, \cdots \cdots l$$

where l is the number of training patterns, C is a parameter which gives a tradeoff between maximum margin and classification error, and ϕ is a mapping from the input space to the feature space.

In this paper there are five emotions to be recognized and the kernel function of SVM is Radial Basis Function (RBF). A binary tree structure illustrated in Fig. 2(d) is adopted in the paper, which is different with the traditional one-versus-rest, one-versus-one, or hierarchy SVM structure show in Fig. 2(a), (b) and (c). The selection of the binary tree structure is based on the emotional recognition accuracies, which is verified in the following experiments.

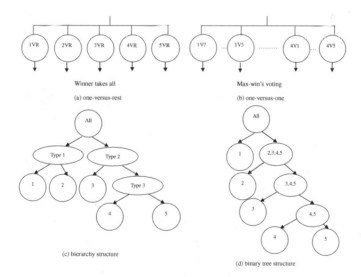

Fig. 2. Structures of multi-class classifiers

5 Experiments and Result Analysis

In this paper, samples in a mandarin emotional corpus are studied. The corpus was developed by Institute of Automation, Chinese Academy of Sciences and the utterances pronounced by native Chinese speaker were sampled at 16 kHz, quantized in 16 bit.

In this paper, we select 250 utterances of 50 different texts pronounced by a female speaker in 5 emotions including happy, angry, fear, sad and neutral. The training set includes 125 samples consisting 25 samples of the 5 emotions respectively, and the rest 125 ones are the testing samples. The procedure of the experiments is illustrated in the Fig. 3.

In Fig. 3, 20 dB Gaussian white noise is added on the original samples to produce the noisy samples. Random Gaussian matrix is used as measurement matrix and Basis Pursuit is adopted in the reconstruction. Acoustic features of the three samples are extracted and selected to train the SVM classifier respectively. The procedures in the dished lined diagram constitute the CS preprocessing. Three types of samples including original, noisy and reconstructed speech are tested. The recognition performances are compared and analyzed in the following experiments.

Except for the recognition accuracies, Root Mean Square Error (RMSE) and Maximum Error (MAXE) are used to evaluate the performance of SER. RMSE and MAXE are calculated by:

$$\text{RMSE} = \sqrt{\frac{1}{N} \sum_{i=1}^{N} e_i^2}, \ \text{MAXE} = \max\{e_i\} \tag{10}$$

where e_i is the recognition error of ith emotion, and N is the size of test set.

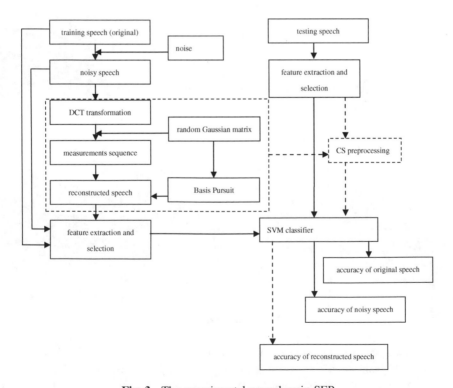

Fig. 3. The experimental procedure in SER

5.1   Classifier Structure Analysis

Recognition accuracies of original samples without feature selection by SVM classifiers of the four different structures are compared in Table 2, and the corresponding RMSE and MAXE are listed in Table 3. In Tables 2 and 3, one-Vs-rest and one-Vs-one is corresponding to the structures in Fig. 2(a) and (b) respectively, H-SVM denotes the hierarchy SVM structure shown in Fig. 2(c), and the binary tree (BT-SVM) classifier structure is illustrated in Fig. 2(d). The emotions are arranged as the followings: fear is "1", angry is "2", happy is "3", neutral is "4" and sad is "5". Through the comparison of data in Tables 2 and 3, we can see that the performance of the binary tree structure, whose total recognition accuracy is the best, is similar with one-Vs-one and better than H-SVM and one-Vs-rest. Meanwhile, the time complexities of different structures are similar because the number of samples is not large. Thus in the following experiments, the binary tree structured classifier is utilized.

Table 2. Recognition performance of classifiers with different structures

Classifier structure	Recognition accuracy (%)					
	Total	Angry	Fear	Happy	Neutral	Sad
H-SVM	74.4	96	0	88	92	96
one-Vs-one	78.4	96	48	100	80	68
one-Vs-rest	72	100	0	96	100	64
BT-SVM	79.2	96	72	96	92	40

Table 3. RMSE and MAXE of classifiers with different structures

Parameters	H-SVM	one-Vs-one	one-Vs-rest	BT-SVM
RMSE	0.4525	0.2879	0.4756	0.2993
MAXE	1	0.52	1	0.6

5.2   Recognition of Noisy Reconstruction

The SER experimental results are illustrated in Figs. 4 and 5. In both Figs. 4 and 5, the horizontal axis describes the different number of the selected features. When the number of selected feature is 45, it means that no feature selection algorithm is performed. Figure 4(a) illustrates the total recognition performance of SVM classifiers on the original speech, noisy speech, the reconstructed speech. The highest recognition accuracies are 80.8% (original), 71.2% (noisy), 81.6% (reconstructed). Figure 4(b) illustrates the accuracy of fear, the most confusable emotion, of original, noisy and reconstructed speech. Figure 5(a) and (b) are the curves of RMSE and MAXE corresponding to the Fig. 4(a).

From Figs. 4 and 5, it is clear that noise plays a negative role on the emotion recognition, because the total recognition accuracy of noisy samples is the worst and has the highest RMSE and MAXE. The reconstruction algorithm of CS theory can reduce the noise influences effectively and improve the recognition accuracy even higher than that of the original speech though the original samples utilized the selected optimal feature subset. This is an impressed action of CS theory in preprocessing procedure.

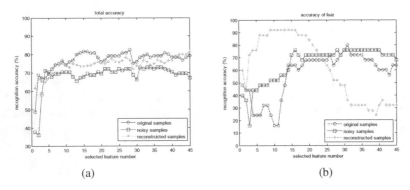

Fig. 4. Recognition accuracies of SER

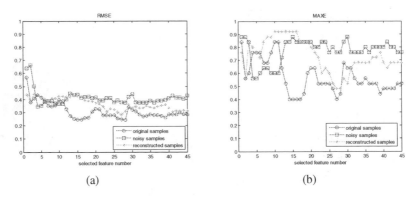

Fig. 5. RMSE and MAXE of SER

Table 4 is the detailed recognition accuracies of 5 emotions in the emotional corpus with the best selected feature numbers. And Table 5 is the confusion matrices with highest recognition accuracies corresponding to the situations in Table 4. Tables 4 and 5 show numerically that the detrimental influence of noisy in SER is reduced by CS effectively and feature selection plays an important role in analysis of the best performance in original and noisy samples.

Table 4. Recognition details of different samples with the best feature number

Type	Feature number	Recognition accuracy (%)				
		Total	Angry	Fear	Happy	Neutral	Sad
Original	16	80.8	96	68	100	88	52
Noisy	37	71.2	96	84	76	88	12
Reconstructed	45	81.6	96	32	88	96	96

Table 5. Confusion matrices corresponding to the highest recognition accuracies

Speech Type	Emotion	Recognition accuracy (%)				
		Angry	Fear	Happy	Neutral	Sad
Original Speech	Angry	96	0	4	0	0
	Fear	0	68	4	12	16
	Happy	0	0	100	0	0
	Neutral	0	0	12	88	0
	Sad	0	44	4	0	52
Noisy Speech	Angry	96	0	4	0	0
	Fear	0	84	0	4	12
	Happy	0	0	76	24	0
	Neutral	0	0	12	88	0
	Sad	0	88	0	0	12
Reconstructed Speech	Angry	96	0	4	0	0
	Fear	0	32	0	4	64
	Happy	0	0	88	12	0
	Neutral	0	0	4	96	0
	Sad	0	4	0	0	96

From the experimental data in Tables 2, 4 and 5, the confusion between fear and sad is the most critical factor influencing the overall recognition accuracies. In noisy environment this type of confusion is serious. For example, 84% testing sad samples are recognized as fear falsely when 37 features are selected, even though the total accuracy is the highest 71.2%. In original or reconstructed samples, the confusion is slightly less but cannot be eliminated by CS preprocessing or the classifier structures, which is the main limitation of the proposed method in this paper.

6   Conclusions

This paper adopts CS theory as a preprocessing procedure in noisy SER. Based on the comparison and analysis of experimental results, we can draw the following conclusions: (1) Compressed Sensing can remove the added noise from speech effectively in the preprocessing procedure; (2) Reconstructed speech samples have better emotional recognizability than noisy speech samples and even original speech samples, because of the high reconstruction quality of the voiced speech; (3) feature selection can improve the accuracies of SER and plays an important role in the analysis of methods and classifiers.

The proposed method also has aspects to be optimized, such as the reduction of confusion among emotions, automatic acquisition of optimal dimension of selected feature subset, the performance of the proposed method in various noisy environments with different SNR, and the better reconstruction of unvoiced speech. These challenging aspects need to be dealt with in our future work.

Acknowledgement. This work was supported by the National Natural Science Foundation of China (No. 61501204, No. 61601198), Shandong Province Natural Science Foundation (No. ZR2015FL010), and Science and Technology Program of University of Jinan (XKY1710).

References

1. Picard, R.W.: Affective Computing. MIT Press, Cambridge (1997)
2. Tao, J.H., Tan, T.N.: Affective computing: a review. In: Proceedings of 1st International Conference on Affective Computing and Intelligent Interaction, vol. 10, pp. 981–995 (2005)
3. Ayadi, M.E., Kamel, M.S., Karray, F.: Survey on speech emotion recognition: features, classification schems, and databases. Pattern Recogn. **44**(3), 572–587 (2011)
4. Schuller, B., Arsic, D., Wallhoff, F., Rigoll, G.: Emotion recognition in the noise applying large acoustic feature sets. Proc. Speech Prosody **5**, 128 (2006)
5. You, M.Y., Chen, C., Bu, J.J., Liu, J., Tao, J.H.: Emotion recognition from noisy speech. Proc. ICME **7**, 1653–1656 (2006)
6. Donoho, D.L.: Compressed sensing. IEEE Trans. Inf. Theory **52**(4), 1289–1306 (2006)
7. Candès, E.J.: The restricted isometry property and its implications for compressed sensing. C.R. Math. **346**(9–10), 589–592 (2008)
8. Sharma, P., Abrol, V., Sao, A.K.: Speech enhancement using compressed sensing. In: Proceeding of INTERSPEECH 2013, vol. 8, pp. 3274–3274 (2013)
9. Chapelle, O., Vapnik, V., Bousquet, O., Mukherjee, S.: Choosing multiple parameters for support vector machines. Mach. Learn. **46**(1), 131–159 (2002)
10. Chen, S., Donoho, D.L., Saunders, M.A.: Atomic decomposition by Basis Pursuit. Siam Rev. **43**(1), 129–159 (2006)
11. Saligrama, V., Zhao, M.: Thresholded basis pursuit: LP algorithm for order-wise optimal support recovery for sparse and approximately sparse signals from noisy random measurements. IEEE Trans. Inf. Theory **57**(3), 1567–1586 (2011)
12. Haupt, J., Nowak, R.: Signal reconstruction from noisy random projections. IEEE Trans. Inf. Theory **52**(9), 4036–4048 (2006)
13. Meyer, P.E., Schretter, C., Bontempi, G.: Information-theoretic feature selection in microarray dada using variable complementarity. IEEE J. Sel. Top. Sign. Proces. **2**(3), 261–274 (2008)

A Novel Adaptive Beamforming with Combinational Algorithm in Wireless Communications

Yue Zhao[1,2,3], Bo Ai[1,2(✉)], and Yiru Liu[1,2]

[1] State Key Laboratory of Rail Traffic Control and Safety,
Beijing Jiaotong University, Beijing 100044, People's Republic of China
aibo@ieee.org
[2] Beijing Engineering Research Center of High-Speed Railway Broadband
Mobile Communications, Beijing 100044, People's Republic of China
[3] School of Information Science and Engineering, University of Jinan,
Jinan 250022, Shandong, China

Abstract. A novel combinational adaptive beamforming algorithm is proposed for wireless communication applications. The significant advantage of the LMS (Least mean square) algorithm is its simplicity. Nevertheless its defect is that it has got relatively slow rate of convergence. The convergence rate of the RLS (Recursive least Squares) algorithm is faster than the LMS algorithm by an order of magnitude. However this advantage is gained at the cost of an increase in computational complexity. Considering the characteristics of two classic adaptive algorithms, a combinational algorithm is investigated in this paper by using combining merits of different algorithms as well as avoiding defects of them. The simulation was carried out and results show that the algorithm has comparable performance compared with above algorithms and faster convergence speed than LMS algorithm.

Keywords: Beamforming · Adaptive · Combinational · LMS · RLS

1 Introduction

Beamforming is widely employed to increase the system capacity and improve the frequency efficiency in wireless communications. It utilizes multiple antenna arrays to regulate the direction of main lobe by continuously weighting the magnitude and phase of each antenna signals [1–3]. Beamforming can be utilized at both transmitting and receiving end to achieve spatial gain respectively. There are two types of beamforming technologies, switched beamforming and adaptive beamforming. In switched beamforming, antenna arrays have plenty of fixed beam patterns from which the most appropriate beam is selected according to the reference signal. In adaptive beamforming, beam pattern can be dynamically steered to the desired direction and null can be controlled in the interference direction by means of adaptive algorithm. Typical adaptive beamforming architecture is shown as Fig. 1, where a uniform linear antenna array is regulated by several complex weights according to an adaptive algorithm. The signal transmitted or received by plurality of antenna elements is multiplied by weight

© Springer International Publishing AG 2017
D.-S. Huang et al. (Eds.): ICIC 2017, Part II, LNCS 10362, pp. 637–646, 2017.
DOI: 10.1007/978-3-319-63312-1_56

vector ω that adjusts the amplitude and phase of the signal accordingly. An adaptive algorithm is applied to minimize the error $e(n)$ between a reference signal $d(n)$ and the array output $y(n)$ and generate the weight coefficients of next iteration. $y(n)$ is the output of a transversal adaptive filter.

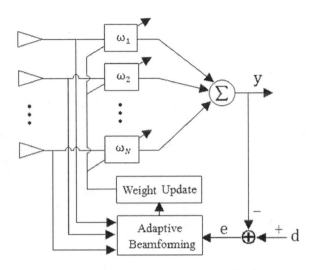

Fig. 1. Architecture of adaptive beamforming.

Godara L. C. summarized a large number of adaptive beamforming algorithm [1, 2]. Chryssomallis M. has also carried on the induction summary to adaptive beamforming algorithms [3]. In [4], a two-dimension direction of arrival estimation was researched and the scheme based on the unitary estimation of signal parameters was simulated. With the significantly increase of wireless communication rate requirement, beamforming in millimeter wave communications has attracted more and more attention for the past few years [5–9]. Kutty S. and Sen D. provided the evolution and improvements of adaptive beamforming technique in millimeter wave communications [5]. In [8], a new hybrid beamformer was designed for downlink multiuser massive MIMO (multiple input multiple output) systems. Cheng and Chen proposed a novel 3D beamforming algorithm combing conventional horizontal beamforming and elevation beamforming [10].The latest paper [11] evaluated adaptive beamforming techniques in mobile communication systems and envisioned future research areas of massive beamforming in 5G. [12] proposed a hybrid beamforming with unified analog beamforming based on the spatial covariance matrix knowledge of all user equipment. Similar as synchronization and other technologies in physical layer, beamforming is a key technology to ensure the quality of the received signal and enhance the capacity of system. Great efforts have been made to research and promote these critical technologies [13–16]. Beamforming in high-speed mobile environment is also one of the focuses of research. The main differences in research field for wireless communications between the high speed railway scenarios and the conventional public scenarios was discussed, as well as the characteristics of channel and the challenging techniques were analyzed in [17]. In addition, multi-stream beamforming was an effective solution to increase the throughput of high speed railway communication systems [18]. Professor Ai Bo etc. proposed mobile communications network architecture in varieties of railway scenarios and analyzed wireless coverage based on massive MIMO for railway communications to improve transmission and spectrum efficiency as well as the resulting technical challenges in [19]. In the existing research results, many scholars focused on the optimal beam selection and precoding technique. In this paper, a novel combinational adaptive beamforming algorithm is investigated and analyzed with comparing to the conventional adaptive algorithms.

2 Adaptive Beamforming

Adaptive beamforming has already played an important role in several areas ranging from radar, sensor arrays, acoustic signal processing, smart antenna and wireless communications [20, 21]. The value of adaptive antenna array which consist of N elements is able to produce N times antenna amplifier and bring $10 \lg N$ signal to noise ratio (SNR) improvement theoretically. The performances of existing adaptive algorithms were in-depth and systematically studied and analyzed in [20]. There are various adaptive beamforming algorithms investigated and improved to satisfy the needs of modern communication. Among them, the LMS algorithm and RLS algorithm are classical algorithms.

2.1 LMS Adaptive Algorithm

The LMS algorithm is a typical stochastic gradient algorithm that utilizes a stochastic gradient in a recursive computation of the Wiener filter for stochastic inputs [20]. A striking feature of LMS algorithm is its simplicity and it does not require calculation of the pertinent correlation functions and matrix inversion. The LMS algorithm consists of a filtering process and an adaptive process. In filtering process, the output of a transversal filter produced by input signal is measured as (1) and the evaluated error $e(n)$ is generated by (2).

$$y(n) = \boldsymbol{\omega}^H(n)\mathbf{u}(n) \tag{1}$$

$$e(n) = d(n) - y(n) \tag{2}$$

where, $y(n)$ is the output of the transversal filter. The signal received/transmitted by multiple antenna arrays is $\mathbf{u}(n)$. The complex vector $\boldsymbol{\omega}$ is the weight coefficient vector.H denotes complex conjugate transpose.

$$\boldsymbol{\omega}(n+1) = \boldsymbol{\omega}(n) + \mu\mathbf{u}(n)e^*(n) \tag{3}$$

An Adaptive process involves the filter self-regulation according to the evaluated error $e(n)$. Equation (3) is the recursive formulation for adaptive updating of the weight coefficient vector, in which μ is the step size factor. It is critical for LMS algorithm to choice an appropriate value of the step size factor. It can be given as

$$0 < \mu < \frac{2}{MS_{\max}} \tag{4}$$

M is the order of the adaptive transversal filter and $S_{\max}$ is the maximum power spectral density of $\mathbf{u}(n)$. Figure 2 depicts a signal-flow diagram representation of the LMS algorithm in the form of a feedback model. It can be seen that the LMS algorithm demands only $2M + 1$ complex multiplications and $2M$ complex additions per iteration. Namely the computational complexity of the LMS algorithm is $O(M)$.

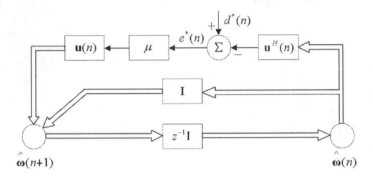

Fig. 2. Signal flow diagram of the LMS algorithm.

Because of the relatively slow speed of convergence, performance of the LMS algorithm will be affected by quick changing of environment. And the quality of the algorithm for covariance matrix eigenvalues spread of array signal is very sensitive. When the spread is large, it is difficult to converge.

2.2 RLS Adaptive Algorithm

The RLS algorithm makes uses of information contained in the input signal and its convergence speed is faster than that of the LMS algorithm by an order of magnitude. However, this advantage is realized at the cost of an increase in computational complexity [20]. The basic formulas of the RLS algorithm are as:

$$\mathbf{k}(n) = \frac{\lambda^{-1}\mathbf{P}(n-1)\mathbf{u}(n)}{1 + \lambda^{-1}\mathbf{u}^H(n)\mathbf{P}(n-1)\mathbf{u}(n)} \tag{5}$$

$$\xi(n) = d(n) - \boldsymbol{\omega}^H(n-1)\mathbf{u}(n) \tag{6}$$

$$\boldsymbol{\omega}(n) = \boldsymbol{\omega}(n-1) + \mathbf{k}(n)\xi^*(n) \tag{7}$$

$$\mathbf{P}(n) = \lambda^{-1}\mathbf{P}(n-1) - \lambda^{-1}\mathbf{k}(n)\mathbf{u}^H(n)\mathbf{P}(n-1) \tag{8}$$

The M-by-M matrix $\mathbf{P}(n)$ is referred to as the inverse correlation matrix and the M-by-1 vector $\mathbf{k}(n)$ is referred to as the gain vector. Equation (6) describes the filtering process of the algorithm, whereby the transversal filter is excited to compute the priori evaluated error $\xi(n)$ which is different from the posteriori evaluated error $e(n)$ in (2).

Equation (7) represents the adaptive process of the algorithm, whereby the weight coefficient vector is adjusted. The step size parameter in the LMS algorithm is replaced by the inverse of the correlation matrix of the input vector $\mathbf{u}(n)$. This alteration results in improvement of convergence speed which is an order of magnitude faster than that of the LMS algorithm. Figure 3 depicts a signal flow diagram presentation of the RLS algorithm in the form of a feedback model. An essential characteristic of the RLS algorithm is that the inversion of the correlation matrix is accomplished by the recursive method. Therefore, convergence speed is faster and it is not sensitive to the eigenvalues spread. It can be achieved the compromise between the convergence speed and computational complexity in the RLS algorithm [3].

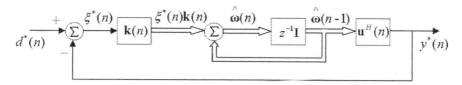

Fig. 3. Signal flow diagram of the RLS algorithm.

2.3   Proposed Combinational Algorithm

Great efforts have been made to improve the performance of adaptive beamforming, such as convergence speed, computation cost and implementation complexity. The combinational algorithm proposed in this paper integrates advantages of different adaptive algorithms as well as avoids defects of them.

The initial weights of the proposed algorithm are valued and updated by the LMS algorithm in the previous iteration. Based on array data information, sampling matrix inversion is calculated by using recursive method. Then the weights are continually updated by using the RLS algorithm.

The relationship between a priori evaluated error $\xi(n)$ and a posteriori evaluated error $e(n)$ is established as:

$$\begin{aligned} e(n) &= d(n) - [\omega(n-1) + \mathbf{k}(n)\xi^*(n)]^H \mathbf{u}(n) \\ &= d(n) - \omega^H(n-1)\mathbf{u}(n) - \mathbf{k}^H(n)\mathbf{u}(n)\xi(n) \\ &= (1 - \mathbf{k}^H(n)\mathbf{u}(n))\xi(n) \end{aligned} \tag{9}$$

The ratio of the posteriori evaluated error $e(n)$ to the priori evaluated error $\xi(n)$ is named the conversion factor $\gamma(n)$

$$\gamma(n) = \frac{e(n)}{\xi(n)}$$

$$= 1 - \mathbf{k}^H(n)\mathbf{u}(n) \tag{10}$$

The value of $\gamma(n)$ is completely determined by the gain vector $\mathbf{k}(n)$ and the input vector $\mathbf{u}(n)$.

3   Results and Discussion

The desired signal has direction of arrival -30^0 and two random interference signals come from the direction of arrival -45^0 and 0^0. Number of elements in the uniform linear array is 8 and the spacing between elements is 0.5λ. The impact of interference noise is taken into account by executing the simulations at -30 dB SNR, where the noise is determined as additive white Gaussian noise. In the RLS algorithm, the forgetting factor $\alpha = 0.9$. In the LMS algorithm, the step factor μ is 0.0114, calculated by

$$0 \le \mu \le \frac{1}{2Tr[R_{xx}]} \tag{11}$$

Beam patterns achieved by using the algorithm proposed in this paper, the LMS algorithm and RLS algorithm respectively are depicted in Fig. 4 in case of $SNR = -30$ dB. The array vector for N elements linear array is given by

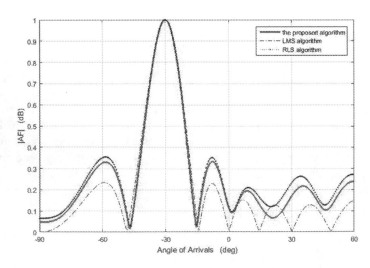

Fig. 4. Beam pattern comparison with the LMS and RLS algorithm, $SNR = -30$ dB.

$$AF = \sum_{n=1}^{N} e^{j(n-1)(kd \sin \theta + \delta)} \tag{12}$$

δ is the phase shift between arrays. It can be seen, the performance of three algorithms were similar.

Figure 5 depicts antenna radiation pattern of the proposed algorithm in case of $SNR = -30$ dB. It can be seen that the algorithm proposed in this paper was able to steer the main lobe in the desired direction of -30^0 and obtain null in the reference direction of -45^0 and 0^0. Figure 6 shows that the mean square error converges to zero after 60 iterations in the LMS algorithm, whereas the algorithm proposed in this paper only needs less than 20 iterations when $SNR = -30$ dB.

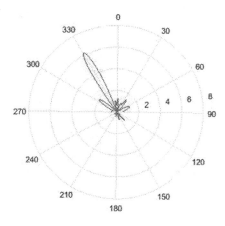

Fig. 5. Antenna radiation pattern of the proposed algorithm, $SNR = -30$ dB.

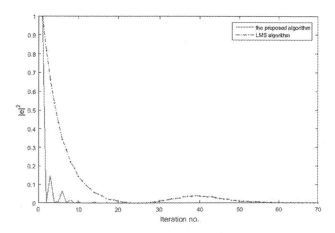

Fig. 6. Mean square error comparison with the LMS algorithm, $SNR = -30$ dB.

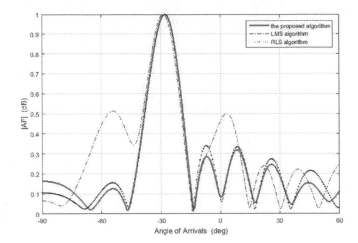

Fig. 7. Beam pattern comparison with the LMS and RLS algorithm, $SNR = -60$ dB

Figure 7 shows beam patterns achieved respectively by using the proposed algorithm, the LMS algorithm and RLS algorithm when $SNR = -60$ dB. Figure 8 shows the mean square errors of the proposed algorithm and the LMS algorithm when $SNR = -60$ dB. It is obvious in the figures when SNR declines to -55 dB ~ -60 dB, performance of the LMS algorithm deteriorates sharply. Especially, it fails in obtaining the null in the direction of -45^0 and 0^0 to suppress interference. Whereas the algorithm proposed in this paper can still work well, which illustrates this algorithm has advantages in robustness.

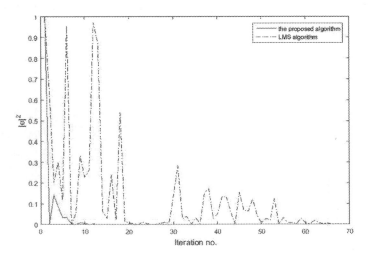

Fig. 8. Mean square error comparison with the LMS algorithm, $SNR = -55$ dB.

4   Conclusion

In this paper, a combinational algorithm taking advantage of merits of different algorithms is proposed for adaptive beamforming in wireless communication systems. The LMS algorithm is simple and easy to implement, but its convergence speed is slow relatively. Based on recursive covariance matrix inversion, the RLS algorithm has better convergence performance at the cost of increase of computation complexity. In proposed algorithm, the weights are valued and updated using the LMS algorithm in first iteration. After the initialization, recursive matrix inversion is calculated according to the RLS algorithm based on array data information. Simulation results of the proposed algorithm are compared with the LMS and RLS algorithm under different SNR conditions, which show that the proposed algorithm works well, even in the case that SNR is low.

Acknowledgements. This work is supported by the National key research and development program under Grant 2016YFB1200102-04, National S&T Major Project 2016ZX03001021-003, and National Natural Science Foundation of China under Grant U1334202).

References

1. Godara, L.C.: Applications of antenna arrays to mobile communications, I. Performance improvement, feasibility, and system considerations. Proc. IEEE **85**(7), 1031–1060 (1997)
2. Godara, L.C.: Applications of antenna arrays to mobile communications, II. Beamforming and direction-of-arrival considerations. Proc. IEEE **85**(8), 1195–1245 (1997)
3. Chryssomallis, M.: Smart antennas. IEEE Antennas Propag. Mag. **42**(3), 129–136 (2000)
4. Wang, T., Ai, B., He, R., Zhong, Z.: Two-dimension direction-of-arrival estimation for massive MIMO systems. IEEE Access **3**, 2122–2128 (2015)
5. Kutty, S., Sen, D.: Beamforming for millimeter wave communications: an inclusive survey. IEEE Commun. Surv. Tutor. **18**(2), 949–973 (2016)
6. Heath, R.W., González-Prelcic, N., Rangan, S., et al.: An overview of signal processing techniques for millimeter wave MIMO systems. IEEE J. Sel. Top. Sig. Process. **10**(3), 436–453 (2016)
7. Bogale, T.E., Le, L.B.: Beamforming for multiuser massive MIMO systems: Digital versus hybrid analog-digital. In: Global Communications Conference, pp. 4066–4071. IEEE (2014)
8. Jiang, J., Kong, D.: Joint user scheduling and MU-MIMO hybrid beamforming algorithm for mmWave FDMA massive MIMO system. Int. J. Antennas Propag. **2016**, 1–10 (2016)
9. Ai, B., et al.: On indoor millimeter wave massive MIMO channels: measurement and simulation. IEEE J. Sel. Areas Commun., TBD, 1–11 (2017)
10. Cheng, Y.S., Chen, C.H.: A novel 3D beamforming scheme for LTE-Advanced system. In: Network Operations and Management Symposium, pp. 1–6. IEEE (2014)
11. Chen, S., Sun, S., Gao, Q., Su, X.: Adaptive beamforming in TDD-based mobile communication systems: state of the art and 5G research directions. IEEE Wirel. Commun. **23**(6), 81–87 (2016)
12. Zhu, D., Li, B., Liang, P.: A novel hybrid beamforming algorithm with unified analog beamforming by subspace construction based on partial CSI for massive MIMO-OFDM systems. IEEE Trans. Commun. **65**(2), 594–607 (2017)

13. Ai, B., Yang, Z.X., et al.: On the synchronization techniques for wireless OFDM systems. IEEE Trans. Broadcast. **52**(2), 236–244 (2006)
14. Ai, B., Jian-Hua, G.E., Yong, W.: Symbol synchronization technique in COFDM systems. IEEE Trans. Broadcast. **50**(1), 56–62 (2004)
15. Ai, B., Jian-Hua, G.E., et al.: Frequency offset estimation for OFDM in wireless communications. IEEE Trans. Consum. Electron. **50**(1), 73–77 (2004)
16. Ai, B., Ge, J.H., et al.: Carrier frequency recovery technique in OFDM systems. Wireless Pers. Commun. **32**(2), 177–188 (2005)
17. Ai, B., et al.: Challenges toward Wireless Communications for High-Speed Railway. IEEE Trans. Intell. Transp. Syst. **15**(5), 2143–2158 (2014)
18. Cui, Y., Fang, X.: A massive MIMO-based adaptive multi-stream beamforming scheme for high-speed railway. EURASIP J. Wirel. Commun. Netw. **1**, 1–8 (2015)
19. Ai, B., Guan, K., Rupp, M., et al.: Future railway traffic services oriented mobile communications network. IEEE Commun. Mag. **53**(10), 78–85 (2015)
20. Haykin, S.: Adaptive Filter Theory, 3rd edn. Prentice-Hall Inc, Englewood Cliffs (1996)
21. Gross, F.: Smart Antennas for Wireless Communications. McGraw-Hill Professional, New York (2005)

Learning Bayesian Networks Structure Based Part Mutual Information for Reconstructing Gene Regulatory Networks

Qingfei Meng[1,2], Yuehui Chen[1,2(✉)], Dong Wang[1,2(✉)],
and Qingfang Meng[1,2]

[1] School of Information Science and Engineering,
University of Jinan, Jinan 250022, China
{yhchen, ise_wangd}@ujn.edu.cn
[2] Shandong Provincial Key Laboratory of Network
Based Intelligent Computing, Jinan 250022, China

Abstract. As a kind of high-precision correlation measurement method, Part Mutual Information (PMI) was firstly introduced into Bayesian Networks (BNs) structure learning algorithm in the paper. Compared to the general search scoring algorithm which set the initial network as an empty network without edge, our training algorithm initialized the network structure as an undirected network. That meant that our initial network identified the genes related to each other. And then the following algorithm only needed to determine the direction of the edges in the network. In the paper, we quoted the classic K2 algorithm based on Bayesian Dirichlet Equivalence (BDE) scoring function to search the direction of the edges. To test the proposed method, We carried out our experiment on two networks: the simulated gene regulatory network and the SOS DNA Repair network of Ecoli bacterium. And via comparison of different methods for SOS DNA Repair network, our proposed method was proved to be effective.

Keywords: Gene regulatory networks · Bayesian Networks · Part Mutual Information · K2 algorithm · BDE scoring function

1 Introduction

The advances on high-throughput technology enabled us to obtain a great deal of gene expression data during the past ten years [1]. Meanwhile, a large number of models and methods emerged for accurately reconstructing gene regulatory networks. For example, the boolean network, the system of differential equations, artificial neural networks, bayesian network, and so on. Among these, the Bayesian Networks become a main model in the research of gene regulatory networks for it's advantages, including handling incomplete data sets, fully combing the prior knowledge of the domain, visual image, and so on.

Generally there were two kinds of algorithms for Bayesian network structure learning: Search scoring method and correlation analysis method. The search scoring method usually introduced a scoring function $S(G, D)$ and then used this function to

© Springer International Publishing AG 2017
D.-S. Huang et al. (Eds.): ICIC 2017, Part II, LNCS 10362, pp. 647–654, 2017.
DOI: 10.1007/978-3-319-63312-1_57

evaluate each possible network structure for finding an optimal solution from all possible network structures. The correlation analysis method captured the dependency between nodes via independence test to study the network structure. In the study, we quoted the classic K2 algorithm [2] based on Bayesian Dirichlet Equivalence (BDE) scoring function to learn Bayesian Networks structure. The innovation of our method lied in introducing Part Mutual Information (PMI) to initialize the Bayesian Network structure.

The method based on the Mutual Information (MI) firstly obtained the mutual information matrix by calculating the mutual information between all possible gene pairs and then constructed the regulatory network through the mutual information matrix [3]. Although the method achieved good results, the limitation of this method was that it could not distinguish the direct and indirect association between genes with a high false positive. So Wang et al. [4] proposed a method based on conditional mutual information (CMI), which could distinguish the direct and indirect relationships between genes, and greatly reduced the false positive rate. However, the model based on conditional mutual information also had some limitations. For example Chen et al. [5] pointed out that there is a false negative problem based on conditional mutual information. When we measured the dependency between variables X and Y given variable Z, CMI could not correctly measure the direct association or dependency if X (or Y) was strongly associated with Z. In order to solve this problem, a method based on partial mutual information [5] was proposed, which solved the false positive problem of mutual information model and the false negative problem of conditional mutual information.

In the paper, we took advantage of partial mutual information to determine the relationship between variables for initialing Bayesian network structure. Then use K2 algorithm based on BDE scoring function to search the optimal network structure. Through two groups of experiments, our proposed method was illustrated to be effective.

2 Method

2.1 Part Mutual Information

Assuming that X and Y were two random variables, MI was defined on the basis of an extended Kullback–Leibler (KL) divergence D [6]:

$$MI(X;Y) = D(p(x,y)\|p(x)p(y)) = \sum_{x,y} p(x,y) \log \frac{p(x,y)}{p(x)p(y)} \tag{1}$$

where $p(x, y)$ was the joint probability distribution of X and Y and $p(x)$ and $p(y)$ were the marginal distributions of X and Y, respectively. Clearly, MI in Eq. 1 was evaluated against the 'mutual independence' of X and Y, which was defined as

$$p(x)p(y) = p(x,y) \tag{2}$$

CMI for variables X and Y given Z could further detect nonlinearly direct association and was defined as:

$$CMI(X;Y|Z) = D(p(x,y,z)\|p(x|z)p(y|z)p(z))$$
$$= \sum_{x,y,z} p(x,y,z) \log \frac{p(x,y|z)}{p(x|z)p(y|z)} \tag{3}$$

So the conditional independence of X and Y given Z, which was defined as:

$$p(x|z)p(y|z) = p(x,y|z) \tag{4}$$

To analogy Eqs. (2), (4), partial independence of X and Y given Z was defined as:

$$p^*(x|z)p^*(y|z) = p(x,y|z) \tag{5}$$

where $p^*(x|z)$ and $p^*(y|z)$ were defined [7] as

$$p^*(x|z) = \sum_y p(x|z,y)p(y), p^*(y|z) = \sum_x p(y|z,x)p(x) \tag{6}$$

So

$$PMI(X;Y|Z) = D(p(x,y,z)\|p^*(x|z)p^*(y|z)p(z)) \tag{7}$$

Or as

$$PMI(X;Y|Z) = \sum_{x,y,z} p(x,y,z) \log \frac{p(x,y|z)}{p^*(x|z)p^*(y|z)} \tag{8}$$

2.2 Bayesian Networks

The Bayesian Networks were a directed acyclic graph $B = (G, \Theta)$ which shown the probabilistic dependency relationship between variables. $G = (V, E)$ was a directed acyclic graph and Θ was the collection of conditional probability table (CPT). $V = (V1, \cdots, Vn)$ was a set of nodes and each node represented a field random variable Xi.

$$E = \{ <Xi, Xk > |Xi, Xk \in V, i \neq k\} \tag{9}$$

Where $<Xi, Xk>$ indicated the directed edge. $Xi \rightarrow Xk$ indicated the probabilistic dependency relationship between variables Xi and Xk. Xi was the father node of Xk. $Pa(Xk)$ indicated the parent node set of variable Xk.

For every variable Xk, it's value xk had the following parameter $\theta xk|pa(Xk) = P(xk|pa(Xk))$, which showed the probability of xk occurrence under the condition of $Pa(Xk)$. So the joint conditional probability distribution on a given set of variables of Bayesian Networks was as:

$$P(X1, X2, \cdots, Xn) = \prod_{i=1}^{n} P(Xi|Pa(Xi))$$ (10)

2.3 K2 Algorithm

Cooper [2] put forward K2 algorithm based on Bayesian Dirichlet Equivalence (BDE) scoring function to learn Bayesian Networks structure. The K2 algorithm used the way of hill-climbing search to learn the network structure.

According to prior knowledge, We first obtained the initial network. Then, Use the search operator including reducing edges, increasing edges and reserving edges to modify the current network to get candidate network. According BDE scoring function, Select the fractional optimal network to replace the current network, and then continue the search until obtaining the best network.

The BDE scoring function an approximation of the marginal likelihood function under the condition of large sample. The BDE scoring function was defined as:

$$S(G, D) = LgP(D|G) + LgP(G) - \frac{d}{2}\log m$$ (11)

Where $P(D|G)$ was an edge distribution of dataset D and was the probability averaging of data from D. $P(G)$ was the prior probability of the network structure G. $\frac{d}{2}\log m$ was a penalty function for sample size m. Because of the addition of a penalty term, the BIC scoring function avoided overfitting.

2.4 Evaluation Index

In our study, five criterions (True Positive Rate (TPR), False Positive Rate (FPR), Positive Predictive (PPV), Accuracy (ACC) and F-score) were used to test the performance of the proposed method. Their definition was given as follows:

$$TPR = \frac{TP}{TP + FN}$$ (12)

$$FPR = \frac{FP}{FP + TN}$$ (13)

$$PPV = \frac{TP}{TP + FP}$$ (14)

$$ACC = \frac{TP + TN}{TP + FP + TN + FN}$$ (15)

$$F - score = \frac{2PPV * TPR}{PPV + TPR} \qquad (16)$$

Where True Positive (TP) meant that edges in real networks were identified as edges in the model. False Positive (FP) meant that edges not in real networks were identified as edges in the model. True Negative (TN) meant that edges not in real networks were not identified as edges in the model. False Negative (FN) meant that edges in real networks were not identified as edges in the model.

3   Experimental Results and Analysis

3.1   Experiment 1: Simulated Gene Regulatory Network

Figure 1 showed a simulated gene regulatory network which was modeled by a S-system model [8]. The model consisted of n non-linear ordinary differential equations and the generic form of equation i is given as follows:

$$X_i'(t) = \alpha_i \prod_{j=1}^{N} X_j^{g_{ij}}(t) - \beta_i \prod_{j=1}^{N} X_j^{h_{ij}}(t) + \varepsilon$$

Where X_i was a vector element of dependent variable, N was the number of variables, α_i and β_i was vector elements of non-negative rate constants, g_{ij} and h_{ij} were matrix elements of kinetic orders and random Gaussian noises (ε) were added to each equation independently.

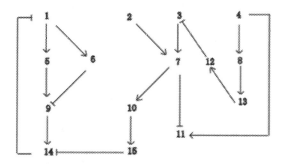

Fig. 1. A simulated gene regulatory network containing 15 genes

Table 1 listed the parameters of S-system containing 15 genes. The initial value is random and we respectively got the simulated data by 20 time points, .30 time points and 50 time points.

Table 1. Parameters of the S-system

α_i	1.0
β_i	1.0
$g_{i,j}$	$g_{1,14} = -0.1$, $g_{3,12} = -0.2$, $g_{5,1} = 1.0$, $g_{6,1} = 1.0$, $g_{7,2} = 0.5$, $g_{7,3} = 0.4$, $g_{8,4} = 0.2$, $g_{9,5} = 1.0$, $g_{9,6} = -0.1$, $g_{10,7} = 0.3$, $g_{11,4} = 0.4$, $g_{11,7} = -0.2$, $g_{12,13} = 0.5$, $g_{13,8} = 0.6$, $g_{14,9} = 1.0$, $g_{14,15} = -0.2$, $g_{15,10} = 0.2$, other $g_{i,j} = 0.0$
$h_{i,j}$	1.0 if $i = j$, 0.0 otherwise

From Table 2, we can see the TPR and FPR obtained by our proposed algorithm.

Table 2. Performances of the proposed model to the simulated data

Number of time points	TPR	FPR
20	0.63	0.57
30	0.77	0.41
50	1.0	0.12

3.2 Experiment 2: SOS DNA Repair Network of E.Coli Bacterium

The datasets from SOS DNA Repair network of E.coli bacterium [9] contained four experiments under various light intensities ($5Jm^{-2}$, $5Jm^{-2}$, $20Jm^{-2}$, $20Jm^{-2}$). Each experiment (http://www.weizmann.ac.il/mcb/UriAlon/ Papers/SOSData) consisted of 50 time points evenly spaced by 6 min and referred to eight genes: uvrD, lexA, umuD, recA, uvrA, uvrY, ruvA and polB. Figure 2 displayed the known real regulation among 8 genes.

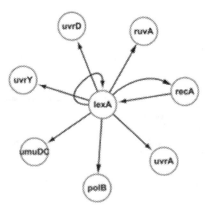

Fig. 2. The true SOS network with eight genes and 9 interactions

In the experiment, we firstly normalized the gene expression data for each gene to the interval [0, 1] using

$$x'_i(t) = \frac{x_i(t) - \min_i}{\max_i - \min_i}$$

Where $x_i(t)$ is the actual measured gene expression level of gene i at time point t. $\min_i$ and $\max_i$ respectively were the minimum and maximum of gene i expression level. Table 3 showed the detail comparison between proposed method and other methods including S-system [10], ODEs [11], RNN [12] and DBN [13]. The results showed that our proposed method performed better than other methods.

Table 3. Comparison of different methods for SOS dataset

Item	S-system	ODEs	RNN	DBN	Proposed Method
TP	5	6	5	4	8
FP	10	15	2	5	7
TPR	0.5556	0.6667	0.55556	0.4444	0.8889
FPR	0.20833	0.3125	0.041667	0.10417	0.2692
PPV	0.3333	0.28571	0.71429	0.44444	0.5333
ACC	0.6667	0.57895	0.80702	0.75439	0.7714
F-score	0.41667	0.4	0.625	0.44444	0.6667

4 Conclusions

In the paper, the Part Mutual Information was introduced into the training algorithm of the Bayesian Networks structure, which meant that the initial network has identified the genes related to each other and the later searching process with K2 algorithm only needed to determine the direction of the edges in the Bayesian Networks. We carried out our experiments on two networks: the simulated gene regulatory network and the SOS DNA Repair network of Ecoli bacterium. And via comparison of different methods for SOS DNA Repair network, our proposed method was proved to be effective.

Acknowledgment. This work was supported by the National Key Research and Development Program of China (2016YFC106000), the National Natural Science Foundation of China (Grant No. 61671220, 61640218, 61201428), the Shandong Distinguished Middle-aged and Young Scientist Encourage and Reward Foundation, China (Grant No. ZR2016FB14), the Project of Shandong Province Higher Educational Science and Technology Program, China (Grant No. J16LN07), the Shandong Province Key Research and Development Program, China (Grant No. 2016GGX101022).

References

1. Cho, K.H., Choo, S.M., Jung, S.H., et al.: Reverse engineering of gene regulatory networks. IET Syst. Biol. **1**(3), 149–163 (2007)
2. Cooper, G.F., Herskovits, E.: A Bayesian method for the induction of probabilistic networks from data. Mach. Learn. **9**(4), 309–347 (1992)
3. Ji, Z., Wu, D., Zhao, W., et al.: Systemic modeling myeloma-osteoclast interactions under normoxic/hypoxic condition using a novel computational approach. Sci. Rep. **5** (2015)
4. Wang, B., Zhang, J., Chen, P., et al.: Prediction of peptide drift time in ion mobility mass spectrometry from sequence-based features. BMC Bioinform. **14**(8), S9 (2013)
5. Altay, G., Emmertstreib, F.: Revealing differences in gene network inference algorithms on the network level by ensemble methods. Bioinformatics **26**(14), 1738–1744 (2010)
6. Bao, W., Chen, Y., Wang, D.: Prediction of protein structure classes with flexible neural tree. Bio-Med. Mater. Eng. **24**(6), 3797–3806 (2014)
7. Zhao, J., Zhou, Y., Zhang, X., et al.: Part mutual information for quantifying direct associations in networks. Proc. Nat. Acad. Sci. **113**(18), 5130–5135 (2016)
8. Schreiber, T.: Measuring information transfer. Phys. Rev. Lett. **85**(2), 461–464 (2000)
9. Anzing, D., Balduzzi, D., Grosse-Wentrup, M., Schölkopf, B.: Quantifying causal influences. Ann. Stat. **41**(5), 2324–2358 (2013)
10. Kimura, S., Ide, K., Kashihara, A.: Inference of S-system models of genetic networks using a cooperative coevolutionary algorithm. Bioinformatics **21**, 1154–1163 (2005)
11. Ronen, M., Rosenberg, R., Shraiman, B.I., et al.: Assigning numbers to the arrows: parameterizing a gene regulation network by using accurate expression kinetics. Proc. Natl. Acad. Sci. U.S.A. **99**(16), 10555 (2002)
12. Noman, N., Iba, H.: Reverse engineering genetic networks using evolutionary computation. In: Genome Informatics International Conference on Genome Informatics, PubMed, pp. 205–214 (2005)
13. Ji, Z., Wang, B., Deng, S.P., et al.: Predicting dynamic deformation of retaining structure by LSSVR-based time series method. Neurocomputing **137**, 165–172 (2014)
14. Kimura, S., Sonoda, K., Yamane, S., et al.: Function approximation approach to the inference of reduced NGnet models of genetic networks. BMC Bioinform. **9**(1), 23 (2008)
15. Xu, R., Wunsch, D.C., Frank, R.L.: Inference of genetic regulatory networks with recurrent neural network models using particle swarm optimization. IEEE/ACM Trans. Comput. Biol. Bioinform. **4**(4), 681–692 (2007)
16. Perrin, B.E., Ralaivola, L., Mazurie, A., Bottani, S., Mallet, J., Buc, D.F.: Gene network inference using dynamic bayesian networks. Bioinformatics **19**(Suppl. 2), 138–148 (2003)

Bilateral Filtering NIN Network for Image Classification

Jiwen Dong[1,2], Yunxing Gao[1,2], Hengjian Li[1,2(✉)], and Tianmei Guo[1,2]

[1] School of Information Science and Engineering,
University of Jinan, Jinan 250022, China
491282999@qq.com
[2] Shandong Provincial Key Laboratory of Network
Based Intelligent Computing, Jinan 250022, China

Abstract. A novel deep architecture bilateral filter NIN for classification tasks is proposed in the paper, in which the input image pixels using the bilateral filter and a multi-path convolution neural network are reconstructed. This network has two input paths, one is the original image and the other is the reconstructed image which independent on and complement each other. Therefore, the loss of foreground object texture and shape information can be reduced during the process of feature extraction from the complex background images. Then, the softmax classifier is employed to classify the extracted features. Experiments are demonstrated on CAFIR-100 dataset, in which some object's feature gradually disappear after pass through a series of convolution layers and average pooling layers. The results show that, Compared with NIN(network in net- work), the classification accuracy rate increased 0.6% on CIFAR-10 database, accuracy rate increased 0.27% on cifar-100 database.

Keywords: Convolutional neural network · Network in Network · Bilateral filter · Image classification

1 Introduction

With the continuous development of the Internet and handheld mobile terminals, they reduce more and more different forms of images. It is unrealistic to identify and sort out these images by artificial methods, so there is an urgent need for a method to automatically identify and automate them. CNN(Convolution Neural Network) [1] is widely used in the field of image classification [2, 3] nowadays because the convolution neural network image classification system has a higher recognition result for many large image datasets such as MINIST(Modified National Institute of Standards and Technology Dataset) [4], CIFAR-10 [5], CIFAR-100 [5] (they are labeled subsets of the dataset),SVHN [6] (The Street View House Numbers Dataset) and ImageNet [7] (is an image dataset organized according to the WordNet hierarchy, in which each node of the hierarchy is depicted by hundreds and thousands of images). In the 2012 ImageNet competition, Ilya Sutskever and Alex Krizhevsky first applied the deep learning model to CNN [1, 7]. They used an 8-layer deep CNN to image classification

© Springer International Publishing AG 2017
D.-S. Huang et al. (Eds.): ICIC 2017, Part II, LNCS 10362, pp. 655–665, 2017.
DOI: 10.1007/978-3-319-63312-1_58

and target location on a dataset which contains millions of web images divided into 1000 classes. This method had achieved very good results and won the championship. In the image classification task, the error rate is 15.3%, and the error rate is reduced by half compared to the 26.3% error rate of the runner-up manual feature design method. Subsequently, Google applied the deep CNN to Google Plus's image annotation and search, and won the ImageNet competition with the GoogleNet model in 2014. This model demonstrates that more convolution times and deeper layers can get a better structure [8]. In 2015, the championship model of ImageNet competition was Deep Residual Learning, it still used the deep convolution model, the deepest model reaching 152 layers. But different from the previous it had the bottleneck block structure in the model, which is a direct connection across several layers [9] (Fig. 1).

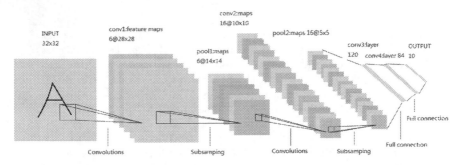

Fig. 1. Lecun et al. proposed a LeNet-5 [4] model based on CNN in 1995, it completed the United States more than 10% of the check recognition, which is one of the successful application of CNN.

CNNs are more likely to extract high-frequency information of images in actual image classification but the object's feature will gradually disappear after pass through a series of convolution layers and average pooling layers, which may lead to the final extraction of the feature vector can't be a good representation of the foreground object and classification effect decreased. Current solutions involve the use of rectified linear units (ReLU) to prevent vanishing gradients [1, 10, 11], because ReLU activates above 0 and its partial derivative is 1. Unfortunately, ReLU has a potential disadvantage. The constant 0 will block the gradient flowing through inactivated ReLUs, such that some units may never activate. Recently, the maxout network provided a remedy to this problem. Maxout Networks [12] is an improvement over the convolution activation function Relu. It has a very strong fitting ability, can be fitted with arbitrary convex function. A more intuitive explanation is that any convex function can be fitted by a piecewise linear function with arbitrary precision, and maxout is to take the maximum value of k hidden nodes, these hidden layer nodes are linear, so in different range of values, the maximum value can also be seen as piecewise linear. But maxout is equivalent to double the parameters of each neuron, this is easy to increase the parameters and increase the complexity of computing [13, 14]. The aforementioned methods are only focus on the improvement of the internal activation function of the

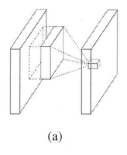 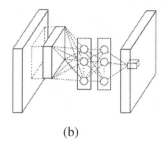

(a)   (b)

Fig. 2. Traditional convolution layer is shown in (a). The mlpconv layer includes a micro network is shown in (b).

network. Motivated by this fact, we directly consider the impact of the original image on the final result (Fig. 2).

In this study, we aimed to reduce the loss of foreground object texture and shape information during the process of feature extraction. Based on the NIN [15] structure, we employ a bilateral filter for feature extraction and refer to the proposed model as bilateral filter NIN. The bilateral filter NIN model uses two input paths become multi-path convolution neural network. First, we analyze the training curves of simple background images and complex background images, the conclusion is that complex background images are more likely to lose texture feature information during feature extraction, reserved is generally high frequency information such as a complex background or irrelevant objects. And then use the bilateral filter to reconstruct the input complex image pixels, this method can suppress the high frequency information of the image, and keep more target object texture information to make the classification accuracy higher (Fig. 3).

(c)   (d)

Fig. 3. Traditional fully connected layer is shown in (c). Global average pooling layer is shown in (d).

2 NIN and Bilateral Filter NIN

At present, there are many CNN deformation structures but their basic structures are very similar. CNN mainly consists of convolutional layer, pooling layer and the fully-connected layer [16].

Similarly, the NIN is also an improvement over the CNN that can learn more abstract and effective non-linear features than CNN. Also NIN can reduce the global over-fitting and the training parameters of fully-connected layer.NIN is a single channel CNN with three convolution layers, six cccp layers (cccp layer is a convolutional layer who has 1×1 kernel size), one maximum pooling layer and two average pooling layer. This section mainly introduces these two layers. NIN has three advantages compare to CNN: (1) Better local abstraction, (2) smaller global overfitting, (3) fewer parameters (no fully-connected layer) (Fig. 4).

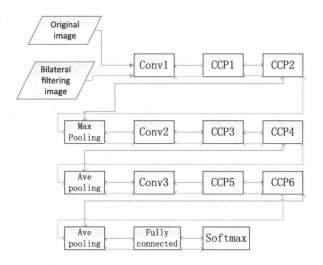

Fig. 4. The bilateral filter NIN network structure.

2.1 MLP Convolutional Layer

In the traditional CNN, the convolution operation is a generalized linear model (GLM). GLM combines the feature of the previous layer into a linear combination and then performs a non-linear activation. But this model has limited abstraction ability. NIN uses a tiny neural network (Mainly Multilayer Perceptron) instead of the linear convolution layer, the implement method is to add a few convolution operations after linear convolution operations [17]. These added convolution operations have very small kernel. This tiny neural network is also weight sharing which is the same weight for a feature layer (Fig. 5).

The traditional convolution is only a linear process, and the deeper layers of the network layer are the integration of the features learned by the shallow network layer. Therefore, it is very necessary to use a tiny neural network for further abstraction before high-level integration of features. The calculation performed by linear convolution layer is shown as follows:

$$f_{i,j,k} = \max(w_k^T x_{i,j}, 0) \tag{1}$$

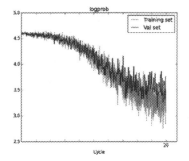 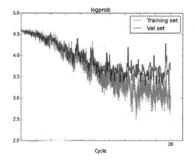

Fig. 5. In (1), group one which have simple images produce relatively smooth learning curve. The blue curve of the validation set by group 1 shows that the degree of overfitting is small. In (2), the learning curve produced by group 4 which have complex images. Learning curve fluctuation is very big. The learning curve of the training set and the learning curve of the validation set are separated that means the overfitting phenomenon is serious.

Here (i,j) is the pixel index in feature map, $x_{i,j}$ stands for the input patch centered at location $x_{i,j}$, and k is used to index the channels of the feature map. The calculation performed by linear convolution layer is shown as follows:

$$f_{i,j,k_1}^1 = \max(w_k^{1\,T} x_{i,j} + b_{k_1}, 0)$$
$$f_{i,j,k_2}^2 = \max(w_k^{2\,T} f_{i,j}^1 + b_{k_2}, 0)$$
$$\vdots \qquad\qquad\qquad\qquad\qquad (2)$$
$$f_{i,j,k_n}^n = \max(w_k^{n\,T} f_{i,j}^{n-1} + b_{k_n}, 0)$$

Here n is the number of layers in the multilayer perceptron. Rectified linear unit is used as the activation function in the multilayer perceptron (Table 1).

There are many advantages to choose a multilayer perceptron (MLP) as a tiny neural network architecture. One is the MLP is also using the BP algorithm for training, so it can be integrated with the CNN. The other is a MLP not only can be used as a deep structure but also contains the feature re-use thought. From the perspective of cross feature map pooling, Eq. (2) is equivalent to cascaded cross feature map parametric pooling on a normal convolution layer. This cascaded cross feature map parametric pooling structure allows complex and learnable interactions of cross feature map information (Table 2).

2.2 Global Average Pooling

NIN uses global average pooling layer replace the traditional fully connected layer. The idea is to generate one feature map for each corresponding category of the classification task in the last mlpconv layer. Instead of adding fully connected layers on top of the feature maps, we take the average of each feature map, and the resulting vector is fed directly into the softmax layer. One advantage of global average pooling over the fully connected layers is that it is more native to the convolution structure by enforcing

Table 1. The detailed configuration of the network parameters

Network configuration	
Conv1	5*5*192 and padding of 2 Followed by relu
cccp1	1*1*160 Followed by relu
cccp2	1*1*96 group of 1 Followed by relu
Max pooling	3*3 and stride of 2 Followed by dropout = 0.5
Conv2	5*5*192 and padding of 2 Followed by relu
cccp3	1*1*192 group of 1 Followed by relu
cccp4	1*1**192 group of 1 Followed by relu
Ave pooling	3*3 and stride of 2 Followed by dropout = 0.5
Conv3	3*3*192 and padding of 1 Followed by relu
cccp5	1*1*192 group of 1 Followed by relu
cccp6	1*1*10 group of 1 Followed by relu
Ave pooling	8*8 and stride of 1

Table 2. Details of the performance comparison on CAFIR-10

Methods	Accuracy %
PCANet	78.67
Convolutional kernel networks	82.18
Maxout networks	90.65
Network in network	91.2
our method	91.7

Table 3. Details of the performance comparison on CAFIR-100

Methods	Accuracy%
Spatial pyramids	54.23
Stochastic pooling	57.49
Maxout networks	61.43
Network in network	64.32
our method	64.59

correspondences between feature maps and categories. Another advantage is that there is no parameter to optimize in the global average pooling thus overfitting is avoided at this layer (Table 3).

2.3   The Bilateral Filter NIN

We reconstruct the input image pixels using the bilateral filter [18]. The bilateral filter consists of two functions, one of the functions determines the filter coefficients by geometric spatial distance another determines the filter coefficients by pixel-value difference, the output pixel of the bilateral filter depends on the weighted combination of neighborhood pixel value.

The calculation performed by bilateral filter is shown as follows:

$$g(i,j) = \frac{\sum_{k,l} f(k,l)w(i,j,k,l)}{\sum_{k,l} w(i,j,k,l)} \tag{3}$$

The weighting coefficient $w(i,j,k,l)$ in the above formula depends on the product of the domain kernel and range kernel:

$$w(i,j,k,l) = \exp\left(-\frac{(i-k)^2 + (j-l)^2}{2\delta_d^2} - \frac{\|f(i,j) - f(k,l)\|^2}{2\delta r^2}\right) \tag{4}$$

The Bilateral Filter NIN has two input paths, one input the original image, another input the reconstructed image. These two paths are used to extract the image features independently, and finally the feature vectors extracted by the two paths are merged after the average pooling layer, then we input it into the softmax classifier to classify the extracted features.

3   Experiments

3.1   Image Complexity and Learning Curve Analysis

We selected the first 50 categories of images in CIFAR-100 [5], each category has 500 training sets. If a category does not have 500 training sets, we will randomly select some images from the existing images to form new images by left or right translation, make sure that the training set has 500 pictures. We use two-dimensional wavelet transform to test the complexity of images because it has fast computation speed and sensitive to different directions. In an image, the larger part of the gradient is also have large wavelet coefficient. The complexity of the image and the number of large coefficients are positively correlated. We convert all 32×32 RGB training sets and test sets into gray scale, perform first-order wavelet transform and standardize (normalize) the wavelet coefficients between 0 to 1.

We define a variable C to represent the complexity of each image:

$$C = \sum_{k} \sum_{x,y} d_{x,y}^{k} \begin{cases} 1, & d_{x,y}^{k} > 0.5 \\ 0, & d_{x,y}^{k} \leq 0.5 \end{cases} \tag{5}$$

In the above equation, we define a variable C to calculate the number of normalized coefficients, d_{xy}^{k} represent the wavelet coefficients of the image horizontal, vertical and diagonal direction, the threshold of the wavelet transform coefficient is set to 0.5, k represents the image index of pixel $x \times y$.

We sort the 500 training sets in each category according to the variable C, and the images in each category is divided into four groups, each group has 125 images. The image complexity gradually increases from the first group. The first group are simple and ideal training samples so they have a small variable C. The fourth group are complex images so they have relatively large variable C. Then we use the above method to set the validation set, the difference is each type of validation set contains 100 pictures, each divided into four groups, each group has 25 images. Each group of the training set and validation set corresponding to each other.

We use the structure and parameter settings in [18]. Group one Simple image data set and group four complex image data set are individually cycled 20 times in the above network structure, each data dictionary training and validation set contains 100 images and 100 category labels. The image sequence is randomly exchanged in each data dictionary. The frequency of the validation set is set to 10 which means verify once when training 10 times. In all groups each category has 125 training images, the total number of categories is 50. So we have 25,000 training images totally. Each group has 1250 images in the verification set.

3.2   Performance Degradation Analysis

The first convolutional layer in the convolutional neural network learns the basic information of the image area, such as color and edge texture and other high-frequency information. The feature map is obtained after the convolution operation then feature map is input to the activation function, if the filter and the image area have high similarity, the feature graph will get a large activation value. So in the first layer, the high-frequency information such as the boundary is more easily extracted and retained by the filter. Max pooling is a subsampling process that selects the maximum value of a sub-region of the feature map. This operation will increase the possibility of that high-frequency information will be remained. After a series of convolution and max pooling operation, the low frequency information of the flat area of the original image is gradually lost.

In Fig. 6, the leftmost column is the original image, the rightmost column of is the last convolution layer of the feature map. Through them, we can see that only the high frequency region, the boundary part and the texture region have the activation value in the feature map, the other area is almost 0.

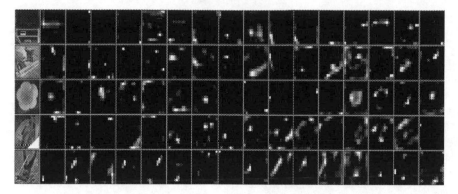

Fig. 6. Random subset feature maps of the last convolutional layer from [1].

The essence of all image classification algorithms is to extract the feature of the target object from the image. However, after our observation and analysis, the convolution neural network is more sensitive to the high frequency information of the image. The reality is that not all images contain a single background and texture complex target object. When we use the convolution neural network to classify images with simple textures in complex backgrounds, the accuracy is not very well. It can be seen from Fig. 7 that our method can better suppress the high frequency information of the image and retain more texture features than the NIN method. Our bilateral filter NIN can effectively prevent this shortcoming, more detail our experiment on CAFIR-10, CAFIR-100 in Sects. 3.3 and 3.4.

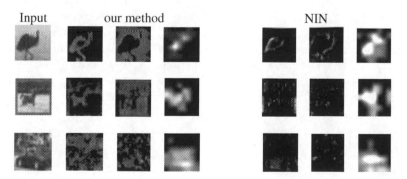

Fig. 7. Visualization of learned feature maps before the first pooling layer obtained using our method and NIN method.

3.3   CAFIR-10

The CIFAR-10 dataset is composed of 10 categories of natural images with 50,000 training images in total, and 10,000 testing images. There is no overlap between categories and categories. Each image is an RGB image of size 32 × 32. This database can be applied not only to multiple classifications but also transfer to the universal natural image. Compared with the already mature face recognition, universal natural image classification is difficult. Cifar-10 database contains a large number of features, noise, increase the difficulty of classification. We re-train the network from scratch with both the training set and the validation set. The resulting model is used for testing. We obtain accuracy rate of 10.41% on CAFIR-10, which improves more than 0.5% percent compared to NIN.

3.4 CAFIR-100

The CIFAR-100 dataset is the same in size and format as the CIFAR-10 dataset, but it contains 100 classes. 100 categories are divided into 20 large categories. Each image has a small class of "fine" tags and a large category "coarse" tags. CIFAR-100 dataset for each type of image less than the number of CIFAR-10 database but the number of images in each class is only one tenth of the CIFAR-10 dataset. So it is more difficult to recognition than CIFAR-10 dataset. An accuracy rate of 64.59% is obtained for CIFAR-100 which surpasses the NIN network of 64.32%.

4 Conclusion

This paper presents a novel deep architecture bilateral filter NIN for classification tasks, we reconstruct the input image pixels using the bilateral filter and construct a multi-path convolution neural network. This network has two input paths, one input the original image, another input the reconstructed image. These two paths, which are independent of each other and complement each other, are used to extract the image features independently, and finally the feature vectors extracted by the two paths are merged after the average pooling layer, then we input it into the softmax classifier to classify the extracted features. During the process of feature extraction, the loss of foreground object texture and shape information can be reduced for the complex background images. The experimental results show that the multi-path convolution neural network based on bilateral filtering is superior to the traditional single-path convolution neural network.

References

1. Krizhevsky, A., Sutskever, I., Hinton, G.E.: Imagenet classification with deep convolutional neural networks. In: Advances in Neural Information Processing Systems, pp. 1097–1105 (2012)
2. Zeiler, Matthew D., Fergus, R.: Visualizing and understanding convolutional networks. In: Fleet, D., Pajdla, T., Schiele, B., Tuytelaars, T. (eds.) ECCV 2014. LNCS, vol. 8689, pp. 818–833. Springer, Cham (2014). doi:10.1007/978-3-319-10590-1_53
3. Hinton, G.E., Srivastava, N., Krizhevsky, A., Sutskever, I., Salakhutdinov, R.R.: Improving neural networks by preventing co-adaptation of feature detectors. arXiv:1207.0580 (2012)

4. LeCun, Y., Bottou, L., Bengio, Y., Haffner, P.: Gradient-based learning applied to document recognition. Proc. IEEE **86**(11), 2278–2324 (1998)

5. Krizhevsky, A., Hinton, G.: Learning multiple layers of features from tiny images (2009)

6. Netzer, Y., Wang, T., Coates, A., Bissacco, A., Wu, B., Ng, A.Y.: Reading digits in natural images with unsupervised feature learning. In: NIPS Workshop on Deep Learning and Unsupervised Feature Learning, vol. 2011, No. 2, p. 5 (2011)

7. Deng, J., Dong, W., Socher, R., Li, L.J., Li, K., Fei-Fei, L.: Imagenet: a large-scale hierarchical image database. In: Proceedings of the IEEE Conference on Computer Vision and Pattern Recognition (2009)

8. Szegedy, C., Liu, W., Jia, Y., Sermanet, et al.: Going deeper with convolutions. In: Proceedings of the IEEE Conference on Computer Vision and Pattern Recognition, pp. 1–9 (2015)

9. He, K., Zhang, X., Ren, S., Sun, J.: Deep residual learning for image recognition. In: Proceedings of the IEEE Conference on Computer Vision and Pattern Recognition, pp. 770–778 (2016)

10. Maas, A.L., Hannun, A.Y., Ng, A.Y.: Rectifier nonlinearities improve neural network acoustic models. In: Proceedings of ICML, vol. 30, No. 1 (2013)

11. Nair, V., Hinton, G.E.: Rectified linear units improve restricted boltzmann machines. In: Proceedings of the 27th International Conference on Machine Learning, pp. 807–814 (2010)

12. Goodfellow, I.J., Warde-Farley, D., Mirza, M., Courville, A.C., Bengio, Y.: Maxout networks. In: ICML-3, vol. 28, pp. 1319–1327 (2013)

13. Le, Q.V., Karpenko, A., Ngiam, J., Ng, A.Y.: ICA with reconstruction cost for efficient overcomplete feature learning. In: Advances in Neural Information Processing Systems, pp. 1017–1025 (2011)

14. Goodfellow, I.J.: Piecewise linear multilayer perceptrons and dropout. Stat **1050**, 22 (2013)

15. Lin, M., Chen, Q., Yan, S.: Network in network. arXiv:1312.4400 (2013)

16. Zeiler, M.D., Fergus, R.: Stochastic pooling for regularization of deep convolutional neural networks. arXiv:1301.3557 (2013)

17. Bengio, Y., Courville, A., Vincent, P.: Representation learning: a review and new perspectives. IEEE Trans. Pattern Anal. Mach. Intell. **35**(8), 1798–1828 (2013)

18. Paris, S., Durand, F.: A fast approximation of the bilateral filter using a signal processing approach. In: Leonardis, A., Bischof, H., Pinz, A. (eds.) ECCV 2006. LNCS, vol. 3954, pp. 568–580. Springer, Heidelberg (2006). doi:10.1007/11744085_44

A High and Efficient Sparse and Compressed Sensing-Based Security Approach for Biometric Protection

Changzhi Yu[1,2], Hengjian Li[1,2(✉)], Ziru Zhao[1,2], and Jiwen Dong[1,2]

[1] School of Information Science and Engineering,
University of Jinan, Jinan 250022, China
ise_lihj@ujn.edu.cn
[2] Shandong Provincial Key Laboratory of Network
Based Intelligent Computing, Jinan 250022, China

Abstract. We propose a highly efficient sparse code with compressive sensing security algorithm based on the Dual-tree Complex Wavelet Transform (DT-CWT) and Hadamard measurement matrix in this paper for biometric protection. Firstly, we use DT-CWT to translate the image into frequency domain and use chaotic systems to encrypt measurement matrices. Also noise shaping is employed in the DT-CWT coefficients to represent the image sparsely. Then, we use compression sensing algorithm to improve the compression rate of encrypted images, and reduce the storage space occupied by images. Finally, in order to improve the algorithm's capability of handle contaminated images, we use the robustness of the double random phase encoding based on 4f optics system algorithm as secondary encryption. In the image decryption, we use the OMP algorithm. Finally, we can see that our proposed algorithm achieves 37.9863 dB in PSNR, 0.0245 in ERROR and 0.9977 in NC.

Keywords: Biometric image encryption · Double - random phase coding compression sensing · Dual tree complex wavelet transform · Noise shape

1 Introduction

Nowadays people pay more and more attention to the confidentiality of biometric information, because biometric images store unique biological information. Whether biometric information can be used safely, how to store and transmit is become more and more important for the research and development of biometric cryptography. Due to biometric images occupy a lot of storage space, they will waste a lot of resources when we store and transfer them, so it is important to compress and encrypt biometric images. How to maintain the privacy of the image while maintaining the compression efficiency has already became a big challenge.

Nowadays the algorithm of scrambling encryption for digital images are as follows: image pixel scrambling encryption algorithm based on Arnold transform [1], image pixel scrambling encryption algorithm based on magic square transform, image pixel scrambling encryption algorithm based on Fibonacci transform [2], and scrambling

© Springer International Publishing AG 2017
D.-S. Huang et al. (Eds.): ICIC 2017, Part II, LNCS 10362, pp. 666–677, 2017.
DOI: 10.1007/978-3-319-63312-1_59

image Pixel gray value or color value, etc., all this models can disturb the image pixel position, rearrange the image pixel position order to achieve the effect of scrambling.

As we all know Compression Sensing has been widely concerned, recently, the Compression Sensing as a new method of signal acquisition is applied to encryption research, CS breaks the traditional Shannon sampling theory and allows random sampling of a signal at a much lower Nyquist sampling rate, it uses a measurement matrix to sense a small amount of measured values and restore the original signal as much as possible without distortion. Initially, Rachlin and Baron [3] demonstrated that compression sensing does not conform to Shannon's definition of confidentiality, but with computational secrecy. In order to improve the image transmission rate in the channel and reduce image redundancy, E.J. Candes [4] et al. proposed Compression Sensing in 2004. In 2010, Ramezani and Mayiami proposed [5] under certain conditions, which is Compression Sensing becomes an important theoretical basis for encryption algorithm and perform a perfect confidentiality. Many scholars have proposed some combination of Compression Sensing and image encryption schemes such as C.W. Deng [6] and others proposed a multi-scale wavelet transform based on the Compression Sensing. Y. Rachlin pointed out: For the security of image encryption, the compression-based encryption scheme is not the most secure, but it has high computational complexity when this scheme is faced with attacks. So, the compression-based encryption scheme has more important application significance. In 1998, J Fridrich [7] proposed an improved method of Symmetric ciphers based on two-dimensional chaotic maps, which is a kind of compression-aware image encryption method. In 2014, Nanrun Zhou [8] proposed a Novel image compression–encryption hybrid algorithm based on key-controlled measurement matrix in compressive sensing.

The encryption system based on optical is widely used, because of encryption system can achieve high-speed parallel encryption of two-dimensional image data. In 1994, B. Javidi and P. Refregier [9, 10] proposed the first Dual- Phase Coding (Double Random Phase Encoding, DRPE) optical system encryption algorithm. They added a random phase plate and spectrum planes to the input images based on the 4-f system, so that the complex amplitude of the output information was stationary and had random white noise. In 2014, Lupe et al. proposed an image encryption scheme based on compression perception and optical theory, which is a highly robust optical application in combination with Compression Sensing and the most typical optical theory application - Dual Random Phase Encoding technique [11, 12]. In this model, the input two-dimensional digital image can output in the form of smooth white noise and has high security, but the image quality after decryption and reconstruction needs to be further improved. Then Hong Liu et al. proposed a dual random phase coding based on chaos for each image block in the fractional Fourier transform domain, which can resist the selection of plaintext attacks and improve the image quality after decryption and reconstruction [13].

In this paper, we aim to design an efficient CS-based image Compression-Encryption coding algorithm. Chaotic system [14, 15] can improve the intensity of the encryption process due to its instability, so in our algorithm, we use randomness chaotic system to encrypt the measurement matrix of CS. Furthermore, we also use DRPE in our algorithm in order to improve the safety of coding process. Experimental results show that our algorithm has a better performance than other algorithms. The detail of our algorithm is given in the Sect. 2. Experimental results (Security and performance evaluation) are described in Sect. 3. Conclusions are presented in Sect. 4.

2   The Proposed Algorithm

We propose an algorithm for image compression and encryption which is based on compression-sensing. In our algorithm, we use Dual-tree Complex Wavelet Transform (DT-CWT) to sparsely represent the image, and then get the sparse coefficient matrix. We chose the Fibonacci scrambling algorithm [2, 11] to scramble the sparse coefficient matrix, so that the human eyes cannot observe the specific image information. We combine the Logistic chaotic system with encrypt the Hadamard matrix, and we use this matrix as the measurement matrix in the compression sensing process. In the process of decrypting and reconstructing, the encrypted image is reconstructed by Orthogonal Matching Pursuit (OMP) algorithm and anti-even tree complex wavelet transform.

2.1   Encryption Process

Our algorithm is based on compression Sensing and Dual Random Phase Coding. The specific steps are shown in Fig. 1.

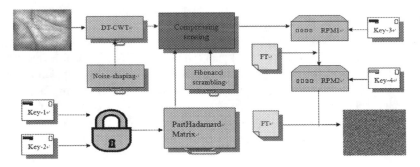

Fig. 1. Encryption process

Firstly, we input an image, which size is M * N. In order to transform the image from the airspace to the frequency domain, we select DT-CWT as the sparse base and obtaining a coefficient matrix to represent the sparse representation of the image. In this way, the characteristic information of this image in multi - direction can be obtained. In order to improve the security of sparse matrices, the Fibonacci scrambling algorithm is used to scramble and diffuse sparse matrices.

At the same time, we use one-dimensional Logistic chaotic mapping to encrypt the Hadamard matrix which is used as a measurement matrix in the compression sensing model. The one-dimensional Logistic model is shown in Eq. (1), where the parameters u and x are defined as $u \in [0, 4], x \in (0, 1)$. The Logistic is in a chaotic state and extremely sensitive to the initial value when it is generated. Two random sequences (used as the key) are used to select a matrix of M * N randomly from a Hadamard matrix of $N_1 * N_1$ ($N_1 > M$, $N_1 > N$). At last this Hadamard matrix of M * N is used as the measurement matrix.

$$x_{n+1} = u \times x_n \times (1 - x_n) \quad u \in [0, 4], x \in (0, 1) \tag{1}$$

The specific steps of DRPE encryption process are as follows. We define a uniformly distributed random phase mask (RPM) function $\theta(x, y)$, as the optical system to encrypt the image's key function and acting on the input image of the airspace. And then we do Fourier transform behind the lens to reach the spectral plane. By defining another RPM function $\varphi(x, y)$ which is evenly distributed in the [0, 1] interval, this RPM is consider as another key function and acts on the frequency domain of the image. And then do Fourier transform behind the lens again, at last we can get encrypted ciphertext which is a kind of smooth white noise. The encryption process can be expressed by the formula (2); decryption process can be expressed by the formula (3) (Fig. 2).

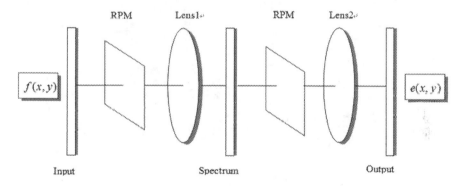

Fig. 2. DRPE encryption model

$$e(x, y) = FT^{-1}\{FT\{f(x, y) \times exp[j2\pi \times \theta(x, y)]\} \times exp[j2\pi \times \varphi(u, v)]\} \tag{2}$$

$$e(x, y) = FT^{-1}\{FT\{f(x, y) \times exp[-j2\pi \times \varphi(x, y)]\} \times exp[-j2\pi \times \theta(u, v)]\} \tag{3}$$

Where (x, y) is the coordinates of the image in the spatial domain and $exp[j2\pi \times \theta(x, y)]$ is the random phase mask function on the spatial domain. FT is the Fourier transform, FT^{-1} is the inverse Fourier transform. At the same time, (u, v) is the frequency domain coordinates of the image after Fourier transform, $exp[j2\pi \times \varphi(u, v)]$ is the random phase mask function in the frequency domain.

2.2   Decoding Process

In the decrypt process, the ciphertext and encrypted image should be prepared. When you decrypt this ciphertext, you need to enter the Key-1 and Key-2 to generate the needed Hadamard measurement matrix, and using the Key-3 and Key-4 to get the Random Phase Mask (RPE). Finally, you need to enter encrypted image into this model. We chose the orthogonal matching pursuit algorithm to reconstruct the original image. Comparing with the results of some existing algorithms, our encryption scheme can not only reconstruct an image with high robustness but also is sensitivity of the key. the step of image decryption is shown in Fig. 3.

In our algorithm, dual-tree Complex Wavelet Transform (DT-CWT) is used to overcome the shortcomings of wavelet transform (DWT) such as translation sensitivity and lacking of direction sensitivity. In image recognition, Image fusion, and other research directions, DT-CWT performance is better than the conventional DWT. Due to the DT-CWT has some redundancy, so we introduce the Noise shaping (NS) algorithm to minimize the number of non-zero coefficients.

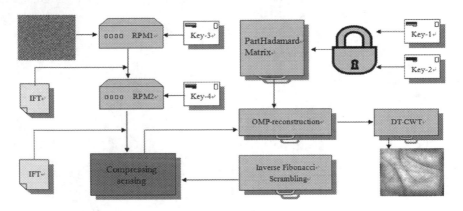

Fig. 3. Decryption process

3   Security and Performance Evaluation

In this section, security and performance evaluation including Recovery, key sensitivity is investigated. In order to determine the effect of the NS model, we also chose n different thresholds (T) to analyze the performance of ours method. The size of our experimental test images is 256×256 pixels. The coefficient matrix of the image is obtained by DT-CWT algorithm, and then the noise shaping algorithm is used to process this matrix. The sparse matrix is obtained by scrambling algorithm. Finally, we identify the scrambled matrix with the encrypted Hadamard measurement matrix. The DRPE technique is used to encrypt the measured value matrix and the results are shown as the Figs. 4, 5 and 6.

Fig. 4. Original image, encrypted image, decrypted image of Iris

Fig. 5. Original image, encrypted image, decrypted image of Palmprints

Fig. 6. Original image, encrypted image, decrypted image of Lena

In order to quantify the performance of the restructured image, we chose Peak Signal to Noise Ratio (PSNR), Error Rate (ERROR) and the Normalized Cross-Correlation Coefficient (NC) as evaluation indexes. The formula shown as (4), (5) and (6), where $f(x, y)$ is the pixel value in the original image and $g(x, y)$ is the image pixel value for decrypting the reconstructed image. In NC, $f(\acute{x}, y) = \sum_{x=0}^{m} \sum_{y=0}^{n} \frac{f(x,y)}{mn}$ and $g(\acute{x}, y) = \sum_{x=0}^{m} \sum_{y=0}^{n} \frac{g(x,y)}{mn}$ is the mean of the original image pixel value and the decryption of the image pixel value.

$$\text{PSNR} = 10 \times \lg\left(\frac{255^2}{MSE}\right) \tag{4}$$

$$\text{NC} = \frac{\sum_{x=0}^{m} \sum_{y=0}^{n} [f(x, y) - f(\acute{x}, y)][g(x, y) - g(\acute{x}, y)]}{\sum_{x=0}^{m} \sum_{y=o}^{n} [f(x, y) - f(\acute{x}, y)]^2 [g(x, y) - g(\acute{x}, y)]^2} \tag{5}$$

$$\text{ERROR} = \frac{\sum_{x=0}^{m} \sum_{y=0}^{n} |f(x,y) - g(x,y)|}{\sum_{x=0}^{m} \sum_{y=o}^{n} f(x,y)} \tag{6}$$

Table 1 shows the Peak Signal-to-Noise Ratio (PSNR), the Error Rate (ERROR), and the Normalized Cross-Correlation Coefficient (NC) of the decrypted image.

Table 1. Test results (PSNR, ERROR, and NC)

Image name	PSNR/dB (CS + DRPE)	PSNR/dB (CS + DRPE + NS)	Error rate (CS + DRPE)	Error rate (CS + DRPE + NS)	NC (CS + DRPE)	NC (CS + DRPE + NS)
Iris	48.8819	49.0447	0.0050	0.0049	0.9999	0.9999
Palmprints	39.0529	39.7280	0.0164	0.0152	0.9975	0.9979
Fingerprint	25.2716	25.8476	0.0619	0.0579	0.9796	0.9823
Lena	37.7374	37.9863	0.0252	0.0245	0.9976	0.9977

As we all know some algorithm's performance may be affected by the impact of encrypted images, which does not really reflect the advantages of the algorithm, so in order to test the applicability of this algorithm, we chose some standard images as the detection object, finally we get the experimental results which are shown in Table 2.

Table 2. Test results (PSNR, ERROR, and NC)

Image name	PSNR/dB (CS + DRPE)	PSNR/dB (CS + DRPE + NS)	Error rate (CS + DRPE)	Error rate (CS + DRPE + NS)	NC (CS + DRPE)	NC (CS + DRPE + NS)
Barbara	35.0735	35.3421	0.0355	0.0344	0.9959	0.9961
Camera	37.9097	38.0333	0.0242	0.0238	0.9975	0.9987
Gold-hill	34.0519	34.7287	0.0413	0.0382	0.9948	0.9956
Lady	42.9313	43.1202	0.0127	0.0124	0.9996	0.9996
Lena	37.7374	37.9863	0.0252	0.0245	0.9976	0.9977
Peppers	38.6650	38.8680	0.0222	0.0217	0.9984	0.9985

In order to facilitate comparison with other research results, Lena is taken as an example to illustrate the advantages of our proposed model. The data in Table 3 shows that the PSNR values of the image encryption algorithm combined with the dual random phase encoding and compression perception are higher than those of the existing literature [16–19], and the NC values are higher with those in the literature.

Table 3. Results comparison (PSNR and NC)

	(CS + DRPE)	(CS + DRPE + NS)	Method [11] (Lena)	Method [16] (Lena)	Method [17] (Peppers)	Method [18] (Peppers)
PSNR	37.7374	37.9863	30.8170	30.8874	29.3684	37.3011
NC	0.9976	0.9977	0.9901	–	0.9870	–

The measurement matrix has a great influence on the compression efficiency of the whole algorithm, as a core of the compression sensing algorithm. A good measurement matrix can greatly improve the compression efficiency of the algorithm. The conventional measurement matrices are Gauss, Bernoulli, PartHadamard, SparseRandom, and so on. In addition, in the noise shaping model, different shaping threshold will have a different influence on the entire algorithm. Therefore, in the compression progress of the Lena map encryption, we chose a different measurement matrix, different threshold of noise shaping to compare their impact on image compression and decryption. The concrete results are shown in Figs. 7 and 8.

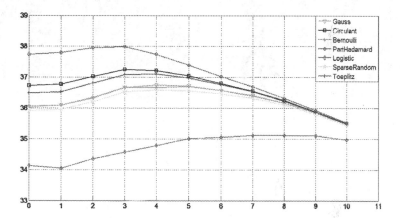

Fig. 7. Different image PSNR, when we use different measurement matrix for CS

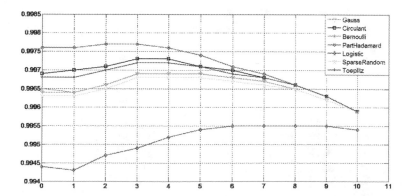

Fig. 8. Different image ERROR, when we use different measurement matrix for CS

From Figs. 7 and 8, we can clearly see that the final recovery of the image PSNR, ERROR is best when we using the Hadamard matrix as a measurement matrix. The performance of the logistics measurement matrix is the worst. The final results of other measurement matrices are similar.

In our model, the initial value of the chaotic system is the key 1 and the key 2. The random phase plates RPM1 and RPM2 in the DRPE process are the key 3 and the key 4 in the encryption algorithm, respectively. Figure 9 shows the decrypted image when the error one of 4 keys is entered. The corresponding PSNR, ERROR and NC are shown as the Table 4. It can be clearly found that if input a key, the image can't be decrypted and can't obtained information related to the original image from the reconstructed image after decryption failure. And only when all the key values are entered correctly, the image can be decrypted correctly.

Fig. 9. Shows the decrypted image when we entered four different error keys

Table 4. Input error keys' results

	KEY1	KEY2	KEY3	KEY4
PSNR/dB	4.9351	4.7791	5.2301	5.2283
ERROR	1.1010	1.1210	1.0642	1.0645
NC	0.0035	0.0360	0.0114	−0.0030

In order to analyze the security of our method, we tested the horizontal correlation of the image pixels and plot grayscale histogram of the original image, encrypted image and the error decrypted image, they are shown in Figs. 10 and 11.

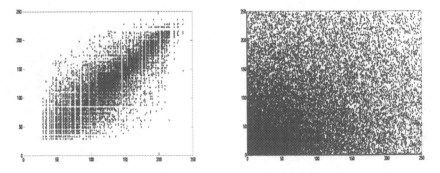

Fig. 10. Original and encrypted image's horizontal correlation of the image pixels

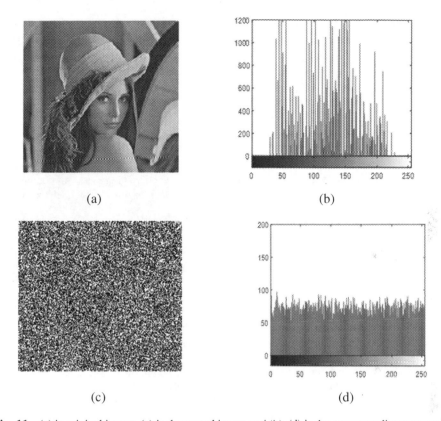

Fig. 11. (a) is original image, (c) is decrypted image and (b), (d) is the corresponding gray scale.

We can see that the horizontal correlation of the original image pixels is strong, but in Fig. 11(c) the pixel correlation of the image is obviously reduced after using our algorithm. Moreover, since some attacks are carried out by counting up the histogram of the image, in order to verify the performance of our algorithm in this area, we do a risk test experiment. We give the gray histogram of original image, the simple scrambling image, the encrypted image by our algorithm. We can see that the encryption of the gray histogram of original image is more uniform. These experimental results are shown in Fig. 11(b) and (d).

4 Conclusion

We propose a new algorithm compression-based and dual-random phase coding techniques for image encryption and make considerable improvement over existing encryption algorithm. In our algorithm, we transform the input images into spare representation through the dual tree complex wavelet, the reconstructed image quality can higher in this way. And we also use the Logistic chaotic system to encrypt the Hadamard measurement matrix, the encryption algorithm includes four keys which have high sensitivity. This can make the key space larger than before. The experimental results show that the algorithm has a good performance in resist the violent attack or cipher text statistics attack. Furthermore, the algorithm can achieve image encryption and image information hidden in the same time, also it can resist noise pollution such as salt, pepper and Gaussian.

References

1. Nitin, R., Byoungho, K., Rajesh, K.: Fast digital image encryption based on compressive sensing using structurally random matrices and Arnold transform technique. Optik **127**, 2282–2286 (2016)
2. Nan, L., Yanhong, S., Jiancheng, Z.: An audio scrambling method based on Fibonacci transformation. J. North China Univ. Technol. **16**(3), 8–11 (2004)
3. Rachlin, Y., Baron, D.: The secrecy of compressed sensing measurements. In: 46th Annual Conference on Communication Control and Computing, Allerton, pp. 813–817. IEEE (2008)
4. Candes, E.J., Romberg, J., Tao, T.: Robust uncertainty principles: exact signal reconstruction from highly incompletely frequency information. IEEE Trans. Inf. Theory **52**(2), 489–509 (2006)
5. Ramezani Mayiami, M., Seyfe, B., Bafghi, H.G.: Perfect secrecy via compressed sensing. In: Communication & Information Theory, pp. 1–5 (2013)
6. Deng, C.W., Lin, W.S., Lee, B.-S.: Robust image compression based on compressive sensing. In: Proceedings of IEEE International Conference on Multimedia and Expo (ICME), Singapore, pp. 462–467. IEEE (2010)
7. Fridrich, J.: Symmetric ciphers based on two-dimensional chaotic maps. Int. J. Bifurc. Chaos **8**(6), 1259–1284 (1998)
8. Zhou, N., Zhang, A.: Novel image compression–encryption hybrid algorithm based on key-controlled measurement matrix in compressive sensing. Optic Laser Technol. **62**, 152–160 (2014)
9. Javidi, B., Horner, J.L.: Optical pattern recognition for validation and security verification. Opt. Eng. **33**, 1752–1756 (1994)
10. Refregier, P., Javidi, B.: Optical image encryption based on input plane and Fourier plane random encoding. Opt. Lett. **20**, 767–769 (1995)
11. Unnikrishnan, G., Joseph, J., Singh, K.: Optical encryption by double-random phase encoding in the fractional Fourier domain. Opt. Lett. **25**(12), 887–889 (2000)
12. Tao, R., Xin, Y., Wang, Y.: Double image encryption based on random phase encoding in the fractional Fourier domain. Opt. Express **15**(24), 16067–16079 (2007)
13. Liu, H., Xiao, D., Liu, Y., Zhang, Y.: Securely compressive sensing using double random phase encoding. Optik **126**, 2663–2670 (2015)
14. Gao, T., Chen, Z.: A new image encryption algorithm based on hyper-chaos. Phys. Lett. A **372**, 394–400 (2008)
15. Gao, H., Zhang, Y., Liang, S., Li, D.: A new chaotic algorithm for image encryption. Chaos Solitons Fractals **29**, 393–399 (2006)
16. Liu, X., Cao, Y., Lu, P.: Study on optical image encryption based on compression sensing. Acta Microbiol. Sinica **34**(3), 1–8 (2014)

17. Lu, P., Xu, Z., Lu, X., Liu, X.: Digital image information encryption based on compressive sensing and double random-phase encoding technique. Optik **124**, 2514–2518 (2013)
18. Liu, X., Cao, Y., Lu, P., Lud, X., Li, Y.: Optical image encryption technique based on compressed sensing and Arnold transformation. Opt. Lett. **18**(5), 6590–6593 (2013)
19. Zhang, Y., Zhou, J., Chen, F., Zhang, L.Y., Wong, K.-W., He, X., Xiao, D.: Embedding cryptographic features in compressive sensing. Neurocomputing **12**(9), 472–480 (2016)

Robust Real-Time Head Detection by Grayscale Template Matching Based on Depth Images

Yun-Xia Liu[1,2(✉)], Yang Yang[3], and Min Li[3]

[1] School of Information Science and Engineering, University of Jinan, Jinan, China
ise_liuyx@ujn.edu.cn
[2] School of Control Science and Engineering, Shandong University, Jinan, China
[3] School of Information Science and Engineering, Shandong University, Jinan, China

Abstract. Head detection conducted on color images has been an active research topic in the computer vision community. Recently, depth sensors have made a new type of data available, which demonstrate good invariance against illumination changes. Head detection based on depth images can be significantly simplified as background subtraction and segmentation are no longer critical issues. In this paper, a robust head detection algorithm is proposed. Firstly, a grayscale template is employed for better modeling and precise detection of human head. Meanwhile, statistical analysis of the correlation coefficients is presented and the optimal threshold is deducted. Secondly, candidate head regions are further examined by seed point selection based on a novel feature taking both correlation and local standard deviation into consideration. Finally, the detected head area is obtained by region-growing and computation efficiency issues are discussed. In order to test the validity of the proposed algorithm, we constructed a Microsoft Kinect depth database with 670 images which includes extreme conditions such as complex background and 180° rotation. Experimental results shows that the proposed algorithm achieves robust real-time head detection.

Keywords: Head detection · Depth image · Template matching · Correlation analysis

1 Introduction

The ability to identify heads of human beings is a critical component of many applications, such as a three-dimension (3D) head reconstruction, human-computer interaction (HCI), people monitoring. Head detection is high challenging due to illumination condition, human pose and background complexity changes in real applications.

In the past few years, head detection and tracking based on color images has been widely explored by researchers and many methods have been proposed [1, 2]. Although great development has been made [3– 5] in recent years, a head detection algorithm with improved robustness that can work in real applications is still demanded. For example, when people being monitored in these studies rotate or otherwise change pose or their environment becomes cluttered, the accuracy of these results can drop markedly, due to

© Springer International Publishing AG 2017
D.-S. Huang et al. (Eds.): ICIC 2017, Part II, LNCS 10362, pp. 678–688, 2017.
DOI: 10.1007/978-3-319-63312-1_60

changes in shape signatures of the head, missing features from the face, or color-changes and the interference of clutter.

An effective way to counter these challenges is to employ depth images. In depth images, pixel values reflects the distance between the camera and object being photographed, thus invariance against illumination changes is achieved. Several studies on head detection or tracking, pose estimation and fall detection are based on the use of depth images [5–14]. Reference [8] adopted a binary head contour and 3D head model to locate the head, but it can't effectively deal with bend and local occlusion. Reference [9] utilized both head and shoulder information to locate the head. However, the robustness to noise and pose variation is not satisfying. Some work [12] use 3D human skeleton for tracking, which suffer from detection failure due to instability. Reference [13, 14] adopted "similarity" methods to detect head portion, where the static scenery assumption could not be guaranteed all the time.

In order to achieve real-time robust head detection based on depth images, a novel detection algorithm is proposed based on grayscale template matching and statistical modeling. Firstly, after preprocessing procedures including denoising, hole filling and edge detection, distance transform is applied for better characterization of local shapes in depth images. Secondly, template matching is conducted with a modified gray-scale template, and the optimal threshold to discriminate candidate head locations and backgrounds is deducted based on statistical modeling of the correlation coefficients. The final detection output is obtained by candidate seed selection and morphological region growing operation. Experimental results are presented to demonstrate the validity of the proposed algorithm.

The remainder of this paper is organized as follows. Section 2 introduces the overall architecture of the proposed algorithm and details about pre-processing. The proposed algorithm is discussed in detail in Sect. 3. Section 4 discusses experimental results. Section 5 presents our conclusion and suggest the possible directions for future research.

2 System Overview and Preprocessing

This section provides an overview of the major modules in our method, which is summarized in Fig. 1. We also present implementation details of the preprocessing pressures in this section.

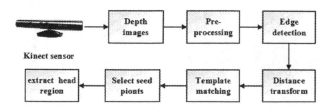

Fig. 1. Overview of the proposed head detection algorithm

2.1 System Overview

In the proposed head detection algorithm as shown in Fig. 1, the depth images are obtained from Mircosoft Kinect sensor. The quality of such depth images inevitably contaminated by noise and often suffer from limited accuracy and stability due to depth holes and inconsistent depth values. Preprocessing procedures are necessary for image quality enhancement.

We follow the paradigm in [8] that head detection is achieved in two steps. Firstly, we conduct edge detection on preprocessed images as we regard shape as the main feature for head detection. However, nearby objects in depth images often demonstrate similar depth values, which adds difficulty in precise head detection. To solve this problem, we utilize Euclidean distance transform to enrich shape related detail information to roughly locate possible head regions.

For the second step of head detection, a grayscale template is proposed and template matching is conducted. Based on statistical analysis of the correlation coefficients of head and background regions, the optimal threshold value is determined. False acceptance rate is further decreased by seed point selection and the final detection output is obtained by region growing.

2.2 Preprocessing

Preprocessing operations are necessary for robust head detection. In order to remove or decrease the effect of noise and enhance the local smooth in the acquired depth image (as shown in Fig. 2(a)), the most commonly used mean filter with 3×3 window is adopted where all weights equal to 1/9. Mean filtering helps to get a smoothed depth image.

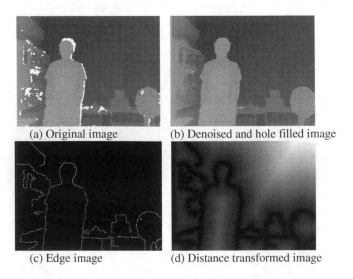

(a) Original image   (b) Denoised and hole filled image

(c) Edge image   (d) Distance transformed image

Fig. 2. Intermediate preprocessing images.

However, there are still holes left that cannot be eliminated after average filtering. Morphological methods are applied to fulfill the holes. Firstly, utilize open operation to erode the small white region, and then adopt reconstruct function to recover the pixels' values. Example depth image after hole filling is shown in Fig. 2(b).

Human heads to be detected in real applications often suffer from different sizes, poses and backgrounds. We assume a ball model for the head and regard this shape information is the origin for robust head detection. The Canny operator is utilized to obtain the edge image (shown in Fig. 2(c)), where edge pixels take value of 1 and background pixels equal to 0.

2.3 Euclidean Distance Transform

In the depth images, depth values of substances locating similar distance range in a relative small interval, which lead to low level of discrimination. Given a binary image (the edge image obtained after preprocessing), distance transform transforms it into a grayscale image by calculating the distance between the pixel in consideration to its nearest target (edge). In this way, it enriches detail information related with the objects shape and contributes to robust head detection.

Distance transform includes Euclidean distance transform, chess-board distance transform and Quasi-Euclidean distance transform. In our work, we adopt the Euclidean distance transform, which does not deform objects and coincides with the ball model for the head.

Denote T as the set of the target points (edge pixels), the distance transformed image $D(i,j)$ is defined as

$$D(i,j) = \underset{(x,y)}{\mathrm{argmin}}(d[(x_i, y_i), (x, y)] | (x, y) \in T)\ 0 \leq i < M, 0 \leq j < N \tag{1}$$

where M and N are the depth and width of the input image, and $d(\bullet, \bullet)$ defines the Euclidean distance between two arbitrary points in the binary edge image:

$$d[(x_i, y_i), (x, y)] = [(x_i - x)^2 + (y_i - y)^2]^{1/2} \tag{2}$$

By means of the Euclidean distance transform, we obtain a distance image, as shown in Fig. 2(d), the value of an arbitrary point represents the distance between that and the nearest non-zero pixel in the edge image.

3   Robust Head Detection by Grayscale Template Matching

In this section, we first introduce the proposed grayscale template, based on witch the correlation could be calculated. In order to reduce the missing detection rate, we set a small threshold on the correlation coefficients that allow more head candidates (denoted as seed points) to get through. Afterwards, seed point selection is performed based on the proposed correlation and standard deviation feature (CSF). Computational efficiency techniques for real-time application is also discussed.

3.1 Grayscale Template Matching with Optimal Threshold

Template matching is a simple yet effective technique that is widely used in detection, tracking and so on. In the study in [8], the authors propose a semi-sphere 3D head model and utilize a binary head template to mark out the candidate regions, whose characterization ability is limited. Due to binary template only utilizing shape information, it requires a relative lower threshold of correlation coefficients in order to detect out all head regions, which will inevitably lead to a high level of false acceptance rate. The 3D head model considers head region as a semi-sphere, which does not conform to the common sense.

In order to overcome this modelling inefficiency, a greyscale head template which taking both shape and depth texture information around the head into consideration is proposed. The proposed head template with size 80×80, shown in Fig. 3(b) is obtained by averaging manually annotated ground truth head region in the distance transformed images after distance correction and mapping. Clearly, the proposed grayscale model reflects head characteristics including both shape and local depth texture compared to Ref. [8].

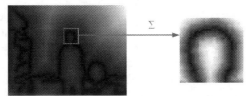

(a) Example distance transformed image (b) Graysclae head template

Fig. 3. Proposed grayscale head template.

Template matching involves translating and positioning the grayscale template T at all possible locations on the distance transformed images. The correlation coefficient between arbitrary region R (with the same size of template) and the grayscale template T could be calculated as:

$$C_R = \frac{\sum\limits_m \sum\limits_n \left(R - \bar{R} \right) \left(T - \bar{T} \right)}{\left[\sum\limits_m \sum\limits_n \left(R - \bar{R} \right)^2 \cdot \sum\limits_m \sum\limits_n \left(T - \bar{T} \right)^2 \right]^{1/2}} \tag{3}$$

where m and n denotes the size of R and T. It serves as an indicator for possible head regions. We expect higher C_R values for head regions, thus when C_R is larger than a predetermined threshold T_C, the region R under consideration is classified as a candidate head region.

Apparently, the detection result is highly dependent on the setting of the threshold T_C. For statistical robustness, we employ the logistic normal distribution (LND)

$$f(x) = \frac{1}{x\sigma\sqrt{2\pi}} e^{-\frac{\ln^{(x-\mu)}}{2\sigma^2}}, x > 0 \tag{4}$$

to fit the correlation coefficients density of the head and background regions, where μ and σ denotes the mean and standard deviation of the variable's logit, which better characterized the high dimensional features of data, such as skewness and kurtosis.

We use the correlation coefficients in manually annotated ground truth head regions and backgrounds to fit two family of LND curves using the maximum likelihood estimator (MLE):

$$\begin{cases} \hat{\mu}_i = \frac{1}{N_i} \sum_{j=1}^{N_i} \ln(C_j) \\ \hat{\sigma}_i^2 = \frac{1}{N_i} \sum_{j=1}^{N_i} \left(\ln(C_j) - \hat{\mu}_i \right)^2 \end{cases}, \quad i = 1, 2 \tag{5}$$

where $i = 1,2$ denotes the head region and background region respectively, N_i denotes number of regions in each class and j indexes all possible regions in each class.

The error probability of system consists of two parts, namely the false acceptance probability P_{FAR} and the false reject probability P_{FRR}:

$$\begin{aligned} P_e &= P_{FAR} + P_{FRR} \\ &= \int_t^{\infty} \frac{1}{\sigma_1\sqrt{2\pi}} e^{-\frac{(x-\mu_1)^2}{2\sigma_1^2}} dx + \int_{-\infty}^{t} \frac{1}{\sigma_1\sqrt{2\pi}} e^{-\frac{(x-\mu_2)^2}{2\sigma_2^2}} dx \end{aligned} \tag{6}$$

depending on the threshold value t. By setting the derivative of the error probability P_e with respect to t equals to zero,

$$\frac{\partial P_e}{\partial t} = \frac{1}{\sqrt{2\pi}\sigma_2} e^{-\frac{(t-\mu_2)^2}{2\sigma_2^2}} - \frac{1}{\sqrt{2\pi}\sigma_1} e^{-\frac{(t-\mu_1)^2}{2\sigma_1^2}} = 0 \tag{7}$$

We can the conclusion that the optimal threshold T_{opt} fulfills

$$\frac{(T_{opt} - \mu_1)^2}{2\sigma_1^2} - \frac{(T_{opt} - \mu_2)^2}{2\sigma_2^2} = \ln\left(\frac{\sigma_1}{\sigma_2}\right) \tag{8}$$

Based on abundant experimental data collected from six different scenes and a loose FAR setting of 2%, the optimal threshold value T_{opt} is predetermined to be 0.38 in our experiments. We find it balances false acceptance as well as false reject and leads to robust head detection in our experiments.

3.2   Seed Points Selection and Region Growing

As mentioned in Sect. 2.1, we intendedly lowered the T_{opt} value for a lower false rejection rate. As a consequence, some less related background regions will survive the thresholding as shown in in Fig. 4(a). Seed point selection is necessary to ease this effect and decrease the false acceptance rate.

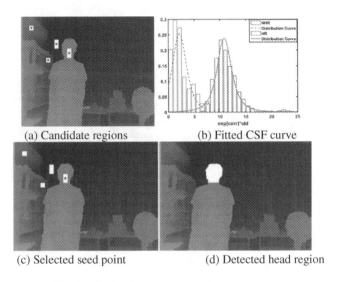

(a) Candidate regions   (b) Fitted CSF curve

(c) Selected seed point   (d) Detected head region

Fig. 4.   Seed point selection and region growing

We find the local variance a good indicator to discriminate whether there involves head. In order to balance the different dynamic range of the correlation coefficients C_R and the local standard deviation Std, a novel correlation standard-deviation feature (CSF) is proposed:

$$CSF_R = \exp\left(C_R\right) \cdot Std \tag{9}$$

When adopting T-location distribution of the CSF feature, we can obtain fitted CSF probability density function curve based on the empirically histogram of head and non-head regions in as shown in Fig. 4(b). After the optima threshold discriminating head and non-head regions T_{CSF} has been determined similarly with T_{opt} described in Sect. 3.1, one can easily select the seed points as shown in Fig. 4(c).

The final detection output can thus be obtained by region growing algorithm based on the selected seed points, which is depicted in Fig. 4(d). Specifically, if the absolute difference between the pixel value near the seed location and the seed depth value is smaller than a threshold, we classify the point as belonging to the head area, or we simply stop the growth process altogether.

3.3   Computational Acceleration for Real-Time Implementation

In order to achieve real-time head detection, several techniques is adopted to accelerate the proposed algorithm.

Firstly, the computational complexity of template matching is known to be largely dependent on the correlation calculation between the template and image. Specifically, the cost is always increasing exponentially with the sizes of the template and image. To improve the efficiency of matching algorithm, we down-sample both the image and the template before running the algorithm.

Furthermore, computational efficiency could be markedly improved if we sample the possible candidate regions every S pixels in both directions of the image, that the time cost will inverse proportional with the square of stride S. However, this acceleration is achieved at the cost of increased miss matching. Stride value of $S = 8$ is adopted in this paper for trade-off between computation efficiency and detection accuracy.

Finally, we observe that heads usually appear in the upper portion of the image, which allows us to adopt the following strategy. We first run the algorithm on the upper half of the image, and this procedure will be terminated if any candidate seed point is detected. Otherwise (i.e. no seed point in the upper half), we continue to run the program on the lower part. This simple approach grants approximately twice speed up as before.

4   Experimental Results and Discussion

We construct a test database consisting of 670 images in our laboratory by Microsoft Kinect, with complex backgrounds including tables, air-conditioner, books, and book-shelves, etc. The database consists of five parts: (1) no light part to simulate night

Fig. 5. Example test images and head detection results.

condition, (2) a light part to simulate the daytime condition, (3) the normal distance part to represent normal condition, (4) the smaller distance part to represent the occasion that human is too close to camera and (5) the extreme part that involves huge pose changes and severe local occlusions (See the corresponding color images in Fig. 5).

Real-time implementation is achieved at 60 frame per second with an Intel Core i5-4570 CPU equipped desktop computer, where all parameters are pre-determined without fine-tuning. As shown in color images in Fig. 5 that the Kinect head scanning program (KHS) fails to detect head in such extreme situations. KHS utilizes depth feature and skeleton information (shoulder joints and head point) to detect human head, thus cannot work well in strong local occlusion and large angle rotation conditions. However, robust detection results are reported by the propose algorithm, which strongly demonstrates the robustness superiority of the propose algorithm.

Figure 6 shows more examples of detection results of the proposed algorithm. We observe 667 true positives and 3 false negatives on the whole database, with detection accuracy equal to 99.5%. From the results we see that the proposed algorithm is highly robust in terms of detecting variation in human pose, illumination change, rotation, local occlusion and the direction between cameras and people in consideration.

Fig. 6. More head detection results on the database.

5   Conclusions

In this study, we propose a novel head detection algorithm with improved grayscale template matching based on Kinect depth images. The optimal threshold value of the correlation coefficients are deducted imposing a logistic normal distribution density. A novel correlation standard deviation feature is proposed for further seed point selection that balances the false acceptance and false reject rate. Several acceleration techniques are adopted for computational efficiency improvement. Experimental results demonstrated real-time and robust head detection results on test database.

Currently, the threshold values depend on characteristics of the database and are predetermined. In the future, we plan to develop adaptive thresholds that lead to improved robustness for real applications. In addition, we plan to expand the work so that it can simultaneously support detection of multiple heads.

Acknowledgement. This work was supported by the National Nature Science Foundation of China (No. 61305015, No. 61203269), the Shandong Province Key Research and Development Program, China (Grant No. 2016GGX101022), the National Key Research And Development Plan (No. 2016YFC0106001), and the Postdoctoral Science Foundation of China (No. 2015M580591).

References

1. Birchfield, S.: Elliptical head tracking using intensity gradients and color histograms. In: Proceedings of IEEE Computer Society Conference on Computer Vision and Pattern Recognition, pp. 232–237. IEEE (1998)
2. Yoon, H., Kim, D., Chi, S., et al.: A robust human head detection method for human tracking. In: 2006 IEEE/RSJ International Conference on Intelligent Robots and Systems, pp. 4558–4563. IEEE (2006)
3. Zhao, M., Sun, D., He, H.: Hair-color modeling and head detection. In: 7th World Congress on Intelligent Control and Automation, WCICA 2008, pp. 7773–7776. IEEE (2008)
4. Ishii, Y., Hongo, H., Yamamoto, K., et al.: Face and head detection for a real-time surveillance system. InL Proceedings of the 17th International Conference on Pattern Recognition, ICPR 2004, vol. 3, pp. 298–301. IEEE (2004)
5. Krotosky, S.J., Cheng, S.Y., Trivedi, M.M.: Real-time stereo-based head detection using size, shape and disparity constraints. In: Proceedings of IEEE Intelligent Vehicles Symposium, pp. 550–556. IEEE (2005)
6. Li, M., Yang, S., Li, X.: A head detection method based on curvature scale space. In: Third International Symposium on Intelligent Information Technology Application Workshops, IITAW 2009, pp. 390–393. IEEE (2009)
7. Rodgers, J., Anguelov, D., Pang, H.C., et al.: Object pose detection in range scan data. IEEE Computer Society Conference on Computer Vision and Pattern Recognition, vol. 2, pp. 2445–2452. IEEE (2006)
8. Zhu, Y., Dariush, B., Fujimura, K.: Controlled human pose estimation from depth image streams. In: IEEE Computer Society Conference on Computer Vision and Pattern Recognition Workshops, CVPRW 2008, pp. 1–8. IEEE (2008)

9. Lee, K., Choo, C.Y., See, H.Q., et al.: Human detection using histogram of oriented gradients and human body ratio estimation. In: 2010 3rd IEEE International Conference on Computer Science and Information Technology (ICCSIT), vol. 4, pp. 18–22. IEEE (2010)
10. Wu, S., Yu, S., Chen, W.: An attempt to pedestrian detection in depth images. 2011 Third Chinese Conference on Intelligent Visual Surveillance (IVS), pp. 97–100. IEEE (2011)
11. Xia, L., Chen, C.C., Aggarwal, J.K.: Human detection using depth information by kinect. In: 2011 IEEE Computer Society Conference on Computer Vision and Pattern Recognition Workshops (CVPRW), pp. 15–22. IEEE (2011)
12. Nghiem, A.T., Auvinet, E., Meunier, J.: Head detection using kinect camera and its application to fall detection. In: 2012 11th International Conference on Information Science, Signal Processing and their Applications (ISSPA), pp. 164–169. IEEE (2012)
13. Ye, M., Yang, R.: Real-time simultaneous pose and shape estimation for articulated objects using a single depth camera. In: Proceedings of the IEEE Conference on Computer Vision and Pattern Recognition, pp. 2345–2352 (2014)
14. Bagautdinov, T., Fleuret, F., Fua, P.: Probability occupancy maps for occluded depth images. In: Proceedings of the IEEE Conference on Computer Vision and Pattern Recognition, pp. 2829–2837 (2015)

A Data Stream Clustering Algorithm Based on Density and Extended Grid

Zheng Hua[1,2], Tao Du[1,2(✉)], Shouning Qu[1,2], and Guodong Mou[1,2]

[1] School of Information Science and Engineering, University of Jinan,
No. 336, West Road of Nan Xinzhuang, Jinan 250022, Shandong, China
ise_dut@ujn.edu.cn
[2] Shandong Provincial Key Laboratory of Network Based Intelligent
Computing, University of Jinan, Jinan 250022, China

Abstract. Based on the traditional grid density clustering algorithm, proposing A Data Stream Clustering Algorithm Based on Density and Extended Grid (DEGDS). The algorithm combines the advantages of grid clustering algorithm and density clustering algorithm, by improving the defects of clustering parameters by artificially set, get any shape of the cluster. The algorithm uses the local density of each sample point and the distance from the other sample points, determining the number of clustering centers in the grid, and realizing the automatic determination of the clustering center, which avoids the influence of improper selection of initial centroid on clustering results. And in the process of combining the Spark parallel framework for partitioning the data to achieve its parallelization. For data points clustered outside the grid, the clustering within the grid has been effectively expanded by extending the grid, to ensure the accuracy of clustering. Introduced density estimation is connected and grid boundaries to merging grid, saving memory consumption. Using the attenuation factor to incremental update grid density, reflect the evolution of spatial data stream. The experimental results show that compared with the traditional clustering algorithm, the DEGDS algorithm has a large performance improvement in accuracy and efficiency, and can be effectively for large data clustering.

Keywords: Density clustering · Grid clustering · Data stream · Spark parallel

1 Introduction

With the rapid development of hardware technology, network communication technology, various sensing devices and various information technologies, in many applications [1], such as social networks, sensor networks, electronic commerce, network monitoring, meteorological and environmental monitoring, financial and retail enterprises, scientific and engineering experiment of dynamic link has produced lots of real-time dynamic data [2]. We call kind of data for the data stream. Different from the traditional data types, streaming data have the characteristics of large amount of data, infinite or unknown length, fast dynamic change, unstable rate, and high cost of accessing historical data. It is very difficult to use the traditional data mining algorithm

© Springer International Publishing AG 2017
D.-S. Huang et al. (Eds.): ICIC 2017, Part II, LNCS 10362, pp. 689–699, 2017.
DOI: 10.1007/978-3-319-63312-1_61

to deal with the streaming data, so the clustering algorithm based on data stream emerge at the right moment [3].

There are a variety of clustering algorithms, including partitioning methods, hierarchical methods, density-based, grid-based and model-based clustering algorithms [4]. Most of the traditional clustering algorithms use distance-based measurement methods. For data with continuous attributes, the cluster prototype is usually centroid, and the clustering results prefer spherical shape, and the clustering quality is poor on non-convex data. When faced with massive data, the more classical clustering algorithm has density-based clustering algorithm and grid-based clustering algorithm [5].

Aggarwal et al. proposed a dynamic CluStream algorithm for data stream, which provides the double-layer architecture for data stream analysis [6, 7]. The online process carries on the elementary clustering of the data stream, the on-line statistical data stream characteristic vector, uses the micro-cluster to store the summary information of the data stream regularly, form the initial clustering and dynamic results of data processing is to update, incremental processing and updating data. The off-line process analyzes the primary clusters obtained online according to the needs of the users [8, 9]. As the framework of the use of the idea of BIRCH, it is only sensitive to the clustering of spherical shape, but it is difficult to achieve good results for other types of data [10]. On the basis the algorithm, Aggarwal et al. in 2004 also proposed the HPStream algorithm. The algorithm uses projection method to solve the data stream difficult to high dimensional clustering problem. To dynamically select the smallest dimension and the clustering correlation, realizes a subspace clustering algorithm and uses the attenuation factor with the time decay of historical data, and too much in the clustering number, delete the first to join the cluster [11, 12]. CluStream and HPStream algorithms are based on K-means clustering ideas, the clustering results are usually spherical, without the ability to get any shape of the cluster [13]. In this paper, based on the improvement of CluStream algorithm, a Clustering Algorithm Based on Density and Extended Grid (DEGDS) algorithm based on data stream is proposed.

2 Clustering Algorithm

Based on the classical clustering algorithm Clustream, the algorithm follows the double layer frame structure of the algorithm, data processing of the form of parallel processing [14]. The algorithm divides the clustering process into two stages: online processing and offline adjustment. First, in the online layer to quickly receive has accumulated a period of data stream, get a sample set on the data stream, each of the data items to be processed is mapped into the corresponding grid cells [15]. The data find in the center of the micro-clusters of each grid according to the corresponding rules, and then form the initial clustering according to the adjacent grid boundaries. Recalculate the grid density after density attenuation on the basis of the incoming data points. Then the off-line adjustment process of the algorithm is relatively independent, after the off-line layer detects the change of the data stream, updates the clustering result, preserve pyramid time frame based on the clustering structure. After each grid cluster adjustment, the algorithm detecting isolated points outside the micro-clusters boundary though use the periodic characteristics of the data stream.

2.1   Online Processing

The DEGDS algorithm adopts distributed cluster computing framework based on memory computation. In the DEGDS algorithm, a Spark Streaming processing system that can deploy for streaming data to perform real-time data processing. Spark will enter the data stream by batch size segmentation, each piece of data convert into a spark RDD, a RDD partition corresponds to a grid unit, each partition node update grid information, adjust the grid density, thus form a new micro-cluster. Spark only needs to do the transformation of RDD. Finally, the RDD manipulate intermediate results and stored into memory.

Definition 1. Assume that the data set $(S_1, S_2, \ldots, S_d)$ has a d-dimensional attributes. And the data space $S = S_1 \times S_2 \times \ldots \times S_d$ is a d-dimensional data space. $x = (x_1, x_2, \ldots, x_d)$ represents the set of data points at the time t on the data space S. Divide each dimension $S_i(1 \leq i \leq d)$ of the data space into p copies which equal length and mutually disjoint, extending the interval length in the positive direction of each dimension so that form a finite super-rectangular area. Each hyper-rectangular area is a grid cell, any data record x can be mapped to a grid cell, denote by $g_{(x)} = (j_1, j_2, \ldots, j_d)$, where $j_i = 1, 2 \ldots p$, there are $N = \prod\limits_{i=1}^{d} p$ grid units. For the sake of generality, it is assumed that any dimension interval is a half-open and half-closed interval $[l_j, r_j)$ $(j = 1, 2, \ldots, d)$, then the point set x can be expressed as $x = [l_1, r_1) * [l_2, r_2) * \ldots l_d, r_d)$.

In order to further improve the computational efficiency, we can draw the grid length δ_j of the j_{th} dimension though the definition of grid unit, and can be expressed by the formula (1).

$$\delta_j = (r_j - l_j)/p \tag{1}$$

Owing to the streaming data has real-time dynamic characteristics, in order to avoid the historical data is too large for the current calculation of the proportion, real-time data on the mining results of the influence is too small, make the density of the grid unit can better reflect the distribution of the current data, the introduction of the density attenuation coefficient of the data weight $\lambda(0 \leq \lambda \leq 1)$. λ base on the weight threshold of the data and the life cycle. Assuming that the weight threshold is ω and the life cycle is τ, so the density attenuation coefficient satisfies $\lambda^\tau \leq \omega$. That is, after the τ cycle of the data, its weight is not greater than ω. The value of λ set by different values of ω and τ.

Definition 2. The grid cell density is the number of data points in the grid. When the new streaming data reaches the grid cell, the density of the grid cells is updated. Assuming that the density value of the data x at a time t is: D (x, t); then the grid density at time t + 1 is:

$$D(x, t + 1) = \lambda D(x, t) \tag{2}$$

Assuming that the grid cell g receives a new streaming data at T_n and the last time of the update is T_i ($T_n > T_i$), the density of the grid cell can be updated as follows:

$$D(g, T_n) = \lambda^{T_n - T_i} D(g, T_i) + 1 \tag{3}$$

When the data stream arrives, Spark Streaming segments the data stream. When each segment of data arrives, each RDD partition processes a stream of data and divides the data stream into the corresponding grid. The density of the clusters in the grid is higher than the density of the cluster outside and the density of the boundary points. There is a clear gap between the density of the clustering isolated point and the density of the cluster center point. Due to the same clustering set containing a connected region that a relatively higher density point set, the different clusters are distinguished by the relatively lower density point set area, so there is a large difference that area density between different clusters. In the grid cell, find the center of the grid, which can determine the number of clusters within the grid.

Definition 3. Since the density of the cluster center point is not less than the density of the point near it, where the distance between the two cluster centers is very far. Define the density value ρ_x of the sample point x:

$$\rho_x = \sum_y \chi(d_{xy} - d_c) \tag{4}$$

$$\chi(x) = \begin{cases} 1 & x \leq 0 \\ 0 & otherwise \end{cases} \tag{5}$$

Where d_{xy} is the distance from point y to point x, d_c is the critical value of distance. And ρ_x is the number of points whose distance from the sample point x is less than d_c.

Definition 4. δ is used to measure the distance that the sample point x and the other local density are larger than their own sample points. A δ_x is denoted by the repulsive point of the sample point x.

$$\delta_x = \min(d_{xy}) \quad y : \rho_y > \rho_x \tag{6}$$

The smaller of the repulsive point, explains the closer the point x and the point y, the greater the dependency of the point x on the point y, the more likely it is that the point x is the clustering of the point y; the larger the repulsive point, explains the farther the distance between the point x and the point y, the smaller the dependence of the point x on the point y, the more likely that the point x is to be an outlier or to another cluster. Data points ρ and δ distribution shown in Fig. 1.

RDD calculates ρ_x and δ_x of each data point in parallel in the current time, according to decision chart of ρ and δ distribution of data point. It can be determined that when the ρ and δ distributions are large, the data points are clustering centers, and there are several such points where there are many cluster centers. When the δ value is large but the ρ value is very small, the data points are likely to be isolated points, because they are far from the cluster center. This makes it possible to determine the

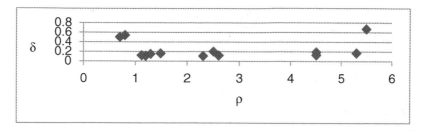

Fig. 1. Data points ρ and δ distribution

number of clusters in the grid. After the clustering center is determined, each of the other data points is divided into the corresponding cluster center that the closest to their own and the highest density, thus forming the clusters in the grid.

Clustering data points that outside the grid by extending the grid. Determining cluster center of inside each grid unit after, making a circle by the center point O as the center of a circle, the distance from O to A that the furthest point in the grid is the radius r. The extended area is shown in Fig. 2.

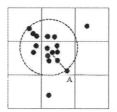

Fig. 2. Extended area

The influence of the data in the extended area on the micro-cluster by Inf ($0 < Inf < 1$) to judge, that $Inf = 1 - \sqrt{\sum_{j=1}^{d} (X_j - cen_j)^2 / \delta_j}$, where j represents the effect of the j_{th} data point on the grid in the extended area, the X_j represents the data points within the extended area, the cen_j represents the centroid of the grid unit, and the δ_j represents the j_{th} grid length. The greater the degree of influence, the greater the degree of agglomeration of clusters in the extended area, the higher the similarity; the smaller the degree of influence, the smaller the degree of agglomeration of clusters in the extended area, the smaller the similarity.

With the constant flow of data, new grid cells will be more and more. In the limited memory space conditions, must be in accordance with certain rules on the grid unit to merge, to save memory consumption. The grid density threshold has a great effect on the clustering results. However, based on the artificial set of grid density threshold has a lot of limitations. When the data points in the grid accumulate to a certain extent, the grid density thresholds can be set in advance to distinguish between different types of grid cells. Suppose the density of the largest grid cell is Den_max and the density of the non-empty minimum grid cell is Den_min, the average density of all the grid cells is denoted by Den_ave. The Den_ave $= \frac{\sum_{i=1}^{n} Di}{n}$, where D_i is the grid cell density and n is the number of non-empty grid cells.

Dense grid threshold is $D_m = \frac{Den\_max + Den\_ave}{2}$. Sparse grid threshold is $D_n = \frac{Den\_min + Den\_ave}{2}$. when $D(g,t) \geq D_m$ is dense grid unit; when $D_n \leq D(g,t) < D_m$ is transition grid unit; when $D(g,t) < D_n$ is sparse grid cell.

Suppose there are any two grid cells g_1 and g_2 in the d-dimensional space, if and only when the two grid cells exist an intersection in any dimension or have the same common plane, so the two grid cells called adjacent grid cell. If the grid cells g_1 and g_2 are adjacent grid cells, and g_2 and grid cell g_3 in the space are also adjacent grid cell, so then g_1 and g_3 are also adjacent grid cell. The adjacent grid cells are shown in the Fig. 3.

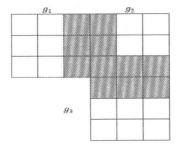

Fig. 3. Adjacent grid cells

Whether or not the grid can be merged is determined by the level of the inter-connection between the grid and the boundary of the neighbouring grid elements. If the two adjacent grids are dense mesh cells, they can be merged into one cluster. If there is a non-dense mesh cell between two adjacent grids, whether the grid can be combined is determined by the boundary of the adjacent grid cell.

When the new data stream is coming, the dense two grid units are transformed from the non-dense grid to the dense grid, which satisfies the condition of the grid unit merging and forms a new grid. According to the new grid for each data point ρ_x and δ_x, recalculation of grid density. For grid cells that do not meet the dense conditions, they are merged according to the adjacent boundaries. The length of the initial mesh unit was reduced, after repeated tests, set the reduced length to $\delta_j/4$. If you want to merge two adjacent non-dense grid units, you only need to find the adjacent boundaries of the two grid units. When the streaming data is updated, RDD only needs to transfer the boundary grid data of each grid unit to each partition, and micro-cluster that formed within the grid cell. Micro-clusters are globally collapsing and finally return the results to the master node.

The process of the algorithm is shown in Table 1:

Table 1. Online processing algorithm flow

Online processing

Step1	Calculate the local density of each point in the grid ρ_x .
Step2	According to the data distribution of ρ_x and δ_x, select the center point of the cluster in the grid.
Step3	According to the center of the cluster, divide the remaining points into clusters in the grid.
Step4	Search the grid g within the distance from the center point O the farthest data point x_j.
Step5	Calculate the distance between x_j and the center point O as the extension radius r.
Step6	According to the influence degree Inf, it is judged whether or not the point in the extended area is a point in the micro-cluster.
Step7	Scan the extended grid, and any adjacent dense mesh cells into a cluster; the non-dense grid cell examines its boundaries.
Step8	Scan all grid cells until all data has been updated.

2.2 Reflections on the Isolated Point

Unlike the isolated points of static data, the isolated points of streaming data are constantly changing. It is divided into two cases, one situation is in the dynamic space, the streaming data point may not always be isolated, but if it is always isolated over a long period of time, then it can be considered that this point is isolated. There is also a situation where the data points appear in the remote area of space, reflected in the space of the chance, and in the static environment similar to the situation. In the formation of clustering, to eliminate the isolated point, because the isolated point is a very strong noise in the data stream, it not only does not reflect the overall characteristics of the data stream, but will be in the final clustering results bring great error. For the above two types of isolated points, should to be deleted.

2.3 Off-Line Process

The off-line adjustment process is relatively independent, and its work is based on the accumulation of intermediate knowledge base. As the time goes on, the continuous data stream is coming, and the result of clustering is updated. The longer the data, the importance will be worse. Therefore, the attenuation coefficient is introduced, and the weight of the historical data is reduced. Using the incremental update of the way, in the original cluster clustering results to update, this method has good shrinkage and high efficiency. When a new data stream arrives, the data space is composed of initial micro-clusters and the new data affects the grid. Get the result of the mining according to pyramid time frame structure to store data.

3   Experimental Results

In the local cluster, the use of real data set for optimized algorithm for comparative experimental analysis. The Spark cluster selection is built on the VMware server virtualization platform. The Spark local cluster consists of a master node and several worker nodes. Each node is a dual-core, 4 GB memory PC. Each node's operating system is Linux Ubuntu 12.04 version. All nodes in the cluster are installed with JDK 1.8.0, Spark-1.2.0, Scala-2.10.4. The program is written by Scala. The algorithm tests the performance of the algorithm in a stand-alone environment and in a cluster environment. Using a thermal power plant during the 2015.11.13–2015.11.23 data as test data, the data sampling period of 60 s, about 15,000 data records. Although the amount of data is small, but as a test has been sufficient, the follow speed-up calculation can be extended to big data situation.

3.1   Clustering Quality Evaluation

In the experiment, the clustering quality is defined as the proportion of the data point clustering correctness in the data set. The horizontal axis represents the size of the data set, and the vertical axis represents the accuracy of the cluster. The algorithm is compared with CluStream algorithm and D-Stream algorithm.

It can be seen from the Fig. 4 that the average clustering accuracy of the algorithm can reach more than 97%, which is obviously better than CluStream algorithm and D-Stream algorithm. Because the DEGDS algorithm divides the sample points into the cluster center, allocates the data points more rationally into the micro-cluster, and improves the clustering accuracy. CluStream algorithm unable to find any shape of the cluster, unable to be a good identification of outliers, making the accuracy of clustering decreased. D-Stream algorithm uses a meshing method, the density of the boundary region of the cluster is relatively low, and many boundary grids contain noise data, which will affect the clustering accuracy.

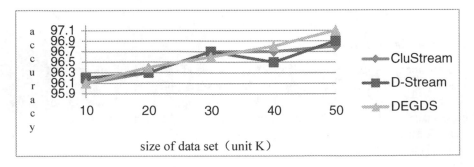

Fig. 4. Comparison of clustering accuracy

3.2   Execute Time Test

In order to test the running speed of the algorithm, the experiment uses a four-dimensional simulation data stream to test data set, the size of 50 K. In the case of the same size of each data set, the algorithm runs several times, each record the running time, statistical average.

It can be seen from the Fig. 5 that when the test data set is small, the algorithm's time is really slower than the execution time of the D-Stream algorithm on a single node. That is because in the Spark framework under the parallel computing convection data partition, the clustering results have some interference. When the data stream is growing, the DEGDS algorithm is basically linearly increasing, which proves that the algorithm has good scalability. The D-Stream algorithm uses a grid map to represent the mesh density in terms of the number of spatial data points, avoiding a large number of distance calculations and reducing the time complexity.

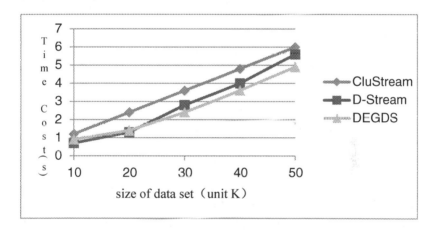

Fig. 5. Execute time test

3.3   Speed-up

Speed-up, usually refers to the completion of the same operation, in the stand-alone environment system running time spent with multiple processors parallel processing time spent by the ratio. The higher the speed-up, the relative time of parallel computing is less, and the higher the efficiency and performance of parallelism. The speed-up $S_{(i)}$ can be defined as the formula (7). In order to test the efficiency of parallel processing operations, the experiment uses different data sets, data sets D1, D2, D3, respectively, with 50,000 records, 100000 records and 400,000 records.

$$S_{(i)} = \frac{T_1}{T_{(i)}} \tag{7}$$

Where T_1 represents the time required for the job to run under a stand-alone processor, $T_{(i)}$ represents the time required to process the job in parallel under i processor.

It can be seen from the Fig. 6 that the speed-up in the Spark framework can maintain a linear increase with the growth of the number of nodes, mainly because the number of nodes increases, which can shorten the time required for clustering, which means that Spark has a great advantage in big data. It is foreseeable that when the number of nodes increases, the performance of the algorithm is more obvious. In the case of the data set remains unchanged, with the increase in the number of nodes, the speed-up also increases, and finally stabilized trend. Because each node in the cluster needs to communicate with each other, and when the network node increases, the network overhead also increases. Plus the data set itself is not very large, unable to fully reflect the Spark ability to deal with data in parallel environment, so the final speed-up tends to slow.

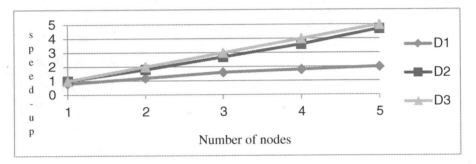

Fig. 6. Speed-up

4 Summary and Future Work

In order to solve the clustering problem of data stream in space, this paper discusses several basic problems of data stream clustering, and gives the corresponding solutions, propose a stream clustering algorithm based on density clustering and grid clustering. The algorithm utilizes the advantages of Spark parallel platform to improve the efficiency of clustering. The experimental results show that the algorithm has better performance in terms of accuracy and efficiency.

In this paper, study several basic problems of data stream clustering analysis, and propose a new solution. But there are still many problems that need further study. There is no way to dig out clustering from a multi-granularity perspective. And how to deal with the problem of isolated points, so reduce the error of clustering results.

References

1. J. Comput. Appl. **36**(12), 3292–3297 (2016)
2. Fiori, A., Mignone, A., Rospo, G.: DeCoClu: density consensus clustering approach for public transport data. Inf. Sci. **328**, 378–388 (2016)
3. Tang, Y.: A distributed data flow clustering algorithm based on grid block. Small Microcomput. Syst. **37**(3), 488–493 (2016)

4. Gao, Y.: A data flow clustering algorithm based on grid and density. Comput. Sci. **35**(2), 134–137 (2008)
5. Ma, C., Hong, S.: A dense peak clustering algorithm based on cluster center point automatic selection strategy. Comput. Sci. **43**(7), 255–258 (2016)
6. Jiang, L.: Optimization of fast clustering algorithm for fast search and discovery density. Appl. Res. Comput. **33**(11), 3251–3254 (2016)
7. Zheng, Y.: Data flow clustering algorithm based on mobile grid and density. Comput. Eng. Appl. **45**(8), 129–131 (2009)
8. Feng, C.: Data Flow Clustering Analysis Algorithm. Fudan University (2006)
9. Chen, J.Y., He, H.H.: A fast density-based data stream clustering algorithm with cluster centers self-determined for mixed data. Inf. Sci. **345**(C), 271–293 (2016)
10. Skála, J., Kolingerová, I.: Dynamic hierarchical triangulation of a clustered data stream. Comput. Geosci. **37**(8), 1092–1101 (2011)
11. Samwel, B., Whipkey, C.: Efficient top-down hierarchical join on a hierarchically clustered data stream (2016)
12. Krawczyk, B., Stefanowski, J., Wozniak, M.: Data stream classification and big data analytics. Neurocomputing **150**, 238–239 (2015)
13. Nguyen, H.L., Woon, Y.K., Ng, W.K.: A survey on data stream clustering and classification. Knowl. Inf. Syst. **45**(3), 1–35 (2015)
14. Xu, S., Wang, J.: Dynamic extreme learning machine for data stream classification. Neurocomputing **238**, 433–449 (2017)
15. Xiaoyun, C., Yufang, M., Yan, Z., et al.: GMDBSCAN: multi-density DBSCAN cluster based on grid. In: IEEE International Conference on E-Business Engineering, pp. 780–783. IEEE (2008)

Credit Risk Assessment Based on Long Short-Term Memory Model

Yishen Zhang[1,2], Dong Wang[1,2(✉)], Yuehui Chen[1,2(✉)],
Huijie Shang[1,2], and Qi Tian[3]

[1] School of Information Science and Engineering,
University of Jinan, Jinan 250022, China
{ise_wangd,yhchen}@ujn.edu.cn
[2] Shandong Provincial Key Laboratory of Network Based Intelligent
Computing, University of Jinan, Jinan 250022, China
[3] University of Toronto Mississauga,
3359 Mississauga Road, Mississauga, Canada

Abstract. At present, with continuously expanding of Chinese credit market, thus large amounts of P2P (person-to-person borrow or lend money in Internet Finance) platform were born and have been in development. Most of P2P platform in China carries out the credit risk evaluation of loan applicant by data mining method. As an emerging data mining tool, the artificial neural network has better classification capability. The improvement of risk assessment capabilities of applicant can effectively reduce the overdue rate of analysis, thus in this paper, a kind of credit risk evaluation model based on the Long Short-Term Memory (LSTM) model is presented. The sample data of overdue and non-overdue credits are provided by Hengxin Investment Consulting Co., Ltd. in Jinan, by which the model is established. After the trial, this model is applied to the aspect of overdue classification of credit evaluation with higher accuracy.

Keywords: Artificial neural network · Credit risk assessment · Long Short-Term memory

1 Introduction

Loan on credit is a kind of unsecured loan. In recent years, the credit market has rapidly expanded in China, on the one hand, the cash conversion cycle is getting shorter and shorter because of the China's emerging economy, and on the other hand, the request for fund of the merchants is getting higher and higher. Thus, large amounts of P2P online inclusive financial platform emerge at the right moment [1]. However, considering that a rounded credit evaluation system has not been formed and P2P appeals different customer group and banks, especially to the customers without securities which will have smaller restraint capability to customers. Thus better risk evaluation results can be obtained by establishing corresponding credit risk evaluation model [2–4], and large amounts of platform are exploring their own method of credit risk evaluation, wherein, majority of platforms adopt data mining method, that is to get a

© Springer International Publishing AG 2017
D.-S. Huang et al. (Eds.): ICIC 2017, Part II, LNCS 10362, pp. 700–712, 2017.
DOI: 10.1007/978-3-319-63312-1_62

better trip on the authenticity and the validity of customer information according to the obtained customer information, to properly evaluate the financial position of customer, to accurately estimate the operating condition, repayment willing and capacity of the borrower.

The establishment of good credit risk evaluation model is the greatest challenge for P2P platform. Therefore, the establishment of credit risk evaluation model will be of great importance in against the occurrence of stagnant and bad debts, driving the circulation speed of fund and safeguarding the security and stabilization of fund [5, 6]. Nowadays, in credit risk evaluation field, we can get some good progress in methods such as the artificial neural network, genomic programming, support vector machine as well as logistic regression and some mixed model in terms of performance and precision. Over the past few years, based on the customer information data, many trials have been carried out by adopting many excellent algorithms and research methods in credit risk evaluation field, wherein, Khashman (2010) carried out a research on the artificial neural network algorithm with a precision rate of 83.6% by using Germanic customers data-sets [7]; Bekhet (2014) carried out a research on the logistic regression algorithm and RBF network with a precision rate of 86.5% respectively by using a dataset from Jordan Commercial Bank [8], since the logical regression algorithm used in Bekhet's paper does not a kind of neural network, we would not compare our method with the logistic regression algorithm; Wang (2016) carried out a research on the Improved Back Propagation (BP) neural network algorithm with a precision rate of 86% [9].

In those above studies, the data sets are from different regions, but the data sets are all from customer data, and the dimensions of the feature descriptions are very similar, so the classification accuracy of the gap is very small, the gap comes from the effectiveness of the algorithm used, there is still much room for improvement. Our method effectively improves the classification accuracy through three aspects of the work. The first is to use a description of the information characteristics of the customer more in line with the geographical characteristics, and objective reality, especially the selected feature dimensions which could effectively verification, this way can be described by the customer through the feature dimensions more accurate. The second is to focus on maintaining the interrelationship between dimensions when choosing to describe the dimensions of the customer, for example, Bekhet used some dimensions like Total income, Loan purpose, Loan amount, interest rate and other dimensions are not related to each other, so it cannot accurately judge the customers' risk situation, and there is a strong correlation between the dimensions chosen in this paper. The third is choosing a classifier with stronger learning ability in the interrelationship between dimensions, and the interconnection between the three gates of multiple cells in Long Short-Term Memory (LSTM) has a strong learning ability for the interrelationship between dimensions, so the LSTM is used as a classifier in this paper.

After the research of this paper, a new representation method of customer information using LSTM model as the classifier is presented. During 10-fold cross validation, it is proved that our method has better sort results.

2 Definition of Data Collection and Dimensionality

Customer information data can be described upon many dimensions, in this paper, 300 positive samples (overdue customer) and 300 negative samples (non-overdue customer), 600 in total, each accounting for 50% of total number, is randomly selected from more than 2000 customers to whom the matching of creditor's right was carried out between 2014 and 2016 by Hengxin Investment Consulting Co., Ltd. in Jinan. After the research, ultimately, 13 dimensions are selected to describe and evaluate the customer information, and the standard of selected dimension is: (1) without containing of customer identity information; (2) from the aspect of human review, the exclusion of subjective information of customer, the dimension with difficulty of verifying and calculating, such as the intended use of the loan, business model, profit etc., only containing the objective information that can be verified by the third party.

Selecting the dimensions according to this standard, we will make the correct classification by maximizing the presentation of the customer objective data that is hard to forge, and excluding the subject-reported data with a real basis, and the name, value, definition, selecting reason as well as the verification method of 13 dimensions selected in this paper are as shown in Tables 1 and 2 shows the examples of datasets, and the abbreviations for dimensions used in Table 2 are listed at the end of each line in the first column of Table 1.

Table 1. Proposed variables for building dataset

Name	Value	Definition	Selecting Reason	Verification Method
Gender, G	0 or 1	0: Female 1: Male	The performance capacity of men and women under the same environment is different, it can affect the performance	Verified by the Sex in ID card
Education, E	1–4	1: Middle School 2: High School 3: Junior School 4: Bachelor degree or above	To some extent, the education can reflect personal quality, which will affect the implementation of the results	Verified by the education background in the individual credit reporting issued by China Higher-education Student Information and the Credit Information Center of People's Bank of China

(continued)

Table 1. (*continued*)

Name	Value	Definition	Selecting Reason	Verification Method
Age, A	Actual value	The number of years that customer have lived	People of different ages have different experiences, strengths and potentials, and the behavioral pattern of their age will affect the performance	Verified by the date of birth in ID card
Marital Status, M	0–2	0: Single 1: Married 2: Divorced	Marital status can reflect the customer's maturity and responsible attitude from the side, the status reflected in this item will affect the performance	Verified by the customer whether to provide the marriage certificate, divorce certificate and the marital status provided in individual credit reporting issued by the Credit Information Center of People's Bank of China
Household Type, HT	0 or 1	0: Local registered permanent residence 1: non-local registered permanent residence	The different nature of the account will have corresponding restraint capability by the platform to the customer, and different sense of urgency of customer will affect the performance	Verified by the Address and Authority in ID card
Establishment period of work unit, EW	Actual value	The number of years that the work unit where customer worked have been set up	Establishment period of work unit reflects the stability and the strength of the work unit from the side, which consequently can affect the future income of customer as well as the performance	Verified by the registration date searched in the National Enterprise Credit Information Publicity System of the People's Republic of China by using the name of the work unit

(*continued*)

Table 1. (*continued*)

Name	Value	Definition	Selecting Reason	Verification Method
Industrial Classification, IC	0–5	The industrial classification that the customer is engaged	This item mainly targets individual businesses, as the different business is with different development potential and level of profits, which consequently can affect the future income of customer as well as the performance	Verified by the Scope of Business in the Business License
Positional Ranks in the Work Unit, PR	0–4	0: General staff 1: Primary administrant 2: Middle-manager 3: Senior management 4: Cooperate person or founder	Different positional ranks can reflect the customer's management ability and coordination ability, then his or her personal quality from the side, and different personal quality will affect the performance	Verified by the work permit, the capital verification report and the registration information of enterprise
Income, I	Actual value	For the last six months, the difference of total amount of input and outlay in the customer's itemized account printed by the bank of deposit of each bank card	Only the objective value that can be queried by the third party can be calculated, which do not conclude the cash income described by the customer but hard to verify, and the earning power of customer should be evaluated from the perspective that can be fully affirmed	Verified by the customer's itemized account printed and sealed by the bank of deposit of each bank card in the last six months
Debt, D	Actual value	The sum of average used credit line in all credit cards of the account without cancellation in the last six months and the total debt	Only the objective value that can be queried by the third party can be calculated, which do not conclude the stock account and the	Verified by the credit extension and the summary of debt information of individual credit reporting issued by the Credit

(*continued*)

Table 1. (*continued*)

Name	Value	Definition	Selecting Reason	Verification Method
		information in the credit information report	fund account, and the debt of customer should be evaluated from the perspective that can be fully affirmed	Information Center of People's Bank of China
House Ownership, HO	0–2	0: No house property 1: Purchased entirely with cash 2: Bullet house-purchase	Different house ownership reflects the financial situation of customer, thus affecting performance	Verified by the Building Ownership Certificate and the Real Property Right Certificate
Ownership of Vehicle, OV	0–2	0: No car 1: Car being purchased entirely with cash 2: Car purchased by loan	Different ownership of vehicle reflects the financial situation of customer, thus affecting performance, but has relatively weak impact compared with the house ownership	Verified by the Motor Vehicle Register Certificate
Overdue Times in Credit Information Reports, OR	Actual value	The sum of overdue times for each item in line items of credit transaction in credit information report	This item reflects the historical performance of the customer, thus affecting the performance	Verified by the Credit Transactions in the individual credit reporting issued by the Credit Information Center of People's Bank of China

Table 2. Examples

No.	G	E	A	M	HT	EW	IC	PR	I	HO	OV	OR	D
1	1	3	48	1	2	4	0	1	150	2	1	3	137.8
2	1	2	36	1	2	10	1	1	30	0	1	0	41.2
3	2	3	49	1	1	2	1	1	48	1	1	2	91.5
4	1	3	54	1	2	11	3	1	102	2	0	4	16.5
5	1	3	36	1	3	7	1	1	20	0	0	3	69.8

The innovative way of describing the characteristics of customer information was proposed by us has the following advantages: (1) integrity, the selected 13 dimensions contain information that all customers can provide; (2) objectivity, all dimensions are objective features, without the subjective description of the customers, without the need for subjective evaluation; (3) accuracy, all dimensions can be verified by official means, and those were shown in Table 1; (4) correlation, the dimensions were selected after thoughtful consideration, different from the traditional scorecard model which each dimension has its own scoring criteria. For example, a 20-year-old young man has multiple sets of real estate and vehicles with a monthly income of more than ¥10,000, after being judged by human beings, would classify as an information fraud customer or no need to require a loan customer, but in above researches, since the selected description of the dimensions were not sufficient or the evaluation criteria were relatively independent, and the algorithms were lack of learning in mining the interrelationships between the dimensions, this case will be recognized as a quality customer. However, the dimensions we selected have a close correlation, and after the learning through all the examples by LSTM model, it can judge the rationality of the relationship between the values of all dimensions, resulting in higher classification accuracy.

The 600 samples are based on the statistics in Table 1. In order to give full play to the advantages of LSTM, each element of all samples in dataset will sent into input unit of LSTM model one by one. In this way, LSTM model can get better performance in mining the relation of each element in the classification samples, thus, LSTM model can get more details and get better classification accuracy from the learning.

3   Classification Method

3.1   Long Short-Term Memory

LSTM model constructs a special memory storage unit with the mutual complex connection of all three gates in each Memory Cell and adept at find and create a dependency between the elements before and after which are from the examples [10]. This model is suitable for mining the correlation of all 13 dimensions in this research. LSTM adds a structure called Memory Cell to remember the past information in the neural nodes of the hidden layer of the recurrent neural network. The purpose for this structure is to record the historical information, LSTM model also adds three gate structure: Input Gate, Forget Gate and Output Gate, these are used to control the usage of the historical information [11]. The structure of Memory Cell in LSTM model is shown in Fig. 1.

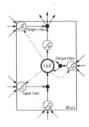

Fig. 1. The structure of Memory Cell in LSTM model

Input Gate determines whether an information from the input layer can be added into the node of the hidden layer and denoted by ι. Output Gate determine whether an output from the current node can be passed on to the next layer and denoted by ω. Forget Gate determine whether the historical information stored inside the node of the current hidden layer will be kept or not and denoted by φ. s^t_c represented the information stored at time t. The input value of an Input Gate consists of three values which include: (1) output vector from the node in the input layer; (2) output vector from the cell in the previous layer; (3) the information kept by the previous cell. We use a^t_l to represent the input vector of the Input Gate of time t, then the output vector which come from the activation function of this Input Gate at time t is calculated by

$$b^t_l = f(a^t_l)$$ (1)

The input value of Forget Gate also consists of three input vector, which is the same as the input value of the Input Gate. We use a^t_φ to represent the input vector of the Forget Gate at time t, then the output vector which come from the activation function of this Forget Gate at time t is calculated by

$$b^t_\varphi = f(a^t_\varphi)$$ (2)

The input of the cell unit consists of two parts: (1) the input vector of the input layer; (2) the output from the Output Gate of the previous hidden layer. We use a^t_c to represent the input vector of cell unit of time t. Forget Gate determine whether the value of the historical information will be kept or not and calculated by

$$s^t_c = b^t_\varphi s^{t-1}_c + b^t_l g(a^t_c)$$ (3)

The input value of Output Gate consists of three parts: (1) the output vector of the input layer; (2) the output vector of the cell from the previous hidden layer; (3) the information which recorded by the current cell unit. We use a^t_ω to represent the input vector of the Output Gate at time t, then the output vector which come from the activation function of this Output Gate at time t is calculated by

$$b^t_\omega = f(a^t_\omega)$$ (4)

As a result the output vector of cell unit is calculated by

$$b^t_c = b^t_\omega h(s^t_c)$$ (5)

In Eq. (5), h represent the activation function.

The output vector of cell unit which is also known as the output vector of the hidden layer as the input vector of the output layer and calculated by

$$a_k^t = \sum_{c=1}^{h} \omega_{ck} b_c^t \tag{6}$$

The final output vector which come from the output layer is calculated by

$$b_k^t = f\left(a_k^t\right) \tag{7}$$

Based on the Back Propagation Algorithm of Time (BPTT), we can conclude the updated weight from node i to node j at time t is calculated by

$$\omega_{ij} = \omega_{ij} - \eta \delta_j^t b_j^t \tag{8}$$

In Eq. (8), δ_j^t represent the residual value of node j at time t.

3.2 Prediction Assessment

In the statistical analysis, two methods can be used to check the effectiveness of classification in practical applications, that is test of independent data set and 10 fold cross validation test. In 10-fold cross validation test, 10% of all samples were treated as test data, and the remaining 90% were used as training data. Each data set is calculated for the overall accuracy (OA). In addition, two indicators, the sensitivity and specificity, are used to assess the accuracy of classification.

4 Discussion and Results

After this research, 10-fold cross validation is carried out on the data set containing all 600 samples by using the LSTM model, which means that 540 training samples and 60 test samples have been used in each trial, and 10 repeated trials have been conducted on each data set, that is 100 trials in total. The results in Table 3 show that the average classification accuracy of the test set is 92.97%. In the Table 3, the terms are abbreviated, "T" means "trail", "D" means "data", "A-acc" means "Average accuracy rate".

Table 3. The results of LSTM model in 10-fold cross validation

		T0	T1	T2	T3	T4	T5	T6	T7	T8	T9	A-acc
D0	miss	5	6	7	1	5	4	1	4	7	6	92
	acc	92	90	88	98	92	83	83	88	93	91	
D1	miss	5	5	4	3	6	8	2	3	5	3	93
	acc	92	92	93	95	90	87	97	95	92	95	
D2	miss	6	5	4	2	6	2	0	3	5	1	94
	acc	90	92	93	97	90	97	100	95	92	98	
D3	miss	3	1	5	2	3	4	4	4	4	6	94
	acc	95	98	92	97	95	93	93	93	93	90	

(continued)

Table 3. (*continued*)

		T0	T1	T2	T3	T4	T5	T6	T7	T8	T9	A-acc
D4	miss	4	5	9	4	3	2	1	7	3	5	93
	acc	93	92	85	93	95	97	98	88	95	92	
D5	miss	7	4	7	5	6	7	5	9	10	8	89
	acc	88	93	88	92	90	88	92	85	83	87	
D6	miss	6	5	0	5	3	4	0	2	2	0	96
	acc	90	92	100	92	95	93	100	97	97	100	
D7	miss	4	4	3	7	4	5	1	3	2	4	94
	acc	93	93	95	88	93	92	98	95	97	93	
D8	miss	3	5	8	4	5	2	5	4	5	0	93
	acc	95	92	87	93	92	97	92	93	92	100	
D9	miss	2	2	4	5	3	9	6	5	6	4	92
	acc	97	97	93	92	95	85	90	92	90	93	
OA												92.97

Since the data sets used in those studies were involved privacy information of customers, we cannot get the data sets used in the previous studies, so we used the dimensions definition methods of the data sets which are mentioned in previous studies in our customer database for the corresponding collection and collation, the dimensions which selected by Wang were similar to ours, but we did not suggest the method of data collation and validation, so we temporarily consider that we use the same data set with Wang. Based on the data set which collated by the Khashman's dimensions definition method, 10-fold cross validation is carried out on the Artificial neural networks, Radial basis function scoring model and Improved BP Neutral Network, the comparison of our method and other methods is shown in Table 4.

Table 4. The results of the experiment based on the Khashman's dimensions definition method

Algorithm	Accuracy (%)	Sens (%)	Spec (%)
Improved BP Neutral Network	80.0	86.7	73.3
Radial basis function scoring model	81.7	80.0	83.3
Artificial neural networks	68.3	73.3	63.3
This method	86.7	88.3	85.0

Based on the Bekhet's dimensions definition method, 10-fold cross validation is also carried out on those methods, and the results are shown in Table 5.

Based on ours dimensions definition method, the results are shown in Table 6.

We can draw conclusions by experiments in the above three sets of comparative: (1) the experiment based on the data set which collated by Khashman's dimensions definition method, the classification accuracy of Artificial neural networks algorithm is very close to the experiment at 900:100 (68.3% vs 68%) of Khashman, the experiment based on the data set which collated by Bekhet's dimensions definition method, the

Table 5. The results of the experiment based on the Bekhet's dimensions definition method

Algorithm	Accuracy (%)	Sens (%)	Spec (%)
Improved BP Neutral Network	85.0	90.0	80.0
Radial basis function scoring model	86.7	85.0	88.3
Artificial neural networks	76.7	80.0	73.3
This method	88.3	88.3	88.3

Table 6. The results of the experiment based on ours method

Algorithm	Accuracy (%)	Sens (%)	Spec (%)
Improved BP Neutral Network	88.3	91.6	85.0
Radial basis function scoring model	88.3	86.7	90.0
Artificial neural networks	78.3	81.7	75.0
This method	92.97	93.15	92.79

classification accuracy of the Radial basis function scoring model is very close to the experiment which down by Bekhet (86.7% vs 86.5%), the experiment based on the data set which collated by ours method, the classification accuracy of the Improved BP Neutral Network is very close to the experiment which down by Wang (88.3% vs 86%). It can be explained that these experiments using the data sets in different regions, but are all the collection of customer information, so there is no difference in the expression of customer information. (2) The dimensions definition method proposed by Khashman is extensive, but it is relatively small in effective and objective. For example, the dimensions like Have telephone or not, Foreign worker and so on have no important impact in judging the credit risk of customers, so it shows that the basis for improving the classification accuracy is the dimensions definition method which has the completely, objectively and accurately. (3) Although the dimensions definition method proposed by Bekhet is more objective and complete, it did not propose a method of verification, so there is room for improvement in accuracy of customer information, and in using the scoring model, each dimension of the model is independent and also lack of contact between each other, so compared to the experiment based on the data set collated by Khashman's method, the classification accuracy of these algorithms increased by nearly 5%, but there is still room for improvement. (4) The dimensions definition method proposed by us is complete, objective, accurate and relevant, it is made up for the shortcoming of insufficient in previous data sets of the customer information, so compared the experiment based on the data set collated by ours method to the experiment based on the data set collated by Bekhet's dimensions definition method, the classification accuracy rate of other three models have a upgrade nearly 2%. This proves the validity in our dimensions selection, not only that, since the relevance of our data set dimensions, LSTM model which has the advantage in learning interrelationships between elements base on this data set plays extremely, so the classification accuracy rate of the LSTM model has increased nearly 5%, this shows the effectiveness of the framework and the tight fit with dimensions and the model which we proposed. Both of an excellent dimensions definition method and a model with specialty are indispensable to get a higher classification accuracy.

5 Conclusion

During this research, the redesigned and redefined characteristic dimension of customer information and LSTM model is presented for the credit risk evaluation field, compared with other methods, the method presented in this research improves all evaluation indicators with different scales, which illustrates the effectiveness of adopting this presentation method of characteristic dimension for customer information and LSTM model. In future, we will continue to search for more effective classifier to obtain better classification accuracy on this data set.

Acknowledgements. This research was supported by Na National Key Research and Development Program of China (No. 2016YFC0106000), National Natural Science Foundation of China (Grant No. 61302128, 61573166, 61572230, 61671220, 61640218), the Youth Science and Technology Star Program of Jinan City (201406003), the Natural Science Foundation of Shandong Province (ZR2013FL002), the Shandong Distinguished Middle-aged and Young Scientist Encourage and Reward Foundation, China (Grant No. ZR2016FB14), the Project of Shandong Province Higher Educational Science and Technology Program, China (Grant No. J16LN07), the Shandong Province Key Research and Development Program, China (Grant No. 2016GGX101022).

References

1. Jiang, D., Li, X.: The study on the credit risk assessment of borrower in P2P network of China. In: Xu, J., Hajiyev, A., Nickel, S., Gen, M. (eds.) Proceedings of the Tenth International Conference on Management Science and Engineering Management. AISC, vol. 502, pp. 1619–1630. Springer, Singapore (2017). doi:10.1007/978-981-10-1837-4_131
2. Guo, Y., Zhou, W., Luo, C., Liu, C., Xiong, H.: Instance-based credit risk assessment for investment decisions in P2P lending. Eur. J. Oper. Res. **249**(2), 417–426 (2016)
3. Blanco, A., Mejias, R., Lara, J., Rayo, S.: Credit scoring models for the microfinance industry using neural networks: evidence from Peru. Exp. Syst. Appl. **40**(1), 356–364 (2013)
4. Heiat, A.: Comparing performance of data mining models for computer credit scoring. J. Int. Fin. Econ. **12**(1), 78–83 (2012)
5. Chen, N., Ribeiro, B., Chen, A.: Financial credit risk assessment: a recent review. Artif. Intell. Rev. **45**(1), 1–23 (2016)
6. Oricchio, G., Lugaresi, S., Crovetto, A., Fontana, S.: Banking crisis and SME credit risk assessment. In: Oricchio, G., Crovetto, A., Lugaresi, S., Fontana, S. (eds.) SME Funding, pp. 1–6. Palgrave Macmillan, London (2017)
7. Khashman, A.: Neural network for credit risk evaluation: investigation of different neural models and learning schemes. Exp. Syst. Appl. **37**(9), 6233–6239 (2010)
8. Bekhet, H., Eletter, S.: Credit risk assessment model for Jordanian commercial banks: neural scoring approach. Rev. Dev. Finance **4**(1), 20–28 (2014)
9. Wang, L., Chen, Y., Zhao, Y., Meng, Q., Zhang, Y.: Credit management based on improved BP neural network. IHMSC **1**, 497–500 (2016)
10. Hochreiter, S., Schmidhuber, J.: Long short-term memory. Neural Comput. **9**(8), 1735–1780 (1997)
11. Bao, W., Chen, Y., Wang, D.: Prediction of protein structure classes with flexible neural tree. Bio-Med. Mater. Eng. **24**(6), 3797–3806 (2014)

12. Ji, Z., et al.: NMFBFS: a NMF-based feature selection method in identifying pivotal clinical symptoms of Hepatocellular carcinoma. Comput. Math. Methods Med. **2015**, 1–12 (2015)

13. Huang, D.S.: Systematic Theory of Neural Networks for Pattern Recognition (in Chinese). Publishing House of Electronic Industry of China, May 1996

14. Ji, Z., Xia, Q., Meng, G.: A review of parameter learning methods in bayesian network. In: Huang, D.-S., Han, K. (eds.) ICIC 2015. LNCS, vol. 9227, pp. 3–12. Springer, Cham (2015). doi:10.1007/978-3-319-22053-6_1

15. Xu, L.-L., et al.: Immune-based rough sets attribute reduction algorithm and its application. Comput. Eng. Des. **30**(22), 5158–5161 (2009)

16. Huang, D.S., Du, J.-X.: A constructive hybrid structure optimization methodology for radial basis probabilistic neural networks. IEEE Trans. Neural Netw. **19**(12), 2099–2115 (2008)

17. Gers, F.A., Schmidhuber, J., Cummins, F.: Learning to forget: Continual prediction with LSTM. Neural Comput. **12**(10), 2451–2471 (2000)

18. Han, S.-Y., Chen, Y.-H., Tang, G.-Y.: Sensor fault and delay tolerant control for networked control systems subject to external disturbances. Sensors **17**, 700 (2017)

19. Han, S.-Y., Zhang, C.-H., Tang, G.-Y.: Approximation optimal vibration for networked nonlinear vehicle active suspension with actuator time delay. Asian J. Control (2017). doi:10.1002/asjc.1419

The Feature Extraction Method of EEG Signals Based on the Loop Coefficient of Transition Network

Mingmin Liu[1,2], Qingfang Meng[1,2(✉)], Qiang Zhang[3],
Hanyong Zhang[1,2], and Dong Wang[1,2]

[1] School of Information Science and Engineering,
University of Jinan, Jinan 250022, China
ise_mengqf@ujn.edu.cn
[2] Shandong Provincial Key Laboratory of Network Based Intelligent
Computing, Jinan 250022, China
[3] Institute of Jinan Semoconductor Elements Experimentation,
Jinan 250014, China

Abstract. High accuracy of epilepsy EEG automatic detection has important clinical research significance. The combination of nonlinear time series analysis and complex network theory made it possible to analyze time series by the statistical characteristics of complex network. In this paper, based on the transition network the feature extraction method of EEG signals was proposed. Based on the complex network, the epileptic EEG data were transformed into the transition network, and the loop coefficient was extracted as the feature to classify the epileptic EEG signals. Experimental results show that the single feature classification based on the extracted feature obtains classification accuracy up to 98.5%, which indicates that the classification accuracy of the single feature based on the transition network was very high.

Keywords: Transition network · The loop coefficient · Epilepsy EEG automatic detection

1 Introduction

Epilepsy is a chronic recurrent transient brain dysfunction syndrome. Its attack, excessive synchronization of the brain neurons discharge causes temporary dysfunction of the central nervous system. So the patient's health has a great deal of harm. At present, complex epilepsy EEG data is enormous in clinical medicine, and the efficiency of artificial classification is low and the accuracy is not high. Therefore, the automatic detection method of epilepsy EEG has important significance for clinical research.

The Hurst exponent of the epileptic EEG was discussed in [1] and the results shown that the normal EEG was uncorrelated whereas the epileptic EEG was long range anti-correlated. Spectral entropy and embedding entropy, which could be used to measure the system complexities, were introduced to epilepsy detection in [2, 3]. Combined with these classification features, the classifiers, such as artificial neural

© Springer International Publishing AG 2017
D.-S. Huang et al. (Eds.): ICIC 2017, Part II, LNCS 10362, pp. 713–719, 2017.
DOI: 10.1007/978-3-319-63312-1_63

network (ANN) and support vector machine (SVM), had also been widely applied into the epilepsy detection algorithm [4–9]. From these literature, we can conclude that an excellent classification feature not only obtains better classification accuracy but also spends less computational complexity because of it does not need combined with classifier. These advantages are significant for the clinical application.

Recently, complex networks theory provided a new perspective for nonlinear time series analysis. Zhang and Small [10] proposed an algorithm that transformed the pseudo-periodic time series into complex networks. A bridge between nonlinear time series analysis and complex networks theory had been built. Lacasa [8] first proposed the visibility graph algorithm, which could convert arbitrary time series into a graph. Time series conversion to complex network algorithm made it possible to the application of complex network theory researching time series. Sun & Small [11] took the Rossler chaotic system as an example to give a concrete algorithm for the conversion of nonlinear time series into transition network. Based on the statistical properties of complex networks, Sun & Small gave a detailed analysis of the different periods of the Rossler system, thus converting the nonlinear time series into transition network to maturity. In the paper [12], an improved method for converting nonlinear time series into transition networks was proposed, and the possibility of transforming any time series into transition networks was proved. In paper [13], the transformation of epileptic EEG to proximity network was proposed. According to the statistical characteristics of complex network, the classification of epileptic EEG could be realized by combining classification method.

In this paper, we improve the method that transform nonlinear time series into transition network mentioned by Sun & Small [11], so the operation rate and classification accuracy are raised. According to the statistical characteristics of complex networks, we extract the loop coefficient to classify the epileptic EEG and the classification accuracy up to 98.5%. So we improve the accuracy of the automatic detection of epilepsy EEG.

2   The Feature Extraction Method

2.1   A Method of Constructing Complex Network by Time Series

Sun & Small [11] gave a concrete algorithm for the conversion of nonlinear time series into transition network. However, the efficiency of this method was low and classification accuracy rate was not high. The value of sliding window length L had a great influence on classification accuracy. So, in this paper, this method had been improved. The time series was segmented by the maximum value, and then the nodes were constructed. This method eliminated the influence of L and improved the efficiency of operation and the accuracy of classification. The improved method as follows:

Given a time series $\{x_n\}_{n=1}^N$. Then, found all the maximum value in the time series. The time series between every two adjacent maximum value was defined as one sub-segment contained a maximum value, labeled $\{X_m\}_{m=1}^L$ and L was the length of each sub-segment. The ordinal pattern of was defined as $\pi_\tau = (\tau_1, \tau_2, \tau_3 \ldots \tau_L)$ of X_m was defined the permutation of $(1, 2, 3 \ldots L)$ satisfying $X_{m+\tau_1} < X_{m+\tau_2} < X_{m+\tau_3} < X_{m+\tau_4} < \cdots \ldots X_{m+\tau_l}$.

To distinguish same ordinal pattern, but amplitude differently (while maintaining the useful features of the ordinal representation), we proposed the simple modification of adding amplitude information. We first found the maximum X_{max} of each sub-segment and the minimum X_{min} of this time series, and then we used rounded operation formula computing the value of M:

$$M = \left[\frac{X_{max} - x_{min}}{Q} \right] \tag{1}$$

Finally each sub-segment was symbolized as $S_i = \{M, \pi_\tau\}_{i=1}^k$.

To investigate the transitions among the different states identified by modification, a weighted and directed network was constructed with fixed Q as follows: S_i represents a node in a complex network, thus we got the node set of complex network; naturally, built the corresponding connections. The building the corresponding connections method employed the method given by Sun & Small. The link starting from the node corresponding to S_i ends at the node corresponding to S_{i+1}. The weight W_{ij} of the link directed from node i to j is given by

$$W_{ij} = \#(S_i \rightarrow S_j) \tag{2}$$

$\#(S_i \rightarrow S_j)$ is the number of times that the transition from S_i to S_j occurs in a given time series. Finally, the adjacency matrix $W_{ij} = (w_{ij})_{M*M}$ was used to represent the generated networks. So we got the improved transition network.

2.2 Feature Extraction

The complex network with adjacency matrix W_{ij} was obtained by the transition networks construction algorithm. The geometric topological structure stored the dynamic characteristic information of the original time series, and the characteristics of the epileptic EEG were extracted by studying the statistical properties of the complex network.

In a directed networks [11], $k_i == \sum_{j=1}^{M} a_{ij} = k_{in}^i + k_{out}^i$, where k_{out}^i is the number of outward links from node i and k_{in}^i is the number of inward ones.

According to our transformation, it was easy to see that the loop structures in our networks have a specific dynamical meaning. Loops corresponded to the system returning to a state similar to a former state (a state recurrence). Hence, in this work, we used the loop coefficient to study the recurrence time distributions from the network structure.

There was node i connecting to the network and we set the remaining part of the network is C. If the network C was virtual into a node, starting from node i, the number of possible paths (excluding duplicate edges) returning to node i through network C was $k_{in}^i * k_{out}^i$. Loop coefficient was defined as H(I), it can be calculated as follows:

$$H(I) = \sqrt{k_{in}^i * k_{out}^i} \tag{3}$$

The greater the value of H(I), the recurrence time was longer and the connection between A and network C was more closely. The epileptic seizure EEG owned lower complexity than that of intermittent EEG, and the chaos was weakened. Hence the loop coefficient based on epileptic seizure EEG was smaller. So it could be used as epileptic EEG signal classification.

3   Experiment Results and Analysis

In this study, we use a clinical epileptic EEG data set from the University of Bonn, Germany. The epileptic EEG data file contains 100 ictal EEG data and 100 interictal EEG data. Every EEG datum was sampled at a rate of 128 Hz, and has 4096 points and EEG data such as manual or eye movement disturbances were removed. In the experiment, we set the sample length of 512 and 1024, respectively, to construct a complex network. Furthermore, we evaluate the performance of EEG feature extraction method and epileptic EEG automatic detection algorithm.

Experiments set the data length of 512 and 1024, respectively, the detailed classification results in Table 1. When the data length is 1024, the classification accuracy is the higher, reaching 98.5%. The classification results with data lengths of 512 and 1024 are shown in Figs. 1 and 2, respectively. The classification threshold shows by the solid line in the figure separates the two types of epileptic EEG. The loop coefficient in the interictal period is significantly higher than that of the ictal period. This conclusion is consistent with one fact that the complexity of ictal EEG data is lower than that of the interictal EEG data.

Table 1. Result of the feature automatic detection of transition network

Method	Data length	ACC
Transition network	512	95.5
	1024	98.5

Table 2 shows the accuracy of epileptic EEG single feature classification based on transition network and other methods. It can be seen from the table that the accuracy of classification by the extracted single feature based on improved transition network is significantly higher than that of single-feature classification of epileptic EEG based on other methods, achieved highest the classification accuracy of the single feature based on the complex network.

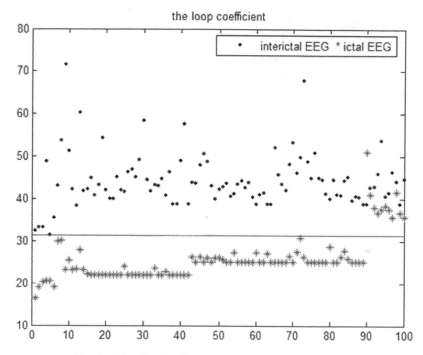

Fig. 1. The classification results with data lengths of 512

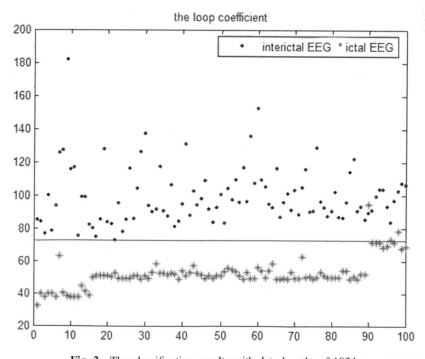

Fig. 2. The classification results with data lengths of 1024

Table 2. Result of the feature automatic detection algorithm

Method	Data length	Feature	ACC
Transition network	1024	H(I)	98.5
Proximity networks [13]	2048	NEED	96.5
Proximity networks [14]	2048	Pclu	94.5
RQA [15]	1024	DET	90.5
Weighted network [16]	1024	wd_r	94.5
Hurst + SVM [1]	****	****	87.5

4 Conclusion

Combined with the theory of complex network, the epilepsy EEG data was constructed as a transition network firstly and the construction method was improved.

We extracted the loop coefficient that was applied to classify the epileptic EEG data set. As the global topological structure of complex network based on nonlinear time series H(I), characterize the non-linear dynamic characteristics of the original nonlinear time series, which could be used to distinguish EEG with different nonlinear dynamic modes. Compared with the single feature classification accuracy of proximity networks, we got higher classification accuracy. The feature extracted in this paper improved the performance of automatic detection classification algorithm of epilepsy effectively.

Acknowledgments. This work was supported by the National Natural Science Foundation of China (Grant No. 61671220, 61640218, 61201428), the National Key Research And Development Plan (No. 2016 YFC0106001), the Shandong Distinguished Middle aged and Young Scientist Encourage and Reward Foundation, China (Grant No. ZR2016FB14), the Project of Shandong Province Higher Educational Science and Technology Program, China (Grant No. J16LN07), the Shandong Province Key Research and Development Program, China (Grant No. 2016 GGX101022).

References

1. Nurujjaman, M., Ramesh, N., Sekar Iyengar, A.N.: Comparative study of nonlinear properties of EEG signals of normal persons and epileptic patients. Nonlin. Biomed. Phys. **3**(1), 6–15 (2009)
2. Acharya, U.R., Molinari, F., Vinitha Sree, S., Chattopadhyay, S.: Kwan-Hoong, Ng., Suri, J. S.: Automated diagnosis of epileptic EEG using entropies. Biomed. Signal Process. Control **7**(4), 401–408 (2012)
3. Kannathal, N., Lim, C.M., Acharya, U.R., Sadasivan, P.K.: Entropies for detection of epilepsy in EEG. Comput. Methods Programs Biomed. **80**(3), 187–194 (2005)
4. Acharya, U., Vinitha Sree, S., Chattopadhyay, S., Wenwei, Y.U., Alvin, A.P.C.: Application of recurrence quantification analysis for the automated identification of epileptic EEG signal. Int. J. Neural Syst. **21**(3), 199–211 (2011)

5. Übeyli, E.D.: Combined neural network model employing wavelet coefficients for EEG signals classification. Digit. Signal Proc. **19**(2), 297–308 (2009)
6. Gandhi, T., Panigrahi, B.K., Gandhi, T., Bhatia, M., Anand, S.: Expert model for detection of epileptic activity in EEG signature. Expert Syst. Appl. **37**(4), 3513–3520 (2010)
7. Song, Y., Liò, P.: A new approach for epileptic seizure detection sample entropy based feature extraction and extreme learning machine. J. Biomed. Sci. Eng. **3**(6), 556–567 (2010)
8. Huang, D.S., Yu, H.-J.: Normalized feature vectors: a novel alignment-free sequence comparison method based on the numbers of adjacent amino acids. IEEE/ACM Trans. Comput. Biol. Bioinf. **10**(2), 457–467 (2013)
9. Huang, D.S., Jiang, W.: A general CPL-AdS methodology for fixing dynamic parameters in dual environments. IEEE Trans. Syst. Man Cybern. Part B **42**(5), 1489–1500 (2012)
10. Ji, Z., Wang, B., Deng, S.P., et al.: Predicting dynamic deformation of retaining structure by LSSVR-based time series method. Neurocomputing **137**, 165–172 (2014)
11. Ji, Z., Wu, G., Hu, M.: Feature selection based on adaptive genetic algorithm and SVM. Comput. Eng. **14**, 072 (2009)
12. Ji, Z., Wu, D., Zhao, W., et al.: Systemic modeling myeloma-osteoclast interactions under normoxic/hypoxic condition using a novel computational approach. Sci. Rep. **5** (2015)
13. Yuan, Q., Zhou, W., Liu, Y.X., Wang, J.W.: Epileptic seizure detection with linear and nonlinear features. Epilepsy Behav. **24**(4), 415–421 (2012)
14. Yuan, Q., Zhou, W., Li, S., Cai, D.M.: Epileptic EEG classification based on extreme learning machine and nonlinear features. Epilepsy Res. **96**, 29–38 (2011)
15. Zhang, J., Small, M.: Complex network from pseudoperiodic time series: topology versus dynamics. Phys. Rev. Lett. **96** (2006)
16. Lacasa, L., Luque, B., Ballesteros, F., Luque, J., Nuno, J.C.: From time series to complex networks: the visibility graph. Proc. Natl. Acad. Sci. U.S.A. **105**, 4972–4975 (2008)
17. Sun, X.R., Small, M., Zhao, Y., Xue, X.P.: Characterizing system dynamics with a weighted and directed network constructed from time series data. Phys. A **24**(2), 1054–1500 (2013)
18. Wang, M., Tian, L.X.: From time series to complex networks: the phase space coarse graining. Phys. A **461**, 456–468 (2016)
19. Wang, F.L., Meng, Q.F., Zhou, W., Chen, S.: The feature extraction method of EEG signals based on degree distribution of complex networks from nonlinear time series. In: Huang, D.-S., Bevilacqua, V., Figueroa, J.C., Premaratne, P. (eds.) ICIC 2013. LNCS, vol. 7995, pp. 354–361. Springer, Heidelberg (2013). doi:10.1007/978-3-642-39479-9_42
20. Wang, F.L., Meng, Q.F., Chen, Y.H., Zhao, Y.Z.: Feature extraction method for epileptic seizure detection based on cluster coefficient distribution of complex network. WSEAS Trans. Comput., 351–360 (2014)
21. Bao, W., Chen, Y., Wang, D.: Prediction of protein structure classes with flexible neural tree. Bio-Med. Mater. Eng. **24**(6), 3797–3806 (2014)
22. Meng, Q.F., Chen, S., Chen, Y.H.: Automatic detection of epileptic EEG based on recursive quantification analysis and support vector machine. Acta Phys. **6**(5) (2014)
23. Wang, F.L., Meng, Q.F., Xie, H.B., Chen, Y.H.: Novel feature extraction method based on weight difference of weighted network for epileptic seizure detection. In: Proceedings of the 36th Annual International IEEE EMBS Conference, Chicago, Illinois, USA (2014)

Safety Inter-vehicle Policy Based on the Longitudinal Dynamics Behaviors

Xiao-Fang Zhong, Ning Yuan, Shi-Yuan Han[⊠], Yue-Hui Chen,
and Dong Wang

Shandong Provincial Key Laboratory of Network Based Intelligent Computing,
University of Jinan, Jinan 250022, China
swu_zhongxf@126.com,
{ise_zhouj,ise_yuann,ise_hansy,yhchen,
ise_wangd}@ujn.edu.cn

Abstract. The analysis of the safety inter-vehicle distance plays important roles for driving assistant system, which can give the warning signal to drivers timely. In order to provide the drivers a warning signal about an impending collision reasonable and timely, safety inter-vehicle policies between host vehicle and preceding vehicle are proposed based on the longitudinal dynamics behaviors in this paper. First, by analyzing the driving force and resistance force generated from the road surface, the basis safety inter-vehicle distance is designed, in which the surface friction coefficient and the air's visibility are considered. Taken the driving state of preceding vehicle into consideration, including braking hard until to a complete stop and braking with constant deceleration, the safety inter-vehicle policies are derived from the basis safety inter-vehicle distance, which is composed of the sliding distance, duration distance and deceleration distance. Finally, by comparing with the classic safety distances, the effectiveness and elasticity of the proposed inter-vehicle policies are illustrated.

Keywords: Safety distance · Longitudinal dynamics behaviors · Driving assistant systems

1 Introduction

Traffic accident and congestion have become serious problems in the modern society. In China alone, the number of the private vehicles reaches 140 million. Meanwhile, the number of the traffic accident reaches around ninety thousand, in which about 60,000 people dead and 200,000 people injured. In the past decades, considerable efforts have been devoted to the intelligent transport systems, which includes the applications of the driving assistance systems [1–3], the active suspension systems [4–6], the distributed wireless networks [7–10] and cooperative adaptive cruised control systems [11–13], aiming at providing new technologies for reducing road accident and congestion. In various recent search works, the vehicle collision warning system has been regarded as one of the effective techniques in the driving assistance systems applications. Obviously, the safety inter-vehicle policy plays an important role in the vehicle collision warning

© Springer International Publishing AG 2017
D.-S. Huang et al. (Eds.): ICIC 2017, Part II, LNCS 10362, pp. 720–729, 2017.
DOI: 10.1007/978-3-319-63312-1_64

system that provides the reasonable occasion for reminding the vehicle driver to adjust the driving states. Here, we will investigate the safety inter-vehicle policy in this paper.

In order to improve the driving safety and traffic efficiency, various inter-vehicle distance policies have been developed, mainly including two types: the constant spacing policy and the variable spacing policy [14]. The constant spacing policy is with the advantages of simple structure, less calculation and easily implementation. However, it cannot handle complex traffic environment. For the variable spacing policy, the constant time-gap policy and variable time-gap policy are the typical spacing policies, in which the host driving velocity is focused. By using the variable spacing policies, the time gap can be reduced in the range of 0 s–0.5 s so that the traffic efficiency can be improved [15]. Meanwhile, a safety spacing policy was proposed in [16], in which the safety distance was described as a nonlinear function involving the driving velocity and braking ability; by considering the influence from the preceding vehicle, the safety distance was viewed as a saturation function with a reasonable range in [17]; taking the relationship among a platooned vehicles into consideration, a safety policy was proposed by introducing the related position of the host vehicle in [12]. In practical driving situations, the driving action of the preceding vehicle has a direct effect on the decision making for the host vehicles'. As the core factors in traffic accidents, the relationship between host vehicle and preceding vehicle must be discussed more detailed. This is the motivation of this paper.

The aim of this paper is to design the safety inter-vehicle distance policies so that the rear-end collision can be avoided under the situations of preceding vehicle with the uniform motion and deceleration motion. First, by analyzing the longitudinal dynamic of host vehicle, the braking distance is divided into three stages, including the driving and vehicle engineering response time, the traveling distance before braking lock, and the traveling distance of the duration time and slipping time. Then the basis safety inter-vehicle policy is designed based on the host vehicle only. After that, by considering the different driving states of preceding vehicle, two inter-vehicle policies are proposed for the uniform motion with the same velocity and uniform deceleration motion of preceding vehicle. As well be shown later in numerical simulations under different situations, the proposed policies can server as an effective and feasible tool for driving assistance systems.

The rest of paper is organized as follows. In Sect. 2, by discussing the basis theories of the longitudinal vehicle dynamics, the research problem is formulated. Then the safety inter-vehicle policies are designed based on the driving states of preceding vehicle in Sect. 3. The simulation results are shown in Sect. 4 to prove the effectiveness and the feasibility of the proposed polices. Finally, the conclusions are given in Sect. 5.

2   Problem Formulation

In this section, the basis theories of longitudinal vehicle dynamics are described first. Then the objective of this paper is given.

2.1   Basis Theory of Longitudinal Vehicle Dynamics

The longitudinal vehicle dynamics can be defined as the longitudinal external forces while the vehicle straight driving on road under different conditions. Based on the model of longitudinal vehicle dynamics, the suitable evaluating performance index can be defined and the vehicle force condition can be analyzed. Then the states of vehicle can be analyzed according to the driving forces and resistance factors.

The driving forces and resistance factors will be used to design the safety inter-vehicle policy. With the normal work of the engine, the driving force can be viewed as the reacting force from driving wheel, which is generated from peripheral force F_2 caused by the wheel torque F_1. The Fig. 1 displays the driving force analysis graphics.

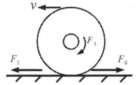

Fig. 1. The driving force analysis graphics

For resistance factors, under the assumptions that the gradient is ignored while the vehicle is driving, the rolling resistance and slipping resistance will be analyzed. The rolling resistance force F_0 can be defined as the reacting force between the wheel and road surface along the tangential direction. The rolling resistance force F_r after locking action of wheel is shown in Fig. 2.

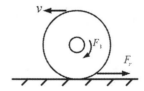

Fig. 2. The rolling resistance force

Meanwhile, the slipping resistance force is defined as the reaction force between the wheel and road surface while the wheel is slipping. The slipping resistance force F_s during locking action of wheel is shown in Fig. 3.

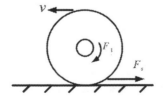

Fig. 3. The slipping resistance force

Besides, the air friction of vehicle body ignores while analyzing the longitudinal vehicle dynamics.

2.2   Problem Formulation

The vehicles drive in a platoon. While the preceding vehicle is emergency braking, the dynamic behaviors of host vehicle and preceding vehicle are shown in Fig. 1. S is the safety inter-vehicle gap, S_0 is the inter-vehicle distance between the head of host vehicle and the rear of preceding vehicle, S_1 is the distance consumption for driver response time and coordinating time of auxiliary brake system, S_2 denotes distance travelled by the braking process of host vehicle.

By analyzing the process of the braking process of host vehicle, the deceleration situation is displayed in Fig. 2 during the braking process. It can be seen that the consumption times can be divided into three parts: the driver's response time t_a, brake response time t_2, and duration time t_b and slipping time t_c after braking action. The duration time t_b denotes the process from reacting of braking system to maximum of brake pedal. The slipping time t_c is the process of remaining the maximum brake pedal. After doing uniform motion, the vehicle will be under the decelerated motion.

Then, the problem can be formulated as how to design the inter-vehicle distance policy so that the rear-end collision between the host vehicle and the preceding vehicle can be avoided and the warning signal can be sent to drivers timely.

3   The Safety Inter-vehicle Policies

In this section, the safety inter-vehicle distance will be discussed under four satiations, including the motion state of host vehicle only, the emergency braking of preceding vehicle to stop, the uniform motion of preceding vehicle with lower velocity than host vehicles', and the deceleration uniform motion of preceding vehicle with lower velocity than host vehicles'. The designed safety inter-vehicle policies are described in the following theorem.

Theorem 1: Consider the surface friction coefficient, the air's visibility and the motion state of host vehicle, the safety inter-vehicle policy can be described as

$$\begin{cases} S = S_a + S_b + S_c + l - l_2 = v_0\left(t_a + \frac{t_b}{2}\right) - \frac{\mu g t_b^2}{24} + \frac{v_0^2 - v_c^2}{2\mu} + l - l_2 \\ S_a + S_b + S_c \leq l_1 + l_2 \end{cases} \quad (1)$$

where S is the total safety distances; t_a is the sum of the driver's response time t_1 and the braking response time t_2; S_a denotes the traveling distance during time gap t_a; S_b denotes the traveling distance before braking lock; S_c is the traveling distance of the duration time t_b and slipping time t_c; l_1 denotes the range of air's visibility; l_2 is the total distance during the driver's response time t_a, the duration time t_b and slipping time t_c; μ denotes the surface friction coefficient; g is the gravitational acceleration; l is the inter-vehicle distance between host vehicle and preceding vehicle after stopping action; v_c denotes the vehicle velocity after braking action.

While the preceding vehicle brakes hard until to a complete stop, the host vehicle must be slowed down and stopped quickly. Based on the Eq. 1, the safety inter-vehicle distance is designed as

$$
\begin{cases}
S = v_0(t_a + \frac{t_b}{2}) - \frac{\mu g t_b^2}{24} + \frac{v_0^2}{2\mu g} + l \\
S_a + S_b + S_c \leq l_1
\end{cases}
\tag{2}
$$

While the preceding vehicle drives with uniform velocity with the lower velocity than the host vehicles', the host vehicle must be slowed down and kept the safety distance. Then the safety inter-vehicle distance is designed as

$$
\begin{cases}
S = v_1(t_a + \frac{t_b}{2}) - \frac{\mu g t_b^2}{24} + \frac{v_0^2}{2\mu g} + l \\
S_a + S_b + S_c \leq l_1 + v_{a0}(t_a + \frac{t_b}{2} + \frac{v_{b0}-v_{a0}}{\mu g})
\end{cases}
\tag{3}
$$

where v_{a0} is the initial velocity of host vehicle, v_2 denote the related velocity between host vehicle and preceding vehicle.

While the preceding vehicle brakes with constant deceleration a_1 and initial velocity v_2, the safety inter-vehicle distance is designed as

$$
S = \frac{v_2^2}{2a_2} - \frac{v_{a0}^2}{2a_1} + l - v_2 t_0
\tag{4}
$$

Proof: First, the first situation that the motion state of host vehicle is considered only is discussed. For the stage of the driver's response time t_1 and the braking response time t_2, the vehicle travels with uniform velocity v_0 and the traveling distance can be calculated by

$$
S_a = v_0 t_a
\tag{5}
$$

For the stage t_b that the driver tread the vehicle dash, the deceleration a can be increased along with a linear function to the maximum value a_{max}. The deceleration $_a$ is described as

$$
a = -a_{max} t / t_b
\tag{6}
$$

Therefore, the velocity t_b can be calculated by

$$
v_b = v_0 - \int \frac{\mu g}{t_b} t dt = v_0 - \frac{\mu g}{2t_b} t^2
\tag{8}
$$

Then the traveling distance S_b before wheel was locked completely can be described

$$
S_b = \int_0^{t_b} (v_0 - \frac{\mu g}{2t_b} t^2) dt = v_0 t_b - \frac{\mu g t_b^2}{6}
\tag{9}
$$

For the stage t_c, the vehicle is with the slipping friction force F_f related to the surface friction coefficient μ. Then the maximum value of deceleration a of stage t_b can be obtained

$$a = \mu g t / t_b \tag{10}$$

Then the traveling distance S_c can be described

$$S_c = \frac{v_c^2 - v_b^2}{2\mu g} \tag{11}$$

where v_c is the velocity while the wheel is locking, which is described as

$$v_c = v_0 - \frac{\mu g}{2} t_b \tag{12}$$

Substituting the Eq. (12) into (11), the Eq. (11) can be rewritten as

$$S_c = \frac{v_c'^2 - v_0^2}{-2\mu g} - \frac{v_0 t_b}{2} + \frac{\mu g t_b^2}{8} \tag{13}$$

By summing the (5), (9) and (13), the first formula in (1) can be obtained.

In order to ensure the safety driving under air's visibility, the inequation formula in (1) is obtained

$$S_a + S_b + S_c \leq l_1 + l_2 \tag{14}$$

While the preceding vehicle brakes hard until to a complete stop, $v_c = 0$ and $l_2 = 0$. Then the safety inter-vehicle distance (2) is obtained.

While the preceding vehicle drives with uniform velocity with the lower velocity than the host vehicles', the difference value between host vehicle and preceding vehicle can be described as

$$v_1 = v_0 - v_{a0} \tag{15}$$

Then the traveling distance the stage of the driver's response time t_1 and the braking response time t_2 can be calculated by

$$S_a = v_{a0} t_a \tag{16}$$

Meanwhile, the traveling distance S_b for the stage t_b can be described

$$S_b = v_{a0} t_b \tag{17}$$

The traveling distance S_c for the stage t_c can be described

$$S_c = v_{a0} \left(\frac{v_0 - v_{a0}}{\mu g} - \frac{t_b}{2} \right) \tag{18}$$

Then the Eq. (3) is obtained.

While the preceding vehicle brakes with constant deceleration a_1 and initial velocity v_2, the traveling distance of host vehicle can be calculated from

$$S_2 = \frac{t_0 + t_2}{2} v_2 \tag{19}$$

Meanwhile, the traveling distance of preceding vehicle can be calculated from

$$S_1 = \frac{v_1 t_1}{2} \tag{20}$$

Therefore, the safety distance can be described as

$$S = S_2 + l - S_1 = \frac{v_1 t_1}{2} + l - \frac{t_0 + t_2}{2} v_2 \tag{21}$$

Then, the safety inter-vehicle distance (4) is obtained. The proof is completed.

4   Simulation

In order to demonstrate the effectiveness of proposed safety inter-vehicle policies, the situations, including the emergency braking of preceding vehicle to stop, the uniform motion of preceding vehicle with lower velocity than host vehicles', and the deceleration uniform motion of preceding vehicle with lower velocity than host vehicles', will be discussed in the following (Fig. 4).

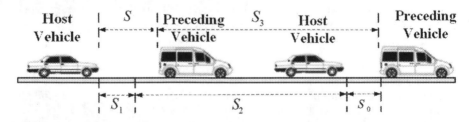

Fig. 4. The driving states after braking of preceding vehicle

The sum t_a of the driver's response time t_1 and the braking response time t_2 is chosen in the range of *0.8 s–1* s. Meanwhile, the consumption time for accelerating to the maximum acceleration a_{max} is in the range of *0.1 s–0.2 s*. The initial distance between host vehicle and preceding vehicle is in the range of *2 m–5 m*. In this case, 2 m is chosen. Meanwhile, the surface friction coefficient μ set as 0.6.

While the preceding vehicle brakes hard until to a complete stop, the simulation results of the safety inter-vehicle distance (2) are shown in Table 1.

Table 1. The safety distances under the case that the preceding vehicle brakes hard until to a complete stop.

Velocity (km/h)	40	50	60	70	80	90	100	110
Safety distances	21.94	30.21	39.79	50.68	62.88	76.40	91.22	107.36

While the preceding vehicle drives with uniform velocity with the lower velocity than the host vehicles', the simulation results of the safety inter-vehicle distance (3) are shown in Table 2.

Table 2. The safety distances under the case that the preceding vehicle drives with uniform velocity with the lower velocity than the host vehicles'.

Relative velocity	50	60	70	80	90	100
Initial velocity of host vehicle	60	70	80	90	100	110
Safety distances		37.43	48.32	60.52	74.04	76.40	105.00

While the preceding vehicle brakes with constant deceleration a_1 and initial velocity v_2, the simulation results of the safety inter-vehicle distance (4) are shown in Table 3.

Table 3. The safety distances under the case that the preceding vehicle brakes with constant deceleration a_1 and initial velocity v_2.

Relative velocity	40	40		40	40
Deceleration of host vehicle (m/s^2)	5.88	5.88	5.88	5.88	5.88
Deceleration of preceding vehicle (m/s^2)	7	7	7	7	7
Velocity of host vehicle (km/h)	60	70	80	90	100
Velocity of preceding vehicle (km/h)	20	30	40	50	60
Safety distances	23.67	29.77	35.67	41.36	46.83

In order to display the effective the proposed safety inter-vehicle distance more clearly, the classic safety distances are shown in Table 4.

Table 4. The classic safety distances.

Initial velocity of host vehicle	30	40	50	60	70
The maximum deceleration of host vehicle	5.88	5.88	5.88	5.88	5.88
The braking time of host vehicle	2	2	2	2	2
Safety distances	27.57	37.72	49.18	61.95	76.04

Based on the above simulation results, it can be seen clearly that the proposed safety inter-vehicle policies can be take effectively under different situations. Meanwhile, by considering the longitudinal dynamic behaviors of preceding vehicle, the host vehicle can response the cases under dangerous moments (Fig. 5).

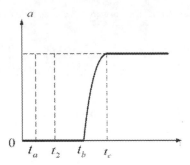

Fig. 5. The deceleration situation after braking action

5   Conclusions

In this paper, the problem of design of the safety inter-vehicle distances was discussed. Based on the longitudinal dynamic of the host and preceding vehicle, the safety inter-vehicle policy was designed first, in which the driving state of host vehicle was considered. Based on this basis policy, two safety inter-vehicle policies were proposed, in which the preceding vehicle under braking hard until to a complete stop and braking with constant deceleration were discussed. Our future study would focus on extending the obtained results in this paper to more complicated applications in intelligent transportation systems.

Acknowledgments. This work was supported by the National Key Research and Development Plan (No. 2016YFC0106000), the National Natural Science Foundation of China (Grant No. 61640218, 61671220, 61573166, 61673357), the Shandong Distinguished Middle-aged and Young Scientist Encourage and Reward Foundation, China (Grant No. BS2014DX015, ZR2016FB14), the Project of Shandong Province Higher Educational Science and Technology Program, China (Grant No. J16LN07), the Shandong Province Key Research and Development Program, China (Grant No. 2016GGX101022).

References

1. Lu, M., Wevers, K.: Forecasting and evaluation of traffic safety impacts: driving assistance systems against road infrastructure measures. IET Intell. Transp. Syst. 1(2), 117–123 (2007)
2. Reina, G., Johnson, D., Underwood, J.: Radar sensing for intelligent vehicles in urban environments. Sensors 15(6), 14661–14678 (2015)
3. Olivares-Mendez, M.A., Sanchez-Lopez, J.L., Jimenez, F., Campoy, P., Sajadi-Alamdari, S. A., Voos, H.: Vision-based steering control, speed assistance and localization for inner-cityvehicles. Sensors 16(3), 362 (2016)
4. Han, S.-Y., Zhang, C.-H., Tang, G.-Y.: Approximation optimal vibration for networked nonlinear vehicle active suspension with actuator time delay. Asian J. Control (2017). doi:10.1002/asjc.1419

5. Han, S.-Y., Tang, G.-Y., Chen, Y.-H., Yang, X.-X., Yang, X.: Optimal vibration control for vehicle active suspension discrete-time systems with actuator time delay. Asian J. Control **15**, 1579–1588 (2013)
6. Han, S.-Y., Wang, D., Chen, Y.-H., Tang, G.-Y., Yang, X.-X.: Optimal tracking control for discrete-time systems with multiple input delays under sinusoidal disturbances. Int. J. Control Autom. Syst. **13**, 292–301 (2015)
7. Zhou, J., Chen, C.L.P., Chen, L., Li, H.X.: A collaborative fuzzy clustering algorithm in distributed network environments. IEEE Trans. Fuzzy Syst. **22**, 1443–1456 (2014)
8. Zhou, J., Chen, C.L.P., Chen, L., Li, H.X.: Fuzzy clustering with the entropy of attribute weights. Neurocomputing **198**, 125–134 (2016)
9. Han, S.-Y., Chen, Y.-H., Tang, G.-Y.: Sensor fault and delay tolerant control for networked control systems subject to external disturbances. Sensors **17**, 700 (2017)
10. Zhou, J., Chen, C.L.P., Chen, L., Li, H.X.: A user-customizable urban traffic information collection method based on wireless sensor networks. IEEE Trans. Intell. Transp. Syst. **14**, 1119–1128 (2013)
11. Liu, B., Kamel, A.E.: V2X-based decentralized cooperative adaptive cruise control in the vicinity of intersections. IEEE Trans. Intell. Transp. Syst. **17**(3), 644–658 (2016)
12. Bao, W., Chen, Y., Wang, D.: Prediction of protein structure classes with flexible neural tree. Bio-med. Materials Eng. **24**(6), 3797–3806 (2014)
13. Han, S.-Y., Chen, Y. H., Wang, L., Abraham, A.: Decentralized longitudinal tracking control for cooperative adaptive cruise control systems in a platoon. In: IEEE International Conference on Systems, Man, and Cybernetics, pp. 2013–2018 (2013)
14. Zheng, Y., Li, S.E., Wang, J., Cao, D., Li, K.: Stability and scalability of homogeneous vehicular platoon: study on the influence of information flow topologies. IEEE Trans. Intell. Transp. Syst. **17**(1), 14–26 (2016)
15. Zhang, J., Ioannou, P.: Adaptive vehicle following control system with variable time headways. In: IEEE Conference on Decision and Control, pp. 12–15 (2009)
16. Dey, K.C., Wang, X., Wang, Y.: A review of communication, driver characteristics, and controls aspects of cooperative adaptive cruise control (CACC). IEEE Trans. Intell. Transp. Syst. **17**(2), 491–509 (2016)
17. Zhao, J., Oya, M.: A safety spacing policy and its impact on highway traffic flow. In: IEEE Conference on Intelligent Vehicle Symposium, pp. 960–965 (2009)
18. Yanakiew, D., Kanellakopoulos, I.: A safety spacing policy and its impact on highway traffic flow. IEEE Trans. Veh. Technol. **47**(4), 1365–1377 (1998)

Global Adaptive and Local Scheduling Control for Smart Isolated Intersection Based on Real-Time Phase Saturability

Shi-Yuan Han[1(✉)], Fan Ping[2], Qian Zhang[3], Yue-Hui Chen[1], Jin Zhou[1], and Dong Wang[1]

[1] Shandong Provincial Key Laboratory of Network Based Intelligent Computing, University of Jinan, Jinan 250022, China
{ise_hansy,yhchen,ise_zhouj,ise_wangd}@ujn.edu.cn
[2] Information Network Center, Qilu Hospital of Shandong University, Jinan 250012, China
pingfan@163.com
[3] School of Physics and Optoelectronic Engineering, Xidian University, Xi'an 710126, China
qianz678@hotmail.com

Abstract. Linking real-time phase saturability directly to the traffic signal control, the global adaptive control scheme for traffic light loop and the local scheduling control strategy for phase green time are proposed in this paper for real-time traffic signal control systems with multiple phases. First, applying the real-time phase saturability of each phase as the time-varying weight factor, an elastic scheduling model is designed to describe the competitive relationship among the different phase in a traffic light loop. Then the traffic green time scheduling problem in a traffic light loop is formulated as a trade-off optimization problem between the green light time and real-time phase saturability for each phase. By solving a quadratic programming problem that seeks the minimum value of the sum of the squared deviation between the green time and the maximum allowable green time in each phase, the allocated green time in next traffic loop is obtained. If the total real-time traffic load exceeds the allocable maximum value or unreaches the allocable minimum value, the proportional global adaptive control schemes are triggered for rearranging traffic light loop time. Undertaking different traffic flow conditions, the effectiveness of proposed adaptive control schemes and scheduling strategies are illustrated compared with the classic average allocating method.

Keywords: Traffic signal control system · Elastic scheduling · Adaptive control · Proportional control · Real-time phase saturability

1 Introduction

The traffic congestion has become serious problem in the urban fields with limited resources of road infrastructure. In the past decades, considerable efforts have been devoted to the Intelligent Transport Systems (ITS) aiming at providing new technologies for reducing traffic congestion and improving traffic capacity, for example, the applications

© Springer International Publishing AG 2017
D.-S. Huang et al. (Eds.): ICIC 2017, Part II, LNCS 10362, pp. 730–739, 2017.
DOI: 10.1007/978-3-319-63312-1_65

of the traffic signal control systems [1–3], the active control technologies [4–7] and the distributed wireless networks [8–11]. Specially, the traffic signal control system has been viewed as one of the effective technique for dredging traffic flow and relieving the traffic congestion. Meanwhile, traffic signal light can be viewed as a controller for releasing the traffic flow, thereby improving the road capacity and reducing the waiting time. Recently, various kinds of traffic signal control schemes have been derived based on the concepts of reinforcement learning method [1], self-organizing control method [12], approximate dynamic programming method [13], and adaptive control schemes [14].

Real-time traffic conditions are the crucial factor for scheduling the traffic signal green time of phases. Due to the easier implement and lower calculation, the clock-based fixed signal scheduling scheme is widely applied in a round-robin manner for traffic signal control systems. However, it has limitations in satisfying both high traffic capacity and good performance requirements. Compared with the fixed signal scheduling scheme, many real-time signal control schemes have been proposed and actually implemented. For example, the real-time traffic signal control problem was formulated as a hybrid Neural-Networks-based multi-agent system to improve the traffic capacity of an intersection in [15]; the architecture of traffic signal control was designed as a cooperative and hierarchical hybrid agents system in [16]; by using vehicular Ad Hoc Networks, an optimal signal control scheme was proposed based on the collecting information of the related vehicles' speed and position in [17]; the traffic-signal duration was optimize by using an efficient dynamic programming formulations of the state space in [18]. The aforementioned traffic signal control schemes are effective tools for scheduling traffic signal green time of phases. It could be pointed that the green time in a traffic signal loop is under the influence from real-time traffic saturability. How to link the real-time traffic saturability to the traffic signal control system is the motivation of this paper.

The traffic signal control system can be viewed as a real-time system. Fortunately, with the development of the quality of control (QoC) for real-time systems, it provides another perspective for discussing the traffic signal control problem. For QoC management of the real-time systems, a general flexible framework was designed to adjust the periods for flexible workload management adaptively in [19]; considering the QoC directly in process of control system, a hierarchical feedback management framework was proposed to local adjust and global adaptive control of periods in [20]; the optimal control frequencies were found by integrating the control performance and scheduling methods [21]; the scheduling method was proposed to manage the QoC under overload situations by using feedback mechanism in [22]. Actually, the relationship among different phases can be described as the competitive relationship for traffic green time in a traffic signal loop. Form the standpoint of improving the utilization of the limited traffic signal loop, the local scheduling control problem will be discussed in this paper.

The aim of this paper is to design the global adaptive control schemes and local scheduling control strategies for traffic signal control systems based on the real-time phase saturability. First, the control problem for traffic green time is formulated as the scheduling problem for real-time systems with limited resource based on the real-time phase saturability. An elastic scheduling model is introduced to describe the competitive relationship between the different phases in an intersection. Defining the real-time phase saturability as the competitive ability of different phases, the phases' allocated green time is obtained from the solution of a minimum optimization problem. For the overload or lightweight traffic flow situation, the event-triggered proportional control schemes are designed to adjust the traffic signal loop period. As well be shown later in numerical simulations under different situations, the presented schemes can server as an effective and feasible tool for traffic signal control systems.

The rest of paper is organized as follows. The traffic signal control problem is described in Sect. 2. For the intersection with normal traffic load, an elastic scheduling model is designed and the local scheduling control scheme is proposed to design the allocating traffic green light for each phase in an intersection in Sect. 3. The proportional control schemes are designed to deal with the intersection with overload or lightweight traffic flow in Sect. 4. The simulation results are shown in Sect. 5 to prove the effectiveness and the feasibility of the proposed polices. Finally, the conclusions are given in Sect. 6.

2   Problem Description

Consider an isolated traffic intersection, the geometric description is shown in Fig. 1, in which three lanes are set in each orientation. The ramp loop and regional loop are installed to count the number of vehicles in each phase.

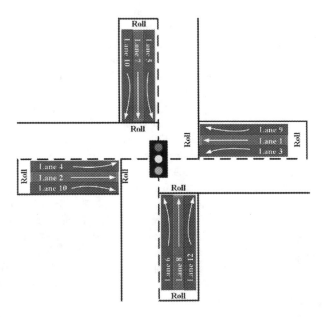

Fig. 1. Geometric illustration of an isolated intersection with multiple phases.

In traffic signal control systems, the phases are defined to avoid the traffic flow collision in different orientations. Then the traffic signal loop period T will be divided into several time slices T_i for i phase, $i = 1, 2, ..., N$, where N denotes the number of phases. Usually, the divided time slice T_i is the traffic green signal time of the ith phase. The phases are displayed in Fig. 2 in a signal intersection.

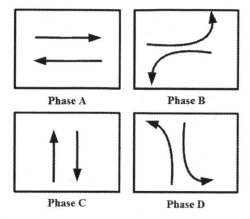

Fig. 2. Traffic phases in an isolated intersection.

In general, the signal conversion order is defined as $a \to b \to c \to d \to a$ for traffic signal control systems with four phases. Meanwhile, another conversion order can be set as $a \to c \to a$ for traffic signal control system with two phases.

In order to improve the capacity of traffic intersection, the one of objective of this paper is to design the local scheduling schemes to allocate the traffic green time T_i for the ith phase in a traffic signal loop T under the normal traffic load. Meanwhile, another objective of this paper is to increase or decrease the traffic signal loop time responding to the overload or lightweight traffic flow.

3   Local Adaptive Control Scheme of Traffic Green Lights Time for Each Phases

3.1   Real-Time Phase Saturability

Traffic saturation x_{ij} is an important parameter for evaluating the real-time traffic load, which is defined as the specific value between the entrance traffic flow q_{ij} from the jth phase and the saturation flow Q_j in the jth lane. The traffic saturation can be described as

$$\begin{cases} x_{ij} = \frac{q_{ij}}{Q_j}, \\ x_i = \max\{x_{ij}\}, \\ X = \sum_{i=1}^{N} x_i, \end{cases} \tag{1}$$

where x_i denotes the maximum traffic saturation in the ith phase; X is the allowable maximum traffic saturation in a traffic light loop.

While the traffic flow is under normal conditions, the allocated green time is directly related to the real-time phase saturability in the ith phase. For an intersection with multiple phases, the greater traffic saturability, the more allocated green time. In order to evaluate the real-time traffic load in the ith phase, the traffic saturability $w_i(t)$ is defined as

$$w_i(t) = \frac{x_i(t)}{X} = \frac{\max\{x_{ij}(t), x_{ji}(t)\}}{X}, \tag{2}$$

Applying the traffic saturability $w_i(t)$ as the weight factor, the reasonable traffic green time of each phase can be allocated based on the real-time traffic load.

3.2   Real-Time Phase Saturability

While the traffic load can release in the current traffic loop T, $\sum_{i=1}^{N} T_i \geq T$ and $\forall x_i \leq 1$, an elastic scheduling model is designed, in which the problem for allocating traffic signal green time of each phase is described as the processes of scheduling the limited traffic loop time based on the different traffic phase saturability.

The elastic scheduling model for allocating the traffic signal green time of each phase is designed as

$$\begin{cases} \min: \quad E(T_1, \ldots, T_N) = \sum_{i=1}^{N} w_i(t)(T_{i_{\max}} - T_i)^2, \\ s.t.: \quad \sum_{i=1}^{N} T_i \leq T, \\ \qquad T_i \leq T_{i_{\max}}, T_i \geq T_{i_{\min}}, \\ \qquad i = 1, 2, \ldots, N, \end{cases} \tag{3}$$

where N is the number of phase in traffic intersection; T_{imin} and T_{imax} are the allowable minimum and maximum signal green time for i phase, respectively; the weight factor $w_i(t)$ reflects the phase saturability in the ith phase.

By designing the above elastic scheduling model, the competitive relationship among phases can be shown clearly. While the traffic load is heavier in the ith phase, the weight factor $w_i(t)$ is greater, corresponding, the allocated green time is closer to the allowable maximum green time T_{imax}. Therefore, the greater traffic load in the ith phase can be released in the next traffic signal loop.

3.3   Local Scheduling Control Scheme for Traffic Signal Green Time

In order to solve the constrain optimization problem (3), the following lemma is introduced first.

Lemma 1. *Under the condition $\sum_{i=1}^{N} T_{i_{\max}} \geq T$, for any solutions T_i, the following conditions could be satisfied*

$$\sum_{i=1}^{N} T_i = T, \quad T_i \neq T_{i_{\max}}, \quad i = 1, 2, \ldots, N, \tag{4}$$

Theorem 1. *Under the conditions* $\sum_{i=1}^{N} T_{i_{\max}} \geq T$ *and* $\sum_{i=1}^{N} T_{i_{\min}} \leq T$, *let*

$$\bar{T} = \sum_{T_i \neq T_{i_{\min}}} T_{i_{\max}} + \sum_{T_i = T_{i_{\min}}} T_{i_{\min}}, \quad i = 1, 2, \ldots, N. \tag{5}$$

By solving the constrain optimization problem (3), the analytical solution T_i^* *of the allocated green time in the ith phase is described as*

$$T_i^* = T_{i_{\max}} - \frac{\bar{T} - T}{w_i \sum_{T_j \neq T_{j_{\max}}} (1/w_j)}, \tag{6}$$

for $\bar{T} > T_{i_{\max}}, T_i^* > T_{i_{\min}}$, *and* $T_i^* = T_{i_{\min}}$ *otherwise.*

Based on the previous elastic model (3), Lemma 1 and Theorem 1, the local scheduling control scheme for traffic signal green time of each phase can be described as the following corollary.

Corollary 1. *Consider a set of N phase where* T_i *is the allocated green time of the ith phase,* $i = 1, 2, \ldots, N$. *Let* T_{imin} *denotes the allowable minimum green time of the ith phase and let* $e_i = 1/w_i(t) > 0$ *be a set of weight parameters,* $i = 1, 2, \ldots, N$. *Let* T_{imax} *be the allowable maximum period for the ith phase and* $\sum_{i=1}^{N} T_{i_{\max}} \geq T$. *The allocated green time* T_i, $i = 1, 2, \ldots, N$, *obtained from the analytical solution in (6) minimize*

$$E(T_1, \ldots, T_N) = \sum_{i=1}^{N} w_i(t)(T_{i_{\max}} - T_i)^2, \tag{7}$$

subject to the inequality constraints $\sum_{i=1}^{N} T_i \leq T, T_i \geq T_{i_{\min}}$ *and* $T_i \leq T_{i_{\max}}$ *for* $i = 1, 2, \ldots, N$.

From the obtained corollary, it reveals that the optimization criterion inherent while allocating the traffic signal green time for different phases. Meanwhile, the allocated traffic signal green time is depended on the real-time traffic saturability. Finally, the allocated traffic signal green time is derived from the analytical solution that the computational efficiency can be improved.

4   Event-Triggering Global Adjustment of Traffic Light Loop

For the case of $\sum_{i=1}^{N} T_i \geq T$ and $\forall x_i > 1$, it means that the sum of the required signal green time for each phase is greater than the traffic lights period T. Then the real time traffic load cannot be released in next traffic light loop. A proportional global adaptive control strategy is designed to enlarge T, which is described as

$$T = \frac{\sum_{i=1}^{N} x_i(t)}{\sum_{i=1}^{N} x_i^{old}} = (X - n \times N) \times T^{old}, \tag{8}$$

where x_i^{old} denotes the last traffic saturation in the ith phase; T^{old} is the last traffic signal light loop. By using the adjusted period (8), the traffic load can be released more effectively in the next period. Also the proportional method is used only in traffic overload conditions.

For the case of $\sum_{i=1}^{N} T_i \le T$, it indices that the sum of the required signal green time for each phase is smaller than the traffic lights period T. Then the traffic light period cannot be fully utilized. The traffic green light loop T_i is set as the minimum value $T_{i_{min}}$. Then the abbreviated period is shown as

$$T = \sum_{i=1}^{N} x_{i_{min}}, \tag{9}$$

From the event-triggering global adaptive adjustment schemes (8) and (9) for traffic light loop, the traffic light loop can be adaptive triggered based on the traffic load, i.e., overload situation or lightweight traffic load situation.

5 Simulation Results

In order to testify the effectiveness of the proposed schemes, the intersection with 4 phases and 12 lanes will be undertaken. The parameters are listed as follows: the initial traffic signal period set as $T = 120$ s; the minimum allowable green time for each phase is 5 s; the saturability abilities of four orientation are set as $Q_{East_{max}} = 10$, $Q_{South_{max}} = 15$, $Q_{West_{max}} = 20$, and $Q_{North_{max}} = 25$, respectively.

To prove the ability of improving road capability of proposed allocating schemes, the classic average allocating method is introduced, which is described as

$$T_i = T/N, \quad i = 1, 2, \ldots N. \tag{10}$$

Three traffic flow conditions are considered in this simulation, including lightweight traffic flow, normal traffic flow and overload traffic flow. Applying the proposed allocating schemes and the class average allocating method (10) to an intersection, the traffic flow conditions and allocating green time for each phase are shown in Tables 1, 2 and 3 under different traffic loads. Meanwhile, the adjustment results of traffic light periods and the road capacity are shown in Table 4.

Table 1. Allocated green time under lightweight traffic flow under the proposed scheduling scheme and average allocated method (10).

Phase	Phase 1		Phase 2		Phase 3		Phase 4	
Lane	L1	L2	L3	L4	L4	L7	L5	L6
Amount of vehicle	3	4	5	3	2	4	3	3
Average green time	30 s		30 s		30 s		30 s	
Proposed scheduling green time	20 s		25 s		20 s		15 s	

Table 2. Allocated green time under normal traffic flow under the proposed scheduling scheme and average allocated method (10).

Phase	Phase 1		Phase 2		Phase 3		Phase 4	
Lane	L1	L2	L3	L4	L4	L7	L5	L6
Amount of vehicle	3	8	9	10	13	12	14	10
Average green time	30 s		30 s		30 s		30 s	
Proposed scheduling green time	22 s		21 s		35 s		42 s	

For the intersection under lightweight traffic load, the simulation results are shown in Tables 1 and 4. It shows that the traffic capability is the same level by using the proposed allocating schemes and the classic average allocating method. However, by applying scheme (9) to adjust the traffic light loop, the loop under proposed allocating schemes is shorter than that of the classic average allocating methods'. It indicates that the consumption time is less to dredge the traffic flow under proposed allocating scheme.

For the case of normal traffic flow, the simulation results are shown in Tables 2 and 4. It can be seen clearly that, although with the same period $T = 120$ s, the road capacity can be improved effectively under proposed allocating schemes than that of the classic average allocating methods'. Meanwhile, the allocating green time are relative to traffic loads and saturability abilities in each phase.

For the case of overload traffic flow, the simulation results are shown in Tables 3 and 4. While diagnosing the overload situation in intersection, the traffic signal period enlarge to $T = 160$ s. By applying the allocating scheme (6), the road capacity can be improved effectively than that of the classic average allocating methods' with the same period. Then the traffic load can be dispersed rapidly.

Table 3. Allocated green time under overload traffic flow under the proposed scheduling scheme and average allocated method (10).

Phase	Phase 1		Phase 2		Phase 3		Phase 4	
Lane	L1	L2	L3	L4	L4	L7	L5	L6
Amount of vehicle	24	13	15	25	20	26	18	28
Average green time	40 s		40 s		40 s		40 s	
Proposed scheduling green time	30 s		46 s		42 s		42 s	

To sum up the above simulation results, by adjusting the traffic light period for the cases of lightweight and overload traffic flow, the proposed global adaptive proportional control schemes can cover the extreme traffic flow. Meanwhile, by applying the local scheduling schemes for the traffic signal green time under the case of normal traffic load, the traffic signal green time for each phase can be allocated based on the real-time saturabilities. Then the ability of road capacity can be improved effectively by using the proposed local scheduling green time control schemes.

Table 4. Traffic signal loop and road capacity under different traffic flow.

Traffic flow	Traffic loop		Road capacity	
	Proposed scheme	Average method	Proposed scheme	Average method
Lightweight flow	80 s	120 s	27 vehicles	27 vehicles
Normal flow	120 s	120 s	58 vehicles	48 vehicles
Overload flow	160 s	160 s	80 vehicles	64 vehicles

6   Conclusions

Integrating the real-time saturabilities into traffic signal control systems, the competitive relationship was described as a trade-off optimization problem in a designed elastic scheduling model. Then an analytical solution was obtained to allocate the green time for each phase with the weight factor of real-time saturability. Meanwhile, an event-triggering global adjustment scheme was proposed to deal with the overload or lightweight traffic flow, in which the traffic signal period was proportional enlarged or decreased. By comparing with the classic average allocating method under different traffic flow situations, the ability of improving road capacity was illustrated for the proposed control schemes.

Acknowledgments. This work was supported by the National Key Research and Development Plan (No. 2016YFC0106000), the National Natural Science Foundation of China (Grant No. 61640218, 61671220, 61573166, 61673357), the Shandong Distinguished Middle-aged and Young Scientist Encourage and Reward Foundation, China (Grant No. BS2014DX015, ZR2016FB14), the Project of Shandong Province Higher Educational Science and Technology Program, China (Grant No. J16LN07), the Shandong Province Key Research and Development Program, China (Grant No. 2016GGX101022).

References

1. Prashanth, L.A., Shalabh, B.: Reinforcement learning with function approximation for traffic signal control. IEEE Trans. Intell. Transp. Syst. **12**, 412–421 (2001)
2. Zhao, D., Dai, Y., Zhang, Z.: Computational intelligence in urban traffic signal control: a survey. IEEE Trans. Syst. Man Cybern. Part C Appl. Rev. **42**, 485–494 (2012)
3. Behrang, A., Ardalan, V.: Predictive cruise control: utilizing upcoming traffic signal information for improving fuel economy and reducing trip time. IEEE Trans. Control Syst. Technol. **19**, 707–714 (2011)
4. Han, S.-Y., Zhang, C.-H., Tang, G.-Y.: Approximation optimal vibration for networked nonlinear vehicle active suspension with actuator time delay. Asian J. Control (2017). doi:10.1002/asjc.1419
5. Han, S.-Y., Tang, G.-Y., Chen, Y.-H., Yang, X.-X., Yang, X.: Optimal vibration control for vehicle active suspension discrete-time systems with actuator time delay. Asian J. Control **15**, 1579–1588 (2013)

6. Han, S.-Y., Wang, D., Chen, Y.-H., Tang, G.-Y., Yang, X.-X.: Optimal tracking control for discrete-time systems with multiple input delays under sinusoidal disturbances. Int. J. Control Autom. Syst. **13**, 292–301 (2015)
7. Han, S.-Y., Chen, Y.-H., Wang, L., Abraham, A.: Decentralized longitudinal tracking control for Cooperative Adaptive Cruise Control Systems in a platoon. In: IEEE International Conference on Systems, Man, and Cybernetics, pp. 2013–2018 (2013)
8. Zhou, J., Chen, C.L.P., Chen, L., Li, H.X.: A collaborative fuzzy clustering algorithm in distributed network environments. IEEE Trans. Fuzzy Syst. **22**, 1443–1456 (2014)
9. Zhou, J., Chen, C.L.P., Chen, L., Li, H.X.: Fuzzy clustering with the entropy of attribute weights. Neurocomputing **198**, 125–134 (2016)
10. Han, S.-Y., Chen, Y.-H., Tang, G.-Y.: Sensor fault and delay tolerant control for networked control systems subject to external disturbances. Sensors **17**, 700 (2017)
11. Zhou, J., Chen, C.L.P., Chen, L., Li, H.X.: A user customizable urban traffic information collection method based on wireless sensor networks. IEEE Trans. Intell. Transp. Syst. **14**, 1119–1128 (2013)
12. Wei, J., Wang, A., Du, N.: Study of self organizing control of traffic signals in an urban network based on cellular automata. IEEE Trans. Veh. Technol. **54**, 744–748 (2005)
13. Yin, B., Dridi, M., Abdellah, E.M.: Traffic network microsimulation model and control algorithm based on approximate dynamic programming. IET Intell. Transp. Syst. **10**, 186–196 (2016)
14. Mirchandani, P.B., Zou, N.: Queuing models for analysis of traffic adaptive signal control. IEEE Trans. Intell. Transp. Syst. **8**, 50–59 (2007)
15. Srinivasan, D., Choy, M.C., Chen, R.L.: Neural networks for real-time traffic signal control. IEEE Trans. Intell. Transp. Syst. **7**, 261–272 (2009)
16. Choy, M.C., Srinivasan, D., Chen, R.L.: Cooperative, hybrid agent architecture for real-time traffic signal control. IEEE Trans. Syst. Man Cybern. Part A Syst. Hum. **33**, 597–604 (2003)
17. Sameh, S., Ahmed, E.-M., Yasutaka, W.: A linear time and spacing algorithm for optimal traffic signal duration at an intersection. IEEE Trans. Intell. Transp. Syst. **16**, 387–395 (2015)
18. Yang, X., Li, X., Xue, K.: A new traffic signal control for modern roundabouts: method and application. IEEE Trans. Intell. Transp. Syst. **5**, 282–297 (2004)
19. Buttazzo, G., Lipari, G., Caccamo, M., Abeni, L.: Elastic scheduling for flexible workload management. IEEE Trans. Comput. **51**, 289–302 (2002)
20. Tian, Y.-C., Jiang, X., Levy, D.C., Agrawala, A.: Local adjustment and global adaption of control periods for QoC management of control systems. IEEE Trans. Control Syst. Technol. **20**, 846–854 (2012)
21. Chandra, R., Liu, X., Sha, L.: On the scheduling of flexible and reliable real-time control systems. Real-Time Syst. **24**, 153–169 (2003)
22. Amirijoo, M., Hansson, J., Son, S.H., Gunnarsson, S.: Experimental evaluation of linear time-invariant models for feedback performance control in real-time systems. Real-Time Syst. **35**, 209–238 (2007)

Classifying DNA Microarray for Cancer Diagnosis via Method Based on Complex Networks

Peng Wu[1,2(✉)], Likai Dong[1], Yuling Fan[1], and Dong Wang[1,2]

[1] School of Information Science and Engineering,
University of Jinan, Jinan, People's Republic of China
{ise_wup,ise_donglk,ise_fanyl,ise_wangd}@ujn.edu.cn
[2] Shandong Provincial Key Laboratory of Network Based Intelligent Computing,
University of Jinan, Jinan, People's Republic of China

Abstract. Performing microarray expression data classification can improve the accuracy of a cancer diagnosis. The varying technique including Support Vector Machines (SVMs), Neuro-Fuzzy models (NF), K-Nearest Neighbor (KNN), Neural Network (NN), and etc. have been applied to analyze microarray expression data. In this investigation, a novel complex network classifier is proposed to do such thing. To build the complex network classifier, we tried a hybrid method based on the Particle Swarm Optimization algorithm (PSO) and Genetic Programming (GP), of which GP aims at finding an optimal structure and PSO accomplishes the fine tuning of the parameters encoded in the proposed classifier. The experimental results conducted on Leukemia and Colon data sets are comparable to the state-of-the-art outcomes.

Keywords: DNA microarray data · Complex network classifier · Particle Swarm Optimization algorithm · Genetic Programming

1 Introduction

As a high-throughput multiplex technology, DNA microarray analysis is widely used in molecular biology and biomedicine. An accurate cancer diagnosis can be achieved by performing microarray expression data classification [4, 11, 13–18]. The varying technique including Support Vector Machines (SVMs) [5], Neuro-Fuzzy models (NF) [12], K-Nearest Neighbor (KNN) [6], Neural Network (NN) [4], and etc. have been applied to analyze microarray expression data.

Artificial neural networks have proven to be a powerful tool in DNA microarray analysis. Nevertheless, a neural network behaves highly dependent on what kind of structure it has. Based on a practice case, it may be appropriate to have more than one hidden layer, feedback or feed-forward connections, or in some circumstances, direct connections between input and output layer.

In this paper, a novel classifier based on Complex Network (CN) was proposed for DNA microarray analysis. This classifier has structures similar to that of scale-free CNs [1, 3]. We use the modified algorithm proposed in [2] to initialize a CN classifier, which allows input variables to select over layer connections and different activation functions

© Springer International Publishing AG 2017
D.-S. Huang et al. (Eds.): ICIC 2017, Part II, LNCS 10362, pp. 740–747, 2017.
DOI: 10.1007/978-3-319-63312-1_66

for different nodes. To construct the CN classifier, we utilized a hybrid method integrating the Genetic Programming (GP) [8] and the Particle Swarm Optimization algorithm (PSO) [7], of which the former aim at finding an optimal structure and the latter accomplishes the fine tuning of the parameters encoded in the model.

The rest of this paper is organized as follows. In Sect. 2, some relative methods are briefly summarized. Section 3 serves to depict the classifier based on the CN. Its application for real DNA microarray analysis is reported in Sect. 4, the whole paper is finally concluded in Sect. 5.

2 The Methodology

This section gives an introduction to the basis of the framework, i.e., CN, GP, and PSO.

2.1 Complex Networks

A CN has the structure which is generally characterized by non-trivial topology and a high number of nodes and connections. CN can be classified into two main classes: scale-free and small-world. Barabasi and Albert proposed a method to generate a series of scale-free networks. Growth and preferential attachment compose the two principal characters in the constructing of the network. The former means that the number of nodes in the network increases with time and the latter stipulates that the probability for a new node to be connected to an existing node depends on the number of links that this node already has [9, 13].

In this investigation, we put emphasis on scale-free networks, and the generation algorithm of which is similar to the algorithm proposed by Mauro Annunziato, Ilaria Bertini et al. in [9]. It should be stressed that the main of the evolution strategy proposed by Mauro et al. is maintained but some modifications have been made in order to evolve our CN classifiers.

2.2 Genetic Programming

GP is a kind of algorithm which can simulate evolution process in nature. In spite of the similar genetic operations between GP and Genetic Algorithms (GA), GP has tree structural individual rather than string structural individual owned by GA. In order to evolve CN classifier, we made some modifications on the genetic operators, i.e., crossover and mutation, which will be described in the next section.

2.3 Particle Swarm Optimization Algorithm

The PSO is a high effective tool for searching the optimal solution in a given space, which simulates the searching food process of birds by using a population of particles. For the details of PSO please refer to [8].

3   The Complex Network Classifier Based on GP and PSO

3.1   The Description of Complex Network Classifier

In this paper, a directed and weighted graph is selected for representing a CN classifier. We use the method introduced in Sect. 2 to generate a CN classifier (see Fig. 1). In order to initialize a CN classifier, for any non-input node, i.e. $Node_i$, several small real numbers are randomly generated and used for representing the connection strength between $Node_i$ and its adjacent nodes. Moreover, the flexible activation function is used to develop the CN classifier, which is shown in Eq. 1:

$$f(a_i, b_i, x) = e^{-\left(\frac{x - a_i}{b_i}\right)^2}. \tag{1}$$

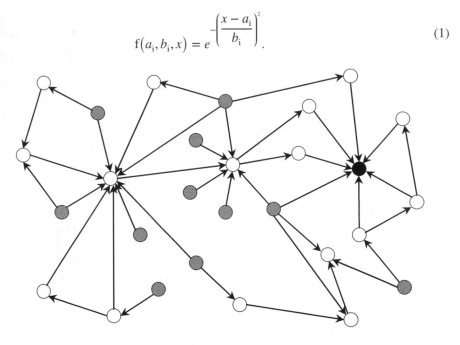

Fig. 1. A sample of CN classifier (the gray nodes denote input nodes and the black one is output node).

Where two control parameters a_i and b_i are randomly created as the flexible activation function parameters.

The output of a non-input node, taken as a flexible neuron model, is calculated by Eq. 2:

$$Output_{node_i} = \sum_{j=1}^{n} w_j * x_j, \tag{2}$$

where x_j ($j = 1, 2, ..., n$) are the inputs to node $Node_i$. The overall output of CN classifier can be calculated by the depth-first method, recursively.

3.2   The Optimization of Complex Network Classifier

Optimizing Structure. To optimize the structure of CN classifier means finding an optimal or near-optimal product of evolution. We use GP with some modification to achieve the goal, which can be summarized as follows:

1. Initialization. A population of individuals with CN structure is randomly generated.
2. Evaluation. Each individual with a CN structure of the current population, i.e. P_{gi}, is evaluated by using the defined fitness function to calculate a fitness value.
3. Crossover. Given a certain probability P_c and the tournament selection mechanism, i.e. m individuals are randomly selected from the current population to go through a competition and the winner is selected. Figure 2 depicts how the crossover works.

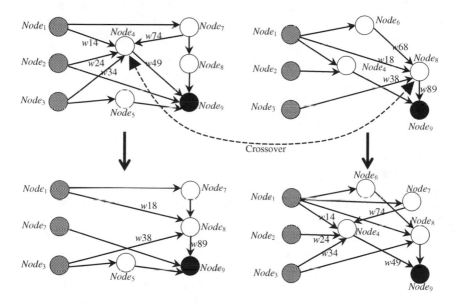

Fig. 2. Crossover operation between $Node_4$ and $Node_8$ in two CN classifiers.

4. Mutation. According to a probability, P_m, an individual with CN structure is selected randomly for mutation. Four kinds of operators can be free to choose under the same condition, the steps of which are described as follows:
    - Adding one connection: randomly select two nodes in the CN classifier and add a connection between them.
    - Adding one node: Insert one new node and connect it to a selected destination node with a preferential-attachment function which defines the probability that a node in the network receive a link from a newly inserted node [11]. The formula of this function is defined as Eq. 3:

$$\prod (k_i) = \frac{k_i}{\sum_{j=1}^{N} k_j} \tag{3}$$

where k_i is the sum of in-degree and out-degree of $Node_i$.

- Deleting one connection: randomly select a connection and delete it.
- Deleting one node: randomly select a node and cut off all of its connections.

5. Selection. Given a union of μ parents and μ offspring, the pairwise comparison will carry out among them. According to the comparison results, select μ individuals out of the union individuals to form the next generation. Repeat Steps 3–5 until a new population is constructed. If the satisfied solution is found, then stop; otherwise go to Step 2.

Tuning Parameters. The basic PSO algorithm is selected for parameter optimization due to its fast converge and ease to implementation.

4   Experimental Study

In this section, the performance of the CN based classifier and comparisons with other intelligent models are reported. Two benchmark DNA microarray datasets, namely the Leukemia and Colon [4, 11], are studied. They consist of binary cancer classification problems. The initial leukemia dataset consists of 72 samples with 7129 genes. These samples are measurements corresponding to acute lymphoblast leukemia (ALL) and acute myeloid leukemia (AML) samples from bone marrow and peripheral blood. There are 25 samples of AML, and 47 samples of ALL. The colon cancer dataset contains the expression of the 2000 genes with highest minimal intensity across the 62 tissues derived from 40 tumors and 22 normal colon tissue samples.

For this experiment, the normalization procedure is firstly used for preprocessing the raw data, and the details for data procedures are described in [4]. Due to small samples with thousands of genes, gene selection approach based on correlation has been applied in this study.

In order to use as many samples as possible, we adopt leave-one-out cross validation (LOOCV) to train and evaluate our models. In LOOCV, one of all samples is evaluated as testing data while the rest are used as training data. To obtain reliable experimental results, we run on 45 different tests for two microarray datasets. Figures 3 and 4 describe the details of results. Tables 1 and 2 compare different classification methods for leukemia and colon. From the tables, we can see that our model obtain excellent accuracy for leukemia dataset and comparable for colon dataset.

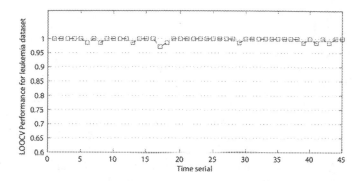

Fig. 3. The classification accuracy rates running on 45 different tests for leukemia data set.

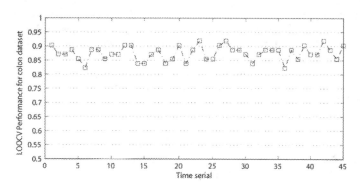

Fig. 4. The classification accuracy rates running on 45 different tests for colon data set.

Table 1. Relative works on leukemia dataset

Common parameters for GP	LOOCV accuracy (%)
Our method	99.69
SVM [4]	94.1
Nero-fuzzy [6]	87.5
NN [4]	86.1
KNN [9]	72.64

Table 2. Relative works on colon dataset

Common parameters for GP	LOOCV accuracy (%)
Our method	87.52
SVM [11]	90.3
Nero-fuzzy [6]	93.55
NN [4]	88.7
KNN [9]	75.81

5   Conclusion

In this paper, we presented an optimized classifier based on CN classifier for DNA microarray data analysis. The experimental results show that our model can obtain competitive results. But there are still serial issues that deserve to study in future research. For DNA microarray data analysis, gene selection methods are always essential; we choose the correlation-based method in this study, which is commonly believed it may not fully extract the information from data corrupted by high-dimensional noise. Therefore, other gene selection methods such as principal components analysis, Fisher-ratio, and t-test, should prove to be ideal candidates. For CN classifier, there are many other evolution algorithms that can be used to construct the structure.

Acknowledgment. This research was supported by the National Key Research and Development Program of China (No. 2016YFC0106000, 2016YFC0106001), the Youth Science and Technology Star Program of Jinan City (201406003), the Natural Science Foundation of Shandong Province (ZR2013FL002).

References

1. Albert, R., Barabasi, A.L.: Statistical mechanics of CNs. Rev. Mod. Phys. **74**(1), 47 (2002)
2. Annunziato, M., Bertini, I., Felice, M., Pizzuti, S.: Evolving complex neural networks. In: Basili, R., Pazienza, M.T. (eds.) AI*IA 2007. LNCS, vol. 4733, pp. 194–205. Springer, Heidelberg (2007). doi:10.1007/978-3-540-74782-6_18
3. Barabasi, A.L., Albert, R.: Emergence of scaling in random networks. Science **286**(5439), 509–512 (1999)
4. Chen, Y., Zhao, Y.: A novel ensemble of classifiers for microarray data classification. Appl. Soft Comput. **8**(4), 1664–1669 (2008)
5. Furey, T.S., Cristianini, N., Duffy, N., Bednarski, D.W., Schummer, M., Haussler, D.: Support vector machine classification and validation of cancer tissue samples using microarray expression data. Bioinformatics **16**(10), 906–914 (2000)
6. Jirapech Umpai, T., Aitken, S.: Feature selection and classification for microarray data analysis: evolutionary methods for identifying predictive genes. BMC Bioinform. **6**(1), 148 (2005)
7. Kennedy, J.: Particle swarm optimization. In: Sammut, C., Webb, G.I. (eds.) Encyclopedia of Machine Learning, pp. 760–766. Springer, US (2011)
8. Koza, J.R.: Genetic Programming: on the Programming of Computers by Means of Natural Selection, vol. 1. MIT Press, Cambridge (1992)
9. Lai, Y.C., Motter, A.E., Nishikawa, T.: Attacks and cascades in complex networks. In: Ben-Naim, E., Frauenfelder, H., Toroczkai, Z. (eds.) Complex Networks. Lecture Notes in Physics, vol. 650, pp. 299–310. Springer, Heidelberg
10. Mehmood, R., et al.: Clustering by fast search and merge of local density peaks for gene expression microarray data. Sci. Rep. **7**, 45602 (2017)
11. Niedzwiecki, D., Frankel, W.L., Venook, A.P., Ye, X., Friedman, P.N., Goldberg, R.M., Mayer, R.J., Colacchio, T.A., Mulligan, J.M., Davison, T.S., et al.: Association between results of a gene expression signature assay and recurrence-free interval in patients with stage II colon cancer in cancer and leukemia group B 9581(alliance). J. Clin. Oncol. **34**(25), 3047–3053 (2016)

12. Wang, Z., Palade, V., Xu, Y.: Neuro-fuzzy ensemble approach for microarray cancer gene expression data analysis. In: 2006 International Symposium on Evolving Fuzzy Systems, pp. 241–246. IEEE (2006)
13. Zanin, M., Papo, D., Sousa, P.A., Menasalvas, E., Nicchi, A., Kubik, E., Boccaletti, S.: Combining complex networks and data mining: why and how. Phys. Rep. **635**, 1–44 (2016)
14. Huang, D.S.: Systematic Theory of Neural Networks for Pattern Recognition (in Chinese), Publishing House of Electronic Industry of China, May 1996
15. Ji, Z., Wu, D., Zhao, W., et al.: Systemic modeling myeloma-osteoclast interactions under normoxic/hypoxic condition using a novel computational approach. Sci. Rep. 5 (2015)
16. Wang, B., Zhang, J., Chen, P., et al.: Prediction of peptide drift time in ion mobility mass spectrometry from sequence-based features. BMC Bioinform. **14**(8), S9 (2013)
17. Huang, D.S., Yu, H.-J.: Normalized feature vectors: a novel alignment-free sequence comparison method based on the numbers of adjacent amino acids. IEEE/ACM Trans. Comput. Biol. Bioinform. **10**(2), 457–467 (2013)
18. Ji, Z., Wu, G., Hu, M.: Feature selection based on adaptive genetic algorithm and SVM. Comput. Eng. **14**, 072 (2009)

Predicting Multisite Protein Sub-cellular Locations Based on Correlation Coefficient

Peng Wu[1,3], Dong Wang[1,3(✉)], Xiao-Fang Zhong[2], and Qing Zhao[1]

[1] School of Information and Engineering, University of Jinan,
Jinan 250022, China
{ise_wup,ise_wangd}@ujn.edu.cn, 798046481@qq.com
[2] Technology, Shandong Womens University, Jinan 250300, China
215046321@qq.com
[3] Shandong Provincial Key Laboratory of Network Based Intelligent
Computing, Shandong Womens University, Jinan 250022, China

Abstract. With the development of proteomics and cell biology, protein sub-cellular location has become a hot topic in bioinformatics. As the time goes on, more and more researchers make great efforts on studying protein sub-cellular location. But they only do research on single-site protein sub-cellular location. However, some proteins can belong to two or more sub-cellulars. So, we should transfer the line of sight to multisite protein sub-cellular location. In this article, we use Virus-mPLoc data set and choose pseudo amino acid composition and correlation coefficient two effective feature extraction methods. Then, putting these features into multi-label k-nearest neighbor classifier to predict protein sub-cellular location. The experiment proves that this method is reasonable and the precision reached 68.65% through the Jack-knife test.

Keywords: Multisite · Pseudo amino acid composition · Correlation coefficient · Multi-label k-nearest neighbor

1 Introduction

Studying protein is important as its special function for life activities. However, protein must locate in correct sub-cellulars if it wants to perform corresponding functions. If it locates in the wrong sub-cellulars, it will do harm to the body, or even cause serious dangerous. The development of proteome projects pushes forward research of protein sub-cellular location. As researchers believe that proteins only locate in one sub-cellular, they do a study on single-site protein sub-cellular location for many years. But, the experiment data show that some proteins can belong to two or more sub-cellulars. The appearance of multisite protein opens a new door of protein sub-cellular location. Therefore, it's necessary for us to predict multisite protein sub-cellular location [11]. Nowadays, although many researchers have studied multisite protein sub-cellular location, there is still a large space to promote the predicting precision [2].

In this study, we tried to use multi-label k-nearest neighbor classifier to predict multisite protein sub-cellular location. To this end, pseudo amino acid composition and a correlation coefficient of coding features were selected.

© Springer International Publishing AG 2017
D.-S. Huang et al. (Eds.): ICIC 2017, Part II, LNCS 10362, pp. 748–756, 2017.
DOI: 10.1007/978-3-319-63312-1_67

In the remaining part of this paper: Sect. 2 gives brief introduction of the feature expression we used; following, the classification algorithm is described in Sect. 3 and a case study is presented in Sect. 4; eventually, we draw a conclusion in Sect. 5.

2 Feature Expression

As we all know, extracting effective protein features is the key step of predicting protein sub-cellular locations. There are many commonly used protein feature extraction models [10], for example Amino Acid composition Model, Physicochemical Properties Model and so on. However, only using one feature extraction method customarily cannot represent protein adequately. It is necessary for researchers to combine some methods to improve the prediction precision [5]. In this article, we choose two feature extraction methods, i.e., pseudo amino acid composition and a correlation coefficient of coding.

2.1 Pseudo Amino Acid Composition (PseAAC)

The traditional amino acid composition only considers the composition of one protein sequence [8]. However, if we want to express protein sequences adequately, the location information must be considered. Then CHOU proposed pseudo amino acid composition which is a modified form [7]. It not only takes composition information into account but also thinks about location information. In the pseudo amino acid composition, protein sequence M which length is L could be expressed as the following feature vector:

$$v = \{m_1, m_2, m_3, \ldots, m_{20}, m_{20+1}, m_{20+2}, m_{20+3}, \ldots m_{20+n}\}^T \qquad (n < L) \qquad (1)$$

The former 20 dimensions of the feature vector are frequency that every amino acid appeared in the protein sequence, it represents the protein composition information. The other n dimensions represent location information and reflect the relationship between every amino acid [3].

The relation between every amino acid can be calculated through the following formula

$$\tau_n = \frac{1}{L-n} \sum_{i=1}^{L-n} J_{i,i+n} \qquad (n < L) \qquad (2)$$

In the above formula, $J_{i,i+n}$ can be calculated by the following method:

$$J_{i,i+n} = \frac{1}{3} \left\{ [H_1(R_{i+n}) - H_1(R_i)]^2 + [H_2(R_{i+n}) - H_2(R_i)]^2 \right.$$
$$\left. + [M(R_{i+n}) - N(R_i)]^2 \right\} \qquad (3)$$

Where $H_1(R_i)$ is the hydrophobicity value of amino acid residue R_i, $H_2(R_i)$ represent the hydrophilic of amino acid residue R_i, $M(R_i)$ is the side chain molecular weight of amino acid residue R_i. In this article, we choose 16 physical and chemical characteristics to express protein sequence (it is shown in Tables 1, 2 and 3).

Table 1. The value of features 1–8

Amid acid	F1	F2	F3	F4	F5	F6	F7	F8
A	−0.4	−0.5	15	8.1	0.046	0.67	1.28	0.3
C	0.17	−1	47	5.5	0.128	0.38	1.77	0.9
D	−1.31	3	59	13	0.105	−1.2	1.6	−0.6
E	−1.22	3	73	12.3	0.151	−0.76	1.56	−0.7
F	1.92	−2.5	91	5.2	0.29	2.3	2.94	0.5
G	−0.67	0	1	9	0	0	0	0.3
H	−0.64	−0.5	82	10.4	0.23	0.64	2.99	−0.1
I	1.25	−1.8	57	5.2	0.186	1.9	4.19	0.7
K	−0.67	3	73	11.3	0.219	−0.57	1.89	−1.8
L	1.22	−1.8	57	4.9	0.186	1.9	2.59	0.5
M	1.02	−1.3	75	5.7	0.221	2.4	2.35	0.4
N	−0.92	0.2	58	11.6	0.134	−0.61	1.6	−0.5
P	−0.49	0	42	8	0.131	1.2	2.67	−0.3
Q	−0.91	0.2	72	10.5	0.18	−0.22	1.56	−0.7
R	−0.59	3	101	10.5	0.291	−2.1	2.34	−1.4
S	−0.55	0.3	31	9.2	0.062	0.01	1.31	−0.1
T	−0.28	−0.4	45	8.6	0.108	0.52	3.03	−0.2
V	0.91	−1.5	43	5.9	0.14	1.5	3.67	0.6
W	0.5	−3.4	130	5.4	0.409	2.6	3.21	0.3
Y	1.67	−2.3	107	6.2	0.298	1.6	2.94	−0.4

As the original values vary widely, it is necessary to process them in advance. This article uses the maximum and minimum standardized treatment. The formula is as following:

$$H(R_i) = \frac{h^0(R_i - \min(h^0(R_i)))}{\max(h^0(R_i)) - \min(h^0(R_i))} \tag{4}$$

After calculating the n related factors, we can get feature vector from the following formula:

$$P_j = \begin{cases} \dfrac{f_j}{\sum_{i=1}^{20} f_j + w\sum_{n=1}^{\lambda}\tau_n} & 1 \leq j \leq 20 \\[2ex] \dfrac{w\tau_{j-20}}{\sum_{i=1}^{20} f_j + w\sum_{n=1}^{\lambda}\tau_n} & 20+1 \leq j \leq 20+\lambda \end{cases} \tag{5}$$

In the formula, w is the weight factor, its value usually is 0.05. f_i represents the frequency of amino acid residue i in the protein sequence. The value of λ is 20 or 30. Then the feature vector from this model is 40 or 50 dimensions [4].

Table 2. The values of features 9–16

Amid acid	F9	F10	F11	F12	F13	F14	F15	F16
A	0	0.687	115	0.28	154.3	27.5	1.18	0.007
C	2.75	0.263	135	0.28	219.8	44.6	1.46	−0.037
D	1.38	0.632	150	0.21	194.9	40	1.59	0.024
E	0.92	0.669	190	0.33	223.2	62	1.86	0.007
F	0	0.577	210	2.18	204.7	115.5	2.23	0.038
G	0.74	0.67	75	0.18	127.9	0	0.88	0.179
H	0.58	0.594	195	0.21	242.5	79	2.03	−0.011
I	0	0.564	175	0.82	233.2	93.5	1.81	0.022
K	0.33	0.407	200	0.09	300	100	2.26	0.018
L	0	0.541	170	1	232.3	93.5	1.63	0.052
M	0	0.328	185	0.74	202.6	94.1	2.03	0.003
N	1.33	0.489	160	0.25	207.9	58.7	1.66	0.005
P	0.39	0.6	145	0.39	179.9	41.9	1.47	0.24
Q	0.9	0.527	183	0.35	235.5	80.7	1.93	0.049
R	0.64	0.591	225	0.1	341	105	2.56	0.044
S	1.41	0.693	116	0.12	174.1	29.3	1.29	0.004
T	0.71	0.713	142	0.21	205.8	51.3	1.52	0.003
V	0	0.529	157	0.6	207.6	71.3	1.65	0.057
W	0.12	0.632	258	5.7	237.1	145.5	2.66	0.038
Y	0.21	0.495	234	1.26	229.2	117.3	2.37	0.024

Table 3. The sixteen kinds of features

Index	F1	F2	F3	F4
Name	Hydrophobic	Hydrophilic	Side chain mass	Polar
Index	F5	F6	F7	F8
Name	Polarization	Solvation free energy	Curve shape index	Transfer free energy
Index	F9	F10	F11	F12
Name	Amino acid composition	Correlation coefficient	Surface residues	Distribution coefficient
Index	F13	F14	F15	F16
Name	Information entropy	Side chain volume	Dissolve ability	Load index

2.2 Correlation Coefficient

In practice, a certain amino acid not only acts on its interfacing amino acid but also connects on some amino acids that away from long distance. The correlation coefficient of coding could consider location information and the interrelationship of amino acids no matter the short distance or long distance [1].

$$cc(\lambda, k) = \frac{1}{L-\lambda} \sum_{i=1}^{L-\lambda} \frac{A_{i,k} \times B_{i+\lambda,k}}{\sqrt{A_{i,k} \times A_{i,k}^T} \times \sqrt{B_{i+\lambda,k} \times B_{i+\lambda,k}^T}} \tag{6}$$

where $A_{i,k}$ and $B_{i+\lambda,k}$ represent the value of feature k on i and $i + \lambda$ place in the protein sequence, respectively. And they can be got from:

$$\begin{cases} A_{i,k} = R_{i,k} - \frac{1}{L}\sum_{i=1}^{L} R_{i,k} \\ B_{i+\lambda,k} = R_{i+\lambda,k} - \frac{1}{L}\sum_{i=1}^{L} R_{i,k} \end{cases} \tag{7}$$

From this method, protein sequence can be represented as $20 + 16 \times \lambda$ dimensions feature vector. And the former 20 dimensions of feature vector represent the protein composition information, the other $16 \times \lambda$ dimensions represent $1 \sim \lambda$ correlation coefficient information of 16 physical and chemical characteristics.

3   Classification Algorithm

Selecting an appropriate classifier is also an important link to predict multisite protein sub-cellular location [6]. In this article, the multi-label k-nearest neighbor algorithm is used to complete classification.

Multi-label k-nearest neighbor algorithm is improved from k-nearest neighbor algorithm, it determines the label information of unknown sample from its k neighbors label information [14].

In the training set, $\mathbf{Y}$ represents Binary label vector. For every label i. X_0^i indicates that training sample does not contain label i and X_1^i indicates that training sample contains label i, C_m^i is the number that k neighbors of unknown sample m contain. Then calculating prior probability $P(X_0^i)$ and $P(X_1^i)$. For sample set $\mathbf{T}$, calculating conditional probability $P(E_j^i/X_0^i)$ and $P(E_j^i/X_1^i)$, E_j^i represent that there are j samples contain label i in sample set $\mathbf{T}$.

According to the principle of maximum posterior probability:

$$y_{(m)}(i) = \arg maxP\left(\frac{X_b^i}{E_{C_m^i}^i}\right) \qquad (b = 0 \; or \; 1) \tag{8}$$

Finally, on the basis of Bayes formula determines the final result,

$$y_m(i) = \arg max \frac{P\left(X_b^i\right)P\left(\frac{E_{C_m^i}^i}{X_b^i}\right)}{P\left(E_{C_m^i}^i\right)} \tag{9}$$

$$= \arg max \; P(X_b^i)P(E_{C_m^i}^i, X_b^i)$$

4   Case Study

4.1   Data Set

Viruses can reproduce their progenies only within a host cell, and their actions depend both on its destructive tendencies toward a specific host cell and on environmental conditions. Therefore, knowledge of the sub-cellular localization of viral proteins in a host cell or virus-infected cell is very useful for in-depth studying their functions and mechanisms as well as designing antiviral drugs [9]. Figure 1 describes the progenies reproduced by a virus. In this article, we use Virus-mPLoc dataset. This dataset concludes 252 protein sequences totally, classified into 6 sub-cellular locations. There are 207 different proteins, 165 single site protein, 39 two sites protein and 3 three sites protein among them.

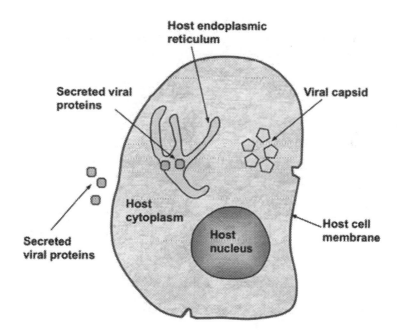

Fig. 1. Six sub-cellular locations of viral proteins [9].

4.2   Results and Analysis

After all, features pass through the ML-kNN classifier, some standards should be used to measure the result be good or bad. In multi-label learning, the commonly used measures [13] are Hamming Loss, One Error, Coverage, Ranking Loss and Average Precision.

Supposing, given multi-label classifier y(.) and the test set $N = \{(x_i, Y_i)|1 < i < m\}$, Y_i is the related label set of x_i. There are m samples and q labels totally in the training set.

(1) Hamming Loss

$$H = \frac{1}{m} \sum_{i=1}^{t} \frac{1}{q} |y(x_i \Delta Y_i)| \tag{10}$$

Where Δ is used to measure the difference between two symmetric set. This index which judging the misclassification of individual labels indicates whether the system is good, if this index is smaller, then the system is better.

(2) One Error

$$O = \frac{1}{m} \sum_{i=1}^{m} \{[\arg \max f(x_i, y)] \notin Y_i\}. \tag{11}$$

Where $f(,)$ is the real valued function. This standard tests the condition that the label which with the highest degree of membership value does not include in the label set. The smaller this standard is, the better system designed.

(3) Coverage

$$C = \frac{1}{m} \sum_{i=1}^{m} \max \text{rank} f(x_i, f) - 1. \tag{12}$$

Where rankf(,) represents sorting function. The standard evaluates how deep should search if all related labels are covered in the collating sequence. It is hoped smaller.

(4) Ranking Loss

$$R = \frac{1}{m} \sum_{i=1}^{m} \frac{1}{|Y_i||\bar{Y}_i|} ||\{(y', y'')|f(x_i, y') \le f(x_i, y''), (y', y'' \in Y_i \times \bar{Y}_i)\}|| \tag{13}$$

Where $\bar{Y}_i$ is the complementary set of Y_i. This value reflects the condition that appearing the wrong sort in the sort sequence. The smaller value proves superior performance of the system.

(5) Average Precision

$$A = \frac{1}{m} \sum_{i=1}^{m} \frac{1}{Y_i} \frac{||\{y'|rankf(x_i, y) \le rankf(x_i, y), y \in Y_i\}||}{rankf(x_i, y)} \tag{14}$$

This index investigates the condition that in front of the related label also is a related label in the sort sequence. The index is hoped larger. As we all know, such as k-nearest neighbor algorithm, different k get different results for multi-label k-nearest neighbor algorithm [12]. The following Table 4 displays the results with different k values.

Table 4. Different k gets different results

k	H	O	C	R	A
1	0.107	0.324	0.998	0.125	0.812
2	0.132	0.376	1.053	0.147	0.779
3	0.148	0.354	1.132	0.156	0.758
4	0.145	0.333	1.13	0.153	0.762

From the above table, we can see that when k take the value of 1, the system works best and get the highest accuracy. And the prediction accuracy can achieve 68.65% through this system.

5 Conclusion

In this paper, we tried to predict multisite protein sub-cellular location. In order to do this, we used two effective feature extraction methods, i.e., pseudo amino acid composition and correlation coefficient. At the same time, we developed multi-label k nearest neighbor algorithm to be the classifier. The experimental results on Virus-mPLoc data set prove that this method is reasonable.

Acknowledgment. This research was supported by the National Key Research And Development Program of China (No. 2016YFC0106000), National Natural Science Foundation of China (Grant No. 61302128, 61573166, 61572230, 61671220, 61640218), the Youth Science and Technology Star Program of Jinan City (201406003), the Natural Science Foundation of Shandong Province (ZR2013FL002), the Shandong Distinguished Middle-aged and Young Scientist Encourage and Reward Foundation, China (Grant No. ZR2016FB14), the Project of Shandong Province Higher Educational Science and Technology Program, China (Grant No. J16LN07), the Shandong Province Key Research and Development Program, China (Grant No. 2016GGX101022).

References

1. Chen, K., Kurgan, L.A., Ruan, J.: Prediction of protein structural class using novel evolutionary collocation-based sequence representation. J. Comput. Chem. **29**(10), 1596–1604 (2008)
2. Du, P., Xu, C.: Predicting multisite protein subcellular locations: progress and challenges. Expert Rev. Proteomics **10**(3), 227–237 (2013)
3. Fan, G.L., Li, Q.Z.: Predict mycobacterial proteins subcellular locations by incorporating pseudo-average chemical shift into the general form of Chous pseudo amino acid composition. J. Theor. Biol. **304**, 88–95 (2012)
4. Huang, C., Yuan, J.Q.: Predicting protein subchloroplast locations with both single and multiple sites via three different modes of Chou's pseudo amino acid compositions. J. Theor. Biol. **335**, 205–212 (2013)
5. Li, L., Yu, S., Xiao, W., Li, Y., Li, M., Huang, L., Zheng, X., Zhou, S., Yang, H.: Prediction of bacterial protein subcellular localization by incorporating various features into Chou's PseAAC and a backward feature selection approach. Biochimie **104**, 100–107 (2014)

6. Lin, W.Z., Fang, J.A., Xiao, X., Chou, K.C.: iLoc-animal: a multi-label learning classifier for predicting subcellular localization of animal proteins. Mol. BioSyst. **9**(4), 634–644 (2013)
7. Mei, S.: Multi-kernel transfer learning based on Chou's PseAAC formulation for protein submitochondria localization. J. Theor. Biol. **293**, 121–130 (2012)
8. Shen, H.B., Chou, K.C.: PseAAC: a flexible web server for generating various kinds of protein pseudo amino acid composition. Anal. Biochem. **373**(2), 386–388 (2008)
9. Shen, H.B., Chou, K.C.: Virus-mPLoc: a fusion classifier for viral protein subcellular location prediction by incorporating multiple sites. J. Biomol. Struct. Dyn. **28**(2), 175–186 (2010)
10. Wang, Z., Zou, Q., Jiang, Y., Ju, Y., Zeng, X.: Review of protein subcellular localization prediction. Curr. Bioinform. **9**(3), 331–342 (2014)
11. Xiao, X., Wu, Z.C., Chou, K.C.: iLoc-virus: a multi-label learning classifier for identifying the subcellular localization of virus proteins with both single and multiple sites. J. Theor. Biol. **284**(1), 42–51 (2011)
12. Huang, D.S.: Systematic Theory of Neural Networks for Pattern Recognition (in Chinese). Publishing House of Electronic Industry of China, Beijing (1996)
13. Ji, Z., Wu, D., Zhao, W., et al.: Systemic modeling myeloma-osteoclast interactions under normoxic/hypoxic condition using a novel computational approach. Sci. Rep. **5**, 13291 (2015)
14. Wang, B., Zhang, J., Chen, P., et al.: Prediction of peptide drift time in ion mobility mass spectrometry from sequence-based features. BMC Bioinform. **14**(8), S9 (2013)

Using a Hierarchical Classification Model to Predict Protein Tertiary Structure

Peng Wu[1,3], Dong Wang[1,3(✉)], Xiao-Fang Zhong[2],
and Fanliang Kong[1]

[1] School of Information and Engineering,
University of Jinan, Jinan 250022, China
{ise_wup,ise_wangd}@ujn.edu.cn, kongfl91@163.com
[2] Technology, University of Jinan, Jinan 250300, China
[3] Shandong Provincial Key Laboratory of Network Based Intelligent
Computing, University of Jinan, Jinan 250022, China

Abstract. To predict protein tertiary structure accurately is helpful for understanding the functions of proteins. In this study, a hierarchical classification method based on flexible neural tree was proposed to predict the structures, in which the tier classifiers were flexible neural trees due to their excellent performances. In order to classify the structures, three types of feature are used, i.e. the tripeptide composed of dimension reduction, the pseudo amino acid composition and the position information of amino acid residues. To evaluate our method, the 640 data set was used in this investigation. The experimental results suggest that our method overwhelms several representative approaches to predicting protein tertiary structure.

Keywords: Protein tertiary structure · Hierarchical classification model · Flexible neural tree

1 Introduction

In the past two decades, researchers devote to figuring out the connections between the function and the structure of a protein. A lot of findings show that the structure of a protein plays a significant role in its function. Levitt and Chothia [6] firstly give the definitions of four types of proteins, i.e., all-α, all-β, $\alpha + \beta$, and α/β. For the types all-α and all-β, there are only α helices and β strands contained respectively. As for the other types, all-α and all-β, they are hybrids of α helices and β strands. The former type comprises anti-parallel β strands and the latter one contains the parallel α strands [2].

It is a challenge to achieve an accuracy result of protein structure prediction. Researchers have been trying to promote the classification accuracy by using different methods. All of these approaches can be categorized to two main directions: finding more suitable features to character the amino acid sequence, and designing better classification models. In this study, we try to use some new feature expression, and at the same time, exploit a powerful classification tool to improve the accuracy of protein structure prediction.

© Springer International Publishing AG 2017
D.-S. Huang et al. (Eds.): ICIC 2017, Part II, LNCS 10362, pp. 757–764, 2017.
DOI: 10.1007/978-3-319-63312-1_68

The rest of this paper is organized as follows. Section 2 servers to give the explanation of features used in our experiments. The description of classification model is given in Sect. 3. The ensemble strategy and experimental analysis are discussed in Sects. 4 and 5, respectively. Finally, the conclusion is draw in the last section.

2 Feature Expression

2.1 Pseudo Amino Acid Composition (PseAAC)

There are 20 amino acids existing in proteins. The sequence forms of amino acids decide the primary structure of proteins. Each one in the 20 amino acids has a single attribute [7–9, 18]. However, these attributes cannot be used directly as the feature expression due to theirs' loss of sequence information of the protein. As an alternative, Chou et al. [10, 11] proposed pseudo amino acid composition, i.e. PseAAC, which comprises $(20 + \lambda)$-dimensional vector and can be formulated as following:

$$v = \left\{ p_1, p_2, p_3, \ldots, p_{20}, p_{20+1}, p_{20+2}, p_{20+3}, \cdots p_{20+\lambda} \right\}^T \tag{1}$$

The first 20 components are the simple amino acid composition and the remaining part can be calculated by the following formula:

$$p_i = \frac{w\tau_{i-20}}{\sum_{k=1}^{20} P_k + w \sum_{k=1}^{\delta} \tau_k} \qquad (20+1) \leq i \leq (20+\lambda) \tag{2}$$

Where w is a weighted coefficient, and its value generally is set to be 0.05. $\tau_k (k = 1, 2, \ldots \lambda)$ is a k-order correlation factor and reflects the sequence order correlation between pairwise of amino acid residues. This paper set τ to 20. As a result, the feature PseAAC derived as a 40-dimensional vector.

2.2 The Position Information of Amino Acid Residues

The position information of amino acid residues refers to the position proportional of the 20 kinds of amino acid residues in the protein sequence [14]. In spite of a low dimensional feature vector, the position feature is distinctive and very useful to distinguish diverse structures of proteins. Any of a protein sequence P, each amino acid residues tagged with increasing Arabic numerals from N to C terminal. The first amino acid residue labeled as 1, the second amino acid residue labeled as 2, and so on, the last amino acid residue labeled as L which is the length of the protein sequence. Protein sequences P, therefore, can be expressed with the corresponding Arabic numerals, these numbers can reflect the position information of 20 kinds of amino acid positions in protein sequence P. Calculate the sum of each amino acid residues tag number, marked as S_k^i, where $i = (A, C, D, E, F, G, H, I, K, L, M, N, P, Q, R, S, T, V, W, Y)$, k represents the k-th protein sequences in the data set.

$$S_k^i = \sum_{j=1}^{L} j e_{ij}, \quad k = 1, 2, \ldots, N, \tag{3}$$

where L is the length of the protein sequences, the value of A can take 1 or 0, when the j-th element of the protein sequences is i, $e_{ij} = 1$; On the contrary, if the j-th element of the protein sequences is not i, $e_{ij} = 0$.

This paper uses the percentage method for data normalization as following:

$$L_i^k = S_i^k, i = 1, 2, \ldots, 20; \quad k = 1, 2, \ldots, N \tag{4}$$

2.3   Tripeptide Composed of Dimension Reduction

The number of triplets (a, b, c) occur in the amino acid sequence labeled as $P_{(a,b,c)}$, the formula of which is as following:

$$P_{(a,b,c)} = \sum_{i=1}^{L-2} W_{a,b,c}(i, i+1, i+2) \qquad a, b, c = 1, 2, 3, \ldots, 20 \tag{5}$$

Where $W_{a,b,c}(i, i+1, i+2)$ is defined: only when the i-th residue is a, the $(i + 1)$-th residue is b and the $(i + 2)$-th residue is c in the amino acid sequence, $W_{a,b,c}(i, i+1, i+2) = 1$; Otherwise, $W_{a,b,c}(i, i+1, i+2) = 0$. L is the total length of the protein sequence.

Tripeptide composed refers to the number of 20 kinds of amino acids in the protein sequence appearing in the form of triplets. It put a protein sequence mapped to an 8000-dimensional feature vector. Next, we need to reduce the dimension on feature vector. In this paper, the 20 kinds of amino acids are divided into six categories based on hydrophobic and hydrophilic. The preserved features are shown in Table 1.

Table 1. Classification of amino acids

Classification	Abbreviation	Amino acids
Strongly hydrophilic or polar	L	R,D,E,N,Q,K,H
Strongly hydrophobic	B	L,I,V,A,M,F
Weakly hydrophilic or weak hydrophobic	W	S,T,Y,W
Proline	P	P
Glycine	G	G
Cysteine	C	C

The protein sequence from the six different letters constituted, therefore, should contain a total of 36 different triplets. According to these dimension reduction methods, a protein sequence can be mapped into a 216-dimensional feature vector.

If two triplets can be expressed as X-O-Y or Y-O-X, they can be called symmetrical triplet. Symmetrical triplet tend to have similar function, we take them as the same kind of triplets. Thus, a protein sequence can be mapped into a 126-dimensional feature Vector.

We connected the three feature vectors introduced above into a totally 186-dimensional vector, and which was used in our experiments.

3   Flexible Neural Tree Model

Flexible neural tree (FNT) has obtained great success in many fields such as stock index prediction [5], breast cancer classification [3], reverse engineering of gene regulatory networks, and foreign exchange rate prediction [4] etc. FNT is an extension of artificial neural network, especially forward feedback ones.

Compared to neural network, FNT has a flexible structure which allows the neurons connect to cross-layers and a smaller set of parameters which deceases the adjusting cost. Consequently, it is very easy to code FNT with flexible tree structures. With structure based evolving algorithm such as genetic programming [1, 16], probabilistic incremental program evolution [20, 21], etc. In this study, we took FNT as the base classifier. A tree-structural based encoding method with specific instruction set is selected to represent a FNT model.

The flexible neuron instructors use a function set $F = \{+_2, +_3, \ldots +_n\}$ and a terminal instruction set $T = \{x_1, x_2, x_3, \ldots x_n\}$ to generate a FNT model, which is described as follows.

$$S = F \cup T \tag{6}$$

Where $+_i(i = 1, 2,\ldots, N)$ denotes the non-leaf nodes instruction and has i arguments, and $x_i(i = 1, 2,\ldots,N)$ is the leaf nodes instruction and has no arguments.

Similar to the neurons embedded in neural networks, each non-leaf node of FNT models has a certain number of inputs and one output, which is shown in Fig. 1. From this aspect, we can look the instruction $+_i$ as a flexible neuron operator with i inputs.

The total outcome of a flexible neuron $+_n$ is calculated as following,

$$out_n = f(a_n, b_n, net_n) = \exp -\left(\frac{net_n - a_n}{b_n}\right)^2, \tag{7}$$

where net_n are the inputs to $+n$, and its output is calculated by following,

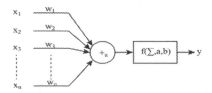

Fig. 1. A neuron embedded in flexible neuron tree.

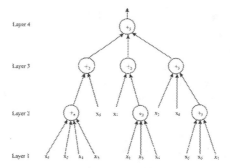

Fig. 2. A FNT model with the non-terminal set $F = \{+_2, +_3, +_4\}$ and the terminal set $T = \{x_1, x_2, x_3, x_4, x_5\}$

$$net_n = \sum_{j=1}^{M} w_j x_j \tag{8}$$

where $w_j (j = 1, 2, \ldots, M)$ is the weighted efficiency.

Figure 2 illustrates a typical FNT model, the overall output of which can be computed from left to right by a depth-first method recursively. The probabilistic incremental program evolution algorithm was used to find the structure of FNT and PSO [13, 15, 19] was used to tune the parameters involved in a FNT model. The general flowchart of generating a FNT model is shown in Fig. 3.

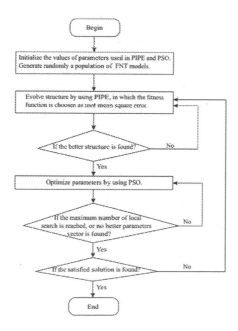

Fig. 3. The general flowchart of generating a FNT model.

4   Hierarchical Classification Model

Protein tertiary structure prediction is a four classification problem; hierarchical classification model constructed in this paper is shown in Fig. 4. Through a large number of experiments show, it's easy to distinguish between α, β and $\alpha + \beta$, α/β. So, the four types can be divided into two categories on the first layer, then separate the two categories one by one on the second layer.

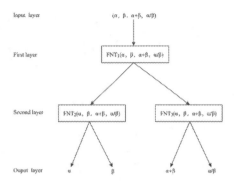

Fig. 4. The hierarchical classification model.

5   Case Study

In statistical analysis, two methods can be used to examine a predictor for its effectiveness in practical application, namely, independent dataset tests and 10-fold cross validation tests. During the process of the 10-fold cross validation tests, 10% of all proteins are treated as test data and others as train data. The overall accuracy (OA) is computed for each dataset.

5.1   Data Set

This paper selected 640 dataset to validate the proposed method. 640 dataset contains 640 proteins with 25% sequence identity, and their classification labels are retrieved from the database SCOP [17]. It includes 138 all-α class, 154 all-β class, 177 $\alpha + \beta$ class and 171 α/β class. 640 dataset can be more objective to value the validity of proposed methods because it contains sequences with low homology.

5.2   Results and Analysis

In this paper, we have used three feature extraction strategies, PseAA composition, position information of amino acid residues and tripeptide composed of dimension reduction. FNT is used as classification algorithm. The experiments are performed on a hierarchical classification model using 10-fold cross validation tests and report the overall accuracy (OA).

As shown in Table 2, although the prediction accuracy of all- class was not the best one, it was relatively high, and was 7.92% and 6.47% higher than Logistic regression and SVM, respectively. Other classes achieved the highest prediction accuracy compared with other methods.

This paper achieved the highest overall accuracy, and was 1.72% higher than SVM which accuracy is the highest among other four methods.

Table 2. The comparison of several methods.

Method	$\alpha(\%)$	$\beta(\%)$	$\alpha + \beta(\%)$	$\alpha = \beta(\%)$	Overall(%)
IB1 [12]	53.62	46.10	68.93	34.50	50.94
Naïve Bayes [12]	55.07	62.34	80.26	19.88	54.38
Logistic regression	69.57	58.44	61.58	29.82	54.06
SVM	73.91	61.04	81.92	33.92	62.34
This paper	61.54	75.00	82.35	44.44	64.06

6 Conclusion

In this paper, the hybrid feature extracted by PseAAC, the position information of amino acid residues and tripeptide composed of dimension reduction were used to present a protein sequence.

Hierarchical classification model based on FNT are proposed to predict the protein sequences with low similarity. Compared with other methods, the proposed method shows higher prediction accuracy. Since the prediction accuracy of all-α class is not ideal, in future work should focus on submitting a better feature extraction method to express the protein sequence.

Acknowledgment. This research was supported by the National Key Research And Development Program of China (No. 2016YFC0106000), National Natural Science Foundation of China (Grant No. 61302128, 61573166, 61572230, 61671220, 61640218), the Youth Science and Technology Star Program of Jinan City (201406003), the Natural Science Foundation of Shandong Province (ZR2013FL002), the Shandong Distinguished Middle-aged and Young Scientist Encourage and Reward Foundation, China (Grant No. ZR2016FB14), the Project of Shandong Province Higher Educational Science and Technology Program, China (Grant No. J16LN07), the Shandong Province Key Research and Development Program, China (Grant No.2016GGX101022).

References

1. Banzhaf, W., Nordin, P., Keller, R.E., Francone, F.D.: Genetic Programming: An Introduction, vol. 1. Morgan Kaufmann, San Francisco (1998)
2. Chen, Y.L., Li, Q.Z.: Prediction of the subcellular location of apoptosis proteins. J. Theor. Biol. **245**(4), 775–783 (2007)
3. Chen, Y., Abraham, A., Yang, B.: Feature selection and classification using flexible neural tree. Neurocomputing **70**(1), 305–313 (2006)

4. Chen, Y., Peng, L., Abraham, A.: Exchange rate forecasting using flexible neural trees. In: Wang, J., Yi, Z., Zurada, Jacek M., Lu, B.-L., Yin, H. (eds.) ISNN 2006. LNCS, vol. 3973, pp. 518–523. Springer, Heidelberg (2006). doi:10.1007/11760191_76

5. Chen, Y., Yang, B., Dong, J., Abraham, A.: Time-series forecasting using flexible neural tree model. Inf. Sci. **174**(3), 219–235 (2005)

6. Michael, L., Chothia, C.: Structural patterns in globular proteins. Nature **261**, 552–558 (1976)

7. Chou, J.J., Zhang, C.T.: A joint prediction of the folding types of 1490 human proteins from their genetic codons. Biochemical and biophysical research communications (1993)

8. Chou, K.C.: A novel approach to predicting protein structural classes in a (20{1)-d amino acid composition space. Proteins: Struct., Funct., Bioinf. **21**(4), 319–344 (1995)

9. Chou, K.C.: A key driving force in determination of protein structural classes. Biochem. Biophys. Res. Commun. **264**(1), 216–224 (1999)

10. Chou, K.C.: Prediction of protein cellular attributes using pseudo-amino acid composition. Proteins: Struct., Funct., Bioinf. **43**(3), 246–255 (2001)

11. Chou, K.C.: Using amphiphilic pseudo amino acid composition to predict enzyme subfamily classes. Bioinformatics **21**(1), 10–19 (2005)

12. Ding, S., Zhang, S., Li, Y., Wang, T.: A novel protein structural classes prediction method based on predicted secondary structure. Biochimie **94**(5), 1166–1171 (2012)

13. Du, K.-L., Swamy, M.N.S.: Particle swarm optimization. In: Du, K.-L., Swamy, M.N.S. (eds.) Search and Optimization by Metaheuristics, pp. 153–173. Springer, Cham (2016). doi:10.1007/978-3-319-41192-7_9

14. Emanuelsson, O., Nielsen, H., Brunak, S., Von Heijne, G.: Predicting subcellular localization of proteins based on their n-terminal amino acid sequence. J. Mol. Biol. **300**(4), 1005–1016 (2000)

15. Kennedy, J.: Particle swarm optimization. In: Sammut, C., Webb, G.I. (eds.) Encyclopedia of Machine Learning, pp. 760–766. Springer, Cham (2011)

16. Koza, J.R.: Genetic Programming ii: Automatic Discovery of Reusable Subprograms. MIT Press, Cambridge (1994)

17. Murzin, A.G., Brenner, S.E., Hubbard, T., Chothia, C.: Scop: a structural classification of proteins database for the investigation of sequences and structures. J. Mol. Biol. **247**(4), 536–540 (1995)

18. Nakashima, H., Nishikawa, K., Tatsuo, O.: The folding type of a protein is relevant to the amino acid composition. J. Biochem. **99**(1), 153–162 (1986)

19. Poli, R., Kennedy, J., Blackwell, T.: Particle swarm optimization. Swarm Intell. **1**(1), 33–57 (2007)

20. Salustowicz, R., Schmidhuber, J.: Probabilistic incremental program evolution. Evol. Comput. **5**(2), 123–141 (1997)

21. Salustowicz, R., Schmidhuber, J.: Probabilistic incremental program evolution: stochastic search through program space. In: ECML, pp. 213–220 (1997)

The Study on Grade Categorization Model of Question Based on on-Line Test Data

YuLing Fan[1], Tao Xu[1,2], Likai Dong[1], and Dong Wang[1,2(✉)]

[1] School of Information Science and Engineering, University of Jinan, Jinan 250022, China
ise_wangd@ujn.edu.cn
[2] Shandong Provincial Key Laboratory of Network Based Intelligent Computing,
University of Jinan, Jinan 250022, China

Abstract. To tackle with the blindness of random questions choosing for exercise and test of the on-line learning system, this paper clusters questions exploiting various feature subsets and parameters via K-means. For the test data of ACM Online Judge system, the features of temporal fluctuations mean of time consumption and repeat submission rate are used to make the question categorization and automatic recommendation come true. The experimental results suggest that the proposed method is simple but effective, and by which an on-line test platform can realize functions such as individuation teaching, intelligently questions choosing, teaching instruction, automatically paper constructing and paper difficult prediction.

Keywords: Online judge · K-means · Difficulty classification

1 Introduction

With the increasing popularity of online learning, both universities, enterprises, and government departments are actively carrying out a variety of Internet-based online learning. It has the advantage of preserving the learners 'learning behavior and the shortcomings of ignoring learners' differences. Participants in the level of good and bad, due to the lack of initial capacity analysis, standardized teaching model makes that learners gradually lost.

At present, the university OJ (Online Judge) system is generally applied to the "C language programming" for all the students of the whole school and the "program design basis" of computer specialty, and the computer professional and non-professional learners are mixed together. Programming levels are uneven. This situation is very similar to the online learning model. How to assign targeted questions to different learners at the different level of practice is a difficult problem in the online learning system.

OJ system database save a large number of evaluation data, from which randomly selected samples, as the original evaluation data; In a lot of learning behavior data, considering the difference between students' understanding degree and completion time, we choose the combination of time variance, average time and repetition rate to form feature set; After the feature normalization process, some of the features are first

© Springer International Publishing AG 2017
D.-S. Huang et al. (Eds.): ICIC 2017, Part II, LNCS 10362, pp. 765–775, 2017.
DOI: 10.1007/978-3-319-63312-1_69

clustered using the K-means clustering algorithm [1]; followed by clustering with different number of centroids; Finally, through the analysis, comparing the different characteristics, and the clustering results of different parameters set, predicting the optimal program, and ultimately getting the difficulty level of each topic.

In the future, to extract the problem of considerable difficulty according to the learners ranking situation, which can avoid the blindness of the previous random topics. To measure whether a learner grasps a certain knowledge point. It no longer simply based on the number of statistical topics, but with the standard of the sum of the difficulty coefficient in completing titles. This is essential to improve learning efficiency and improve the efficiency of online education. However, traditional e-learning methods for providing testing paper are usually based on the questions with difficulty level set manually, which is not suitable for all students. In this paper, the difficulty of questions is exploited based on various features via K-means. The temporal fluctuations mean of time consumption and repeat submission rate are used to classify the questions into different categories, which will be used for automatic recommendation.

The first part of this paper introduces an overview of the classification model; the second part introduces the algorithm and the key technology, and the third introduces the experimental setup and the result analysis.

2   An Overview of the Classification Model

In this paper, it takes evaluation data as analysis of the object, analyzing its clustering results, in order to achieve the purpose of online problem's difficulty classification, the classification model is shown in Fig. 1, which is including the following steps:

(1) Data collection [2]: In order to make the sample more representative, the sample title's data is required by the different professional students to complete, and cover knowledge points of all chapters. These data are collected as the original evaluation data.

(2) Data cleaning: detecting and cleaning the original data for null data and error value, removing the isolated point [3], minimizing the impact of the noise data, and preprocessing to form a data set.

(3) Data normalization [4]: To extract the standard deviation of completing time, average completing time and repeating submission rate as a feature sets from the data. After the normalization of the features, improving the comparability of data indicators

(4) Cluster analysis: Selecting different feature subsets, different parameters of the value to do experiments and analyze, different clustering results are compared.

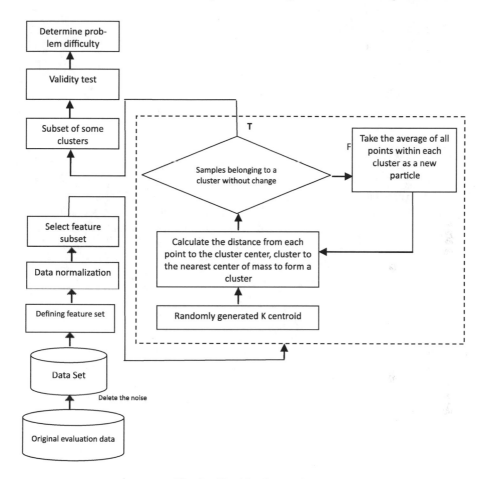

Fig. 1. Classification model

3 Algorithm and Key Technology

3.1 Feature Extraction

Before clustering the algorithm, the feature is extracted based on the original data, and the three-dimensional feature,

(σ_j, T_j, S_j) is used for the topic P_j.

In the following algorithm description, n represents the number of students completing the title P_j; t_i indicates the time at which the i-th student completes the question; S_i shows whether the i-th student is repeated, the one-time commit is counted as 0, and the multiple is written as 1.

3.1.1   Time Fluctuation

For a knowledge point, because the students understanding situation is quite different, resulting in a large range of time with the problem. In order to measure the degree of dispersion when the title P_j is used, the time variance of the problem is used as the first dimension, marking as σ_j, as shown in Eq. (1).

$$\sigma_j = \sqrt{\frac{\sum_{i=1}^{n} (t_i - \bar{t})^2}{n}} \tag{1}$$

3.1.2   Average Time

The longer the student takes the time to complete the problem, the more difficult the description of the problem, so the average use of the selected topic P_j as a second dimension, marking as T_j, as shown in Eq. (2).

$$T_j = \frac{\sum_{i=1}^{n} t_i}{n} \tag{2}$$

3.1.3   Repeated Submission Rate

When the student's job can not correctly handle the test data given by the title, it will not be submitted normally. So the number of repeated submissions representing that students repeatedly modify the job, the number of repetitions is proportional to the difficulty of the title. The repetition rate is taken as the third dimension, as is shown in Eq. (3).

$$S_j = \frac{\sum_{i=1}^{n} s_i}{n} \tag{3}$$

3.2   Data Normalization

Different features often have different dimensions and dimension units. In order to eliminate the influence of dimensions between features, it is necessary to standardize the data to solve the comparability between data features.

For the sake of comparison, this paper uses the difference normalization method to simulate the linear transformation of the data, and map the result to [0–1], marking as W_1, W_2, W_3.

(1) Time fluctuation normalization

$$W_1 = \frac{\sigma_j - \sigma_{\min}}{\sigma_{\max} - \sigma_{\min}} \tag{4}$$

Where j is the title number, σ_j is the standard deviation of the title, $\sigma_{\min} = \min(\sigma_j)$ $(j = 1, 2, \ldots, m)$, $\sigma_{\max} = \max(\sigma_j)$ $(j = 1, 2, \ldots, m)$, and m is the total number of questions.

(2) The average time normalization

$$W_2 = \frac{T_j - T_{min}}{T_{max} - T_{min}} \tag{5}$$

Where j is the title number, T_j is the average time of the title P_j, where j is the title number and the average time of the title, $T_{min} = \min(T_j)$ ($j = 1, 2, ...,m$), $T_{max} = \max(T_j)$ ($j = 1, 2, ..., m$), and m is the total number of questions.

(3) Repeated submission rate normalization

$$W_3 = \frac{S_j - S_{min}}{S_{max} - S_{min}} \tag{6}$$

Where j is the title number, S_j is the repetition rate of the title P_j, $S_{min} = \min(S_j)$ ($j = 1$, $2, ..., m$), $S_{max} = \max(S_j)$ ($j = 1, 2, ..., m$), and m is the total number of questions.

3.3 K-Means Clustering Algorithm

In this paper, K-means algorithm is used for data unsupervised clustering, the specific steps are as follows:

(1) For the point to be clustered to initialize the cluster center, this paper takes a random selection of the center, as is shown in Eq. (7), selected k centroids.

$$\mu^{(0)} = \mu_1^{(0)}, ..., \mu_k^{(0)} \tag{7}$$

(2) Calculate the distance from each point to the cluster center, and cluster each point to the nearest cluster, as is shown in Eq. (8).

$$C_{(j)}^{(t)} \leftarrow \arg\min_t \left\| \mu_i - x_j \right\|^2 \tag{8}$$

(3) Calculate the average of the coordinates of all points in each cluster and use this average as the new cluster center, as is shown in Eq. (9).

$$\mu_i^{(t+1)} \leftarrow \arg\min_\mu \sum_{j:C(j)=i} \left\| \mu - x_j \right\|^2 \tag{9}$$

Repeated (2), (3). Until the cluster, the center is no longer a large range of mobile or clustering times to meet the requirements.

4 Experimental Setup and Result Analysis

4.1 Evaluation of Data Processing

The use of a university OJ system background C language test questions, there are more than 2,000 questions in the question bank. Each subject has saved the student rankings, cumulative learning time, the number of completed questions and other learning behavior. Evaluation of the data selected 160 professional students' 300 questions, about 50000 data as a sample.

4.1.1 Raw Data Set

Table 1 shows the original data about some students on the subject. This assignment has a total of 6 questions. Where Rank is the ranking of this job, User is the username, Solved is the number of questions submitted in this exercise, Penalty represents the time spent on this exercise, Time is the time selected for the topic, Submit represents the number of times of submission, Score is scored for this question.

Table 1. Raw data set

Rank	User	Solved	Penalty	Time	Submit	Score
1	20161222080	6	4:33:47	0:39:58	(−1)	10
24	20161222041	2	3:17:45		(−7)	0
153	20161222134	4	356:12	153:20:57	0	10

As the 20161222041 learners repeated submits the subject 7 times unsuccessful, Time has no value. 20161222134 learner use 153 h, and isolated points will affect the entire process of clustering. So these two types of data will be cleaned. After data preprocessing, there are 256 questions in the data set.

4.1.2 The Extraction of Feature Set

In the classical test theory [5], the difficulty of the test questions is usually used $P = 1-S/F$, where P is the difficulty of the examination, S is the average of the scores on the subject, and F is the full score of the problem.

To judge the difficulty of title from the score is not comprehensive, although some subjects score, but students spent a lot of time, and also affect the completion of other topics. In addition, OJ system provides the subject of test data and the results of data as a basis, that students submit the code in the practice are basically correct, so this algorithm has not used the score as a feature.

Select the completion time variance, average completion time, and repeated submission rate for each topic as a feature. As most students can successfully have a one-time submission, very few students will have to submit many times. When dealing with data can not do a simple sum of times, so marking a one-time submission as 0, repeated as 1, as is shown in Table 2.

Table 2. Features set of P_1

Topic	Time variance	Average time	Repeated submission rate
P1	28.13344	60.8312	0.03247

4.2   Normalized Results

In this paper, we choose the difference method to limit the eigenvalue in [0,1], as is shown in Table 3.

Table 3. Normalized table

Topic	Time fluctuation (W_1)	Average time (W_2)	Repeated submission rate (W_3)
P1	0.39251	0.17667	0.03247

Time fluctuation represents the distance between the original time and the average time and is standardized on the basis of the standard deviation. It represents the degree of discrepancy and difference of the topic P_j. The greater the difference in time fluctuation, the greater the difficulty of the problem. The average time and repeated submission rate are also proportional to the difficulty of the problem.

4.3   Clustering Results Analysis

4.3.1   Analysis of Features

Taking difficult, middle and easy topics as examples, we select different feature subsets and analyze the clustering results.

(1) Time fluctuation W_1 and average time W_2 are related to the time factor, so only select W_2 and W_3 two features, and carry out clustering algorithm, the results are shown in Table 4:

Table 4. Two characteristics clustering results

Category	Class center (W_2, W_3)	Number of questions
1	(0.1789,0.1566)	166
2	(0.3451,0.1515)	44
3	(0.4912,0.1656)	46

From the class center can be seen, the category 1 of the problem is simple, about 65% of the total number of topics. Category 2 of the problem is moderate, about 17%. Category 3 of the problem is the most difficult, about 18%.

(2) Select three characteristics W_1 W_2 and W_3.Clustering results are shown in Table 5, and clustered hash chart is shown in Fig. 2.

Table 5. Three characteristics clustering results

Category	Class center (W_1, W_2, W_3)	Number of questions
1	(0.3515,0.2758, 0.1476)	174
2	(0.7596,0.605,0.1901)	44
3	(0.4659,0.4877,0.5207)	38

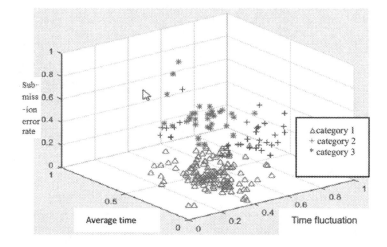

Fig. 2. Three characteristics hash graph

From the analysis of Table 5, category 1 is easy, accounting for 68%, category 2 is moderate, about 17%, and category 3 is the most difficult, about 15%. From the proportion of various topics, the two ways are similar, The proportion of category 1 is comparatively high.

From the analysis of Fig. 2, which category 1 topic marking as "△", the point of the distribution is more intensive, the eigenvalues are close illustrating that the learners are of the same degree about mastering the question; Category 2 using " + ", the distribution are scattered, from the X-axis, which shows that the time fluctuation was significantly higher than the other two categories, indicating the gap that learners master the knowledge of this point is very large, takes the most time, and times to submit is moderate; category 3 with "*", indicates that the situation is between Category 1 and Category 2.

From the analysis of the specific problem, in the topics which two characteristics clustering results are different, randomly selecting a number of topics, to compare the rationality of the two clusters. Such as the function of title P58 is to determine the leap year, the characteristic value is (0.2964, 0.3293, 0.3492), two feature clustering result is category 3, which is difficult. Three characteristics clustering result is category 1, which is a simple question. The classification difference is due to time fluctuation W_1, although the repetition rate is higher, the time fluctuation is smaller, only 0.2964. It is clear that the clustering method of selecting three eigenvalues is better.

4.3.2   The Analysis of Classification Numbers

Clustering is an unsupervised learning [6, 7]. The sample data set a different number of the class center to have different effects on the clustering results. For this experiment, it set the center of the class are k = 3 and k = 4. Clustering results are shown in Tables 5 and 6.

Table 6. Four characteristics clustering results

Category	Class center (W_1, W_2, W_3)	Number of questions
1	(0.3829,0.1789, 0.1566)	77
2	(0.7596,0.605,0.1901)	46
3	(0.3246,0.36,0.1463)	95
4	(0.4737,0.4969,0.5296)	38

From the analysis of Table 6, the title can be classified into four categories. Category 1 is easy, category 2 is difficult, and category 3 is moderate, category 4 is comparatively difficult.

Figure 3 is a four-category hash chart, where the category 1 is denoted by "△", and the category 3 topic is represented by "*", and the two types of topics are basically equivalent to the category 1 in Fig. 2, that is "△"; category 2 with "*" form, from display of the X-axis, the time fluctuations were significantly higher than 1,3 categories, saying the gap between the learners to master knowledge points is large, takes the most time, and times to submit is moderate; category 4 with "○" said that the repetition rate of such topics is very high, the average time is not high.

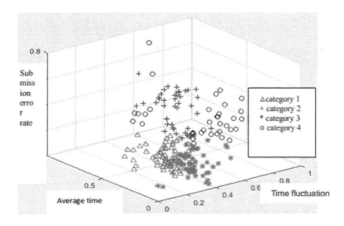

Fig. 3. Four characteristics hash graph

In the measurement of the results of the two clusters, as is shown in Table 7, k = 3 clusters in the summary of the total number of topics accounted for 68% of the total, obviously, the proportion of k = 4 clustering of various types is more reasonable.

Table 7. Comparison of two clustering results

K = 3	The proportion of classification topics	K = 4	The proportion of classification topics
Category 1	68%	Category 1	30%
Category 2	17%	Category 2	18%
Category 3	15%	Category 3	37%
		Category 4	15%

5   Conclusion

This paper proposes a method to determine the difficulty of the problem by using the data mining algorithm. The experimental results are used to determine the eigenvalues (time fluctuation, time variance, and repetition rate). The difficulty of the topic is divided into four categories: easy, medium, comparatively difficult and difficult. Compared with the traditional method of difficulty coefficient determination, this method considers many factors, so it is more reasonable and accurate to measure of the difficulty of the subject. This method can be applied to the difficulty setting of the online problem, which can provide the learners with the level matching problem, and provide the basis for predicting the difficulty of the test paper.

The following work can use the improved algorithm [8, 9] to further study the learning behavior [10], the online learning courses for objective analysis. Through the use of data of network behavior, dynamically generating teaching guidance strategy.

Acknowledgments.   This research was supported by National Natural Science Foundation of China (No. 61302128), the Youth Science and Technology Star Program of Jinan City (201406003), the Teaching Reform Research Project in Undergraduate College of Shandong Province (2016), Industry-University Cooperative Education Project of Ministry of Education (No. 201601023018), the Scientific Research Fund of Jinan University (No. XKY1622) and Teaching Research Project of Jinan University (No. J1638)

References

1. Chonghui, G., Fengzhan, T.: Data Mining Tutorial. Tsinghua University Press, pp. 107–121 (2012)
2. Duda, R.O., Hart, P.E., Stork, D.G.: Pattern Classification, 2nd edn, pp. 11–12. Machine Press, Beijing (2003)
3. Aiwu, Z., Baolou, C., Yan, W.: Study and improve on k-means algorithm. Comput. Technol. Dev. **22**(10), 101–104 (2012)
4. Jigui, S., Jie, L., Lianyu, Z.: Clustering algorithms research. J. Softw. **19**, 48–61 (2008)

5. Jiaxia, S., Xueyong, L.: The algorithm and design of the test difficulty coefficient determined by classical test theory. China Sci. Technol. Inf. **19**(1), 44–45 (2009)
6. Zhijie, L., Yuanxiang, L., Feng, W., Li, K.: Accelerated multi task online learning algorithm for big data stream. J. Comput. Res. Dev. **52**(11), 25–45 (2015)
7. Zhexue, H.: Extensions to the k-means algorithm for clustering large data sets with categorical values. Data Min. Knowl. Discov. **2**, 283–304 (1998)
8. Yiling, H., Xiaoqing, G., Chun, Z.: Modeling and mining of online learning behavior analysis. Open Educ. Res. **20**(2), 102 (2014)
9. Barbara, D.: Using Self-similarity to cluster large data sets. Data Min. Knowl. Disc. **7**, 123–152 (2003)
10. Modha, D.S., Spangler, W.S.: Feature weighting. k-means clustering. Mach. Learn. **52**, 217–237 (2003)

2D Human Parsing with Deep Skin Model and Part-Based Model Inference

Tao Xu[1,2], Zhiquan Feng[1,2(✉)], Likai Dong[1,2], and Xiaohui Yang[1,2]

[1] Shandong Provincial Key Laboratory of Network Based
Intelligent Computing, University of Jinan, Jinan 250022, China
ise_fengzq@ujn.edu.cn
[2] School of Information Science and Engineering,
University of Jinan, Jinan 250022, China

Abstract. Human parsing plays an important role in action understanding, clothing recommendation and human-computer interaction, etc. However, variations of human pose, clothes, viewpoint and cluttered background make the segmentation and pose estimation of body parts more difficult. In this paper, a human parsing framework is proposed based on a combination of deep skin model and part based model inference. First, a deep skin model is trained via deep belief networks, which will be used to reduce the pose searching spaces and enhance the efficiency of model inference. Secondly, pictorial structure model parses human body more accurate with the fusion maps of skin detection and HOG based part detectors. The experimental results demonstrate that the fusion of skin detection improves the detection and pose estimation of human body parts, especially for the parts such as head, arms and legs.

Keywords: Human parsing · Deep skin model · Model inference

1 Introduction

Human parsing plays an important role in computer vision which takes images or videos as the input data, parses the body structure via segmentation of body parts and model inference(such as pictorial structure model inference), present a reasonable spatial human body comprised of detected parts [1].

With the detection of important body parts and physical structure inference, human parsing achieves attributes and poses of human, which is the key step of video understanding, action recognition and marker-less body motion capture [2]. Additionally, action recognition and fashion classification also benefits from human parsing [3]. For example, Yamaguchi et al. utilized a trained body parts detector for producing cloth style and achieving accurate fashion classification. They also proposed an online fashion retrieval and recommendation system based on a unified inference with fusion of human parsing and cloth attributes classification [4].

In recent years, human parsing, especially the 2D human parsing has achieved good performance in some databases [5]. However, variations of human pose, clothes, view

© Springer International Publishing AG 2017
D.-S. Huang et al. (Eds.): ICIC 2017, Part II, LNCS 10362, pp. 776–787, 2017.
DOI: 10.1007/978-3-319-63312-1_70

point and cluttered background make the segmentation and pose estimation of body parts more difficult. Some challenges are as follows:

– Body parts, such as arms and legs are hard to detect due to variation and diversity of body apparent, different clothing, skin color and occlusion. For example, if human body is occluded by objects, it will be failed in the estimation of the occluded parts. How to improve the accuracy of part detector is one key issue of human parsing.
– Many traditional methods used bottom-up framework for structure parsing. For example, pictorial structure model [6] treat the body as a assemble of parts which selects reasonable local parts for inferencing the global body structure. But, inaccurate parts detection is disadvantageous for parsing human pose. How to enhance the parsing performance by using the advantage of both two categories is another issue still need to be resolved.

According to our previous work [7, 8], we argue that skin color is an important cue in visual detection. For its attractive properties of high processing speed, invariance against rotation, partial occlusion and pose change, skin color is now widely used in face detection and hand tracking. Skin color detection could provide region of interest (ROI) candidates for advanced image processing, such as human parsing. In this paper, we propose a human parsing framework based on combination of deep skin model and part based model inference. First, a patch-wise deep skin model is trained via deep belief networks with a skin patch dataset. The image patch is the basic unit in skin region detection instead of pixels. Patches exploit the spatial correlations of pixels in the training and testing stages, which leads the skin detection immune to the false positive skin pixels. The trained deep skin model will be used to reduce the pose searching spaces and enhance the efficiency of model inference. Secondly, pictorial structure model is utilized to parse the body structure based on the fusion map of skin detection and HOG based part detectors. The experiments results show that fusion of skin detection and part detectors improves the pose estimation, especially for the parts such as head, arms and legs.

The rest of the paper is organized as follows. In Sect. 2, we discuss related work. The deep skin model and part-based inference with skin detector is presented in Sects. 3 and 4. In Sect. 5, the experimental results are shown with analysis, and conclusions are given in Sect. 6.

2 Related Work

Human parsing approaches have been studied for decades. Several approaches have been reported, and significant improvements have been obtained in both feature representation and model design [9, 10]. Human parsing is typically formulated by estimating the posterior distribution, $p(\mathbf{x}|\mathbf{z})$, where $\mathbf{x}$ is the pose of the body and $\mathbf{z}$ is a feature set derived from the image [11]. It is widely accepted that describing the human body as an ensemble of parts improves the recognition of human body in complex poses, despite of an increasing of computational time. In this case, the configuration of the human body can be represented as a set of parts, $\mathbf{x} = \{\mathbf{x}_1, \mathbf{x}_2,..., \mathbf{x}_N\}$, each with its own position and orientation in space, $\mathbf{x}_i = \{\tau_i, \theta_i\}$, τ_i is the position of body parts and θ_i represent the orientations of body parts with respect to their parents. Each parts representation is often augmented with an additional variable, s_i, that accounts for uniform scaling of the body part in the image, i.e., $x_i = \{\tau_i, \theta_i, s_i\}$, $\tau_i \in R^2$, $\theta_i \in R^1$, $s_i \in R^1$.

Pictorial structure (PS) [12] is a tree part-based model which is popular used in pose inferencing. The most common features in PS for part detection include, silhouettes, color, edges and gradients for modeling the texture over the body parts. To reduce dimensionality and increase robustness to noise, these raw features are often encapsulated in image descriptors, such as shape context, SIFT and histogram of oriented gradients (HOG). The effectiveness of different feature types on pose estimation has been studied in the context of several inference architectures [13]. Body pose estimation is a challenging problem because of the many degrees of freedom to be estimated. Moreover, limbs vary greatly in appearance due to changes in clothing and body shape, as well as changes in viewpoint.

According to our previous work, skin color showed great ability for segmentation of hand, arm and legs. Fusion of skin detection map and parts detection map will be used to improve the parsing of skin parts. There are a number of studies of pixel based skin detection methods [7, 14, 15]. Jones and Rehg [14] created the Compaq dataset and used Bayesian classifier with the histogram technique (HB) for skin detection. Phung et al. [15] introduced the ECU dataset and presented a study of skin segmentation on eight color spaces using HB, Multilayer perception and Gaussian classifiers, etc. Region-based skin detection methods try to take this spatial arrangement of skin pixels into account to enhance the performance. Xu et al. [8] presented a patch-based skin detection method via deep learning. Image patches with different sizes are treated as the processing units, which consider the spatial arrangement of skin pixels in the patch classification and alleviate the disturbance problem of discrete pixels.

3   Deep Skin Model

Deep learning has induced great attention of human parsing [16, 17]. The proposed method is inspired by the success applications of deep models.

3.1   RBM and DBN

The Restricted Boltzmann Machines (RBM) could be stacked and trained to form DBN [18]. A graphical depiction of an RBM is shown in Fig. 1(a). Each RBM consists of two layers, namely visible layer $\mathbf{x}$ and hidden layer $\mathbf{h}$. The connections between the nodes within each layer are restricted and $\mathbf{w}$ denotes the weights between the visible layer and the hidden layer. The joint probability distribution over $\mathbf{x}$ and $\mathbf{h}$ in RBM is as

$$p(\mathbf{x},\mathbf{h}) \propto e^{b^T x + c^T h + h^T w_x} \tag{1}$$

where $\mathbf{b}$ and $\mathbf{c}$ are the biases of the visible layer and the hidden layer respectively.

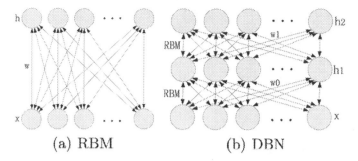

Fig. 1. Graphical model of RBM is showed in (a), and (b) shows a DBN model with input vector **x** and two hidden layers h1 and h2.

The trainable RBMs will be putted together to form DBNs. An example of DBN structure is shown in Fig. 1(b), which consists of two stacked RBMs. The first RBM is formed between **x** (visible layer) and **h1** (hidden layer 1). Layer 2 (**h1**) and layer 3 (**h2**) forms the second RBM. DBN models the joint distribution between observed vector **x** and the ℓ hidden layers $\mathbf{h^k}$ as

$$p(x, h^1, \ldots, h^\ell) = (\prod_{k=0}^{\ell-2} p(h^k|h^{k+1}))p(h^{\ell-1}, h^\ell) \tag{2}$$

where $x = h^0$, $p(h^k|h^{k+1})$ is a visible-given-hidden conditional distribution in an RBM associated with level k of the DBN. $p(h^{\ell-1}, h^\ell)$ is the joint distribution in the top-level RBM. The training of DBN is in a greedy layer-wise fashion. The idea is to get h1 from the first-level RBM, then take **h1** as input for the second-level RBM to computer **h2**, etc., until the last layer. This procedure will generate parameters $\mathbf{w_i}$ and $\mathbf{c_i}$ for each layer which can be used to initialize a multi-layer neural network. These parameters then will be fine-tuned with a supervised learning criterion.

3.2 Deep Skin Modeling

For patch based skin detection, skin deep model is constructed based on DBN. Figure 2 (a) shows the graphical model of the used DBN which consists of four layers and three RBMs. Each RBM is trained in a layer-wise greedy manner with contrastive divergence. Having pre-trained RBMs, the weights and biases will be used to initialize the skin deep model. For skin patch classification, a label layer is added to form a feed-forward neural network with five layers (see Fig. 2(b)). Then the entire model will be fine-tuned as feed-forward, back-propagate neural network. This procedure is equivalent to initializing the parameters of a deep neural network with the weights and hidden layer biases obtained by the unsupervised training strategy.

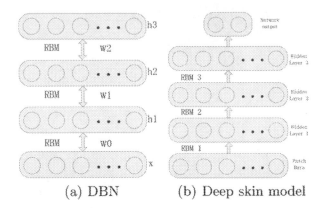

(a) DBN   (b) Deep skin model

Fig. 2. DBN model showed in (a) will be used to form the skin deep model. (b) depicted the skin deep model which has a label layer, and the whole model will be trained as a feed-forward, back-propagate neural network.

There are two stages in the training of the skin deep model: (1) a layer-wise pre-training and (2) a fine-tuning stage. The process is as follows:

1. Train the first layer **h1** as an RBM with input x = $h(0)$ as the visible layer.
2. Use **h1** as data for the second layer.
3. Train the second layer **h2** as an RBM, taking **h1** as training examples.
4. Iterate (2) and (3) for the desired number of layers, each time propagating upward samples.
5. Fine-tune all the parameters with back-propagate networks in supervised learning strategy.

4 Part-Based Model Inference with Skin Map

In part-based methods, part detectors plays key role for the parsing results [19]. Pishchulin et al. [19] improved the performance of state of the art methods in two ways, i.e., investigating various features for part detectors, and using groups of features and model inference to achieve better performance on LSP dataset [20]. For many scenarios of human parsing, skin color is a powerful clue of detecting skin part regions, such as faces, arms, legs and hands. Figure 3 shows skin parts in LSP dataset.

Fig. 3. Skin parts in LSP.

4.1   Part Based Model Inference

Part-based model could represents the body using a Markov Random Field (MRF). the posterior $p(x|z)$ can be expressed as:

$$p(x|z) \propto p(z|x)p(x) = p(z|\{x_1, \ldots, x_M\}) \tag{3}$$

$$\approx \prod_{i=1}^{M} p(z|x_i)p(x_1) \prod_{(i,j) \in E} p(x_i, x_j) \tag{4}$$

With an additional assumption of pair-wise potentials that account for kinematic constraints, the model forms a tree-structured graph known as the Tree-structured Pictorial Structures (PS) model. An approximate inference with continuous variables is also possible.

$$m_{i \rightarrow j}(l_j) = \sum_{l_i} p(l_i, l_j)p(z|l_i) \prod_{k \in A(i) \backslash j} m_{k \rightarrow i}(l_i), \tag{5}$$

where $m_{i \rightarrow j}$ is the message from part i to part j, with $p(l_i, l_j)$ measuring the compatibility of poses for the two parts and $p(z|l_i)$ is the likelihood, and $A(i) \backslash j$ is the set of parts in the graph adjacent to i except for j.

Once all of the message updates are complete, the marginal posteriors for all of the parts can be estimated as:

$$p(l_i|z) \propto p(z|l_i) \prod_{j \in A(i)} m_{i \rightarrow j}(l_i) \tag{6}$$

Then, the most likely configuration can be obtained as a MAP estimate:

$$l_{i,\text{MAP}} = \arg \max_{l_i} p(l_i|z) \tag{7}$$

4.2   Human Parsing Framework

Skin color can not only segment the skin region of body parts, but be used to improve the performance of human parsing by fused with part detectors(such as HOG based detectors).

Figure 4 shows the flowchart of our method. First, part detectors will be used to obtain HOG map by HOG filtering in 24 directions. The HOG based part detectors can be pre-trained on LSP dataset. Secondly, deep skin models are applied on original image with a sliding window framework to achieve the skin map. Thirdly, the HOG map and skin map will be fused to generate a fusion map, which could enhance the probability of some body parts (arms, legs or hands) that HOG descriptors failed in accurate detecting. Last, human parsing will present the body configuration using a trained PS model.

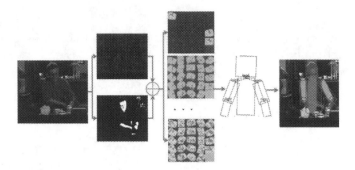

Fig. 4. Human parsing framework via fusion of skin map and HOG map.

5   Experimental Results and Analysis

5.1   Data Set

For training and testing of deep skin model, a patch skin dataset is proposed based on ECU dataset [15] and Compaq dataset [14]. The skin patches are randomly cropped from the skin regions in ECU and non-skin patches are selected from non-skin images of Compaq.

5.2   Skin Segmentation of Deep Skin Model

In the construction of skin deep model, the DBN net consists of three RBMs each with 100 hidden neurons. Each RBM will be trained on the full patch training set with 40 epochs. Having pre-trained each RBM, the weights and biases are used to initialize a feed-forward neural net with 4 layers of sizes 100-100-100-2, the last 2 neurons being the output label units. The skin deep model will be trained using mini-batches of size 100 for 200 epochs using back-propagation. Three components of RGB, YCbCr, HSV and CIELab color spaces in one patch and their combinations are catenated to form the input vector. Skin segmentation based on skin deep model is implemented via a sliding window framework. The size of sliding window equals to the patch size. The sub-window will slide from top to bottom in the image with overlap, in which the input vector will be extracted. If the patch corresponding to the sub-window is classified as skin patch, pixels in this sub-window will be classified as skin ones. The role of overlap is also studied in the experiments.

Based on patch skin dataset, deep skin models are trained and tested on the four color spaces and their combinations. The classification rates and the ROC curves are shown in Table 1 and Fig. 5, respectively.

For the experiments on 20 × 20 dataset, skin deep model in RGB color space achieved 93.5645% ACC with 5.3% FPR. The model in YCbCr has a higher TPR of 94.8%, while its TPR is 18.3%, which is higher than other models. If Y component is discarded, both the TPR and FPR of the model in CbCr are decreased while the ACC increased. For 10 × 10 dataset, the performance of models in RGB, YCbCr, HSV and

Table 1. Testing results on single color spaces

	20 × 20 patch			20 × 20 patch		
	TPR	FPR	ACC	TPR	FPR	ACC
RGB	91.3%	5.3%	93.6%	91.0%	6.9%	92.4%
YCbCr	94.8%	18.3%	86.1%	87.9%	6.4%	91.7%
HSV	79.2%	2.6%	91.4%	91.5%	7.1%	92.4%
CIELab	87.4%	5.5%	92.2%	88.7%	6.8%	91.7%
CbCr	90.4%	13.7%	87.6%	12.6%	89.4%	88.1%

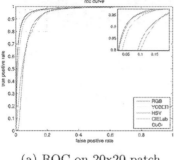
(a) ROC on 20x20 patch

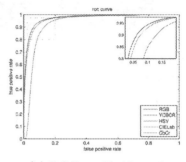
(b) ROC on 10x10 patch

Fig. 5. The ROC curves on single color spaces. (a) shows the ROC curves on 20 × 20 patch set and (b) is the ROC curves on 10 × 10 patch set. The embedded sub-figures show the details of ROC curves.

CIELab color spaces are very similar. The model in YCbCr gained 91.7% ACC, which is better than model in CbCr. If the luminance is discarded in YCbCr space, it can not conclude that the performance will be improved.

5.3 Skin Region Detection via Sliding Window

To get the skin ROI candidates in images, the trained patch based skin deep models are integrated into a sliding window framework. A sub-window with size equal to the patch size slides from top to bottom and left to right in an image. In the experiments, two skin deep models, i.e., the model trained on 10 × 10 dataset in RGB and the model trained on 20 × 20 dataset in RGB, are used to detect skin color with 2, 5 and 10 steps.

Examples of skin detection results are illustrated in Fig. 6. As showed in r1, pixel based methods present the rough face regions by discrete skin pixels, while our method directly obtains the face regions. The results of 10 × 10 skin deep models agree quite well with the ground truth. The proposed patch based skin deep models are more robust in skin ROI detection than pixel based methods.

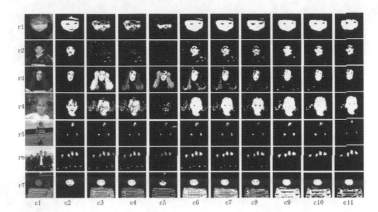

Fig. 6. Samples of skin detection results. c1 shows the original images in ECU dataset and the corresponding ground truth are showed in c2. Pixel based detection results with GMM [15], HB [14] and FNT [7] are showed in c3, c4 and c5. c6 through c8 show the results of 10×10 skin deep model with steps of 2, 5 and 10, respectively. c9 through c11 depict the results of 20×20 skin deep model with steps of 2, 5 and 10, respectively. ri denotes ith row and cj is the jth column.

5.4 Human Parsing with Model Inference

Experimental results of human parsing are showed in Table 2 with the flowchart of proposed in Sect. 4.2. In the experiment, deep skin model (DSM) is used to generate the skin map, which is fused with HOG (Histogram of Oriented Gradient) based part detectors. Pictorial Structure Model (PS) is utilized to parsing the human with a trained body model.

Table 2. Human parsing results

PCP(%)	Torso	Head	Upper leg	Lower leg	Upper arm	Lower arm	Total
HOG + PS	67.1	63.0	36.2	35.2	18.6	10.6	33.1
DSM + HOG + PS	69.0	79.8	66.1	67.2	49.1	31.7	60.5
FMP	84.1	77.1	69.5	65.6	52.5	35.9	60.8
DSM + FMP	81.4	80.7	65.5	69.3	54.7	38.2	64.9

In Table 2, the proposed methods (DSM + HOG + PS, DSM + FMP) achieves better performance than HOG based method (HOG + PS) [19] and Flexible mixture of part (FMP) [21] method using PCP evaluation. For example, the method with HOG feature and pictorial structure model inference(HOG + PS) obtain 35.2% and 10.6% on the estimation of lower leg and lower arm, respectively, while the proposed method (DSM + HOG + PS) get an accuracy of 67.2% and 31.7% on these two parts. And the same is true for the results of FMP and FMP with DSM.

Figure 7 shows the skin segmentation results using deep skin models, and Fig. 8 demonstrates the parsing results with pictorial structure model inference on LSP. According to the results in Fig. 7, deep skin models present approximate explicit skin regions of body parts. Although there are false detected regions (see images in last row), parsing processing with model inference will remove these false regions and generate accurate pose estimation of these parts (see Fig. 8).

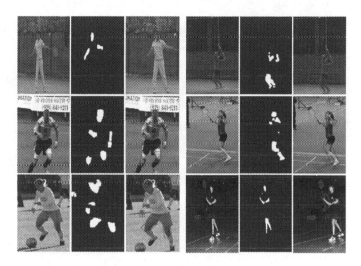

Fig. 7. Skin parts segmentation by deep skin model

Fig. 8. Human parsing result based on map fusion and model inference.

6 Conclusions

A 2D human parsing method is proposed in this paper. With fusion of skin map generated by deep skin models and feature map of HOG based part detectors, the pose searching spaces are reduced, and the performance of body parsing is improved. The experimental results demonstrates that skin color is an important clue for detection of some body parts, especially for face, arms and legs. The parsing result also proves that model inference can be benefit from fusion of feature maps, even using basic pictorial structure model.

Acknowledgments. This work is supported by the National Natural Science Foundation of China (Nos. 61472163, 61603151), the National Key Research & Development Plan of China (No. 2016YFB1001403), the Science and Technology Project of Shandong Province (No. 2015GGX101025), and Doctoral Foundation of University of Jinan (Nos. XBS1653, XBS1621).

References

1. Sigal, L.: Human pose estimation. Comput. Vis. **157**(10), 362–370 (2014). Springer US
2. Eichner, M., Marin-Jimenez, M., Zisserman, A., et al.: 2D articulated human pose estimation and retrieval in (almost) unconstrained still images. Int. J. Comput. Vis. **99**(2), 190–214 (2012)
3. Guo, G., Lai, A.: A survey on still image based human action recognition. Pattern Recogn. **47**(10), 3343–3361 (2014)
4. Yamaguchi, K., Kiapour, M.H., Ortiz, L.E., et al.: Retrieving similar styles to parse clothing. IEEE Trans. Pattern Anal. Mach. Intell. **37**(5), 1028–1040 (2015)
5. Andriluka, M., Pishchulin, L., Gehler, P., et al.: 2D human pose estimation: new benchmark and state of the art analysis. In: Proceedings of the IEEE Conference on Computer Vision and Pattern Recognition, pp. 3686–3693 (2014)
6. Hernndez-Vela, A., Sclaroff, S., Escalera, S.: Poselet-based contextual rescoring for human pose estimation via pictorial structures. Int. J. Comput. Vis. **118**(1), 49–64 (2016)
7. Xu, T., Wang, Y., Zhang, Z.: Pixel-wise skin colour detection based on flexible neural tree. IET Image Proc. **7**(8), 751–761 (2013)
8. Xu, T., Zhang, Z., Wang, Y.: Patch-wise skin segmentation of human body parts via deep neural networks. J. Electron. Imaging **24**(4), 043009 (2015)
9. Andriluka, M., Pishchulin, L., Gehler, P., et al.: MPII Human Pose Dataset (2016). http://human-pose.mpi-inf.mpg.de
10. Carreira, J., Agrawal, P., Fragkiadaki, K., et al.: Human pose estimation with iterative error feedback. In: Proceedings of the IEEE Conference on Computer Vision and Pattern Recognition, pp. 4733–4742 (2016)
11. Perez-Sala, X., Escalera, S., Angulo, C., et al.: A survey on model based approaches for 2D and 3D visual human pose recovery. Sensors **14**(3), 4189–4210 (2014)
12. Felzenszwalb, P.F., Huttenlocher, D.P.: Pictorial structures for object recognition. Int. J. Comput. Vis. **61**(1), 55–79 (2005)
13. Sapp, B., Jordan, C., Taskar, B.: Adaptive pose priors for pictorial structures. In: IEEE Conference on Vision and Pattern Recognition (CVPR 2010), pp. 422–429. IEEE (2010)

14. Jones, M., Rehg, J.: Statistical color models with application to skin detection. Int. J. Comput. Vis. **46**(1), 81–96 (2002)
15. Phung, S., Bouzerdoum, A., Chai, D.: Skin segmentation using color pixel classification: analysis and comparison. IEEE Trans. Pattern Anal. Mach. Intell. **27**(1), 148–154 (2005)
16. Toshev, A., Szegedy, C.: Deeppose: human pose estimation via deep neural networks. In: Proceedings of the IEEE Conference on Computer Vision and Pattern Recognition, pp. 1653–1660 (2014)
17. Tompson, J.J., Jain, A., LeCun, Y., et al.: Joint training of a convolutional network and a graphical model for human pose estimation. In: Advances in Neural Information Processing Systems, pp. 1799–1807 (2014)
18. LeCun, Y., Bengio, Y., Hinton, G.: Deep learning. Nature **521**(7553), 436–444 (2015)
19. Pishchulin, L., Andriluka, M., Gehler, P., et al.: Strong appearance and expressive spatial models for human pose estimation. In: IEEE International Conference on Computer Vision (ICCV 2013), Sydney, Australia, pp. 3487–3494. IEEE (2013)
20. Johnson, S., Everingham, M.: Clustered pose and nonlinear appearance models for human poseestimation. In: British Machine Vision Conference (BMVC 2010), Aberystwyth, UK, pp. 1–11. BMVA Press (2010)
21. Yang, Y., Ramanan, D.: Articulated human detection with flexible mixtures of parts. IEEE Trans. Pattern Anal. Mach. Intell. **35**(12), 2878–2890 (2013)

Terrain Visualization Based on Viewpoint Movement

Ping Yu$^{(\boxtimes)}$ and Songjiang Wang

Department of Information Science and Technology,
Zibo Normal College, Zibo, China
yuxue1980@126.com

Abstract. In real-time rendering of large-scale terrain, often using frame coherence to optimize terrain visualization algorithm, to reduce the amount of calculation of each frame, to improve the efficiency of the algorithm. But using frame coherence optimization algorithm has a premise, it is between two consecutive frames, the viewpoint movement is small. Therefore, you can make use of the changes between the two frames, only modify the part of the display node can get the next frame to display the terrain node. However, in practical applications, however, there is a large difference between the two consecutive frames, which is caused by the discontinuity of the viewpoint and the fast movement speed. In this paper, it is considering the viewpoint of jumping movement, using the threshold of viewpoint movement to optimize the terrain visualization algorithm. To improve the effectiveness and generality of terrain visualization algorithm based on frame coherence. The experimental results show that the use of threshold of viewpoint movement revised visualization algorithm, can effectively reduce the amount of calculation per frame, to enhance the efficiency of the algorithm.

Keywords: Viewpoint movement · LOD · Frame coherence

1 Introduction

Terrain visualization is a pivotal field on computer graphics, and also is a indispensability part in virtual terrain environment. Dynamic terrain is a pivotal content on terrain visualization, and it has been widely used in GIS, VR, battlefield simulation, computer games, etc. On one hand, in order to get natural effect of visual terrain, there are lager amount triangles in rendering terrain scene; on the other hand, in order to meet real-time rendering, requests computer generate more than 15 frame images in each second. At present, the research of terrain rendering is mainly based on (Level of Detail, LOD [1, 2]) model technology.

LOD technology of convenience to choose the right means of data organization LOD model generation. And the error metric is used to select the triangle to render in dynamic terrain visualization. In the real-time rendering of terrain, the performance optimization method is usually used to improve the efficiency of the algorithm. One of the most effective and common techniques is frame coherence [3, 4].

© Springer International Publishing AG 2017
D.-S. Huang et al. (Eds.): ICIC 2017, Part II, LNCS 10362, pp. 788–799, 2017.
DOI: 10.1007/978-3-319-63312-1_71

As in the literature [5–8], the algorithm in the real-time rendering phase of each frame from the root of the binary tree node to re-calculate the split to generate a display of the binary tree. Therefore, the real-time rendering algorithms in large amount of calculation and more time-consuming. Considering the viewpoint is in continuous movement, the level of detail model which between adjoining frames the terrain has great consistency. That is, when the viewpoint movement too little change, the triangle need to draw changed little. Therefore, it is possible to start from a frame on triangle binary tree, according to the point of view of the current space using the error metric of nodes, dynamic adjustment of binary tree nodes get need to draw a frame of the triangle. So, for the frame coherence can greatly reduce the amount of calculation of each frame, so as to improve the efficiency of algorithm.

But there in the scene roaming, 3D scene roaming prone to disorientation questions [9, 10]. 3D virtual scene roaming disorientation refers to, users often can only see in the three-dimensional virtual scene roaming view within the scope of the object, lead to the user when roaming the lack of unity, often get lost. And this kind of 3D virtual scene roaming disorientation in terrain visualization. This needs us provide users can manually choose viewpoint position, or at a high speed movement out of the terrain, to get rid of the confusion of topography and lost. In manual change the viewpoint position or in the process of the high speed movement, a viewpoint is not keep moving continuously. In this time, the frame coherence was used to optimize the terrain visualization algorithm whether to continue to be valid and worthy of study and discussion. In this paper, we discuss the effectiveness of the frame coherence algorithm after the jump. Viewpoints, the author of this paper after jumping movement, the frame coherence between the effectiveness of the algorithm are discussed. Presents a massive terrain visualization algorithm, using the viewpoints jumping factor, the optimization of interframe coherence algorithm, thus improve the effectiveness and generality of the algorithm.

2 Terrain Visualization Algorithm Based on Frame Coherence

Algorithm based on DAG triangle binary tree to storage structure, using the constraint error metric calculation error, so as to produce less redundant triangles at the same time avoid the cracks of T-connection and effectively. Continuous mobile stage of real-time rendering based on the viewpoint, the use of frame coherence reduces the cast time of each frame drawing algorithm efficiency.

2.1 Triangle Binary Tree Based on DAG

In this paper, we use the triangle bin-tree [2, 11] definition. In this context, the initial plane square area is divided into two an isosceles right-triangle and put them in a bin-tree is defined as the top. To the top triangle, for example, connections hypotenuse right angle vertex and the mid-point of its triangle. Then, the top triangle is divided into two sub-triangles. Defines the two sub-triangles is located in binary tree top. Recursively perform the operation, you can generate a triangle binary tree hierarchy. In the process of constructing the bin-tree, the fact also implies the introduction of a strict order of the vertices. This sequence can be described by the Directed Acyclic Graph of vertex (DAG) [5, 11]. The introduction of the vertical's order determines the division of the parent node before the child node can not be divided. This paper is the dependence of the vertices implicitly included in the error metric, to ensure that in every child node requirements have split his father node split.

2.2 The Error Metric with Constraint

First of all, to better reflect the local details of terrain, using local roughness tectonic domain in (0, 1) interval vertex constraint factor k_i, computation formula is as follows:

$$k_i = \left| 1 - \frac{1}{f_i + 1 + \xi} \right| \tag{1}$$

The f_i for local coarse error function formula is as follows, ξ is deduced for the tiny number close to zero.

$$f_i = \frac{1}{n} \sum_{x=1}^{M} \sum_{y=1}^{N} [Z(x, y) - Z_i]^2 \tag{2}$$

The $Z(x, y)$ as the elevation of local terrain vertex (x, y) value, Z_i as the vertex of the elevation values, $n = M \times N$.

The error of the second viewpoint related calculation as the delta vertex produced by $\triangle ABC$ after introducing the hypotenuse AD space error, expressed in elevation of the absolute value of difference before and after the introduction of bevel edge delta $\delta D = |ZD - (ZB + ZC)/2|$. At a given screen space error threshold τ, reuse after local terrain vertex constraint factor k_i, error calculation formula is:

$$r_i = \|p_i - e\| = \frac{\delta_i \times d \times \lambda}{\tau} \times k_i \tag{3}$$

here λ is the numbers of the pixel on the screen projected by the unit-length in space; δ_i, ρ are the object space error and screen space error of the vertex i respectively; P_i, e are the position of the vertex i and viewpoint in space respectively; d is the distance from the viewpoint to the projection plane; τ is the given specified screen-space error threshold. In terrain rendering process, d, λ are constant, δ_i is fixed value. According to introducing the order of the vertex by DAG, the relationship among the vertices implicitly included in the error criterion. So, the radius of nested errors sphere [8, 9] can be expressed as formula (4)

$$R_i = \begin{cases} r_i, & i = \text{leaf node} \\ \max(r_i, \max(\|p_i - p_j\| + R_j)), \\ & i = \text{non} - \text{leafnode} \end{cases} \tag{4}$$

According to the nested errors sphere, when the viewpoint in the internal sphere, $P_i > \tau$, shows that the vertex i is active, and need to split to display; and when the viewpoint is located in the external sphere, $P_i < \tau$, t shows that the vertex i is inactive, no further operation. Splitting of the vertex is determined according to the testing viewpoint and internal and external relations of the spherical surface.

2.3 Frame Coherence

The level of detail model which between adjoining frames the terrain has great consistency. That is, when the viewpoint too little change, the triangle need to draw changed little. Therefore, this paper use frames coherence process the terrain of real-time display. According to the changes of the current view, render the need nodes in the next frame, using the rendering of current frame. Merging the nodes that the viewpoint leave their nest error sphere, and splitting the nodes that the viewpoint enter into their nest error sphere, and directly displaying the other nodes that display in current frame. By merging and splitting the part of nodes displaying in the current frame, it can reduce the calculated amount between the viewpoint and the nested error sphere. Maintaining continuity of the grid in operation, while reducing operating time, to improve the efficiency and speed of rendering terrain to meet the requirements of real-time terrain rendering (Fig. 1).

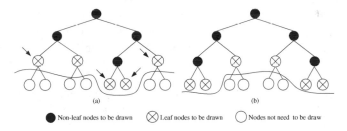

(a)   (b)

● Non-leaf nodes to be drawn   ⊗ Leaf nodes to be drawn   ○ Nodes not need to be draw

Fig. 1. Diagram of frame coherence

Shown in Fig. 2, (a) shows the bin-tree in the current frame, only the arrow leaf nodes requires a change in the next frame. Therefore, using frame coherence the nodes need to compute those changed in the adjoining frames (arrow leaves node). Through the local node merging and splitting operation can get the need to draw a binary tree of the terrain in current frame. It can be seen, the use of frame coherence to been partial adjustment can reduce the number of the node which need to be compute error, compared to merging from the root node of bin-tree every frame. Thus, it can reduce the number of estimation, and calculated amount and the time cost of error calculation each frame, and improves efficiency of the algorithm.

3 The Influence and Research of Viewpoints Jumping Movement

In using the interframe continuity when handling terrain, considering the point of view of mobile under a state of continuous and more simplified mesh model between two adjacent frames change is very small. Therefore, the use of a frame on the grid model appropriate modification can be mapped the grid model in the frame. It can effectively reduce the amount of calculation per frame, to enhance the efficiency of the algorithm. But there is also a point of view the view jumping movement between the two frames, or directly to the specific location of viewpoints, here referred to as viewpoints jumping movement. In viewpoints jumping movement case, frame coherence between technique can effectively improve the efficiency of the algorithm, is worth to explore and research. Based on the analysis of the performance of the algorithm under the viewpoint of continuous and jumping movement, a terrain visualization algorithm is proposed.

3.1 Performance Analysis

In using the frame coherence optimization algorithm has a premise, it is between two consecutive frames, the viewpoint movement is small. Whether or not the frame coherence optimization algorithm is based on the number of vertices between adjacent frames after the viewpoints jumping movement. The larger the number of splitting and merging between adjacent frames, the less the effect of frame coherence optimization, and even more time consuming than re-generated from the root node.

Definition 1 if vertex i appears in the triangular mesh model M, the vertex i is active in M [32], denoted as *active(i)* or i in M. If the vertex i is not in the M, the vertex i is not active in M, denoted as i not in M.

Definition 2 if the vertex i is active, and the i is fully split, then the vertex i is valid in M (*enable*), denoted as *enable(i)*. Otherwise, the vertex i is not a valid point.

Definition 3 efficient active vertex set (the enable active vertex set: EAVS) $\Re = \{V_i \mid enable\ (V_i)$, and $\exists\ V_j$ not in M, $V_j \in Child\ (Vi)$, or the *Child (V_i)* = $\Phi\}$.

$\Re$ is marks the DAG which in no child vertex or at least one child is not active and effective vertices.

Defining 4 remove elements operation "-":

$$\Re - \{p_i\} = \begin{cases} \Re & p_i\ not\ active \\ \{p|p \in \Re \wedge p \neq p_i \wedge p \notin Des(p_i)\} \cup Parent(p_i) & p_i\ active \end{cases} \quad (5)$$

Des (PI) is marks all the descendants vertex of p_i, Parent (p_i) represents the vertex p_i father vertex.

Unlike traditional collection operation "-", delete a vertex p_i from $\Re$ PI, if the p_i does not active, it doesn't change the original collection. If p_i is active, then delete all descendants of the element is the result of the vertex and become inactive, the vertex and its descendants are removed from the $\mathfrak{R}$, also need to add the father of the p_i. Here need to pay attention to several questions:

(1) Whether condition is *active*, not p_i whether in $\mathfrak{R}$, because according to the definition of $\mathfrak{R}$, active vertex is not necessarily in the p_i
(2) After removed p_i from $\mathfrak{R}$, the child of the father is not all active, all father should clearly in $\mathfrak{R}$.
(3) After operation performed vertex p_i, all of its descendants is not active.

Defining 5 all split vertex set $\eta = \{V_i | enable(V_i)\}$. η is recorded in all valid vertex of DAG.

Define 6 data update, means each frame in the frame coherence calculation combined split between vertex number, from the root node determine the ratio of the number of vertices. As shown in the formula 6, including "$| \ |$" said the number of vertices in the set

$$\Psi_{m+1} = \frac{\|\mathfrak{R}_{m+1} - \mathfrak{R}_m\|}{\|\eta_{m+1}\|} \tag{6}$$

$\|\mathfrak{R}_{m+1} - \mathfrak{R}_m\|$ is recorded the number of vertices that need to be re-judged between two frames using the frame coherence principle. $\|\eta_{m+1}\|$ is recorded the number of vertices that need to be re-judged from the root of the terrain. Data update Ψ_{m+1} can reflect the use of the effectiveness of using frame coherence principle. When the $\Psi_{m+1} < 1$, it is shown that the number of vertices that need to be re-judged between two frames is less than the number of vertices that need to be recalculated from the root of the terrain; On the other hand, when the $\Psi_{m+1} > 1$, it is shown that the number of vertices that need to be re-judged between two frames is greater than the number of vertices that need to be recalculated from the root of the terrain and the use of frame coherence optimization algorithm is invalid.

Define 7 change distance γ, $\gamma = \frac{v}{s} \times t$. v is the speed of viewpoint movement, namely distance per second (m/s); s is the sampling interval of terrain (m); t is the time interval of each frame (s).

The frame coherence was used to optimize the terrain visualization algorithm, can well improve the efficiency of the algorithm, but the premise is the terrain changes smaller between the two frames. When change distance γ gradually increase to viewpoints jumping movement occurs, there will be data update $\Psi_{m+1} > 1$, the frame coherence was used to optimize topography algorithm fails, even will reduce the efficiency of the algorithm.

3.2 Research on Terrain Rendering Based on Viewpoints Jumping Movement

Using frame coherence optimization algorithm, improve the efficiency of algorithm. It is necessary to deal with the situation of viewpoints jumping movement, to ensure the effectiveness and generality of the algorithm. Through the section on the definition and analysis, it can be seen between two frames of data update $\Psi_{m+1} > 1$, the frame coherence was used to optimize terrain rendering failure; when the change distance γ increases to viewpoints jumping movement, using the frame coherence optimization of terrain rendering is invalid. So, given a threshold of viewpoint movement δ_E, the algorithm is optimized for:

$$\begin{cases} \gamma_E \leq \delta_E & and \quad \Psi_{m+1} \leq 1, & \text{frame coherence} \\ \gamma_E > \delta_E & or \quad \Psi_{m+1} > 1, & \text{recalculated from the root of the terrain} \end{cases} \tag{7}$$

In the frame coherence effect, still use terrain frame coherence between processing, improve the efficiency of real-time rendering. When the viewpoint jump moves, continue to use the interframe continuity cannot guarantee the validity of the algorithm, the judging from binary tree root node began to generate binary tree. Therefore using the viewpoints jumping movement threshold and data updates than judgement, to ensure that whatever viewpoints or jumping movement has higher efficiency and better realistic real-time rendering.

When the change distance $\gamma_E \leq \delta_E$, namely mobile distance between two frames within the scope of the point of view of mobile threshold, and the data update to $\Psi_{m+1} \leq 1$, this is less than to change the number of vertices between two frames generated when the number of vertices, using frame coherence between thoughts, only on a frame to split, merge all the rendering in the triangle nodes judgment, get this need to render the triangle nodes in a frame. And when the frame change $\gamma_E > \delta_E$, namely mobile distance between two frames outside the scope of view mobile threshold, or data update ratio $\Psi_{m+1} > 1$, which changes the number of vertices between frames than to regenerate the vertices, no longer use the frame coherence between optimization. But directly from the root node of the binary tree began to according to the current point of view of space position determine its division, is the need to render the triangle nodes in the frame. At the same time, there is a definite relation between change distance γ_E and data update Ψ_{m+1}. When change distance γ_E increase, data updates Ψ_{m+1} than will grow bigger. Terrain steep and gentle into degrees of change of distance between different frames of data update than influence will also be different.

4   Experimental Results

Algorithm about large-scale terrain based on this paper has been implemented on the computer. Hardware environment: Intel(R) Core(TM) i3-5-5u, 4G RAM, NVIDI2 A GeForce2400 graphics card; software environment: Windows 10, VS2010, and OpenGL. We used a data [12] of crater and island_1k.zip. As shown in Fig. 2, crater terrain is steep, terrain sampling interval of 1 m/s. Island terrain is more gentle, terrain sampling interval of 30 m/s.

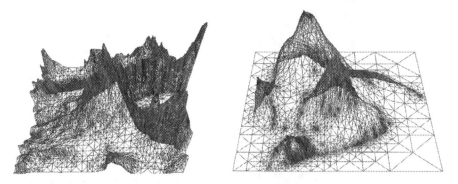

Fig. 2. DEM of crater and island

4.1 The Influence of Frame Change Distance on Data Update

It can be seen from the previous section that the frame change distance and the data update between two consecutive frames affect the performance of the frame coherence optimization algorithm. That is, the larger the change distance of the viewpoint, the larger the data update, the more limited the algorithm of frame coherence. On the contrary, the smaller the change distance is, the smaller the data update is, the more

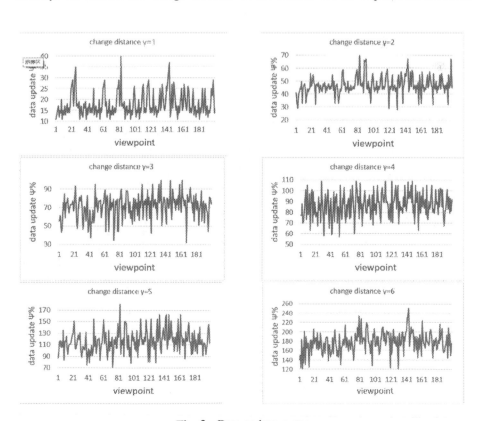

Fig. 3. Data update: crater

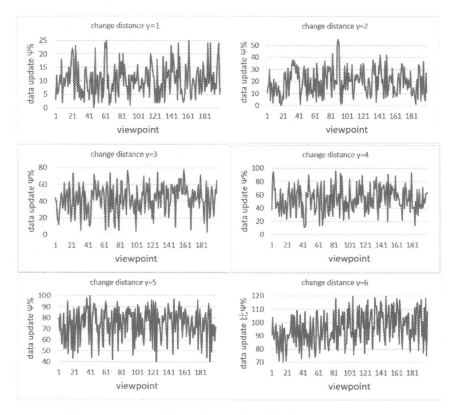

Fig. 4. Data update: Island

effective the inter frame coherence optimization algorithm is. At the same time, the larger the frame change, the greater the data update. Therefore. Using two terrain data of crater and island, under the same trajectory, the influence of the change distance between 200 different positions on the data update is recorded.

In Fig. 3, the crater terrain is used, and the data update of the 200 viewpoint positions is taken in the case of γ_E values from 1 to 6; In Fig. 4, the Island terrain is used, and the data update of the 200 viewpoint positions is taken in the case of γ_E values from 1 to 6. Can be seen from Figs. 3 and 4, when the viewpoints of mobile distance γ_E between two frames is small, data update Ψ_{m+1} is also small. In this case, the two frames need to calculate and determine the number of vertices are less, so using the continuity between frames can greatly reduce the amount of calculation for each frame, so as to achieve to enhance the efficiency of the algorithm. As the point of view between the two frames mobile distance γ_E is more and more big, the data update Ψ_{m+1} is bigger, appear even more than 100% of the time. In this case, the two frames need to calculate and determine the number of vertices has more than double root nodes to calculate and determine the number of vertices, the use of interframe continuity has not reached to enhance the efficiency of the algorithm.

Using different area, under the average of Ψ_{m+1} available more clearly see two different terrain, to influence. As Table 1:

Table 1. Average of Ψ_{m+1}.

Ψ_{m+1}	$\gamma_E = 1$	$\gamma_E = 2$	$\gamma_E = 3$	$\gamma_E = 4$	$\gamma_E = 5$	$\gamma_E = 6$
Ψ_{m+1} of crater	17.875	46.865	71.01	86.815	116.285	177.85
Ψ_{m+1} of island	10.23	19.72	43.075	53.765	74.16	96.075

Can be seen from Table 1, for relatively steep crater topography, Ψ_{m+1} is close to 100% when $\gamma_E = 4$. This means for relatively steep crater topography in $\gamma_E > 4$, using the frame coherence between optimization algorithm will fail. For the relatively gentle Island terrain, Ψ_{m+1} close to 100% in $\gamma_E = 6$. This means for relatively gentle Island terrain in $\gamma_E > 6$, using the frame coherence between optimization algorithm that expire.

4.2 Threshold of Viewpoint Movement

Using frame coherence optimization algorithm, can be very good to improve the efficiency of the algorithm, but along with the rising of the change distance γ_E, the effectiveness of the algorithm is very difficult to guarantee. For this, the set point of threshold of viewpoint movement δ_E. When change distance $\gamma_E \leq \delta_E$, it will still use the frame coherence optimization algorithm; But when frame $\gamma_E > \delta_E$, it will be from the root node to regenerate the terrain, to ensure the effectiveness and generality of the algorithm.

Using the crater and island terrain, in between the frames change distance is the same situation, contrast using frame coherence and from the root node average frame rate of the two algorithms is to regenerate the terrain, as shown in Fig. 3.

Can be seen from Fig. 5, the relatively steep crater topography, change distance γ_E between frame around 6, using the frame coherence between optimization algorithm has no advantage, even when the change distance γ_E between frames after more than 7, from the root node redraw algorithm is more effective. For island terrain is relatively flat, inflection point also appeared around 7, when change distance γ_E between frames after more than 7, also reflected from the root node redraw algorithm is more effective.

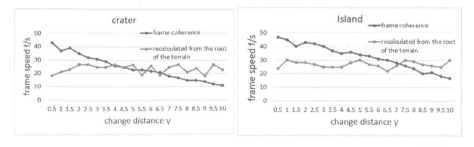

Fig. 5. Frame speed of crater and island

4.3   Analysis of the Efficiency of Frame Coherence Algorithm

Through research and analysis, and the viewpoint of data update than the top mobile change threshold can be seen: for the steep terrain, the inter frame change distance $\gamma_E > 4$, data update than to reflect the use of interframe coherence optimization algorithm will fail, and when the inter frame change distance greater than 7, also will appear by frame coherence optimization algorithm will failure; for a more gentle terrain, the inter frame change distance $\gamma_E > 6$, data update than to reflect the use of interframe coherence optimization algorithm will fail, and when the distance change between frames is greater than 7, also will appear by inter frame coherence optimization algorithm will fail the situation. Therefore, combined with the formula (7) gives the definition of the threshold $\delta_E = 5$, when $\gamma_E > 5$ finds frame coherence algorithm failure from the root node to re generate the terrain topography, to ensure that the terrain visualization algorithm has higher efficiency, achieve the purpose of real-time rendering of terrain.

5   Conclusion

In this paper, based on the special situation of the viewpoint jumping movement, we propose to improve the effectiveness and generality of the terrain visualization algorithm based on the change of the viewpoint motion. This method can guarantee the frame coherence effectively, can use frame coherence to optimize terrain rendering algorithm, and when the failure of coherence between the frames, to generate the terrain from the root node, which can reduce the amount of computation per frame, to improve the overall efficiency of the algorithm. The experimental results show that the proposed method can make up for the deficiency of the inter frame coherence optimization algorithm, and can reduce the amount of computation per frame and improve the efficiency and effectiveness of the algorithm.

Acknowledgements. This work is supported by colleges and universities of Shandong province science and technology plan projects for The research on Large-scale dynamic terrain visualization and key technology No. J15LN66. Thanks are specially given to the anonymous reviewers for their helpful suggestions.

References

1. Duchaineau, M., Wolinsky, M., Sigeri, D.E., et al.: Roaming terrain: real-time optimally adapting meshes. In: Proceedings of IEEE Visualization 1997, Phoenix, pp. 81–88 (1997)
2. Wang, J., Wang, L., Cao, W., Zhao, X.: A 'Drift' algorithm for integrating vector polyline and DEM based on the spherical DQG. IOP Conf. Ser. Earth Environ. Sci. **17**(1), 012203 (2014)
3. Bao, X., Pajarola, R., Shafae, M.: SMART: an efficient technique for massive terrain visualization from out-of-core. In: Proceedings of Vision, Modeling and Visualization (VMV), pp. 413–420 (2004)

4. He, Y., Cremer, J., Papelis, Y.: Real-time extendible-resolution display of on-line dynamic terrain. In: Proceedings of Graphics Interface, Calgary, Alberta, pp. 27–29 (2002)
5. Zhang, J., Chen, G.: Exploiting frame-to-frame coherence for rendering terrain using continuous LOD. In: Pan, Z., Cheok, A., Haller, M., Lau, R.W.H., Saito, H., Liang, R. (eds.) ICAT 2006. LNCS, vol. 4282, pp. 695–704. Springer, Heidelberg (2006). doi:10.1007/11941354_72
6. Cai, X., Li, J., Sun, H., Li, J.: Multi-samples texture synthesis for dynamic terrain based on constraint conditions. In: Pan, Z., Cheok, A.D., Müller, W., Chang, M., Zhang, M. (eds.) Transactions on Edutainment VII. LNCS, vol. 7145, pp. 188–196. Springer, Heidelberg (2012). doi:10.1007/978-3-642-29050-3_17
7. Zhao, X., Xie, B., Wan, D., Wang, Q.: Mix-subdivision dynamic terrain visualization algorithm. Adv. Mater. Res. **756–759**, 3372–3377 (2013)
8. Wang, D., Zhu, Q., Xia, Y.: Real-time multiresolution rendering for dynamic terrain. J. Soft. **9**(4), 889–894 (2014)
9. Huang, J., Guo, L., Long, Y., Wu, H.: Design and implementation of dynamic response mechanism between 2D digital map and 3D visualization scene. J. Geomat. **1**, 33–35 (2003)
10. Yuliang, L., Yingchun, S., Wei, C., Juan, C.: Integration system design and implementation of 2D situation and 3D visualization scene. Ship Electron. Eng. **5**, 2–8 (2005)
11. Ping, Y.U., Zhang, C.-M.: Constraint based error metric for dynamic terrain visualization. J. Gr. **2** (2013)
12. http://www.vterrain.org/BT/index.html

Distributed Processing of Continuous Range Queries Over Moving Objects

Jin Zhou, Hao Teng, Ziqiang Yu[✉], Dong Wang, and Jiaqi Wang

Shandong Provincial Key Laboratory of Network Based Intelligent Computing,
University of Jinan, Jinan 250022, China
ise_yuzq@ujn.edu.cn

Abstract. With the widespread usage of wireless network and mobile devices, the scale of spatial-temporal data is dramatically increasing and a good deal of real world applications can be formulated as processing continuous queries over moving objects. Most existing works investigating this problem mainly concern about the centralized search algorithm for dealing with range queries over a limited volume of objects, but these approaches hardly can scale well in a cluster of servers. Additionally, the existing approaches seldom process the situation that the locations of objects and queries are simultaneously changing. To address this challenge, we propose a distributed grid index and a distributed incremental search approach to handle concurrent continuous range queries over an ocean of moving objects. As to the distributed grid index, it can be deployed on a distributed computing framework to well support the real-time maintenance of moving objects. Further, we take fully into account the condition that locations of objects and queries are both changing at the same time, and put forward a parallel search approach based on the publish/subscribe mechanism to achieve incrementally searching results of each continuous range queries with a cluster of servers. Finally, we conduct extensive experiments to sufficiently evaluate the performance of our proposal.

Keywords: Continuous range queries · Distributed processing · Incremental search · Moving objects

1 Introduction

With the popularity of mobile devices and wireless network, many things such as sharing bicycles can be deemed as moving objects. The problem of processing continuous range queries over moving objects has attracted extensive attentions because many location based services can be formulated as this problem. The semantic of the continuous range query in our work refers to the locations of queries and objects are both continuously changing. Although this semantic enhances the difficulty of processing this type of queries but it accords with the reality and there are some cases that can illustrate this point. To capture suspects, polices will monitor the vehicles passing into or out of a specified region and the polices will probably adjust the search region frequently, which

© Springer International Publishing AG 2017
D.-S. Huang et al. (Eds.): ICIC 2017, Part II, LNCS 10362, pp. 800–810, 2017.
DOI: 10.1007/978-3-319-63312-1_72

is indeed a continuous query processing because the results need to be consecutively updated with vehicles moving and the query scope is also changing constantly.

In another example, a comprehensive service in cap-hailing applications is that the data center needs to continuously seek for nearby taxies for a user who is walking, and this service is also a continuous range query with locations of queries and objects are all dynamically varying. Since the continuous range query has broad applications, so we concentrate on devising an distributed framework with efficiency to address this problem.

In this work, a continuous range query (*CRQ* for short) over moving objects refers to returning moving objects inside a user-defined region in real-time and continuously monitoring the change of query results as the query region constantly changes over a certain time period. In the big data background, processing *CRQ*s over moving objects is faced with unprecedented challenges. First, we have to face the tremendous volumes of moving objects and queries, which are far beyond the computing and storage capacities of one single server. Second, with the ubiquitous mobile internet, most range queries arc online and they desire to be responded in real-time. Last but not the least, it is necessary to constantly monitor the results of *CRQ*s as objects and queries are all moving, which probably involves a gigantic computing cost.

For the sake of challenges, most of existing works are not suitable for dealing with the continuous range queries. This is because many works [7, 10] do not consider the situation that locations of queries and objects are both changing simultaneously. Another critical issue is that most existing proposals [2, 4] always investigate the central algorithms to improve the search efficiency, but they cannot scale well to handle extensive concurrent range queries over a tremendous set of moving objects. Due to the limitations of existing methods, this work explores a distributed framework to search and monitor the results of givcn continuous range queries based on a cluster of servers in real-time. Specifically, we construct a distributed grid index structure that can be seamlessly deployed in a master-slaves model to support the maintenance of moving objects and the parallel processing of range queries. We further design a distributed incremental search method that can update the results of each *CRQ* with only computing the incremental result.

Our main contributions can be summarized as follows:

- This work process *CRQ*s in the scenario where queries and objects are simultaneously moving, which is seldom involved in other existing works.
- We propose DGI, a distributed grid index, for supporting *CRQ*s over moving objects in a distributed setting.
- We design DIS, a comprehensive distributed incremental search approach that can take full advantages of a cluster of servers to continuously monitor the results of *CRQ*s in real-time.
- DGI and DIS are implemented on top of S4, and we conduct extensive experiments to evaluate the performance of DGI and DIS, which confirm its superiority over existing approaches.

2   Related Work

Continuous range query processing is very important due to its broad application base, and it has been extensively studied by existing proposals. Some works [7, 9] investigate processing range queries on road network. Stojanovic et al. [7] propose a framework for continuous range query processing for objects moving on network paths, and introduces an additional pre-refinement step which generates main-memory data structures to support reevaluation of continuous range queries. The work [9] studies the problem of processing range queries on road networks and proposes Voronoi Range Search based on the Voronoi diagram. Due to determining the Voronoi diagram for each object is very expensive, so this approach hardly can support range queries on frequent moving objects. Additionally, some works introduce the concept of safe region to reduce the reevaluation cost. The proposal [4] points out that the cost of monitoring and keeping the location of a moving query updated is very high, and it investigates an efficient technique by adopting the concept of a safe region. That is, as long as the query remains inside its specified safe region, expensive re-computation is not required.

Cheema et al. [2] also adopt the methodology that utilizes he concept of a safe zone to monitor moving circular range queries and propose powerful pruning rules improve the query efficiency. The above works all focus on searching the exact results of queries, but the work [5] pay more attention to approximate range search and proposes approximate static range search (ARS). The work [8] coins a term "Region Queries" to indicate a broad category of spatial range queries, and focuses on showing a complete picture of region queries. These proposals present some excellent search algorithms for monitoring (continuous) range queries, but methods are centralized and their scalability is restricted. Moreover, most of them does not consider the situation that every object is constantly moving as processing continuous range queries.

It is imperative that utilizing the distributed computing model to deal with concurrent range queries over moving objects, and some works [1, 3, 6, 10, 11] have explored this problem. In fact, the distributed framework in these approaches consist of a central server and extensive moving objects and they all require the moving devices to have considerable computational capabilities, which restricts their applicability. In contrast, our approach does not assume any computation capabilities at the mobile objects other than reporting their positions (e.g., the sharing bicycles can be a simple GPS tracking device), and thus has wider applicability.

3   Distributed Grid Index

Given an interest of region covering a large number of moving objects, a grid index partitions this region into four-square cells with the same size. For any cell c_i, it is regarded as an index unit that records the locations of moving objects residing in the cell as well as the queries involving it. Grid index structure has been extensively utilized for processing spatial-temporary queries, and we also utilize it to support the processing of CRQs over moving objects. But unlike the existing approaches utilizing the grid index on a single server, we construct a Distributed Grid Index (DGI), namely, deploying the grid index on a master-slaves model, which consists of one master and multiple servers. Next, we will expound the structure of DGI as well as the deployment of DGI on the master-salves model. The symbols will be used in later text is summarized in Table 1.

Table 1. Summary of symbols

Symbols	Meanings
$o \prec B$	An object o is covered by the region B
$o \nrightarrow B$	An object o is not covered by the region B
$C \sqcap D$	The region C is partially covered by the region D
$C \sqcap\!\!\!= D$	The region C is fully covered by the region D
$H \setminus Z$	The elements belong to t H but do not belong to Z

DGI consists of a Global Schema Index (GSI) and extensive cell indexes. In particular, DGI refers to deploying the grid index on a *master-salves* cluster with a general streaming data processing model. In this model, we use the conception of PE to represent a logical processing element, and the data will be encapsulated as an *event* that can be transferred from one PE to another. Every *event* is formulated as a triple (*type, key, value*). When a PE consumes a received *event*, it will encapsulate the intermediate results as a new event that can be routed to another PE. Here, each PE can receive their desired *events* by specifying their *types* and *keys*, which indeed forms the *events* routing rule. When this model assigns an *event* to a corresponding PE based on the routing rule, if the PE does not exist, a new PE will be automatically created to process this *event*.

In DGI framework, a unique EntrancePE on the *master* maintains the GSI and every CellPE takes charge of one or more cell indexes in different slaves. GSI is maintained by the *master* and cell indexes are distributed in different *slaves*. For GSI, it needs to contain identifiers and boundaries of every cell index. In addition to this, it also has to *master* the relationships between cells and *slaves*. To achieve above purposes, the GSI seems to be necessary to store much information that is apt to make it be a bottleneck. But in fact, the GSI designed by us only needs to record the bottom left cell as the reference cell and a naming rule that can be used to identify the identifier of each cell rapidly. Since every cell is a four-square, then the identifier and boundaries of every cell can be deduced instantly. As to the relationships between cells and *slaves*, we introduce an hash function $f(x)$ that can map the cells to different slaves on the basis of the identifiers of cells, that is, $s_i = f(c_i)$, where s_i and c_i are the identifiers of slaves and cells.

Each cell index is charged of recording the objects covered by itself and the queries with search scopes intersecting with this cell. For any cell c_i, it has three major components, i.e., OL_i, FL_i, and PL_i. The list OL_i is used to record the locations of moving objects covered by itself. FL_i stores the identifiers of queries whose search regions fully cover c_i, while the queries with the search scope partially overlapping with c_i are maintained by PL_i.

4   Distributed Incremental Search Approach

In this section, we propose DIS, a comprehensive approach to address the challenges of incrementally searching the results of extensive *CRQ*s in real-time.

4.1   Search Initial Results of *CRQ*s

The DRQS framework (as shown in Fig. 1) can be deemed as the fundamental computing architecture, based on which we design search algorithms to process *CRQ*s. Now we discuss SIR, an algorithm to search initial results of *CRQ*s.

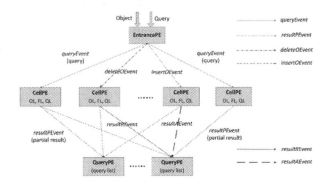

Fig. 1. The framework of DRQS

When EntracePE receives a *CRQ* (q_i) with the search scope sq_i, it first determines the cells intersecting with sq_i based on *GSI* and these cells are called *candidate cells*. We use $\mathscr{F}$ to label the set of *candidate cells*. Since the boundaries of cells are static, the cells in $\mathscr{F}$ can be determined with a brute-force comparison of boundaries of each cell with sq_i, but which will greatly waste computing costs. To settle this issue, we will firstly find the cell c_h that covers the center point of the search scope sq_i, and then detect whether each adjacent cell c_i of c_h has an intersection with sq_i. This process will iteratively enlarge the search region until a group of cells surrounding c_h do not overlap with sq_i. This strategy can quickly find the candidate cells of q_i even if the shape of sq_i is irregular.

After determining the set $\mathscr{F}$ for q, EntrancePE will send a *queryEvent* carrying q to the CellPEs corresponding to cells in $\mathscr{F}$. To facility presentation, we suppose that each CellPE is charged of one cell index. Hence, when a CellPE receives the *queryEvent*, it will instantly find the objects covered by sq_i and these objects form a partial result of q_i. Every CellPE encapsulates the partial result as a *resultPEvent* and sends it to the QueryPE. In DRQS framework, the final result of every query is calculated and maintained by a unique QueryPE, and the *resultPEvent* carrying partial results of q_i will be routed to the same QueryPE with identifier of q_i as the key of *resultPEvent*. Therefore, QueryPE can obtain the final result of q_i by merging all partial results.

4.2 Incrementally Computing Results of *CRQs*

After obtaining the initial results of *CRQs*, we still need to monitor the results of every query. We first design IOS, an algorithm for incrementally searching results of *CRQs* as object are moving based on a publish/subscribe mechanism to constantly update the results for each query in real-time. In IOS algorithm, every CellPE and QueryPE are regarded as a data publisher and query subscriber respectively. We use cp_i to label the CellPE matching the cell c_i. As to the CellPE cp_i, it maintains two registered query lists FL_i and PL_i. For a given query q_i, its candidate cells forms the set $\mathscr{F}$. If $c_i \in \mathscr{F}$, q_i will be registered in c_i. Specifically, q_i will be inserted into FL_i if $sq_i \simeq c_i$ or inserted into PL_i if $s_q \sim c_i$. The query q_i has to be registered in every candidate cell in this way. Once an object in the cell c_i moves, the corresponding CellPE only needs to detect whether the registered queries can be influenced rather than recompute the results of all existed queries.

In fact, we further observe that not all registered queries in c_i will be influenced by the movement of every object. Since the result of q_i just concerns about the objects covered by sq_i rather than their exact positions, so we can deduce that if a CRQ covers the cell ci, it will not be influenced by the movements of objects in

Theorem 1. *For a given cell c_i, $\forall\, o_i \in OL_i$, its location is $\left(x_i',\, y_i'\right)$ at the time point t_i and the location becomes $(x_i,\, y_i)$ at the time point t_{i+1}. If $\left(x_i',\, y_i'\right)$ and $(x_i,\, y_i)$ are both covered by c_i, then $\forall\, q_j \in FL_i$, its result will not be influenced by the changed location of o_i with the condition that $t_{q_j}^s$ is smaller than t_i.*

With the help of Theorem 1, if the location of o_i in a cell c_i changes, we should update the results of the registered queries in c_i by handling the following two cases.

- The first case is that $\left(x_i',\, y_i'\right) \prec c_i$ and $(x_i,\, y_i) \prec c_j$ $(i \neq j)$. In this case, the results of all registered queries of c_i and c_j will be probably influenced. For each registered query q_i of c_i, if $\left(x_i',\, y_i'\right) \prec sq_i$, then the CellPE corresponding to c_i will notify the QueryPE maintaining q_i to remove $\left(x_i',\, y_i'\right)$ from the result of q_i by sending a *resultREvent*; meanwhile, for every registered query q_j of c_j, if $(x_i,\, y_i) \prec sq_j$, then the CellPE matching c_j will send a *resultAEvent* to the QueryPE maintaining qi, which aims to notify this QueryPE to insert o_i $(x_i,\, y_i)$ into results of q_j.

- Another case is that $\left(x_i',\, y_i'\right) \prec c_i$ and $(x_i,\, y_i) \prec c_i$. According to Theorem 1, $\forall\, q_i \in FL_i$, it cannot be affected by the object o_i. Hence, only the results of queries in PL_i need to be updated. In this case, the CellPE matching c_i can directly identify the registered queries affected by oi, and then notify the corresponding QueryPEs to update the query results.

As above, every CellPE corresponds to a publisher and each QueryPE serves as a subscriber. Only if a *CRQ* q is registered in the CellPEs corresponding to its candidate cells, these CellPEs will continuously monitor the results of q in a parallel way only if their matching cells belong to the set $\mathscr{F}$ of q. In this process, different CellPEs will send the moving objects involved with q to a unique QueryPE, which can be deemed a subscriber. This QueryPE will process the received locations of objects as soon as possible to guarantee the exact result of q all the time. Moreover, the QueryPE as a subscriber will take charge of maintaining the result of the query q all through the life-cycle of q. In this scenario, we organize extensive CellPEs and QueryPEs in different servers as a publish/subscribe mechanism that can well support incrementally searching the results of $CRQs$ in a distributed environment.

As to a query q_i, if its search scope is sq_i at time point t_j and sq_i becomes sq_i' at time point t_{j+1}, then q_i is required to be resubmitted to EntrancePE. When receiving q_i, EntrancePE will compute $\mathscr{F}$, the set of candidate cells for q_i, based on sq_i', and then send q_i to each cell c_k ($c_k \in \mathscr{F}$). After receiving q_i, the cell c_k has to update its two lists of registered queries. Meanwhile, the matching CellPE will instantly find the objects covered by sq'_i from the cell c_k. In this case, CellPE indeed employs an incremental search strategy to only compute the incremental result of q_i at time point t_{j+1} based on its existing result, and this incremental search strategy includes the following steps.

(1) If q_i is a new registered query for c_k, the following cases need to be considered.

- If $sq_i' \simeq c_k$, then q_i will be inserted into FL_k and all objects of c_k form a part of the result of q_i.
- If $sq_i' \sim c_k$, then q_i will be inserted into PL_k and c_k has to find the objects covered by sq'_i.

(2) If q_i is an existed registered query of c_k, it will be handled with next steps.

- If $(sq_i \simeq c_k)$ && $(sq_i' \simeq c_k)$, which means q_i has been in FL_k and we have no need to insert q_i into FL_k again. In this case, q_i still covers all objects in c_k though it moves. As to the cell c_k, it does not need to update the result of q_i;
- If $(sq_i \simeq c_k)$ && $(sq_i' \simeq c_k)$, we need to remove q_i from PL_k and insert it into QL_k. In this case, the incremental result of q_i is the set of objects covered by the region $\left(sq_i' - sq_i\right)$, which can be rapidly determined.
- If $(sq_i \simeq c_k)$ && $(sq_i' \sim c_k)$, we has to remove q_i from QL_k and insert it into PL_k. At this moment, we only need to remove the objects residing in the scope $\left(sq_i - sq_i'\right)$ from the result of q_i.
- If $(sq_i \sim c_k)$ && $(sq_i' \sim c_k)$, q_i is not necessary to be added into PL_k again but c_k needs to update the information of q_i. Now, we need to search all objects covered by the region $(sq_i' - (sq_i' \cap sq_i))$ and these objects belong to the new result of q_i.

5   Experiments

We conduct experiments to evaluate the proposed DGI index and DSI approach. To better evaluate the performance of DSI, we introduce two other distributed algorithms as baseline methods. The first method, NS, is a naive search algorithm which does not use any index. For any object, NS uses a hash function to determine which server should store it. Processing a *CRQ* thus involves scanning all objects maintained in all servers at any time point. The second method, S-DIS, is a simplified DIS method that handles a *CRQ* as a new query at any time point, that is, it does not utilize the incremental search strategy when processing *CRQs*.

We use the German road network data to simulate two different datasets for our experiments. In the datasets, all objects appear on the roads only. In the first dataset (UD), the objects follow a uniform distribution. In the second dataset (GD), 70% of the objects follow the Gaussian distribution, and the other objects are uniformly distributed. In these two datasets, the whole area is normalized to a unit square and this square is partitioned into small cells with edge length being 0.01. Moreover, all objects move along the road network, with the velocity uniformly distributed in [0, 0.002] unless otherwise specified. We use V_p and V_q to represent the velocities of an object and a query. The experiments are conducted on a cluster of 8 Dell R210 servers with Gigabit Ethernet interconnect.

We first evaluate the performance of DGI. Figure 2 demonstrates the time of building DGI with the number of objects varying. Based on the results, we observe that the time cost is in proportion to the number of objects and it is almost not influenced by the distribution of objects. When objects are moving, the velocity of objects will exert an impact on the build time of DGI. In Fig. 3, we evaluate the maintaining time of DGI by processing a set of objects in a specified period of time and find that the maintaining time is slightly affected by the velocity of objects. This is because DGI always needs to remove the obsolete location and insert the new position for processing the movement of an object regardless of the distance it moves one time.

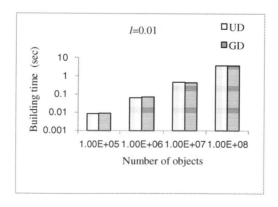

Fig. 2. Building time of DGI

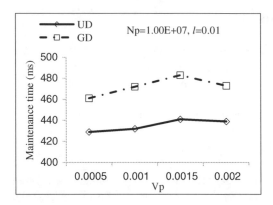

Fig. 3. Influence of V_p on maintenance time of DGI

Next, we conduct an evaluation on the performance of DIS approach. In Fig. 4, we first test the time of three approaches when processing the same set of queries. In this group of experiments, we make 10000 queries continuously move and test the time of processing these queries at five consecutive time points. As can be observed, DIS performs better than other two approaches especially at the last four time points. The reason why the time cost of DIS decreases is that only incremental result of queries need to be calculated after the first time point, which can greatly reduce the computing time, while other two methods always process each query as a new one.

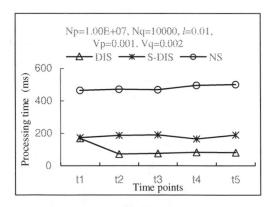

Fig. 4. Comparison of three approaches

Figure 5 demonstrates the influence of number of queries on the processing time of each method. Here, we observe that the response time of each approach increases obviously as more and more queries are processed, but the consuming time of NS approach grows more sharply. Due to this set of experiments does not involve continuous queries, so the performances of DIS and S-DIS are almost identical.

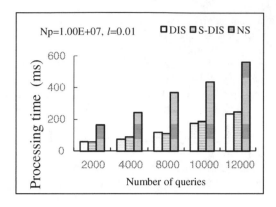

Fig. 5. Processing time of three approaches

6 Conclusion

With the dramatic increase of mobile devices and the advances in wireless network, the efficient processing of *CRQ*s has been of increasing interest. This work propose a distributed incremental search approach that sufficiently considers the situation that queries and objects are both moving and only needs to reevaluate the incremental result of every *CRQ* to cut down the computing costs as well as communication expenses between CellPEs and QueryPEs. Finally extensive experiments are conducted to verify the performance of our proposal.

Acknowledgement. This work was supported in part by the Shandong Provincial Natural Science Foundation (ZR2016FB14) Science and Technology Program of University of Jinan (XKY1737), the National Natural Science Foundation of China under Grants with No. 61640218, and the Project of Shandong Province Higher Educational Science and Technology Program under Grant with No. J16LN07.

References

1. Cai, Y., Hua, K.A., Cao, G.: Processing range-monitoring queries on heterogeneous mobile objects. In: 2004 IEEE International Conference on Mobile Data Management, pp. 27–38 (2004)
2. Cheema, M.A., Brankovic, L., Lin, X., Zhang, W., Wang, W.: Multi-guarded safe zone: An effective technique to monitor moving circular range queries. In: ICDE 2010, pp. 189–200 (2010)
3. Gedik, B., Liu, L.: MobiEyes: Distributed processing of continuously moving queries on moving objects in a mobile system. In: Bertino, E., Christodoulakis, S., Plexousakis, D., Christophides, V., Koubarakis, M., Böhm, K., Ferrari, E. (eds.) EDBT 2004. LNCS, vol. 2992, pp. 67–87. Springer, Heidelberg (2004). doi:10.1007/978-3-540-24741-8_6
4. Haidar, A.-K., Taniar, D., Betts, J., Alamri, S.: On finding safe regions for moving range queries. Math. Comput. Model. **58**(5), 1449–1458 (2013)

5. Haidar, A.-K., Taniar, D., Safar, M.: Approximate algorithms for static and continuous range queries in mobile navigation. Computing **95**(10–11), 949–976 (2013)
6. Hu, H., Xu, J., Lee, D.L.: A generic framework for monitoring continuous spatial queries over moving objects. In: Proceedings of the 2005 ACM SIGMOD International Conference on Management of Data, pp. 479–490 (2005)
7. Stojanovic, D., Papadopoulos, A.N., Predic, B., Djordjevic-Kajan, S., Nanopoulos, A.: Continuous range monitoring of mobile objects in road networks. Data Knowl. Eng. **64**(1), 77–100 (2008)
8. Taniar, D., Rahayu, W.: A taxonomy for region queries in spatial databases. J. Comput. Syst. Sci. **81**(8), 1508–1531 (2015)
9. Xuan, K., Zhao, G., Taniar, D., Rahayu, W., Safar, M., Srinivasan, B.: Voronoi-based range and continuous range query processing in mobile databases. J. Comput. Syst. Sci. **77**(4), 637–651 (2011)
10. Zhou, J., Chen, L., Chen, C.L.P., Zhang, Y., Li, H.: Fuzzy clustering with the entropy of attribute weights. Neurocomputing **198**, 125–134 (2016)
11. Zhou, J., Chen, C.L.P., Chen, L., Li, H.: A collaborative fuzzy clustering algorithm in distributed network environments. IEEE Trans. Fuzzy Syst. **22**(6), 1443–1456 (2014)

Identity Authentication Technology of Mobile Terminal Based on Cloud Face Recognition

Likai Dong[1], Zhikang Ma[1], Tao Xu[1,2], and Dong Wang[1,2(✉)]

[1] School of Information Science and Engineering, University of Jinan, Jinan 250022, China
[2] Shandong Provincial Key Laboratory of Network Based Intelligent Computing,
University of Jinan, Jinan 250022, China
ise_wangd@ujn.edu.cn

Abstract. The face recognition of mobile terminal plays an important role in the identity authentication technology. But there are some problems such as long detection time and low recognition rate due to the performance of mobile devices. This paper presents a mobile terminal identity authentication technology based on cloud face recognition. The whole system consists of cloud server, check-in and checked-in. Checked-in part includes simple face detection and face image acquisition, which sends the acquired face image to the cloud server for face recognition. The wifi hotspot based on mobile phone can determine whether the sign-in end is near the check-in end to achieve the purpose of distance authentication. The face recognition module, which is heavily influenced by hardware performance, is deployed on the cloud server. The information of each part exchange through the network, and human can efficiently authenticate the entire process. Experiment result shows that the method proposed in this paper can effectively improve the speed and accuracy of face recognition on the mobile side, and achieves an average accuracy rate of 92%.

Keywords: Identity authentication · Cloud face recognition · Mobile

1 Introduction

At present, the application of identity authentication technology is very extensive, people need to authenticate their identity all the time [1]. No matter the business card punch or college attendance are inseparable from the identity authentication technology. Traditional identity authentication equipment usually requires high costs for purchase and maintenance, and the queuing time usually too long when a large number of users need authentication at the same time [2]. These drawbacks make the actual use of the process inconvenient. Since there are so many drawbacks of the Traditional identity authentication technology, so the study of a kind of efficient, convenient and easy to use identity authentication technology becomes so important.

Face is the basic biological characteristics of human, which will be very convenient and efficient if it can be used as a channel for personal identity [3, 4]. Because of the large amount of consumption of system resources, it needs special equipment to run. It is difficult to operate on the mobile terminal and the scenes to be used will be greatly

© Springer International Publishing AG 2017
D.-S. Huang et al. (Eds.): ICIC 2017, Part II, LNCS 10362, pp. 811–822, 2017.
DOI: 10.1007/978-3-319-63312-1_73

limited. With the rise of cloud face recognition technology, it is possible to apply the face recognition technology to the mobile terminal [5].

This paper introduces a mobile terminal identity authentication technology based on cloud face recognition, and relying on the technology to develop a set of deployed to Android device on the face sign system. Due to the low system resource consumption and high recognition rate, it can be widely applied to business meetings attendance, college class attendance and other scenes.

2   Key Technical Analysis

Due to the high share occupancy of the current Android system in the mobile terminal system, the system will be deployed first on Android. How to carry out face detection and face recognition on Android system will be the key problem to be solved. And there will be a large number of users at the same time applying for identity authentication during the actual sign-in behavior. In order to ensure that the authentication speed of a large number of users at the same time authentication to minimize the user's "queuing" time, the business is divided into check-in side, was signed and server-side three parts and the face detection and face recognition of identity authentication will be carried out in the sign-in side. The judgment of whether the sign-side is near the sign-side has become the key issues that need to be addressed by the system (Fig. 1).

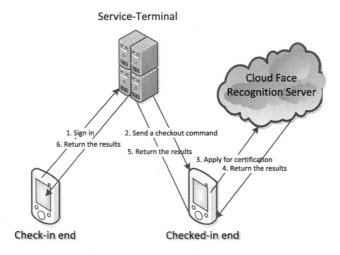

Fig. 1. System call diagram

2.1   Face Detection

Android system built-in API allows developers easily using the video or image for face detection and positioning [6]. You can use the front camera of the mobile terminal for face photo collection. The face detection can be performed after photo collecting during which use the Surface View in the Android system for real-time preview of the camera.

The system uses the FaceDetector class provided by the developer for use by the developer for face detection. Because the image captured by the android mobile phone front camera is rotated after 90°, so the picture needs to be turned counterclockwise 90° firstly before face detection.

As the Android system provided FaceDetector class face detection is based on the eyes of the people, after the detection of the face can be provided by the API method to obtain the center of the eyes and the distance between the eyes. Assuming that the center position of the eyes be (x, y) and the eyes spacing is d, so the left eye coordinates are $(x - dd/2, y)$. Take a point from the left side of the left eye $d/2$ and at the top of the d and The position is $(x - d, y + d)$. As a vertex of the face intercepting rectangle, the vertical direction of the length of 3d, and the horizontal direction of the length of 2d length to intercept the facial area.

2.2 Face Recognition

Face recognition [5, 6] plays an important role in identity authentication technology [7–9]. This technology is very convenient, fast and efficient because it don't needs other users' participating only make judgments through the user's face photo or video stream. It is very easy to apply the face recognition technology to the identity authentication of the mobile terminal because the current mobile terminals are all equipped with camera.

Due to the mobile terminal's own computing speed and storage capacity constraints, the face recognition accuracy of the mobile terminal cannot achieve the desired target. Taking into account the current mobile network speed and the development of cloud face recognition, this paper considers the use of cloud face recognition technology for face recognition. The mobile terminal sends the human face picture acquired by the face detection to the cloud server for face recognition, and the result data is returned to the mobile terminal after the recognition is completed. Data transfer using json data will save a lots of data traffic. This model will greatly reduce the operating costs of mobile terminals. And the recognition rate depends on the recognition rate of the cloud server and will not be limited by the performance of the mobile terminal hardware. The running of the system will consume very little resources.

Remind the users to change the expression automatically according to the prompts to get three different photos during validating. Send the photo to the computing server to verify the result set of the validation. Let the result set returned is $R_{i(i=0,1,2)}$. So the final verification result is $R = \sum_{i=0}^{2} \frac{R_i}{3}$, $i = 0, 1, 2$. If $R \geq 99.5\%$, it shows that the three pictures are almost identical. It is suspected that the user uses the photo cheating system for identity authentication.

2.3 Distance Certification

This paper considers the face detection and face recognition module to be checked at the end to ensure that the system to verify the speed and solve the "queuing" problem when there are a large number of users at the same time using the "queuing". The problem that the system needs to be solved is the judgement that whether the check-in end is near the check-in location. As the using of the system is mostly indoors, the use of mobile devices GPS chip will not be able to reach the accuracy requirements of the system.

Considering that the current mobile terminal devices are all equipped with WiFi module, and the feature that the use of mobile terminal WiFi coverage radius limit and MAC address is unique, the system use the WiFi technology to realize the judgement of the location of this system, so that to judge whether the sign side is near the sign side.

2.4   System Real-Time Analysis

Since the system is deployed on mobile terminals, there is still a shortage of system resources for mobile terminals. In order to ensure system real-time and reduce the system's resource consumption, the system has taken the following points to improve the real-time of the system which will reduce the system's resource consumption:

1. The face picture will be cut during face detection and then will be uploaded to the cloud server. Each photo is guaranteed to be within 100 K. According to each identity authentication need to upload three photos and 6 json data transmission to calculate, each time the identity of all data consumption is also guaranteed within 320 K. It saves a lot of user's data traffic.
2. Due to the use of cloud face recognition, the face recognition business logic ported will be carried out to the cloud. It is very economical to save the mobile terminal running resources. In the 10M bandwidth of the local area network for testing, the average recognition speed is 2.06 ms.
3. The identity system is divided into three parts. It reduces the "queuing" time for multiple users to apply for certification at the same time. In the case of about 200 people in parallel, the time to complete the certification of all people in general circumstances within 30 s.
4. When performing distance certification, the check-in end will only search for WiFi information in the WiFi list. Such design reduces the amount of time of disconnecting the hotspots and avoid the limitation of the number of mobile terminal hotspots.

3   Functional Structure Design

The entire system is divided into check-in side, signed side and server-side. Each part has its certain function. Check-in side as the starting side of the sign, it's main function is to check the user's registration and landing, start the sign and sign information. And with the above functions to increase the sign-in information export and notification of the release and other functions. Signed side is mainly used for sign together with check-in side. It is mainly to achieve the function of the user registration and landing, check the module and with the above features to increase the check-in information to view and notify the view and other functions. The server-side main implementation is maintained with a long link with the checked-in side, so that commands and messages can be delivered at any time as a bridge between the check-in and the sign-in. It is also responsible for saving and maintaining the information of check-in and the user. The specific function chart is as follows (Fig. 2):

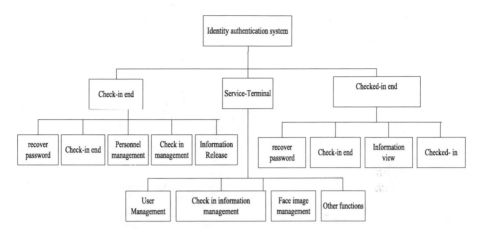

Fig. 2. Functional structure design

The whole system is divided into several modules based on logical structure, such as face authentication, distance authentication, registration, query, notification and mail. The following is description of specific process of face authentication module and distance authentication module in the identity authentication technology. Face authentication module

Face authentication module is the most important and most complex module of the system [8, 10]. The entire module runs in conjunction with the check-in, check-in, server and cloud face recognition.

The request for sign-in is initiated by the check-in end. The sign-in server initiates a check-in request service to the server at the request check-in. After checking the permissions and identity of the sign-in end, the server sends a command to start the check-in to the check-in end of the sign-in request.

Once the check-in is received, the face detection module is turned on. In the check-in side to detect the user's face information after taking three different facial photos of the face. It preprocesses the photos and send them to the cloud face recognition for authentication. The user has been checked by the end of the face to accept the recognition of the results of the face after the data sent to the server to judge. After the server judgment is completed, the authentication result and the reason are returned to the check-in end and the checked-in end. Save the data to the local server and replace the face recognition server's face photos based on the authentication results. The specific face certification flow chart is as follows (Fig. 3):

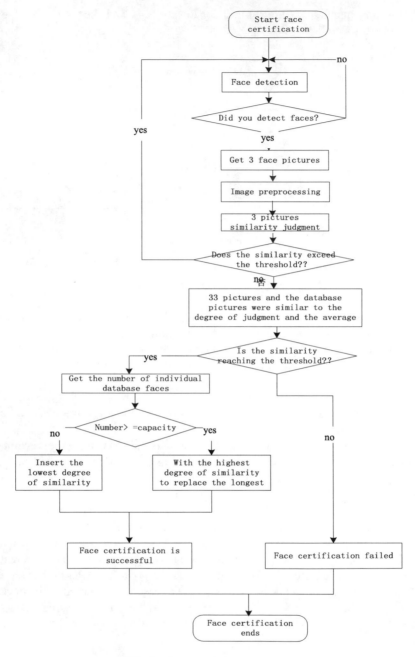

Fig. 3. Face recognition flowchart

3.1 Distance Authentication Module

When you sign up at the sign-inside, Check-in side will start hot spots, and randomly generate WiFi hotspots of ssid and password. They send the SSID of the mobile terminal hotspot and the MAC address of the terminal network card with the sign request. The server receives a check-in request and sends it to the checked-in terminal to encapsulate the SSID of the check-in. After signing the face authentication, it will start the WiFi device on the mobile side to search for the corresponding MAC address of the specified SSID in the WiFi list. And return all queries to the MAC address with the results of face authentication to the server. The server determines whether there is a checkpoint MAC address in the list returned by the check-in end. The specific distance from the certification flow chart is as follows (Fig. 4):

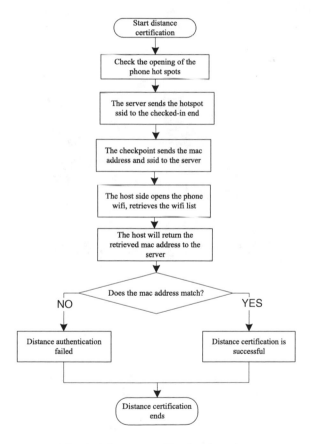

Fig. 4. Distance certification flow chart

4   Experimental Results and Analysis

4.1   System Operation Results

The server uses Tencent Cloud's cloud server, server CPU with 1 core, memory 1G, bandwidth 1M and windows server 2012 operating system, running in jdk1.8 and Tomcat 8.0 environment.

The use of the experimental terminal in the use of Guangdong Oupo company produced OPPO R1c mobile phone. The mobile phone processor for the high-pass Xiao Long 615 (MSM8939), eight nuclear, memory 2 GB, front camera 500 million pixels, the system uses Android 4.4.

The test environment is a college classroom with an area of about 70 m², with 82 seats in the classroom. After testing, in normal light, test facial expression normal, no obvious rotation of the head, the operating system of the experimental terminal system is as follows (Fig. 5):

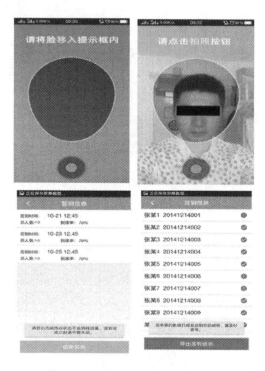

Fig. 5. System running screenshots

4.1.1 System Availability Test

Through a large number of users several tests, after the system is fully deployed, the modules are running normally, system functions are working properly. System installation and uninstallation are very easy. It is easy to use, compatible with Android 4.4–6.0 between the various versions. It has stronger compatibility and meets the basic design requirements.

4.1.2 System Performance Test

10 users were chosen for registration and authentication test, and there are no similar appearance such as twins looks. Three tests per person at the time of registration. The first test was a legitimate test. The last two are repeated tests. The identity is done 5 times per person. Respectively, the number of 10 users registered the number of initial registration, the number of repeated registration, certification and pass the specific time. The specific test results are shown in Table 1.

Table 1. System performance test table

Test items	Test results	Average timeconsuming/ms
Initial registration pass rate	1	2850
Repeat registration pass rate	0	1600
Certification rejection rate	0.06	2560
Certification recognition rate	0.92	2680
Certification misjudgment rate	0.08	2970

The system test performs 50 times. There are 4 miscarriage of justice and all the same user authentication failed. And there were no miscarriage of justice between different users. The emergence of four miscarriage of justice are added in a face recognition to determine the normal. There are three cases of misjudgment. The reason is that no face is detected. It judged normal after adjusting the face angle. Because the recognition range is limited to the wireframe, the impact of other faces on the certification site can be reduced to very low. It can be said that almost no other people's face affected.

According to the experimental results, the system can identification a single person within 3 s. The recognition rate is within the acceptable range of the user. In the case of identification of abnormal circumstances only need to identify the second time. By default, the system will alert the user to re-authentication. The system can basically achieve the actual use requirements.

4.2 Three Platform Data Analysis

Taking into account the actual use of the user's degree of freedom will be higher, the recognition rate of the distance is affected by the user's environment, the user's expression action and so on. This paper on the current cloud face recognition rate of good reputation in the three platforms (TencentYouToLab [11], face++ [12] and EyeKey [13]), carried out more demanding tests. The test uses the ideal test face database. This paper selected the actual situation of the light often appear dark, whether to wear glasses, facial expressions and head movements and other 13 special circumstances plus 2 normal conditions were tested. The specific test results are as follows.

4.2.1   Normal Test

At the first, this paper examines two situations that occur in the vast majority of cases in actual use. One is the same person different photo recognition rate and different people mistaken for the same person's chance. The specific test results are shown in Table 2.

Table 2. System performance test table

Test items	TencentYouTo Lab [11]		Face++ [12]		EyeKey [13]	
	Recognition rate	Timeconsuming (ms)	Recognition rate	Timeconsuming (ms)	Recognition rate	Timeconsuming (ms)
The same person	1	2740	1	3650	1	526
Different people	0	2810	0	3540	0	498

From the test results, for the normal situation, the accuracy of the three platforms is very high, and they all can meet the actual use of the requirements. The test speed is also within the user's acceptable range. It can be said that the three platforms can be selected in actual use.

4.2.2   Special Case Analysis

In order to test out the three platforms in the face of some special circumstances of the processing power, I also selected 13 kinds of life often appear in the special circumstances of a separate test. The specific test results are shown in Table 3.

From the test results, when conducting special case tests, three platforms have the ability to identify and identify the speed of the three platforms in the direction of the focus is also different. Simply from the test results, EyeKey is better. Whether it's recognition rate or time-consuming are more obvious priority over the other two platforms. In terms of recognition rate, Face++ poor EyeKey is not too much. But in terms of recognition speed, Face++ has some gap from the EyeKey. TencentYouTo Lab in the identification of these special circumstances when the other two are somewhat lacking. But TencentYouTo Lab is the basic business completely free of charge. On the above test results, it is enough to solve the actual use of most of the situation.

Table 3. System performance test table

Test items	TencentYouTo Lab		Face++		EyeKey	
	Recognition rate	Timeconsuming (ms)	Recognition rate	Timeconsuming (ms)	Recognition rate	Timeconsuming (ms)
Glasses	0.123	2800	0.805	3740	0.943	530
Laughing	0.504	2060	0.806	3260	1.000	240
Left 30 side face	0.512	2150	0.902	3040	0.992	860
Left 45 side face	0.268	2170	0.642	3210	0.976	870
Left 90 side face	0.000	2160	0.089	2780	0.626	580
Left 45 twist the head	0.350	2140	0.670	3680	0.959	810
Look up	0.090	2130	0.504	2960	0.991	820
Bow down	0.268	2140	0.707	3550	0.967	800
Eyes closed with glasses	0.113	3030	0.578	3820	0.943	330
Right 30 side face	0.626	3490	0.772	3100	0.991	240
Right 45 side face	0.081	3490	0.228	2790	0.943	310
Right 45 twist the head	0.439	2350	0.618	2780	0.959	300
Bow down	0.008	4070	0.114	1300	0.919	270

5   Conclusion

In this paper, we proposed a mobile terminal authentication technology based on the cloud face recognition which is fast, convenient, efficient and accurate for authentication with a large number of users at the same time. It can solve the problem of the cumbersome and inefficient process of identity authentication in the past. The whole system consists of three parts, i.e., cloud server; check-in and checked-in end. The information of each part exchange through the network, and authentication can be efficiently achieved. Experiment result shows that the proposed method improved the speed and accuracy of face recognition on the mobile side.

Acknowledgments. This paper is supported by the National Natural Science Foundation of China (No. 61302128), the Youth Science and Technology Star Program of Jinan City (201406003) and Doctoral Foundation of University of Jinan (XBS1653), Teaching Research Project of University of Jinan (No. J1638).

References

1. Takabi, H., Joshi, J.B.D., Ahn, G.J.: Security and privacy challenges in cloud computing environments. IEEE Secur. Priv. **8**(6), 24–31 (2010)
2. Suo, H., Liu, Z., Wan, J., et al.: Security and privacy in mobile cloud computing. In: Wireless Communications and Mobile Computing Conference (IWCMC), 2013 9th International, pp. 655–659. IEEE (2013)
3. Darwish, A.A., et al.: Human authentication using face and fingerprint biometrics. In: 2010 Second International Conference on Computational Intelligence, Communication Systems and Networks (CICSyN), pp. 274–278. IEEE (2010)
4. Soltane, M., Doghmane, N., Guersi, N.: Face and speech based multi-modal biometric authentication. Int. J. Adv. Sci. Technol. **21**(6), 41–56 (2010)
5. Zhou, H., Sadka, A.H.: Combining perceptual features with diffusion distance for face recognition. IEEE Trans. Syst. Man Cybern. Part C **41**(5), 577–588 (2011)
6. Zhou, H., Yuan, Y., Sadka, A.H.: Application of semantic features in face recognition. Pattern Recogn. **41**(10), 3251–3256 (2008)
7. Kremic, E., Subasi, A., Hajdarevic, K.: Face recognition implementation for client server mobile application using PCA. In: Proceedings of the ITI 2012 34th International Conference on Information Technology Interfaces (ITI), pp. 435–440. IEEE (2012)
8. Kremic, E., Subasi, A.: The implementation of face security for authentication implemented on mobile phone (2011)
9. Pawle, A.A., Pawar, V.P.: Face recognition system (FRS) on cloud computing for user authentication. Int. J. Soft Comput. Eng. (IJSCE) **3**(4) (2013)
10. Celesti, A., Tusa, F., Villari, M., et al.: Security and cloud computing: intercloud identity management infrastructure. In: 2010 19th IEEE International Workshop on Enabling Technologies: Infrastructures for Collaborative Enterprises (WETICE), pp. 263–265. IEEE (2010)
11. Tencent AI Lab. http://ai.tencent.com/ailab/
12. Face++. https://www.faceplusplus.com.cn/face-detection/
13. EyeKey. http://www.eyekey.com/

Prediction and Analysis of Mature microRNA with Flexible Neural Tree Model

Rongbin Xu[1], Huijie Shang[1], Dong Wang[1,2(✉)], Gaoqiang Yu[1], and Yunguang Lin[1]

[1] School of Information Science and Engineering, University of Jinan, Jinan 250022, China
ise_wangd@ujn.edu.cn
[2] Shandong Provincial Key Laboratory of Network Based Intelligent Computing,
University of Jinan, Jinan 250022, China

Abstract. miRNA is a class of small non-coding RNA molecules, length of about 20–24 nucleotides. It combines with mRNA by the principle of complementary base pairing to achieve the objective of cracking or suppressing mRNA, which has the function of gene regulation. Therefore, study on the prediction of miRNA is always the hot topic in bioinformatics. In this paper, we drew on a new method of feature extraction and combined the flexible neural tree (FNT) to predict miRNA. For comparison, we adopted XUE dataset, used the training dataset to train the classifier, and then used the classifier to test on testing dataset. The final average accuracy rate of our experiment that is 93.7% is higher than the prediction method of XUE triple-SVM. So our method achieves a better classification effect.

Keywords: miRNA · pre-miRNA · Flexible neural tree · PSO · Couplet-syntax

1 Introduction

miRNA is a class of small non-coding RNA molecules, length of about 20–24 nucleotides, it combines with mRNA by the principle of complementary base pairing to achieve the objective of cracking or suppressing mRNA, which create huge effects in gene regulation [1]. Because of the important role of miRNA, miRNA was listed as the first of world's ten major breakthroughs of science and technology in 2002 "science". Study on the prediction of miRNA will also help us to gain a deeper understanding of the gene regulatory network. It has been proved that miRNAs is a new biomarker of many diseases, and it is closely related to the occurrence and development of disease. For example, miRNA has become the marker and target for cardiovascular disease diagnosis. It is important to guide the treatment of cancer as well.

At present, the study found that the known miRNA is far less than the number of actual miRNA. On the other hand, miRNA sequence length is too short; it also causes a great difficulty for predicting miRNA. So we use the pre-miRNA that has a longer sequence to predict.

Biological experiment and computation intelligence are two main methods for predicting miRNA. Biological experiment method is mainly used in the early prediction of miRNA, this method has the advantages of accuracy, rapidity, but the cost is high, efficiency is low, and it is difficult to predict the expression of miRNA in a particular

© Springer International Publishing AG 2017
D.-S. Huang et al. (Eds.): ICIC 2017, Part II, LNCS 10362, pp. 823–833, 2017.
DOI: 10.1007/978-3-319-63312-1_74

time or in a particular tissue or cells [2]. Computation intelligence is not affected by the impact of the specific expression of time and organization, and high efficiency, low cost, so it has become the main method in the current miRNA prediction [3].

2   Data

2.1   Composition of Dataset

Human positive pre-miRNA dataset. We downloaded 207 reported human pre-miRNAs from the database miRBASE [4] (Release 5), removed the pre-miRNAs with multi-branch stems and loops in their secondary structure. Finally, we got 193 human pre-miRNAs that account for about 93% of the original.

Negative pre-miRNA dataset. We constructed two negative dataset that are similar to pre-miRNA, but not confirmed. First dataset was extracted from the protein coding region. We extracted the protein coding sequences from the UCSC database [5], then selected no overlapping fragments, let the length of the sequence consistent with human pre-miRNA, and predicted secondary structure of pseudo pre-miRNA by RNAfold [6]. Further sample selection according to the following standards:

Pre-miRNA stem contains at least 18 paired bases;
Minimum free energy of Pre-miRNA cannot be greater than −15 kcal/mol;
Removed the pre-miRNAs with multi-branch stems and loops in their secondary structure.
Finally 8494 pseudo pre-miRNAs were gained from this data.

Another negative dataset is the human chromosome 19 gene sequences [7] that are from the database UCSC. After screening, we received a total of 2444 pseudo pre-miRNAs.

2.2   Constructed the Training Dataset and Testing Dataset

We need to construct the training datasets and testing dataset to train neural network and evaluate the classified performance of neural network. Training dataset (TR-ANN) are made up of 168 pseudo pre-miRNAs that were randomly selected from first negative dataset and 163 pre-miRNAs that were randomly selected from positive dataset. Testing datasets are made up of four parts, as follows:

Te-30set: remaining pre-miRNAs from positive pre-miRNA dataset;
Te-updateset: latest discovery of human pre-miRNA;
Te-1000set: 1000 pseudo pre-miRNA randomly selected from first negative dataset(Does not include training datasets);
Te-2444set: second negative dataset.

3   Method

In this paper, we use Couplet-syntax to exact the feature of pre-miRNA and predict the category by FNT which was optimized by PIPE and PSO. The following flowchart can generalize our approach (Fig. 1).

Fig. 1. The flowchart of approach

3.1   Feature Extraction

Structure and sequence of pre-miRNA was handled by Couplet-syntax [8] for extracting more concrete and representative feature of pre-miRNA. The research has proved that inner bulge and symmetric loop on pre-miRNA stem has great difference with other similar pseudo pre-miRNA, therefore, the principle of Couplet-syntax is more detailed description of pre-miRNA structure and sequence through the different symbols. Different symbols include: "^" represents inner bulge, "*" represents inner symmetric loop, "N" represents empty base (instable base) (Fig. 2).

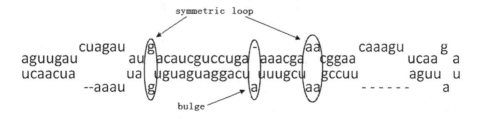

Fig. 2. pre-miRNA.

Specific operations of Couplet-syntax as follows:

Dealt with secondary structure of pre-miRNA: "^" represents inner bulge, "*" represents inner symmetric loop;
Dealt with base sequence of pre-miRNA: used "N" to replace the base on tail ring and stem expansion part;
Calculated frequency of couplet group: a couplet group is made up of a base ("A", "U", "G", "C", "N") on base sequence with corresponding a structure symbol("(", "^", "*") and the next structure symbol. Finally we got a 45 dimensional feature vector (Fig. 3).

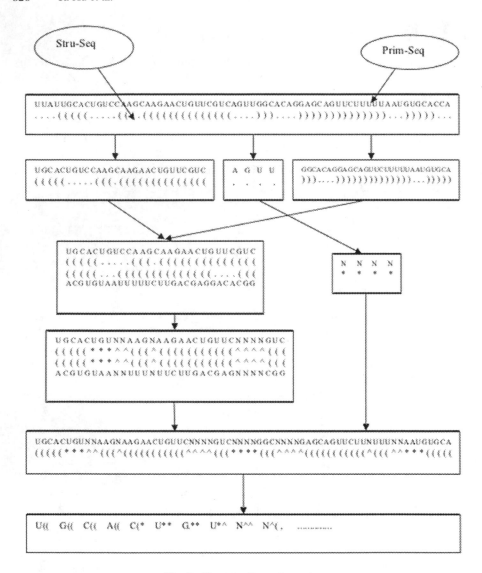

Fig. 3. Process of couple-syntax.

3.2   Flexible Neural Tree

It is a special kind of multilayer neural network that Flexible Neural Tree (FNT) [9, 10] uses a tree structure and don't have a fixed input layer, hidden layer and output layer. The network model can adjust its structure through exchanging information with surrounding environment. In addition, the network model can achieve dimensionality reduction through automatically selecting inputted information.

The construction of flexible neural tree includes: optimization of tree structure and optimization of network parameters. We used Probabilistic Incremental Program Evolution (PIPE) algorithm to optimize tree structure and used Particle Swarm Optimization (PSO) to optimize network parameters.

3.3 Structure of Flexible Neural Tree

FNT is composed by function instruction set (F) and terminal instruction set (T). F represents non terminal node, T represents terminal node (Fig. 4).

$S = F \cup T = \{+2, +3, \ldots \ldots, +N\} \cup \{x_1, x_2 \ldots \ldots x_n\} + i$ (i = 2, 3, ... , n) represents that non terminal node has i branches. xi represents terminal node. We can regard every non terminal node as a neuron operator. As shown below:

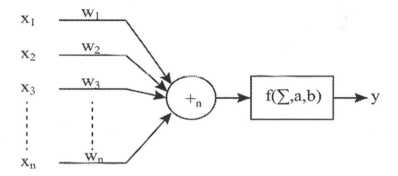

Fig. 4. Neuron operators.

In the process of creating neural tree, if a non terminal node +i is selected, it will randomly generate i parameters ($\omega1, \omega2 \ldots \ldots \omega i$) that will be as connection weights between the node and its child nodes and randomly generate two adjustable parameters a_i and b_i at the same time. Therefore, we can get input of neuron operator by sigmoid function.

$$f(a_i, b_i; x) = \exp\left(-\left(\frac{x - a_i}{b_i}\right)^2\right) \tag{1}$$

$$X = \sum_1^n \omega_i * x_i \tag{2}$$

A FNT structure model can be constituted after all of the neurons are generated. The picture below is a typical FNT. Structure model that is composed of six neuron operators and three feature inputs (x1, x2, x3). Output can be gotten through from left to right Depth-First Traversal of the tree (Fig. 5).

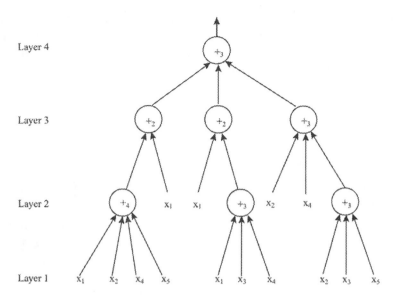

Fig. 5. Flexible neural tree.

3.4 Probabilistic Incremental Program Evolution Algorithm

Probabilistic incremental program evolution (PIPE) [11–14] repeatedly generate probabilistic prototype tree (PPT) according to probability vector and modify probability distribution according to the best individual information, so the best individual information are stored in PPT. PIPE adopts encoded method of tree structure and combines with probability vector coding and probability incremental learning based on population (Fig. 6).

PIPE algorithm procedure is as follows:

First initialize the probabilistic prototype tree;

Repeat the following operation until the condition is satisfied:

Creating a population;
Growing PPT;
Evaluating individual;
Individual learning;
Updating and varying PPT;
Pruning PPT.

Probabilistic prototype tree. Firstly, PIPE needs to produce PPT that produces final structure of FNT. PPT is generated by four processes: initialize PPT, grow PPT, update and vary PPT, prune PPT. process of generating PPT starts from the root node. PPT adopts depth-first traversal and selects a node instruction or a constant according to the probability that has been set (Fig. 7).

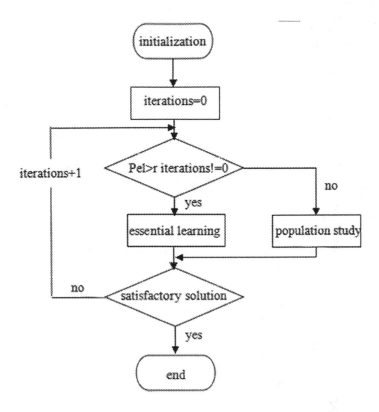

Fig. 6. PIPE algorithm flow chart.

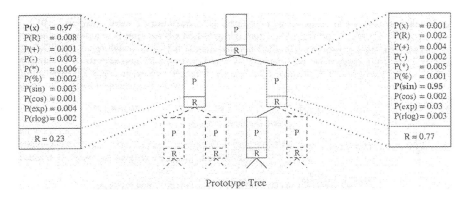

Fig. 7. Probabilistic prototype tree.

PPT is a complete binary tree; each node of PPT contains a lot of information, as follows:

a random constant R;
l + k instructions;

variable probability vector;
P(i): probability of selection i instructions.

The initialization of probabilistic prototype tree

generate a random constant R between 0–1;
let Pt denote the probability of terminal node;
instruction probability of all the terminal leaf node:

$$P(i) = P_t/l \tag{3}$$

Instruction probability of all the function node:

$$P(i) = (1 - P_t)/k \tag{4}$$

Update and variation of probabilistic prototype tree. Expected probability of the optimal tree:

$$P_T = P_b + (1 - P_b)\lambda\left(\frac{\varepsilon + f_{el}}{\varepsilon + f_b}\right), \quad P_b = \prod_{d,w} P_{d,w}(i) \tag{5}$$

Probability of updated:

$$P_{d,w}(i) = P_{d,w}(i) + \lambda(1 - P_{d,w}(i)) \tag{6}$$

Probability of variation:

$$P_{d,w}(i) = P_{d,w}(i) + \lambda(1 - P_{d,w}(i)) \tag{7}$$

Prune probabilistic prototype tree. If the probability of a node is greater than the value that has been setted already, then the PPT that corresponds to the node will be deleted (Fig. 8).

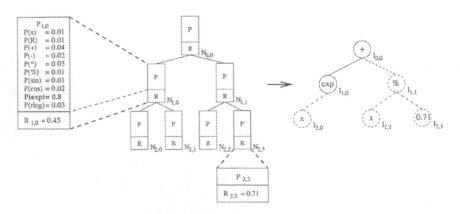

Fig. 8. Prune probabilistic prototype tree.

3.5   Particle Swarm Optimization

PSO [15] that imitates foraged behavior of birds and is based on group cooperation is a global random search algorithm. Firstly, we randomly generate a set of particles that is a set of answers. Each particle has an own position vector xi, velocity vector vi and fitness value that is calculated from fitness function. Iterative process develops towards the direction of optimal fitness value.

Process of PSO algorithm is as follows:

Particles are initialized by an arbitrary position and velocity;
Calculate fitness value of each particle by using the fitness function;
If fitness value of particle is better than pbest (regional optimal solution), then update the value and position of pbest;
If fitness value of particle is better than gbest (global optimal solution), then update the value and position of gbest;

Change velocity and position of particle according to two formulas, as follows:

$$V_{id} = w \times V_{id} + c_1 \times rand() \times (P_{id} - X_{id}) + c_2 \times Rand() \times (P_{gd} - X_{id}) \tag{8}$$

$$X_{id} = X_{id} + V_{id} \tag{9}$$

Repeat these steps from second step until stop condition is satisfied. Stop condition is usually defined as reaching the maximum execution times or reaching the expected fitness value.

3.6   Process of Flexible Neural Tree Algorithm

The generated process of FNT model is as follows:

Generating a random PPT;
Getting a better tree structure through PIPE;
Optimizing parameters through PSO;

Structure and parameters are repeatedly optimized until the terminate condition is satisfied.

4   Analysis of Experimental Results

We adopted datasets from XUE [16] that has gained a better acknowledged prediction result. We trained the constructed FNT model by training dataset and used trained FNT model to test on testing datasets. The comparison results are as follows:

The prediction method that is used in this paper is better than triplet-SVM. Average prediction accuracy of triplet-SVM is 90.6% while our method is 97.3%, therefore method that we propose in this paper can effectively improve the prediction accuracy of miRNA (Table 1).

Table 1. Comparison results

Testing dataset	Type	Size	Classification results (%)	
			Triplet-SVM	FNT
Te-30set	Positive	30	93.3	93.3
Te-updateset	Positive	39	92.3	97.4
Te-1000set	Negative	1000	88.1	90.9
Te-2444set	Negative	2444	89.0	93.2
Te-updateset	Positive	39	92.3	97.4

5   Conclusion

SVM is a main method in current prediction of miRNA. SVM solves support vector by quadratic programming, while solution of quadratic programming will involve calculation of m (m is the number of samples) order matrix. So it is difficult to predict massive pre-miRNA [17]. We proposed the prediction method that is based on FNT. FNT can optimize own structure, deal with massive data and select representative features to achieve dimension reduction. The final test results also prove that our method has greatly improved the prediction accuracy. On the other hand, our experiments provide effective data information for function and mechanism of gene regulation.

Acknowledgment. This research was supported by the National Key Research And Development Program of China (No. 2016YFC0106000), National Natural Science Foundation of China (Grant No. 61302128, 61573166, 61572230, 61671220, 61640218), the Youth Science and Technology Star Program of Jinan City (201406003), the Natural Science Foundation of Shandong Province (ZR2013FL002), the Shandong Distinguished Middle-aged and Young Scientist Encourage and Reward Foundation, China (Grant No. ZR2016FB14), the Project of Shandong Province Higher Educational Science and Technology Program, China (Grant No. J16LN07), the Shandong Province Key Research and Development Program, China (Grant No. 2016GGX101022).

References

1. Bartel, D.P.: MicroRNAs: Genomics, Biogenesis, Mechanism, and Function. Cell **116**(2), 281–297 (2004). ISSN 0092-8674
2. Berezikov, E., Cuppen, E., Plasterk, R.H.A.: Approaches to mi-croRNA discovery. Nat. Genet. **38**(Suppl.), S2–S7 (2006)
3. Kim, V.N., Nam, J.W.: Genomics of microRNA. Trends Genet. **22**(3), 165–173 (2006)
4. Griffiths-Jones, S.: The microRNA registry. Nucleic Acids Res. **32**(1), 109–111 (2004)
5. Karolchik, D., Baertsc, H.R., Diekhans, M., et al.: The UCSC genome browser database. Nucleic Acids Res. **31**(1), 51–54 (2003)
6. Hofacker, I.L.: Vienna RNA secondary structure server. Nucleic Acids Res. **31**(13), 3429–3431 (2003)
7. Pruitt, K.D., Maglot, T.D.R.: RefSeq and LocusLink:NCBI genecentered resources. Nucleic Acids Res. **29**(1), 137–140 (2001)

8. Wang, M.: Identifying pre-miRNA using couplet-syntax for local sequence-structure information. Nanjing University of Aeronautics and Astronautics (2010)
9. Chen, Y.H., Yang, B., Dong, J.W.: Nonlinear systems modeling via optimal design of neural trees. Int. J. Neural Syst. **14**(2), 125–138 (2004)
10. Chen, Y.H., Yang, B., Dong, J.W., et al.: Time-series forecasting using flexible neural tree model. Inf. Sci. **174**(3), 219–235 (2005)
11. Ding, Y.: Computational Intelligence. Science Press, Beingjing (2004)
12. Salustowicz, R.P., Schmidhuber, J.: Probabilistic incremental program evolution. Evol. Comput. **2**(5), 123–141 (1997)
13. Chen, Y., Kawaji, S.: System identification and control using probabilistic incremental program evolution algorithm. J. Robot. Mechatron. **12**(6), 657–681 (2000)
14. Chen, Y., Kawaji, S.: Evolving ANNs by hybrid approaches of PIPE and random search algorithm for system identification. In: Proceedings of the 3rd Asian Control Conference, 2206–2211 (2000)
15. Kennedy, J., Eberhart, R.: Partical swarm optimization. In: Proceedings of the 1995 IEEE International Conference on Neural Networks. [s. I.]: [s. n.], pp. 1942–1948 (1995)
16. Xue, C., Li, F., He, T., et al.: Classification of real and pseudo microRNA precursors using local structure-sequence features and support vector machine. BMC Bioinf. **6**, 310–316 (2005)
17. Griffiths-Jones, S , Grocock, R.J., van Dongen, S., et al.: miRBase: microRNA sequences, targets and gene nomenclature. Nucleic Acids Res. **34**(Database issue), D140–D144 (2006)

Author Index

Printed in the United States
By Bookmasters